TRANSPORTATION SYSTEMS ANALYSIS

Springer Optimization and Its Applications

VOLUME 29

Aims and Scope

Optimization has been expanding in all directions at an astonishing rate during the last few decades. New algorithmic and theoretical techniques have been developed, the diffusion into other disciplines has proceeded at a rapid pace, and our knowledge of all aspects of the field has grown even more profound. At the same time, one of the most striking trends in optimization is the constantly increasing emphasis on the interdisciplinary nature of the field. Optimization has been a basic tool in all areas of applied mathematics, engineering, medicine, economics and other sciences.

The *Springer Optimization and Its Applications* series publishes undergraduate and graduate textbooks, monographs and state-of-the-art expository works that focus on algorithms for solving optimization problems and also study applications involving such problems. Some of the topics covered include nonlinear optimization (convex and nonconvex), network flow problems, stochastic optimization, optimal control, discrete optimization, multi-objective programming, description of software packages, approximation techniques and heuristic approaches.

Ennio Cascetta

TRANSPORTATION SYSTEMS ANALYSIS

Models and Applications

Second Edition

Springer

Ennio Cascetta
Dipartimento di Ingegneria dei Trasporti
Università degli Studi di Napoli Federico II
Via Claudio, 21
80125 Napoli
Italy
cascetta@unina.it

ISBN 978-0-387-75856-5 e-ISBN 978-0-387-75857-2
DOI 10.1007/978-0-387-75857-2
Springer New York Dordrecht Heidelberg London

Library of Congress Control Number: 2009926028

Mathematics Subject Classification (2000): 90-XX, 90B06

Printed on acid-free paper

9 8 7 6 5 4 3 2 1

springer.com

Introduction

> Science is made of facts just as a house is made of bricks, but a collection of facts is no more science than a pile of bricks is a house.
>
> Henri Poincaré

> The aim of the disciplines of praxis is not theoretical knowledge.... It is to change the forms of action....
>
> Aristotle

Transportation systems consist not only of the physical and organizational elements that interact with each other to produce transportation opportunities, but also of the demand that takes advantage of such opportunities to travel from one place to another. This travel demand, in turn, is the result of interactions among the various economic and social activities located in a given area. Mathematical models of transportation systems represent, for a real or hypothetical transportation system, the demand flows, the functioning of the physical and organizational elements, the interactions between them, and their effects on the external world. Mathematical models and the methods involved in their application to real, large-scale systems are thus fundamental tools for evaluating and/or designing actions affecting the physical elements (e.g., a new railway) and/or organizational components (e.g., a new timetable) of transportation systems.

This book discusses the mathematical models that are used to analyze transportation systems, presenting them as the result of a limited number of general assumptions (theory). It also deals with the methods needed to make these models operational, and with their application to transportation system project design and evaluation. This field of knowledge is known as transportation systems engineering.

The development of a transportation system project may involve functional design of new infrastructure facilities such as roads, railways, airports, and car parks; assessment of long-term investment programs; evaluation of project financing schemes; determination of schedules and pricing policies for transportation services; definition of circulation and regulation schemes for urban road networks; and design of strategies for new advanced traffic control and information systems. Physical elements of the system are designed and/or selected from among those available to provide the characteristics and performance that are required of the transportation services to be provided. A transportation system project must of course be technically feasible; but it is equally important that its definition reflects a quantitative assessment of its characteristics and impacts against the objectives and constraints that the project is intended to satisfy.

The difficulty, but also the fascination, of this field derives from the intrinsic complexity of transportation systems. They are, indeed, internally complex systems, made up of many elements influencing each other both directly and indirectly, often nonlinearly, and with many feedback cycles. Furthermore, only some elements in the system are "technical" in nature (vehicles, infrastructure, etc.), governed by the laws of physics and, as such, traditionally studied by engineers. In contrast, the number of travelers or quantity of goods that use these physical elements and, through congestion, the performance of these elements and the impacts of their use, are strictly connected to travel demand and users' behavior. Thus, the analysis of travel demand plays a key role in understanding and designing transportation systems. However, travel demand analysis requires a different kind of approach, one that draws on concepts traditionally used more in social and economic sciences than in engineering.

Apart from their internal complexity, transportation systems are closely interrelated with other systems that are external to them. Transportation projects may have implications for the economy, the location and intensity of the activities in a given area, the environment, the quality of life, and social cohesion. In short, they have a bearing on many, often conflicting, interests, as can easily be seen from the heated debates that accompany almost all decisions concerning transportation at all scales. Both the intensity of these impacts and our sensitivity to them have grown considerably in recent decades due to continued economic and social development, and they have to be addressed in the design and evaluation of transportation projects.

For all these reasons, the consequences of a project cannot be predicted using only experience and intuition. Although they are prerequisites for good design, experience and intuition do not allow quantitative evaluation of the effects of a project, and they may be seriously misleading for complex systems. Modeling supported by empirical evidence sometimes produces unexpected and seemingly paradoxical results: a capacity addition that increases congestion on existing facilities; local projects whose effects propagate to remote parts of the system; price increases that lead to revenue reductions; measures meant to reduce car usage that result in an overall increase in air pollution and energy consumption; and so on. Furthermore, due to the large number of design variables and the complexity of their interactions, modeling the effects of multiple variables requires powerful mathematical tools to help the designer find satisfactory combinations. Finally, social equity issues can only be objectively addressed using a quantitative approach.

The mathematical theory of transportation systems that is presented in this book has been developed over recent decades to develop solutions to these problems. This discipline is based on a systems engineering approach. It is concerned with the relationships among the elements making up a transportation system and with their performance. It possesses a theoretical core that is unique to transportation systems, and also draws on the theory and methods of many other disciplines, especially economics, econometrics, and operations research, in addition to those that are traditionally more directly relevant to transportation engineers, such as traffic engineering, transportation infrastructure engineering, and vehicle mechanics.

The discipline's theoretical foundation is, in my opinion, a "topological–behavioral" paradigm consisting of a set of assumptions and a limited number of functional relationships. This paradigm is an abstract representation of transportation services and their functioning (supply or performance models), of travel demand and users' behavior (demand models), and of the interactions of the two (demand/supply interaction or assignment models).

Over the years, these assumptions and relationships have been extended and formalized. The general mathematical properties of the resulting models have been investigated, producing a wide and internally consistent system of results with a certain degree of formal elegance. This does not preclude the possibility of significant new theoretical and methodological developments in the future. Indeed, transportation systems engineering is probably one of the areas of applied systems engineering in which research is most active, most able to generate extensions and generalizations within the accepted assumptions, and most able to widen and even replace the assumptions on which it is based. Examples can be seen in research on the interactions of transportation with land-use and activity systems, in models of supply design and in the analysis of within-day dynamic systems.

Transportation systems theory would, however, be of little use for addressing practical problems without a set of methods to make it operational. This allows us to specify systems of mathematical models that are consistent with the theory and able to represent the relevant aspects of different transportation systems in the real world. Such methods range from rules for defining a network model to techniques for estimating travel demand and algorithms for solving large-scale computational problems. These methods use the results of a variety of disciplines and, taken as a whole, make up the technical tools and resources of transportation system engineers and analysts.

This book extends and generalizes the contents of my previous book *Transportation Systems Engineering: Theory and Methods* published in 2001, updating both the theory and the application methods. In its attempt to address both general theory and practical methods, the book should be useful to readers with different needs and backgrounds. The various topics are presented, wherever possible, with a gradually increasing level of detail and complexity. Some sections can be used as the basis for beginning and advanced courses in transportation systems engineering and other disciplines, such as economics and regional science. Some sections deal with topics that are mainly of interest for specific applications or are still subjects of research; exclusion of such sections, which are marked with an asterisk, should not limit the understanding of later sections and chapters. The book is made up of ten chapters and an appendix.

Chapter 1 defines a transportation system, and identifies its components and the assumptions on which the theory described in later chapters is developed. It also introduces some application areas of transportation systems engineering, as well as the decision-making process and the role of quantitative methods in this process.

Chapters 2 to 6 explore the theory of transportation systems under the traditional assumption of intraperiod stationarity of the relevant variables. More specifically, Chap. 2 deals with mathematical models that represent transportation supply systems. These models combine traffic flow theory and network flow theory models. The chapter introduces an abstract model that links network flow theory models with the mathematical relationships between transportation costs and flows. The chapter then presents general guidelines concerning the applications of network models and specific models for transportation systems for both continuous and scheduled service. Chapter 3 describes the theoretical basis and mathematical properties of random utility models; these are the general tools most widely adopted to model the travel behavior of transportation system users. Chapter 4 then describes specific mathematical models that represent different aspects of passenger and freight travel demand, introducing their theoretical formulations and providing several examples.

Chapters 5, 6, and 7 describe and analyze assignment models, which predict the outcome of transportation demand/supply interactions; these outcomes include user flows and travel conditions (times, costs, etc.) on the different components of the supply system. Some solution methods are also presented.

Chapter 5 concerns models (and simple algorithms) for within-day static network equilibrium, assuming (fully) pre-trip path choice (either deterministic or stochastic), fixed demand, one transportation mode, a single user class. Shortest path computation as well as assignment to uncongested networks are also addressed.

Chapter 6 extends the results of Chap. 5 to within-day static network equilibrium with combined pre-trip en-route path choice (such as hyperpath assignment), variable demand, several user classes, several transportation modes. Some references are also made to recent inter-period (day-to-day) dynamic models, including both deterministic and stochastic process approaches.

Chapter 7 extends the results of the previous chapters to intra-period (within-day) dynamic systems. In particular, it describes supply, demand and supply/demand interaction (assignment) models for within-day dynamic systems, considering both continuous and scheduled service systems.

Chapter 8 explores methods for estimating travel demand. Methods derived from statistics and econometrics are applied to survey data to estimate existing travel demand in a given area, and to specify and calibrate travel demand models. The chapter also discusses techniques for estimating existing demand flows and model parameters from aggregate data, specifically traffic counts.

Chapter 9 briefly describes several supply design models and algorithms. It considers design problems for road and transit networks that relate to network topology, performance characteristics, and pricing. The design models and algorithms can be used to determine the values of variables that define the design problem at hand by optimizing different types of objective functions under various constraints.

Finally, Chap. 10 describes methods for evaluating and comparing alternative transportation projects. Cost-benefit analysis is presented as an example of economic analysis, cost-revenue analysis as an example of financial analysis, and different multicriteria analysis approaches as examples of quantitative methods for comparing different projects.

For full appreciation and understanding of the book, the reader should have a basic knowledge of calculus, mathematical analysis, optimization techniques, graph and network theory, probability theory, and statistics. Appendix A provides an overview of additional relevant mathematics.

Different reading paths can be followed according to the reader's interests. For example, a path focusing on demand analysis could consist of Chaps. 3, 4, and 8, whereas one focusing on transportation systems design and planning could consist of Chaps. 2, 5, 6, 7, 9, and 10.

A book of this scope and magnitude cannot be completed without the help and the assistance of several individuals. Giulio Erberto Cantarella took part in the entire decision process that underlies the structure of the book and the choice of its contents. He also contributed directly to it, co-authoring Sect. 2.2 of Chap. 2 and Chaps. 5 and 6.

Francesca Pagliara contributed to Chaps. 1 and 10, Vincenzo Punzo contributed to Chaps. 2 and 7, Andrea Papola and Vittorio Marzano to Chaps. 3, 4, and 8, Armando Cartenì to Chaps. 5 and 6, and Mariano Gallo to Chap. 9. Natale Papola and Guido Gentile are the authors of Sect. 7.5 and Appendix 7.A. I would also like to thank Paolo Ferrari and Pietro Rostirolla for their advice and contributions to the preparation of Chap. 10. Almost all topics covered in this book were discussed over the years with Agostino Nuzzolo, who also co-authored Sect. 7.6 of Chap. 7 on scheduled service transportation systems.

I would like also to thank Jon Bottom for revising the English of the whole book as well as for several comments and suggestions.

Despite such extensive contributions and input from others, I take sole responsibility for any mistakes.

Contents

Chapter 1
Modeling Transportation Systems: Preliminary Concepts and Application Areas

1.1 Introduction

Transportation systems consist not only of the physical and organizational elements that interact with each other to produce transportation opportunities, but also of the demand that takes advantage of such opportunities to travel from one place to another. This travel demand, in turn, is the result of interactions among the various economic and social activities located in a given area. Mathematical models of transportation systems represent, for a real or hypothetical transportation system, the demand flows, the functioning of the physical and organizational elements, the interactions between them, and their effects on the external world. Mathematical models and the methods involved in their application to real, large-scale systems are thus fundamental tools for evaluating and/or designing actions affecting the physical elements (e.g., a new railway) and/or organizational components (e.g., a new timetable) of transportation systems.

This book discusses the mathematical models that are used to analyze transportation systems, presenting them as the result of a limited number of general assumptions (theory). It also deals with the methods needed to make these models operational, and with their application to transportation system project design and evaluation. This field of knowledge is known as *transportation systems engineering*. This chapter defines transportation systems and identifies their main elements and the interactions between them (Sect. 1.2). Transportation systems are presented, and the main components and their interactions are defined; the basic assumptions made to analyze these systems are described in Sect. 1.3 through mathematical models that are briefly introduced in Sect. 1.4 and are described at length in later chapters. Finally, Sect. 1.5 describes the "mission" of transportation systems engineering: its role in the wider and more complex decision-making process, as well as some of its typical application areas.

1.2 Transportation Systems

A *transportation system* can be defined as a set of elements and the interactions between them that produce both the demand for travel within a given area and the provision of transportation services to satisfy this demand. Almost all of the components of a social and economic system in a given geographical area interact at some level of intensity. However, in practice it is impossible to take into account

E. Cascetta, *Transportation Systems Analysis*,
Springer Optimization and Its Applications 29,
DOI 10.1007/978-0-387-75857-2_1,

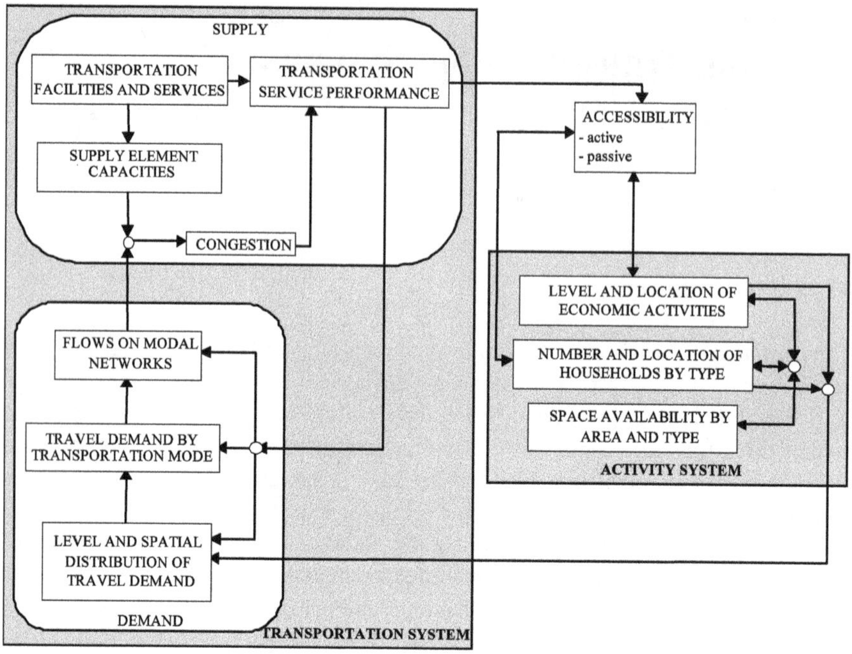

Fig. 1.1 Relationships between the transportation system and the activity system

every interacting element when addressing a given transportation engineering problem. The general approach of systems engineering is to isolate the elements most relevant to a problem at hand, and to group these elements and the relationships between them within the analysis system. The remaining elements are assigned to the external environment; they are taken into account only in terms of their interactions with the analysis system. In general, the analysis system includes the elements and interactions that an action under consideration may significantly affect. Hence there is a strong interdependence between the identification of the analysis system and the problem to be solved. The transportation system of a given area can also be seen as a subsystem of a wider territorial system with which it strongly interacts. The details of the specific problem determine the extent to which these interactions are included either in the analysis system or the external environment.

These concepts can be clarified by some examples. Consider an urban area consisting of a set of households, workplaces, services, transportation facilities, government organizations, regulations, and so on. This system has a hierarchical structure and, within it, several subsystems can be identified (see Fig. 1.1).

One of the subsystems – the *activity system* – represents the set of individual, social, and economic behaviors and interactions that give rise to travel demand. To describe the geographic distribution of activity system features, the urban area is typically subdivided into geographic units called *zones*. The activity system can be further broken down into three subsystems consisting of:

- The households living in each zone, categorized by factors such as income level, life-cycle, composition, and the like
- The economic activities located in each zone, categorized by a variety of socio-economic indicators (e.g., sector of activity; value added; number of employees)
- The real estate system, characterized by the floor space available in each zone for various uses (industrial production, offices, building areas, etc.) and the associated market prices

The different components of the activity system interact in many ways. For example, the number and types of households living in the various zones depend in part on employment opportunities and their distribution, and therefore on the economic activity subsystem. Furthermore, the location of some types of economic activities (retail, social services such as education and welfare, etc.) depends on the geographic distribution of the households. Finally, the number of households and the intensity of economic activities in each zone depend on the availability of specific types of floor space (houses, shops, etc.) and on their relative prices. Detailed analysis of the mechanisms underlying each subsystem of the activity system lies beyond the scope of this book. However, it should be noted that the relative accessibility of the different zones is extremely relevant to many of these mechanisms.

Another subsystem – the *transportation system* – consists of two main components: demand and supply.

Travel demand derives from the need to access urban functions and services in different places and is determined by the distribution of households and activities in the area. Household members make long-term "mobility choices" (holding a driving license, owning a car, etc.) and short-term "travel choices" (trip frequency, time, destination, mode, path,[1] etc.), and use the transportation network and services so that they can undertake different activities (work, study, shopping, etc.) in different locations. These choices result in travel demand flows, that is, the trips made by people between the different zones of the city, for different purposes, in different periods of the day, by means of the different available transportation modes. Similarly, economic activities require the transportation of goods that are consumed by other activities or by households. Goods are moved between production plants, retail locations, and houses or other "final consumption" sites. Their movements make up freight travel demand and corresponding flows.

Both mobility and travel choices are influenced by the characteristics of the transportation services offered by the available travel modes (such as private vehicles, transit, walking). These characteristics are known as *level of service* or performance attributes; they include travel times, monetary costs, service reliability, riding comfort, and the like. For instance, the choice of destination may be influenced by the travel time and cost needed to reach each alternative destination; the choice of departure time depends on the travel time to the destination and the desired arrival time; and the choice of transportation mode is influenced by the time, cost and reliability of the available modes.

[1]The term *path* is used in the book to define both a choice alternative and a path in a graph. The term route is also used in the literature with either or both of these meanings.

The *transportation supply* component is made up of the facilities (roads, parking spaces, railway lines, etc.), services (transit lines and timetables), regulations (road circulation and parking regulations), and prices (transit fares, parking prices, road tolls, etc.) that produce travel opportunities. Travel from one location to another frequently involves the successive use of several connected facilities or services. Transportation facilities generally have a finite capacity, that is, a maximum number of units that may use them in a given time interval. Transportation facilities also generally exhibit *congestion*; that is, the number of their users in a time unit affects their performance. When the flow approaches the capacity of a given facility (e.g., a road section), interactions among users significantly increase and congestion effects can become important. Congestion on a facility can significantly affect the level of service received by its users; for example, travel time, service delay, and fuel consumption all increase with the level of congestion.

Finally, the performance of the transportation system influences the relative accessibility of different zones of the urban area by determining, for each zone, the generalized cost (disutility) of reaching other zones (active accessibility), or of being reached from other zones (passive accessibility). As has been noted, both these types of accessibilities influence the location of households and economic activities and ultimately the real estate market. For example, in choosing their residence zone, households take account of active accessibility to the workplace and other services (commerce, education, etc.). Similarly, economic activities are located to take into account passive accessibility on behalf of their potential clients; public services should be located to allow for passive accessibility by their users, and so on.

Several feedback cycles can be identified in an urban transportation system. These are cycles of interdependence between the various elements and subsystems, as shown in Fig. 1.1. The innermost cycle, the one that involves the least number of elements and that usually shows the shortest reaction time to perturbations, is the interaction between facility flows, the performance due to congestion and transportation costs, in particular those connected with road transportation. The trips made by a given mode (e.g., car) choose from among the available paths and use traffic elements of the transportation network (e.g., road sections). Due to congestion, these flows affect the level of service on the different paths and so, in turn, influence user path choices.

There are also outer cycles, cycles that influence multiple choice dimensions and that involve changes occurring over longer time periods. These cycles affect the split of trips among the alternative modes and the distribution of these trips among the possible destinations. Finally, there are cycles spanning even longer time spans, in which interactions between activity location choices and travel demand are important. Again, through congestion, travel demand influences accessibility of the different areas of the city and hence the location choices of households and firms.

It is clear from the above that a transportation system is a *complex system*, that is, a system made up of multiple elements with nonlinear interactions and multiple feedback cycles. Furthermore, the inherent unpredictability of many features of the system, such as the time needed to traverse a road section or the particular choice made by a user, may require the system state to be represented by random variables.

As a first approximation, these random variables are often represented by their expected values.

Transportation systems engineering has traditionally focused on modeling and analysis of the elements and relationships that make up the transportation system, considering the activity system as exogenously given. More specifically, it has typically considered the influence of the activity system on the transportation system (in particular on travel demand), whereas the inverse influence of accessibility on activity location and level has usually been neglected. However, this divide is rapidly vanishing and transportation system analysis increasingly studies the whole activity–transportation system, albeit at different levels of detail than do disciplines such as regional science and spatial economics.

The aim of transportation systems engineering, as shown in greater detail below, is to design transportation systems using quantitative methods such as those described in the following chapters. Transportation projects may have very different scales and impacts, and consequently the boundaries between the analysis system and the external environment may vary considerably.

If the problem at hand is long-term planning of the whole urban transportation system, including the construction of new motorways, railway lines, parking facilities, and the like, the analysis has to include the entire multimode transportation system and possibly its relationships with the urban activity system. Indeed, the resulting modifications in the transportation network and service performance characteristics and the time needed to implement the plan are such that all components of the transportation and activity systems will likely be affected.

There are cases, however, in which the problem is more limited. If, for example, the aim is to design the service characteristics of an urban transit system without building new facilities (and without implementing new policies affecting other modes, such as car use restrictions), it is common practice to include in the analysis system only those elements (demand, services, prices, vehicles, etc.) related to public transportation. The rest of the transportation system is included in the external environment interacting with the public transportation system.

As shown in the following chapters, the above examples can be generalized to areas of different size (a region, a whole country, etc.) and extended to cover freight transportation.

1.3 Transportation System Identification

Transportation system identification is the definition of the elements and relationships that make up the system to be analyzed. It includes the following steps.

- Identification of relevant spatial dimensions
- Identification of relevant temporal dimensions
- Definition of relevant components of travel demand

Some comments on the different steps are given below. However, it should be stated at the outset that system identification cannot be reduced to the mere application of a set of rigid rules. Rather, it requires the application of professional expertise, which is acquired by combining experience with a thorough knowledge of the methods of transportation systems engineering.

1.3.1 Relevant Spatial Dimensions

The identification of relevant spatial dimensions consists of three phases:

- Definition of the study area
- Subdivision of the area into traffic zones (zoning)
- Identification of the basic network

These three phases necessarily precede the building of any model of the transportation system because they define the spatial extent of the system and its level of spatial aggregation.

Study Area

This phase delineates the geographical area that includes the transportation system under analysis and encompasses most of the project effects. First, the analyst must consider the decision-making context and the type of relevant trips: commuting, leisure, and so on (see Sect. 1.3.3). Most trips of interest should have their origin and destination inside the study area. Similarly, the study area should include transportation facilities and services that are likely to be affected by the transportation project. As one example, the study area for a new traffic scheme should include possible alternative roads for rerouting; as another, the study area for a new infrastructure project should include locations where the number of trips starting or ending may change due to variations in accessibility. The limit of the study area is the *area boundary*. Outside this boundary is the external area, which is only considered through its connections with the analysis system. For instance, the study area might be a whole country if the transportation project is at a national level; alternatively, it may be a specific urban area, or part of an urban area for a traffic management project.

Zoning

In principle, the trips undertaken in a given area may start and end at a large number of points. To model the system, it is necessary to subdivide the study area (and possibly portions of the external area) into a number of discrete geographic units called *traffic analysis zones (TAZs)*. Trips between two different traffic zones are known as *interzonal* trips, whereas *intrazonal* trips are those that start and end within the same zone.

In most transportation models, all trips that start or end within a zone are represented as if their terminal points were at a single fictitious node called the *zone*

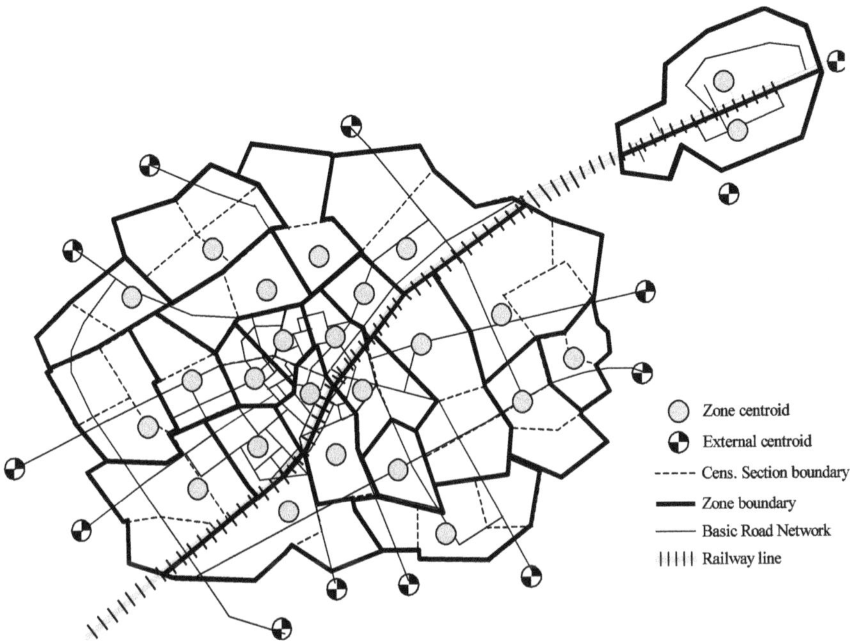

Fig. 1.2 Zoning and basic network

centroid, located in the zone near the geographic "center of gravity" of the full set of actual trip terminal points that it represents. In this representation, intrazonal trips both start and end at the same centroid location, so their effects on the network cannot be modeled.

Zoning can have different levels of detail, that is, a coarser or finer grain. For example, traffic zones may consist of entire cities or groups of cities in a regional or national model, or of one or a few blocks in urban traffic model.

For a given model, the density of zoning should approximately correspond to the density of the relevant network elements: a denser set of network elements corresponds to a finer zoning and vice versa (see Fig. 1.2). For example, if the urban system includes public transportation, it is common practice to consider smaller traffic zones than for a system including only individual cars. This allows walking access to transit stops and/or stations to be realistically represented in terms of the distance from the zone centroid.

The external area is usually subdivided into larger traffic zones. External zones represent trips that use the study area's transportation system but start or end outside of the study area itself. External zones are also represented by zone centroids sometimes called *stations*.

For a given study area and analysis problem, there may be several possible zoning systems. However, some general guidelines are usually followed.

- Physical geographic separators (e.g., rivers, railway lines, etc.) are conventionally used as zone boundaries because they prevent "diffuse" connections between

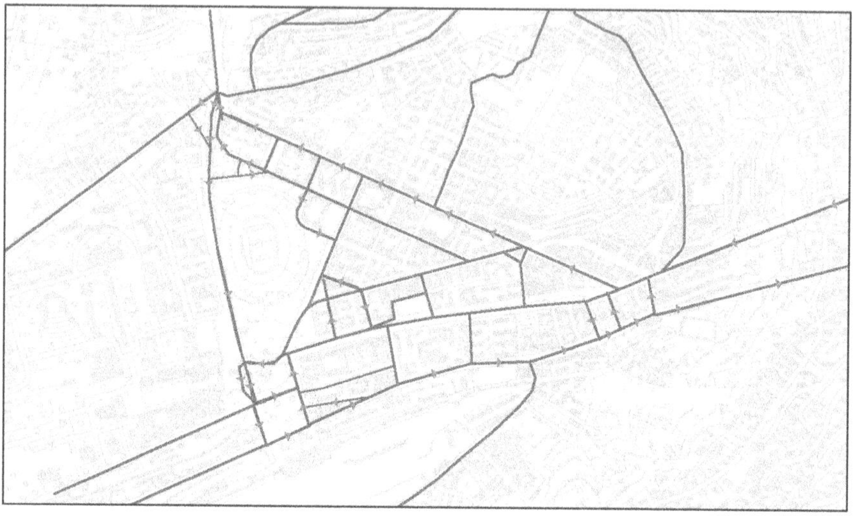

Fig. 1.3 Basic road network for a portion of urban area

adjacent areas and therefore usually imply different access conditions to transportation facilities and services.

- Traffic zones are often defined as aggregations of official administrative areas (e.g., census geographic units, municipalities, or provinces). This allows each zone to be associated with the statistical data (population, employment, etc.) usually available for such areas.
- A different level of zoning detail may be adopted for different parts of the study area depending on the precision needed. For example, smaller zones may be used in the vicinity of a specific facility (e.g., a new road, railway, etc.) for which traffic flows and impacts must be predicted more precisely.
- A traffic zone should group connected portions of the study area that are relatively homogeneous with respect both to their land use (e.g., residential or commercial uses in urban areas; industrial or rural uses in outlying areas) and to their accessibility to transportation facilities and services.

Basic Network

The set of physical elements represented for a given application is called the *basic network.* For example, in urban road systems, the road sections and their main traffic regulations such as one-way, no turn, and the like are indicated (see Fig. 1.3). For scheduled service systems, the infrastructure over which the service is operated (road sections, railways, etc.) will be indicated, together with the main stops or stations, the lines operating along the physical sections, and so on.

The facilities and services included in the network might relate to one or to several transportation modes. The former is referred to as a single mode system and the latter as a multimodal system.

Relevant facilities and services are identified based on their role in connecting the traffic zones in the study area and the external zones. This implies a close interdependence between the identification of the basic network and zone systems. Facilities and services may also be included according to their relationship to the transportation alternatives under consideration.

Because the flows on network elements resulting from intrazonal trips are not modeled, very fine zoning with a coarse basic network will probably cause overestimation of the traffic flows on the included network elements. Conversely, a very detailed basic network with coarse zoning may lead to underestimation of some traffic flows.

Identification of the relevant elements is obviously easier when all the services and facilities play a role in connecting traffic zones, as may be the case, for example, for a national airways network. In the case of road networks, only a subset of roads is relevant in connecting the different zones. In urban areas, for example, local roads are usually excluded from the basic network of the whole area, although they may be included in the basic networks of spatially limited subsystems (a neighborhood or part of it). Similarly, when dealing with a whole region, most of the roads within each city will not be included in the basic network.[2]

1.3.2 Relevant Temporal Dimensions

A transportation system operates and evolves over time, with the characteristics of both travel demand and supply varying at different time scales. For example, the number of trips undertaken in an urban area and the frequency of transit services vary by time of day, by day of the week, and so on. Although space has always been recognized as a fundamental dimension of transportation systems, the time dimension has often been overlooked. However, determination of the relevant analysis time intervals as well as assumptions about system variability within those intervals are crucial modeling decisions.

The main assumptions related to the temporal dimensions of a particular study include the following.

- Definition of the analysis time horizon, and assumptions regarding long-term trends in the exogenous variables
- Selection of reference periods to account for variations in travel demand and supply
- Assumptions about the variability of system parameters within each selected reference period
- Procedures to infer overall system attributes by combining the results obtained from the modeling and analysis of each reference period

[2]Recent developments in databases and Geographic Information Systems (GIS) allow geographically referenced data about the physical elements of the basic network of a given area to be readily stored, retrieved, and represented.

Design and evaluation of transportation projects typically involve two distinct time scales. Design (e.g., determining the required number of road lanes, the settings of a traffic signal at an intersection or the service frequency of a transit line) usually requires information on short *maximum-load periods* such as the peak hour. This information is obtained from a transportation model by analyzing conditions in a particular *reference* or *model period* (see Chap. 9). On the other hand, economic or financial evaluations usually require information about a project's performance over a time span comparable to its technical life (see Chap. 10). The *analysis period* is the entire time duration relevant to the study of a given system.

Depending on the application, the analysis period may include one or more model periods. For major infrastructure projects, for example, the analysis period may span several years or even decades, but the system is typically modeled for only a limited number of reference periods (e.g., one average day per year); the results obtained for the model periods are then expanded to the whole analysis period. By contrast, applications such as traffic signal setting, for example, may only require the modeling of a single reference period (e.g., the A.M. peak period on an average weekday).

If both demand and supply remained approximately constant over the whole analysis period, then any shorter interval could be adopted as a reference period, and the results obtained from modeling the reference period could validly be extrapolated to the whole analysis period. However, because transportation system characteristics change over time, a selected reference period will only be representative of a portion of the analysis period. Thus, the latter is typically subdivided into several model periods, corresponding to different representative situations.[3] Figure 1.4 shows the variation of urban travel demand by trip purpose within an average weekday. In this case, inasmuch as the hypothesis of constancy within the day would clearly be unrealistic, the day would typically be subdivided into shorter model periods (e.g., morning peak, off-peak, evening peak).

One approach is to assume that all relevant transportation characteristics are constant on average during the reference period, and independent of the particular instant at which they are modeled: this is the assumption of *within-period stationarity*. Traditional mathematical models of transportation systems assume that demand and supply remain constant over a period of time long enough to allow the system to reach a stationary or steady-state condition.

The other approach explicitly models the variations in demand and supply within the reference period; this is the assumption of *within-period dynamics*. It should be noted that, in practice, within-period dynamic models typically assume that some elements of the system (e.g., activity-system variables or global travel demand) remain constant within the model period.

In general, three kinds of time variations of system characteristics are important.

---

[3]It might be thought that analysis intervals that include several stationary subperiods (e.g., an average day with several homogeneous peak and off-peak periods) could be dealt with by considering a single reference period with average parameter values (e.g., travel demand or supply). However, this approach could lead to serious errors, especially for congested systems (see Chap. 2). Congestion and demand phenomena are typically highly nonlinear, and average flows and service levels can differ significantly from flows and service levels computed using average parameter values.

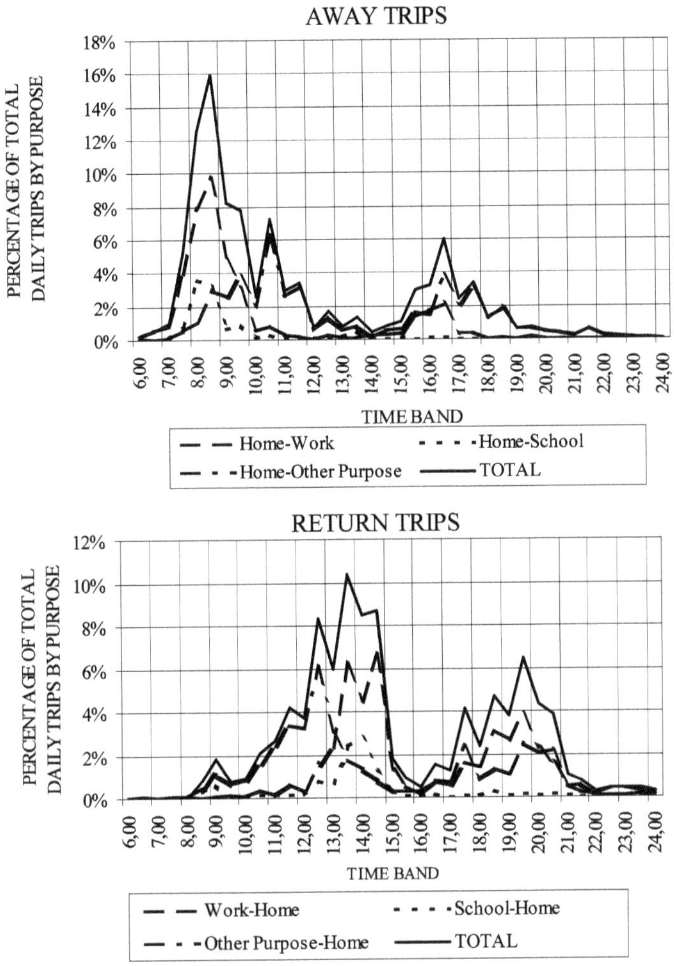

Fig. 1.4 Breakdown of urban travel demand by time of day and purpose

(a) *Long-term variations* or *trends* at the global level and/or systematic variations that can be identified by averaging over multiple reference periods. For example, if reference intervals are single days, a trend consists of variations in the total level and/or in the structure of the average annual demand, observed over several years. In this case, the daily demand is averaged over 365 elementary periods. Long-period variations are often the result of structural changes in the socioeconomic variables underlying travel demand, or in transportation supply. For example, variations in the level of economic activity, production technologies, household income, individual vehicle ownership, sociodemographic population characteristics, lifestyles, urban migration, and the stock of transportation facilities and services have significantly modified the level and structure of passenger and freight travel demand over the years (see Fig. 1.5).

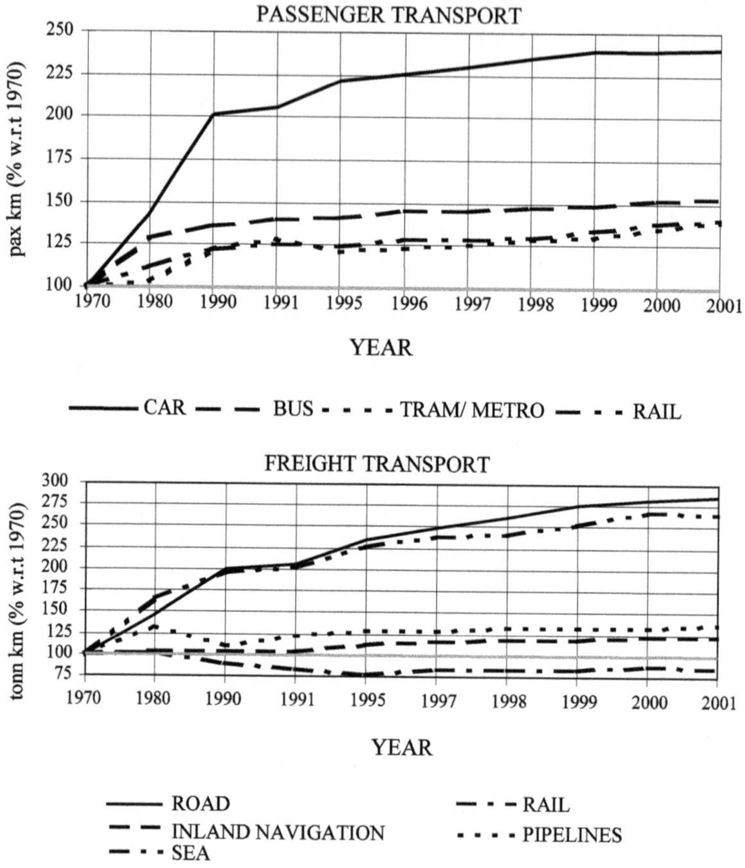

Fig. 1.5 Average long-term trends in European passenger and freight demand

(b) *Cyclical (seasonal) variations* occurring within the analysis period and involving several reference periods. These variations repeat themselves cyclically and can be observed by averaging over a number of cycles. This is the case, for example, with variations in daily demand on different days of the week, or with variations at different times within a typical day. For instance, the fluctuations of urban travel demand by time of day, shown in Fig. 1.4, repeat cyclically over successive workdays. In an analysis period, several cyclic variations with different cycle lengths may occur and overlap with long-term variations. For example, demand and supply change over an analysis period of several years (long-term variation), but they also vary cyclically over the different months of the year, the days of the week and the hours of each day.

(c) *Between-period variations* are variations in demand and supply over reference periods with otherwise identical characteristics, after accounting for the trend and cyclic variations. This is the case with demand variations during morning peak hours of different days with similar characteristics. These fluctuations can be considered random because they cannot be associated with specific events. Travel demand results from the choices made by a large number of decision makers; its actual value in a period therefore depends both on the unpredictable behavioral elements connected with these choices, and on the influence of choices made in previous periods. Similarly, the actual values of some key supply parameters, such as road capacities or travel times, may vary due to unpredictable events, such as an accident. Variations in demand and supply between successive reference periods, for example, the same hours within typical days, are called *between-period* (or *period-to-period*) *dynamics*.

As already mentioned, in reality the three types of dynamics overlap and their identification depends to a great extent on the perspective adopted. In addition, the length of the reference period depends on the modeling approach followed. Some models can *endogenously* represent the variations in relevant parameters within a typical day, which in this case may be taken as the model period. Other models may require the analyst to explicitly specify different exogenous input variables in order to represent variations over different reference periods of the day; in this case, single hours may be the best model periods. Moreover, different applications usually require different assumptions on the relevant temporal dimensions.

Consider, for example, a freight system project for which no significant congestion is expected. This project might require an analysis period several years long. Furthermore, it might be appropriate to consider long-term variations of the system over a number of years, and to account for seasonal variations by considering one or a few typical months as model periods, while ignoring cyclic variations within each month.

For a project with a short-term horizon, such as the traffic plan of an urban area, the long-term trend of daily demand (say over several years) can be ignored. The analysis period might consist of one or more typical days (e.g., average week and weekend days).

Cyclic variations could be modeled as hourly variations within the typical day. Model periods may encompass the morning and evening peak and off-peak hours, with traffic conditions during each period assumed to be stationary. Alternatively, the analyst may consider a different perspective, by which the analysis period is an entire week, cyclic variations are relative to both days of the week and hours of the day, and reference periods encompass full days. In this case, the models would explicitly represent the distribution of demand and supply performances over subintervals of each day, following a within-period dynamic approach (see Fig. 1.6).

1.3.3 Relevant Components of Travel Demand

Passengers and goods moving in a given area demand the transportation services supplied by the system. Travel demand clearly plays a central role in the analysis and modeling of transportation systems because most transportation projects attempt to satisfy this demand (although some projects, such as travel-demand management policies, attempt instead to modify some of its characteristics). In turn,

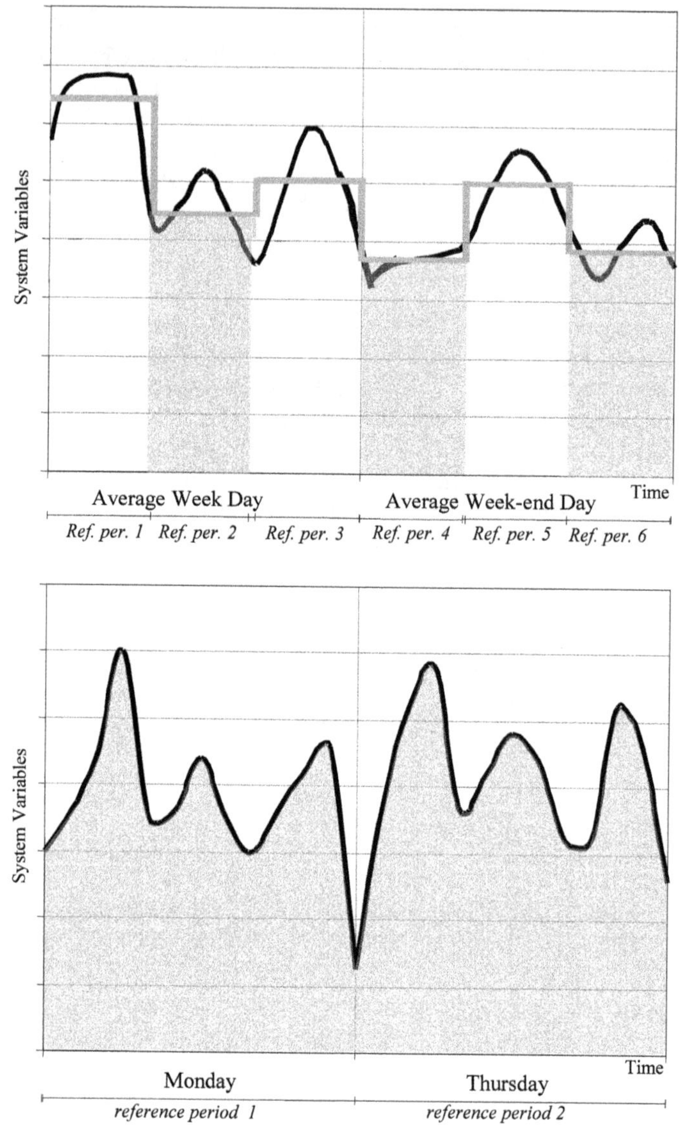

Fig. 1.6 Alternative reference periods

traveler choices can significantly affect the performance of transportation supply elements through congestion (see Chaps. 2 and 7).

Travel does not generally provide utility in itself, but is rather an auxiliary activity necessary for other activities carried out in different locations. Travelers make work-, school-, and shopping-related trips. Goods are shipped from production sites to markets. Travel demand is therefore a derived demand, the result of the interactions between the activity system and the transportation services and facilities, as was seen in Sect. 1.1, as well as of the habits underlying travel behavior in a given area.

A *travel-demand flow* can formally be defined as the number of users with given characteristics consuming particular services offered by a transportation system in a given time period. It is clear that travel demand flows result from the aggregation of individual trips made in the study area during the reference period. A *trip* is defined as the act of moving from one place (origin) to another (destination) using one or more modes of transportation, in order to carry out one or more activities. A sequence of trips, following each other in such a way that the destination of one trip coincides with the origin of the next, is referred to as a *journey* or *trip chain*. With passenger travel, trip chains usually start and end at home; for example, a home–work–shopping–home chain consists of three distinct trips. For freight, individual movements of goods from one place to another are usually referred to as *shipments* or *consignments*. The sequence of manipulations (e.g., packaging) and storage activities applied to shipments is often referred to as the *logistic* or *supply chain*.

Transportation system users, and the trips they undertake, can be characterized in a variety of ways in addition to the temporal characterization described in the previous section. In the following chapters, h stands for the reference period, describing the average weekday, the morning or evening peak hours, the winter or summer seasons, and so on. Some of these ways are described here.

The *spatial characterization* of trips is made by grouping them by place (zone or centroid) of *origin* and *destination*, and demand flows can be arranged in tables, called origin–destination matrices (*O-D matrices*), whose rows and columns correspond to the different origin and destination zones, respectively (see Fig. 1.7). Matrix entry d_{od} gives the number of trips made in the reference period from origin zone o to destination zone d (the *O-D flow*). Some aggregations of the O-D matrix elements are also useful. The sum of the elements of row o:

$$d_{o\cdot} = \sum_d d_{od} \tag{1.3.1}$$

accumulates the total number of trips leaving zone o in the reference period and is known as the flow *produced* or *generated* by zone o. The sum of the elements of column d accumulates the number of trips arriving in zone d in the reference period:

$$d_{\cdot d} = \sum_o d_{od} \tag{1.3.2}$$

and is known as the flow *attracted* by zone d. The total number of trips made in the study area in the reference interval is indicated by $d_{\cdot\cdot}$:

$$d_{\cdot\cdot} = \sum_o \sum_d d_{od} \tag{1.3.3}$$

Trips can be characterized by whether their endpoints are located within or outside of the study area. For *internal (I-I) trips*, the origin and the destination are both within the study area. For *exchange (I-E* or *E-I) trips*, the origin is within the

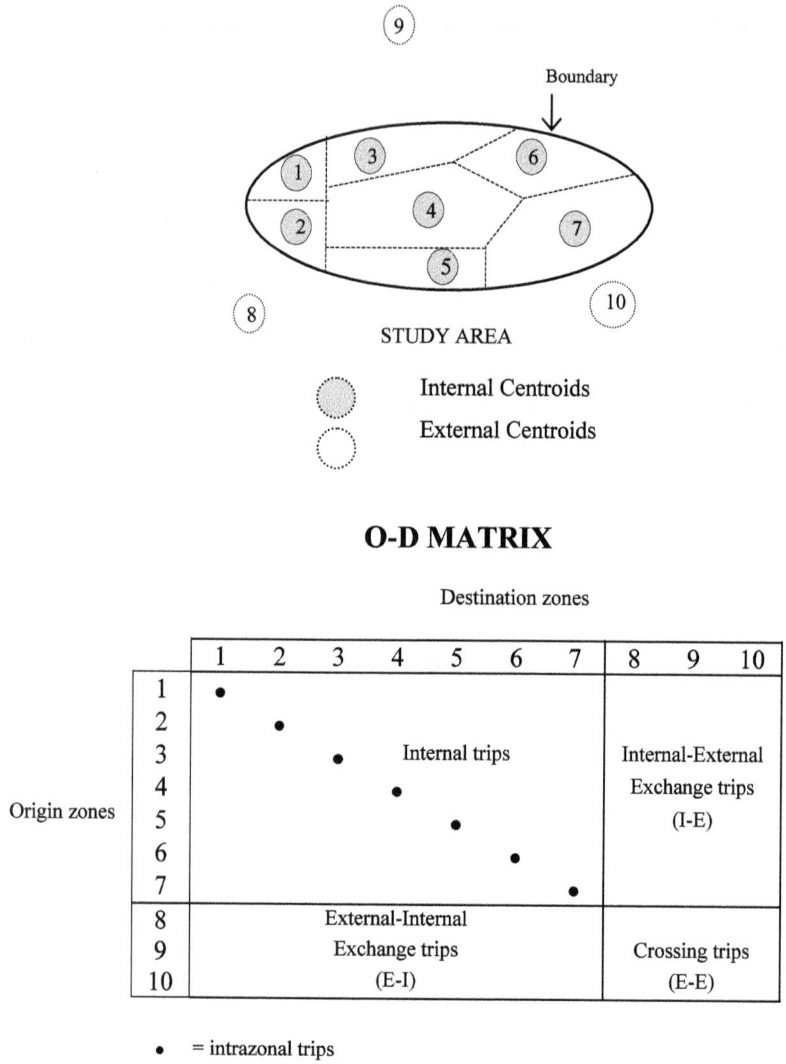

Fig. 1.7 Trip types and their identification in the origin–destination matrix

study area and the destination outside, or vice versa. Finally, *crossing (E-E) trips* have both their origin and their destination external to the study area, but traverse the study area, that is, use the transportation system under study. Figure 1.7 is a schematic representation of the three types of trips and their position in the O-D matrix.

Travel demand can also be classified in terms of *user* and *trip characteristics.* In the case of person trips, user characteristics of interest usually relate to the trip-maker's socioeconomic attributes, such as income level or possession of a driver's license. Groups of users who are homogeneous with respect to a particular set of socioeconomic characteristics are referred to as *market segments*. In a study of different pricing policies, for example, market segments might be defined according to personal or household income. In the case of goods movements, the user characteristics of interest typically relate to attributes of the shipping firm, such as sector of economic activity, firm size, type of plant, production cycle, and so on. In the following chapters, market segments are indicated by i.

Characteristics of individual trips are also of interest. Person trips are often described in terms of the general activities carried out at the origin and destination ends. The pair of activities defines the *trip purpose*: home-based work trips, work-based shopping trips, and so on. A whole sequence of purposes (activities) can be associated with a trip chain. The trip purpose is indicated by s.

Other trip characteristics of interest in a particular analysis may include desired arrival or departure times, and mode, among others, for person trips; and consignment size, type of goods (time sensitivity, value, etc.) and mode for freight trips.

1.4 Modeling Transportation Systems

Design and evaluation require the quantification of interactions among the elements of existing and potential future transportation systems. Values of some elements of existing transportation systems may be obtained from direct measurement, however, it is usually very costly to extend such measurements to all the elements involved. Moreover, proposed future transportation systems obviously cannot be measured. Hence modeling plays a central role in the design and evaluation of transportation systems.

The mathematical models that are described in the following chapters allow representation and analysis of the interactions among the various elements of a transportation system. It is worth giving an overview here of the various classes of models that make up the system of models used to analyze an actual transportation system. The models and their relationships are described in Fig. 1.8; they should be compared with the physical components of the system that they represent, shown in Fig. 1.1.

Supply models, described in Chap. 2, represent the transportation service provided to travel between the different zones; network flow models are frequently used for this purpose. More specifically, supply models represent the performance of transportation facilities and services for the users, and also determine the external impacts (pollution, energy consumption, accidents) of this use (these are sometimes called impact models). The resulting level of service attributes, such as travel time and cost, are input variables for demand models. To predict the performance of single elements (facilities) and the effects of congestion, especially for road systems, supply models often use the results of traffic flow theory, which is briefly described in Chap. 2. Moreover, network models are used to represent the travel opportunities between different locations, and/or the relationships between different trip phases.

Demand models predict the relevant aspects of travel demand as a function of the activity system and of the level of service provided by the supply system. Demand

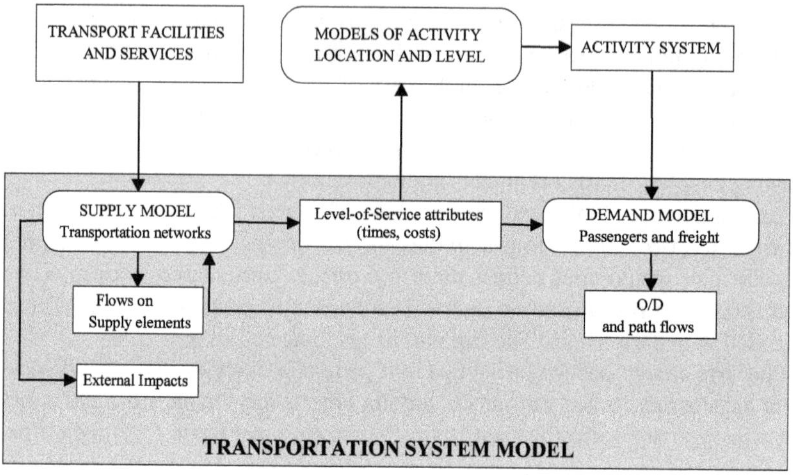

Fig. 1.8 Structure of transportation system models

characteristics typically predicted include the number of trips in the reference period (demand level) and their distribution between different time intervals within the reference period, among different points, different transportation modes, and possible paths. Demand models, described in Chap. 4, can be applied to passenger as well as to freight demand. Travel demand models are usually derived from random utility theory, described in Chap. 3.

Analysis and design of transportation systems require the estimation of present demand and the forecasting of future demand. These estimates and forecasts can be obtained using different sources of information and statistical procedures. To estimate present demand, surveys can be conducted, typically by interviewing a sample of users. From such surveys, *direct estimates* of the demand can be derived using results from sampling theory. Alternatively, the demand (present or future) can be estimated using models similar to those that are described in Chap. 4. *Model-based estimates* require that models be specified (i.e., the functional form and the variables are defined), calibrated (i.e., the unknown model coefficients are determined), and validated (i.e., the ability to reproduce available data is verified). Model estimation procedures are presented in Chap. 8.

Assignment models (or network demand–supply interaction models), studied in Chaps. 5 and 6, predict how O-D demand and path flows will use the various elements of the supply system. Assignment models allow the calculation of link flows, that is, the number of users using each link of the network that represents transportation supply in the reference period. Furthermore, link flows may affect the performance of particular transportation facilities through congestion, and therefore may affect the input to demand models. The mutual interdependencies of demand, flows, and costs are captured by assignment models and are addressed in Chaps. 5, 6 and 7. The models described in this book are based on general assumptions already introduced in the previous sections of this chapter. They are summarized below.

- *Physical and functional delineation of the system.* The transportation system is contained within a defined region (*study area*) and the external area is considered only through its relationships with the analysis system. These relationships are related to both demand (exchange and crossing trips) and supply (transportation infrastructure and services connecting the external area with the analysis system).
- *Spatial discretization (zoning).* The geographic area is subdivided into discrete subareas (traffic analysis zones) to which the socioeconomic variables are related. Departure and arrival points of all the trips traveling to or from a zone are assumed to originate from or go to an arbitrary location in the zone known as the *zone centroid.*
- *Identification of relevant transportation services.* Only those facilities and/or services that connect study area traffic zones together, or that connect them with external traffic zones, are explicitly represented and modeled.

Further assumptions about the representation of time include the following.

- *Identification of relevant model periods.* This refers to the definition of the length of the analysis period, selection of the significant cyclic variations to be modeled, and identification of the corresponding reference or model periods.
- A*ssumptions about within-period variability.* The within-period stationary approach, adopted in Chaps. 2, 4, 5, and 6, assumes that travel demand and supply have constant average characteristics over a period of time long enough to allow stationary conditions to be reached. Under this assumption, the significant variables assume values that are independent of the reference time. Alternatively, within-period dynamic models explicitly represent the variation of supply and some demand dimensions within each reference period. Within-period dynamic models are still at a relatively early stage of development and are discussed in Chap. 7.
- *Type of demand–supply interaction.* In the equilibrium approach, it is assumed that the system is in an *equilibrium configuration* in which demand, flows, and costs are mutually consistent. Equilibrium assignment models have been extensively studied and are described in Chaps. 6 and 7. Alternatively, it is possible to adopt a *between-period dynamic* approach to modeling demand–supply interaction by explicitly representing system evolution over different reference periods. Models of this type are considered in Chap. 6.

Finally, traditional transportation models are sometimes integrated with models that predict *activity location* and *production levels*. These models differ according to the size of the study area (urban, regional, and national) and the type of activities that are considered as endogenous. For example, they may relate to household location in an urban area or to production levels in different sectors of the economy at a multiregional level. Models that jointly analyze the transportation and activity systems are referred to as land use–transportation interaction models. This class of model is less widely used than transportation system models, and their systematic analysis goes beyond the scope of this book. An example of a model that analyzes various interactions among production levels, economic activity location, and transportation is described in Chap. 4, in the context of freight demand models.

1.5 Model Applications and Transportation Systems Engineering

Transportation systems engineering – the design of physical components or policy interventions intended to affect transportation supply and/or demand – is one of the main applications of transportation system modeling.

A set of coordinated, internally consistent actions is referred to as a *project* or *plan*. Projects might relate to transportation facilities, control systems, services, or fares. They can be designed and evaluated from the perspective of the community served by the transportation system under analysis, or from that of the service providers and/or facility operators. Design and decision-making are two interdependent activities. Decision-making for transportation systems is often more complex than for the systems considered in other sectors of engineering. This is especially true when the decision maker must consider, directly or indirectly, the effects of proposed actions on the overall community. Projects concerning decisions and/or typical points of view of a transportation operator, such as the organization of freight distribution or the design of a traffic signal control system, usually undergo a simpler and more straightforward decision-making process. However, even projects that might appear to be of internal concern to a company or public agency, such as the reorganization of transit lines, often produce external impacts that may influence the final decisions.

1.5.1 Transportation Systems Design and the Decision-Making Process

Changes in transportation systems may affect a community and its members in a variety of ways. Building a new facility, for example, may not only change the service experienced by network users, but also produce economic, financial, social, and environmental impacts on groups or individuals who are not system users. These nonusers may be single individuals as well as businesses, landowners, operators, and institutions responsible for the transportation system and the area in which it operates.

Project decisions can be made in many different ways. The "rational" approach to decision-making is based on evaluation of the impacts of the projects under consideration on the various affected parties. This approach, which is commonly adopted for private decisions, is even more necessary when the decisions are made on behalf of a community. The natural dynamics of society, changes in individuals' and decision-makers' attitudes, the occurrence of particular events, and variations in resource availability are all such that decisions and their implementation evolve over time. Increasing recognition over the years of the importance of such long-term dynamic effects has resulted in changes in the very concept of planning. Planning is no longer seen as an activity that leads to the preparation of a single "master" plan identifying a set of projects to be implemented over a long period of time. Rather, planning is now viewed as a process rather than an activity. A planning process results in a sequence of decisions (plans or projects) taken at different, not necessarily predefined, moments in time, with each decision accounting for the effects of previous decisions and exogenous factors. In this framework, the role of quantitative methods for the definition and evaluation of alternative projects is even more relevant as they ensure a sort of "dynamic rationality" for the whole process.

The decision-making process described above is often considered a gross simplification of actual public decision-making processes in the real world. Despite this criticism, it should be seen as a reference paradigm that, with necessary adaptations, can in principle be applied to very different problems and decision contexts. The theoretical analyses that have led to "planning theory" as a theory of collective decision-making are beyond the scope of this book. However, identification of the role and limits of transportation systems analysis and design within the broader decision-making process is extremely relevant. To this end, it is useful to consider the main activities of the decision-making process as shown in Fig. 1.9. The right-hand side of the figure shows schematically the decision process, and the left-hand side shows the phases of analysis and modeling that support its activities.

In the *objectives and constraints identification* phase, the objectives of the decision-maker (or decision-makers) and the relevant constraints for the project are defined. Objectives and constraints may be either explicit or, at least partly, implicit. They depend on the perspective of the decision-maker and, in one way or another, define the type of actions that can be included in the project (e.g., creation of new facilities over the long term or reorganization of existing facilities in the short term).

Modifications to the transportation system can be designed and evaluated from different points of view. Objectives of a private operator, for example, would typically include profit maximization. Constraints might include existing regulations, the available budget, service or fare obligations, the technical limits on the production capacity of the factors employed, and so on. In the case of public decision-makers, the project objectives are numerous, often not clearly defined and frequently conflicting with each other, as, indeed, are the interests of a "complex" society. A public decision-maker may be interested in increasing safety, reducing the generalized transportation cost borne by the users, increasing equity in the distribution of transportation benefits, improving accessibility to economic and social activities, fostering new land development, protecting environmental resources, and reducing the public deficit. Objectives and constraints, explicit or implicit, synthesize the values and attitudes of the firm or of society.

The increasing importance of energy consumption and environmental conservation in recent decades is a clear example of this point. Both objectives and constraints influence the successive phases of the process, especially the analysis of the present situation and the actions that can be included in alternative projects. From the modeling perspective, these factors have an impact on the definition of the analysis system, that is, identification of the elements and their relationships, which are included in the representation of the system in order to evaluate correctly the effects of planned actions. In the *analysis of the present situation* phase, data on the transportation and activity systems are collected. Data are used to analyze the present system state and identify its main deficiencies or "critical points" with

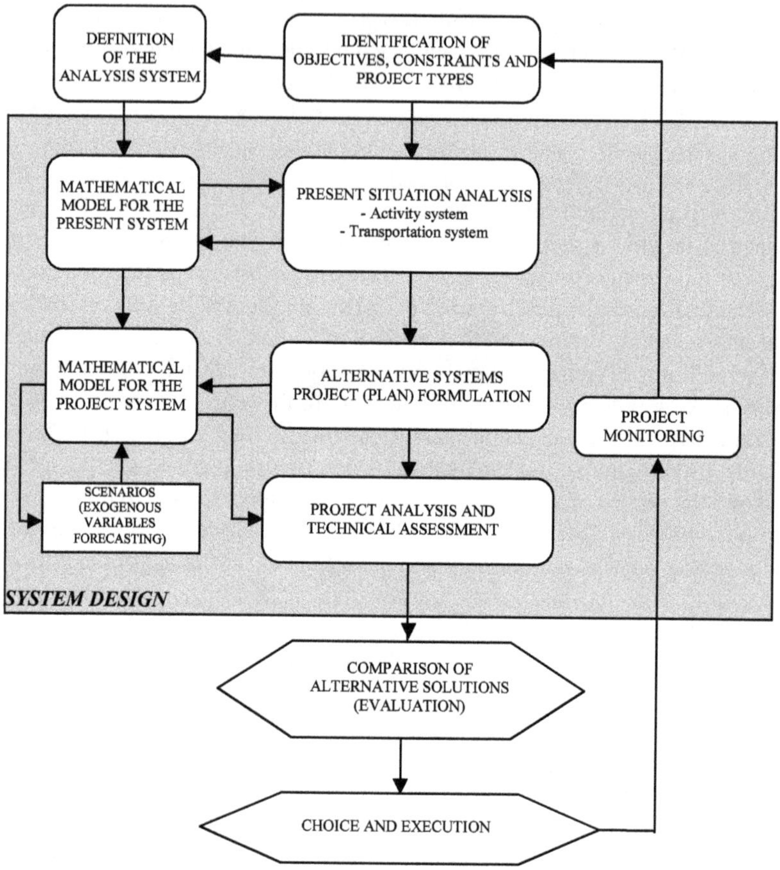

Fig. 1.9 Transportation systems design and the planning process

respect to the project objectives and constraints. These critical aspects should be corrected or mitigated by the planned actions. This phase is also linked to the *building of a mathematical model of the present system*, because it provides the input data for the models (supply, demand, land use). Furthermore, the models often provide some system performance indicators (e.g., flows, saturation levels, generalized transportation costs by the O-D pair) that would be impossible or too costly to measure directly.

The next step is the *formulation of system projects (or plans)*, that is, sets of complementary and/or integrated actions that are internally consistent and technically feasible.[4] The strict interdependence among the elements of a transportation system generally requires that a project be designed taking into account the other system

---

[4]Complementary projects have mutually reinforcing positive effects (e.g., building park-and-ride facilities and improving railway services), whereas integrated projects aim at reducing possible negative interactions (e.g., upgrading public transportation and increasing parking prices).

components that may be significantly influenced by it. A new subway line, for example, requires a reorganization of the surface transit lines to increase the catchment area of the stations (complementary action). Restricting the access of cars to parts of an urban area requires the design of appropriate parking areas, transit lines, pricing policies, and so on in order to alleviate its potentially negative effects (integrated actions). System design is usually limited to the definition of the functional characteristics of the elements composing the system; their physical design, if required, pertains to other branches of engineering.

In general, several alternative projects can be proposed in response to predefined objectives. One alternative is the nonintervention (*do nothing*) option. More realistically, the *do minimum* option involves implementing *committed* decisions (those that, for political or other reasons, cannot be reversed) as well as carrying out basic activities required to keep the system state from deteriorating unacceptably. When a complex project involves multiple actions that cannot be implemented simultaneously, alternative time sequences can be generated, with each sequence considered as an alternative project. Indeed, the impacts of such projects may be significantly influenced by the specific sequence of actions undertaken for their implementation.

Assessment and evaluation of alternative projects require the *prediction of the relevant impacts* of their implementation. Most of the impacts can be forecast quantitatively using the mathematical models and their application methods that are described later in this book. If evaluation of a project requires prediction of its main impacts over a sufficiently long time horizon, assumptions are needed regarding the anticipated future structure of the activity system, or rather the values of the variables that are exogenous to the model. A set of consistent assumptions on the activity system is usually known as a *socioeconomic scenario*. The evolution of exogenous variables over long time periods depends on complex phenomena related to the demographic, social, and economic evolution of the area and on the related external environment. It is very difficult, and perhaps impossible, to forecast these phenomena with precision. Thus, the usual practice is to consider a number of different future scenarios to assess the range of variation of the predicted impacts, and to check the robustness of the alternative projects with respect to the different scenarios.

Technical assessment of the projects concludes the system design phase. This activity verifies that the elements of the supply system will function within their ranges of economic validity and technical feasibility (e.g., that the forecast user flows are not too low or too high with respect to their technical capacity). Moreover, the technical feasibility of the assumed performance of system components and the consistency of this performance with the forecast system state are ascertained. Technical assessment is based on predicted project impacts. Modeling studies can (and often do) influence the high-level design of projects as, indeed, is usually the case in engineering systems design.[5]

---

[5]This assumes that potential projects are exogenously specified prior to analysis; this is the approach most commonly used in applications. However, mathematical models can also be used as supply design tools, as discussed in Chap. 9. As stressed in that chapter, supply design models

Activities related to the analysis of the present situation, formulation of alternative projects, prediction of relevant impacts, and technical assessment can together be defined as the *system design phase*.

The predicted impacts of alternative projects can be further processed to facilitate their comparison. There are many techniques for the analysis and comparison of alternative projects with different levels of aggregation. However, it should be stressed that these techniques cannot and should not replace the actual decision-making process, which is based on compromises among conflicting interests and objectives. Rather, they should be considered as tools to support decision-making.

After a project, or part of it, is implemented, one can compare forecast and actual effects, note the occurrence of unexpected developments and new problems, and evaluate social consent or dissent. These observations may modify some elements of the project or alter its future development. Project *monitoring*[6] is the systematic checking of the main "state variables" of the transportation system using these checks for the a posteriori evaluation of project impacts and the identification of new problems. Monitoring can also identify deficiencies in modeling and analysis, and suggest areas needing improvement. In practice, monitoring transportation systems and projects is often neglected or carried out nonsystematically, although it should play a much more important role in the planning process.

The complexity of the decision-making processes for transportation systems is clear from what has been said so far. The analyst has a technical role in the phases of analysis, design, and forecasting. It should also be recognized that in general the transportation systems engineer does not have all the technical skills required for all the tasks involved. Interaction with specialists from other disciplines (other branches of engineering, economics, urban and regional planning, and social sciences) is needed, particularly if the projects are likely to have significant effects on external systems. On the other hand, understanding the "inner working" of transportation systems, and therefore their design and quantitative modeling, lies at the core of the professional competence of transportation systems engineers.

1.5.2 Some Areas of Application

Some examples of transportation system engineering applications are discussed below, together with their implications for the mathematical models and evaluation methods discussed later in the book.

---

generally pertain to particular types of project (e.g., traffic signal control or transit line frequencies) that are components of wider system projects. In most cases, supply design models should be seen as generators of alternative supply configurations rather than as tools to get the "optimal" solution. For these reasons, supply design models can be included, at least conceptually, in the overall system of mathematical models.

[6]Monitoring has a conceptual function analogous to that of feedback in closed-loop control systems. Closed-loop systems usually prove to be more efficient than open-loop systems, which lack such feedback.

Strategic Planning

Strategic or investment planning involves decisions about long-term (10–20 year) capital investment programs involving the construction of new facilities (e.g., roads, railways, ports) and/or the acquisition of vehicles and technologies (e.g., rolling stock and traffic control systems). In this case, projects usually include transportation services, pricing policies, and, in some cases, travel demand management policies (e.g., access or parking restrictions). Public projects are included in urban, regional, national, or transnational transportation plans, depending on the extent of the area; the projects of agencies or companies are part of their strategic development (or business) plans.

For strategic plans, the study generally encompasses the entire transportation system because substantial changes, even for a single mode, may influence the structure and functioning of the whole system. Returning to the example of an urban transportation plan for a new subway line, the design elements will also include the surface transit lines, parking policy, fare policy, and so on. Evaluation of the line's effects cannot be limited to the public transportation system because the demand split among modes may well change, producing significant effects on road congestion, parking availability, and so on. The time horizon for this level of design requires forecasts of alternative activity system scenarios, and the reverse interactions between the transportation system and the activity system need to be considered as well. Continuing with the same example, it is reasonable to expect that construction of a new subway line may affect, to some extent, the pattern of land use and therefore of travel demand. This broad view of the design system usually entails a less detailed level of representation. Indeed, it is pointless to model extremely detailed effects, such as turning movements at intersections or flows on minor roads, because they are not significant for the evaluation of the project under study.

Feasibility Studies

Feasibility studies are assessments of the technical possibility, economic worth, priority level, and execution mode of individual transportation projects. Project definition is generally derived from a higher-level reference scheme, such as a strategic plan, that identifies new connections needed in the transportation network.

Technical and economic feasibility studies of transportation projects usually require the formulation of project alternatives in terms of their performance and functional characteristics (such as layout, connections, capacity, service performance, type and characteristics of vehicles and technologies, and prices). Alternative projects, including the do-nothing or reference solution, are then evaluated from the functional, economic, and financial points of view, in the context of different transportation and activity system scenarios. The analysis time horizon in this case is usually long-term and the geographic scale varies from urban to regional or national according to the kind of project to be assessed. The definition and functional characteristics of the larger system can be analyzed and modeled at levels of detail that vary according to the intensity of the interactions with the project being studied. For example, a denser zoning system can be adopted around the proposed alignment of a new railway. Whatever the case, the system must be modeled considering the travel demand for and supply of all transportation modes.

There are many examples of feasibility studies. Some studies are aimed at assessing the financial worth of private capital investments in facilities and/or transportation services (project financing). In this case, forecasts of travel demand, user flows, and revenues are of special interest, as are the external conditions under which expected demand and financial returns can be obtained.

Tactical Planning

Short- or medium-term tactical planning involves decisions about projects requiring limited resources, usually assuming minor or no changes in existing facilities. Urban traffic plans or public transportation plans are examples of tactical plans undertaken by public agencies. The design of scheduling or pricing policies for air or rail services are examples of tactical plans carried out from the operators' point of view.

Of primary interest in this context are evaluations of the technical and functional impacts of the project, as well as analysis of its financial performance in terms of operating costs and traffic revenues. These analyses might be accompanied by an economic appraisal, although this is often simplified. For these applications, the socioeconomic scenario is usually taken as given. In practice, it is also assumed that the level and spatial distribution of travel demand are unaffected by the projects, whereas variations in modal split and flows on the project networks are explicitly modeled. In some cases, a single transportation mode is examined in the context of the overall system; the effects of intermodal competition are then considered only through the level of demand of the mode considered (elasticity analysis), without explicit representation of the network and service characteristics of the competing modes.

Operations Management Programs

Short-term operations management programs generally focus on particular aspects of the operations of individual transportation modes, optimizing the use of available resources usually from a company or agency point of view. The design of traffic signal control plans, preparation of transit timetables, and organization of factors necessary for producing transportation services (e.g., assignment of vehicles to lines and travel staff to work shifts) are examples of operations management programs.

In this case, the study is usually limited to a single mode and assumes that the modal demand is fixed. For example, only the road subsystem (network and demand) is considered in designing a traffic-signal control scheme. If necessary, network and assignment models described in later chapters can be integrated with detailed microsimulation models. Furthermore, the design phase can be carried out with the support of supply design models similar to those described in Chap. 9.

Reference Notes

The definition of a transportation system and its elements can be found in most textbooks covering transportation systems analysis and modeling, though with slightly different interpretations. Descriptions of this kind can be found in Manheim (1979), Sheffi (1985), and Ortuzar and Willumsen (2001), among others. The integrated transportation system is described by Cascetta (1995).

Definitions of travel demand and its characteristics can be found in textbooks such as Wilson (1974), Hutchinson (1974), Manheim (1979), Meyer and Miller (2001), Ortuzar and Willumsen (2001), and Train (2003).

Descriptive analyses of the structure of travel demand and its development over time may be found in the European Conference of Ministries of Transport (ECMT, 2001) study of passenger transportation, and the Organization for Cooperation and Economic Development (OCED, 2001) study of freight transportation. An overview of travel demand trends in some transportation markets are provided in Boyer (1998).

A clear and concise description of the different approaches to the general problem of planning and public decision-making, with special reference to town planning, is given in Alexander (1997), which contains a vast bibliography. Many textbooks deal with the process of transportation planning from different viewpoints. Contributions that present differing and sometimes contrasting positions are Hutchinson (1974), Manheim (1979), and Meyer and Miller (2001). Wachs (1985) and Bianco (1986) contain annotated bibliographies of the theoretical developments of the concept of transportation systems planning. The different levels of planning are classified in Florian et al. (1988). Detailed description of the different types of projects and a general outline of the evaluation process is provided in Cascetta (1993). The work by de Luca (2000) deals with the general structure and contents of the different levels of transportation planning for an Italian case study. Finally, the book edited by Cascetta (2005) covers many applications of the main principles of transportation planning and transportation systems engineering applied to Campania regional case studies.

Chapter 2
Transportation Supply Models

2.1 Introduction

This chapter deals with the mathematical models simulating transportation supply systems. In broad terms a transportation supply model can be defined as a model, or rather a system of models, simulating the performances and flows resulting from user demand and the technical and organizational aspects of the physical transportation supply.

Transportation supply models combine traffic flow theory and network flow theory models. The former are used to analyze and simulate the performances of the main supply elements, the latter to represent the topological and functional structure of the system. Therefore, in Sect. 2.2 we present some of the basic results of traffic flow theory. Section 2.3 covers the constituent elements of a transportation network supply model: such elements form an abstract model of transportation supply (transportation network) which combines network flow theory with the functions that express dependence between transportation flows and costs on the network. This is followed by some general indications on the applications of network models in Sect. 2.4. Specific models for transportation systems with *continuous* services (such as road systems) are described in Sect. 2.4.1; models for *discrete* or *scheduled services* (such as bus, train, or airplane) are described in Sect. 2.4.2. Throughout this chapter, as stated in Chap. 1, it is assumed that the transportation system is intraperiod (within-day) stationary (unless otherwise stated); extensions of supply models to intraperiod dynamic systems are dealt with in Chap. 7.

2.2 Fundamentals of Traffic Flow Theory[1]

Models derived from traffic flow theory simulate the effects of interactions between vehicles using the same transportation facility (or the same service) at the same time. For simplicity's sake, the models presented refer to vehicle flow, although most of them can be applied to other types of users, such as trains, planes, and pedestrians. In the sections below we describe stationary uninterrupted flow models (nonstationary models are introduced in Chap. 7), followed by models of interrupted flow, derived from queuing theory.

[1] Giulio Erberto Cantarella is co-author of this section.

E. Cascetta, *Transportation Systems Analysis*,
Springer Optimization and Its Applications 29,
DOI 10.1007/978-0-387-75857-2_2,

2.2.1 Uninterrupted Flows

Multiple vehicles using the same facility may interact with each other and the effect of their interaction will increase with the number of vehicles. This phenomenon, called *congestion*, occurs in most transportation systems, generally worsening the overall performances of the facility, such as the mean speed or travel time. Indeed, it may happen that a vehicle is forced to move at less than its desired speed if it encounters a slower vehicle. The higher the number of vehicles on the infrastructure, the more likely this condition is to happen. This circumstance may also occur in transportation systems with scheduled services: the higher the number of vehicles on the infrastructure, the more likely out-of-schedule vehicles are to cause a delay to other vehicles.

In general, stochastic models may be used to characterize in a probabilistic sense an interaction event that causes a delay. For congested systems with continuous services it is very often sufficient to adopt the aggregate deterministic models described below; they may be applied in areas far away from interruptions such as intersections and toll booths.

2.2.1.1 Fundamental Variables

Several variables can be observed in a *traffic stream*, that is, a sequence of cars moving along a road segment referred to as a link, a. In principle, although all variables should be related to link a, to simplify the notation the subscript a may be implied. The fundamental variables are as follows (see Fig. 2.1).

τ The time at which the traffic is observed
L_a The length of road segment corresponding to link a
s A point along a link, or rather, its abscissa increasing (from a given origin, usually located at the beginning of the link) along the traffic direction ($s \in [0, L_a]$)
i An index denoting an observed vehicle
$v_i(s, \tau)$ The speed of vehicle i at time τ while traversing point (abscissa) s

For traffic observed at point s during time interval $[\tau, \tau + \Delta\tau]$, several variables can be defined (see Fig. 2.1) as follows.

$h_i(s)$ The headway between vehicles i and $i-1$ crossing point s
$m(s \mid \tau, \tau + \Delta\tau)$ The number of vehicles traversing point s during time interval $[\tau, \tau + \Delta\tau]$
$\bar{h}(s) = \sum_{i=1,\ldots,m} h_i(s)/m(s \mid \tau, \tau + \Delta\tau)$ The mean headway, among all vehicles crossing point s during time interval $[\tau, \tau + \Delta\tau]$
$\bar{v}_\tau(s) = \sum_{i=1,\ldots,m} v_i(s)/m(s \mid \tau, \tau + \Delta\tau)$ The time mean speed, among all vehicles crossing point s during time interval $[\tau, \tau + \Delta\tau]$

Similarly, for traffic observed at time τ between points s and $s + \Delta s$, the following variables can be defined.

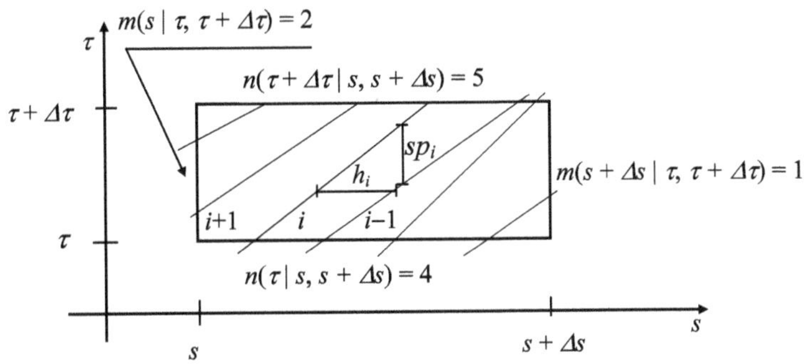

Fig. 2.1 Vehicle trajectories and traffic variables

$sp_i(\tau)$ The spacing between vehicles i and $I - 1$ at time τ

$n(\tau \mid s, s + \Delta s)$ The number of vehicles at time τ between points s and $s + \Delta s$

$\bar{sp}(\tau) = \sum_{i=1,\ldots,n} sp_i(\tau)/n(\tau \mid s, s + \Delta s)$ The mean spacing, among all vehicles between points s and $s + \Delta s$ at time τ

$\bar{v}_s(\tau) = \sum_{i=1,\ldots,n} v_i/n(\tau \mid s, s + \Delta s)$ The space mean speed, among all vehicles between points s and $s + \Delta s$ at time τ

During time interval $[\tau, \tau + \Delta\tau]$ between points s and $s + \Delta s$, a general flow conservation equation can be written:

$$\Delta n(s, s + \Delta s, \tau, \tau + \Delta\tau) + \Delta m(s, s + \Delta s, \tau, \tau + \Delta\tau)$$
$$= \Delta z(s, s + \Delta s, \tau, \tau + \Delta\tau) \tag{2.2.1}$$

where

$\Delta n(s, s + \Delta s, \tau, \tau + \Delta\tau) = n(\tau + \Delta\tau \mid s, s + \Delta s) - n(\tau \mid s, s + \Delta s)$ is the variation in the number of vehicles between points s and $s + \Delta s$ during $\Delta\tau$

$\Delta m(s, s + \Delta s, \tau, \tau + \Delta\tau) = m(s + \Delta s \mid \tau, \tau + \Delta\tau) - m(s \mid \tau, \tau + \Delta\tau)$ is the variation in the number of vehicles during time interval $[\tau, \tau + \Delta\tau]$ over space Δs

$\Delta z(s, s + \Delta s, \tau, \tau + \Delta\tau)$ is the number of entering minus exiting vehicles (if any) during time interval $[\tau, \tau + \Delta\tau]$, due to entry/exit points (e.g., on/off ramps), between points s and $s + \Delta s$

In the example of Fig. 2.1 there are no vehicles entering/exiting in the segment Δs; then $\Delta z = 0$ (Δn is equal to 1 and Δm is equal to -1).

With the observed quantities two relevant variables, *flow* and *density*, can be introduced:

$f(s \mid \tau, \tau + \Delta\tau) = m(s \mid \tau, \tau + \Delta\tau)/\Delta\tau$ is the flow of vehicles crossing point s during time interval $[\tau, \tau + \Delta\tau]$, measured in vehicles per unit of time

$k(\tau \mid s, s + \Delta s) = n(\tau \mid s, s + \Delta s)/\Delta s$ is the density between points s and $s + \Delta s$ at time τ, measured in vehicles per unit of length

Flow and density are related to mean headway and mean spacing through the following relations.

$$f(s \mid \tau, \tau + \Delta\tau) \cong 1/h(s)$$

$$k(\tau \mid s, s + \Delta s) \cong 1/sp(\tau)$$

Note that if observations are perfectly synchronized with vehicles, the near-equality in the previous two equations becomes a proper equality.

Moreover, if the general flow conservation equation (2.2.1) is divided by $\Delta\tau$, the following equation is obtained.

$$\Delta n/\Delta\tau + \Delta f = \Delta e \tag{2.2.2}$$

where

$\Delta f(s, s + \Delta s, \tau, \tau + \Delta\tau) = \Delta m(s, s + \Delta s, \tau, \tau + \Delta\tau)/\Delta\tau$ is the variation of the flow over space

$\Delta e(s, s + \Delta s, \tau, \tau + \Delta\tau) = \Delta z(s, s + \Delta s, \tau, \tau + \Delta\tau)/\Delta\tau$ is the (net) entering/exiting flow

Finally, dividing by Δs, we obtain a further formulation of (2.2.1) (useful for comparisons with nonstationary models based on the fluid-dynamic analogy described in Chap. 7) that expresses the role of variation in density:

$$\Delta k/\Delta\tau + \Delta f/\Delta s = \Delta e/\Delta s \tag{2.2.3}$$

where

$\Delta k(s, s + \Delta s, \tau, \tau + \Delta\tau) = \Delta n(s, s + \Delta s, \tau, \tau + \Delta\tau)/\Delta s$ is the variation of the density over time

2.2.1.2 Model Formulation

In this subsection we describe several deterministic models developed under the assumption of stationarity, formally introduced below. Extensions to nonstationarity conditions are reported in Chap. 7 (some information on stochastic models is reported in the bibliographical note). In formulating such models it is assumed that a traffic stream (a discrete sequence of vehicles) is represented as a continuous (one-dimensional) fluid.

Traffic flow is called *stationary* during a time interval $[\tau, \tau + \Delta\tau]$ between points s and $s + \Delta s$ if flow is (on average) independent of point s, and density is independent of time τ (other definitions are possible):

$$f(s \mid \tau, \tau + \Delta\tau) = f$$

$$k(\tau \mid s, s + \Delta s) = k$$

Note that this condition is chiefly theoretical and in practice can be observed only approximately for mean values in space or time. It is nevertheless useful in that it

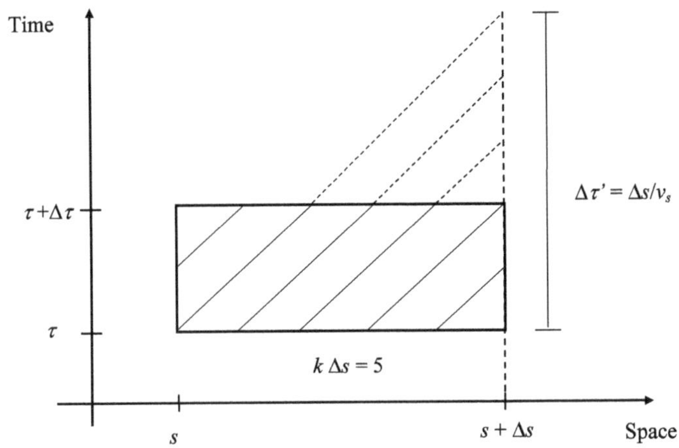

Fig. 2.2 Vehicle trajectories and traffic variables for stationary (deterministic) flows

allows effective analysis of the phenomenon. In this case, the time mean speed is independent of location and the space mean speed is independent of time:

$$\bar{v}_\tau(s) = \bar{v}_\tau$$

$$\bar{v}_s(\tau) = \bar{v}_s$$

In the case of stationarity, both terms in the left side of the conservation equation (2.2.3) are identically null, anyhow other flow conservation conditions may be formulated. Hence, let $n = k \cdot \Delta s$ be the number, time-independent due to the assumption of stationarity, of vehicles on the stretch of road between cross-sections s and $s + \Delta s$, and let $\bar{v}_s$ be the space mean speed of these vehicles. The vehicle that at time τ is at the start of the stretch of road, cross-section s, will reach the end, cross-section $s + \Delta s$, on average at time $\tau + \Delta\tau'$, with $\Delta\tau' = \Delta s / v_s$. Due to the assumption of stationarity, the number of vehicles crossing each cross-section during time $\Delta\tau$ is equal to $f \cdot \Delta\tau$. Thus the number of vehicles contained at time τ on section $[s, s+\Delta s]$ is equal to the number of vehicles traversing cross-section $s+\Delta s$ during the time interval $[\tau, \tau + \Delta\tau']$ (see Fig. 2.2); that is, $k\Delta s = f\Delta\tau' = f\Delta s/v_s$. Hence, under stationary conditions, flow, density, and space mean speed must satisfy the *stationary flow conservation equation*:

$$f = kv \tag{2.2.4}$$

where

$v = \bar{v}_s$ is the space mean speed, simply called speed for further analysis of stationary conditions.[2]

[2]It is worth noting that the time mean speed is not less than the space mean speed, as can be shown because the two speeds are related by the equation $\bar{v}_\tau = \bar{v}_s + \sigma^2/\bar{v}_s$, where σ^2 is the variance of speed among vehicles. In Fig. 2.2 $\sigma^2 = 0$, hence $\bar{v}_\tau = \bar{v}_s$.

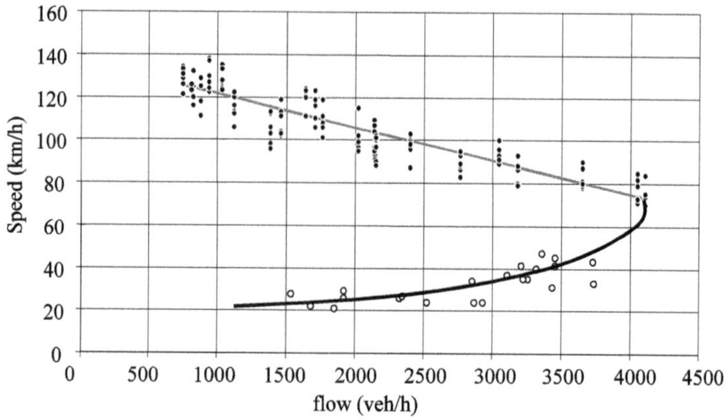

Fig. 2.3 Relationship between speed and flow

In stationary conditions, empirical relationships can be observed between each pair of variables: flow, density, and speed. In general, observations are rather scattered (see Fig. 2.3 for an example of a speed–flow empirical relationship) and various models may be adopted to describe such empirical relationships. These models are generally given the name *fundamental diagram (of traffic flow)* (see Fig. 2.4) and are specified by the following relations.

$$v = V(k) \tag{2.2.5}$$

$$f = f(k) \tag{2.2.6}$$

$$f = f(v) \tag{2.2.7}$$

Although only a model representation of empirical observations, this diagram permits some useful considerations to be made. It shows that flow may be zero under two conditions: when density is zero (no vehicles on the road) or when speed is zero (vehicles are not moving). The latter corresponds in reality to a stop-and-go condition.

In the first case the speed assumes the theoretical maximum value, *free-flow speed* v_0, whereas in the second the density assumes the theoretical maximum value *jam density*, k_{jam}. Therefore, a traffic stream may be modeled through a *partially compressible fluid*, that is, a fluid that can be compressed up to a maximum value.

The peak of the *speed–flow* (and *density–flow*) curve occurs at the theoretical maximum flow, *capacity* Q of the facility; the corresponding speed v_c and density k_c are referred to as the *critical speed* and the *critical density*. Thus any value of flow (except the capacity) may occur under two different conditions: low speed and high density and high speed and low density. The first condition represents an unstable state for the traffic stream, where any increase in density will cause a decrease in speed and thus in flow. This action produces another increase in density and so on until traffic becomes jammed. Conversely, the second condition is a stable state because any increase in density will cause a decrease in speed and an increase in

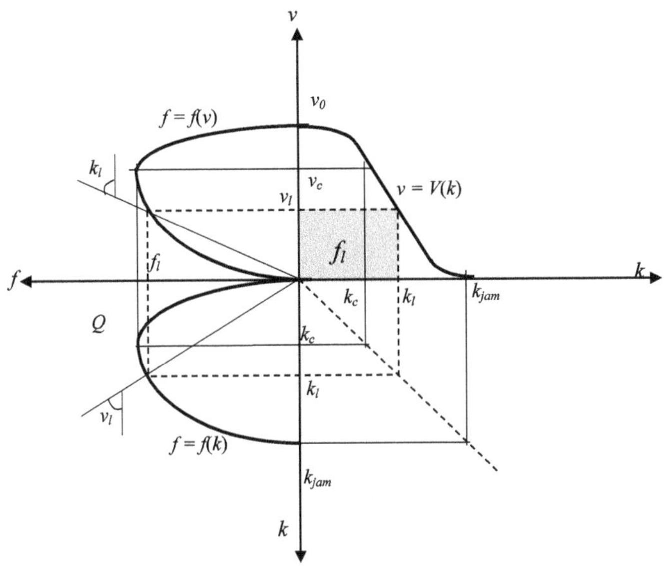

Fig. 2.4 Fundamental diagram of traffic flow

flow. At capacity (or at critical speed or density) the stream is nonstable, this being a boundary condition between the other two.

These results show that flow cannot be used as the unique parameter describing the state of a traffic stream; speed and density, instead, can univocally identify the prevailing traffic condition. For this reason the relation $v = V(k)$ is preferred to study traffic stream characteristics.

Mathematical formulations have been widely proposed for the fundamental diagram, based on single regime or multiregime functions. An example of a single regime function is Greenshields' linear model:

$$V(k) = v_0(1 - k/k_{\text{jam}})$$

or Underwood's exponential model (useful for low densities):

$$V(k) = v_0 e^{-k/k_c}.$$

An example of a multiregime function is Greenberg's model:

$$V(k) = a_1 \ln(a_2/k) \quad \text{for } k > k_{\min}$$
$$V(k) = a_1 \ln(a_2/k_{\min}) \quad \text{for } k \le k_{\min}$$

where a_1, a_2 and $k_{\min} \le k_{\text{jam}}$ are constants to be calibrated.

Starting from the speed–density relationship, the flow–density relationship, $f = f(k)$, may be easily derived by using the flow conservation equation under stationary conditions, or fundamental conservation equation (2.2.4):

$$f(k) = V(k)k$$

Greenshields' linear model yields:

$$f(k) = v_0(k - k^2/k_{\text{jam}})$$

In this case the capacity is given by

$$Q = v_0 k_{\text{jam}}/4$$

Moreover the flow–speed relationship can be obtained by introducing the inverse speed–density relationship: $k = V^{-1}(v)$, thus

$$f(v) = V\big(k = V^{-1}(v)\big) \cdot V^{-1}(v) = v \cdot V^{-1}(v)$$

For example, Greenshields' linear model yields: $V^{-1}(v) = k_{\text{jam}}(1 - v/v_0)$ thus

$$f(v) = k_{\text{jam}}(v - v^2/v_0)$$

In general, the flow–speed relationship may be inverted by only considering two different relationships, one in a stable regime, $v \in [v_c, v_o]$, and the other in an unstable regime, $v \in [0, v_c]$. Greenshield's linear model leads to:

$$v_{\text{stable}}(f) = \frac{v_0}{2}\Big(1 + \sqrt{1 - 4f/(v_0 k_{\text{jam}})}\Big) = \frac{v_0}{2}\Big(1 + \sqrt{1 - f/Q}\Big)$$

$$v_{\text{unstable}}(f) = \frac{v_0}{2}\Big(1 - \sqrt{1 - f/Q}\Big)$$

In the particular case that one can assume the flow regime is always stable, with reference to relation $v = v_{\text{stable}}(f)$ the corresponding relationship between travel time t and flow may be defined (some examples of this type of empirical relationship may be found in Sect. 2.4):

$$t = t(f) = L/v_{\text{stable}}(f) \tag{2.2.8}$$

2.2.2 *Queuing Models*

The average delay experienced by vehicles that queue to cross a flow interruption point (intersections, toll barriers, merging sections, etc.) is affected by the number of vehicles waiting. This phenomenon may be analyzed with models derived from queuing theory, developed to simulate any waiting or user queue formation at a server (administrative counter, bank counter, etc.). The subject is treated below with reference to generic users, at the same time highlighting the similarities with uninterrupted flow.

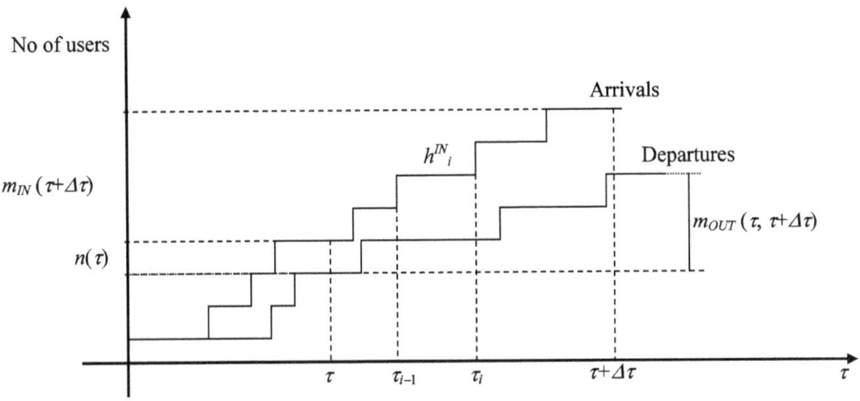

Fig. 2.5 Fundamental variables for queuing systems

2.2.2.1 Fundamental Variables

The main variables that describe queuing phenomena are:

τ The time at which the system is observed
τ_i The arrival time of user i
$h_i = \tau_i - \tau_{i-1}$ The headway between successive users i and $i-1$ joining the queue at times τ_i and τ_{i-1}
$m_{\text{IN}}(\tau, \tau + \Delta\tau)$ Number of users joining the queue during $[\tau, \tau + \Delta\tau]$
$m_{\text{OUT}}(\tau, \tau + \Delta\tau)$ Number of users leaving the queue during $[\tau, \tau + \Delta\tau]$
$h(\tau, \tau + \Delta\tau) = \sum_{i=1,\ldots,m} h_i / m_{\text{IN}}(\tau, \tau + \Delta\tau)$ Mean headway between all vehicles joining the queue in the time interval $[\tau, \tau + \Delta\tau]$
$n(\tau)$ Number of users waiting to exit (queue length) at time τ

With reference to observable quantities, flow variables may be introduced.

$u(\tau, \tau + \Delta\tau) = m_{\text{IN}}(\tau, \tau + \Delta\tau)/\Delta\tau$ arrival (entering) flow during $[\tau, \tau + \Delta\tau]$
$w(\tau, \tau + \Delta\tau) = m_{\text{OUT}}(\tau, \tau + \Delta\tau)/\Delta\tau$ exiting flow during $[\tau, \tau + \Delta\tau]$

Note that the main difference with the basic variables of running links is that space $(s, \Delta s)$ is no longer explicitly referred to because it is irrelevant. Some of the above variables are shown in Fig. 2.5.

With reference to the service activity, let:

$t_{s,i}$ Be service time of user i
$t_s(\tau, \tau + \Delta\tau)$ Average service time among all users joining the queue in time interval $[\tau, \tau + \Delta\tau]$
tw_i Total waiting time (pure waiting plus service time) of user i
$tw(\tau, \tau + \Delta\tau)$ Average total waiting time among all users joining the queue in time interval $[\tau, \tau + \Delta\tau]$

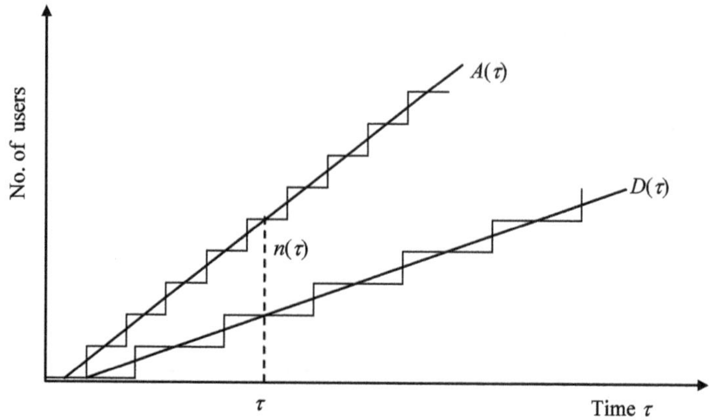

Fig. 2.6 Fluid approximation of deterministic queuing systems

$Q(\tau,\tau+\Delta\tau)=1/t_s(\tau,\tau+\Delta\tau)$ the (transversal[3]) capacity or maximum exit flow, that is, the maximum number of users that may be served in the time unit, assumed constant during $[\tau,\tau+\Delta\tau]$ for simplicity's sake (otherwise $\Delta\tau$ can be redefined)

The capacity constraint on exiting flow is expressed by

$$w \le Q.$$

A general conservation equation, similar to (2.2.1) and (2.2.2) introduced for uninterrupted flow, holds in this case:

$$n(\tau)+m_{\mathrm{IN}}(\tau,\tau+\Delta\tau)=m_{\mathrm{OUT}}(\tau,\tau+\Delta\tau)+n(\tau+\Delta\tau). \tag{2.2.9}$$

Moreover, dividing by $\Delta\tau$ we obtain:

$$\Delta n/\Delta\tau+\left[w(\tau,\tau+\Delta\tau)-u(\tau,\tau+\Delta\tau)\right]=0. \tag{2.2.10}$$

In the following subsection we describe several deterministic models developed under the assumption that the headway between two consecutive vehicles and the service time are represented by deterministic variables. This is followed by a subsection on stochastic models developed using random variables. In formulating such models, as in the case of uninterrupted flow models, we assume arrival at the queue is represented as a continuous (one-dimensional) fluid.

[3]In some cases it is also necessary to introduce longitudinal capacity, that is, the maximum number of users that may form the queue.

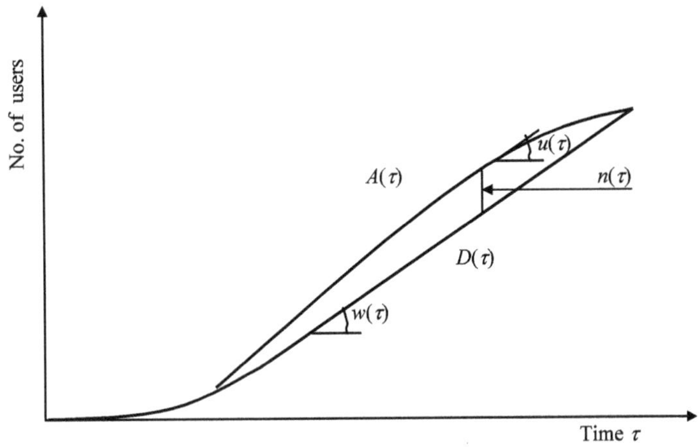

Fig. 2.7 Cumulative arrival and departure curves

2.2.2.2 Deterministic Models

Deterministic models are based on the assumptions that arrival and departure times are deterministic variables. According to the fluid approximation introduced above, the conservation equation (2.2.10) for $\Delta\tau \to 0$ becomes (see Fig. 2.6):

$$\frac{dn(\tau)}{dt} = u(\tau) - w(\tau)$$

Deterministic queuing systems can also be analyzed through the cumulative number of users that have arrived at the *server* by time τ, and the cumulative number of users that have departed from the *server* (leaving the queue) at time τ, as expressed by two functions termed *arrival curve* $A(\tau)$, and *departure curve* $D(\tau) \le A(\tau)$, respectively; see Fig. 2.7. Queue length $n(\tau)$ at any time τ is given by:

$$n(\tau) = A(\tau) - D(\tau) \tag{2.2.11}$$

provided that the queue at time 0 is given by $n(0) = A(0) \ge 0$ with $D(0) = 0$. The arrival and departure functions are linked to entering and exiting users by the following relationships.

$$m_{\text{IN}}(\tau, \tau + \Delta\tau) = A(\tau + \Delta\tau) - A(\tau) \tag{2.2.12}$$

$$m_{\text{OUT}}(\tau, \tau + \Delta\tau) = D(\tau + \Delta\tau) - D(\tau) \tag{2.2.13}$$

The *flow conservation equation* (2.2.9) can also be obtained by subtracting member by member the relationships (2.2.12) and (2.2.13) and taking into account (2.2.11). The limit for $\Delta\tau \to 0$ of (2.2.12) and (2.2.13) leads to (see Fig. 2.7):

$$u(\tau) = \frac{dA(\tau)}{d\tau}$$

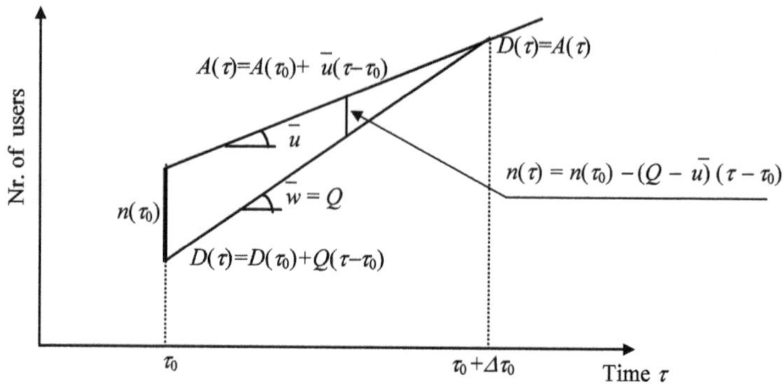

Fig. 2.8 Undersaturated queuing system

$$w(\tau)=\frac{\mathrm{d}D(\tau)}{\mathrm{d}\tau}$$

If during time interval $[\tau_0, \tau_0+\Delta\tau]$ the entering flow is constant over time, $u(\tau)=\bar{u}$, then the queuing system is named *(flow-)stationary* and the arrival function $A(\tau)$ is linear with slope given by $\bar{u}$:

$$A(\tau)=A(\tau_0)+\bar{u}\cdot(\tau-\tau_0) \quad \tau\in[\tau_0, \tau_0+\Delta\tau]$$

The exit flow may be equal to the entering flow $\bar{u}$, or to the capacity Q as described below.[4]

(a) Undersaturation When the arrival flow is less than capacity ($\bar{u}<Q$) the system is *undersaturated.* In this case, if there is a queue at time τ_0, its length decreases with time and vanishes after a time $\Delta\tau_0$ defined as (see Fig. 2.8)

$$\Delta\tau_0=n(\tau_0)/(Q-\bar{u}) \tag{2.2.14}$$

Before time $\tau_0+\Delta\tau_0$, the queue length is linearly decreasing with τ and the exiting flow $\bar{w}$ is equal to capacity Q:

$$\begin{aligned} n(\tau) &= n(\tau_0)-(Q-\bar{u})(\tau-\tau_0)\\ \bar{w} &= Q\\ D(\tau) &= D(\tau_0)+Q(\tau-\tau_0) \end{aligned} \tag{2.2.15}$$

After time $\tau_0+\Delta\tau_0$ the queue length is zero and the exiting flow $\bar{w}$ is equal to the arrival flow $\bar{u}$:

$$n(\tau_0+\Delta\tau_0)=0$$

[4]In stationary queuing models used on transportation networks, the inflow $\bar{u}$ can be substituted with the flow f_a of the link representing the queuing system.

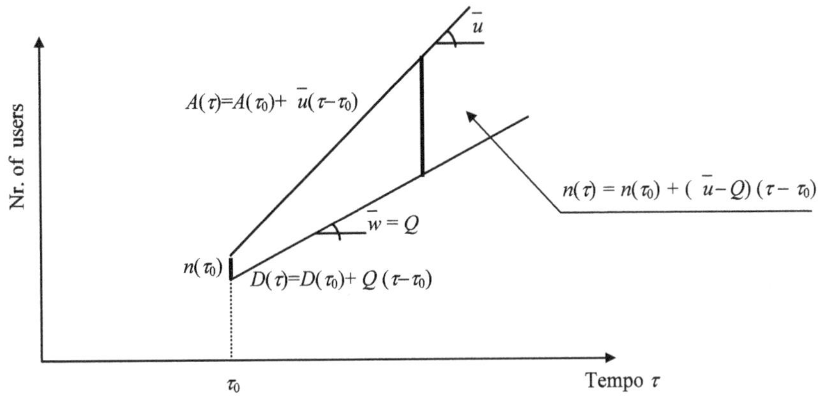

Fig. 2.9 Oversaturated queuing system

$$\begin{aligned}\bar{w} &= \bar{u} \\ D(\tau) &= A(\tau) = A(\tau_0) + \bar{u}(\tau - \tau_0)\end{aligned} \tag{2.2.16}$$

(b) Oversaturation When the arrival flow rate is larger than capacity, $\bar{u} \geq Q$, the system is *oversaturated*. In this case queue length linearly increases with time τ and the exiting flow is equal to the capacity Q (see Fig. 2.9):

$$\begin{aligned}n(\tau_0) &= n(\tau_0) + (\bar{u} - Q)(\tau - \tau_0) \\ \bar{w} &= Q \\ D(\tau) &= D(\tau_0) + Q(\tau - \tau_0)\end{aligned} \tag{2.2.17}$$

(c) General Condition By comparing (2.2.15) through (2.2.17) it is possible to formulate this general equation for calculating the queue length at generic time instant τ:

$$n(\tau) = \max\{0, (n(\tau_0) + (\bar{u} - Q)(\tau - \tau_0))\} \tag{2.2.18}$$

With the above results, any general case can be analyzed by modeling a sequence of periods during which arrival flow and capacity are constant. An important case is that of the queuing system at traffic lights which may be considered a sequence of undersaturated (green) and oversaturated (red) periods with zero capacity (see p. 73: *Application of Queuing Models*).

The *delay* can be defined as the time needed for a user to leave the system (passing the server), accounting for the time spent queuing (pure waiting). Thus the delay is the sum of two terms:

$$tw = t_s + tw_q$$

where

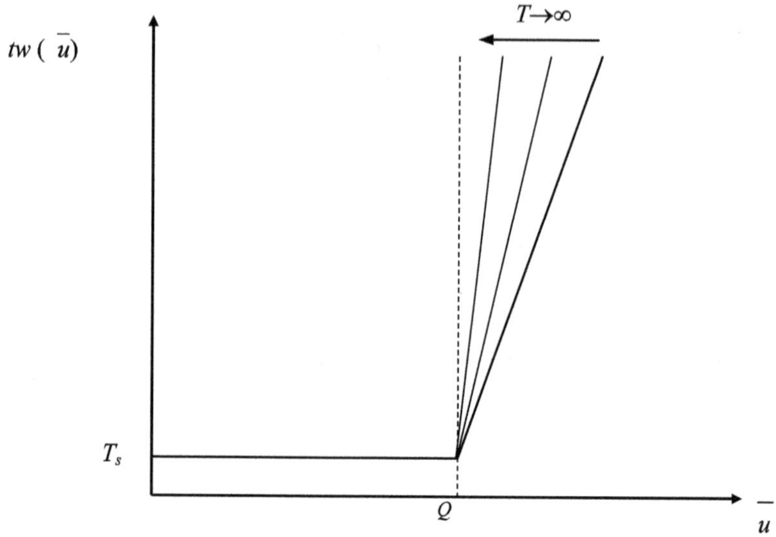

Fig. 2.10 Deterministic delay function at a server

tw is the total delay
$t_s = 1/Q$ is the average service time (time spent at the server)
tw_q is the queuing delay (time spent in the queue)

In undersaturated conditions ($\bar{u} < Q$) if the queue length at the beginning of period is zero (it remains equal to zero), the queuing delay is equal to zero, $tw_q(u) = 0$, and the total delay is equal to the average service time:

$$tw(\bar{u}) = t_s$$

In oversaturated conditions ($\bar{u} \geq Q$), the queue length, and respective delay, would tend to infinity in the theoretical case of a stationary phenomenon lasting for an infinite time. In practice, however, oversaturated conditions last only for a finite period T. If the queue length is equal to zero at the beginning of the period, it will reach a value $(\bar{u} - Q) \cdot T$ at the end of the period. Thus, the average queue over the whole period T is:

$$\bar{n} = \frac{(\bar{u} - Q)T}{2}$$

In this case the average queuing delay is $\bar{x}/Q$, and average total delay is (see Fig. 2.10):

$$tw(\bar{u}) = t_s + \frac{(\bar{u} - Q)T}{2Q} \qquad (2.2.19)$$

2.2.2.3 Stochastic Models

Stochastic models arise when the variables of the problem (e.g., user arrivals, service times of the server, etc.) cannot be assumed deterministic, due to the observed fluctuations, as is often the case, especially in transportation systems. If the system is undersaturated, it can be analyzed through (stochastic) queuing theory which includes the particular case of the deterministic models illustrated above. Some of the results of this theory are briefly reported below, without any claim to being exhaustive.

It is particularly necessary to specify the stochastic process describing the sequence of user arrivals (arrival pattern), the stochastic process describing the sequence of service times (service pattern) and the queue discipline. Arrival and service processes are usually assumed to be stationary renewal processes, in other words with stable characteristics in time that are independent of the past: that is, headways between successive arrivals and successive service times are independently distributed random variables with time-constant parameters. Let N be a random variable describing the queue length, and n the realization of N. The characteristics of a queuing phenomenon can be redefined in the following concise notation,

$$a/b/c(d,e)$$

where

a denotes the type of arrival pattern, that is, the variable which describes time intervals between two successive arrivals:

- $D =$ Deterministic variable
- $M =$ Negative exponential random variable
- $E =$ Erlang random variable
- $G =$ General distribution random variable

b denotes the type of service pattern, such as a

c is the number of service channels: $\{1, 2, \ldots\}$

d is the queue storage limit: $\{\infty, n_{\max}\}$ or longitudinal capacity

e denotes the queuing discipline:

- *FIFO* = First In–First Out (i.e., service in order of arrival)
- *LIFO* = Last In–First Out (i.e., the last user is the first served)
- *SIRO* = Service In Random Order
- *HIFO* = High In–First Out (i.e., the user with the maximum value of an *indicator* is the first served)

Fields d and e, if defined respectively by ∞ (no constraint on maximum queue length) and by *FIFO*, are generally omitted. In the following we report the main results for the *M/M/1* (∞, *FIFO*) and the *M/G/1* (∞, *FIFO*) queuing systems, which are commonly used for simulating transportation facilities, such as signalized intersections.

Some definitions or notation differ from those traditionally adopted in dealing with queuing theory (the relative symbols are in brackets) so as to be consistent with those adopted above. The parameters defining the phenomenon are as follows.

$u, (\lambda)$ The arrival rate or the expected value of the arrival flow
$Q = 1/t_s, (\mu)$ The service rate (or capacity) of the system, the inverse of the expected service time
$u/Q, (\rho)$ The traffic intensity ratio or utilization factor
n A value of the random variable N, number of users present in the system, consisting of the number of users queuing plus the user present at the server, if any (the significance of the symbol n is thus slightly different)
tw A value of the random variable TW, the time spent in the system or overall delay, consisting of queuing time plus service time

(a) *M/M/1* (∞, *FIFO*) Systems In *undersaturated* conditions ($u/Q < 1$):

$$E[N] = \frac{\frac{u}{Q}}{1 - \frac{u}{Q}} = \frac{u}{Q - u} \tag{2.2.20}$$

$$\text{VAR}[N] = \frac{\frac{u}{Q}}{(1 - \frac{u}{Q})^2}$$

According to Little's formula, the expected number of users in the system $E[N]$ is the product of the average time in the system (expected value of delay) $E[TW]$ multiplied by arrival rate u:

$$E[N] = uE[TW] \tag{2.2.21}$$

from which:

$$E[TW] = \frac{1}{Q - u} \tag{2.2.22}$$

The expected time spent in the queue $E[tw_q]$ (or queuing delay) is given by the difference between the expected delay $E[tw]$ and the average service time $t_s = 1/Q$:

$$E[TW_q] = \frac{1}{Q - u} - \frac{1}{Q} = \frac{u}{Q(Q - u)}. \tag{2.2.23}$$

According to Little's second formula, the expected value of the number of users in the queue $E[N_q]$ is the product of the expected queuing delay $E[TW_q]$ multiplied by the arrival rate u:

$$E[N_q] = uE[TW_q] \tag{2.2.24}$$

and then:

$$E[N_q] = \frac{u^2}{Q(Q - u)} \tag{2.2.25}$$

(b) *M/G/1* (∞, *FIFO*) Systems In this case the main results are the following.

$$E[N] = \frac{u}{Q}\left[1 + \frac{u}{2(Q-u)}\right]$$

$$E[TW] = \frac{1}{Q}\left[1 + \frac{u}{2(Q-u)}\right]$$

$$E[TW_q] = \frac{u}{2Q(Q-u)}$$

2.3 Congested Network Models

This section provides a general mathematical formulation of transportation supply models, based on congested network flow models. The bases for these models are *graph models*. Next, *network models*, including link performances and costs, and *network flow models*, including link flows, are introduced. Finally, *congested network (flow) models*, modeling relationships among performances, costs, and flows, are developed.

2.3.1 Network Structure

The network structure is represented by a *graph*. The latter is defined by a set N of elements called *nodes* and by a set of pairs of nodes belonging to N, $L \subseteq N \times N$, called *links*. The graphs used to represent transportation services are generally oriented; that is, the links have a direction and the node pairs defining them are ordered pairs. A link connecting the node pair (i, j) can also be denoted by a single index, say a.

The links in a graph modeling a transportation system represent phases and/or activities of possible trips between different traffic zones. Thus, a link can represent an activity connected to a physical movement (e.g., covering a road) or an activity not connected to a physical movement (such as waiting for a train at a station). Links are chosen in such a way that physical and functional characteristics can be assumed to be homogeneous for the whole link (e.g., the same average speed). In this sense, links can be seen as the partition of trips into segments, each of which has certain characteristics; the level of detail of such a partition can clearly be very different for the same physical system according to the objectives of the analysis.

Nodes correspond to significant events delimiting the trip phases (links), that is, to the space and/or time coordinates in which events occur that they represent. In *synchronic networks*, nodes are not identified by a specific time coordinate, and the same node represents events occurring at different moments (instants) of time. For example, the different entry or exit times in a road segment, an intersection, or a station, may be associated with a single node, representing all the entry/exit events.

Centroid nodes, introduced in Sect. 1.3.1, represent the beginning or end of individual trips. In *diachronic networks*, on the other hand, nodes may have an explicit time coordinate and therefore represent an event occurring at a given instant. The graphs considered in this chapter are synchronic, because diachronic networks assume a within-period system representation; diachronic graphs for scheduled services are introduced in Chap. 7.

A trip is a sequence of several phases and, in a graph that represents transportation supply, it consists of a *path* k, defined as a succession of consecutive links connecting an initial node (path origin) to a final node (path destination). Usually, only paths connecting centroid nodes are considered in transportation graphs. On this basis, each path is unambiguously associated with one, and only one, O-D pair, whereas several paths can connect the same O-D pair. An example of a graph with different paths connecting the centroid nodes is depicted in Fig. 2.11.

A binary matrix called the *link–path incidence matrix* $\boldsymbol{\Delta}$, can represent the relationship between links and paths. This matrix has a number of rows equal to the number of links n_L and a number of columns equal to the number of paths n_P. The generic element δ_{ak} of the binary matrix $\boldsymbol{\Delta}$ is equal to one if link a belongs to path $k, a \in k$, and zero, otherwise, $a \notin k$ (see Fig. 2.11). The row of the link–path incidence matrix corresponding to the generic link a identifies all the paths including that link (columns k for which $\delta_{ak} = 1$). Moreover, the elements of a column corresponding to the generic path k identify all the links that make it up (rows a for which $\delta_{ak} = 1$).

2.3.2 Flows

A *link flow* f_a can be associated with each link a. Link flow is the average number of homogeneous units using link a (i.e., carrying out the trip phase represented by the link) in a time unit. In other words, the link flow is a random variable of mean f_a. Several link flows can be associated with a given link depending on the homogeneous unit considered. *User flows* relate to users, such as travelers or goods, possibly of different classes. *Vehicle flows* relate to the number of vehicles, perhaps of different types such as automobiles, buses, trains, and so on.

For individual modes, such as automobiles or trucks, user flows can be transformed quite straightforwardly into vehicle flows through average occupancy coefficients. For scheduled modes, such as trains, vehicle flows derive from the service schedule and are often treated as an input to the supply model.

The link flow of the generic user class or vehicle type i is denoted by f_a^i. In accordance with the results of traffic flow theory (see Sect. 2.2), link performance and cost variables are affected by user or vehicle flow. To allow for this dependence it is often worth homogenizing the various classes of users or various types of vehicles by defining *equivalent flows* associated with links. In this case the flows of different user classes or vehicle types are homogenized to a reference class or type:

$$f_a = \sum_i w_i f_a^i$$

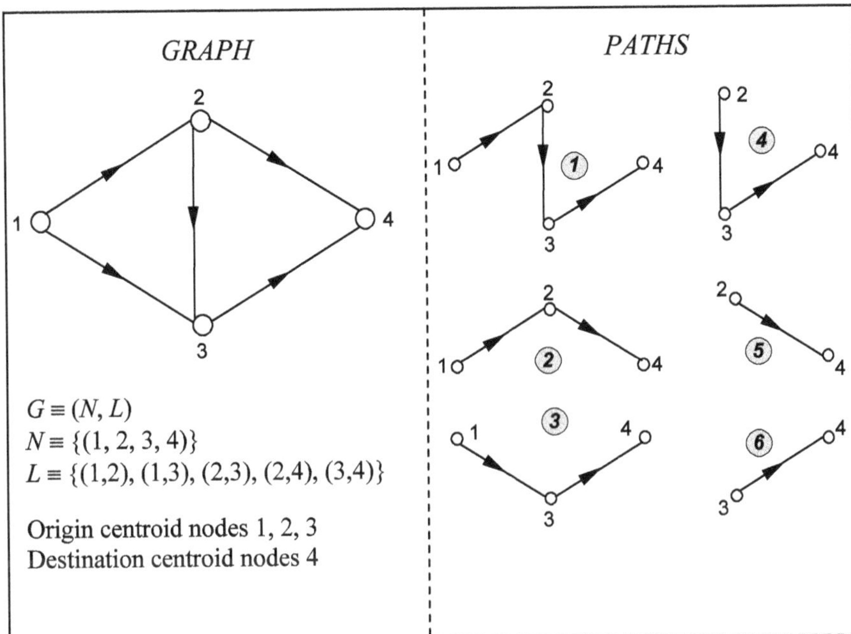

O-D Pairs		1-4			2-4		3-4
Links \ Paths		1	2	3	4	5	6
1		1	1	0	0	0	0
2		0	0	1	0	0	0
3		1	0	0	1	0	0
4		0	1	0	0	1	0
5		1	0	1	1	0	1

Fig. 2.11 Example of a graph and link–path incidence matrix

where w_i is the homogenization coefficient of the users of class i with respect to their influence on link performances. For example, for road flows, automobiles are usually the reference vehicle type ($w_i = 1$) and the other vehicle flows are transformed into equivalent auto flows with coefficients w_i. The latter are greater than one if the contribution to congestion of these vehicles is greater than that of cars (buses, heavy vehicles, etc.), less than one in the opposite case (motorcycles, bicycles, etc.).

The *vector of link flows* $\boldsymbol{f}$ has, as a generic component, the flow on link a, f_a, for each $a \in L$ (see Fig. 2.12).

Flow variables can also be associated with paths. Under the within-day stationarity hypothesis, the average number of users, who in each subinterval travel along each path, is constant. The average number of users, who in a time unit follow path k, is called the *path flow* h_k. If the users have different characteristics (i.e., they belong to different classes), path flows per class i, h_k^i, can be introduced. Path flows of different user classes or vehicle types can be homogenized by means of coefficients w_i similar to those introduced for link flows; the equivalent path flow is obtained as:

$$h_k = \sum_i w_i \cdot h_k^i$$

There is clearly a relationship between link and path flows. Indeed, the flow on each link a can be obtained as the sum of the flows on the various paths containing that link. This relationship can be expressed by using the elements δ_{ak} of the link–path incidence matrix as

$$f_a = \sum_k \delta_{ak} \cdot h_k \tag{2.3.1}$$

or in matrix terms:

$$\boldsymbol{f} = \boldsymbol{\Delta h} \tag{2.3.2}$$

where $\boldsymbol{h}$ is the path flow vector.

Equation (2.3.1) or (2.3.2) expresses the way in which path flows induce flows on individual links. For this reason it is referred to as the (*static*) *Network Flow Propagation* (*NFP*) model (see Fig. 2.11). Note that the linear algebraic structure of (2.3.1) depends crucially on the assumption of intraperiod stationarity (within-day static model); if this assumption is removed, the model loses its algebraic-linear nature as shown in Chap. 7.

2.3.3 Performance Variables and Transportation Costs

Some variables perceived by users can be associated with individual trip phases. Examples of such variables are travel times (transversal and/or waiting), monetary cost, and discomfort. These variables are referred to as *level-of-service* or *performance attributes*. In general, performance variables correspond to disutilities or costs for the users (i.e., users would be better off if the values of performance variables were reduced). The average value of the nth performance variable, related to link a, is denoted by r_{na}. The *average generalized transportation link cost*, or simply the

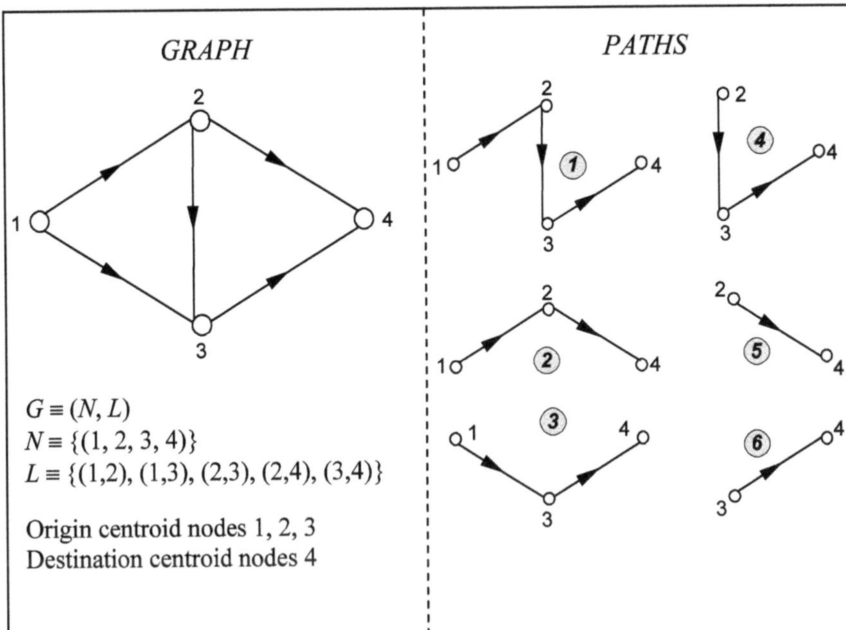

Link flow vector

$$\boldsymbol{f} = \begin{bmatrix} f_{12} \\ f_{13} \\ f_{23} \\ f_{24} \\ f_{34} \end{bmatrix}$$

Path flow vector

$$\boldsymbol{h} = \begin{bmatrix} h_1 \\ h_2 \\ h_3 \\ h_4 \\ h_5 \end{bmatrix}$$

Link-path incidence matrix

$\Delta =$

	1	2	3	4	5	6
1,2	1	1	0	0	0	0
1,3	0	0	1	0	0	0
2,3	1	0	0	1	0	0
2,4	0	1	0	0	1	0
3,4	1	0	1	1	0	1

Link flow vector

$$\boldsymbol{f} = \begin{bmatrix} f_{12} \\ f_{13} \\ f_{23} \\ f_{24} \\ f_{34} \end{bmatrix} = \Delta \boldsymbol{h} = \begin{bmatrix} 1 & 1 & 0 & 0 & 0 & 0 \\ 0 & 0 & 1 & 0 & 0 & 0 \\ 1 & 0 & 0 & 1 & 0 & 0 \\ 0 & 1 & 0 & 0 & 1 & 0 \\ 1 & 0 & 1 & 1 & 0 & 1 \end{bmatrix} \begin{bmatrix} h_1 \\ h_2 \\ h_3 \\ h_4 \\ h_5 \\ h_6 \end{bmatrix} = \begin{bmatrix} h_1 + h_2 \\ h_3 \\ h_1 + h_4 \\ h_2 + h_5 \\ h_1 + h_3 + h_4 + h_6 \end{bmatrix}$$

Fig. 2.12 Transportation network with link and path flows

transportation link cost, is a variable *synthesizing* (the average value of) the different performance variables borne and perceived by the users in travel-related choice and, more particularly, in path choices (see Sect. 4.3.3). Thus, the transportation link cost reflects the average users' disutility for carrying out the activity represented by the link. Other performance variables and costs, which cannot be associated with individual links but rather to the whole trip (path), are introduced shortly.

Performance variables making up the transportation cost are usually nonhomogeneous quantities. In order to reduce the cost to a single scalar quantity, the different components can be homogenized into a generalized cost applying reciprocal substitution coefficients β, whose value can be estimated by calibrating the path choice model (see Sect. 4.3.3). For example, the generalized transportation cost c_a relative to the link a can be formulated as

$$c_a = \beta_1 \cdot t_a + \beta_2 \cdot mc_a$$

where t_a is the travel time and mc_a is the monetary cost (e.g., the toll) connected with the crossing of the link. More generally, the link transportation cost can be expressed as a function of several link performance variables as

$$c_a = \sum_n \beta_n \cdot r_{na}$$

Different users may experience and/or perceive transportation costs, which differ for the same link. For example, the travel time of a certain road section generally differs for each vehicle that covers it, even under similar external conditions. Furthermore, two users experiencing the same travel time may have different perceptions of its disutility. If we then add the fact that the analyst cannot have perfect knowledge of such costs, we realize that the perceived link cost is well represented by a random variable distributed among users, whose average value is link transportation cost c_a. There may be other "costs" both for users (e.g., accident risks or tire consumption) and for society (e.g., noise and air pollution) associated with a link. It is usually assumed that these costs are not taken into account by users in their travel-related choices and are not included in the perceived transportation cost. The transportation cost is, therefore, an internal cost, used to simulate the transportation system and, in particular, travelers' choices. The other cost items are external costs, used for project design and assessment. External costs are sometimes referred to as impacts; they are dealt with in Sect. 2.3.5.

Different groups (or classes) of users may have different average transportation costs. This may be due to different performance variables (e.g., their speeds and travel times are different or they pay different fares) or to differences in the homogenization coefficients β_n (e.g., different time/money substitution rates corresponding to different incomes). In this case a link cost c_l^i can be associated with each user class i. In what follows, for simplicity of notation, the class index i is taken as understood unless otherwise stated. Other considerations relative to users belonging to different classes are made in Chap. 6.

Link performance variables and transportation costs can be arranged in vectors. The *performance vector* $\boldsymbol{r}_a$ is made up by the nth performance variable for each link, its components being r_{na}. Analogously, the vector $\boldsymbol{c}$, whose generic component c_a is the generalized transport cost on link a, is known as the *link cost vector*.

The concepts of performance variables and generalized transportation cost can be extended from links to paths. The *average performance variable* of a path k, z_{nk}, is the average value of that variable associated to a whole origin–destination trip, represented by a path in the graph. Some path performance variables are *linkwise additive*; that is, their path value can be obtained as the sum of link values for all links making up the path.

Examples of additive path variables are travel times (the total travel time of a path is the sum of travel times over individual links) or some monetary costs, which can be associated with some or all individual links. An *additive path performance variable* can be expressed as the sum of link performance variables as

$$z_{nk}^{\text{ADD}} = \sum_{a \in k} r_{na} = \sum_{a} \delta_{ak} r_{na}$$

or in vector notation

$$\boldsymbol{z}_n^{\text{ADD}} = \boldsymbol{\Delta}^T \boldsymbol{r}_n$$

Other path performance variables are *nonadditive*; that is, they cannot be obtained as the sum of link specific values. These variables are denoted by z_{nk}^{NA}. Examples of nonadditive performance variables are monetary cost in the case of tolls that are nonlinearly proportional to the distance covered or the waiting time at stops for high-frequency transit systems, as shown below.

The *average generalized transportation cost* of a path k, g_k, is defined as a scalar quantity homogenizing in disutility units the different performance variables perceived by the users (of a given category) in making trip-related choices and, in particular, path choices.

The path cost in the most general case is made up of two parts: linkwise additive cost g_k^{ADD} and nonadditive cos, g_k^{NA}, assuming that they are homogeneous:

$$g_k = g_k^{\text{ADD}} + g_k^{\text{NA}} \tag{2.3.3}$$

The *additive path cost* is defined as the sum of the linkwise additive path performance variables:

$$g_k^{\text{ADD}} = \sum_{n} \beta_n \cdot z_{nk}^{\text{ADD}}$$

Under the assumption that the generalized cost depends linearly on performance variables, the additive path cost can be expressed as the sum of generalized link costs. The relationship between additive path cost and link costs can be expressed by combining all the equations previously presented:

$$g_k^{\text{ADD}} = \sum_{n} \beta_n z_{nk}^{\text{ADD}} = \sum_{n} \beta_n \sum_{a} \delta_{ak} r_{na} = \sum_{a} \delta_{lk} \sum_{n} \beta_n r_{na} = \sum_{a} \delta_{ak} c_a$$

or

$$g_k^{\text{ADD}} = \sum_{a} \delta_{ak} c_a \tag{2.3.4}$$

GRAPH

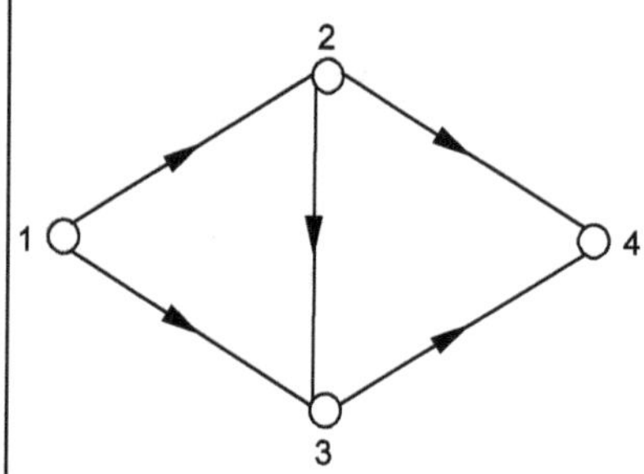

$G \equiv (N, L)$
$N \equiv \{(1, 2, 3, 4)\}$
$L \equiv \{(1,2), (1,3), (2,3), (2,4), (3,4)\}$

Origin centroid nodes 1, 2, 3
Destination centroid nodes 4

PATHS

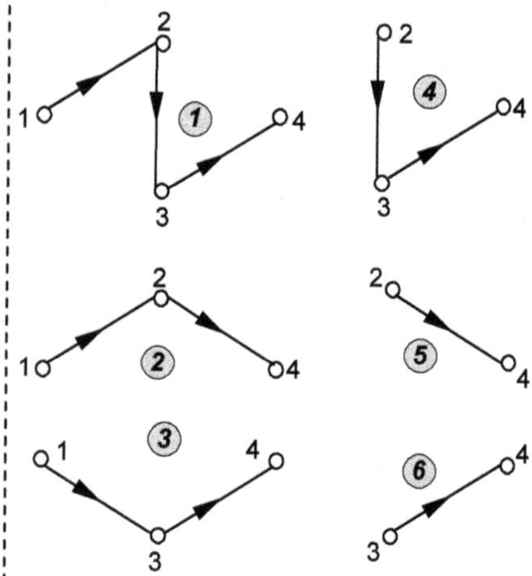

Link cost vector

$$\boldsymbol{c} = \begin{bmatrix} c_{12} \\ c_{13} \\ c_{23} \\ c_{24} \\ c_{34} \end{bmatrix}$$

Link-path incidence matrix

$\Delta =$

	1	2	3	4	5	6
1,2	1	1	0	0	0	0
1,3	0	0	1	0	0	0
2,3	1	0	0	1	0	0
2,4	0	1	0	0	1	0
3,4	1	0	1	1	0	1

Path cost vector

$$\boldsymbol{g}^{ADD} = \Delta^T \boldsymbol{c} = \begin{bmatrix} 1 & 0 & 1 & 0 & 1 \\ 1 & 0 & 0 & 1 & 0 \\ 0 & 1 & 0 & 0 & 1 \\ 0 & 0 & 1 & 0 & 1 \\ 0 & 0 & 0 & 1 & 0 \\ 0 & 0 & 0 & 0 & 1 \end{bmatrix} \begin{bmatrix} c_{12} \\ c_{13} \\ c_{23} \\ c_{24} \\ c_{34} \end{bmatrix} = \begin{bmatrix} c_{12} + c_{23} + c_{34} \\ c_{12} + c_{24} \\ c_{13} + c_{34} \\ c_{23} + c_{34} \\ c_{24} \\ c_{34} \end{bmatrix}$$

Fig. 2.13 Transportation network with link and path costs

The expression (2.3.4) can also be formulated in vector format by introducing the vector of additive path costs $\boldsymbol{g}^{\text{ADD}}$ (see Fig. 2.13):

$$\boldsymbol{g}^{\text{ADD}} = \boldsymbol{\Delta}^T \boldsymbol{c} \tag{2.3.5}$$

The nonadditive path cost g_k^{NA} includes nonadditive path performance variables:

$$g_k^{\mathrm{NA}} = \sum_n \beta_n z_{nk}^{\mathrm{NA}}$$

Finally, the path cost vector $\boldsymbol{g}$, of dimensions $(n_P \times 1)$, can be expressed as

$$\boldsymbol{g} = \boldsymbol{\Delta}^T \boldsymbol{c} + \boldsymbol{g}^{\mathrm{NA}} \tag{2.3.6}$$

where $\boldsymbol{g}^{\mathrm{NA}}$ is the nonadditive path cost vector.

In many applications, the nonadditive path cost vector is, or is assumed to be, null. This affects the efficiency of the calculation algorithm for assignment models, as shown in Chaps. 5 and 6.

2.3.4 Link Performance and Cost Functions

Link performance attributes generally depend on the physical and functional characteristics of the facility and/or the service involved in the trip phase represented by the link itself. Typical examples are the travel time on a road section depending on its length, alignment, allowed speed, or the waiting time at a bus stop depending on the headway between successive bus arrivals. When several travelers or vehicles use the same facility, they may interact with each other, thereby influencing link performance. This phenomenon is known as *congestion* and was introduced in Sect. 2.2.1. Typically, the effects of congestion on link performance increase as the flow increases. For instance, the larger the flow of vehicles traveling along a road section, the more likely faster vehicles will be slowed by slower ones, thus increasing the average travel time. Moreover, the larger the flow arriving at an intersection, the longer is the average waiting time; the larger the number of users on the same train, the lower is the riding comfort.

In general, congestion effects are such that the performance attributes of a given link may be influenced by the flow on the link itself and by flows on other links.

Link performance functions relate the generic link performance attribute r_{na} to physical and functional characteristics of the link, arranged in a vector $\boldsymbol{b}_{na}$, and to the equivalent flow on the same link and, possibly, on other links, arranged in the vector $\boldsymbol{f}$:

$$r_{na} = r_{na}(\boldsymbol{f}; \boldsymbol{b}_{na}, \boldsymbol{\gamma}_{na})$$

where $\boldsymbol{\gamma}_{na}$ is a vector of parameters used in the function.

Because the generalized transportation cost of a link c_a is a linear combination of link performance attributes, *link cost functions*[5] can be expressed as functions of

[5]A distinction should be made between cost functions in microeconomics and in transportation systems theory. In the first case, the cost function is a relationship connecting the production cost of a good or service to the quantity produced and the costs of individual production factors. Cost

the same parameters:

$$c_a = c_a(\boldsymbol{f}; \boldsymbol{b}_a, \boldsymbol{\gamma}_a) \tag{2.3.7}$$

where vectors $\boldsymbol{b}_a$ and $\boldsymbol{\gamma}_a$ have the same meaning as above.

Link performance and cost functions may have some mathematical properties, which are used in Chaps. 5 and 6 to study the properties of supply–demand interaction models and to analyze the convergence of their solution algorithms.

Performance and cost functions can be classified as *separable* and *nonseparable* across a link. In the former case, the performances and cost variables of a link depend exclusively on the (equivalent) flow on the link itself:

$$c_a(\boldsymbol{f}) = c_a(f_a)$$

In the latter case, they also depend on the flow on other links. Examples of both types of function are given in the following sections.

The *cost function vector* $\mathbf{c}(\boldsymbol{f})$ is obtained by ordering the n_L functions of the individual network links:

$$\mathbf{c} = \mathbf{c}(\boldsymbol{f}) \tag{2.3.8}$$

Under the assumption that the first partial derivative of $\boldsymbol{c}(\boldsymbol{f})$ exists and is finite, the Jacobian matrix, $\mathbf{Jac}[\boldsymbol{c}(\boldsymbol{f})]$, may be defined:

$$\mathbf{Jac}\big[\boldsymbol{c}(\boldsymbol{f})\big] = \begin{vmatrix} \frac{\partial c_1}{\partial f_1} & \cdots & \frac{\partial c_1}{\partial f_{n_L}} \\ & \frac{\partial c_i}{\partial f_i} & \cdots \\ \frac{\partial c_{n_L}}{\partial f_1} & \cdots & \frac{\partial c_{n_L}}{\partial f_{n_L}} \end{vmatrix}$$

The cost functions generally have an asymmetric Jacobian. In some cases, they may have a symmetric Jacobian: $\partial c_i/\partial f_j = \partial c_j/\partial f_i,\ \forall i, j$; that is, the cost variation on link a, due to a flow variation on link j, is equal to the cost variation on link j, due to a flow variation on link i. Separable cost functions are clearly a special case, the Jacobian being a diagonal matrix: $\partial c_i/\partial f_j = 0,\ \forall i \neq j$.

In the case of uncongested networks the cost functions are independent of the flows, so the partial derivatives are all equal to zero and the Jacobian is null.

2.3.5 Impacts and Impact Functions

Design and evaluation of transportation systems, in addition to performance variables perceived by the users, require the modeling of impacts borne by the users, but not perceived in their mobility choices, and of impacts on nonusers. Examples

---

functions in transportation systems provide the cost perceived by users in their trips. Transportation cost is therefore a cost of use rather than of production. The cost of producing transportation services is usually indicated as the service production cost, and similarly the functions correlating it to the relevant quantities are called production cost functions.

of the first type include indirect vehicle costs (e.g., tire or lubricant, vehicle depreciation, etc.) and accident risks with their consequences (death, injury, material damage). The impacts for nonusers include those for other subjects directly involved in the transportation system, such as costs and revenues for the producers of transportation services, and impacts "external" to the transportation system (or market). Examples of externalities are the impacts on the real estate market, urban structure, or on the environment such as noise and air pollution. The mathematical functions relating these impacts to physical and functional parameters of the specific transportation systems and, in some cases, to link flows are called *impact functions*. Often these functions are named with respect to the specific impact they simulate (e.g., fuel consumption functions or pollutant emission functions). Some impacts can be associated with individual network links and depend on the flows, $e_l(f)$. Link-based impact functions are usually included in transportation supply models; see Fig. 2.1. Some impact functions may be quite elementary whereas others may require complex systems of mathematical models. Examples of link-based impact functions are those related to air and noise pollution due to vehicular traffic. Some impact functions are discussed in Chap. 10 in the context of evaluation of transportation system projects.

2.3.6 General Formulation

To summarize the above points, a *transportation network* consists of the set of nodes N, the set of links L, the vector of link costs $\boldsymbol{c}$, which depend on the vector $\boldsymbol{r}$ of link performances, the vector $\boldsymbol{g}^{\mathrm{NA}}$ of nonadditive path costs and the vector $\boldsymbol{e}$ of relevant impact variables: $(N, L, \boldsymbol{c}, \boldsymbol{g}^{\mathrm{NA}}, \boldsymbol{e})$. For congested networks, the link cost vector is substituted by the flow-dependent cost functions $\boldsymbol{c}(\boldsymbol{f})$; the same holds for flow-dependent internal and external impacts $\boldsymbol{e}(\boldsymbol{f})$, whereas the nonadditive costs vector $\boldsymbol{g}^{\mathrm{NA}}$ is usually assumed to be independent of the flows. In this case the abstract transportation network model can be expressed as $(N, L, \boldsymbol{c}(\boldsymbol{f}), \boldsymbol{g}^{\mathrm{NA}}, \boldsymbol{e}(\boldsymbol{f}))$. Performance variables and functions are not explicitly mentioned, as they are included in the generalized transportation cost functions.

The set of relationships connecting path costs to path flows is known as the *supply model*. The supply model can therefore be formally expressed combining (2.3.2), (2.3.6), and (2.3.8) into a relationship connecting path flows to path costs:

$$\boldsymbol{g}(\boldsymbol{h}) = \boldsymbol{\Delta}^T \boldsymbol{c}(\boldsymbol{\Delta h}) + \boldsymbol{g}^{\mathrm{NA}} \tag{2.3.9}$$

where it is assumed that nonadditive path costs, if any, are not affected by congestion. Link characteristics can be obtained through performance, cost and impact functions for the link flows corresponding to the path flow vector. Clearly the model (2.3.9) expresses the abstract congested network model described in the previous sections. The same type of models can be used to describe other systems such as electrical or hydraulic networks.

The general structure of a supply model is depicted in Fig. 2.14. The graph defines the topology of the connections allowed by the transportation system under

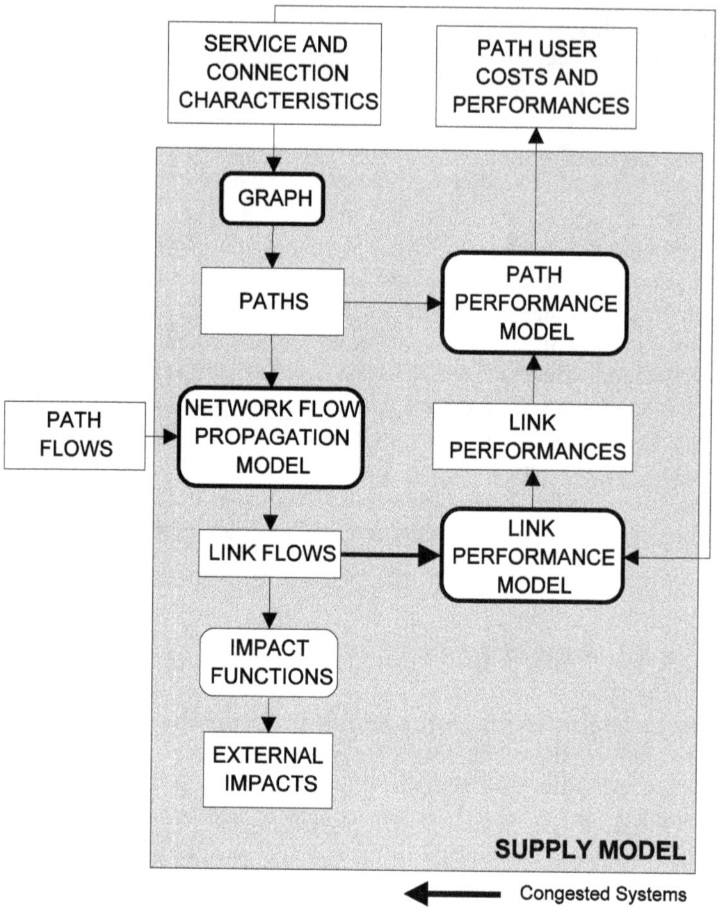

Fig. 2.14 Schematic representation of supply models

study, and the flow propagation model defines the relationship among path and link flows. The link performance model expresses for each element (link) the relationships among performances, physical and functional characteristics, and flow of users. The impact model simulates the main external impacts of the supply system. Finally, the path performance model defines the relationship between the performances of single elements (links) and those of a whole trip (path) between any origin–destination pair.

2.4 Applications of Transportation Supply Models

Network models and related algorithms are powerful tools for modeling transportation systems. A network model is a simplified mathematical description of the phys-

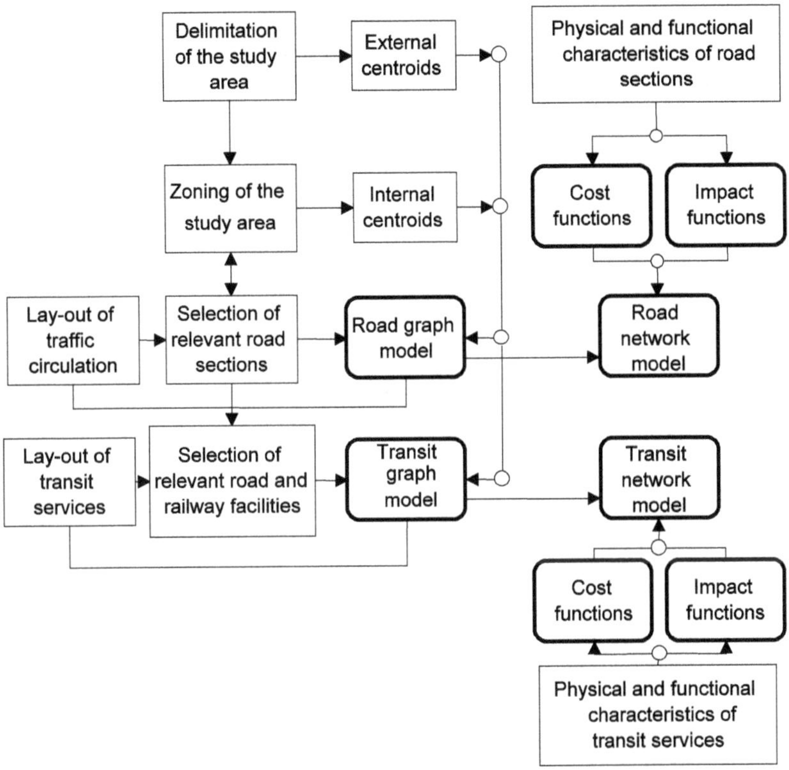

Fig. 2.15 Functional phases for the construction of an urban bimodal network model

ical phenomena relevant to the analysis, design, and evaluation of a given system. Thus transportation network models depend on the purpose for which they are used.

Building a network model usually requires a sequence of operations whose general criteria are described in the following. A schematic representation of the main activities in the case of a bimodal supply system (road and transit urban systems) is depicted in Fig. 2.15.

In the most general case, a supply network model is built through the following phases.

(a) Delimitation of the study area
(b) Zoning
(c) Selection of relevant supply elements (basic network)
(d) Graph construction
(e) Identification of performance and cost functions
(f) Identification of impact functions

Phases (a), (b), and (c) relate to the relevant supply system definition. They are described, respectively, in Sect. 1.3.1 of Chap. 1 and are not repeated here. The rest of this section introduces some general considerations related to phases (d), (e), and

(f) for a generic system. Specific models are described separately for two different types of transportation systems: continuous services (such as road), in Sect. 2.4.1, and scheduled services (such as train or buses), in Sect. 2.4.2.

The construction of a transportation graph requires the definition of the relevant trip phases and events (links and nodes) that depend on the physical system to be represented. Important nodes in transportation graphs are the so-called *centroid nodes*. They correspond to the events of beginning and ending a trip in a given zone. As was seen in Sect. 1.3.1, the centroids can approximate the internal points within a traffic zone. In general, the zone centroid is a *fictitious node*, that is, a node which does not correspond to any specific location but which represents the set of points of the zone where a trip can start or end. Therefore, a zone centroid is placed "barycentrically" with respect to such points or to some proxy variables (e.g., the number of households or workplaces). In principle, different centroid nodes may be associated to different trip types (e.g., origin and destination centroids). In other cases, centroids represent the places of entry into or exit from the study area for the trips, which are partly undertaken within the system (*cordon centroids*). In this case they are usually associated with physical locations (road sections, airports, railway stations, etc.).

A graph usually includes *links* of different types: *real links* and *connectors*. Real links represent trip phases corresponding to "physical" components (infrastructures or services), such as traversing a road section or riding a train between two successive stations. When centroid nodes do not correspond to a physical element, *connector links* are introduced into the graph. These links represent the trip phase between the terminal point (zone centroid) and a physical element of the network. In the remainder of this section, links are referred to according to the trip phase (activity) or the infrastructure or service which allows that activity. For example, there are road links, transit line links, and waiting links at stops.

A transportation graph will have different levels of complexity, depending on the system being represented and the details required to do so. In general, short-term or operational projects, such as a road circulation plan or the design of transit lines, require a very detailed representation of the real system. By contrast, strategic or long-term projects usually require less detailed, larger-scale graphs both because of the geographical size of the area and the number of elements included in the system.

As shown shortly, different graphs can be associated with the same basic network, depending on the aim of the model. Graphs can also represent transportation infrastructures; in general, infrastructure graphs are not used directly for system models, but rather they are referred to during the construction of service graphs. User flows and supply performances depend on the transportation services using the infrastructures rather than on the infrastructures themselves.

Specification of link performance and cost functions for a transportation network requires the study of the functioning of the individual elements that comprise it. In practice, performance functions used at times derive from explicit assumptions on system behavior, following a "deductive" approach, as for queuing models for barrier systems such as motorway toll booths, road intersections, air and sea terminals, and the like (see Sect. 2.2.2). When this approach, albeit based on simplifying assumptions, proves particularly complex, we use "descriptive" models developed according to an "inductive" approach, as in most stationary traffic flow models (see Sect. 2.2.1). Such models are made up of statistical relationships between performance attributes and the explicative variables of the phenomenon. Examples of both types of performance functions are given in the next two sections.

Both approaches use unknown parameters, vectors $\boldsymbol{\gamma}_n$ and $\boldsymbol{\gamma}$, respectively, in expressions (2.4.11) and (2.4.12), which should be calibrated for each specific supply model. To estimate behavioral model parameters or to specify the functional form and estimate nonbehavioral model parameters, the usual methods of inferential statistics may be used. However, in many applications the cost functions calibrated in similar contexts are transferred to the system in question to save application time and costs.

2.4.1 Supply Models for Continuous Service Transportation Systems

Continuous and simultaneous services are available at every instant and can be accessed from a very large number of points. Typical examples are individual modes such as cars and pedestrians using road systems.

2.4.1.1 Graph Models

In graphs representing road systems, nodes are usually located at the intersections between road segments included in the supply model. Nodes can also be located where significant variations occur in the geometric and/or functional characteristics of a single segment (such as changes in a road cross-section and lateral friction). Intersections with secondary roads not included in the "base network," however, are not represented by nodes. Links usually correspond to connections between nodes allowed by the circulation scheme. Therefore, a two-way road is represented by two links going in opposite directions, whereas a one-way road has a single link going in the allowed direction. Figure 2.16 shows the graph representing part of the urban road network shown in Fig. 1.3.

In applications two distinct types of links are considered: *running links*, which represent the vehicle's real movement as the trip along a motorway or urban road section; and *waiting* or *queuing links*, representing queuing at intersections, toll barriers, and so on (see Fig. 2.17).

The level of detail of the road system depends on the purpose of the model. This is especially true for road intersections. In a coarse representation, a road intersection is usually represented by a single node where the access links converge. Alternatively, we can adopt a more detailed representation that distinguishes different turning movements and excludes nonpermitted turns (if any). Such a representation can be obtained by using a larger number of nodes and links. Figure 2.18 shows the two possible representations of a four-arm road intersection. Note that in the single-node representation, paths requiring a left turn (4-5-2) cannot be excluded if

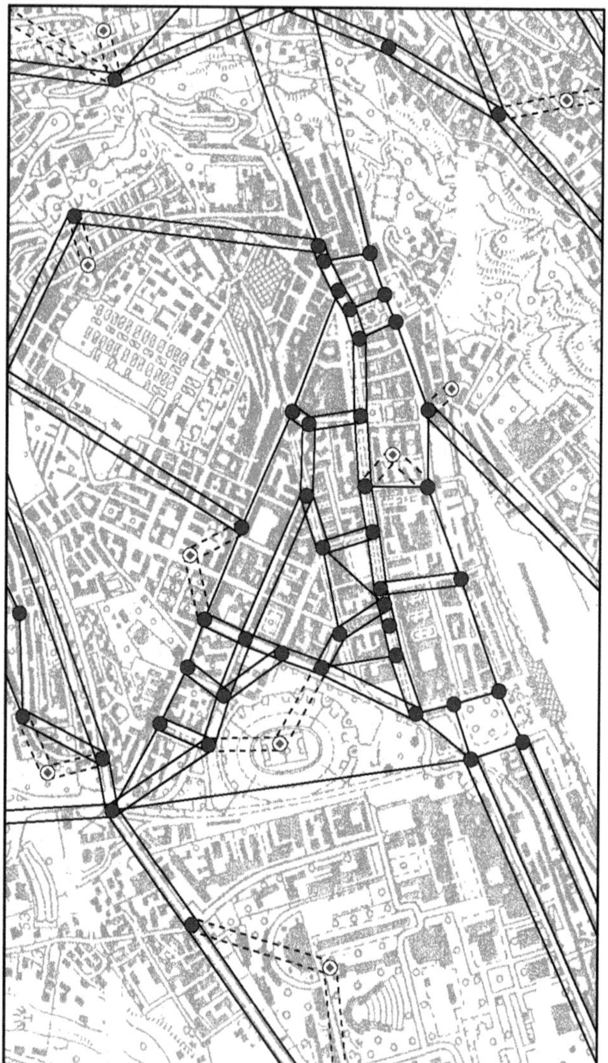

Fig. 2.16 Example of a graph representing part of an urban road system

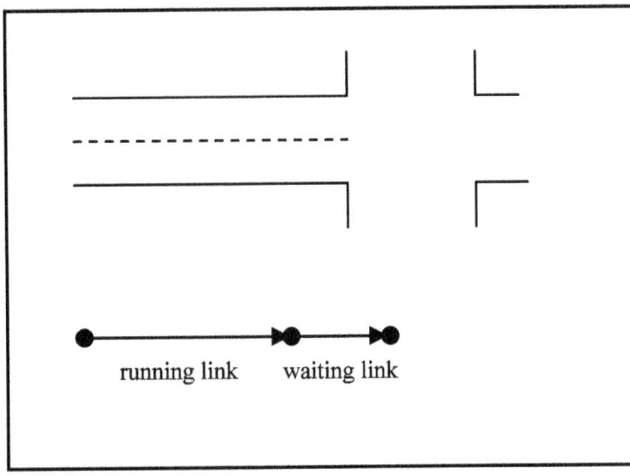

Fig. 2.17 Representation of a road intersection with running and waiting links

this turning movement is not allowed; furthermore, different waiting times cannot be assigned to maneuvers with different green phase durations, such as right turns (4-5-3). Both of these possibilities are allowed by the detailed representation.

Parking is another element of a road system that can be represented with different levels of detail. In detailed road graphs, trip phases corresponding to parking can be represented with different links for different parking facilities available in a given zone (see Fig. 2.19). *Parking links* can be connected through pedestrian links to the centroid of the zone where they are located, and to the centroids of traffic zones within walking distance. In less detailed graphs, parking is included in connector links; in this case, however, congestion and different parking policies cannot be simulated.

2.4.1.2 Link Performance and Cost Functions

The *generalized transportation cost* of a road link is usually made up by several performance attributes. For example, three attributes can be selected: travel time along the section, waiting time (e.g., at the final intersection, at the tollbooth, etc.), and monetary cost. In this case, the cost function can be obtained as the sum of three performance functions:

$$c_a(\boldsymbol{f}) = \beta_1 tr_a(\boldsymbol{f}) + \beta_2 tw_a(\boldsymbol{f}) + \beta_3 mc_a(\boldsymbol{f}) \tag{2.4.1}$$

where

$tr_a(\boldsymbol{f})$ is the function relating the running time on link a to the flow vector
$tw_a(\boldsymbol{f})$ is the function relating the waiting time on link a to the flow vector
$mc_a(\boldsymbol{f})$ is the function relating the monetary cost on link a to the flow vector

The dependence on physical and functional variables $\boldsymbol{b}_a$, and parameters $\boldsymbol{\gamma}$, has been omitted for simplicity's sake. Note that in (2.4.1) it has been assumed that homogenization coefficients may differ for the different time components. Furthermore, not all of the components in (2.4.1) are present for each link; for example,

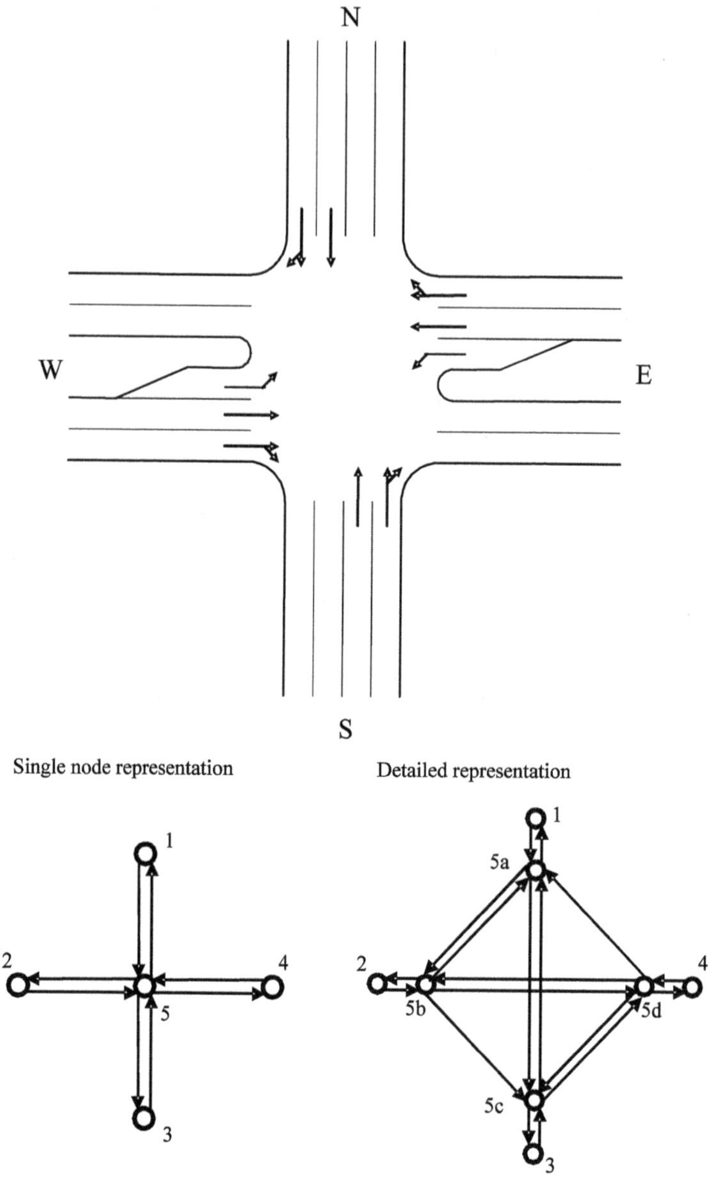

Fig. 2.18 Graphs for a road intersection

if the link represents only the waiting time for a maneuver, tr_a and mc_a are zero, and the same consideration is true for monetary costs and waiting times on most pedestrian links. If an individual link represents both the trip along a road section and queuing at the intersection, its cost function will include both travel time tr_a and queuing time tw_a.

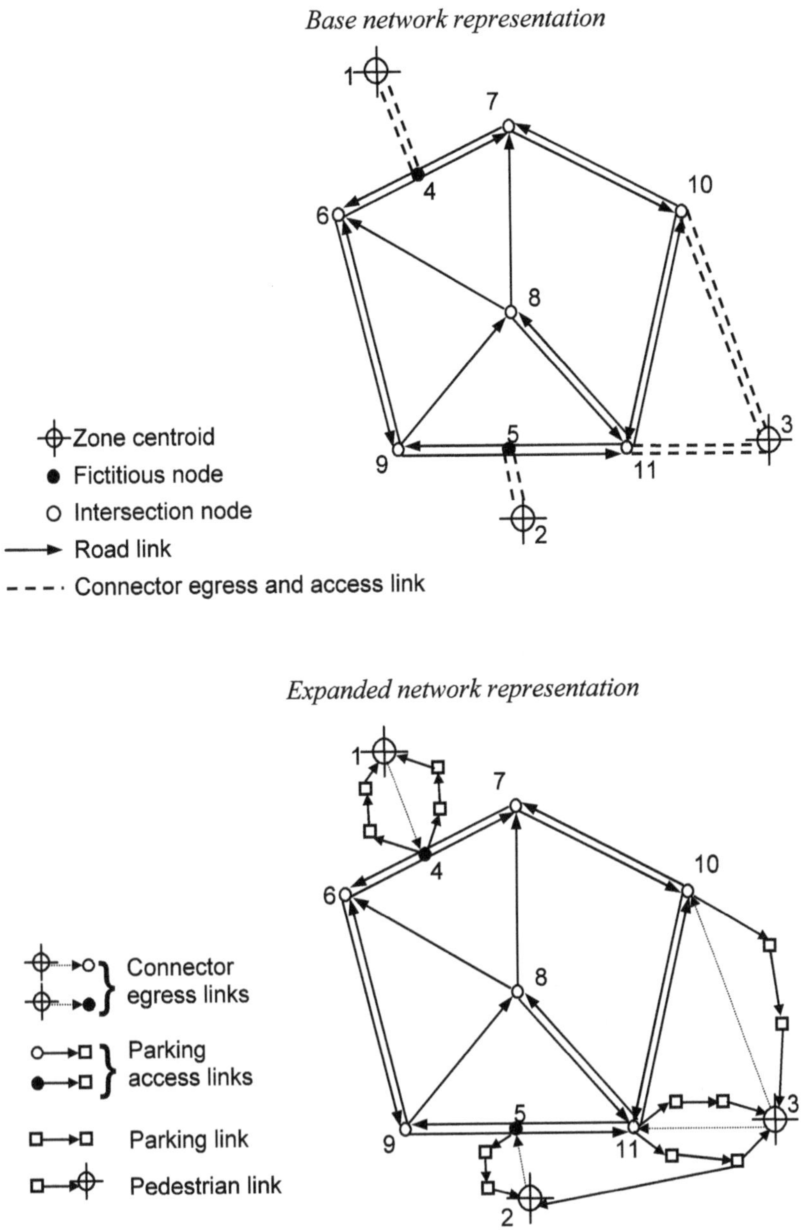

Fig. 2.19 Explicit representation of parking supply

In the most general case, the monetary cost term mc_a includes the cost items that are perceived by the user. Because users do not usually perceive other consumption (motor oil, tires, etc.), in applications monetary costs are usually identified as the toll (if any) and fuel consumption:

$$mc_a = mc_{\text{toll}} + mc_{\text{fuel}}(f).$$

The latter depends on the specific consumption (liters/km), which can vary in relation to the average speed and hence to the congestion level. In practice, these variations are sometimes ignored and the monetary cost is calculated as a function of the toll and the average unit consumption.

Performance functions for travel time and queuing time attributes are derived by following both a behavioral (deductive) and experimental (inductive) approach. For the waiting links, for example, the results of queuing theory are generally used (see Sect. 2.2.2). However, their mere implementation has not always permitted proper coverage of all situations in practice, which is why such relations often include approximated adjustment terms obtained from empirical observations.

Listing all the performance functions that can be adopted for the elements of different continuous service systems is beyond the scope of this book. In the following, we therefore present some examples of performance functions both for travel links and waiting links, following the two approaches mentioned. It should also be stressed that, consistently with the assumption of intraperiod stationarity, stationary traffic flow variables and results are used.

Running Links Starting from the (stable regime) speed–flow relationship, the (stable regime) travel time of a running link a can be calculated as a function of the flow:

$$tr_a = L_a / v_a(f_a) \tag{2.4.2}$$

where

tr_a	is the running time on link a
f_a	is the flow on link a
L_a	is the length of the running link a
v_a	is the mean speed on link a assuming a stable regime

Below we introduce the relationships between travel time tr_a and flow f_a for uninterrupted flow conditions, for various types of road infrastructures: motorways and urban and extraurban roads.

(a) Motorway Links On motorway links flow conditions are typically uninterrupted and it is assumed that the waiting time component is negligible because it occurs on those sections (ramps, tollbooths, etc.) that are usually represented by different links.

Link travel time is usually obtained through empirical statistical relationships. One of the most popular expressions, referred to as the BPR cost function, has the following specification.

$$tr_a(f_a) = \frac{L_a}{v_{oa}} + \left(\frac{L_a}{v_{ca}} - \frac{L_a}{v_{oa}}\right)\left(\frac{f_a}{Q_a}\right)^4 \tag{2.4.3}$$

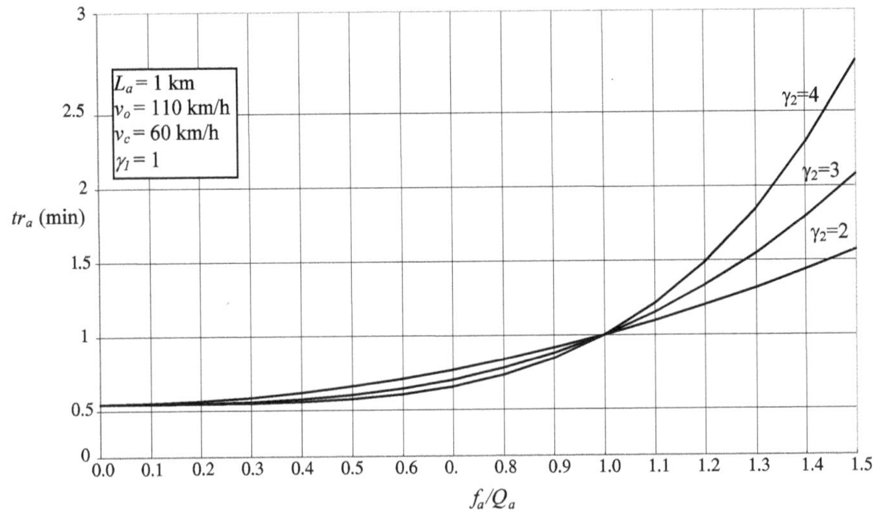

Fig. 2.20 Motorway travel time function (2.4.3) for different values of some parameters

where

L_a is the length of link a
v_{0a} is the free-flow average speed
v_{ca} is the average speed with flow equal to capacity
Q_a is link capacity, that is, the average maximum number of equivalent vehicles that can travel along the road section in a time unit. Capacity is usually obtained as the product of the number of lanes on the link a, N_a, and lane capacity, Q_{ua}

From (2.4.3) it can be noted that, in the case of motorways, cost functions are separable. The influence of flows on the performances of other links (e.g., the opposite direction or entrance/exit ramps) is significantly reduced by the characteristics of the infrastructure (divided carriageways, grade-separated intersections, etc.).

The values of v_{oa}, v_{ca}, and Q_a depend on the geometric and functional characteristics of the section (width of lanes, shoulders, and median strips; bend radiuses; longitudinal slopes; etc.). Typical values can be found in different sources; the Highway Capacity Manual (HCM) is the most complete and systematic (see Reference Notes). Parameters γ_1 and γ_2 are typically estimated on empirical data.

Figure 2.20 shows a diagram of (2.4.3) for different parameter values. Note that this function associates a travel time with the link also when flows are above link capacity (oversaturation), even though such flows are not possible in reality. However, in applications oversaturation is often allowed for reasons connected with mathematical properties and solution algorithms of static equilibrium assignment models (see Chap. 5). From a computational point of view, the oversaturation assumption should not influence the results significantly if the value of parameter γ_2, that is, the delay penalty due to capacity overloading, is large enough.

Values of γ_2 are typically much larger than one; that is, the function is more-than-linear in flow/capacity ratios. This phenomenon is rather frequent in congested systems. It should also be noted that, if the flow is close to capacity, resulting instability challenges the within-day stationarity assumptions and the cost functions adopted. In this sense, delay functions should be considered as "penalty" functions preventing major oversaturation, rather than estimates of actual travel times.

(b) Extraurban Road Links Users traveling on an extraurban road behave differently according to the number of lanes available for each direction: single lane (two-lane arterial) or two or more lanes (four-lane arterial, six-lane arterial, etc.).

In the former case, the capacity and travel conditions in each direction are not influenced by the flow in the opposite direction. For this type of road, the same formula (2.4.3) described for motorway links can be used, although with different parameters. These can again be deduced from capacity manuals, such as the HCM, or from other specific empirical studies.

In the case of roads with one lane in each direction, link performances depend on the flow in both directions: because overtaking is not always possible, vehicles may reduce the average speed. In practice, it is often assumed that link capacity has a value common to both directions, and the travel time function is modified as follows.

$$tr_a(f_a, f_{a^*}) = \frac{L_a}{v_{0a}} + \gamma_a \left(\frac{L_a}{v_{ca}} - \frac{L_a}{v_{0a}} \right) \left(\frac{f_a + f_{a^*}}{Q_{aa^*}} \right)^{\gamma_2} \tag{2.4.4}$$

where, apart from the symbols introduced previously, the link in the opposite direction is denoted by a^* and overall capacity in both directions by Q_{aa^*}.

(c) Urban Road Links In an urban context, given the relatively short lengths of road sections, travel speed is more dependent upon road physical and functional characteristics than upon the flow traveling on them. The higher the dependence is on factors such as section bendiness or roadside parking, the lower the impact of flow.

As an example, we report the empirical relation for estimating travel speed calibrated on survey sample data from the Napoli (Italy) urban area, integrated with microscopic simulation data (see the bibliographical note):

$$\begin{aligned} v_a = {} & 29.9 + 3.6Lu_a - 0.6P_a - 13.9T_a - 10.8D_a - 6.4S_a + 4.7Pv_a \\ & - 1.0\text{E}{-}04 \frac{(f_a/Lu_a)^2}{1 + T_a + D_a + S_a} \end{aligned} \tag{2.4.5}$$

where

- Lu_a is the useful width in meters of link a
- P_a is the nonnegative slope in % of link a
- T_a is the tortuosity of link a, in values in the interval [0, 1]
- D_a is an index of disturbance to traffic from external factors (entry from sideroads, irregular parking, pedestrian crossings, etc.) in values in the interval [0, 1]

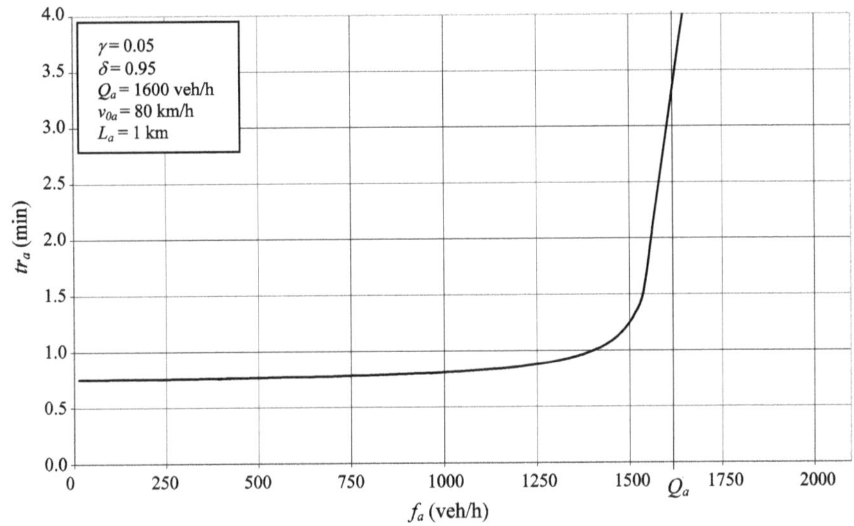

Fig. 2.21 Hyperbolic travel time cost function

S_a	is the percentage of length of a occupied by parking
Pv_a	is a dummy variable of 1 if the pavement of link a is asphalt, 0 otherwise
f_a	is the equivalent flow on link a in equiv. vehicles/hour

The travel time on link a may thus be calculated by multiplying the time obtainable from (2.4.5) by a corrective factor $c(L_a)$, which makes allowance for the effect of transient motions at the ends of the link (in the case of stopping at intersections):

$$tr_a = \frac{L_a}{v_a} \cdot c(L_a) = \frac{L_a}{v_a} \cdot \frac{1}{1 - \exp(-0.47 - 0.48\text{E}{-}02 \cdot L_a)} \tag{2.4.6}$$

where L_a is the road section length in km.

A further example of link travel time function is the hyperbolic expression given by Davidson, which also holds for interrupted flow (delays at intersections are thus included):

$$\begin{cases} tr_a = (L_a/v_{0a})(1 + \gamma f_a/(Q_a - f_a)) & \text{for } f_a \leq \delta Q_a \\ tr_a = \text{tangent approximation} & \text{for } f_a > \delta Q_a \end{cases} \tag{2.4.7}$$

with $\delta < 1$ and $Q_a =$ link capacity. Also see Fig. 2.21.

In this last case the tangent approximation is necessary because tr_a tends to ∞ for f_a going to Q_a. This condition is unrealistic because the oversaturated period has a finite duration.

Waiting Links

(a) Toll-Barrier Links In the case of links representing queuing systems, it is assumed that average waiting time is the only significant time performance variable. In

simple cases (e.g., a link corresponds to all toll lanes), the average undersaturation waiting time can be obtained by using a stochastic queuing model:

$$tw_a^u(f_a) = T_s + \left(T_s^2 + \sigma_s^2\right) \cdot \frac{f_a}{2} \cdot \frac{1}{1 - f_a/Q_a} \tag{2.4.8}$$

where

T_s is the average service time for each toll lane
σ_s^2 is the variance of the service time at the pay-point
$Q_a = N_a/T_s$ is the link (toll-barrier) capacity equal to the product of the number of lanes (N_a) by the capacity of each lane ($1/T_s$)

Expression (2.4.8) is derived from the assumption of a queuing system $M/G/1$ (∞, *FIFO*) with Poisson arrivals and general service time (see Sect. 2.2.2.3).

The values of T_s and σ_s^2 depend on various factors such as the tolling structure (fixed, variable) and the payment method (manual, automatic, etc.). Note that the average waiting time obtained through (2.4.8) is larger than the average service time T_s even though the arriving flow is lower than the system's capacity. This effect derives from the presence of random fluctuations in the headways between user arrivals and service times. Hence the delay expressed by (2.4.8) is known as "stochastic delay."

Moreover, the average delay computed with (2.4.8) tends to infinity as the flow f_a tends to capacity (i.e., if f_a/Q_a tends to one). This would be the case if the arrivals flow f_a remained equal to capacity for an infinite time, which does not occur in reality. In order to avoid unrealistic waiting times and for reasons of theoretical and computational convenience, two different methods can be adopted. The first, and less precise, method assumes that (2.4.8) holds for flow values up to a fraction α of the capacity, for example, $f_a \leq 0.95Q_a$. For higher values, the curve is extended following its *linear approximation*, that is, in a straight line passing through the point of coordinates $\alpha Q_a, tw(\alpha Q_a)$ with angular coefficient equal to the derivative of (2.4.8) computed at this point:

$$tw_a(f_a) = tw_a(\alpha Q_a) + K(f_a - \alpha Q_a) \tag{2.4.9}$$

with

$$K = \frac{T_s^2 + \sigma_s^2}{2} \cdot \frac{1}{(1-\alpha)^2}$$

Figure 2.22 shows the relationships (2.4.8) and (2.4.9) for some values of the parameters.

A more rigorous method is based on calculating oversaturation delay using a deterministic queuing model with an arrival rate equal to f_a, deterministic service times equal to T_s and an oversaturation period equal to the reference period duration T (see Sect. 2.2.2.2). The deterministic average (oversaturation) delay tw_a^d is then equal to:

$$tw_a^d = T_s + \left(\frac{f_a}{Q_a} - 1\right)\frac{T}{2} \tag{2.4.10}$$

which, for a *given* capacity, is a linear function of the arrivals flow f_a.

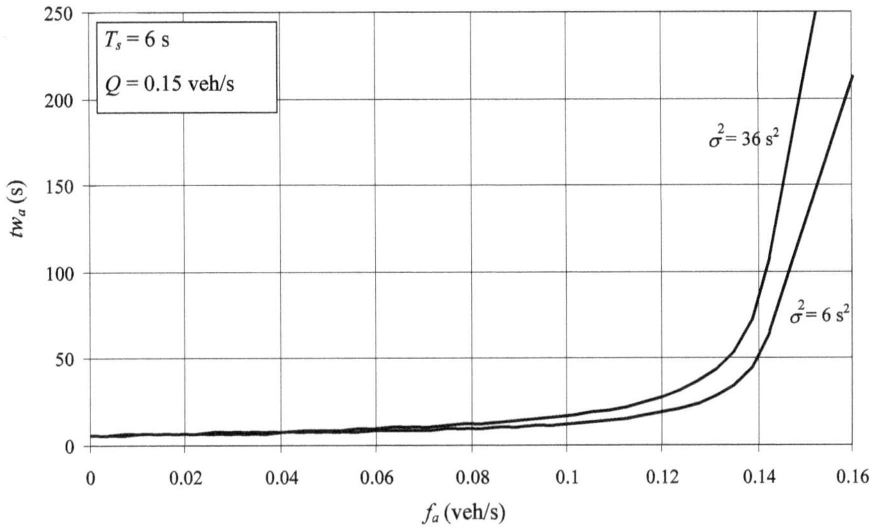

Fig. 2.22 Waiting time functions (2.4.8) and (2.4.9) at toll-barrier links

Note that in this case the assumption of intraperiod stationarity is challenged because even if the arrivals flow rate f_a and capacity $1/T_s$ are constant over the whole reference period T, the waiting time is different for users arriving in different instants of the reference period. In static models it is assumed that users perceive the average waiting time. Intraperiod dynamic models, discussed in Chap. 7, remove this assumption.

The average delay tw_a can be calculated by combining the stochastic undersaturation average delay tw_a^u expressed by (2.4.8) with the deterministic average oversaturation delay tw_a^d, expressed by (2.4.10). The combined delay function is such that the deterministic delay function is its oblique asymptote (see Fig. 2.23). The following equation results.

$$tw_a(f_a) = T_s + \left(T_s^2 + \sigma^2\right)\frac{f_a}{2} + \frac{T}{4}\left\{\frac{f_a}{Q_a} - 1 + \left[\left(\frac{f_a}{Q} - 1\right)^2 + \frac{4(f_a/Q_a)}{Q_a T}\right]^{1/2}\right\} \tag{2.4.11}$$

(b) Signal-Controlled Intersection Links Queuing and delay phenomena at signalized intersections can be obtained from the queuing theory results reported in Sect. 2.2.2. In fact, signalized intersections are a particular case of servers for which capacity is periodically equal to zero (when the signal is *red*). During such times the system is necessarily oversaturated.

The simplest case is that of a *signal-controlled intersection* not interacting with adjacent ones (*isolated intersection*), without lanes reserved for right or left turns.

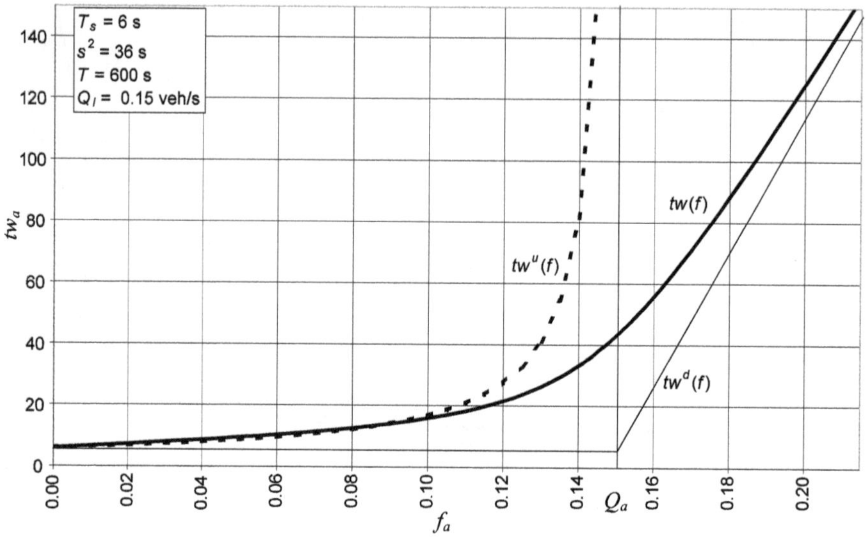

Fig. 2.23 Under- and oversaturation waiting time functions for toll barrier links

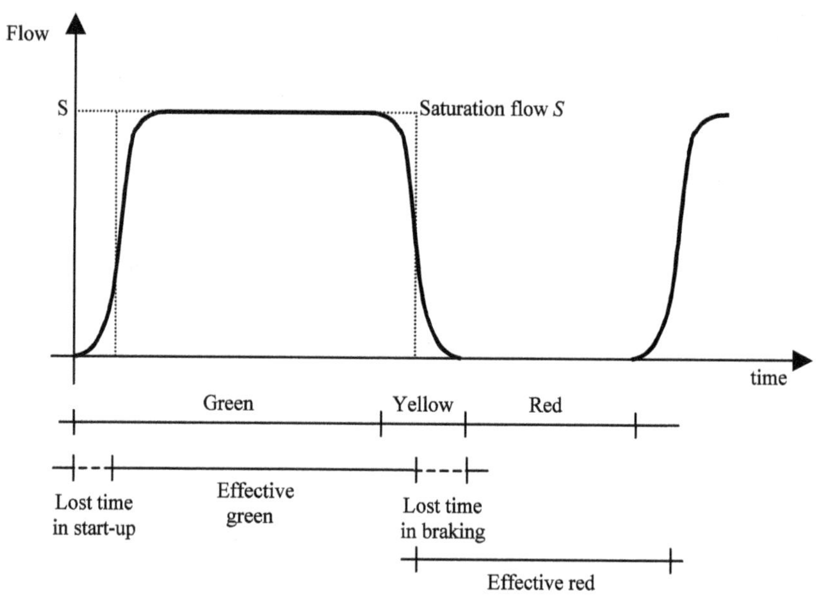

Fig. 2.24 Discharge flow from signal-controlled intersection in relation to cycle phases

Below we first introduce the assumptions and variables for each access as well as the most widely used calculation method. We then present the various models for calculating delays at intersections.

It is common to divide the cycle length into two time intervals (Fig. 2.24 illustrates the quantities associated with a traffic-light cycle). The effective green time equals the green plus yellow time minus the lost time, during which departures occur at a constant service rate, given by the inverse of saturation flow. The effective red time is the difference between cycle length and the effective green time, during which no departures occur.

Below, to simplify the notation, we omit the index of link a. Moreover, to facilitate application of the results in Sect. 2.2.2, the symbol $\bar{u}$ instead of f is used for the arrivals flow. Let:

T_c be the cycle length for the whole intersection
G be the effective green time for an approach
$R = T_c - G$ be the effective red time for the approach
$\mu = G/T_c$ be the effective green/cycle ratio for the approach

The number of vehicles arriving at the approach during time interval T_c is given by the following equation.

$$m_{\mathrm{IN}}(\tau, \tau + T_c) = \bar{u} \cdot T_c$$

The maximum number of users that may leave the approach, during time interval T_c, is given by:

$$S \cdot G = \mu \cdot S \cdot T_c$$

where S is the saturation flow of the intersection approach, that is, the maximum number of equivalent vehicles which in the time unit could cross the intersection if the traffic lights were always green ($\mu = 1$). Alternatively, the saturation flow may be defined as the maximum discharge rate that may be sustained by a queue during the green–amber time.

Hence the actual capacity of the approach is given by:

$$Q = \frac{S \cdot G}{T_c} = \mu \cdot S$$

Thus, the approach can be defined *undersaturated* if:

$$\bar{u} \cdot T_c < \mu \cdot S \cdot T_c$$

that is:

$$\bar{u} < \mu \cdot S \tag{2.4.12}$$

On the other hand the approach is defined *oversaturated* if:

$$\bar{u} \geq \mu \cdot S \tag{2.4.13}$$

The saturation flow rate of an intersection can in principle be obtained through specific traffic surveys; in practice, however, empirical models based on average results are often used. The Highway Capacity Manual (HCM) describes one of the

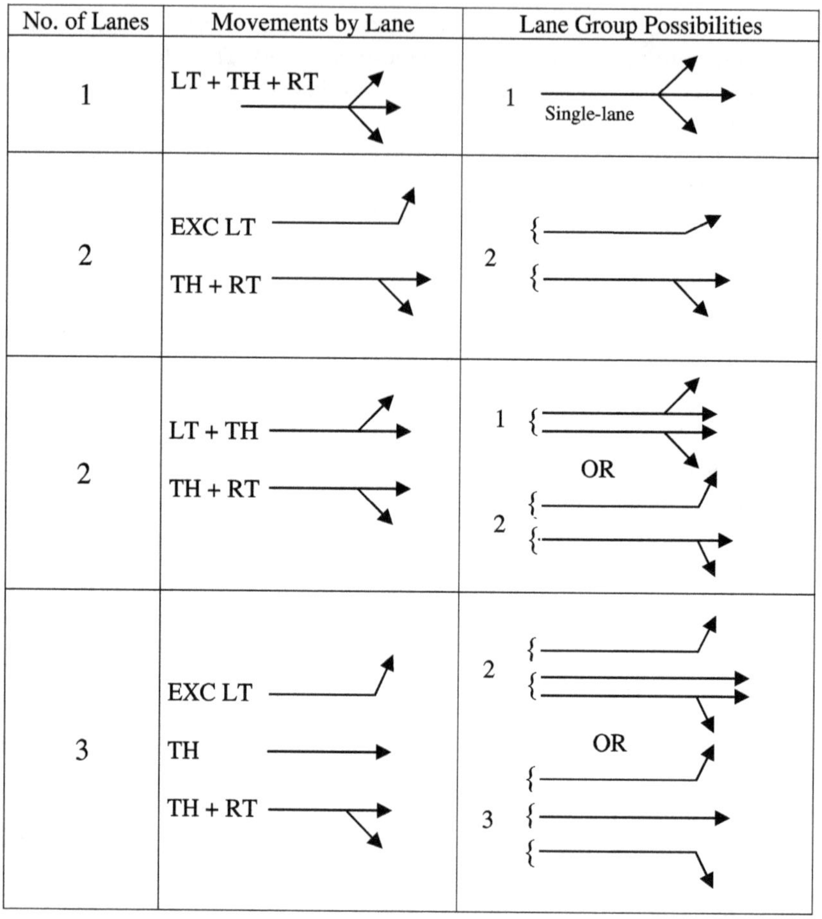

Fig. 2.25 Typical lane groups for the HCM method for calculating saturation flow

most popular methods. To apply this method, it is necessary to determine appropriate lane groups. A lane group is defined as one or more lanes of an intersection approach serving one or more traffic movements with which a single value of saturation flow, capacity, and delay can be associated. Both the geometry of the intersection and the distribution of traffic movements are taken into account to segment the intersection into lane groups. In general, the smallest number of lane groups that adequately describes the operation of the intersection is used. Figure 2.25 shows some common lane group schemes suggested by the HCM. The saturation flow rate of an intersection is computed from an "ideal" saturation flow rate, usually 1900 equivalent passenger cars per hour of green time per lane (pcphgpl), adjusted for a variety of prevailing conditions that are not ideal. The method can be summarized by the following expression,

$$S = S_0 \cdot N \cdot F_w \cdot F_{\mathrm{HV}} \cdot F_g \cdot F_p \cdot F_{\mathrm{bb}} \cdot F_a \cdot F_{\mathrm{RT}} \cdot F_{\mathrm{LT}}$$

where

S	is the saturation flow rate for the specific lane group, expressed as a total for all lanes in the lane group under prevailing conditions, in vphg
S_0	is the ideal saturation flow rate per lane, usually 1900 pcphgpl
N	is the number of lanes in the lane group
F_w	is the adjustment factor for lane width (12 ft or 3.66 m lanes are standard)
F_{HV}	is the adjustment factor for heavy vehicles in the traffic flow
F_g	is the adjustment factor for approach grade
F_p	is the adjustment factor for the existence of a parking lane adjacent to the lane group and the parking activity in that lane
F_{bb}	is the adjustment factor for the blocking effect of local buses that stop within the intersection area
F_a	is the adjustment factor for the area type
F_{RT}	is the adjustment factor for right turns in the lane group
F_{LT}	is the adjustment factor for left turns in the lane group

The first six adjustment factors not connected with the type of turning maneuvers are reported in Fig. 2.26.

Once the approach capacity $Q_l = \mu S$ is known, we may calculate the queue length and mean waiting time tw_a, using models derived from different approaches.

Application of Queuing Models From (2.4.12) and (2.4.13) it is clear that the results discussed in Sect. 2.2.2 hold for a queuing system representing a signalized intersection approach. In this context, the server's capacity Q coincides with the actual capacity of access: $Q = \mu \cdot S$. The latter is the weighted mean between the zero value of the "red" period and that equal to S for the "green" period, with $\mu = G/T_c$.

In the case in which access occurs in undersaturation conditions, the queue length may be calculated using (2.2.18) in which capacity assumes alternatively a value of zero, in intervals of length R (intervals of effective red), and a value of S, in intervals of length G (intervals of effective green) (see Fig. 2.27). As the system is undersaturated, at the end of each interval of effective green the queue is zero: $n_u(I \cdot T_c) = 0\ \forall i$, where i stands for the progressive number of cycles. Thus, for each interval of effective red we have $n(\tau_0) = 0$ with $\tau_0 = I \cdot T_c$ and, setting $Q = 0$ in (2.2.18), the queue length is equal to:

$$n_u^R(\tau) = \bar{u}(\tau - I \cdot T_c) \quad I \cdot T_c \le \tau \le I \cdot T_c + R \tag{2.4.14}$$

The queue length reaches a maximum value at the end of the red-time, equal to:

$$n_u^R(I \cdot T_c + R) = \bar{u}R = \bar{u}(1 - \mu)T_c$$

Thus, at the beginning of the interval of effective green we have $n(\tau_0) = \bar{u}(1 - \mu)T_c$ with $\tau_0 = I \cdot T_c + R$, and the queue length in a certain instant τ of the interval is given by (2.2.18) with $Q = S$:

$$n_u^G(\tau) = \max\{0, \bar{u}(1 - \mu)T_c - (S - \bar{u})(\tau - I \cdot T_c - R)\}$$

ADJUSTMENT FACTOR FOR AVERAGE LANE WIDTH F_w									
Average lane width, W (FT)	8	9	10	11	12	13	14	15	16
F_w	0.867	0.900	0.933	0.967	1.000	1.033	0.067	1.100	1.133

ADJUSTMENT FACTOR FOR HEAVY VEHICLES F_{HV}								
Percentage of heavy vehicles (%)	0	2	4	6	8	10	15	20
F_{HW}	1.000	0.980	0.962	0.943	0.926	0.909	0.870	0.833
Percentage of heavy vehicles (%)	25	30	35	40	45	50	75	100
F_{HW}	0.800	0.769	0.741	0.714	0.690	0.667	0.571	0.500

ADJUSTMENT FACTOR FOR APPROACH GRADE F_g									
Grade (%)	−6	−4	−2	0	+2	+4	+6	+8	≥ 10
F_g	1.030	1.020	1.010	1.000	0.990	0.980	0.970	0.960	0.950

ADJUSTMENT FACTOR FOR PARKING F_p						
F_p	No. of parking maneuvers per hour					
No. of lanes in lane group	No parking	0	10	20	30	≥ 40
1	1.000	0.900	0.850	0.800	0.750	0.700
2	1.000	0.950	0.925	0.900	0.875	0.850
3 or more	1.000	0.967	0.950	0.933	0.917	0.900

ADJUSTMENT FACTOR FOR BUS BLOCKAGE F_{bb}					
F_{bb}	No. of buses stopping per hour				
No. of lanes in lane group	0	10	20	30	≥ 40
1	1.000	0.960	0.920	0.880	0.840
2	1.000	0.980	0.960	0.940	0.920
3 or more	1.000	0.987	0.973	0.960	0.947

ADJUSTMENT FACTOR FOR AREA TYPE F_a	
Type of area	F_a
CBD (Center Business District)	0.900
All other areas	1.000

Fig. 2.26 Adjustment factors in the HCM method for saturation flow

$$I \cdot T_c + R \leq \tau \leq I \cdot T_c + R + G \tag{2.4.15}$$

The time period (within the green) in which the queue is exhausted is (see (2.4.15)):

$$\Delta\tau_0 = \frac{\bar{u}(1-\mu)T_c}{(S-\bar{u})}$$

The queue in undersaturation conditions therefore shows a periodic time trend, with zero values at the end of effective green time (i.e., at the beginning of the red interval) and maximum values at the end of the effective red interval (see Fig. 2.27).

However, if the system is in oversaturation conditions ($\bar{u} \geq \mu \cdot S$), the *total queue length* is obtained by summing the queue length in undersaturation to the queue length in oversaturation (see Fig. 2.28). The *queue length in undersaturation, $n_u(\tau)$*, is obtained once again by (2.4.14) and (2.4.15), for an arrivals rate equal to capacity

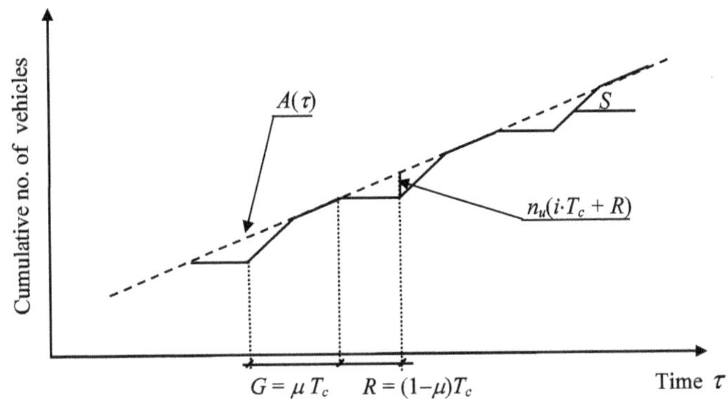

Fig. 2.27 Deterministic queuing model for signalized intersections, undersaturated conditions

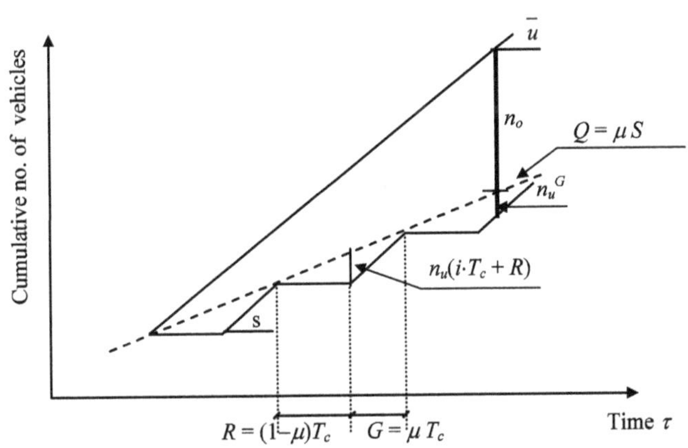

Fig. 2.28 Deterministic queuing model for signalized intersections, oversaturated conditions

$(\bar{u} = \mu \cdot S)$:

$$n_u^R(\tau) = \mu \cdot S(\tau - I \cdot T_c) \quad I \cdot T_c \leq \tau \leq I \cdot T_c + R \tag{2.4.16}$$

$$n_u^G(\tau) = \mu \cdot S(1-\mu)T_c - S(1-\mu)(\tau - I \cdot T_c - R)$$
$$I \cdot T_c + R \leq \tau \leq I \cdot T_c + R + G \tag{2.4.17}$$

The *oversaturated queue length* can be computed with the queue obtained from (2.2.18) with $Q = \mu \cdot S$, $\tau_0 = 0$ and $n(\tau_0) = 0$ (see Fig. 2.28):

$$n_0(\tau) = (\bar{u} - \mu \cdot S)\tau \tag{2.4.18}$$

The expressions of queue length allow us to determine the deterministic delay at intersections, as described below.

For undersaturated conditions $\bar{u} < \mu S$, the average individual delay tw_{US} can easily be obtained from the evolution over time of the queue length, as described

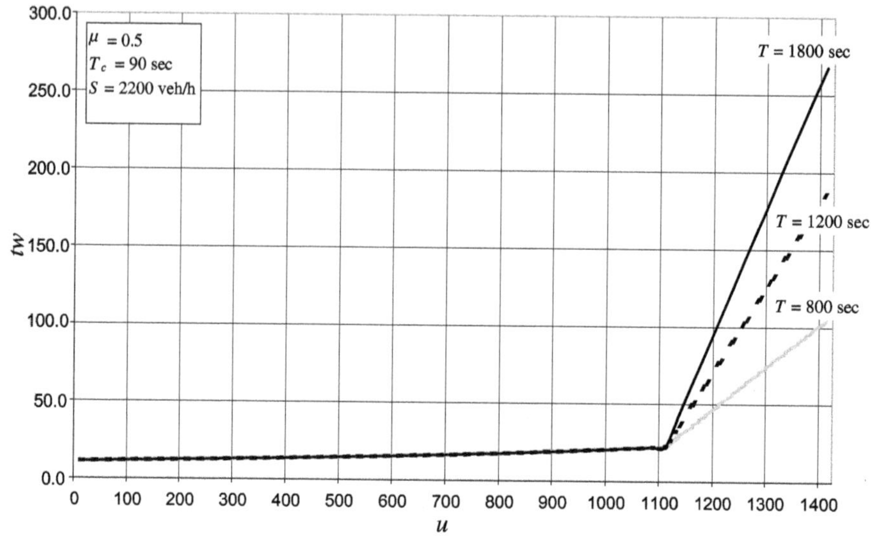

Fig. 2.29 Deterministic delay function at a signalized intersection

by (2.4.14) and (2.4.15):

$$tw_{\text{US}} = \frac{T_c[1-\mu]^2}{2[1-\bar{u}/S]} \tag{2.4.19}$$

In oversaturated conditions, $\bar{u} > \boldsymbol{\mu} S$, for the deterministic case, the queue length, and respective delay, would tend theoretically to infinity. In practice, however, oversaturation lasts only for a finite period of time T, and the average delay tw_{OS} can be calculated from the evolution over time of queue length as described by (2.4.16) through (2.4.18):

$$tw_{\text{OS}} = \frac{T_c[1-\mu]}{2} + \frac{T}{2}\big[(\bar{u}/\mu S) - 1\big] \tag{2.4.20}$$

Note that the first term is the value of (2.4.19) for $\bar{u} = \mu \cdot S$. The delay for the arrival flows can be computed through (2.4.19) for $\bar{u} < \mu \cdot S$, and through (2.4.20) for $\bar{u} \geq \mu \cdot S$, as depicted in Fig. 2.29. Note that the diagram depicted in Fig. 2.29 shows an increase in average delay also for flows below the capacity. This is due to the increase in the undersaturated delay expressed by (2.4.19).

Stochastic delay models are based on the results of queuing theory. More precisely, a signalized intersection is considered to be a *M/G/1* (∞, *FIFO*) system. Therefore, the average delay is (see Sect. 2.2.2.3):

$$tw_q^{st}(u) = \frac{(\bar{u}/\mu S)^2}{2\bar{u}(1-\bar{u}/\mu S)} \tag{2.4.21}$$

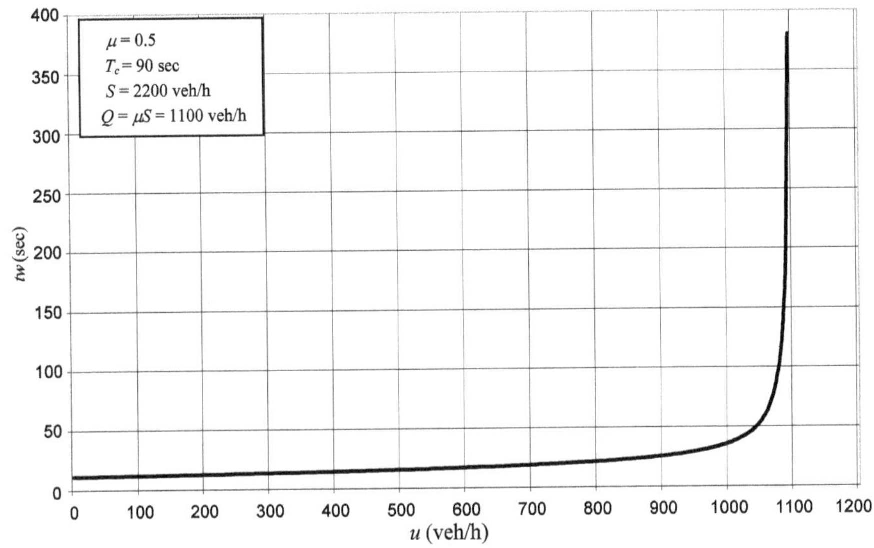

Fig. 2.30 The Webster delay model

Overall Delay Models The total (mean individual) delay equals the sum of the deterministic and the stochastic terms (introduced in the previous section), and sometimes, of terms calibrated through experimental observations.

One of the best known expressions is *Webster's three-term* formula, proposed for an isolated intersection under the assumption of random (Poisson) arrivals and undersaturation conditions ($f_a/Q_a < 1$) (see Fig. 2.30):

$$tw_a(f_a) = \frac{T_c(1-\mu)^2}{2(1-f_a/S_a)} + \frac{(f_a/Q_a)^2}{2f_a(1-f_a/Q_a)} - 0.65\left(Q_a/f_a^2\right)^{1/3}(f_a/Q_a)^{2+\mu} \tag{2.4.22a}$$

where

T_c is the cycle length
μ is the effective green to cycle length ratio for the lane group represented by link a
Q_a is the capacity of the lane group represented by link a

The first term expresses the deterministic delay (see (2.4.19)), the second is the stochastic delay due to the randomness of the arrivals (see (2.4.21)), and the third term is an adjustment term obtained by simulation results. This term amounts to about 10% of the sum of the other two, hence its established use in practical applications of Webster's two-term formula:

$$tw_a(f_a) = 0.9\left[\frac{T_c(1-\mu)^2}{2(1-f_a/S_a)} + \frac{(f_a/Q_a)^2}{2f_a(1-f_a/Q_a)}\right] \tag{2.4.22b}$$

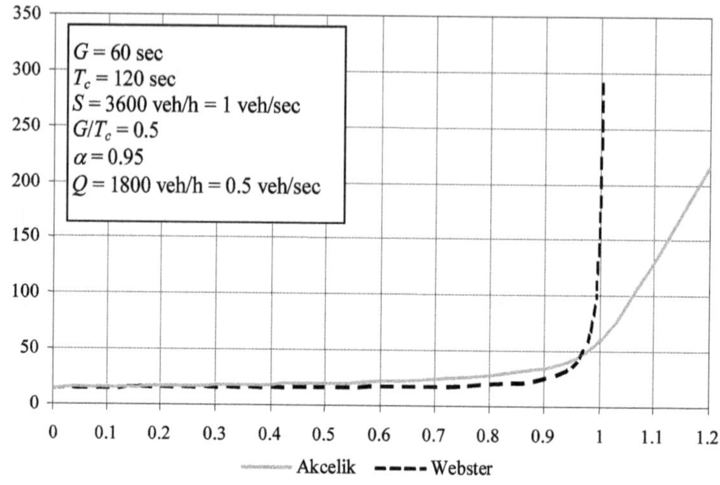

f	f/Q	Akcelik	Webster
0.00	0.00	15.00	15.00
0.10	0.20	16.67	16.87
0.20	0.40	18.75	19.26
0.25	0.50	20.00	20.77
0.30	0.60	21.93	22.61
0.40	0.80	27.95	28.45
0.50	1.00	60.00	
0.60	1.20	216.75	

Fig. 2.31 Waiting time functions at a signalized intersection

The delay given by (2.4.22) tends to infinity for an arrivals flow f_a, which tends to capacity $Q = \mu \cdot$ S (see Fig. 2.30). Thus Webster's formula cannot be used to simulate delays at oversaturated signalized intersections. To overcome this limit, it is possible to apply the two heuristic methods described for (2.4.8).

The first method applies (2.4.22) for values of f_a up to a percentage α of the capacity whereas for higher values a linear approximation of the function is used, thereby ensuring the continuity of the function and its first derivative:

$$tw_a(f_a) = tw_a(\alpha Q_a) + \frac{\mathrm{d}}{\mathrm{d}f} tw_a(f)\bigg|_{f_a=\alpha Q_a} \cdot (f_a - \alpha Q_a) \quad f_a \geq \alpha Q_a \qquad (2.4.23)$$

The second method computes the oversaturation delay combined with the stochastic delay, deforming the stochastic delay function so that it has an oblique asymptote defined by the deterministic delay. Based on these considerations, *Akcelik's* formula was proposed:

$$
\begin{aligned}
tw_a(f_a) &= \frac{0.5T_c(1-\mu_a)^2}{1-\mu_a X_a} \qquad X_a \le 0.50 \\
tw_a(f_a) &= \frac{0.5T_c(1-\mu_a)^2}{1-\mu_a X_a} + 900 \cdot T \cdot \left\{ X_a - 1 \right. \\
&\quad \left. + \left[(X_a-1)^2 + \frac{8(X_a-0.5)}{\mu_a S_a T} \right]^{1/2} \right\} \qquad 0.50 \le X_a \le 1 \\
tw_a(f_a) &= 0.5T_c(1-\mu_a) + 900 \cdot T \cdot \left\{ X_a - 1 \right. \\
&\quad \left. + \left[(X_a-1)^2 + \frac{8(X_a-0.5)}{\mu_a S_a T} \right]^{1/2} \right\} \qquad X_a > 1
\end{aligned}
\tag{2.4.24}
$$

where $X_a = f_a/Q_a$ is the flow/capacity ratio, the times tw_a and T_c are expressed in seconds, S_a in pcph, and T is the duration of the oversaturation period in hours. Equation (2.4.24) is compared with the Webster formula in Fig. 2.31 for a value of $T = 0.5$ h.

Note that application of the previous formulae for calculating saturation flows, capacities, and average waiting times (delays) in the case of multiple lane groups requires an "exploded" representation of the intersection with several links corresponding to the relevant lane groups and their maneuvers (see Fig. 2.18). For example, in the case of an exclusive right-turn lane a single link can represent such a movement and the associated delay. Sometimes, to simplify the representation, fewer links than lane groups are used; in this case the total capacity of all lane groups is associated with the single link and the resulting delay is associated with the whole flow.

From a mathematical point of view the delay functions discussed so far are separable only if the traffic-signal regulation (assumed known) is such as to exclude interference between maneuvers represented by different links. For example, this is the case for the three-phase regulation scheme of a T-shaped intersection shown in Fig. 2.32. However, if the phases allow conflict points, for example, left turns from the opposite direction with through flows during the same phase, nonseparable cost functions may be necessary, which take account of the reciprocal reduction in saturation flow for maneuvers in conflict, such as for the two-phase scheme for the X intersection in Fig. 2.33.

In general, if a single node represents the entire intersection, the effects of individual maneuvers and lane groups are impossible to distinguish and separable functions are adopted, with a single value of saturation flow, reduced to account for the interfering turns.

In the case of control systems at signalized intersections, the control parameters (cycle length T_c, ratio μ of green time to cycle length) depend on flows arriving at the access roads which converge at the intersection. In this case the delay functions are different and definitely nonseparable.

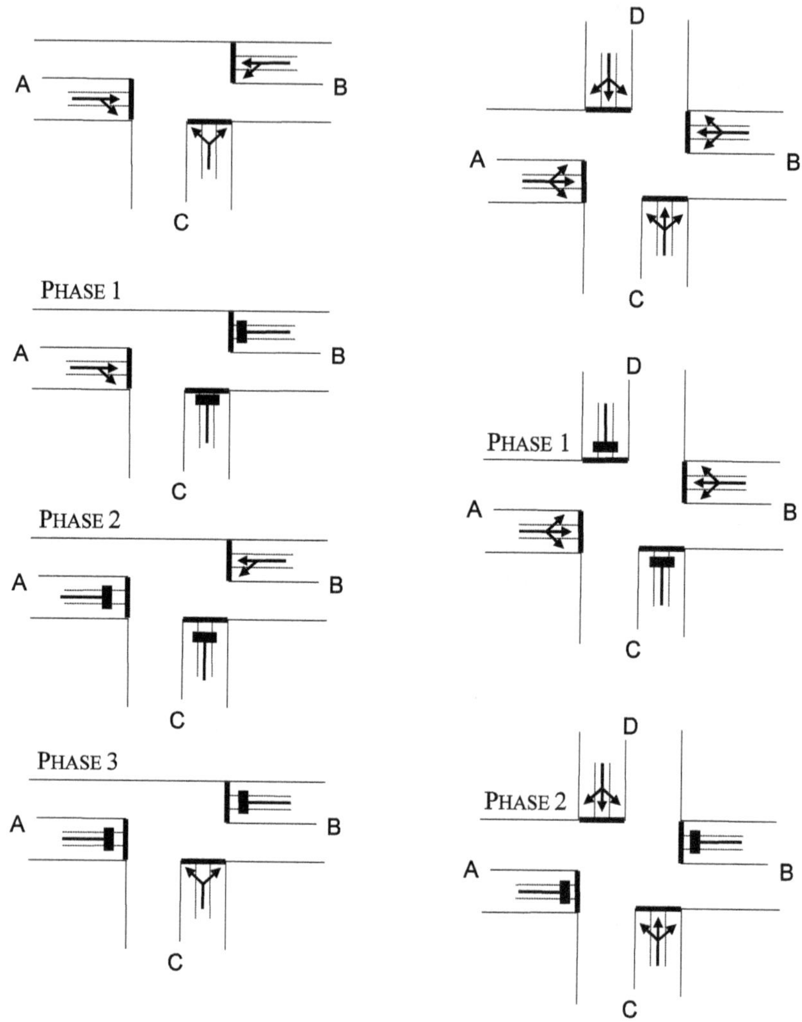

Fig. 2.32 Examples of traffic light phases for 3- and 4-arm intersections

Finally, in the case of networks of interacting intersections (i.e., so close as to affect one another), further regulation parameters must be introduced; hence, calculation of the delay cannot be performed with the formulae presented, but requires more detailed flow simulation models along the road sections joining a pair of adjacent intersections.

(c) Priority Intersections To complete the survey of the delay functions, *priority intersections* (i.e., intersections regulated by give-way rules rather than traffic lights) need to be considered. Empirical functions are often used to express average delays; these functions are nonseparable in that right-of-way rules cause delays due to con-

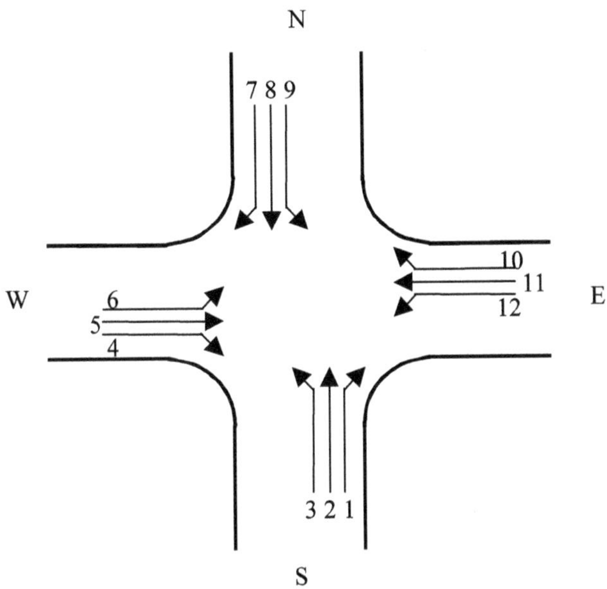

Maneuver		Flows influencing the delay
Direction South-North, right-hand turn	1	1,2,3,5,9
Direction South-North, crossing	2	1,2,3,5,6,9,10,11,12
Direction South-North, left-hand turn	3	1,2,3,5,6,7,8,11,12

Fig. 2.33 Flow conflicts for computing delays at a priority intersection

flicts between flows. As an example, the delay corresponding to the maneuvers at a 4-arm intersection can be calculated by means of the following HCM function.

$$tw_a(\boldsymbol{f}) = \exp\big(-0.2664 + 0.3967\ln(f_{\text{conf}}) + 3.959A\big(\ln(f_{\text{conf}}) - 6.92\big)\big) \quad (2.4.25)$$

where

$tw_a(\boldsymbol{f})$ is the waiting time expressed in seconds
f_{conf} is the total conflicting flow, which varies according to the maneuver as shown in Fig. 2.33
$A = 1$ if $f_{\text{conf}} > 1062$ vehicles/h, 0 otherwise

(d) Parking Links Monetary cost (fares) and search time are the most important performance attributes connected to links representing parking in a given area. In general, these attributes differ for links representing different parking types (facilities). The more sophisticated models of search time take into account the congestion effect through the ratio between the average occupancy of the parking facilities of type p, represented by link a, and the parking capacity Q_l.

The average search time can be calculated through a model assuming that available parking spaces of type p are uniformly distributed along a circuit, possibly mixed with parking spaces of different types (e.g., free and priced parking). If occupancy of a given parking type at the beginning and end of the reference period is inferior to capacity, the following expression can be obtained.

$$ts_a(f_a) = \frac{L_p}{v_s}\frac{1}{occ_2(f_a) - occ_1} \cdot \frac{Q_{\text{tot}} \cdot (Q_p + 1)}{Q_p} \cdot \ln\left(\frac{1 + Q_p - occ_1}{1 + Q_p - occ_2(f_a)}\right) - \frac{(Q_{\text{tot}} - Q_p)}{Q_p} \tag{2.4.26}$$

where

$ts_a(f_a)$	is the search time in minutes
f_a	is the flow on parking link a
L_p	is the average length of a parking space
v_s	is the average search speed for a free parking space
occ_1	is the parking occupancy at the beginning of the reference period
occ_2	is the parking occupancy at the end of the reference period, depending on flow assigned to the parking link and the turnover rate
Q_p	is the parking capacity of type p corresponding to link a
Q_{tot}	is the total capacity of all parking types mixed with type p in the zone

If one or both *occ* are above capacity, similar but formally more complicated formulas can be obtained. These expressions are not reported here.

2.4.2 *Supply Models for Scheduled Service Transportation Systems*

Discontinuous and nonsimultaneous transportation services can be accessed only at given points and are available only at given instants. Typical examples are scheduled services (buses, trains, airplanes, etc.), which can be used only between terminals (bus stops, stations, airports, etc.) and are available only at certain instants (departure times). Scheduled services can be represented by different supply models according to their characteristics and to the consequent assumptions on users' behavior (see Sect. 4.3.3.2). The approach followed in this chapter is based upon the modeling of *service lines*, that is, a set of scheduled runs with equal characteristics. This approach is consistent with the assumption of intraperiod stationarity and with path choice behavior, typical of high frequency and irregular urban transit systems.

If service frequency is low and/or it is assumed that the users choose specific runs, it is necessary to represent the service with a different graph known as a *run graph* or *diachronic graph*. This is usually the case with extraurban transportation services (airplanes, trains, etc.), which have low service frequencies and are largely punctual. In this case, however, the assumption of within-day stationarity does not hold. Indeed, the supply characteristics are often nonuniform within the reference period (arrival and departure times of single runs may be nonuniformly spaced). Furthermore, in order to simulate the traveler's behavior *desired departure* or *arrival times* should be introduced. For these reasons run-based supply models are described in Chap. 7 dealing with intraperiod dynamic systems.

2.4.2.1 Line-based Graph Models

If the scheduled services have high frequencies (e.g., one run every 5–15 min) and low regularity, it is usually assumed that the users do not choose an individual run, but rather a service line or a group of lines. A *service line* is a set of runs sharing the same terminals, the same intermediate stops, and the same performance characteristics, as in the case of an urban bus or underground lines. In this case a *line graph* is typically used. In this graph, nodes correspond to stops, more precisely to the relevant events occurring at the stops. *Access nodes* represent the arrival of the user at the stop, the *stop node*, or *diversion node*, represents the boarding of a vehicle, and the *line nodes* represent the arrival and departure of vehicles of a given line at a given stop. The links represent activities or phases of a trip: access trips between access nodes (*access links*), waiting at the stop (*waiting links*), boarding and alighting from the vehicles of a line (*boarding* and *alighting links*), the trip from one stop to another of the same line (*line links*), and vehicle dwelling at the stop (*dwelling links*).

Essentially, each stop is represented by a subgraph such as that shown in Fig. 2.34. The graph representing an entire public transportation system can be built by combining the *line graph* and the *access graph* through the stop subgraphs. Access links may represent different access modes depending on the system modeled. In urban areas, they may represent pedestrian connections or, sometimes, undifferentiated "access modes" including local transit lines to the main network of bus and rail services. The line graph is completed by adding nodes and links allowing entry/exit from the centroids to the stops; in the urban context this usually occurs through pedestrian nodes and links or through road links connected to park-and-ride facilities (nodes).

2.4.2.2 Link Performance and Cost Functions

The typical performance attributes used in line-based supply models are travel time components related to different trip phases and monetary costs. Travel times can be decomposed into on-board travel times T_b, dwelling times at stops T_d, waiting times T_w, boarding times T_{br}, alighting times T_{al}, and access/egress times T_a, which may correspond to walking or driving time for urban transit networks. In general, a single time component is associated to each link and the coefficients β, homogenizing travel times into costs (disutilities) are different. In fact, several empirical studies have shown that waiting and walking times have coefficients two to three times larger than that of on-board time for urban transit systems.

Performance functions used in many applications do not take congestion into account, at least with respect to flows of transit users, as it is assumed that services are designed with some extra capacity with respect to maximum user flows.

On-board travel time of a transit link can be obtained through a very simple expression:

$$Tb_a = \frac{L_a}{v_a(\boldsymbol{b}_a, \boldsymbol{\gamma}_a)} \tag{2.4.27}$$

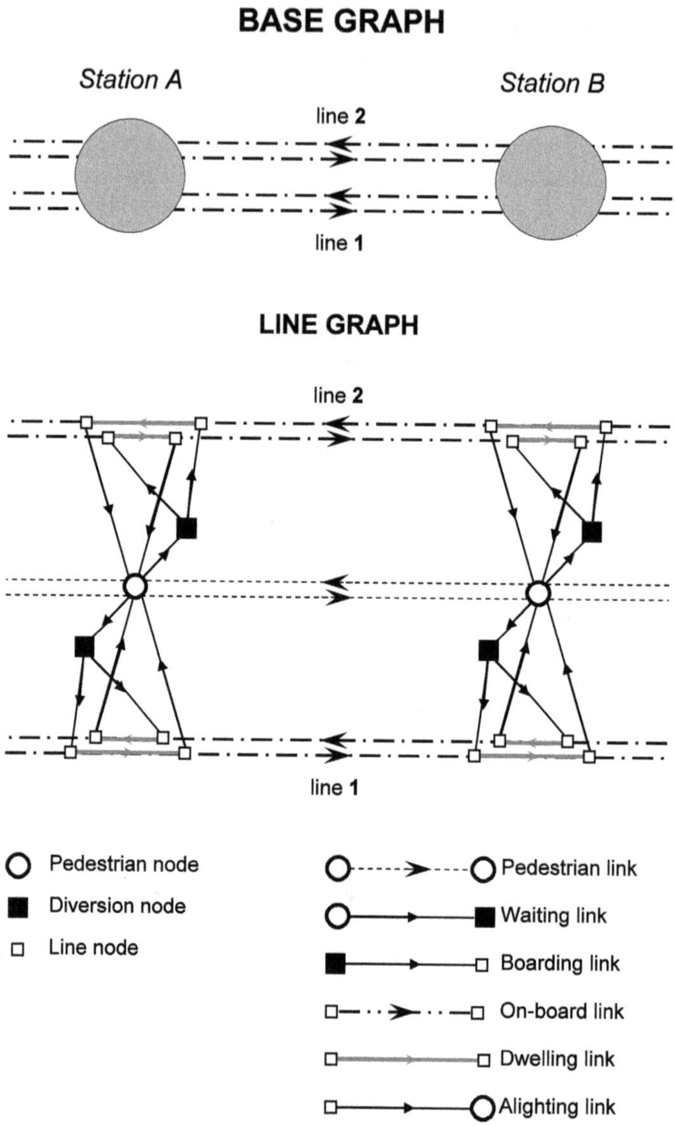

Fig. 2.34 Line-based graph for urban transit systems

where vector $\boldsymbol{b}_a$ includes the relevant characteristics of the transit system represented by link a, and vector $\boldsymbol{\gamma}_a$ comprises a set of parameters. The average speed is strongly dependent on the type of right-of-way. For exclusive right-of-way systems, such as trains, the average speed v_a can be expressed as a function of the characteristics of the vehicles (weight, power, etc.), of the infrastructure (slope, radius of bends, etc.), of the circulation regulations on the physical section and the type of service represented. Relationships of this type can be deduced from mechanics for which specialized texts should be referred to. For partial right-of-way systems, such as surface buses, the average speed depends on the level of protection (e.g., reserved bus lane) and the vehicle flows on the links corresponding to interfering movements. Performance functions of this type typically derive from descriptive models.

The *waiting time* is the average time that users spend between their arrival at the stop/station and the arrival of the line (or lines) they board. Waiting time is usually expressed as a function of the line *frequency* φ_{ln}, that is, the average number of runs of line ln in the reference period. When only one line is available the average waiting time Tw_{ln} will depend on the regularity of vehicle arrivals and the pattern of users' arrivals at the stop. It can be shown that, under the assumption that users arrive at the stop according to a Poisson process with a constant arrival rate[6] (consistent with the within-day stationarity assumption), the average waiting time is:

$$Tw_{ln} = \frac{\theta}{\varphi_{ln}} \tag{2.4.28}$$

where θ is equal to 0.5 if the line is perfectly regular (i.e., the headways between successive vehicle arrivals are constant), and it is equal to 1 if the line is "completely irregular" (i.e., the headways between successive arrivals are distributed according to a negative exponential random variable); see Fig. 2.35.

In the case of several "*attractive lines*," that is, when the user waits at a diversion node m for the first vehicle among those belonging to a set of lines Ln_m, the average waiting time can again be calculated with expression (2.4.28) by using the *cumulated frequency* Φ_m of the set of attractive lines:

$$Tw_{ln} = \frac{\theta}{\Phi_m} \quad \text{with} \quad \Phi_m = \sum_{ln \in Ln_m} \varphi_{ln} \tag{2.4.29}$$

Expression (2.4.29) holds in principle when vehicle arrivals of all lines are completely irregular. In this case cumulated headways can still be modeled as a negative exponential random variable, with a parameter equal to the inverse of the sum of line frequencies. In practice, however, expression (2.4.29) is often used also for intermediate values of θ.

These expressions of average waiting times are revisited in Sect. 4.3.3.2 on path choice models for transit systems.

Access/egress times are also usually modeled through very simple performance functions analogous to expression (2.4.27):

$$Ta_{ln} = \frac{L_{ln}}{v_{al}(\boldsymbol{b}_{ln}, \gamma_{ln})}$$

where v_{al} represents the average speed of the access/egress mode. Also in the case of pedestrian systems, it is possible to introduce congestion phenomena and correlate

---

[6]To be precise, it is assumed that users' arrival is a Poisson process; that is, the intervals between two successive arrivals are distributed according to a negative exponential variable.

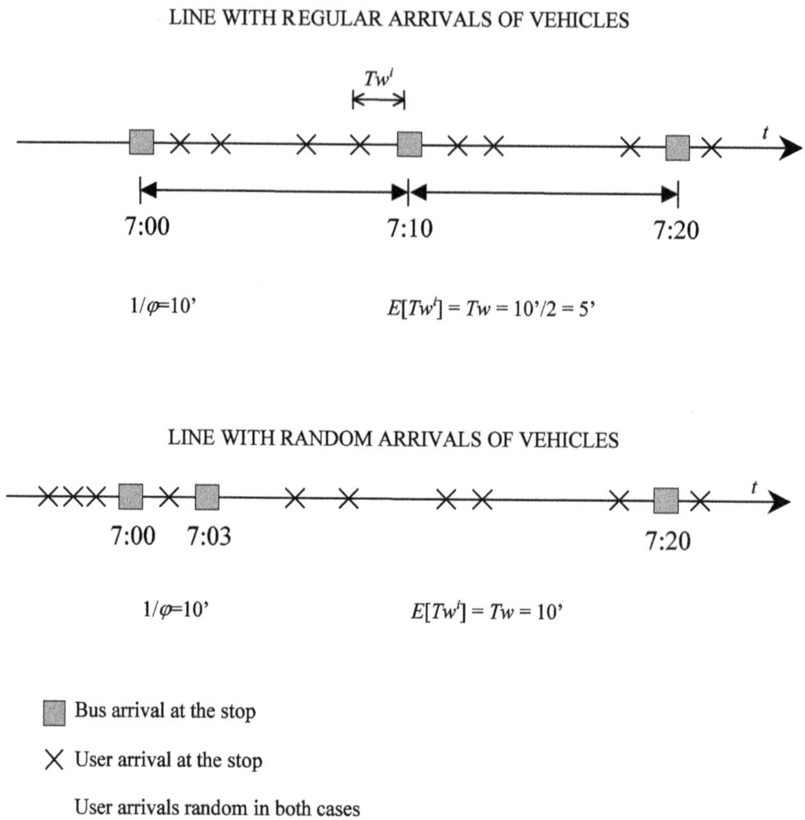

Fig. 2.35 Arrivals and waiting times at a bus stop

the generalized transportation cost with the pedestrian density on each section by using empirical expressions described in the literature.

More detailed performance models introduce congestion effects with respect to user flows both on travel times and on comfort performance attributes. An example of the first type of function is that relating the *dwelling time* at a stop Td_{ln} to the user flows boarding and alighting the vehicles of each line:

$$Td_{ln} = \gamma_1 + \gamma_2 \left(\frac{f_{al(a)} + f_{br(a)}}{Q_D} \right)^{\gamma_3} \tag{2.4.30}$$

where

$f_{al(a)}$ is the user flow on the alighting link
$f_{br(a)}$ is the user flow on the boarding link
Q_D is the door capacity of the vehicle
$\gamma_1, \gamma_2, \gamma_3$ are parameters of the function

Another example is the function relating the average waiting time to the flow of users staying on board and those waiting to board a single line. This function takes into account the "refusal" probability, that is, the probability that some users may not be able to get on the first arriving run of a given line because it is too crowded and have to wait longer for a subsequent one. In the case of a single attractive line l the waiting time function can be formally expressed as

$$Tw_{ln} = \frac{\theta}{\varphi_{ln(.)}}\left(\frac{f_{b(.)} + f_{w(.)}}{Q_{ln}}\right) \tag{2.4.31}$$

where $\varphi_{ln}(.)$ is the actual available frequency of line ln, that is, the average number of runs of the line for which there are available places. It depends on the ratio between the demand for places – sum of the user flow staying on board $f_{b(.)}$ and the user flow willing to board, $f_{w(.)}$ – and the line capacity Q_{ln}. This formula is valid only for $f_{b(.)} + f_{w(.)} > Q_{ln}$.

Note that both performance functions (2.4.30) and (2.4.31) are nonseparable, in that they depend on flows on links other than the one to which they refer.

Discomfort functions relate the average riding discomfort on a given line section represented by link a, dc_a, to the ratio between the flow on the link (average number of users on board) and the available line capacity Q_a:

$$dc_a = \gamma_3 f_a + \gamma_4\left(\frac{f_a}{Q_a}\right)^{\gamma_5} \tag{2.4.32}$$

where, as usual, γ_3, γ_4, and γ_5 are positive parameters, usually with γ_5 larger than one expressing more-than-linear effect of crowding.

Reference Notes

The application of network theory to the modeling of transportation supply systems can be found in most texts dealing with mathematical models of transportation systems, such as Potts and Oliver (1972), Newell (1980), Sheffi (1985), Cascetta (1998), Ferrari (1996), and Ortuzar and Willumsen (2001). All of these, however, deal primarily or exclusively with road networks. The presentation of a general transportation supply model and its decomposition into submodels as described in Fig. 2.14 is original.

Performance models and the traffic flow theory are dealt with in several books and scientific papers. The former include Pignataro (1973), the ITE manual (1982), May (1990), McShane and Roess (1990), the Highway Capacity Manual (2000), and the relevant entries in the encyclopaedia edited by Papageorgiou (1991). Among the latter, the pioneering work of Webster (1958), later expanded in Webster and Cobbe (1966) and those of Catling (1977), Kimber et al. (1977), Kimber and Hollis (1978), Robertson (1979), and Akcelik (1988) on waiting times at signalized intersections. In-depth examinations of some aspects of traffic flow theory can be found in Daganzo (1997).

For a theoretical analysis of queuing theory, reference can be made to Newell (1971) and Kleinrock (1975).

The work of Drake et al. (1967) reviews the main speed–flow–density relationships, and gives an example of their calibration. The linear model was proposed by Greenshields (1934). References to nonstationary traffic flow models are in part reported in the bibliographical note to Chap. 7.

A review of the road network cost functions can be found in Branston (1976), Hurdle (1984), and Lupi (1996). The study of Cartenì and Punzo (2007) contains experimental speed–flow relationships for urban roadways, reported in the text (2.4.5) and updates the work by Festa and Nuzzolo (1989). The cost function for parking links (2.4.26) was proposed by Bifulco (1993).

Supply models for scheduled services have traditionally received less attention in the scientific community. The line representations of scheduled systems are described in Ferrari (1996) and in Nuzzolo and Russo (1997).

Several authors, such as Seddon and Day (1974), Jolliffe and Hutchinson (1975), Montella and Cascetta (1978), and Cascetta and Montella (1979), have studied the relationships between waiting times and service regularity in urban transit systems. Congested performance models discussed in Sect. 2.4.2 have been proposed by Nuzzolo and Russo (1993), and other models for waiting time at congested bus stops are quoted in Bouzaiene-Ayari et al. (1998). Mechanics of motion is treated in detail in several classical books. For an updated bibliographical note see Cantarella (2001).

Chapter 3
Random Utility Theory

3.1 Introduction

In Chap. 1 it was stated that transport flows result from the aggregation of individual trips. Each trip is the result of a number of choices made by transport system users: by travelers in the case of personal transport or by operators (manufacturers, shippers, carriers) in goods transport. Some choices are made infrequently, such as where to reside and work and whether to own a vehicle. Other choices are made for each trip; these include whether to make the trip, at what time, to what destination or destinations, by what mode, and using what path. Each choice context, defined by the available alternatives, evaluation factors, and decision procedures, is known as a "choice dimension." In most cases, travel choices are made among a finite number of discrete alternatives.

Starting from these assumptions, many travel demand models, such as those described in the next chapter, attempt to reproduce users' choice behavior,[1] and so are called behavioral models. The present chapter describes the behavioral models derived from random utility theory, which is the richest and by far the most widely used[2] theoretical paradigm for modeling transport-related choices and more generally, choices among discrete alternatives. Within this paradigm, it is possible to specify a number of models, having various functional forms, and applicable to a wide variety of contexts. It is also possible to study their mathematical properties and estimate their parameters using well-established statistical methods.

It should be said that random utility models are not the only behavioral models that can be used to represent transport-related choices. Other models proposed in the literature are based on choice mechanisms that violate one or more of the general hypotheses described in Sect. 3.2. These models are often called "noncompensatory," because they do not allow the compensation of negative attributes with positive ones or, more generally, trading off one attribute against another. Noncompensatory models are at present mostly research tools and are not widely used in

---

[1]Behavioral models, like all microeconomic demand models, attempt to reproduce the results of choice behavior "as if" decision-makers behaved in accordance with certain hypotheses; they do not claim to represent the actual psychological mechanisms leading to decisions.

[2]Discrete choice models in general, and random utility models in particular, can be considered one of the most significant contributions of the transport field to economics and econometrics. From the theoretical point of view, they represent a development of classical microeconomic demand models. In fact, discrete choice models represent choices made among discrete alternatives whereas classical microeconomic demand models represent the choice of a (continuous) quantity of "commodities" to be consumed. Discrete choice models, originally developed for travel-demand modeling, are used in many applications of econometrics, from the choice of insurance policy types and investment portfolios to the choice of car models.

E. Cascetta, *Transportation Systems Analysis*,
Springer Optimization and Its Applications 29,
DOI 10.1007/978-0-387-75857-2_3,

practice. Furthermore, it has been shown that a properly specified random utility model can very often satisfactorily approximate the choice probabilities obtained with noncompensatory models.

In this chapter, random utility models are discussed in terms of personal mobility choices. The same kinds of model can be applied to freight transport-related choices, as shown in Sect. 4.7. The chapter does not consider the statistical estimation of model parameters, except where particular estimation issues are relevant to the discussion; estimation is discussed in Chap. 8.

Section 3.2 introduces the general hypotheses underlying random utility models, and Sect. 3.3 describes their most widely used functional forms. Section 3.4 defines the expected maximum perceived utility variable and analyzes the mathematical properties of this variable and of random utility models. Section 3.5 considers the problem of choice set modeling. Section 3.6 introduces the concept of elasticity of random utility models. Finally, Sect. 3.7 analyzes various aggregation procedures that allow the estimation of aggregate demand from models that represent individual choices.

3.2 Basic Assumptions

Random utility theory is based on the hypothesis that every individual is a rational decision-maker, maximizing utility relative to his or her choices. Specifically, the theory is based on the following assumptions.

(a) The generic decision-maker i, in making a choice, considers m_i mutually exclusive alternatives that constitute her *choice set* I^i. The choice set may differ according to the decision-maker (e.g., in the choice of transport mode, the choice set of an individual without a driver's license or car obviously should not include the alternative "car as a driver");
(b) Decision-maker i assigns to each alternative j in his choice set a *perceived utility* or "attractiveness" U^i_j and selects the alternative that maximizes this utility;
(c) The utility assigned to each choice alternative depends on a number of measurable characteristics, or *attributes*, of the alternative itself and of the decision-maker: $U^i_j = U^i(\boldsymbol{X}^i_j)$, where $\boldsymbol{X}^i_j$ is the vector of attributes relative to alternative j and to decision-maker i;
(d) Because of various factors described later, the utility assigned by decision-maker i to alternative j is not known with certainty by an external observer (analyst) wishing to model the decision-maker's choice behavior, thus U^i_j must be represented in general by a random variable.

From the above assumptions, it is not usually possible to predict with certainty the alternative that the generic decision-maker will select. However, it is possible to express the probability that the decision-maker will select alternative j conditional on her choice set I^i; this is the probability that the perceived utility of alternative j is greater than that of all the other available alternatives:

$$p^i(j/I^i) = Pr\left[U^i_j > U^i_k \;\forall k \neq j,\; k \in I^i\right] \tag{3.2.1}$$

The perceived utility U_j^i can be expressed as the sum of two terms: a systematic utility and a random residual. The *systematic utility* V_j^i represents the mean (expected value) utility perceived by all decision-makers having the same choice context (alternatives and attributes) as decision-maker i. The *random residual* ε_j^i is the (unknown) deviation of the utility perceived by user i from this mean value; it captures the combined effects of the various factors that introduce uncertainty into choice modeling:

$$U_j^i = V_j^i + \varepsilon_j^i \quad \forall j \in I^i \tag{3.2.2a}$$

with

$$V_j^i = E\left[U_j^i\right], \qquad \sigma_{i,j}^2 = \mathrm{Var}\left[U_j^i\right]$$

and therefore

$$E\left[V_j^i\right] = V_j^i, \qquad \mathrm{Var}\left[V_j^i\right] = 0$$

$$E\left[\varepsilon_j^i\right] = 0, \qquad \mathrm{Var}\left[\varepsilon_j^i\right] = \sigma_{i,j}^2$$

Replacing expression (3.2.2a) in (3.2.1) yields:

$$p^i[j/I^i] = Pr\left[V_j^i - V_k^i > \varepsilon_k^i - \varepsilon_j^i \ \forall k \neq j,\ k \in I^i\right] \tag{3.2.3a}$$

From (3.2.3a) it follows that the choice probability of an alternative depends on the systematic utilities of all competing (available) alternatives, and on the joint probability law of the random residuals ε_j.

Random utility models and the variables they involve can be compactly represented using vector notation. Let

- $\boldsymbol{p}^i$ be the vector of choice probabilities, of dimension $(m_i \times 1)$, with elements $p^i[j]$
- $\boldsymbol{U}^i$ be the vector of perceived utilities, of dimension $(m_i \times 1)$, with elements U_j^i
- $\boldsymbol{V}^i$ be the vector of systematic utility values, of dimension $(m_i \times 1)$, with elements V_j^i
- $\boldsymbol{\varepsilon}^i$ be the vector of random residuals, of dimension $(m_i \times 1)$, with elements ε_j^i
- $f(\boldsymbol{\varepsilon})$ be the joint probability density function of the random residuals
- $F(\boldsymbol{\varepsilon})$ be the joint probability distribution function of the random residuals

Expression (3.2.2a) can therefore be written in vector notation as:

$$\boldsymbol{U}^i = \boldsymbol{V}^i + \boldsymbol{\varepsilon}^i \tag{3.2.2b}$$

In general, the choice model (3.2.3a) can be viewed as a function, known as a *choice function*, that associates a vector of choice probabilities to each vector $\boldsymbol{V}^i$ of systematic utilities for a given probability law of random residuals:

$$\boldsymbol{p}^i = \boldsymbol{p}^i(\boldsymbol{V}^i) \quad \forall \boldsymbol{V}^i \in E^{m_i} \tag{3.2.3b}$$

A random utility model is said to be *invariant* (or *additive*) if neither the form nor the parameters of the joint probability density function of the random residuals, $f(\boldsymbol{\varepsilon})$, depends on the vector $\boldsymbol{V}$ of systematic utilities:

$$f(\boldsymbol{\varepsilon}/\boldsymbol{V}) = f(\boldsymbol{\varepsilon}) \quad \forall \boldsymbol{\varepsilon} \in E^{m_i}$$

It follows immediately from expression (3.2.3a) that, for invariant models, the choice probabilities of the alternatives do not change if a constant V_0 is added to the systematic utility of each of them:

$$\begin{aligned} p^i[j/I^i] &= Pr\left[V_j^i + V_0 - V_k^i - V_0 > \varepsilon_k^i - \varepsilon_j^i\right] \\ &= Pr\left[V_j^i - V_k^i > \varepsilon_k^i - \varepsilon_j^i\right] \quad \forall k \neq j;\ j,k \in I^i \end{aligned} \tag{3.2.4}$$

From the previous expression it also follows that, in the case of invariant models, choice probabilities depend on the differences between the systematic utility of each alternative and that of a reference alternative h; these differences $V_j - V_h$ are known as relative systematic utilities.

Before describing some of the random utility models derived from particular assumptions on the random residual joint probability functions, some further general remarks on the implications of the hypotheses introduced so far should be made.

The variance–covariance matrix of random residuals. In general, a variance–covariance matrix $\boldsymbol{\Sigma}$ is symmetric and positive semidefinite (see Appendix 3.B). When the variance of each random residual ε_k is zero, $\sigma_{kk} = 0$, all the covariances must also be zero, $\sigma_{kh} = 0\ \forall h$, and therefore the variance–covariance matrix is itself zero, $\boldsymbol{\Sigma} = 0$; this case yields the *deterministic* choice model whose properties are described in Sect. 3.4. If the variance–covariance matrix is not zero, $\boldsymbol{\Sigma} \neq 0$, a *non-deterministic* choice model is obtained. In this case, it is usually assumed that the variance $\sigma_{kk} = \sigma_k^2$ of each random residual ε_k is strictly positive, $\sigma_{kk} > 0$, and that the random residuals are imperfectly correlated, $(\sigma_{kh})^2 < \sigma_k^2\sigma_h^2$; that is, the rows (or columns) of $\boldsymbol{\Sigma}$ are pairwise linearly independent. These conditions are equivalent to assuming that the variance–covariance matrix is not singular, $|\boldsymbol{\Sigma}| \neq 0$, in addition to being nonzero, $\boldsymbol{\Sigma} \neq 0$. In this case the models are called *probabilistic*,[3] and the choice function $p = p(\boldsymbol{V})$ can be shown to be continuous with continuous first partial derivatives.

The set of available alternatives I^i, or choice set, significantly influences the choice probabilities, as can be seen from (3.2.1) and (3.2.3a). If a particular decision-maker's choice set I^i is known, the definition of choice probability (3.2.1) can be applied directly. However, it often happens that the analyst has no exact knowledge of the generic decision-maker's choice set. In this case, the problem can be handled with different levels of approximation, as shown in Sect. 3.5.

[3]The case in which the variance–covariance matrix is nonzero, $\boldsymbol{\Sigma} \neq 0$, but singular, $|\boldsymbol{\Sigma}| = 0$, because the variance of a random residual is zero and/or two random residuals are perfectly correlated, is of limited practical interest and is not given further attention.

The expression for the systematic utility. Systematic utility is the mean perceived utility among all individuals who have the same attributes; it is expressed as a function $V_j^i(X_{kj}^i)$ of attributes X_{kj}^i relative to the alternatives and the decision-maker. Although in principle the function $V_j^i(\mathbf{X}_j^i)$ may be of any type, it is usually assumed for analytical and statistical convenience that the systematic utility V_j^i is a linear function, with coefficients β_k, of the attributes X_{kj}^i or of functional transformations $f_k(X_{kj}^i)$ of them:

$$V_j^i(\mathbf{X}_j^i) = \sum_k \beta_k X_{kj}^i = \boldsymbol{\beta}^{\mathrm{T}} \mathbf{X}_j^i \tag{3.2.5a}$$

or

$$V_j^i(\mathbf{X}_j^i) = \sum_k \beta_k f_k(X_{kj}^i) = \boldsymbol{\beta}^{\mathrm{T}} \boldsymbol{f}(\mathbf{X}_j^i) \tag{3.2.5b}$$

Further details on the specification of the systematic utility are given in Chap. 8.

The attributes included in the vector X_j^i can be classified in different ways. Those related to the service offered by the transport system are known as *level-of-service* or *performance attributes* (times, costs, service frequency, comfort, etc.). Those related to the land-use characteristics of the study area (e.g., the number of shops or schools in each zone) are known as *activity system attributes*. Those related to the decision-maker or to his household (income, holding a driver's license, number of cars in the household, etc.) are referred to as *socioeconomic attributes*.

Attributes of any type are called *generic* if they are included in the systematic utility of more than one alternative in the same form and with the same coefficient β_k. They are called *specific* if they are included with different functional forms and/or coefficients in the systematic utilities of different alternatives. A dummy variable is usually included in the systematic utility of the generic alternative j; its value is one for alternative j and zero for the others. This variable is usually denoted the *Alternative Specific Attribute* (ASA) or "modal preference" attribute,[4] and its coefficient β_k is known as the Alternative Specific Constant (ASC). The ASA is a kind of constant term in the systematic utility; it can be viewed as the difference between the mean utility of an alternative and the portion that is explained by its other attributes X_{kj}.

From expression (3.2.4), it can be seen that the choice probabilities of invariant models depend in part on the differences between the ASC of each alternative j and that of a reference alternative h. If alternative specific attributes were included in the systematic utilities of all alternatives, any combination of coefficients β that led to the same ASC differences between alternatives would result in the same choice probability values, so the ASCs could not be statistically estimated. For this reason, when specifying invariant models, the ASC of at least one of the alternatives must

[4]This term derives from early applications of random utility models to the choice among different transport modes.

$$V_{\text{walking}} = \beta_1 t_{wl}$$

$$V_{\text{auto}} = \beta_1 t_{wla} + \beta_2 t_{ba} + \beta_3 mc_a + \beta_4 AVAIL + \beta_5 INC + \beta_6 AUTO$$

$$V_{\text{bus}} = \beta_1 t_{wlb} + \beta_2 t_{bb} + \beta_3 mc_b + \beta_7 t_{wb} + \beta_8 BUS$$

Alternative specific attributes (ASA)	Level of service attributes	Socioeconomic attributes

AUTO
BUS
	t_b = Time on board (generic)
	t_w = Waiting time at stop (specific)
	t_{wl} = Walking time (generic)
	mc = Monetary cost (generic)
		AVAIL = # Auto/# licenses
		INC = Disposable household income

Fig. 3.1 Specification of systematic utilities and classification of attributes

be set to zero; equivalently, ASAs may be included in the systematic utilities of at most all the alternatives except one.

An elementary example of systematic utilities related to transport mode choice is given in Fig. 3.1. Many other examples are given in the following chapters.

Utilities are merely a way of capturing the preference ordering among alternatives, and so have no intrinsic units of measurement; alternatively, they can be expressed in arbitrary dimensionless units, sometimes called *utils*. From expression (3.2.5) it can be seen that, in order to sum attributes expressed in different units (e.g., time and cost), their respective coefficients β_k have to be expressed in units that are inverses of those of the attributes themselves (e.g., time^{-1} and cost^{-1}). The coefficients β are sometimes called reciprocal substitution coefficients because they make it possible to evaluate reciprocal "exchange rates" (rates of substitution) between attributes. This point is developed in Chap. 4.

The randomness of perceived utilities. Various factors account for the difference between the utility perceived by an individual decision-maker and the systematic utility common to all decision-makers with equal values of the attributes. These factors are related both to the model (factors a, b, and c below) and to the decision-maker (factors d and e). They include:

(a) Errors in measuring the attributes that are included in the systematic utility. For example, level-of-service attributes are often obtained from a network model and so are subject to modeling and aggregation (zoning) errors; other attributes are intrinsically variable and only their average value can be considered.
(b) Attributes that are not included in the systematic utility because they are not directly observable or are difficult to evaluate (e.g., travel comfort or total travel time reliability).
(c) Instrumental attributes that are included in the systematic utility specification but only imperfectly represent the actual attributes that influence the alternatives' perceived utility (e.g., modal preference attributes replacing variables such as the comfort, privacy, image, etc. of a certain transport mode; the total number of commercial establishments in a given zone replacing the number and variety of shops).

(d) Variability among decision-makers, or variations in tastes and preferences among different decision-makers and, for an individual decision-maker, over time. Different decision-makers with otherwise identical attributes might have different utility functions or different values of the reciprocal substitution coefficients β_k according to personal preferences (e.g., walking distance is more or less disagreeable to different people). The same decision-maker might weigh an attribute differently in different decision contexts (e.g., according to different physical or psychological conditions).

(e) Errors in the evaluation of attributes by the decision-maker (e.g., erroneous estimation of travel time).

From the above discussion, it follows that the more accurate a model is (the greater the number of relevant attributes included in the systematic utilities, the more precise their determination, etc.), the lower should be the variance of its random residuals ε_j. Experimental evidence generally confirms this conjecture.

3.3 Some Random Utility Models

Given the general hypotheses presented in the previous section, different random utility model forms can be derived by assuming different joint probability distribution functions for the perceived utility random residuals ε_j[5] (expression (3.2.3a)). This section describes the random utility models that are the most widely used in travel-demand modeling. Models are introduced in order of increasing generality and analytical complexity. Section 3.3.1 describes the Multinomial Logit (or MNL) model, which is the simplest functional form. Subsequently, progressive generalizations of the MNL to the single-level hierarchical or nested logit model (Sect. 3.3.2), to the multilevel hierarchical or tree logit model (Sect. 3.3.3), to the cross-nested logit model (Sect. 3.3.4), and to the Generalized Extreme Value (GEV) model (Sect. 3.3.5) are described. Each of these models includes the MNL as a special case, and each can be obtained in turn from the GEV model. Finally, Sect. 3.3.6 describes the probit model and Sect. 3.3.7 introduces the mixed logit model.

3.3.1 *The Multinomial Logit Model*

The multinomial logit model is the simplest random utility model. It is based on the assumption that the random residuals ε_j are independently and identically distributed (i.i.d.) as Gumbel random variables (r.v.) with zero mean and scale para-

[5] In this section, for the sake of simplicity, the symbol i indicating the generic decision-maker is systematically taken as understood.

meter θ.[6] The marginal probability distribution function of each random residual is given by:

$$F_{\varepsilon_j}(x) = Pr[\varepsilon_j \leq x] = \exp\big[-\exp(-x/\theta - \Phi)\big] \tag{3.3.1}$$

where Φ is Euler's constant ($\Phi \approx 0.577$). In particular, the mean and variance of the Gumbel variable expressed by (3.3.1) are, respectively,

$$\begin{aligned} &E[\varepsilon_j] = 0 \quad \forall j \\ &\mathrm{Var}(\varepsilon_j) = \sigma_\varepsilon^2 = \frac{\pi^2}{6}\theta^2 \quad \forall j \end{aligned} \tag{3.3.2}$$

Further characteristics of the Gumbel r.v. are given in Appendix 3.B.

The independence of the random residuals implies that the covariance between any pair of residuals is zero:

$$\mathrm{Cov}[\varepsilon_j, \varepsilon_h] = 0 \quad \forall j, h \in I \tag{3.3.3}$$

From this it can be deduced that alternative js perceived utility U_j, which is the sum of its systematic utility V_j (a constant) and the random ε_j, is also a Gumbel random variable with probability distribution function, mean and variance given by:

$$\begin{aligned} &F_{U_j}(x) = Pr[U_j \leq x] = Pr[\varepsilon_j \leq x - V_j] = \exp\big[-\exp\big(-(x - V_j)/\theta - \Phi\big)\big] \\ &E[U_j] = V_j, \qquad \mathrm{Var}[U_j] = \frac{\pi^2\theta^2}{6} \end{aligned} \tag{3.3.4}$$

Based on the above assumptions about the residuals ε_j (and therefore about the perceived utilities U_j), the variance–covariance matrix of the residuals $\mathbf{\Sigma}_\varepsilon$ is a scalar multiple (by σ_ε^2) of an identity matrix having the same number of rows and columns as the number of alternatives. Figure 3.2 shows, for a multinomial logit model involving four choice alternatives, a graphic representation of the assumptions made regarding the distribution of the random residuals and their variance–covariance matrix. This representation, known as a choice tree, should be compared to those of the hierarchical logit models described in the following sections.

The Gumbel variable has an important property known as *stability with respect to maximization*: the maximum of a set of independent Gumbel variables, all with scale parameter θ, is also a Gumbel variable with parameter θ. More specifically, if $\{U_j\}$ is a set of independent Gumbel variables having equal parameter θ but different means V_j, the variable U_M:

$$U_M = \max_j \{U_j\}$$

---

[6]Some texts define the Gumbel distribution scale parameter to be the reciprocal of θ; that is, $\alpha = 1/\theta$. In the text, the θ notation is normally used because of its analytical convenience in the specification of hierarchical logit models. Clearly, it is possible to express all results in terms of the parameter α with a simple variable substitution.

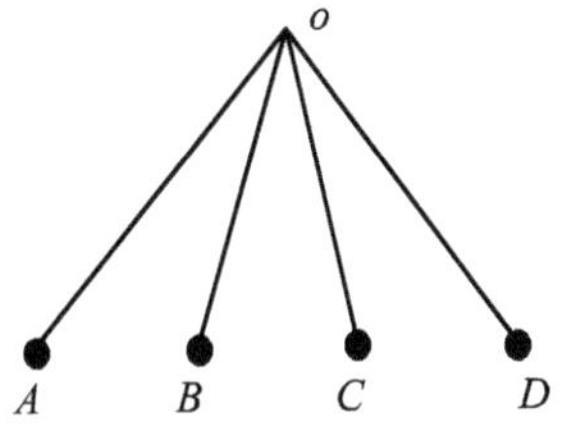

$$\sum_{\varepsilon} = \sigma_{\varepsilon}^{2} \boldsymbol{I} = \frac{\pi^2\theta^2}{6} \begin{array}{c} \begin{array}{cccc} A & B & C & D \end{array} \\ \left[\begin{array}{cccc} 1 & 0 & 0 & 0 \\ 0 & 1 & 0 & 0 \\ 0 & 0 & 1 & 0 \\ 0 & 0 & 0 & 1 \end{array}\right] \end{array} \begin{array}{c} \\ A \\ B \\ C \\ D \end{array}$$

Fig. 3.2 Choice tree and variance–covariance matrix of a multinomial logit model

is again a Gumbel variable with parameter θ and with mean V_M given by

$$V_M = E[U_M] = \theta \ln \sum_j \exp(V_j/\theta) \tag{3.3.5}$$

The variable V_M is called the *Expected Maximum Perceived Utility* (*EMPU*)[7] or the *inclusive utility*. The variable Y

$$Y = \ln \sum_j \exp(V_j/\theta)$$

which is proportional to it, is called the *logsum* because of its analytical form.

Because of the property of stability with respect to maximization, the assumption of Gumbel-distributed residuals is particularly convenient in random utility models. In fact, under the assumptions made here, the probability of choosing alternative j from among those available $(1, 2, \ldots, m) \in I$, given by (3.2.4), can be expressed[8] in closed form as

$$p[j] = \frac{\exp(V_j/\theta)}{\sum_{i=1}^{m} \exp(V_i/\theta)} \tag{3.3.6}$$

Expression (3.3.6) defines the multinomial logit model, which is the simplest and one of the most widely used random utility models. Under the common assumption that the parameter θ is independent of the systematic utility, the MNL model is invariant (see Sect. 3.4) and has a number of important properties that are described in the following.

Dependence on the differences among systematic utilities.[9] In the case of only two alternatives (A and B), the MNL model (3.3.6) is called binomial logit and can

---

[7]The Expected Maximum Perceived Utility variable is dealt with extensively in Sect. 3.4.

[8]A proof of the Gumbel random variable's stability with respect to maximization and a derivation of the multinomial logit model from the general expression (3.2.3) are presented in Appendix 3.B.

[9]This property and its implications hold for the entire class of invariant models, as was stated in Sect. 3.2. In the following, the general results are particularized for the logit model, where they can be obtained analytically.

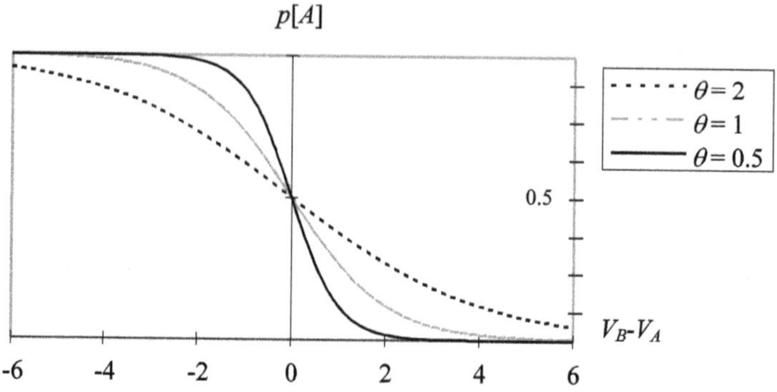

Fig. 3.3 Diagram of choice probability $p[A]$ of a binomial logit model

be expressed as

$$p[A] = \frac{\exp(V_A/\theta)}{\exp(V_A/\theta) + \exp(V_B/\theta)} = \frac{1}{1 + \exp[(V_B - V_A)/\theta]}$$

As can be seen, the choice probability of alternative A depends on the difference between the systematic utilities. Furthermore, as shown in Fig. 3.3, this choice probability is equal to 0.5 if the two alternatives have equal systematic utilities ($V_B - V_A = 0$). It has an S-shaped semisymmetric graph for positive and negative values of $V_B - V_A$. In addition, it tends to one as $V_B - V_A$ tends to $-\infty$ (as the systematic utility of alternative A becomes infinitely greater than that of B) and it tends to zero as $V_B - V_A$ tends to $+\infty$. The rate of variation of the choice probability of A with respect to variations of $V_B - V_A$ is larger for values of $V_B - V_A$ close to zero, where it is almost linear, and increases as the variance of the random residuals (parameter θ) decreases. As the absolute value of $V_B - V_A$ increases, the slope of $p[A]$ approaches the horizontal; for large differences $V_B - V_A$ the variations of choice probability have low sensitivity to the variations of $V_B - V_A$.

Similar considerations apply to the more general case of the multinomial logit model with m alternatives. From expression (3.3.6) it can be seen that:

$$p[j] = \frac{1}{1 + \sum_{h \neq j} \exp[(V_h - V_j)/\theta]}$$

Influence of residual variance. From (3.3.6) it can be seen that a smaller random residual variance (smaller parameter θ) leads to a larger choice probability for the alternative with maximum systematic utility. This probability tends to one (a deterministic utility model) as the variance tends to zero. Conversely, as the variance of the residuals increases, the exponents V_j/θ tend to the same value (zero) and the choice probabilities of the different alternatives tend to the same value, equal to $1/m$. The effect of the random residual variance is graphically illustrated in Fig. 3.2 and numerically in Fig. 3.4 for two choice alternatives cor-

$$p[A] = \frac{\exp[(-0.1 \cdot t_A - 1 \cdot mc_A)/\theta]}{\exp[(-0.1 \cdot t_A - 1 \cdot mc_A)/\theta] + \exp[(-0.1 \cdot t_B - 1 \cdot mc_B)/\theta]}$$

$t_A = 20$ min $\quad c_A = 3.6$ unit $\quad V_A = -5.6$

$t_B = 40$ min $\quad c_B = 0.6$ unit $\quad V_B = -4.6$

	$\theta = 10$	$\theta = 1$	$\theta = 0.5$
p_A	0.48	0.27	0.12
p_B	0.52	0.73	0.88

Fig. 3.4 Effect of the variance of random residuals on choice probabilities for a binomial logit model

responding to two paths with attributes given by travel time (t) and monetary cost (mc).

Independence from irrelevant alternatives. From expression (3.3.6), another general property of the logit model can easily be deduced. Choice probability ratios between any two alternatives depend only on the systematic utilities of the two alternatives and, in particular, are independent of the number and systematic utilities of other choice alternatives:

$$p[j]/p[h] = \exp(V_j/\theta)/\exp(V_h/\theta) \tag{3.3.7}$$

This property, known in the literature as Independence from Irrelevant Alternatives (IIA), can sometimes lead to unrealistic results.

Consider, for example, the choice between two alternatives A and B having equal systematic utility. In this case, the logit model probability (3.3.6) of choosing each alternative is 0.50 and the ratio between the probabilities of choosing A and B is equal to one:

$$p[A]/p[B] = \exp(V_A/\theta)/\exp(V_B/\theta) = 1$$

Suppose now that a third alternative C is added to the choice set. Alternative C has the same systematic utility as the other two, but is otherwise very similar to alternative B. To give a specific example, suppose that the choice is between transport modes, where alternative A is a car and alternative B is a bus. Suppose further that the systematic utilities of the two are the same so they have the same choice probability. A third alternative C is introduced, consisting of a new bus line that runs on the same timetable, makes the same stops, and is generally perceived the same as B. Alternatives B and C would have the same choice probabilities. Moreover, because of the IIA property, the ratio between the probabilities of choosing car A and bus B remains equal to one. Therefore, each of the three alternatives would have a probability of 1/3 of being chosen. Thus, the probability of choosing the car would change from 0.50 to 0.33 simply because of the illusory increase in the number of choice alternatives. This result is clearly paradoxical and derives from the lack of realism of the basic assumptions of the logit model in the case described: namely, that the decision-maker perceives the alternatives as completely distinct, and therefore that their random residuals are independent. A more realistic choice model can be obtained by introducing a covariance between the random residuals of alternatives B and C, as shown in the following sections. In general, as shown below, a multinomial logit model has the property that any variation in the choice probability of one alternative (resulting from a change in its attributes) leads to proportional variations in the choice probabilities of all other alternatives.

In applications, the multinomial logit model should be used with choice alternatives that are sufficiently distinct for the assumption of independent random residuals to be plausible.

3.3.2 The Single-Level Hierarchical Logit Model

The hierarchical logit model[10] partially overcomes the assumption of independent random residuals that underlies the multinomial logit model although retaining a closed-form analytical expression for the choice probabilities.

To simplify the exposition, this section deals with the case of a single level of choice hierarchy, with equal choice model parameters. Furthermore, the presentation of the model relies on a graphic representation of the choice process and a particular decomposition scheme of the random residuals. These simplifications are not necessary and are relaxed in the next section dealing with general hierarchical logit models and in Sect. 3.3.5 dealing with generalized extreme value models.

Suppose that the decision-maker's choice set I is subdivided into nonoverlapping subsets $I_1, I_2, \ldots, I_k, \ldots$, called *groups* or *nests*. Suppose also that the utility function of the generic alternative j, belonging to the subset I_k, can be expressed[11] as

$$U_j = V_j + \varepsilon_j = V_j + \eta_k + \tau_{j/k} \quad \forall j \in I_k,\ \forall k \tag{3.3.8}$$

with

$$E[\varepsilon_j] = E[\eta_k] = E[\tau_{j/k}] = 0$$

$$\text{Cov}[\eta_k, \eta_h] = \text{Cov}[\eta_k, \tau_{j/k}] = \text{Cov}[\tau_{j/k}, \tau_{i/k}] = 0$$

As can be seen, it is assumed that the overall random residual ε_j is decomposed into the sum of two zero-mean random variables. The first, η_k, takes on one value for all the alternatives belonging to the same group, although it can assume different values for different groups. The second, $\tau_{j/k}$, takes on different values for each alternative. It is also assumed that the variables η_k and $\tau_{j/k}$ are statistically independent.

---

[10]The hierarchical logit model is also known in the international literature as the nested logit model.

[11]The hierarchical logit model can be obtained in a different and more rigorous way, as a special case of the GEV model described in Sect. 3.A.2.

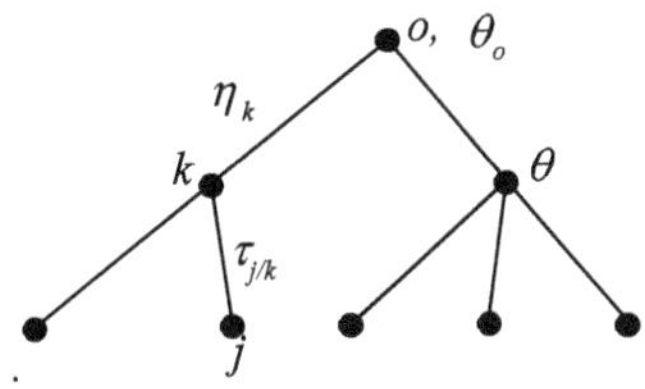

Fig. 3.5 Choice tree of a single-level hierarchical logit model

These assumptions imply that the decision-maker perceives alternatives belonging to the same group as similar; the similarity is captured by the covariance among the overall random residuals of these alternatives. In a mode choice situation, for example, the available modes can be divided into two groups: public modes (bus and train) and private modes (car and motorbike). Assumption (3.3.8) implies that the decision-maker perceives the modes belonging to the same group to be similar inasmuch as they share a number of attributes (flexibility, privacy, etc.).

The utility structure and the choice mechanism corresponding to a single-level hierarchical logit model can be represented by a choice tree, as shown in Fig. 3.5. In the choice tree, "elementary" choice alternatives (e.g., transport modes) correspond to nodes with no exit links ("leaves" of the tree), whereas the root node o has no entering links. The intermediate nodes k, one for each group, represent compound alternatives: groups of elementary alternatives. The random residuals η_k and $\tau_{j/k}$ are associated with the branches that correspond to groups and to elementary alternatives, respectively.

The choice tree can be viewed as the representation of a hypothetical choice process. Starting from the root node, the decision-maker first chooses group k from the available groups (represented by nodes linked to the root); she then chooses elementary alternative j from those belonging to group k (represented by the leaves connected to the node k). The expression for the overall choice probability of an alternative j, $p[j]$, is obtained as the product of the conditional probability $p[j/k]$ of choosing elementary alternative j within group k (lower level), multiplied by the probability $p[k]$ of choosing group k (upper level):

$$p[j] = p[j/k] \cdot p[k] \tag{3.3.9}$$

The name of the model is derived, in fact, from this probability structure.

To specify the probabilities in (3.3.9), further assumptions on the distribution of random residuals must be introduced. For the single-level hierarchical logit model, it is assumed that the random residuals relative to the alternatives available at each decision node (the root and the intermediate nodes) are identically and independently distributed Gumbel random variables.

Considering first the lower-level nodes (elementary alternatives), the residuals $\tau_{j/k}$ are assumed to be i.i.d. Gumbel variables with zero mean and the same parameter θ for all groups k and all alternatives j. In the choice among alternatives belonging to group k, the perceived utility associated with alternative j, $U_{j/k}$, can be expressed as

$$
\begin{aligned}
&U_{j/k} = V_j + \tau_{j/k} \quad \forall j \in I_k,\ \forall k \\
&E[\tau_{j/k}] = 0 \quad \forall j \in I_k,\ \forall k \\
&\mathrm{Var}[\tau_{j/k}] = \pi^2\theta^2/6 \quad \forall j \in I_k,\ \forall k
\end{aligned}
\tag{3.3.10}
$$

Under these assumptions, the conditional choice probability of the elementary alternative j can be expressed as

$$
p[j/k] = Pr[U_{j/k} > U_{i/k}] = Pr[V_j - V_i > \tau_{i/k} - \tau_{j/k}] \quad \forall i \in I_k,\ i \neq j \tag{3.3.11}
$$

and, given the assumptions on the distribution of the residuals $\tau_{j/k}$, probability (3.3.11) can be expressed as a multinomial logit model:

$$
p[j/k] = \frac{\exp(V_j/\theta)}{\sum_{i \in I_k} \exp(V_i/\theta)} \tag{3.3.12}
$$

At the upper level, the choice is made among groups of alternatives, with each group k being considered as a compound alternative. Group k will be chosen if any one of the elementary alternatives belonging to it is chosen. Because the perceived utilities of elementary alternatives are random, the probability $p[k]$ that group k is chosen is the same as the probability that one of its elementary alternatives has the maximum perceived utility among all elementary alternatives in the choice set. Equivalently, probability $p[k]$ can be obtained by assigning to group k an inclusive perceived utility U_k^* equal to the utility of its most attractive alternative, that is, the maximum utility of all the elementary alternatives belonging to the group

$$
U_k^* = \max_{j \in I_k}\{U_j\} = \max_{j \in I_k}\{V_j + \tau_{j/k}\} + \eta_k \tag{3.3.13}
$$

which is again a random variable. The probability that group k is chosen is then the probability that its inclusive perceived utility U_k^* is greatest among the different groups.

The perceived utilities $U_j = V_j + \tau_{j/k}$ of the various alternatives j in group k are, by assumption (3.3.8), independently distributed Gumbel variables with the same scale parameter θ. As stated earlier, the maximum of a set of such random variables is also distributed as a Gumbel variable with parameter θ and with mean equal to:

$$
V_k^* = E[U_k^*] = E\left[\max_{j \in I_k}\{V_j + \tau_{j/k}\}\right] = \theta \ln \sum_{j \in I_k} \exp(V_j/\theta) = \theta Y_k \tag{3.3.14}
$$

where V_k^* is the Expected Maximum Perceived Utility (EMPU) or inclusive systematic utility and Y_k is the corresponding logsum variable. In the expression for the inclusive perceived utility (3.3.13), the r.v. $\max(V_j + \tau_{j/k})$ can be replaced by its expected value plus a deviation τ_k^*[12] from this value, which is another zero-mean Gumbel variable with parameter θ. Then:

$$U_k^* = \theta Y_k + \tau_k^* + \eta_k = \theta Y_k + \varepsilon_k^* \tag{3.3.15}$$

Thus, the perceived utility of group k has a mean value θY_k and a deviation ε_k^*, which is the sum of the two zero-mean random variables τ_k^* and η_k. The basic assumption of the hierarchical logit model is that at each choice level the random residuals of the available alternatives are i.i.d. Gumbel variables; that is, it is assumed that the ε_k^* are i.i.d. Gumbel variables with zero mean and parameter θ_o, with η_k distributed in a way that makes this so:

$$\begin{aligned} E\left[\varepsilon_k^*\right] &= 0 \quad \forall k \\ \mathrm{Var}\left[\varepsilon_k^*\right] &= \pi^2\theta_o^2/6 \quad \forall k \end{aligned} \tag{3.3.16}$$

In accordance with this assumption, the choice probability of group k is expressed by a multinomial logit model. In fact:

$$p[k] = Pr\left[U_k^* > U_h^*\right] = Pr\left[\theta Y_k - \theta Y_h > \varepsilon_h^* - \varepsilon_k^*\right] \quad \forall h \neq k$$

and, given the results of the previous section:

$$p[k] = \frac{\exp(\theta Y_k/\theta_o)}{\sum_h \exp(\theta Y_h/\theta_o)} = \frac{\exp(\delta Y_k)}{\sum_h \exp(\delta Y_h)} \tag{3.3.17}$$

where δ is the ratio of parameters θ and θ_o associated with the two choice levels:

$$\delta = \theta/\theta_o \tag{3.3.18}$$

Replacing expressions (3.3.12) and (3.3.17) in (3.3.9), the choice probability of the generic elementary alternative j is obtained:

$$p[j] = p[j/k] \cdot p[k] = \frac{\exp(V_j/\theta)}{\sum_{i\in I_k} \exp(V_i/\theta)} \cdot \frac{\exp(\delta Y_k)}{\sum_h \exp(\delta Y_h)} \tag{3.3.19}$$

Variances and covariances of the random residuals ε_j of the elementary alternatives' overall perceived utility (3.3.8) can also be derived. The variance of ε_j coincides with that of the random residual ε_k^* because the two variables are the sum of the same variable (η_k) and another independent Gumbel variable (τ_k^* and $\tau_{j/k}$, respectively) with zero mean and the same parameter θ. Therefore:

$$\mathrm{Var}[\varepsilon_j] = \mathrm{Var}\left[\varepsilon_k^*\right] = \pi^2\theta_o^2/6 \quad \forall j \tag{3.3.20}$$

---

[12] From the Gumbel variable's property of stability with respect to maximization, the r.v. τ_k^* is distributed like the variable $\tau_{j/k}$ associated with each alternative j belonging to group k, that is, as a Gumbel variable with zero mean and parameter θ.

The variance of the random residual ε_j is identical for all elementary alternatives. There is also a positive covariance between the random residuals of any pair of alternatives i and j belonging to the same group. In fact:

$$\begin{aligned}\text{Cov}[\varepsilon_i, \varepsilon_j] &= E\big[(\eta_k + \tau_{i/k}) \cdot (\eta_k + \tau_{j/k})\big] \\ &= E\big[\eta_k^2\big] + E[\eta_k \tau_{j/k}] + E[\eta_k \tau_{i/k}] + E[\tau_{i/k}\tau_{j/k}] \quad \forall i, j \in I_k\end{aligned}$$

Because all the variables η_k, $\tau_{i/k}$, and $\tau_{j/k}$ have zero mean and are mutually independent, the first term is equal to the variance of η_k and the others are zero, because they are the covariances of independent random variables:

$$\text{Cov}(\varepsilon_i, \varepsilon_j)\,\text{Var}(\eta_k) \quad \forall i, j \in I_k \tag{3.3.21}$$

However, if two elementary alternatives i and j belong to different groups, all the terms are zero and so also is the covariance between ε_i and ε_j.

The variance of η_k can be expressed as a function of the two parameters θ and θ_o:

$$\text{Var}[\eta_k] = \text{Var}[\varepsilon_j] - \text{Var}[\tau_{j/k}] = \frac{\pi^2(\theta_o^2 - \theta^2)}{6} \quad \forall k \tag{3.3.22}$$

From the previous results, the structure of the random residual variance–covariance matrix can be determined. The elements of the main diagonal are all equal to the variance of the residuals ε_j, expressed by (3.3.20). The covariance between each pair of alternatives belonging to the same group is the same and equal to the value given by (3.3.21) and (3.3.22), whereas the covariance between alternatives belonging to different groups is zero. Therefore, if the alternatives of each group are ordered sequentially, the resulting variance–covariance matrix has a block diagonal structure. Figure 3.6 shows a choice tree and the corresponding variance–covariance matrix.

It is also possible to express the coefficient of correlation between the perceived utilities of two alternatives i and j as a function of the basic model parameters:

$$\rho_{ij} = \begin{cases} \frac{\text{Cov}[\varepsilon_i \varepsilon_j]}{\text{Var}[\varepsilon_i]^{1/2}\,\text{Var}[\varepsilon_j]^{1/2}} = \frac{\theta_o^2 - \theta^2}{\theta_o^2} = 1 - \delta^2 & \text{if } i, j \in I_k \\ 0 & \text{otherwise} \end{cases} \tag{3.3.23}$$

The parameters θ, θ_o, and δ play a major role in the structure of the hierarchical logit model and in determining the choice probabilities.

First, parameter δ defined by (3.3.18) must take on values in the interval [0, 1]. It is defined by the ratio between two nonnegative quantities and, because the variance of $\varepsilon_j(\pi^2\theta_o^2/6)$ cannot be less than that of one of its components $\tau_{j/k}(\pi^2\theta^2/6)$, the following must hold.

$$\theta_o \geq \theta \;\rightarrow\; 0 \leq \delta \leq 1$$

As the variance of $\tau_{j/k}$ tends to that of ε_j (i.e., as θ tends to θ_o), parameter δ tends to one. In this case, the variance of η_k (3.3.22) and the covariance between two alter-

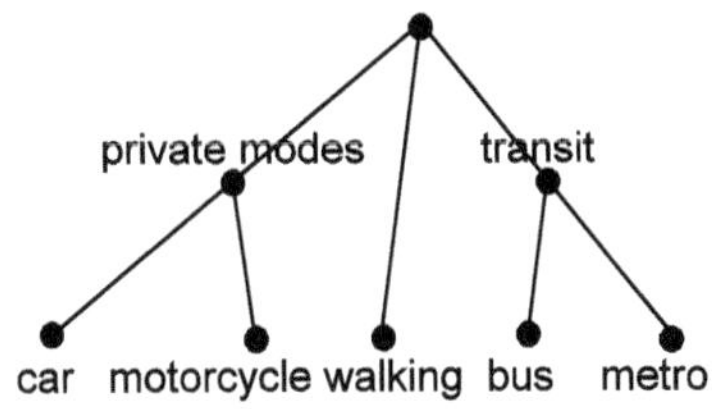

$$
\begin{array}{l}
\\
\text{car} \\
\text{motorcycle} \\
\text{walking} \\
\text{bus} \\
\text{metro}
\end{array}
\quad
\begin{array}{c}
\\
\\
\\
\dfrac{\pi^2}{6} \\
\\
\\
\end{array}
\begin{array}{ccccc}
\text{car} & \text{motorcycle} & \text{walking} & \text{bus} & \text{metro} \\
\end{array}
\left[
\begin{array}{ccccc}
\theta_o^2 & \theta_o^2-\theta^2 & 0 & 0 & 0 \\
\theta_o^2-\theta^2 & \theta_o^2 & 0 & 0 & 0 \\
0 & 0 & \theta_o^2 & 0 & 0 \\
0 & 0 & 0 & \theta_o^2 & \theta_o^2-\theta^2 \\
0 & 0 & 0 & \theta_o^2-\theta^2 & \theta_o^2
\end{array}
\right]
$$

Fig. 3.6 Choice tree and variance–covariance matrix of a single-level hierarchical logit model

natives belonging to the same group (3.3.21) both tend to zero, and the hierarchical logit model (3.3.19) reduces to the multinomial logit model.

This can be seen by substituting $\delta = 1$ in (3.3.19), yielding:

$$
\begin{aligned}
p[j] &= \frac{\exp(V_j/\theta)}{\sum_{i\in I_k}\exp(V_i/\theta)} \cdot \frac{\exp[\ln\sum_{i\in I_k}\exp(V_i/\theta)]}{\sum_h \exp[\ln\sum_{i\in I_h}\exp(V_i/\theta)]} \\
&= \frac{\exp(V_j/\theta)}{\sum_h \sum_{i\in I_k}\exp(V_i/\theta)} \qquad (3.3.24)
\end{aligned}
$$

which is a multinomial logit model with a different expression for the summation in the denominator.

If the variance of $\tau_{j/k}$ tends to zero (i.e., θ tends to zero), parameter δ will also tend to zero. In this case, the two probabilities in the model (3.3.19) will be modified as follows.

- The conditional choice of an elementary alternative within a group degenerates into a deterministic choice of the alternative with maximum systematic utility:

$$
\lim_{\theta\to 0} p[j/k] = \lim_{\theta\to 0} \frac{\exp(V_j/\theta)}{\sum_{i\in I_k}\exp(V_i/\theta)} = \begin{cases} 1 & \text{if } V_j = \max_{i\in I_k}(V_i) \\ 0 & \text{otherwise} \end{cases} \qquad (3.3.25)
$$

- The systematic utilities of alternative groups, equal to θY_k, assume the value of the maximum systematic utility among the elementary alternatives in each group:

$$
\lim_{\theta\to 0} \theta Y_k = \lim_{\theta\to 0} \theta \ln \sum_{i\in I_k} \exp(V_i/\theta) = \max_{i\in I_k}(V_i)
$$

The choice probability of the group therefore becomes

$$p[k] = \frac{\exp[\max_{i \in I_k}\{V_i\}/\theta_o]}{\sum_h \exp[\max_{i \in I_h}\{V_i\}/\theta_o]} \tag{3.3.26}$$

Thus, if parameter δ is zero, the random residuals associated with the conditional utilities of elementary alternatives within a group are zero ($\mathrm{Var}[\tau_{j/k}] = 0$). In this case, the choice between groups is modeled by comparing, using a probabilistic logit model, the alternatives having maximum systematic utility within each group: a random residual at the group level still exists, and the maximum utility alternative is deterministically chosen within each group.

Some special cases of the model presented can be analyzed. If a group k consists of a single alternative j, then $p[j/k] = 1$ and the general expression (3.3.19) for this alternative becomes

$$p[j] = \frac{\exp(V_j/\theta_o)}{\exp(V_j/\theta_o) + \sum_{h \neq k} \exp(\delta Y_h)} \tag{3.3.27}$$

In some applications of the single-level hierarchical logit model, and in particular for systems of partial share models covered in the next chapter, the systematic utility V_j of alternative j in group k is decomposed into two parts: one part, V_k, associated with group k itself; and a second part, $V_{j/k}$, associated with the alternative within the group:

$$V_j = V_k + V_{j/k} \tag{3.3.28}$$

This decomposition leads to an alternative formulation of the choice probabilities $p[j/k]$ and $p[k]$. By replacing (3.3.28) in (3.3.12) and (3.3.17), respectively, it follows that

$$\begin{aligned} p[j/k] &= \frac{\exp(V_j/\theta)}{\sum_{i \in I_k} \exp(V_i/\theta)} = \frac{\exp[(V_k + V_{j/k})/\theta]}{\exp(V_k/\theta) \cdot \sum_{i \in I_k} \exp(V_{i/k}/\theta)} \\ &= \frac{\exp(V_{j/k}/\theta)}{\sum_{i \in I_k} \exp(V_{i/k}/\theta)} \end{aligned} \tag{3.3.29}$$

and

$$p[k] = \frac{\exp(V_k/\theta_o + \delta Y_k')}{\sum_h \exp(V_h/\theta_o + \delta Y_h')} \tag{3.3.30}$$

because

$$\begin{aligned} \delta Y_k &= \delta \ln \sum_{j \in I_k} \exp(V_j/\theta) = \delta \ln \sum_{j \in I_k} \exp\big[(V_k + V_{j/k})/\theta\big] \\ &= \delta \ln \left[\exp(V_k/\theta) \cdot \sum_{j \in I_k} \exp(V_{j/k}/\theta) \right] = \delta V_k/\theta + \delta \ln \sum_{j \in I_k} \exp(V_{j/k}/\theta) \\ &= V_k/\theta_o + \delta Y_k' \end{aligned}$$

where Y_k' is the logsum variable of group k obtained with the alternative specific systematic utilities $V_{j/k}$.

3.3.3 The Multilevel Hierarchical Logit Model*

The single-level hierarchical logit model described in the previous section is a first generalization of the multinomial logit model. However, it retains many simplifying features of the multinomial logit model, such as the assumption of identical covariance between the alternatives belonging to each group and the representation of a single level of nesting, or correlation, of alternatives. These assumptions can be generalized considerably, as described in the following.

The starting point is once again the representation of the choice process and of the covariance between the perceived utilities by means of a general choice tree; the name "tree logit," sometimes given to these models, derives from this approach. The leaves, or terminal nodes, of the tree correspond to elementary choice alternatives (e.g., different transport modes). Nodes i, j, l in Fig. 3.7 are elementary alternatives belonging to the total choice set I. Each intermediate node r can be seen as representing a conditional choice situation in which the decision-maker has available a set of elementary and/or compound alternatives corresponding to the leaves and/or intermediate nodes directly linked to node r. Thus, each intermediate node represents a compound alternative, that is, the set of elementary alternatives that can be reached by the intermediate node itself. At each intermediate node, the choice is made among all the elementary alternatives that can be reached, either directly or indirectly through other intermediate nodes, from the node itself. In the example in Fig. 3.7, the choice represented by node r is made between alternatives i, j, l, with the elementary alternatives i and l grouped in the compound alternative f. More formally, the following elements in Fig. 3.7 can be defined on the choice tree.

- o is the *root or initial node*, the beginning of the decision process
- i, j, l are the *terminal nodes or leaves*, the elementary choice alternatives
- r is a *generic node* of the tree; if this is an *intermediate* (or *structural*) node, it represents both a group of alternatives (compound alternative) and an intermediate choice
- I is the set of elementary alternatives or choice set
- I_r is the set of *descendant nodes* (*children*) *of* r; the set of nodes that can be reached directly from r; it represents the set of elementary or compound choice alternatives available for the conditional choice at r; $I_r = \emptyset$ if $r \in I$
- $a(r)$ is the *predecessor node* (*parent* or *first ancestor*) of node r, a node linked to r by the single directed link $(a(r), r)$ belonging to the graph; $a(o) = \emptyset$
- A_r is the set of all ancestor nodes of r, the set of nodes belonging to the unique branch linking the root o and r, but excluding both node r and the root o, $A_r \equiv \{a(r), a(a(r))\ldots\}$
- $p(r, s)$ is the first common ancestor node of the pair of nodes r and s

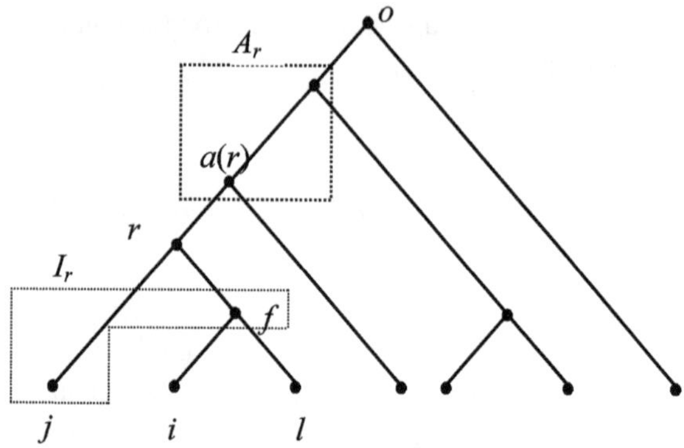

Fig. 3.7 Choice tree of multilevel hierarchical logit models

In this notation, single nodes are indicated with lowercase letters (o, i, j, l, r, s), groups of nodes with capital letters (A, I), and nodes related in some structural way to particular other nodes as lowercase letter functions of those other nodes $(a(r), p(r, s))$.

At each choice node, whether intermediate or initial, it is assumed that a conditional choice is made among all the available alternatives. These alternatives are represented by nodes r, and may be either elementary alternatives (leaves of the tree) or compound alternatives (intermediate nodes). For any such alternative, the node that represents the choice situation directly involving it is $a(r)$, and the full set of alternatives in the choice situation is $I_{a(r)}$.

To model the conditional choice, a perceived utility $U_{r/a(r)}$ is assigned to each node (alternative) r. This is a random variable that, as usual, is decomposed into the sum of its mean, V_r, and a random residual, $\varepsilon_{r/a(r)}$, with the following properties.

- If r is a leaf of the tree, V_r is the expected value of its perceived utility $U_{r/a(r)}$. If r is an intermediate node, V_r is the expected value of the maximum perceived utility (EMPU or inclusive value) of the alternatives, whether elementary or not, belonging to I_r;
- The random residuals $\varepsilon_{r/a(r)}$ of all nodes r that are descendants of $a(r)$ are assumed to be i.i.d. Gumbel variables with zero mean and parameter $\theta_{a(r)}$. Therefore, the variance $\mathrm{Var}[\varepsilon_{r/a(r)}] = \pi^2\theta^2_{a(r)}/6$ is associated with the conditional choice made at node $a(r)$ from all the elementary alternatives directly or indirectly reached from $a(r)$.

From the above assumptions, it follows that

$$
\begin{aligned}
&U_{r/a(r)} = V_r + \varepsilon_{r/a(r)} \quad \forall r \in I_{a(r)} \\
&E[\varepsilon_{r/a(r)}] = 0 \\
&\mathrm{Var}[\varepsilon_{r/a(r)}] = \frac{\pi^2\theta^2_{a(r)}}{6}
\end{aligned}
\tag{3.3.31}
$$

From the results on the expected value of the maximum of Gumbel variables referred to in Sect. 3.3.1, the systematic utility assigned to any node can be determined recursively by starting from the choice tree leaves as

$$V_r = \begin{cases} E[U_{r/a(r)}] & \text{if } r \in I \\ \theta_r \ln \sum_{h \in I_r} \exp(V_h/\theta_r) = \theta_r Y_r & \text{if } r \notin I \end{cases} \tag{3.3.32}$$

Under the above hypotheses, the conditional probability of choosing alternative r at the choice node $a(r)$ is expressed by a multinomial logit model:

$$p\big[r/a(r)\big] = \frac{\exp(V_r/\theta_{a(r)})}{\sum_{r' \in I_{a(r)}} \exp(V_{r'}/\theta_{a(r)})} \tag{3.3.33}$$

and also, from (3.3.32):

$$p\big[r/a(r)\big] = \frac{\exp(V_r/\theta_{a(r)})}{\exp(Y_{a(r)})} \tag{3.3.34}$$

If the alternative r is a compound alternative (i.e., r is an intermediate node) in (3.3.32), the numerator of (3.3.33) becomes:

$$\exp(V_r/\theta_{a(r)}) = \exp\left(\frac{\theta_r}{\theta_{a(r)}} Y_r\right) = \exp(\delta_r Y_r)$$

where δ_r is the ratio of coefficients θ_r and $\theta_{a(r)}$. It is analogous to the coefficient δ introduced in the previous section (see (3.3.18)) and, as such, must be in the interval [0, 1]. In this case, expressions (3.3.33) and (3.3.34) can be reformulated as

$$p\big[r/a(r)\big] = \frac{\exp(\delta_r Y_r)}{\sum_{r'} \exp(V_{r'}/\theta_{a(r)})} = \frac{\exp(\delta_r Y_r)}{\exp(Y_{a(r)})} \tag{3.3.35}$$

Finally, the absolute (unconditional) probability of choosing the elementary alternative $j \in I$ can be obtained from the definition of conditional probability and from the assumptions made on the tree choice mechanism:

$$p[j] = p\big[j/a(j)\big] \cdot p\big[a(j)/a\big(a(j)\big)\big] \cdots \quad j \in I$$

or

$$p[j] = p\big[j/a(j)\big] \prod_{r \in A_j} p\big[r/a(r)\big] \quad j \in I \tag{3.3.36}$$

Replacing expressions (3.3.34) and (3.3.35) in (3.3.36) yields:

$$p[j] = \frac{\exp(V_j/\theta_{a(j)})}{\exp(Y_{a(j)})} \cdot \prod_{r \in A_j} \frac{\exp(\delta_r Y_r)}{\exp(Y_{a(r)})} \quad j \in I \tag{3.3.37}$$

and also

$$p[j] = \frac{\exp(V_j/\theta_{a(j)})}{\exp(Y_o)} \cdot \prod_{r \in A_j} \frac{\exp(\delta_r Y_r)}{\exp(Y_r)}$$

$$= \frac{\exp(V_j/\theta_{a(j)})}{\exp(Y_o)} \cdot \prod_{r \in A_j} \exp\left[(\delta_r - 1)Y_r\right] \quad j \in I \qquad (3.3.38)$$

Absolute choice probabilities $p[j]$ can therefore be computed recursively through the following steps.

- Calculate $\delta_r = \theta_r/\theta_{a(r)}$ for each node r.
- Recursively calculate values Y_r, with expression (3.3.32).
- Calculate probabilities $p[j]$, $j \in I$, with expression (3.3.38).

$$\begin{array}{llll} \text{Given:} & \theta_r & r \notin I & \text{with } \theta_r = 0 \text{ if } r \in I \\ & I_r & r \notin I & \text{with } I_r = \emptyset \text{ if } r \in I \\ & V_j & \forall j \in I & \end{array}$$

The model described can be demonstrated with the choice tree in Fig. 3.8. The leaves of the tree (*AI*, *CD*, *CP*, *BS*, *ST*, *FT*) represent the elementary choice alternatives that, in this example, are the transport modes available for an intercity trip: air (*AI*), car driver (*CD*), car passenger (*CP*), bus (*BS*), slow train (*ST*), and fast train (*FT*). The intermediate nodes represent groups of alternatives, or compound alternatives. Node *CR* represents the car, combining the two alternatives of car driver and car passenger, node *LT* the public land transport modes (bus, slow train, and fast train), and node *RW* combines the railway alternatives. Finally, the respective values of parameters θ and δ are assigned to each intermediate node and to the root.

Following expression (3.3.36), the choice probability of fast train (*FT*) can be written as

$$p[FT] = p[FT/RW].p[RW/LT].p[LT/o]$$

where

$$p[FT/RW] = \frac{\exp(V_{FT}/\theta_{RW})}{[\exp(V_{ST}/\theta_{RW}) + \exp(V_{FT}/\theta_{RW})]} = \frac{\exp(V_{FT}/\theta_{RW})}{\exp(Y_{RW})}$$

with

$$Y_{RW} = \ln\left[\exp(V_{ST}/\theta_{RW}) + \exp(V_{FT}/\theta_{RW})\right]$$

$$p[RW/LT] = \frac{\exp(\theta_{RW} Y_{RW}/\theta_{LT})}{\exp(\theta_{RW} Y_{RW}/\theta_{LT}) + \exp(V_{BS}/\theta_{LT})}$$

$$= \frac{\exp(\delta_{RW} Y_{RW})}{\exp(\delta_{RW} Y_{RW}) + \exp(V_{BS}/\theta_{LT})} = \frac{\exp(\delta_{RW} Y_{RW})}{\exp(Y_{LT})}$$

with

$$Y_{LT} = \ln\left[\exp(\delta_{RW} Y_{RW}) + \exp(V_{BS}/\theta_{LT})\right] \qquad (3.3.39)$$

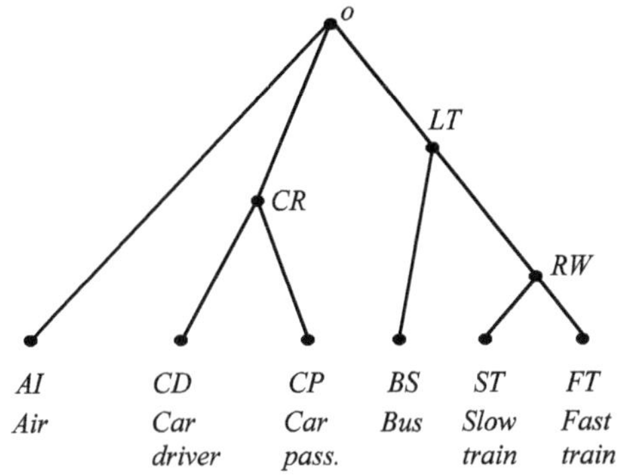

$$\Sigma = \frac{\pi^2}{6}\begin{array}{c} \\ AI \\ CD \\ CP \\ BS \\ ST \\ FT \end{array}\begin{array}{cccccc} AI & CD & CP & BS & ST & FT \end{array}$$

$$\Sigma = \frac{\pi^2}{6}\left[\begin{array}{cccccc} \theta_o^2 & & & & & \\ & \theta_o^2 & \theta_o^2-\theta_{CR}^2 & & & \\ & \theta_o^2-\theta_{CR}^2 & \theta_o^2 & & & \\ & & & \theta_o^2 & \theta_o^2-\theta_{LT}^2 & \theta_o^2-\theta_{LT}^2 \\ & & & \theta_o^2-\theta_{LT}^2 & \theta_o^2 & \theta_o^2-\theta_{RW}^2 \\ & & & \theta_o^2-\theta_{LT}^2 & \theta_o^2-\theta_{RW}^2 & \theta_o^2 \end{array}\right]$$

Fig. 3.8 Choice tree and variance–covariance matrix for a multilevel hierarchical logit model

$$p[LT/o] = \frac{\exp(\theta_{LT}Y_{LT}/\theta_o)}{\exp(\theta_{LT}Y_{LT}/\theta_o) + \exp(\theta_{CR}Y_{CR}/\theta_o) + \exp(V_{AI}/\theta_o)}$$

$$= \frac{\exp(\delta_{LT}Y_{LT})}{[\exp(\delta_{LT}Y_{LT}) + \exp(\delta_{CR}Y_{CR}) + \exp(V_{AI}/\theta_o)]} = \frac{\exp(\delta_{LT}Y_{LT})}{\exp(Y_o)}$$

with

$$Y_{CR} = \ln\big[\exp(V_{CD}/\theta_{CR}) + \exp(V_{CP}/\theta_{CR})\big]$$

$$Y_o = \ln\big[\exp(\delta_{LT}Y_{LT}) + \exp(\delta_{CR}Y_{CR}) + \exp(V_{AI}/\theta_o)\big]$$

The absolute choice probability can be written in the form (3.3.38) as follows.

$$P[FW] = \frac{\exp(V_{FT}/\theta_{RW})}{\exp(Y_o)} \cdot \exp\big[(\delta_{LT}-1)Y_{LT}\big] \cdot \exp\big[(\delta_{RW}-1)Y_{RW}\big]$$

This choice probability can be thought of as resulting from a choice process in which the decision-maker first chooses the compound alternative "public land transport" from the available alternatives, which in this case are air, the compound alternative "car" and the compound alternative "public land transport". Subsequently, she chooses the group "train" from the alternatives available within the land transport group (bus and train), and finally fast train from the two elementary alternatives (fast and slow train) that make up the train group.

Returning to the general model, it is possible to express the variances and covariances of the random residuals as functions of the parameters θ_r. Rigorous demonstration of these results involves the use of GEV models described in Sect. 3.3.5. The same results can be obtained in a less rigorous way using the total variance decomposition method described for the single-level hierarchical logit model in the previous section. It is assumed that the total variance of each of the elementary alternatives j is identical and equal to:

$$\mathrm{Var}[\varepsilon_j] = \pi^2\theta_o^2/6 \tag{3.3.40}$$

The overall random residual of each elementary alternative ε_j is decomposed into the sum of independent zero-mean random variables $\tau_{a(r),r}$ associated with each link of the choice tree. The total variance of an elementary alternative is equal to the sum of the variances corresponding to the links of the (single) branch connecting the root to the leaf that represents the alternative. Furthermore, it is assumed that the random residual variance of each elementary alternative j that can be reached from an intermediate node r, and that is associated to the conditional choice represented by node r itself, is equal to $\pi^2\theta_r^2/6$. It follows that, for all these alternatives, the sum of the contributions of the variances associated with the links that connect r to j must be identical and equal to $\pi^2\theta_r^2/6$:

$$\mathrm{Var}[\varepsilon_{j/r}] = \pi^2\theta_r^2/6 = \mathrm{Var}[\tau_{a(j),j}] + \mathrm{Var}[\tau_{a(a(j)),a(j)}] + \cdots + \mathrm{Var}[\tau_{r,f(r,j)}]$$

where $f(r, j)$ is the only descendant of r that is on the path from r to j.

In Fig. 3.8, for example, the variances of the elementary alternatives *BS*, *ST*, and *FT*, corresponding to the conditional choice between public land transport modes represented by intermediate node *LT*, are all equal to $\pi^2\theta_{LT}^2/6$. This variance will correspond to the fraction of variance associated to the link (*LT*, *BS*) and to the sum of the variances associated with links (*LT*, *RW*) and (*RW*, *ST*) or to the links (*LT*, *RW*) and (*RW*, *FT*).

The random residual variance of the elementary alternatives relative to the conditional choice represented by node $a(r)$, the predecessor of r, is in turn the sum of the variance corresponding to r and the nonnegative term $\mathrm{Var}[\tau_{a(r),r}]$, associated with link $(a(r), r)$; this variance will therefore not be less than that associated with r, or:

$$\theta_{a(r)} \geq \theta_r \tag{3.3.41}$$

The variance contribution associated with each link $(a(r), r)$ of the graph can be expressed as

$$\mathrm{Var}[\tau_{a(r),r}] = \frac{\pi^2}{6}\left(\theta_{a(r)}^2 - \theta_r^2\right) \tag{3.3.42}$$

Inequality (3.3.41) can be generalized, assigning zero variance and $\theta_j = 0$ to the leaves of the graph, thus yielding:

$$\theta_j \leq \theta_{a(j)} \leq \cdots \leq \theta_o \tag{3.3.43}$$

From the preceding expression and the definition of the coefficients $\delta_r = \theta_r/\theta_{a(r)}$, it follows that these coefficients must belong to the interval [0, 1].

Continuing with the example in Fig. 3.8, the variance of alternatives *ST* and *FT* involved in the conditional choice between railway services (node *RW*) will be $\pi^2\theta_{RW}^2/6$, whereas the variance of alternatives involved in the choice between public land transport modes (node *LT*) will be $\pi^2\theta_{LT}^2/6$, with $\theta_{LT} \geq \theta_{RW}$. The variance contribution assigned to link (*LT*, *RW*) will therefore be $\pi^2(\theta_{LT}^2 - \theta_{RW}^2)/6$.

The variance decomposition model described here allows one to derive the covariances between the perceived utilities of any two elementary alternatives i and j. This covariance will correspond to the sum of the variances of the random residuals $\tau_{a(r),r}$ (which are independent with zero mean) associated with the links common to the two branches connecting the root to leaves i and j. Because of the tree structure, these branches can have in common only links from the root to the node where they separate, which is their last node in common. By repeatedly applying (3.3.42), the covariance of ε_i and ε_j is found to be:

$$\mathrm{Cov}[\varepsilon_i; \varepsilon_j] = \frac{\pi^2(\theta_o^2 - \theta_{p(i,j)}^2)}{6} \quad \forall i, j \in I \tag{3.3.44}$$

where $p(i, j)$ is the first common ancestor of elementary nodes i and j.

If two alternatives have the root node as their first common ancestor, that is, if they do not belong to any intermediate compound alternative, their covariance is zero. The correlation coefficient between two elementary alternatives can be deduced from expression (3.3.40) and (3.3.44) as follows.

$$\rho[i, j] = \frac{\mathrm{Cov}[\varepsilon_i; \varepsilon_j]}{[\mathrm{Var}[\varepsilon_i] \cdot \mathrm{Var}[\varepsilon_j]]^{1/2}} = \frac{\theta_o^2 - \theta_{p(i,j)}^2}{\theta_o^2} = 1 - \frac{\theta_{p(i,j)}^2}{\theta_o^2} \tag{3.3.45}$$

For the tree in Fig. 3.8, the covariance between alternatives *ST* and *FT* is given by $\pi^2(\theta_o^2 - \theta_{RW}^2)/6$, the sum of the variances relative to links (*o*, *LT*) and (*LT*, *RW*). The covariance between *ST* and *BS* will be $\pi^2(\theta_o^2 - \theta_{LT}^2)/6$ which, as stated before, is less than or equal to the covariance between *FT* and *ST*.

In the literature, the parameter θ_o is sometimes taken to be equal to one because, as shown in Chap. 8 on travel-demand estimation, only the parameters δ_r can be statistically estimated. Because all the parameters θ_r but one can be obtained from the coefficients δ_r, specifying one of the θ_rs immediately allows the others to be determined. Setting $\theta_o = 1$ leads to a simple expression for the other parameters. In this case, the covariance and the correlation coefficient between any two elementary alternatives become, respectively,

$$\mathrm{Cov}[\varepsilon_i, \varepsilon_j] = \frac{\pi^2(1 - \theta_{p(i,j)}^2)}{6}$$

$$\rho[\varepsilon_i, \varepsilon_j] = 1 - \theta_{p(i,j)}^2$$

In conclusion, the structure of the choice tree is also the structure of the covariances between the perceived utilities of the elementary alternatives. Two alternatives that have no nodes in common along the branches connecting them to the root o are independent. On the other hand, the covariance between elementary alternatives i and j belonging to the same group (their branches meet at an intermediate node) increases with greater "distance" of their first common ancestor from the root node and with smaller values of the parameter $\theta_{p(i,j)}$ associated with this node. Furthermore, the covariance between the perceived utilities of two alternatives i and j whose first common ancestor is their mutual parent $(p(i,j) = a(i) = a(j))$ is not less than the covariance between either of them and any other alternative. Continuing with the example of Fig. 3.8, the covariance between *ST* and *FT* will be greater than or equal to that of either of the two elementary alternatives with any other elementary alternative.

Choice probabilities are significantly affected by the values of parameters θ_r, and therefore by the levels of correlation between alternatives. Figure 3.9 shows the values of choice probabilities for the alternatives in Fig. 3.8, for different parameters θ_r and assuming that all systematic utilities have the same value: $V_{AI} = V_{CD} = V_{CP} = V_{BS} = V_{ST} = V_{FT}$. If the alternatives are independent (specification 1: $\theta_r/\theta_o = 1\ \forall r$), the model becomes a multinomial logit and all the alternatives have equal choice probabilities. As the correlation increases, that is, as parameters θ_{CR}, θ_{LT}, and θ_{RW} decrease, the choice probability of the most correlated alternatives tends to decrease. For example, in specification 3, the alternatives belonging to the two groups car (*CD*, *CP*) and public land transport (*BS*, *ST*, *FT*) are strongly correlated with a correlation coefficient $\rho = 0.9775$. They tend to be seen as a single alternative and their choice probabilities tend to be equal shares of the probability of a single alternative associated with each group. For the same reasons, the choice probability of alternative *AI*, which is not correlated with any other alternative, increases with increases in the correlation of the alternatives belonging to the various groups (specifications 2 and 3).

From the previous results, it can easily be demonstrated that multinomial logit and single-level hierarchical logit models are special cases of the multilevel hierarchical logit. Two different approaches can be used to show this for the multinomial logit model. In the first approach, the tree is that of the multinomial logit model described in Fig. 3.1. In this case, there are no intermediate nodes and the ancestor $a(j)$ of every leaf $j \in I$ is the root o. It then follows that $\theta_{a(j)} = \theta_o$, $A_j = \emptyset$ and, by applying expression (3.3.38), that

$$p[j] = \frac{\exp(V_j/\theta_o)}{\exp(Y_o)}$$

which, by developing the term $\exp(Y_o)$, gives rise to expression (3.3.6) for the multinomial logit.

Alternatively the multinomial logit model can be obtained from a tree of any form in which the parameters θ_r of all the intermediate nodes are the same and equal to θ_o. In this case, it follows from (3.3.44) that the covariance between any pair of alternatives is equal to zero (the residuals are independent), the coefficients $\delta_r = \theta_r/\theta_{a(r)}$ are all equal to one, and (3.3.38) reduces to the MNL expression.

Specification No.	1	2	3	4	5	6	7
θ_{LT}/θ_o	1.000	0.900	0.150	1.000	1.000	0.800	0.400
θ_{CR}/θ_o	1.000	0.900	0.150	0.800	0.800	0.600	0.200
θ_{RW}/θ_o	1.000	0.900	0.150	0.600	0.200	0.600	0.200
$p[AI]$	0.166	0.180	0.304	0.190	0.205	0.212	0.280
$p[CD]$	0.166	0.168	0.169	0.166	0.178	0.161	0.161
$p[CP]$	0.166	0.168	0.169	0.166	0.178	0.161	0.161
$p[BS]$	0.166	0.161	0.120	0.190	0.205	0.174	0.165
$p[FT]$	0.166	0.161	0.120	0.144	0.117	0.146	0.117
$p[ST]$	0.166	0.161	0.120	0.144	0.117	0.146	0.117

Fig. 3.9 Choice probabilities of the multilevel hierarchical logit model of Fig. 3.8 for varying parameters

The single-level hierarchical logit model described in the previous section can be considered as a special case of a tree with only one level of intermediate nodes

$$a\big(a(j)\big) = o \quad \forall j \in I$$

Furthermore, the parameters θ_r are all equal to θ whereas the parameter associated with the root is still indicated by θ_o. It can easily be demonstrated that the choice probability (3.3.19) obtained for the single-level hierarchical logit model results as a special case of expression (3.3.38).

Finally, as in the case of single-level hierarchical logit model, a systematic utility can be assigned to structural or intermediate nodes. This could be the part of the systematic utility common to all the alternatives connected by an intermediate node. In this case, if r is a structural node and V_r the systematic utility assigned to it, (3.3.35) becomes

$$p\big[r/a(r)\big] = \frac{\exp(V_r/\theta_{a(r)} + \delta_r Y_r')}{\exp(Y_{a(r)})}$$

where Y_r' is the logsum variable associated with a node r calculated without the systematic utility V_r, "transferred" to the structural node. Specifications of this type are used in Chap. 4.

*3.3.4 The Cross-nested Logit Model**

The single-level and multilevel hierarchical logit models described above allow us to reproduce only covariance matrices among alternatives' perceived utilities with a "block-diagonal" structure. Therefore, in order to reproduce choice contexts underlying more general covariance matrix structures,[13] the cross-nested logit model has

[13] It is worth mentioning that the Cross-Nested Logit model, as all other GEV models, is *homoskedastic* since it allows equal-variance across random residuals.

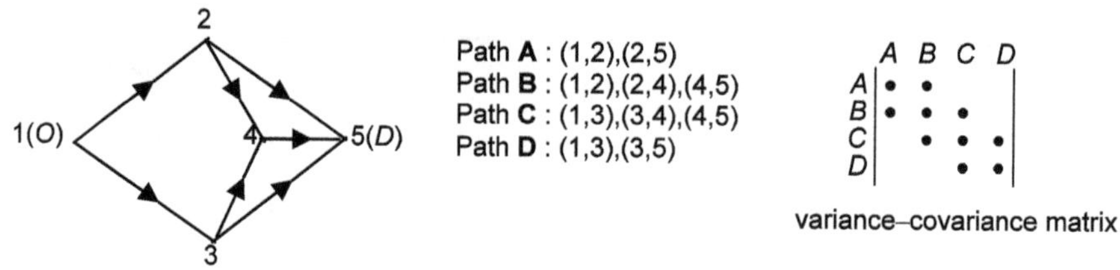

Fig. 3.10 Example of path choice and its variance–covariance matrix

Fig. 3.11 Cross-nested correlation structure for the path choice example in Fig. 3.10

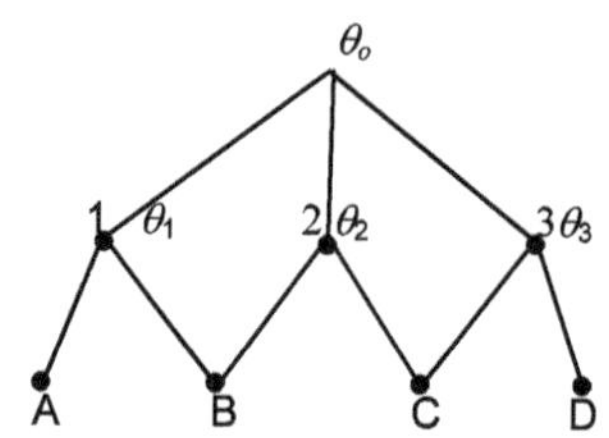

been proposed in the literature as a generalization of the hierarchical logit model, wherein an alternative may belong to more than one group, or nest, with different degrees of membership.

Consider, as an example, the path choice context reported in Fig. 3.10. There are four alternatives (paths A, B, C, D). It can be assumed that there is covariance between the perceived utilities of paths A and B (having link $(1,2)$ in common), between paths B and C (link $(4,5)$ in common) and between paths C and D (sharing link $(1,3)$). Such a covariance structure cannot be represented by a tree and, indeed, the variance–covariance matrix does not in general have a block-diagonal structure. Using a cross-nested structure, on the other hand, three "cross" nests can be specified corresponding to the three assumed binary correlations, with alternative B belonging to nests 1 and 2 and alternative C belonging to nests 2 and 3 (see Fig. 3.11).

It should be noted that, in the case of cross-nested models, the graph representing the correlation structure should be referred to as a choice graph (it is no longer a tree) even though there is no immediate interpretation in terms of a choice process. In the choice graph, intermediate nodes correspond to a group of alternatives (a nest).

With these assumptions, by adapting the formulation of the single-level hierarchical logit model, the choice probability of the generic alternative j can be expressed as

$$p[j] = \sum_k p[j/k] \cdot p[k] \tag{3.3.46}$$

where k represents the generic nest in the single-level nesting structure. The degree of membership of an alternative j in a nest k is denoted by α_{jk} and is included in the interval [0, 1]. Degrees of membership have to satisfy the following normalizing equation.

$$\sum_k \alpha_{jk} = 1 \quad \forall j \tag{3.3.47}$$

The analytical expressions for $p[j/k]$ and $p[k]$ are as follows.

$$p[j/k] = \frac{\alpha_{jk}^{1/\delta_k} e^{V_j/\theta_k}}{\sum_{i\in I_k} \alpha_{ik}^{1/\delta_k} e^{V_i/\theta_k}}; \qquad p[k] = \frac{(\sum_{i\in I_k} \alpha_{ik}^{1/\delta_k} e^{V_i/\theta_k})^{\delta_k}}{\sum_{k'}(\sum_{i\in I_{k'}} \alpha_{ik'}^{1/\delta_{k'}} e^{V_i/\theta_{k'}})^{\delta_{k'}}} \tag{3.3.48}$$

where I_k is the set of alternatives belonging to nest k, θ_k is the parameter associated with an intermediate node, θ_o is the parameter associated with the root and δ_k is the ratio θ_k/θ_o. Combining (3.3.46) and (3.3.48) gives

$$p[j] = \frac{\sum_k [\alpha_{jk}^{1/\delta_k} e^{V_j/\theta_k} \cdot (\sum_{i\in I_k} \alpha_{ik}^{1/\delta_k} e^{V_i/\theta_k})^{\delta_k - 1}]}{\sum_k (\sum_{i\in I_k} \alpha_{ik}^{1/\delta_k} e^{V_i/\theta_k})^{\delta_k}} \tag{3.3.49}$$

Analogously to the hierarchical logit model, the parameters δ_k determine the correlation among the alternatives and, for $\delta_k = 1$ (i.e., $\theta_k = \theta_o$) $\forall k$, the multinomial logit model (3.3.6) is obtained from (3.3.49):

$$p[j] = \frac{\sum_k \alpha_{jk} e^{V_j/\theta_o}}{\sum_k \sum_{i\in I_k} \alpha_{ik} e^{V_i/\theta_o}} = \frac{e^{V_j/\theta_o} \cdot \sum_k \alpha_{jk}}{\sum_i e^{V_i/\theta_o} \cdot \sum_k \alpha_{ik}} = \frac{e^{V_j/\theta_o}}{\sum_i e^{V_i/\theta_o}}$$

The cross-nested logit model can be derived from the general assumptions of random utility theory as a special case of the Generalized Extreme Value (GEV) model, as shown in Appendix 3.A.

Unlike the hierarchical logit models presented in the previous sections, the relationship between cross-nested logit model parameters and corresponding covariances cannot be expressed in a closed-form expression. Therefore, CNL covariances should be calculated through a numerical procedure based on the expression of the joint distribution of random residuals derivable from the formulation of the CNL model as a GEV model.

Interestingly, an empirical expression of CNL covariances, incorporating as specific cases the hierarchical logit covariances, is available in the literature:

$$\begin{aligned} \mathrm{Cov}[\varepsilon_i, \varepsilon_j] &= \frac{\pi^2 \theta_o^2}{6} \cdot \sum_k (\alpha_{ik})^{1/2} \cdot (\alpha_{jk})^{1/2} \cdot (1 - \delta_k^2) \\ \mathrm{Var}[\varepsilon_i] &= \frac{\pi^2 \theta_o^2}{6} \cdot \sum_k (\alpha_{ik})^{1/2} \cdot (\alpha_{ik})^{1/2} = \frac{\pi^2 \theta_o^2}{6} \cdot \sum_k \alpha_{ik} = \frac{\pi^2 \theta_o^2}{6} \end{aligned} \tag{3.3.50}$$

Numerical tests show that conjecture (3.3.50) provides a satisfactory approximation of the actual covariances when the degrees of membership tend to the $0/1$ limit bounds, whereas a slight overestimation is observed in other cases.

α_{B1}	1	0.75	0.5	0.25	0
α_{B2}	0	0.25	0.5	0.75	1
α_{C2}	0	0.25	0.5	0.75	1
α_{C3}	1	0.75	0.5	0.25	0
$\delta = 0.5$					
$p(A)$	0.25	0.2804	0.3039	0.3107	0.2929
$p(B)$	0.25	0.2196	0.1961	0.1893	0.2071
$p(C)$	0.25	0.2196	0.1961	0.1893	0.2071
$p(D)$	0.25	0.2804	0.3039	0.3107	0.2929

Fig. 3.12 Choice probabilities for the example in Fig. 3.10

Figure 3.12 reports choice probabilities for the example in Fig. 3.10 using various values of the vector $\boldsymbol{\alpha}$; equal systematic utilities are assumed.

From these results, it can be observed that an alternative that belongs to several nests has a lower choice probability than another alternative with the same systematic utility but that belongs to only one nest.

3.3.5 The Generalized Extreme Value (GEV) Model*

Generalized Extreme Value models, also known as *GEV* models, are a further generalization of logit, hierarchical logit, and cross-nested logit models. Rather than being a single model, GEV models are a whole class of random utility models. They are defined by a general mathematical formulation involving a characteristic function that has certain properties; different specifications of the characteristic function give rise to different models, including the various logit family models described in previous sections.

GEV models are consistent with the behavioral hypotheses on which random utility theory is based, that is, that the generic decision-maker associates a perceived utility to each alternative j belonging to his choice set. This perceived utility is decomposed into a deterministic part V_j (the systematic utility) and a random residual ε_j. The random residual joint distribution function implied by GEV models is such that the residuals have the same variance and, in general, non-negative covariances.

A GEV model is defined by means of a continuous and differentiable function $G(y_1, y_2, \ldots, y_m)$ of m nonnegative variables $y_1, y_2, \ldots, y_m \geq 0$ (m being the number of choice alternatives) that has the following properties.

(1) $G(\cdot)$ is nonnegative, $G(\cdot) \geq 0$.
(2) $G(\cdot)$ is homogeneous of order $\mu > 0$; that is,

$$G(\alpha y_1, \alpha y_2, \ldots, \alpha y_m) = \alpha^{\mu} G(y_1, y_2, \ldots, y_m)$$

(3) $G(\cdot)$ tends asymptotically to infinity for each y_j tending to infinity:

$$\lim_{y_j \to \infty} G(y_1, y_2, \ldots, y_m) = \infty \quad j = 1, 2, \ldots, m$$

(4) The kth partial derivative of $G(\cdot)$ (or the order k derivative of $G(\cdot)$) with respect to a generic combination of k variables y_j, for $j = 1, 2, \ldots, m$, is nonnegative if k is odd and nonpositive if k is even.

Recall that, by Euler's theorem, if $G(\cdot)$ is homogeneous of order μ, the first partial derivative of $G(\cdot)$ with respect to one of its variables y_j, $\partial G/\partial y_j = G_j(y_1, y_2, \ldots, y_m)$, is homogeneous of order $\mu - 1$.

By substituting y_i with $\exp(V_i)$ (therefore satisfying the nonnegativity of the y_i), the GEV model can be derived from the *random utility theory* hypotheses. Indeed, if the function $G(\cdot)$ meets the above conditions (1) to (4), it may be proved that the function:

$$F(\varepsilon_1, \varepsilon_2, \ldots, \varepsilon_m) = \exp\left[-G(e^{-\varepsilon_1}, e^{-\varepsilon_2}, \ldots, e^{-\varepsilon_m})\right] \tag{3.3.51}$$

is a multivariate extreme value distribution, whose marginals are homoskedastic Gumbel random variables. Moreover, the probabilistic choice model

$$p[j] = \frac{e^{V_j}}{\mu} \cdot \frac{G_j(e^{V_1}, e^{V_2}, \ldots, e^{V_m})}{G(e^{V_1}, e^{V_2}, \ldots, e^{V_m})} \tag{3.3.52}$$

is a random utility model (GEV model).

In fact, as was seen in Sect. 3.2, the probability of choosing alternative j is equal to:

$$p[j/I] = Pr[V_j - V_k > \varepsilon_k - \varepsilon_j \ \forall k \neq j,\ k \in I] \tag{3.3.53}$$

that is, the probability that, for each alternative $k \neq j$, $\varepsilon_k < \varepsilon_j + V_j - V_k$ as ε_j assumes any value between $-\infty$ and $+\infty$. Introducing the joint probability density function of the random residuals ε_j, $f(\varepsilon_1, \varepsilon_2, \ldots, \varepsilon_m)$, this probability can also be expressed as

$$p[j] = \int_{\varepsilon_1=-\infty}^{V_j-V_1+\varepsilon_j} \int_{\varepsilon_2=-\infty}^{V_j-V_2+\varepsilon_j} \cdots \int_{\varepsilon_j=-\infty}^{+\infty} \times \cdots \times \int_{\varepsilon_m=-\infty}^{V_j-V_m+\varepsilon_j} f(\varepsilon_1, \ldots, \varepsilon_m)\, d\varepsilon_1 \ldots d\varepsilon_m \tag{3.3.54}$$

Alternatively, if $F(\varepsilon_1, \varepsilon_2, \ldots, \varepsilon_m)$, is the cumulative distribution function of the random residuals, the partial derivative of F with respect to ε_j, F_j, is equal to the product of the probability density function of ε_j and the joint distribution function for all ε_k with $k \neq j$. The latter, evaluated at $\varepsilon_k = V_j - V_k + \varepsilon_j$, gives the probability that each $\varepsilon_k \neq \varepsilon_j$ is less than $V_j - V_k + \varepsilon_j$, for a given value of ε_j. Consequently, (3.3.54) can be expressed more synthetically as

$$p[j] = \int_{\varepsilon_j=-\infty}^{+\infty} F_j(V_j - V_1 + \varepsilon_j, \ldots, \varepsilon_j, \ldots, V_j - V_m + \varepsilon_j)\, d\varepsilon_j \tag{3.3.55}$$

All the formulations obtained by specifying the joint probability density function $f(\varepsilon_1, \varepsilon_2, \ldots, \varepsilon_m)$, or alternatively the joint probability distribution function $F(\varepsilon_1, \varepsilon_2, \ldots, \varepsilon_m)$, are consistent with the behavioral assumptions of random utility theory expressed by (3.3.53).

In particular, the function (3.3.51) where $G(\cdot)$ satisfies the properties (1) through (4) mentioned above, is a cumulative distribution function in that it has the following three properties.

- $F(\cdot)$ is nondecreasing in the ε_j over the whole range of definition and has nonnegative mixed partial derivatives up to mth order.
- $F(\cdot)$ tends asymptotically to zero if at least one of its variables tends to minus infinity; it tends asymptotically to one as all its variables tend to infinity:

$$\lim_{\varepsilon_j \to -\infty} F(\varepsilon_1, \ldots, \varepsilon_m) = 0$$

$$\lim_{\varepsilon_1, \ldots, \varepsilon_m \to +\infty} F(\varepsilon_1, \ldots, \varepsilon_m) = 1$$

- $F(\cdot)$ is continuous from the right.

To demonstrate the first property, it is sufficient to show that the function $G(e^{-\varepsilon_1}, e^{-\varepsilon_2}, \ldots, e^{-\varepsilon_m})$ defined earlier is nonincreasing in ε_j. Indeed, from condition (4) on the mixed partial derivatives of $G(\cdot)$, it follows that:

$$G_j(\cdot) \geq 0 \quad j = 1, 2, m \tag{3.3.56}$$

that is, $G(.)$ is nondecreasing with respect to the variables $e^{-\varepsilon_j}$. Hence:

$$\partial G(.)/\partial \varepsilon_j = \partial G(.)/\partial e^{-\varepsilon_j} \cdot \partial e^{-\varepsilon_j}/\partial \varepsilon_j = G_j(.) \cdot (-e^{-\varepsilon_j}) \leq 0$$

The function $G(e^{-\varepsilon_1}, e^{-\varepsilon_2}, \ldots, e^{-\varepsilon_m})$ is therefore nondecreasing in $e^{-\varepsilon_j}$ but nonincreasing in ε_j. Starting from this result, it can be proved through a recursive approach that the condition on the sign of the partial mixed derivatives of the G function implies for the F function nonnegative mixed partial derivatives up to mth order.

As for the second property, from (3.3.51) and condition (3) required for $G(\cdot)$, it follows that

$$\begin{aligned}\lim_{\varepsilon_i \to -\infty} F(\varepsilon_1, \ldots, \varepsilon_j, \ldots, \varepsilon_m) &= \lim_{\varepsilon_j \to -\infty} \exp\left[-G(e^{-\varepsilon_1}, \ldots, e^{-\varepsilon_j}, \ldots, e^{-\varepsilon_m})\right] \\ &= \exp\left[-G(e^{-\varepsilon_1}, \ldots, \infty, \ldots, e^{-\varepsilon_m})\right] \\ &= \exp[-\infty] = 0\end{aligned}$$

which is the first of the two limits. The second limit results from the homogeneity of $G(\cdot)$ (condition (2)), which implies that $G(0, 0, \ldots, 0) = 0$. Therefore from (3.3.51) it follows that

$$\lim_{\varepsilon_1,\ldots,\varepsilon_m \to +\infty} F(\varepsilon_1,\ldots,\varepsilon_m) = \lim_{\varepsilon_1,\ldots,\varepsilon_m \to +\infty} \exp\big[-G(e^{-\varepsilon_1},\ldots,e^{-\varepsilon_m})\big]$$
$$= \exp\big[-G(0,\ldots,0)\big] = \exp[-0] = 1$$

The third property is easily verified, because $F(\cdot)$ is defined by (3.3.51), a continuous function.

Furthermore, it can be demonstrated that the solution of (3.3.55), with F defined as in (3.3.51), actually gives expression (3.3.52) for the choice probabilities defining a GEV model.

Indeed, substituting (3.3.51) in expression (3.3.55), and from the homogeneity of $G(\cdot)$ and $G_j(\cdot)$, it follows that

$$p[j] = \int_{\varepsilon_j=-\infty}^{+\infty} \exp\big[-G(e^{V_1-V_j-\varepsilon_j},\ldots,e^{V_m-V_j-\varepsilon_j})\big]$$
$$\cdot G_j(e^{V_1-V_j-\varepsilon_j},\ldots,e^{V_m-V_j-\varepsilon_j}) \cdot e^{-\varepsilon_j}\, d\varepsilon_j$$
$$= \int_{\varepsilon_j=-\infty}^{+\infty} \exp\big\{-[e^{-(V_j+\varepsilon_j)}]^{\mu} \cdot G(e^{V_1},\ldots,e^{V_m})\big\} \cdot [e^{-(V_j+\varepsilon_j)}]^{\mu-1}$$
$$\cdot G_j(e^{V_1},\ldots,e^{V_m}) \cdot e^{-\varepsilon_j} d\varepsilon_j$$
$$= \int_{\varepsilon_j=-\infty}^{+\infty} \big\{\exp -[e^{-(V_j+\varepsilon_j)}]^{\mu}\big\}^{G(e^{V_1},\ldots,e^{V_m})} \cdot [e^{-(V_j+\varepsilon_j)}]^{\mu-1}$$
$$\cdot G_j(e^{V_1},\ldots,e^{V_m}) \cdot e^{-\varepsilon_j} d\varepsilon_j$$
$$= \frac{e^{V_j} \cdot G_i(e^{V_1},\ldots,e^{V_m})}{\mu \cdot G(e^{V_1},\ldots,e^{V_m})} \cdot \Big|\big\{\exp -[e^{-(V_j+\varepsilon_j)}]^{\mu}\big\}^{G(e^{V_1},\ldots,e^{V_m})}\Big|_{-\infty}^{+\infty}$$
$$= \frac{e^{V_j} \cdot G_j(e^{V_1},\ldots,e^{V_m})}{\mu \cdot G(e^{V_1},\ldots,e^{V_m})}$$

which is (3.3.52).

Multinomial logit, single-level hierarchical logit, multilevel hierarchical logit, and cross-nested logit models can be obtained as special cases of the GEV model by appropriately specifying the function $G(\cdot)$, as shown in Appendix 3.A.

3.3.6 The Probit Model

The probit model overcomes most of the drawbacks of the logit model and its generalizations, although at the cost of analytical tractability. It is based on the hypothesis that the perceived utility residuals ε_j are MultiVariate Normal (MVN) r.v. with zero mean and fully general variances and covariances:

$$\begin{aligned} E[\varepsilon_j] &= 0 \\ \mathrm{Var}[\varepsilon_j] &= \sigma_j^2 \\ \mathrm{Cov}[\varepsilon_j, \varepsilon_h] &= \sigma_{jh} \end{aligned} \tag{3.3.57}$$

Further characteristics of the multivariate normal r.v. are given in Appendix 3.B. Variances and covariances of the random residual vector $\boldsymbol{\varepsilon}$ are elements of the $m \times m$ dispersion matrix $\boldsymbol{\Sigma}$, where m is the number of alternatives. The multivariate normal probability density of the residual vector $\boldsymbol{\varepsilon}$ is given by

$$f(\boldsymbol{\varepsilon}) = \left[(2\pi)^m \det(\boldsymbol{\Sigma})\right]^{-1/2} \exp[-1/2\boldsymbol{\varepsilon}^{\mathrm{T}}\boldsymbol{\Sigma}^{-1}\boldsymbol{\varepsilon}] \tag{3.3.58}$$

Perceived utilities U_j are also jointly distributed according to a multivariate normal distribution with mean vector $\boldsymbol{V}$ and variances and covariances equal to those of the residuals ε_j; $\boldsymbol{U} \sim \mathrm{MVN}(\boldsymbol{V}, \boldsymbol{\Sigma})$.

The choice probability of alternative j, $p[j]$, can be formally expressed in terms of the joint probability that utility U_j will assume a value within an infinitesimal interval and that the utilities of the other alternatives will have lower values. This probability element must then be integrated over all possible values of U_j to obtain $p[j]$ (see (3.3.54)):

$$p[j] = \int_{U_1<U_j} \cdots \int_{U_j=-\infty}^{+\infty} \cdots \int_{U_m<U_j} \frac{\exp[-1/2(\boldsymbol{U}-\boldsymbol{V})^T\boldsymbol{\Sigma}^{-1}(\boldsymbol{U}-\boldsymbol{V})]}{[(2\pi)^m \det(\boldsymbol{\Sigma})]^{1/2}} dU_1 \ldots dU_m \tag{3.3.59}$$

The probit model is invariant (see Sect. 3.2) if the matrix $\boldsymbol{\Sigma}$ does not depend on the vector of systematic utilities $\boldsymbol{V}$. In this case, the choice probability of a generic alternative depends only on systematic utility differences. Thus, Alternative Specific Attributes (ASA) and their coefficients (ASC) can be replaced by their differences with respect to the value of a reference alternative.

To illustrate the effect of variances and covariances on choice probabilities, consider the case of three alternatives ($m = 3$), with systematic utilities equal to zero ($V_A = V_B = V_C = 0$) and the following variance–covariance matrix.

$$\boldsymbol{\Sigma} = \begin{bmatrix} 1 & \sigma_{AB} & 0 \\ \sigma_{AB} & 1 & 0 \\ 0 & 0 & \sigma_C^2 \end{bmatrix}$$

Figure 3.13 charts the probability $p[C]$ obtained with the probit model (3.3.59) for varying values of the parameters σ_{AB} and σ_C. As the variance of U_C increases compared with those of the other alternatives, the choice probability of C also increases. The value of the random residual ε_C eventually dominates the value of V_C,

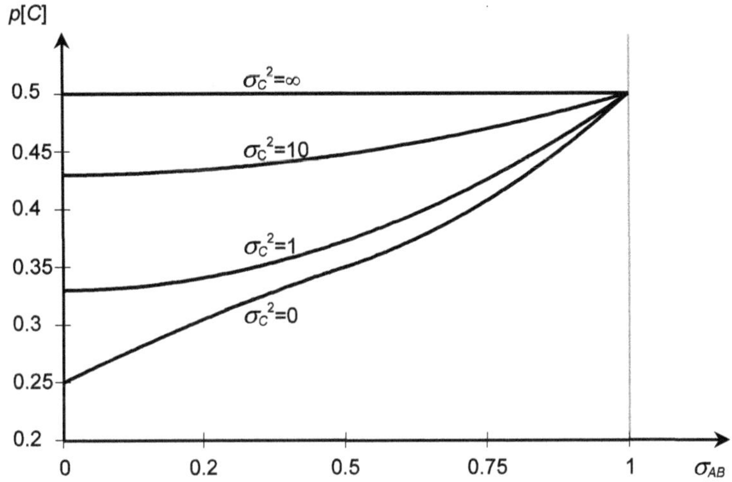

Fig. 3.13 Influence of the variance and covariance of residuals on probit choice probabilities

and the perceived utility U_C is, with high probability, either much higher or much lower than the perceived utilities U_A and U_B ($\lim_{\sigma_C \to \infty} p[C] = 0.5$). Moreover, as the covariance (in this case identical with the correlation coefficient) between the residuals of alternatives A and B increases, the choice probability of alternative C also increases, because A and B are increasingly perceived as a single alternative. The same effect was shown in Sects. 3.3.2 and 3.3.3 for the hierarchical logit model.

In general, the probit model yields choice probabilities similar to those obtained from logit and hierarchical logit models if the same variance–covariance matrix is assumed. Moreover, as mentioned above, it allows for greater flexibility in the specification of the covariance matrix, whose elements can assume whatever value and can be "directly" specified, unlike the logit-type models whose covariance matrix is indirectly defined through the choice network and model parameters.

The flexibility of the variance–covariance matrix can in fact be a problem in the practical use of the probit model. A variance–covariance matrix can contain up to $(m(m+1))/2$ distinct values, as noted in Sect. 8.3.2, where m is the number of choice alternatives. When m is large, specification and calibration of all the possible values can be problematic. Different methods have been proposed to reduce the number of unknown variance–covariance matrix elements requiring estimation. All of these methods assume some structure underlying the random residuals. The parameters of this structure determine the elements of the variance–covariance matrix but are fewer in number than the total number of possible unknowns of such a matrix.

A first method, known as *Factor Analytic Probit*, expresses the vector of random residuals as a linear function of a vector ζ of independent standard normal variables:

$$\varepsilon_j = \sum_{k=1}^{n} f_{jk}\zeta_k, \tag{3.3.60a}$$

$$\boldsymbol{\varepsilon} = \boldsymbol{F}\boldsymbol{\zeta} \tag{3.3.60b}$$

where

$\boldsymbol{\varepsilon}$ is the $(m \times 1)$ vector of multivariate normal random variables (factors) with elements $\varepsilon_j : \boldsymbol{\varepsilon} \sim \text{MVN}(\mathbf{0}, \boldsymbol{\Sigma})$

$\boldsymbol{F}$ is the $(m \times n)$ matrix of factor "loadings" with elements f_{jk}, mapping the vector $\boldsymbol{\zeta}$ of standard normal random variables to the vector $\boldsymbol{\varepsilon}$ of random residuals

$\boldsymbol{\zeta}$ is the $(n \times 1)$ vector of identical and independent standard normal random variables with elements $\zeta_k : \boldsymbol{\zeta} \sim \text{MVN}(\mathbf{0}, \boldsymbol{I})$

From (3.3.60a), the elements of the variance–covariance matrix $\boldsymbol{\Sigma}$ of the random residuals ε_j can be expressed as a function of the elements f_{jk} of matrix $\boldsymbol{F}$:

$$\text{Var}[\varepsilon_j] = E\left[\varepsilon_j^2\right] = E\left[\sum_{k=1}^{n} f_{jk}^2 \zeta_k^2\right] = \sum_{k=1}^{n} f_{jk}^2 \cdot E\left[\zeta_k^2\right] = \sum_{k=1}^{n} f_{jk}^2, \tag{3.3.62}$$

$$\text{Cov}[\varepsilon_j, \varepsilon_h] = E[\varepsilon_j \varepsilon_h] = E\left[\sum_{k=1}^{n} f_{jk}\zeta_k \cdot \sum_{k=1}^{n} f_{hk}\zeta_k\right]$$

$$= \sum_{k=1}^{n} f_{jk} f_{hk} \cdot E\left[\zeta_k^2\right] = \sum_{k=1}^{n} f_{jk} f_{hk} \tag{3.3.63}$$

or in vectorial form:

$$\boldsymbol{\Sigma} = E[\boldsymbol{\varepsilon}\boldsymbol{\varepsilon}^{\mathrm{T}}] = E[\boldsymbol{F}\boldsymbol{\zeta}\boldsymbol{\zeta}^{\mathrm{T}}\boldsymbol{F}] = \boldsymbol{F}E[\boldsymbol{\zeta}\boldsymbol{\zeta}^{\mathrm{T}}]\boldsymbol{F}^{\mathrm{T}} = \boldsymbol{F}\boldsymbol{I}\boldsymbol{F}^{\mathrm{T}} = \boldsymbol{F}\boldsymbol{F}^{\mathrm{T}} \tag{3.3.64}$$

Because typically $n \ll m$, the number of unknown elements is reduced from $m(m+1)/2$ in the matrix $\boldsymbol{\Sigma}$ to $m \cdot n$ in the matrix $\boldsymbol{F}$. In the extreme case $(m = n)$, the matrix $\boldsymbol{F}$ is low triangular and univocally determined through the Cholesky factorization of the matrix $\boldsymbol{\Sigma}$. A relevant application of the factor analytic representation of the probit model is in path choice, as shown in Sect. 4.3.3.1. Another relevant application based on a particular specification of (3.3.60a) is known in the literature as the *random coefficient probit*. It is based on the assumption that the random residual ε_j derives from the variability of utility function coefficients β_k over the population of decision makers. In particular, for each individual i, coefficient β_k^i is assumed equal to an average value β_k plus a random residual η_k^i:

$$\beta_k^i = \beta_k + \eta_k^i \quad k = 1, 2, \ldots, K$$

where K is the total number of coefficients used to define the systematic utilities of the m alternatives. By assuming that the η_k^i are independently distributed normal variables with zero mean and variance σ_k^2,

$$\eta_k^i \sim N\left(0, \sigma_k^2\right) \quad \forall i, k$$

$$\text{Cov}\left[\eta_k^i, \eta_h^i\right] = 0 \quad \forall i, k, h$$

it follows that

$$U_j^i = V_j^i + \varepsilon_j^i = \sum_k \beta_k^i X_{kj}^i = \sum_k \beta_k X_{kj}^i + \eta_k^i X_{kj}^i$$

with:

$$V_j^i = \sum_k \beta_k X_{kj}^i; \qquad \varepsilon_j^i = \sum_k \eta_k^i X_{kj}^i; \qquad \boldsymbol{\varepsilon}^i \sim \text{MVN}(\mathbf{0}, \boldsymbol{\Sigma}_\varepsilon) \tag{3.3.61}$$

where X_{kj} is the value of attribute k in alternative j; it is equal to zero if attribute X_k does not appear in the systematic utility of alternative j.

From comparison between (3.3.60a) and (3.3.61) it follows that

$$f_{jk}^i = \sigma_k X_{jk}^i$$

and by substituting into (3.3.62) and (3.3.63) then:

$$\text{Var}\left[\varepsilon_j^i\right] = \sum_k \left(X_{kj}^i \sigma_k\right)^2 \tag{3.3.65}$$

$$\text{Cov}\left[\varepsilon_j^i, \varepsilon_h^i\right] = \sum_k X_{kj}^i X_{kh}^i \sigma_k^2 \tag{3.3.66}$$

that is, in vectorial form:

$$\boldsymbol{F} = \boldsymbol{X} \boldsymbol{\Sigma}_\eta^{1/2}$$

and from (3.3.64):

$$\boldsymbol{\Sigma}_\varepsilon = \boldsymbol{X} \boldsymbol{\Sigma}_\eta \boldsymbol{X}^{\text{T}}$$

Using this approach, the number of unknown elements of the variance–covariance matrix is reduced from a possible maximum of $(m(m+1))/2$ to the K of the matrix $\boldsymbol{\Sigma}_\eta$.

The flexibility of the probit model is achieved at the cost of computational complexity. The probit model does not possess analytical expressions for its choice probabilities inasmuch as there is no known closed-form solution of the integral (3.3.59). Numerical integration methods are computationally burdensome when there are more than about five alternatives. Calculation of probit choice probabilities with several alternatives is typically carried out by approximation methods. In the following, three traditional approximate methods are described: the so-called Monte

Carlo or Acceptance–Reject (AR) method, the GHK method, and the Clark approximation. However, it should be said that the last is computationally inefficient and is rarely used in practice.

The Monte Carlo method generates a sample of perceived utilities for the alternatives (these can be thought of as the utilities perceived for each alternative by a sample of decision-makers) and estimates the choice probability of each alternative j as the fraction of times that j is the alternative with maximum perceived utility.

More specifically, at the kth iteration, the method generates:

- A vector $\boldsymbol{\varepsilon}^k = (\varepsilon_1^k, \ldots, \varepsilon_m^k)^{\mathrm{T}}$ of random residuals drawn from a zero-mean multivariate normal distribution with dispersion matrix $\boldsymbol{\Sigma}$.
- A vector $\boldsymbol{U}^k$ of perceived utilities: $\boldsymbol{U}^k = \boldsymbol{V} + \boldsymbol{\varepsilon}^k$.
- A vector $\boldsymbol{p}^k$ of deterministic alternative choice probabilities: $\boldsymbol{p}^k = (0, \ldots, 1, \ldots, 0)$ where the value one is associated to the largest component of $\boldsymbol{U}^k$ (the alternative with maximum perceived utility).

Consequently, after n iterations, the sample estimate $\hat{p}[j]$ of the probability $p[j]$ is:

$$\hat{p}[j] = \frac{1}{n} \sum_{k=1}^{n} p[j/\boldsymbol{\varepsilon}^k] = \frac{n_j}{n} \tag{3.3.67}$$

where $\boldsymbol{\varepsilon}^k$ denotes the kth draw of vector $\boldsymbol{\varepsilon}$ from an MVN($\mathbf{0}$, $\boldsymbol{\Sigma}$) distribution, and n_j is the number of times that alternative j is the maximum perceived utility alternative in the sample. It can be shown that the estimator (3.3.67) is unbiased and efficient. With the Monte Carlo method, each extraction can be considered as the execution of a generalized Bernoulli trial with m possible outcomes, where outcome j corresponds to alternative j with maximum perceived utility, and occurs with probability $p[j]$. The joint sample frequency of the results is thus multinomially distributed and the sample variance of the estimate $\hat{p}[j]$ is:

$$\mathrm{Var}\big[\hat{p}[j]\big] = \frac{1}{n} \hat{p}[j]\big(1 - \hat{p}[j]\big) \tag{3.3.68}$$

For large enough values of n, a confidence interval for $p[j]$ can be obtained by assuming that $p[j]$ is approximately distributed as a normal r.v. with mean $\hat{p}[j]$ given by (3.3.67) and variance given by (3.3.68).

In applications, drawing a random m-vector $\boldsymbol{\varepsilon}$ from an MVN($\mathbf{0}$, $\boldsymbol{\Sigma}$) distribution can be accomplished indirectly by drawing m independent values from a standard normal $N(0, 1)$ distribution by means of (3.3.60b). In practice, at the generic iteration k the vector $\boldsymbol{\varepsilon}^k$ of pseudorandom draws from a normal multivariate distribution MVN($\mathbf{0}$, $\boldsymbol{\Sigma}$) can be obtained through:

- Drawing a vector $\boldsymbol{z}^k$ of m normal standard independent variables.
- Calculating $\boldsymbol{\varepsilon}^k = \boldsymbol{F}\boldsymbol{z}^k$ where $\boldsymbol{F}$ is known within a factor analytic approach or through a Cholesky factorization of the matrix $\boldsymbol{\Sigma}$.

The Monte Carlo method, albeit simple to interpret and apply, exhibits some theoretical drawbacks that can be overcome by using different procedures for calculating probit probabilities. As described in Chap. 8, methods for random utility model estimation are based on specific theoretical properties of the function $p[\boldsymbol{\beta}]$, that is, on how choice probabilities change with respect to model parameters. Namely, $p[\boldsymbol{\beta}]$ is required to be doubly differentiable and strictly positive. Because $p[\boldsymbol{\beta}]$ does not exhibit a closed form for the probit model, these properties depend on how choice probabilities are simulated. Notably, when applying the Monte Carlo method, $p[\boldsymbol{\beta}]$ is a step function (i.e., not continuous) and, in the presence of alternatives with low systematic utilities, it is not guaranteed to be strictly positive.

A possible solution is represented by the smoothed Monte Carlo method, according to which the choice probability vector p^k at the generic iteration k is given by a θ-parameter multinomial logit probability vector $p^k = (p_1^k, \ldots, p_m^k)$ rather than a deterministic vector. This leads to a continuous, doubly differentiable (3.3.67) function, given as the average of strictly positive logit probabilities rather than 0/1 values. Obviously, probit choice probabilities provided by a smoothed Monte Carlo represent an approximation of actual probit probabilities, proportional to the value of the variance parameter θ. In other words, θ should be chosen so as to provide a satisfactory compromise between speed and stability of convergence, increasing with θ, and reliability in simulated choice probabilities, decreasing with θ. Those concepts are extended in Sect. 3.3.7 when describing the mixed logit model.

Another possible solution to the operative problems of the Monte Carlo method lies in the GHK method, considered in the literature one of the most stable and accurate. Unlike the Monte Carlo method which supplies contemporaneously an estimate for the choice probabilities of all the alternatives, the GHK method determines the probability of choosing a single alternative on each occasion. This makes it naturally burdensome if the number of alternatives is very high. So as to illustrate the mechanism, let us consider initially the case of a choice set consisting of three alternatives, and let us suppose we wish to determine the probability of choosing alternative 1. Allowing for (3.2.2a) and the theoretical properties of invariant random utility models, the perceived utility of the other two alternatives may be expressed in differential terms with respect to the utility of the considered alternative:

$$U_2 - U_1 = (V_2 - V_1) + (\varepsilon_2 - \varepsilon_1) \to U_{21} = V_{21} + \varepsilon_{21}$$
$$U_3 - U_1 = (V_3 - V_1) + (\varepsilon_3 - \varepsilon_1) \to U_{31} = V_{31} + \varepsilon_{31}$$

The covariance matrix $\boldsymbol{\Sigma}_1$ of random residuals ε_{21} and ε_{31} may be derived directly from matrix $\boldsymbol{\Sigma}$ of residuals $\varepsilon_1 \ldots \varepsilon_3$. Then, because we are dealing with a symmetric and positive definite matrix, it may be expressed by Choleski factorization as $\boldsymbol{\Sigma}_1 = \boldsymbol{C}\boldsymbol{C}^{\mathrm{T}}$, given that:

$$\boldsymbol{C} = \begin{bmatrix} c_{11} & 0 \\ c_{21} & c_{22} \end{bmatrix}$$

Recalling what was stated above concerning the Monte Carlo method, if z_1 and z_2 are two standardized normal r.v. then we may write:

$$\varepsilon_{21} = c_{11} \cdot z_1 \ \to \ U_{21} = V_{21} + c_{11} \cdot z_1$$
$$\varepsilon_{31} = c_{21} \cdot z_1 + c_{22} \cdot z_2 \ \to \ U_{31} = V_{31} + c_{21} \cdot z_1 + c_{22} \cdot z_2$$

and the probability of choosing the first alternative may be reformulated as follows.

$$
\begin{aligned}
p[1] &= Pr\big[(U_{21} < 0) \cap (U_{31} < 0)\big] \\
&= Pr\big[(V_{21} + c_{11}z_1 < 0) \cap (V_{31} + c_{21}z_1 + c_{22}z_2 < 0)\big] \\
&= Pr[V_{21} + c_{11}z_1 < 0] \cdot Pr\big[(V_{31} + c_{21}z_1 + c_{22}z_2 < 0)/(V_{21} + c_{11}z_1 < 0)\big] \\
&= Pr\left[z_1 < -\frac{V_{21}}{c_{11}}\right] \cdot Pr\left[\left(z_2 < -\frac{V_{31} + c_{21}z_1}{c_{22}}\right) \Big/ \left(z_1 < -\frac{V_{21}}{c_{11}}\right)\right]
\end{aligned}
$$

If F stands for the distribution law of normal cumulative probability, the probability product previously written becomes

$$
p[1] = F\left(-\frac{V_{21}}{c_{11}}\right) \cdot \int_{-\infty}^{-V_{21}/c_{11}} F\left(-\frac{V_{31} + c_{21}z_1}{c_{22}}\right) f(z_1)\, dz_1 \qquad (3.3.69)
$$

The first factor of (3.3.69) may be directly obtained from probability tables of standard normal random variables, and the integral may be calculated numerically by performing at the generic iteration k the following steps.

- A draw z_1^k is generated of the standard normal random variable z_1 truncated at $-V_{21}/c_{11}$ (to generate a z_1 truncated at $-V_{21}/c_{11}$ it is enough to generate a standard normal z and calculate $z_1 = F^{-1}(zF(-V_{21}/c_{11}))$.
- From the standard normal probability tables we calculate the value

$$
i^k = F\left(-\frac{V_{31} + c_{21}z_1^k}{c_{22}}\right)
$$

It may be demonstrated that a correct and efficient estimate of the integral of (3.3.69) is obtained by calculating the average of values i^k on a certain number of iterations. The product of the two factors thus calculated, inserted into (3.3.69), supplies a correct and efficient estimate of the choice probability $p[1]$ sought.

Generalization of the procedure to the case of m alternatives is immediate. In this regard, suffice it to think that for a generic alternative j (with $j > 3$) we obtain:

$$
\begin{aligned}
p[j] = Pr[U_{ij} < 0\ \forall i \neq j] = Pr\left[z_1 < -\frac{V_{1j}}{c_{11}}\right] \cdot Pr\bigg[\left(z_2 < -\frac{V_{2j} + c_{21}z_1}{c_{22}}\right) \\
\Big/ \left(z_1 < -\frac{V_{1j}}{c_{11}}\right)\bigg] \\
\cdot Pr\bigg[\left(z_3 < -\frac{V_{3j} + c_{31}z_1 + c_{32}z_2}{c_{33}}\right) \\
\Big/ \left(z_2 < -\frac{V_{2j} + c_{21}z_1}{c_{22}}\right) \cap \left(z_1 < -\frac{V_{2j}}{c_{11}}\right)\bigg] \ldots
\end{aligned}
$$

The *Clark approximation*, another traditional method for calculating probit choice probabilities, is based on an approximation for the maximum of a set of normal random variables (the maximum is of course itself a random variable). The procedure is first illustrated by referring to a choice among three alternatives. In this case, perceived utilities U_1, U_2, and U_3 are distributed according to a multivariate normal distribution with mean vector $\boldsymbol{V} = (V_1, V_2, V_3)^{\mathrm{T}}$ and the following variance–covariance matrix.

$$\boldsymbol{\Sigma} = \begin{bmatrix} \sigma_1^2 & \sigma_{12} & \sigma_{13} \\ \sigma_{21} & \sigma_2^2 & \sigma_{23} \\ \sigma_{31} & \sigma_{32} & \sigma_3^2 \end{bmatrix}$$

Suppose the choice probability of alternative 3, $p[3]$, is to be computed. Clark's results express the mean V_{12} and the variance S_{12}^2 of the random variable $U_{12} = \max(U_1, U_2)$ as

$$V_{12} = V_2 + (V_1 - V_2)F(\alpha) + \gamma f(\alpha) \qquad (3.3.70)$$

$$S_{12}^2 = \mathrm{var}[U_{12}] = m_{12} - V_{12}^2$$

where m_{12} is the second moment around zero of the variable U_{12}, and is given by

$$m_{12} = V_2^2 + \sigma_2^2 + \left(V_1^2 + \sigma_1^2 - V_2^2 - \sigma_2^2\right)F(\alpha) + (V_1 + V_2)\gamma f(\alpha) \qquad (3.3.71)$$

The constants γ and α in expressions (3.3.70) and (3.3.71) are, respectively, the standard deviation of the random variable $(U_1 - U_2)$:

$$\gamma = \left[\sigma_1^2 + \sigma_2^2 - -2\sigma_{12}\right]^{1/2}$$

and the mean standardized value of the random variable $(U_1 - U_2)$:

$$\alpha = (V_1 - V_2)/\gamma$$

The symbols $f(\alpha)$ and $F(\alpha)$ denote, respectively, the value of the probability density function and probability distribution function of a standard normal r.v. $N(0, 1)$ evaluated at α:

$$f(\alpha) = (2\pi)^{-1/2}\exp(-\alpha^2/2)$$

$$F(\alpha) = \int_{-\infty}^{\alpha} f(x)\,dx$$

Clark's formulas also give the covariance between variables U_j and $U_{12\ldots j-1}$ as:

$$S_{j.12\ldots i} = \mathrm{cov}(U_j, U_{12\ldots i}) = \sigma_{ij} + (S_{j.12\ldots i-1} - \sigma_{ij})F(\alpha)$$

where $i = j - 1$. Thus the covariance between variables U_3 and U_{12} is:

$$S_{3.12} = \mathrm{cov}(U_3, U_{12}) = \sigma_{23} + (\sigma_{13} - \sigma_{23})F(\alpha)$$

The probability of choosing alternative 3 is:

$$p[3] = Pr[U_3 \geq U_{12}] = Pr[U_{12} - U_3 \leq 0] \qquad (3.3.72)$$

Although U_{12} is not in fact normally distributed, Clark's method assumes that it can be satisfactorily approximated by a normal r.v. having mean V_3, variance S_{12} and covariance $S_{3.12}$ with the normal r.v. U_3. Thus, the choice probability (3.3.72) can be evaluated using standard results on the distribution of the difference of two normal variables (Appendix 3.B.2):

$$p[3] = F\left[\frac{V_3 - V_{12}}{(\sigma_3^2 + S_{12}^2 - 2S_{3.12})^{1/2}}\right] \tag{3.3.73}$$

Choice probabilities for more than three alternatives can be calculated by sequentially applying the procedure described above. The probability of choosing the generic alternative j can be obtained by computing sequentially the mean, variance, and covariance of nested pairs of perceived utilities ordered in such a way that j is the last alternative. For example, the mean and variance of $U_{12} = \max(U_1, U_2)$ as well as its covariance with U_3 are computed first. Subsequently the mean and variance of the variable $U_{123} = \max(U_3, U_{12})$ are computed together with its covariance with U_4, and so on until the comparison is made between:

$$U_{12\ldots j-1} = \max\big(U_{j-1}, \max\big(U_{j-2} \ldots \max(U_1, U_2)\big)\big)$$

and U_j. At this point, the probability $p[j]$ is obtained by applying expression (3.3.73). The entire sequence has to be repeated to calculate the probability of each alternative.

*3.3.7 The Mixed Logit Model**

The probit model described in the previous section is a tool able to reproduce choice contexts characterized by any covariance matrix whatsoever. Contrasting with this flexibility some applicative problems are encountered, including the already mentioned need to simulate choice probabilities numerically. Analysis of this problem and identification of possible solutions laid the basis for developing a new class of models that are briefly described below.

First of all, we saw previously that one of the possible simulation methods of Probit choice probabilities is the smoothed Monte Carlo, at whose generic iteration the vector of choice probability $\boldsymbol{p}^k$ of (3.3.67) is not deterministic but is calculated by using a multinomial logit of parameter θ, whose value conditions the dichotomy between convergence velocity and approximation of probit probabilities. As regards the latter question, it is immediately recognized that the random utility model to which the choice probabilities calculated with the smoothed Monte Carlo actually correspond, assumes in practice that the overall residual ε_j of (3.2.2a) may be broken down into the sum of two independent terms λ_j and τ_j:

$$U_j = V_j + \varepsilon_j = V_j + \lambda_j + \tau_j \tag{3.3.74}$$

where λ_j are normal r.v. with zero mean and any covariance matrix $\boldsymbol{\Sigma}$ and τ_j are the independent Gumbel r.v. with zero mean and parameter of variance θ. The model (3.3.74) is known in the literature as multinomial probit with logit kernel.[14] In general, the hypotheses made on the distribution of residuals λ_j and τ_j are not restrictive, insofar as they may follow any distribution whatsoever, and this generates a class of random utility models called mixed models. In particular, when residuals τ_j are i.i.d. Gumbel random variables, the model (3.3.74) is known in the literature as mixed logit.

To appreciate that the model (3.3.74) is perfectly consistent with the probabilities simulated by means of a smoothed Monte Carlo, suffice it to consider that, with $\boldsymbol{\lambda}^*$ a vector of pseudorandom draws λ_j^* of normal residuals λ_j, (3.3.74) may be reconsidered in the form $U_j = (V_j + \lambda_j^*) + \tau_j$, and by virtue of the assumption made on the distribution of τ_j the corresponding choice probabilities $p^{MNL}[j/\boldsymbol{\lambda}^*]$ may be calculated by using a multinomial logit of parameter θ:

$$p^{MNL}[j/\boldsymbol{\lambda}^*] = \frac{\exp[V_j + \lambda_j^*]}{\sum_{h \in \boldsymbol{I}} \exp[V_h + \lambda_h^*]} \tag{3.3.75}$$

Clearly, with $f(\boldsymbol{\lambda})$ the joint probability density law of normal residuals λ_j, the choice probabilities $p[j]$ supplied by the mixed logit are given by

$$p[j] = \int p^{MNL}[j/\boldsymbol{\lambda}^*] \cdot f(\boldsymbol{\lambda})\, d\boldsymbol{\lambda} = \int \frac{\exp[V_j + \lambda_j^*]}{\sum_{h \in \boldsymbol{I}} \exp[V_h + \lambda_h^*]} \cdot f(\boldsymbol{\lambda})\, d\boldsymbol{\lambda} \tag{3.3.76}$$

and a correct estimate of the integral (3.3.76) is obtained immediately with an estimator of type (3.3.67) in which the $p[j/\boldsymbol{\varepsilon}^k]$ are given by (3.3.75).

Downstream of these theoretical considerations it thus appears evident that the mixed logit model allows us to simulate the choice probabilities with more efficient methods compared with use of the probit. The same considerations can also be extended to model (3.3.76) regarding the possibility of reducing the number of unknowns.

In particular, by applying factor analysis (3.3.60a) to the normal component λ_j of the overall residual ε_j of the perceived utility of alternative j one obtains:

$$\varepsilon_j = \sum_{k=1}^{n} f_{jk} \zeta_k + \tau_j \tag{3.3.77}$$

[14]The mixed logit model imposes an upper bound on the correlation of any pair of random residuals due to the positive variance σ^2 of the i.i.d. Gumbel residuals. In fact the maximum correlation between two alternatives is:

$$\mathrm{Corr}(\varepsilon_j, \varepsilon_h) = \frac{\mathrm{Cov}[\varepsilon_j, \varepsilon_h]}{[\mathrm{Var}[\varepsilon_j]]^{1/2} \cdot [\mathrm{Var}[\varepsilon_h]]^{1/2}} = \frac{[\mathrm{Var}[\xi_j]]^{1/2} \cdot [\mathrm{Var}[\xi_h]]^{1/2}}{[\mathrm{Var}[\xi_j] + \sigma^2]^{1/2} \cdot [\mathrm{Var}[\xi_h] + \sigma^2]^{1/2}}$$

of choice alternatives), that has the following properties.

or in vector form:

$$\boldsymbol{\varepsilon} = \boldsymbol{F}\boldsymbol{\xi} + \boldsymbol{\tau}$$

with $\boldsymbol{\zeta}$ standing for the vector of dimension $(k \times 1)$ of standard normals ζ_i. The perceived utility U_j of alternative j may then be expressed as:

$$U_j = V_j + \sum_{k=1}^{n} f_{jk}\zeta_k + \tau_j \tag{3.3.78}$$

Hence, the variance of the single residual may be written, in analogy with (3.3.62), in the following way,

$$\mathrm{Var}[\varepsilon_j] = E\left[\sum_{k=1}^{n} f_{jk}^2\zeta_k^2\right] + E\left[\tau_j^2\right] = \sum_{k=1}^{n} f_{jk}^2 + \frac{\pi^2\theta^2}{6} \tag{3.3.79}$$

and the covariance between the residuals of two alternatives is still expressed by (3.3.63). Note that it is possible to specify the matrix $\boldsymbol{F}$ appropriately in order to reproduce specific covariance structures; for example, if $\boldsymbol{F} = \boldsymbol{0}$ the model (3.3.78) degenerates into a multinomial logit, whereas to obtain a covariance matrix similar to that of a single-level hierarchical logit model with n groups, it is sufficient to assume $\boldsymbol{F}$ of dimension equal to the number of alternatives for the number of groups and so that the generic element f_{ji} is equal to 1 if j belongs to the ith group and 0 otherwise.

The model (3.3.78) is known in the literature as the mixed logit error component. As with the random coefficient probit, a random coefficient specification of component λ_j may be adopted:

$$U_j = \sum_k \beta_k X^i_{jk} + \eta^i_k X^i_{jk} + \tau_j \tag{3.3.80}$$

having adopted the same notation as in (3.3.61).

Models (3.3.75) and (3.3.76) may then be rewritten in the following way, leading to the explicit emergence of the joint probability density $f(\boldsymbol{\beta})$ of the coefficients:

$$p^{MNL}[j/\boldsymbol{\beta}] = \frac{\exp[V_j(\boldsymbol{\beta})]}{\sum_{h\in \boldsymbol{I}} \exp[V_h(\boldsymbol{\beta})]} \tag{3.3.81}$$

$$p[j] = \int p^{MNL}[j/\boldsymbol{\beta}] \cdot f(\boldsymbol{\beta})\, d\boldsymbol{\beta}$$

$$= \int \frac{\exp[V_j(\boldsymbol{\beta})]}{\sum_{h\in \boldsymbol{I}} \exp[V_h(\boldsymbol{\beta})]} \cdot f(\boldsymbol{\beta})\, d\boldsymbol{\beta} \tag{3.3.82}$$

In this sense, the choice probabilities supplied by (3.3.81) should be interpreted as choice probabilities calculated for a specific draw of the multivariate random variable $\boldsymbol{\beta}$, and the absolute choice probabilities are obtained as the average of probabilities (3.3.81) weighted by means of $f(\boldsymbol{\beta})$, that is (3.3.82).

3.4 Expected Maximum Perceived Utility and Mathematical Properties of Random Utility Models

The Expected Maximum Perceived Utility (EMPU) is an important variable associated with each choice context. As was seen in Sect. 3.2, random utility models are based on the assumption that the ith decision-maker chooses from the available choice set the alternative $j(i)$ with maximum perceived utility $U_{j(i)}$, where the perceived utilities are modeled as random variables:

$$U_{j(i)} = \max_j \{U_j^i\} = \max(\boldsymbol{U}^i) \quad j \in I^i \tag{3.4.1}$$

The variable $U_{j(i)}$ therefore denotes the perceived utility "obtained" by the decision-maker in the choice context. This variable is not observed by the analyst because it is the maximum value of unobserved perceived utilities. Therefore $U_{j(i)}$ can also be modeled as a random variable.

The expected maximum perceived utility s^i associated with a given choice context is defined as the expected value of $U_{j(i)}$ over the alternatives available in the choice set:

$$\begin{aligned} s^i = s^i(\boldsymbol{V}) &= E[U_{j(i)}] = E\big[\max_j(\boldsymbol{U}^i)\big] = E\big[\max(\boldsymbol{V}^i + \boldsymbol{\varepsilon}^i)\big] \\ &= \int \cdots \int \cdots \int \max(\boldsymbol{V}^i + \boldsymbol{\varepsilon}^i) f(\boldsymbol{\varepsilon})\, d\boldsymbol{\varepsilon} \end{aligned} \tag{3.4.2}$$

From (3.4.2) it can be deduced that the EMPU is a function of the systematic utilities of all the alternatives, vector $\boldsymbol{V}^i$, and that it depends on the joint probability density function of the random residuals $f(\boldsymbol{\varepsilon})$, as well as on the composition of the choice set I^i.[15]

A number of mathematical properties of random utility models can be demonstrated using the EMPU variable. These properties are useful for the construction of travel-demand model systems (see Chap. 4), for the analysis of assignment models (see Chap. 5), and for the evaluation of transport system projects (see Chap. 10).

In the following, probabilistic ($\boldsymbol{\varepsilon} \neq \boldsymbol{0}$) and deterministic ($\boldsymbol{\varepsilon} = \boldsymbol{0}$) choice models are addressed separately.

Mathematical properties of probabilistic choice models. The EMPU associated with a particular choice context is always greater than or equal to the maximum systematic utility:

$$s(\boldsymbol{V}) \geq \max(\boldsymbol{V}) \tag{3.4.3}$$

[15]In what follows, for the sake of simplicity, the dependence of the Expected Maximum Perceived Utility on the joint density function $f(\boldsymbol{\varepsilon})$ and on the choice set I^i is not explicitly expressed. When the choice set is not observed, the Expected Maximum Perceived Utility should be calculated by averaging over the various choice sets with their respective probabilities. The index i denoting the generic decision-maker will also be taken as understood.

By definition,

$$s(\boldsymbol{V}) = \int_{\varepsilon_1=-\infty}^{\infty} \cdots \int_{\varepsilon_m=-\infty}^{\infty} \max(\boldsymbol{V}+\boldsymbol{\varepsilon}) f(\boldsymbol{\varepsilon})\, d\boldsymbol{\varepsilon}$$

and because $f(\boldsymbol{\varepsilon}) \geq 0$ and $\max(\boldsymbol{V}+\boldsymbol{\varepsilon}) \geq V_k + \varepsilon_k,\ \forall k \in \boldsymbol{I}$, it follows that

$$\begin{aligned}
s(\boldsymbol{V}) &= \int_{\varepsilon_1=-\infty}^{\infty} \cdots \int_{\varepsilon_m=-\infty}^{\infty} \max(\boldsymbol{V}+\boldsymbol{\varepsilon}) f(\boldsymbol{\varepsilon})\, d\boldsymbol{\varepsilon} \\
&\geq \int_{\varepsilon_1=-\infty}^{\infty} \cdots \int_{\varepsilon_m=-\infty}^{\infty} V_k f(\boldsymbol{\varepsilon})\, d\boldsymbol{\varepsilon} + \int_{\varepsilon_1=-\infty}^{\infty} \cdots \int_{\varepsilon_m=-\infty}^{\infty} \varepsilon_k f(\boldsymbol{\varepsilon})\, d\boldsymbol{\varepsilon} \\
&= V_k \int_{\varepsilon_1=-\infty}^{\infty} \cdots \int_{\varepsilon_m=-\infty}^{\infty} f(\boldsymbol{\varepsilon})\, d\boldsymbol{\varepsilon} + \int_{\varepsilon_1=-\infty}^{\infty} \cdots \int_{\varepsilon_m=-\infty}^{\infty} \varepsilon_k f(\boldsymbol{\varepsilon})\, d\boldsymbol{\varepsilon} \\
&= V_k + E[\varepsilon_k] = V_k \quad \forall k \in \boldsymbol{I}
\end{aligned}$$

Therefore $s(\boldsymbol{V})$ is greater than or equal to the largest systematic utility, $s(\boldsymbol{V}) \geq V_k,\ \forall k \in \boldsymbol{I}$.

In addition, the mean systematic utility, calculated by weighing the systematic utility of each alternative k by its respective choice probability $p_k(\boldsymbol{V})$, is less than or equal to the EMPU variable. From expression (3.4.3), it follows that

$$p(V)^T V = \sum_k p_k(V) V_k \leq \sum_k p_k(V) \max(V) = \max(V) \leq s(V)$$

In order to analyze the EMPU variable in more detail, consider first a multinomial logit model with constant parameter θ. For this model, $s(\boldsymbol{V})$ can be expressed in closed form. Referring to the results reported for the maximization of Gumbel variables,[16] the EMPU is given by expression (3.3.5), repeated here:

$$s(\boldsymbol{V}) = \theta \ln \sum_j \exp(V_j/\theta) \tag{3.4.4}$$

It can easily be shown that expression (3.4.4) satisfies condition (3.4.3); Fig. 3.14 illustrates this result. From expression (3.4.4) it can also be deduced that the EMPU of a multinomial logit model increases if the systematic utility of one or more alternatives increases because the functions $\ln(\cdot)$ and $\exp(\cdot)$ are both monotonic increasing. Furthermore, because of the nonnegativity of the exponential function, the EMPU increases with the number of available alternatives. In fact, the addition of a new alternative to the choice set results in an increase in the EMPU even if the new alternative has a systematic utility less than that of the alternatives already available. This is because of the randomness of perceived utilities: there is a positive probability that the new alternative will be perceived as having a utility greater than that of

[16]The maximum of i.i.d. Gumbel variables having scale parameter θ is also a Gumbel variable with the same scale parameter. See also Appendix 3.B.

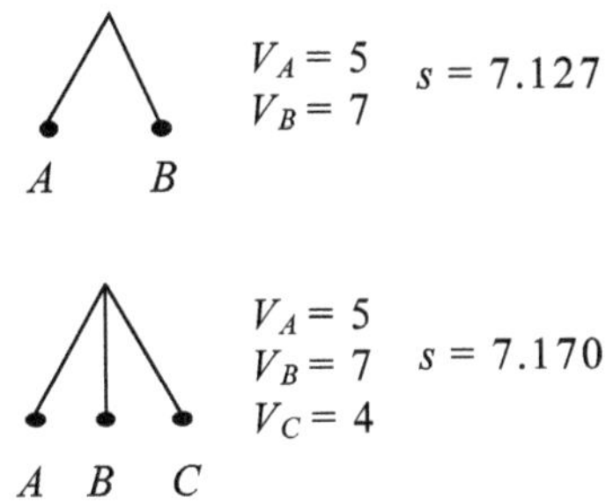

Fig. 3.14 Example of calculation of the expected maximum perceived utility (EMPU)

any other alternative. In this case, the $\max_j(\boldsymbol{U}^i)$ will clearly increase, and this will lead to a general increase in the mean value of $\max_j(\boldsymbol{U}^i)$, which is the EMPU.

The example in Fig. 3.14 also illustrates this point.

These properties of EMPU, directly derived here for the multinomial logit model, also apply to the larger class of invariant random utility models. Recall that, for these models, the density function of the random residuals does not depend on $\boldsymbol{V}$.

$$f(\boldsymbol{\varepsilon}/\boldsymbol{V}) = f(\boldsymbol{\varepsilon}) \quad \forall \boldsymbol{\varepsilon} \in E^m \tag{3.4.5}$$

All of the random utility models described in Sect. 3.3 are invariant if the parameters of $f(\boldsymbol{\varepsilon})$ do not depend on the vector $\boldsymbol{V}$. If the joint density function of the random residuals $f(\boldsymbol{\varepsilon})$ is continuous with continuous first derivatives, the choice probabilities $p(\boldsymbol{V})$ and the EMPU $s(\boldsymbol{V})$ are also continuous functions of $\boldsymbol{V}$ with continuous first derivatives. All random utility models described in Sect. 3.3 satisfy these continuity requirements. Under these assumptions, invariant random utility models share a number of general mathematical properties that are connected with the expected maximum perceived utility.

(1) The *partial derivative* of the EMPU with respect to the systematic utility V_k is equal to the choice probability of alternative k:

$$\frac{\partial s(\boldsymbol{V})}{\partial V_k} = p[k](\boldsymbol{V}) \tag{3.4.6}$$

The gradient of the EMPU is thus equal to the vector of choice probabilities:

$$\nabla s(\boldsymbol{V}) = \boldsymbol{p}(\boldsymbol{V}) \tag{3.4.7a}$$

and its Hessian is equal to the Jacobian of choice probabilities:

$$\boldsymbol{Hess}\big[s(\boldsymbol{V})\big] = \boldsymbol{Jac}\big[p(\boldsymbol{V})\big] \tag{3.4.7b}$$

For a continuous function with continuous first derivatives, the integration and differentiation operators can be exchanged:

$$\frac{\partial s(\boldsymbol{V})}{\partial V_k} = \frac{\partial}{\partial V_k}\int_{\varepsilon_1=-\infty}^{\infty}\cdots\int_{\varepsilon_m=-\infty}^{\infty}\max(\boldsymbol{V}+\boldsymbol{\varepsilon})f(\boldsymbol{\varepsilon})\,d\boldsymbol{\varepsilon}$$
$$= \int_{\varepsilon_1=-\infty}^{\infty}\cdots\int_{\varepsilon_m=-\infty}^{\infty}\frac{\partial\max(\boldsymbol{V}+\boldsymbol{\varepsilon})}{\partial V_k}f(\boldsymbol{\varepsilon})\,d\boldsymbol{\varepsilon} \qquad (3.4.8)$$

Because

$$\frac{\partial\max(\boldsymbol{V}+\boldsymbol{\varepsilon})}{\partial V_k} = \begin{cases}1 & \text{for } k \text{ such that } V_k+\varepsilon_k = \max(\boldsymbol{V}+\boldsymbol{\varepsilon})\\ 0 & \text{otherwise}\end{cases}$$

the integral (3.4.8) is equal to the probability that the perceived utility of alternative k, $V_k+\varepsilon_k$, is the largest among all the m alternatives available, from which expression (3.4.6) derives.

This result can be checked immediately for the multinomial logit model, for which the EMPU, expressed by (3.4.4), can be differentiated analytically:

$$\frac{\partial}{\partial V_k}\left[\theta \ln \sum_j \exp(V_j/\theta)\right] = \frac{\exp(V_k/\theta)}{\sum_j \exp(V_j/\theta)} = p[k](V) \qquad (3.4.9)$$

Furthermore, because the choice probability $p[k]$ is always greater than or equal to zero (3.4.6) shows that the derivative of the EMPU with respect to the systematic utility is always nonnegative: the EMPU increases (or does not decrease) as the systematic utility of each alternative increases and, by extension, as the number of available alternatives increases.[17]

(2) The EMPU function is *convex*[18] with respect to $\boldsymbol{V}$, the vector of systematic utilities.

In fact, for each $\boldsymbol{\varepsilon}$, $f(\boldsymbol{\varepsilon}) \geq 0$ and $\max(\boldsymbol{V}+\boldsymbol{\varepsilon})$ is a convex function of $\boldsymbol{V}$; it follows that the expected maximum perceived utility function $s(\boldsymbol{V})$, expressed by (3.4.2), is a linear combination with nonnegative coefficients of convex functions, and therefore is convex too.

Note that by virtue of property (2) the EMPU function has a Hessian matrix, $\boldsymbol{Hess}(s(\boldsymbol{V}))$, which is (symmetric and) positive semidefinite. Consequently, the Jacobian of choice probabilities, $\boldsymbol{Jac}(\boldsymbol{p}(\boldsymbol{V}))$, is (symmetric and) positive semidefinite (see (3.4.7b)).

(3) If the EMPU function is continuous and differentiable then:

$$s(\boldsymbol{V}') \geq s(\boldsymbol{V}'') + p(\boldsymbol{V}'')^T(\boldsymbol{V}'-\boldsymbol{V}'') \quad \forall \boldsymbol{V}', \boldsymbol{V}'' \qquad (3.4.10a)$$

and the choice probabilities are monotonic increasing functions of the systematic utilities.

$$\left(p(\boldsymbol{V}') - p(\boldsymbol{V}'')\right)^T(\boldsymbol{V}'-\boldsymbol{V}'') \geq 0 \quad \forall \boldsymbol{V}', \boldsymbol{V}'' \qquad (3.4.10b)$$

---

[17]The availability of a new alternative can be seen, in fact, as a change in the systematic utility of that alternative from minus infinity to a finite value.

[18]Convexity of a scalar-valued function of a vector is defined in Appendix A.

Because the EMPU function is convex and differentiable, it follows that

$$s(\boldsymbol{V}') \geq s(\boldsymbol{V}'') + \nabla s(\boldsymbol{V}'')^T(\boldsymbol{V}' - \boldsymbol{V}'') \quad \forall \boldsymbol{V}', \boldsymbol{V}''$$

and its gradient must be an increasing monotonic function (see Appendix A):

$$\left(\nabla s(\boldsymbol{V}') - \nabla s(\boldsymbol{V}'')\right)^T(\boldsymbol{V}' - \boldsymbol{V}'') \geq 0 \quad \forall \boldsymbol{V}', \boldsymbol{V}''$$

Applying (3.4.7a), the two preceding expressions can be formulated in terms of the vector of choice probabilities as in (3.4.10a) and (3.4.10b). Moreover, from (3.4.10a) it follows that:

$$s(\boldsymbol{V}') - s(\boldsymbol{V}'') \geq p(\boldsymbol{V}'')^T(\boldsymbol{V}' - \boldsymbol{V}'') \quad \forall \boldsymbol{V}', \boldsymbol{V}''$$
$$s(\boldsymbol{V}'') - s(\boldsymbol{V}') \geq p(\boldsymbol{V}')^T(\boldsymbol{V}'' - \boldsymbol{V}') \quad \forall \boldsymbol{V}', \boldsymbol{V}''$$

Summing the last two inequalities yields:

$$0 \geq p(\boldsymbol{V}'')^T(\boldsymbol{V}' - \boldsymbol{V}'') + p(\boldsymbol{V}')^T(\boldsymbol{V}'' - \boldsymbol{V}') \quad \forall \boldsymbol{V}', \boldsymbol{V}''$$

from which (3.4.10b) is easily obtained.

In particular, (3.4.10b) can be expressed for a single alternative, assuming that the systematic utilities of all other choice alternatives are constant:

$$p_k(V_k') \geq p_k(V_k'') \quad \text{if } V_k' \geq V_k''$$

In other words, the choice probability of a generic alternative does not decrease as its systematic utility increases, if all the other systematic utilities remain unchanged. Using an analogous argument it can be demonstrated that, as V_k tends to minus infinity, the choice probability of alternative k tends to zero:

$$\lim_{V_k \to -\infty} p[k] = 0$$

Mathematical properties of the deterministic choice model. The deterministic choice model[19] is obtained if the random residuals are all equal to zero. In this case, the perceived utility coincides with the systematic utility and only the alternative(s) having maximum utility can be chosen:

$$p[k] > 0 \quad \Rightarrow \quad V_k = \max(\boldsymbol{V})$$

and

$$V_k = \max(\boldsymbol{V}) \quad \Rightarrow \quad p[k] \in [0, 1], \qquad V_k < \max(\boldsymbol{V}) \quad \Rightarrow \quad p[k] = 0$$

[19]Deterministic utility models and their properties are mainly used in Sect. 4.3.3 on path choice models and in Chap. 5 on assignment models.

Note that the deterministic choice model satisfies condition (3.4.5) and can therefore be considered an invariant model. If there are two or more alternatives with (equal) maximum systematic utility, there are infinitely many choice probability vectors satisfying the above conditions. In this case, the relation $\boldsymbol{p}(\boldsymbol{V})$ is not a function, but a one-to-many map. Let $\boldsymbol{p}_{\mathrm{DET}}(\boldsymbol{V})$ be one of the possible choice probability vectors corresponding to vector $\boldsymbol{V}$ through the deterministic choice map.

The following necessary and sufficient condition guarantees that a probability vector $\boldsymbol{p}^*$ (with $\boldsymbol{p}^* \geq \boldsymbol{0}$ and $\boldsymbol{1}^T \boldsymbol{p}^* = 1$) is a deterministic choice probability vector:

$$\boldsymbol{p}^* = \boldsymbol{p}_{\mathrm{DET}}(\boldsymbol{V}) \quad \Leftrightarrow \quad \boldsymbol{V}^T \boldsymbol{p}^* = \max(\boldsymbol{V})\boldsymbol{1}^T \boldsymbol{p}^* = \max(\boldsymbol{V}) \tag{3.4.11a}$$

Given a vector of deterministic probabilities $\boldsymbol{p}^* = \boldsymbol{p}_{\mathrm{DET}}(\boldsymbol{V})$, it follows that $\boldsymbol{V}^T \boldsymbol{p}^* = \max(\boldsymbol{V})$ because p_k^* can be positive only for an alternative k having maximum systematic utility, and conversely. Furthermore, the condition $\boldsymbol{1}^T \boldsymbol{p}^* = 1$ implies that $\max(\boldsymbol{V})\boldsymbol{1}^T \boldsymbol{p}^* = \max(\boldsymbol{V})$.

In general, for any vector of choice probabilities $\boldsymbol{p}$, because $\boldsymbol{1}^T \boldsymbol{p} = 1$ then, as observed earlier:

$$\boldsymbol{V}^T \boldsymbol{p} \leq \max(\boldsymbol{V})\boldsymbol{1}^T \boldsymbol{p} = \max(\boldsymbol{V}) \quad \forall \boldsymbol{p} : \boldsymbol{p} \geq \boldsymbol{0},\ \boldsymbol{1}^T \boldsymbol{p} = 1$$

Consistent with (3.4.11a), equality holds in the above relationship only for a vector of deterministic probabilities. Combining the above relationship with (3.4.11a), the following basic relationship can be obtained.

$$\left(\boldsymbol{V} - \max(\boldsymbol{V})\boldsymbol{1}\right)^T \left(\boldsymbol{p} - \boldsymbol{p}_{\mathrm{DET}}(\boldsymbol{V})\right) \leq 0 \quad \forall \boldsymbol{p} : \boldsymbol{p} \geq \boldsymbol{0},\ \boldsymbol{1}^T \boldsymbol{p} = 1 \tag{3.4.11b}$$

This is applied in the analysis of deterministic assignment models in Chap. 5.

The deterministic utility model has properties (2) and (3) described above for probabilistic and invariant models.[20] Regarding property (2), the expected maximum perceived utility of a deterministic model is a convex function of systematic utilities and is equal to the maximum systematic utility:

$$s(\boldsymbol{V}) = \max(\boldsymbol{V}) = \boldsymbol{p}_{\mathrm{DET}}(\boldsymbol{V})^T \boldsymbol{V} \tag{3.4.12}$$

This condition and result (3.4.3) imply that, for a given vector of systematic utilities $\boldsymbol{V}$, the EMPU of a deterministic choice model is less than or equal to that of any probabilistic choice model involving the same systematic utility. A behavioral interpretation of this result suggests that the presence of random residuals makes the perceived utility for the chosen alternative, on average, larger than the alternative's systematic utility, which is the perceived utility in a deterministic choice model.

Regarding property (3), the deterministic choice map is monotone nondecreasing with respect to systematic utilities, just as are invariant probabilistic choice functions:

$$s(\boldsymbol{V}') \geq s(\boldsymbol{V}'') + \boldsymbol{p}_{\mathrm{DET}}(\boldsymbol{V}'')^T (\boldsymbol{V}' - \boldsymbol{V}'') \quad \forall \boldsymbol{V}', \boldsymbol{V}'' \tag{3.4.13a}$$

---

[20] Property (1) requires the introduction of the concept of subgradients of a convex function.

or

$$\left(p_{\mathrm{DET}}(\boldsymbol{V}') - p_{\mathrm{DET}}(\boldsymbol{V}'')\right)^T (\boldsymbol{V}' - \boldsymbol{V}'') \geq 0 \quad \forall \boldsymbol{V}', \boldsymbol{V}'' \tag{3.4.13b}$$

in perfect formal analogy with expressions (3.4.10).

In fact, from (3.4.11a) it follows that:

$$\max(\boldsymbol{V}') = (\boldsymbol{V}')^T \boldsymbol{p}_{\mathrm{DET}}(\boldsymbol{V}'),$$
$$\max(\boldsymbol{V}'') = (\boldsymbol{V})^T \boldsymbol{p}_{\mathrm{DET}}(\boldsymbol{V}'')$$

Subtracting the last two equations term by term gives:

$$\max(\boldsymbol{V}') - \max(\boldsymbol{V}'') = (\boldsymbol{V}')^T \boldsymbol{p}_{\mathrm{DET}}(\boldsymbol{V}') - (\boldsymbol{V}'')^T \boldsymbol{p}_{\mathrm{DET}}(\boldsymbol{V}'') \tag{i}$$

Because

$$(\boldsymbol{V}')^T \boldsymbol{p}_{\mathrm{DET}}(\boldsymbol{V}') = \max(\boldsymbol{V}') \geq (\boldsymbol{V}')^T \boldsymbol{p} \quad \forall \boldsymbol{p}$$

for $\boldsymbol{p} = \boldsymbol{p}_{\mathrm{DET}}(\boldsymbol{V}'')$ it follows that:

$$(\boldsymbol{V}')^T \boldsymbol{p}_{\mathrm{DET}}(\boldsymbol{V}') \geq (\boldsymbol{V}')^T \boldsymbol{p}_{\mathrm{DET}}(\boldsymbol{V}'')$$

from which

$$(\boldsymbol{V}')^T \boldsymbol{p}_{\mathrm{DET}}(\boldsymbol{V}') - (\boldsymbol{V}'')^T \boldsymbol{p}_{\mathrm{DET}}(\boldsymbol{V}'') \geq (\boldsymbol{V}')^T \boldsymbol{p}_{\mathrm{DET}}(\boldsymbol{V}'') - (\boldsymbol{V}'')^T \boldsymbol{p}_{\mathrm{DET}}(\boldsymbol{V}'') \tag{ii}$$

Therefore, combining (i) and (ii) yields

$$\max(\boldsymbol{V}') - \max(\boldsymbol{V}'') \geq (\boldsymbol{V}' - \boldsymbol{V}'')^T \boldsymbol{p}_{\mathrm{DET}}(\boldsymbol{V}'')$$

which is expression (3.4.13a), because $s(\boldsymbol{V}) = \max(\boldsymbol{V})$.

3.5 Choice Set Modeling*

Random utility models represent the choice made by a generic individual i from the set of alternatives that make up her choice set I^i, under the hypothesis that the modeler is able to specify this set correctly. When this hypothesis is not acceptable, it is necessary to model explicitly the composition of the generic decision-maker's choice set. This problem has been tackled in two fundamentally different ways. The *implicit approach* incorporates within the choice model itself attributes related to an alternative's actual or perceived availability. The *explicit approach* uses a distinct model to explicitly represent the choice set generation.

The first approach has been adopted in many specifications of random utility models proposed in the literature. Some attributes in the systematic utility function of an alternative play the role of "proxy" variables, representing the availability or perception of that alternative. For example, a variable equal to the number of cars divided by the number of licensed drivers in a household is often used to represent car availability in mode choice models. Attributes with this interpretation can be easily identified in a number of the random utility models described in the next chapter. The implicit approach is undoubtedly simpler from the application point of view, although there is a noticeable lack of consistency because "utility" attributes are mixed with "availability" attributes.

In the explicit approach, the choice probability of an alternative j for decision-maker i is usually expressed through a two-stage choice model:

$$p^i[j] = \sum_{I^i \in G^i} p^i[j, I^i] = \sum_{I^i \in G^i} p^i[j/I^i] p^i[I^i] \tag{3.5.1}$$

where

- I^i is the generic choice set of decision-maker i
- G^i is the set made up of all possible nonempty choice sets for decision-maker i (nonempty subsets of the set of all the possible alternatives)
- $p^i[j, I^i]$ is the joint probability that decision-maker i will choose alternative j and that I^i is his choice set
- $p^i[j/I^i]$ is the probability that decision-maker i will choose alternative j, her choice set being I^i
- $p^i[I^i]$ is the probability that I^i is the choice set of individual i

The choice probability conditional on set I^i, $p^i[j/I^i]$, can be represented with one of the random utility models described in Sect. 3.3.

An example of an explicit choice set generation model can be obtained, starting from the general model (3.5.1), by assuming that the probabilities that each single alternative belongs to the choice set are independent of each other:

$$Pr[j \in I^i / h \in I^i] = Pr[j \in I^i] \quad \forall j, h \tag{3.5.2}$$

In this case, the probability $p[I^i]$ can be expressed as

$$p[I^i] = \frac{\prod_{h \in I^i} p[h \in I^i] \cdot \prod_{k \notin I^i} [1 - p[k \in I^i]]}{1 - p[I^i \equiv \emptyset]} \tag{3.5.3}$$

where the first product is extended to all the alternatives included in I^i and the second to all those not included in I^i. The denominator of expression (3.5.3) normalizes the probabilities $p[I^i]$ to take into account the fact that an empty choice set ($I^i \equiv \emptyset$) is usually excluded, under the assumption that the decision-maker's choice set includes at least one alternative; the probability that the choice set is empty is given by

$$p[I^i \equiv \emptyset] = \prod_j [1 - p[j \in I^i]] \tag{3.5.4}$$

Replacing expressions (3.5.3) and (3.5.4) in (3.5.1), the choice probability of the generic alternative is:

$$p^i[j] = \frac{\sum_{I^i \in G^i} \{\prod_{h \in I^i} p^i[h \in I^i] \cdot \prod_{k \notin I^i} [1 - p^i[k \in I^i]] \cdot p^i[j/I^i]\}}{1 - \prod_j [1 - p^i[j \in I^i]]} \quad (3.5.5)$$

Specification of model (3.5.5) requires a model to represent the probability $p[j \in I^i]$ that generic alternative j belongs to the choice set. Various authors have proposed a binomial logit model[21]:

$$p[j \in I^i] = \frac{1}{1 + \exp(\sum_k \gamma_k Y^i_{kj})} \quad (3.5.6)$$

where the Y_k are "availability/perception" variables mentioned above and the γ_k are their coefficients.

The explicit approach, although very interesting and consistent from a theoretical point of view, poses some computational problems. The number of all possible choice sets (i.e., the cardinality of G^i) grows exponentially with the number of alternatives. This complicates the calculation of choice probabilities (3.5.1), and therefore the joint calibration of the parameters β_k in the systematic utility and γ_k in the choice set model.

An intermediate approach, named Implicit Availability Perception (IAP), accounts for the availability and perception of an alternative by modifying its systematic utility in the random utility model. This approach is based on a generalization of the conventional concepts of availability and choice set membership. Instead of assuming that an alternative is either available or not, the approach considers that an alternative may have intermediate levels of availability and perception to a decision-maker. The decision-maker's choice set is then viewed as a "fuzzy set"; it is no longer represented as a set of [0/1] Boolean variables (1 if the alternative is available or perceived, 0 otherwise), but rather as a set of continuous variables $\mu_I(j)$ defined on the interval [0, 1]. This representation could apply, for example, to an alternative that is theoretically available but not completely perceived as such for a particular journey, due to factors that may be either subjective (lack of information, time constraints, state of health, etc.) or objective (weather conditions, etc.) Obviously, extreme values of $\mu_I(j)$ are still possible, corresponding respectively to the nonavailability and the complete availability and perception of alternative j.

The model accounts for different levels of availability and perception of an alternative by directly introducing an appropriate functional transformation of $\mu_I(j)$ into the alternative's utility function:

$$U^i_j = V^i_j + \ln \mu^i_I(j) + \varepsilon^i_j \quad (3.5.7)$$

where

---

[21]In this application, the Binomial Logit model (3.5.6) should be seen as a convenient functional relationship rather than a random utility model since it does not represent any "choice".

U^i_j is the perceived utility of alternative j for decision-maker i
V^i_j is the systematic utility of alternative j for decision-maker i
ε^i_j is the random residual of alternative j for decision-maker i
$\mu^i_I(j)$ is the level of membership of alternative j in the choice set I^i of decision-maker i $(0 \leq \mu \leq 1)$

In this way, all the alternatives can be considered as theoretically available. If alternative j is not available ($\mu^i_I(j) = 0$), the term $\ln \mu^i_I(j)$ forces its perceived utility U^i_j to minus infinity and the probability of choosing it to zero, regardless of the value of V^i_j. Furthermore, choice probabilities of all the other alternatives are no longer influenced by alternative j. If, on the other hand, an alternative j is definitely available and taken into consideration ($\mu^i_I(j) = 1$), the additional term is equal to zero and the perceived utility has the conventional expression. Intermediate values of $\mu^i_I(j)$ reduce the utility of the alternative according to its level of availability.

For a generic individual i, the true value of the availability and perception level, and therefore of the term $\ln \mu^i_I(j)$, is unknown to the analyst. It can therefore be modeled as a random variable, which in turn can be expressed as the sum of its mean value, $E[\ln \mu^i_I(j)]$, and a random residual, η^i_j, defined by the difference $\ln \mu^i_I(j) - E[\ln \mu^i_I(j)]$. Expression (3.5.7) then becomes:

$$U^i_j = V^i_j + E\big[\ln \mu^i_I(j)\big] + \eta^i_j + \varepsilon^i_j \tag{3.5.8}$$

In order to make expression (3.5.8) more tractable, $E[\ln \mu^i_I(j)]$ can be approximated by its second-order Taylor series expression around the point $\bar{\mu}^i_I(j) = E[\mu^i_I(j)]$. Substituting this approximation in (3.5.8) yields:

$$U^i_j \cong V^i_j + \ln \bar{\mu}^i_I(j) - \frac{1 - \bar{\mu}^i_I(j)}{2\bar{\mu}^i_I(j)} + \sigma^i_j \quad \text{with } \sigma^i_j = \varepsilon^i_j + \eta^i_j \tag{3.5.9}$$

The choice probability of alternative j can therefore be calculated using the random utility models described in Sect. 3.3; it will depend on the systematic utility of each alternative, on the mean availability and perception of each alternative and on the joint distribution of the random variables σ^i_j. For example, if the latter are assumed to be i.i.d. Gumbel $(0, \theta)$ variables, a new multinomial logit model is obtained:

$$p^i[j] = \frac{\exp\big[\frac{1}{\theta} \cdot \big(V^i_j + \ln \bar{\mu}^i_I(j) - \frac{1-\bar{\mu}^i_I(j)}{2\bar{\mu}^i_I(j)}\big)\big]}{\sum_h \exp\big[\frac{1}{\theta} \cdot \big(V^i_h + \ln \bar{\mu}^i_I(h) - \frac{1-\bar{\mu}^i_I(h)}{2\bar{\mu}^i_I(h)}\big)\big]} \tag{3.5.10}$$

where the sum in the denominator is extended to all the alternatives theoretically available to decision-maker i. From the above expression, it can be deduced that, everything else being equal, the choice probability of a generic alternative increases with increases in its mean availability/perception.[22]

Other functional specifications of choice models can be obtained from expression (3.5.9). For example, if the perception/availability of two alternatives j and h are similar (i.e., they are both likely either to be perceived or not to be perceived), a positive covariance between the residuals η_j and η_h can be assumed.

To specify completely the model (3.5.10) (or a similar model with a different functional form for the choice probabilities), the mean availability/perception $\bar{\mu}^i_I(j)$ must be expressed as a function of the availability and perception attributes using, for example, a binomial logit model of the form given by (3.5.6):

$$\mu^i_j(j) = \frac{1}{1+\exp(\sum_{k=1}^{K_j} \gamma_k Y^i_{kj})} \tag{3.5.11}$$

Note the different interpretation of the two expressions (3.5.6) and (3.5.11). Expression (3.5.6) gives the probability that alternative j belongs to the choice set of a given decision-maker, whereas expression (3.5.11) gives the average degree of availability and perception of the alternative for decision-makers with the same attributes Y^i_{kj}.

3.6 Direct and Cross-elasticities of Random Utility Models*

Random utility models can be considered econometric demand functions in every respect. Choice probabilities can be viewed as mean values of the fractions of a market segment (a group of decision-makers with the same characteristics) that select the different available alternatives.[23] Furthermore, random utility models express these fractions as functions of the available alternatives' attributes. In the context of this interpretation, it is possible to extend to random utility models the microeconomic concepts of direct and cross-elasticities of demand functions with respect to infinitesimal or discrete variations of the variables in the utility function.

Recall that *direct elasticity* is defined as the percentage variation in the demand for a certain commodity (in the discussion here, the "demand" for a commodity j refers to the choice probability of an alternative j) divided by the percentage variation in the value of an attribute k of the same commodity X_{kj}:

$$E^{p[j]}_{kj} = \frac{\Delta p[j]}{p[j]} \Big/ \frac{\Delta X_{kj}}{X_{kj}}$$

[22] This consideration clarifies the importance of information on the availability of alternatives.

[23] The actual number of decision-makers with the same attributes who actually choose alternative j is a random variable, so the ratio between this number and the total number of decision-makers is random as well. The mean of this r.v. is equal to choice probability $p[j]$ given by the model.

Analogously, *cross-elasticity* is defined as the percentage variation in the demand for a certain commodity j divided by the percentage variation in the value of an attribute k of another commodity h, X_{kh}:

$$E_{kh}^{p[j]} = \frac{\Delta p[j]}{p[j]} \Big/ \frac{\Delta X_{kh}}{X_{kh}}$$

In the above definitions, the variations in the values of attributes and demand are assumed to be finite. This case defines the arc elasticity, which is calculated as the ratio of incremental ratios over an "arc" of the demand curve. Point elasticities are defined for infinitesimal variations and can be expressed analytically.

The point direct elasticity of the choice probability for alternative j with respect to an infinitesimal variation in the kth attribute X_{kj} of its own utility function is defined as

$$E_{kj}^{p[j]} = \frac{\partial p[j](\boldsymbol{X})}{\partial X_{kj}} \frac{X_{kj}}{p[j]} = \frac{\partial \ln p[j](\boldsymbol{X})}{\partial \ln X_{kj}} \tag{3.6.1}$$

where $\boldsymbol{X}$ includes the vectors of attributes for all alternatives.

Similarly the point cross-elasticity of the choice probability of alternative j with respect to an infinitesimal variation of the kth attribute, X_{kh}, of the utility function of alternative h is defined as

$$E_{kh}^{p[j]} = \frac{\partial p[j](\boldsymbol{X})}{\partial X_{kh}} \frac{X_{kh}}{p[j]} = \frac{\partial \ln p[j](\boldsymbol{X})}{\partial \ln X_{kh}} \tag{3.6.2}$$

Both direct and cross-elasticities[24] are useful measures of the model's sensitivity to variations in the attributes. It is evident from (3.6.1) and (3.6.2) that elasticities depend on the functional form of the model as well as on the values of attributes and parameters in the systematic utilities.

Analytic and compact expressions for direct and cross-elasticities (3.6.1) and (3.6.2) can be obtained for the multinomial logit model with a linear systematic

[24]The elasticities discussed in this section are disaggregate, i.e. related to variations in the probabilities of a single decision maker or of a group of decision makers sharing the same attribute values. Aggregate elasticities refer to variations in the average choice fraction:

$$\bar{p}(j) = \sum_{i=1}^{n} p^i(j)$$

of a group of decision makers with different attributes. Variations are computed with respect to a uniform infinitesimal variation of a given attribute. In this case, it is possible to express the aggregate elasticity as a weighted average of individual elasticities. For instance the direct point elasticity is:

$$E_{kj}^{\bar{p}[j]} = \frac{\sum_{i=1}^{n} p^i[j] E_{kh}^{p^i[j]}}{\sum_{i=1}^{n} p^i[j]}$$

utility function $V_j = \beta^T \boldsymbol{X}_j$. In this case:

$$E_{kj}^{p[j]} = \big(1 - p[j]\big)\beta_k X_{kj}/\theta \tag{3.6.3}$$

$$E_{kh}^{p[j]} = -p[k]\beta_k X_{kh}/\theta \tag{3.6.4}$$

From (3.6.3) it can be deduced that the direct elasticity is positive if attribute X_{kj} is positive (as is usually the case) and if its coefficient β_k is positive. In other words, the choice probability of an alternative increases if the value of an attribute that contributes to its utility (β positive) increases.[25] The increase will be higher for higher values of coefficient β_k and attribute X_{kj}, and for lower values of the alternative j choice probability. Thus, in a mode choice model, direct elasticities of the probability of choosing a car with respect to travel time and cost will be negative because the coefficients β_k of these attributes are negative; these elasticities will be larger, in absolute terms, for an origin–destination pair with relatively large time and cost values. Lastly, if the probability of choosing the car is low, its elasticity will be larger, for given values of parameter β_k and attribute X_{kj}.

Similar considerations, although with inverted signs, hold for cross-elasticities, which will be positive if β_k or X_{kh} are negative, and will be larger for larger absolute values of β_k, X_k, and $p[h]$. Continuing with the above example, the cross-elasticities of the probability of using a car with respect to the travel time and cost of another mode will be positive (because $\beta_k < 0$).

Qualitatively similar conclusions apply to elasticities of random utility models other than *MNL*.

Note that the cross-elasticity (3.6.4) of the multinomial logit model is identical for all alternatives because a variation in the value of one alternative's attribute produces the same percentage variation in the choice probabilities of all other alternatives. This result can be considered as a different manifestation of the logit model independence from irrelevant alternatives property described in Sect. 3.3.1.

Expressions (3.6.3) and (3.6.4) also show that, for given values of coefficients and attributes, direct and cross-elasticities are higher in absolute terms when the variance of the random residuals (directly related to the scale parameter θ) is lower. Conversely, as the random residual variances tend to infinity, the elasticities tend to zero. Figure 3.15 shows the values of direct and cross-elasticities with respect to a generic attribute in a multinomial logit model.

For more complex random utility models it is not easy, or even possible, to derive analytic expressions for direct and cross-elasticities. However, it is useful to discuss elasticities for a single-level hierarchical logit model inasmuch as they provide some insight into the influence of random residual covariances on direct and cross-elasticities.

[25]The result that multinomial logit choice probabilities increase monotonically with respect to systematic utilities is obtained again. It holds, more generally, for all invariant models described in previous sections.

	X_{kA}	X_{kB}	X_{kC}	X_{kD}
$E_{p[A]}$	0.75	−0.25	−0.25	−0.25
$E_{p[B]}$	−0.25	0.75	−0.25	−0.25
$E_{p[C]}$	−0.25	−0.25	0.75	−0.25
$E_{p[D]}$	−0.25	−0.25	−0.25	0.75

Fig. 3.15 Direct and cross-elasticities for a multinomial logit model

Consider the single-level hierarchical logit model in Fig. 3.16; it contains one nest whose only component is the elementary alternative A, and another nest G containing elementary alternatives B, C, and D.

It is possible to obtain in closed form the elasticities of the choice probability of alternative A with respect to a generic attribute X_k that is included in the systematic utility of all the alternatives. Applying the definitions of elasticity (3.6.1) and (3.6.2) to the single-level hierarchical logit model in expression (3.3.19) with parameter $\theta_o = 1$, the direct elasticity (variation of attribute X_{kA}) and the cross-elasticity with respect to alternative B (variation of attribute X_{kB}) are, respectively,

$$E_{kA}^{p[A]} = \big(1 - p[A]\big)\beta_k X_{kA}/\theta \tag{3.6.5}$$

$$E_{kB}^{p[A]} = -p[B]\beta_k X_{kB}/\theta \tag{3.6.6}$$

The elasticities in this case are completely analogous to those obtained for the multinomial logit model, expressed by (3.6.3) and (3.6.4). Things are different, however, for the choice probability elasticities of alternative B in nest G. Its direct elasticity (variation of attribute X_{kB}) is

$$E_{kB}^{p[B]} = \big\{\big(1 - p[G]\big) \cdot p[B/G] + \big(1 - p[B/G]\big)/\theta\big\}\beta_k X_{kB} \tag{3.6.7}$$

If the hierarchical logit were reduced to a multinomial logit model, that is, if $\theta = 1$, the direct elasticity (3.6.7) would become analogous to (3.6.3) or (3.6.5). On the other hand, if θ is less than one, the hierarchical logit elasticity is larger than that of a multinomial logit model with the same parameters, attributes, and residual variance.

The cross-elasticities of $p[B]$ with respect to variations in attribute X_{kA} of the "isolated" alternative A, and in attribute X_{kC} of alternative C in the same nest G are, respectively,

$$E_{X_{kA}}^{p[B]} = -p[A]\beta_k X_{kA}/\theta \tag{3.6.8}$$

$$E_{X_{kC}}^{p[B]} = -\left[p[C] + \frac{1-\theta}{\theta} p[C/G]\right]\beta_k X_{kC} \tag{3.6.9}$$

Equation (3.6.8) shows that the cross-elasticity of B's choice probability with respect to an attribute of alternative A not belonging to B's nest G is equivalent

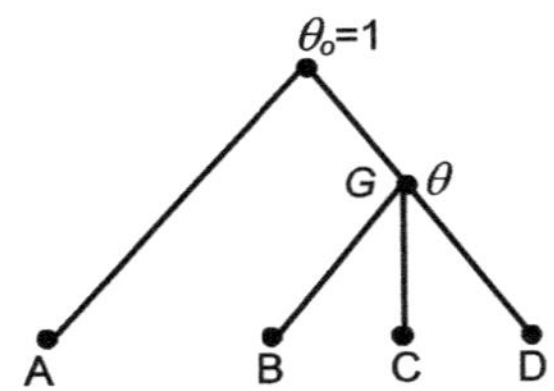

Hierarchical Logit

$V_A = V_B = V_C = V_D$

$\beta_k = 1 \quad \forall k$

$X_{kj} = 1 \quad \forall k, j$

	$\theta = 1$				$\theta = 0.8$				$\theta = 0.4$				$\theta = 0.1$			
	X_{kA}	X_{kB}	X_{kC}	X_{kD}	X_{kA}	X_{kB}	X_{kC}	X_{kD}	X_{kA}	X_{kB}	X_{kC}	X_{kD}	X_{kA}	X_{kB}	X_{kC}	X_{kD}
$E_{p[A]}$	0.75	−0.25	−0.25	−0.25	0.71	−0.24	−0.24	−0.24	0.61	−0.20	−0.20	−0.20	0.52	−0.18	−0.18	−0.18
$E_{p[B]}$	−0.25	0.75	−0.25	−0.25	−0.29	0.93	−0.32	−0.32	−0.39	1.80	−0.70	−0.70	−0.47	6.82	−3.18	−3.18
$E_{p[C]}$	−0.25	−0.25	0.75	−0.25	−0.29	−0.32	0.93	−0.32	−0.39	−0.70	1.80	−0.70	−0.47	−3.18	6.82	−3.18
$E_{p[D]}$	−0.25	−0.25	−0.25	0.75	−0.29	−0.32	−0.32	0.93	−0.39	−0.70	−0.70	1.80	−0.47	−3.18	−3.18	6.82

Fig. 3.16 Direct and cross-elasticities for a hierarchical logit model

to that of the corresponding multinomial logit model. On the other hand, the cross-elasticity with respect to an attribute of an alternative belonging to B's nest G (and so correlated with B) is larger for smaller values of parameter θ, that is, for larger covariance between the two alternatives. If two alternatives are perceived as being very similar (i.e., their respective random residuals are highly correlated), the probability of choosing one of them is very sensitive to variations of the attributes of the other. From (3.6.9) it also follows that if $\theta = 1$ the hierarchical logit model becomes a multinomial logit model and the cross-elasticity is analogous to (3.6.8).

Direct and cross-elasticities of the hierarchical logit model, for different values of parameter θ, are shown in Fig. 3.16. For $\theta = 1$, the elasticities reported in Fig. 3.15 are obtained.

The general conclusion from the above example is that, given equal attributes and coefficients, the more an alternative is perceived as "similar" to other alternatives, the higher are its direct and cross-elasticities. Thus, for any random utility model, variations in the attributes of an alternative will have the greatest effects on the choice probabilities of alternatives that are perceived as close substitutes to it.

3.7 Aggregation Methods for Random Utility Models

Random utility models described in the previous sections express the probability that a decision-maker i chooses an alternative j as a function of the attributes of all available alternatives. To highlight the dependence of choice probabilities on the individual decision-maker, expression (3.2.3a) can be reformulated as

$$p^i[j/V(X^i)] = Pr[V_j(X^i_j) + \varepsilon^i_j \geq V_k(X^i_k) + \varepsilon^i_k \ \forall k \in I^i] \tag{3.7.1}$$

where X^i_j is the vector of attributes of alternative j for decision-maker i, and X^i the vector of the attributes of all alternatives. For convenience of notation, (3.7.1) will be represented more compactly as $p[j/X^i]$ below.

Applications of random utility models for travel-demand modeling often require the mean value of total demand flows, that is, the mean number of decision-makers choosing each alternative. Aggregation techniques allow passage from individual choice probabilities to group, or aggregate, probabilities. To introduce these techniques, it is useful to describe the theoretical aggregation process. Suppose that the vector X^i of attributes, the functional form, and the coefficients of the random utility model are known for each individual i of the population. Suppose also that there are N_T individuals in the population and that they choose independently of each other. Under these assumptions, the number of decision-makers who actually choose the generic alternative j is a random variable, the sum of N_T independent Bernoulli random variables y^i_j, each of which is equal to one if individual i chooses alternative j and zero otherwise. The mean number of individuals choosing alternative j, D_j, is therefore the sum of the means, $p[j/X^i]$, of the N_T Bernoulli random variables:

$$D_j = \sum_{i=1}^{N_T} E[y^i_j] = \sum_{i=1}^{N_T} p[j/X^i] \tag{3.7.2}$$

The average fraction P_j of the population choosing alternative j can be estimated as

$$P_j = \frac{1}{N_T} \sum_{i=1}^{N_T} p[j/\boldsymbol{X}^i] = \frac{D_j}{N_T} \tag{3.7.3}$$

For populations large enough to replace the sum with an integral, (3.7.3) can be rewritten as

$$P_j = \int_X p[j/\boldsymbol{X}] g(\boldsymbol{X})\, d\boldsymbol{X} \tag{3.7.4}$$

where $g(\boldsymbol{X})$ represents the joint probability density function of the vector of attributes over the whole population, a measure of the frequency with which the different values of $\boldsymbol{X}$ occur in the population. In practice, the distribution $g(\boldsymbol{X})$ is not known and, to calculate the percentage P_j, aggregation techniques that estimate $\hat{P}_j$ using information on a limited number of individuals must be used.

In the literature, various aggregation methods have been proposed; these can be seen as approximate techniques for integrating (3.7.4).

The methods most frequently applied are:

(1) Average individual
(2) Classification
(3) Sample enumeration
(4) Classification/enumeration

(1) In the first method, an "*average individual*" is considered, whose attributes $\bar{\boldsymbol{X}}$ are the average population values calculated from the density $g(\boldsymbol{X})$. The aggregated choice percentage is determined as a function of these attributes:

$$\hat{P}_j = p[j/\bar{\boldsymbol{X}}] \tag{3.7.5}$$

This method is acceptable only if the relationship between the vector of attributes and the choice probabilities $p[j/\boldsymbol{X}]$ is linear or almost linear. Should the probability function be convex or concave, the method would, respectively, underestimate or overestimate the actual value of the fraction of the population choosing alternative j (see Fig. 3.17). It can also be shown that the deviation of linear estimate $\hat{P}_j$ from its true value is larger for greater dispersion of the values of X in the population, that is, for larger variances in the marginal distributions of $g(\boldsymbol{X})$.

(2) The *classification* method can be seen as an extension of the average individual method described above. In order to reduce the variance of $g(\boldsymbol{X})$, the population is divided into homogeneous and mutually exclusive classes. Let i represent a generic class with N_i members. The average individual technique is then applied to each such class, and the estimated fraction of the population choosing alternative j becomes:

$$\hat{P}_j = \sum_{i=1}^{I} \frac{N_i}{N_T} P[j/\bar{\boldsymbol{X}}^i] \tag{3.7.6}$$

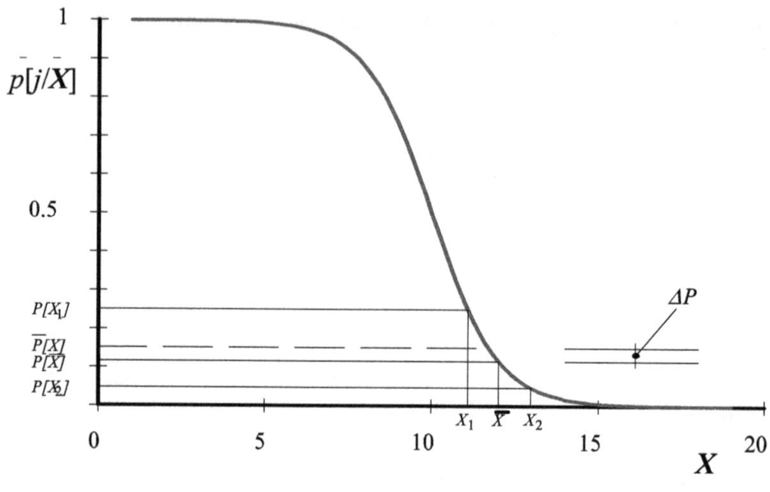

Fig. 3.17 Bias of average individual estimates of population fractions

where $\bar{X}^i$ is the vector of attributes for the average individual of the ith class.

In applications, classes are defined on the basis of a few criteria that are expected to have the greatest effect on systematic utilities. Variables influencing the distribution of the attributes are often adopted as classification criteria, for example, professional status or income. The number N_i of individuals belonging to each class should be available from statistical sources. The classification technique gives satisfactory results when the number of classes is limited and the individual classes are relatively homogeneous with respect to the attributes included in the model.

(3) With the *sample enumeration* method, it is assumed that the whole population can be represented by a random sample of individuals (decision-makers) extracted from it. The average fraction of individuals choosing alternative j in the overall population is estimated from the probability that j is chosen by the individuals belonging to the random sample. If N_s is the number of individuals in the sample, then:

$$\hat{P}_j = \frac{1}{N_s} \sum_{h=1}^{N_s} p[j/\boldsymbol{X}^h] \tag{3.7.7}$$

where $\boldsymbol{X}^h$ is the vector of the attributes relative to the hth individual in the sample. Expression (3.7.7) applies to the estimation of the mean population choice fraction when the individuals are chosen using simple random sampling.[26]

(4) *Sample enumeration and classification* methods can be combined; this is equivalent to assuming a stratified random sample of decision-makers. A random sample of individuals is extracted from each of the I strata (the homogeneous and

[26]Further elements of sample theory are discussed in Chap. 8 on demand estimation and its bibliography.

mutually exclusive classes) into which the population is divided. If N_i is the number of individuals belonging to stratum i and N_{si} is the number of sample individuals extracted from stratum i, the fraction $\hat{P}_j$ can be estimated as

$$\hat{P}_j = \sum_{i=1}^{I} W_i \frac{1}{N_{si}} \sum_{h=1}^{N_{si}} p[j/\boldsymbol{X}^h] \tag{3.7.8}$$

where the ratio $W_i = N_i/N_T$ is the weight of stratum i in the population.

The total number of decision-makers choosing each alternative j (the aggregate demand for alternative j) can be calculated by multiplying expressions (3.7.6), (3.7.7) and (3.7.8) by N_T. The ratio between the number of individuals in the population (or a class) and the number of individuals in the sample, N_T/N_s or N_i/N_{si}, is called the "expansion factor" of individuals from the sample to the population.

A number of extensions to these basic methods have been proposed to overcome difficulties sometimes encountered in their application.

The sample enumeration method allows significant flexibility in the use of random utility models, because the attributes considered in vector $\boldsymbol{X}$ might include variables relating to the individual for which it is difficult, if not impossible, to obtain mean values over the whole population or subpopulations (classes). This flexibility is achieved at the cost of greater computational complexity. However, this drawback is becoming less important with the steady increase in available computing power. Another problem associated with the sample enumeration method relates to the availability of samples of decision-makers for each class i and each choice context (e.g., each traffic zone in the study area). The samples should be large enough to guarantee adequate coverage of the distribution of attributes $\boldsymbol{X}$. This would require large samples of decision-makers for each zone. The *prototypical sample* method overcomes this problem by using the same sample of N_{si} decision-makers of class i for different traffic zones, but applying different weights W_i^z to each class i in each zone z ($W_i^z = N_i^z/N_T$). This method requires knowledge of the number, N_i^z, of individuals of class i in each zone, which can be obtained from statistical sources (present scenario), or from sociodemographic forecasts (future scenarios).

In methods based on sample enumeration, estimation of the average number of individuals choosing alternative j in zone z, D_j^z, requires the expansion factors g_i^z of each class in each zone:

$$D_j^z = \sum_{i=1}^{I} g_i^z \sum_{h=1}^{N_{si}} p[j/\boldsymbol{X}^h] \tag{3.7.9}$$

where these expansion factors can be formally expressed as

$$g_i^z = \frac{N_i^z}{N_{si}}$$

Sometimes the number N_i^z of individuals of class i in zone z is unknown, especially when several classes have been defined. In this case, it is not possible to estimate either the weights of the individual classes ($W_i^z = N_i^z/N_T$) and the average choice percentages by (3.7.8), or the expansion factors g_i^z and the total number of individuals choosing alternative j, D_j, by (3.7.9). To overcome this problem, the *target variable method* can be adopted. This method is described here in reference to the calculation of expansion factors; once these are known, the weights W_i^z can easily be calculated. The expansion factors are calculated so that, when the prototypical sample is rescaled to its universe, it reproduces the zonal values of selected aggregated variables, known as target variables T_t^z. Typical target variables are the number of residents by professional status, age, sex, income group, and so on. Formally, the expansion factors g_i^z must satisfy the following equations.

$$\sum_i g_i^z \sum_{h=1}^{N_{si}} K(t,h) = T_t^z \tag{3.7.10}$$

where $K(t,h)$ is the contribution to the tth target variable of the hth component of the prototypical sample belonging to category i. For example, if the tth target variable is the number of workers in the zone, individual h of class i will contribute one if employed, zero otherwise. In general, the number of unknown expansion factors (i.e., of classes in each zone) is larger than the number N_t of target variables, so the system of equations (3.7.10) does not have a unique solution. In this case, the vector $\mathbf{g}^z$ of expansion factors for the classes in each zone can be obtained by solving a least squares problem that minimizes the weighted distance from a vector of reference expansion factors $\hat{\mathbf{g}}$ while, at the same time, satisfying as closely as possible the system of equations (3.7.10):

$$\mathbf{g}^z = \underset{\mathbf{g}^z \geq 0}{\operatorname{argmin}} \left[\sum_i \left(g_i^z - \hat{g}_i\right)^2 + \alpha \sum_{t=1}^{N_t} \left(\sum_i g_i^z \sum_{h=1}^{N_{si}} K(t,h) - T_t^z \right)^2 \right] \tag{3.7.11}$$

Reference expansion factors can be obtained as sample estimates of the fraction of users belonging to each class. The parameter α is the relative weight of the two parts of the objective function in (3.7.11), that is, the relative weight that the analyst associates with the target variables (3.7.10) and to the initial estimates $\hat{\mathbf{g}}$ in the solution of problem (3.7.11).

Note that this least squares problem imposes nonnegativity constraints on the variables (3.7.11). It is similar in structure to the problem of estimating O-D demand flows from traffic count data that is formulated and discussed in Chap. 8, and can be solved by using the projected gradient algorithm described in Appendix A.

3.A. Derivation of Logit Models from the GEV Model

As stated in Sect. 3.3.5, the choice probability of a GEV model can be expressed as (see (3.3.52)):

$$p[j] = \frac{e^{V_j} \cdot G_j(e^{V_1}, \ldots, e^{V_j}, \ldots, e^{V_m})}{\mu \cdot G(e^{V_1}, \ldots, e^{V_j}, \ldots, e^{V_m})} \tag{3.A.1}$$

where $G_j(y_1, y_2, \ldots, y_m) = \partial G / \partial y_j$.

In the same section, it was also stated that multinomial logit, hierarchical logit and cross-nested logit models can be derived as GEV models. For the multinomial logit and the hierarchical logit this is possible by specifying the function $G(\cdot)$ as

$$G(e^{V_1}, \ldots, e^{V_m}) = e^{Y_o} \tag{3.A.2}$$

where Y_o is the logsum variable relative to the root node of the choice tree for the model under study. The following sections carry out these derivations.

3.A.1 Derivation of the Multinomial Logit Model

In the case of the multinomial logit model, the choice tree has the root node o directly connected to all the elementary alternatives j (see Fig. 3.2).

In this case the variable Y_o can be expressed as

$$Y_o = \ln \sum_{i=1}^{m} e^{V_i/\theta}$$

and (3.A.2) becomes:

$$G(e^{V_1}, \ldots, e^{V_m}) = \sum_{i=1}^{m} e^{V_i/\theta} \tag{3.A.3}$$

It can easily be verified that this function satisfies the four properties mentioned in Sect. 3.3.5, given some restrictions on parameter θ.

In fact:

(1) $G \geq 0$ for any value θ and V_i $(i = 1, \ldots, m)$.
(2)

$$G(\alpha e^{V_1}, \ldots, \alpha e^{V_m}) = \sum_{i=1}^{m} (\alpha e^{V_i})^{1/\theta} = \alpha^{1/\theta} \sum_{i=1}^{m} (e^{V_i})^{1/\theta}$$
$$= \alpha^{1/\theta} G(e^{V_1}, \ldots, e^{V_m});$$

that is, $G(.)$ is homogeneous of degree $1/\theta$, which is positive if $\theta > 0$.

(3)

$$\lim_{e^{V_i} \to \infty} G(e^{V_1}, \ldots, e^{V_m}) = \lim_{e^{V_i} \to \infty} \sum_{i=1}^{m} e^{V_i/\theta} = \infty, \quad \text{for } i = 1, 2, \ldots, m.$$

(4) The first derivative of $G(\cdot)$ with respect to any e^{V_j} is equal to

$$G_k = \partial G(.)/\partial e^{V_j} = \frac{e^{V_j[(1/\theta)-1]}}{\theta}$$

which is nonnegative for any $\theta \geq 0$. Furthermore, higher-order mixed derivatives are all zero, and therefore both nonnegative and nonpositive. Condition (4) is therefore certainly verified if Condition (2) on the positivity of the coefficient θ is verified.

Substituting expression (3.A.3) in (3.A.1), it follows that

$$p[j] = \frac{e^{V_j}}{1/\theta} \cdot \frac{1/\theta \cdot e^{V_j(1/\theta)-1}}{\sum_{i=1}^{m} e^{V_i/\theta}} = \frac{e^{V_j/\theta}}{\sum_{i=1}^{m} e^{V_i/\theta}}$$

which is the expression of the multinomial logit model of parameter θ.

To complete the demonstration, the joint probability distribution of the random residuals can be derived. In fact, substituting expression (3.A.3) in the joint probability distribution function (3.3.51), the product of m Gumbel probability distribution functions with parameter θ is obtained:

$$F(\varepsilon_1, \ldots, \varepsilon_m) = \exp\left[-\sum_{i=1}^{m} e^{-\varepsilon_i/\theta}\right] = \prod_{i=1}^{m} \exp[-e^{-\varepsilon_i/\theta}]$$

Thus expression (3.A.3) for the function $G(\cdot)$ implies that the random residuals are identically and independently distributed as Gumbel variables with parameter θ and therefore with variances and covariances defined by expressions (3.3.2) and (3.3.3). Note that the inclusion of Euler's constant Φ in the systematic utilities V_i entails no loss of generality because, as stated in Sect. 3.3.1, MNL choice probabilities are invariant with respect to the addition of a constant to all utilities.

3.A.2 Derivation of the Single-Level Hierarchical Logit Model

In the single-level hierarchical logit model with equal covariances, the choice tree has the root node o connected to intermediate nodes k to which elementary alternatives j are connected (see Fig. 3.5). The parameters θ associated with all intermediate nodes k are equal.

With this tree structure, the variable Y_o becomes:

$$Y_o = \ln \sum_k \exp\left(\frac{\theta}{\theta_o} \cdot Y_k\right) = \ln \sum_k \left(\sum_{i \in I_k} e^{V_i/\theta}\right)^{\theta/\theta_o}$$

with

$$Y_k = \ln \sum_{i \in I_k} e^{V_i/\theta}$$

Consequently (3.A.2) becomes:

$$G(e^{V_1}, \ldots, e^{V_m}) = \sum_k \left(\sum_{i \in I_k} e^{V_i/\theta}\right)^{\theta/\theta_o} \tag{3.A.4}$$

In this case, it can again be shown that $G(.)$ satisfies the four properties mentioned above, given some restrictions on the parameters θ and θ_o.

In fact:

(1) $G \geq 0$ for any value of θ, θ_o, and V_k, for $k = 1, \ldots, m$.

(2)

$$G(\alpha e^{V_1}, \ldots, \alpha e^{V_m}) = \sum_k \left[\sum_{i \in I_k} (\alpha e^{V_i})^{1/\theta} \right]^{\theta/\theta_o} = \sum_k \left[(\alpha)^{1/\theta} \sum_{i \in I_k} (e^{V_i})^{1/\theta} \right]^{\theta/\theta_o}$$

$$= \sum_k (\alpha)^{1/\theta_o} \cdot \left[\sum_{i \in I_k} (e^{V_i})^{1/\theta} \right]^{\theta/\theta_o}$$

$$= (\alpha)^{1/\theta_o} \cdot \sum_k \left[\sum_{i \in I_k} (e^{V_i})^{1/\theta} \right]^{\theta/\theta_o}$$

$$= (\alpha)^{1/\theta_o} \cdot G(e^{V_1}, \ldots, e^{V_m});$$

that is, G is homogeneous of degree $1/\theta_o$, which is positive if $\theta_o > 0$.

(3) $\lim_{e^{V_k} \to \infty} G(e^{V_1}, \ldots, e^{V_m}) = \infty$, for $k = 1, 2, \ldots, m$.

(4) The first-order partial derivative of $G(.)$ with respect to any e^{V_h} is equal to:

$$G_h = \partial G(.)/\partial e^{V_h} = \theta/\theta_o \cdot \left(\sum_{i \in I_k} e^{V_i/\theta} \right)^{(\theta/\theta_o)-1} \cdot 1/\theta \cdot e^{V_h[(1/\theta)-1]} \quad \text{with } h \in I_k$$

which is nonnegative if:

$$\theta_o \geq 0 \tag{3.A.5}$$

Inequality (3.A.5) is implied by Condition (2) on the positivity of the homogeneity coefficient.

Moreover, second-order mixed derivatives are equal to:

$$\partial^2 G(.)/\partial e^{V_j} \partial e^{V_h}$$
$$= \begin{cases} \frac{1}{\theta_o} \cdot e^{V_j[(1/\theta)-1]} \cdot (\frac{\theta}{\theta_o} - 1) \cdot (\sum_{i \in I_k} e^{V_i/\theta})^{(\theta/\theta_o)-2} \frac{1}{\theta} e^{V_h[(1/\theta)-1]} \\ \quad \text{for } j, h \in I_k \; \forall k \\ 0, \quad \text{otherwise} \end{cases}$$

which, given (3.A.5), are nonpositive if:

$$0 \leq \theta \leq \theta_o \tag{3.A.6}$$

It can be easily shown that if (3.A.6) holds, Condition (4) is always satisfied for higher-order mixed derivatives.

Also in this case, therefore, Conditions (2) and (4) impose restrictions on the two parameters θ and θ_o ($0 < \theta \leq \theta_o$) analogous to those described in Sect. 3.3.2.

Fig. 3.A.1 Choice tree for a multilevel hierarchical logit model

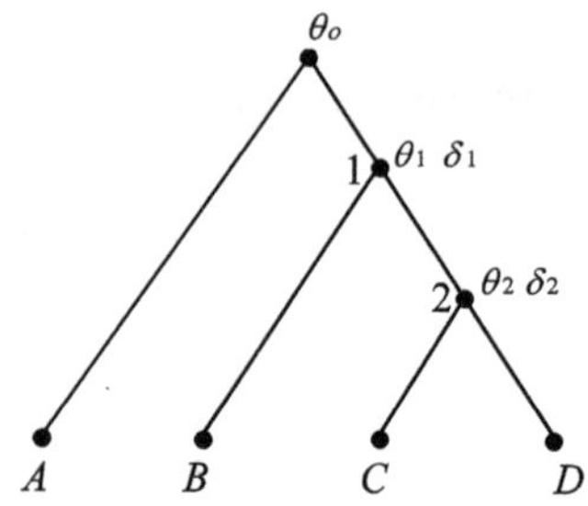

Choice probabilities can be obtained by substituting function (3.A.4) in (3.A.1):

$$p[j] = \frac{e^{V_j}}{\frac{1}{\theta_o}} \cdot \frac{\frac{\theta}{\theta_o} \cdot (\sum_{i \in I_h} e^{V_i/\theta})^{\frac{\theta}{\theta_o}-1} \cdot \frac{1}{\theta} \cdot e^{V_j[(1/\theta)-1]}}{\sum_k (\sum_{i \in I_k} e^{V_i/\theta})^{\frac{\theta}{\theta_o}}}$$

$$= \frac{e^{V_j/\theta}}{\sum_{i \in I_h} e^{V_i/\theta}} \cdot \frac{(\sum_{i \in I_h} e^{V_i/\theta})^{\frac{\theta}{\theta_o}}}{\sum_k (\sum_{i \in I_k} e^{V_i/\theta})^{\frac{\theta}{\theta_o}}} \tag{3.A.7}$$

which is the expression of the single-level hierarchical logit model with parameters θ_o and θ. Introducing the parameter $\delta = \theta/\theta_o$ and the logsum variable Y_k:

$$Y_k = \ln \sum_{i \in I_k} \exp(V_i/\theta)$$

(3.A.7) becomes:

$$p[j] = \frac{e^{V_j/\theta}}{\sum_{i \in I_k} e^{V_i/\theta}} \cdot \frac{e^{\delta Y_h}}{\sum_k e^{\delta Y_k}}$$

which is the expression of the single-level hierarchical logit model (see (3.3.19)).

3.A.3 Derivation of the Multilevel Hierarchical Logit Model

The demonstration that the multilevel hierarchical logit (tree-logit) can be derived from function (3.A.2) satisfying the four properties mentioned cannot be easily generalized, because it is difficult to express the choice tree structure in a general form. To demonstrate the statement that the multilevel hierarchical logit model is a GEV model, reference to an easily generalizable example is made.

Consider the structure of the choice tree in Fig. 3.A.1.

There are two intermediate levels and three parameters: $\theta_o, \theta_1, \theta_2$. Let V_A, V_B, V_C, and V_D be the systematic utilities of the four elementary nodes. According to what was stated in Sect. 3.3.3, it follows that

$$\delta_1 = \theta_1/\theta_o$$

$$\delta_2 = \theta_2/\theta_1$$

$$Y_2 = \ln(e^{V_C/\theta_2} + e^{V_D/\theta_2}) \tag{3.A.8}$$

$$Y_1 = \ln(e^{V_B/\theta_1} + e^{\delta_2 Y_2}) = \ln\left[e^{V_B/\theta_1} + (e^{V_C/\theta_2} + e^{V_D/\theta_2})^{\theta_2/\theta_1}\right]$$

$$Y_o = \ln(e^{V_A/\theta_o} + e^{\delta_1 Y_1}) = \ln\left\{e^{V_A/\theta_o} + \left[e^{V_B/\theta_1} + (e^{V_C/\theta_2} + e^{V_D/\theta_2})^{\theta_2/\theta_1}\right]^{\theta_1/\theta_o}\right\}$$

$$p[A] = \frac{e^{V_A}}{e^{Y_o}}$$

$$p[B] = \frac{e^{V_B/\theta_1}}{e^{Y_o}} \cdot e^{(\delta_1 - 1)Y_1} \tag{3.A.9}$$

$$p[C] = \frac{e^{V_C/\theta_2}}{e^{Y_o}} \cdot e^{(\delta_1 - 1)Y_1} \cdot e^{(\delta_2 - 1)Y_2}$$

Substituting in (3.A.2) the expression for Y_o given by (3.A.8) yields:

$$G(e^{V_A}, \ldots, e^{V_D}) = e^{V_A/\theta_o} + \left[e^{V_B/\theta_1} + (e^{V_C/\theta_2} + e^{V_D/\theta_2})^{\theta_2/\theta_1}\right]^{\theta_1/\theta_o} \tag{3.A.10}$$

It can be verified that, given some restrictions on the parameters θ, this function satisfies the four properties required of $G(\cdot)$.

In fact:

(1) $G \geq 0$ for any value of θ_j, $(j = o, 1, 2)$, V_i $(i = A, B, C, D)$.

(2)

$$\begin{aligned}
G(\alpha e^{V_A}, \ldots, \alpha e^{V_D}) &= (\alpha e^{V_A})^{1/\theta_o} + \{(\alpha e^{V_B})^{1/\theta_1} \\
&\quad + \left[(\alpha e^{V_C})^{1/\theta_2} + (\alpha e^{V_D})^{1/\theta_2}\right]^{\theta_2/\theta_1}\}^{\theta_1/\theta_o} \\
&= (\alpha)^{1/\theta_o} \cdot (e^{V_A})^{1/\theta_o} + \{(\alpha)^{1/\theta_1} \cdot (e^{V_B})^{1/\theta_1} \\
&\quad + \left[(\alpha)^{1/\theta_2} \cdot (e^{V_C})^{1/\theta_2} + (\alpha)^{1/\theta_2} \cdot (e^{V_D})^{1/\theta_2}\right]^{\theta_2/\theta_1}\}^{\theta_1/\theta_o} \\
&= (\alpha)^{1/\theta_o} \cdot (e^{V_A})^{1/\theta_o} + \{(\alpha)^{1/\theta_1} \cdot (e^{V_B})^{1/\theta_1} \\
&\quad + (\alpha)^{1/\theta_1} \cdot \left[(e^{V_C})^{1/\theta_2} + (e^{V_D})^{1/\theta_2}\right]^{\theta_2/\theta_1}\}^{\theta_1/\theta_o} \\
&= (\alpha)^{1/\theta_o} \cdot (e^{V_A})^{1/\theta_o} + (\alpha)^{1/\theta_o} \cdot \{(e^{V_B})^{1/\theta_1} \\
&\quad + \left[(e^{V_C})^{1/\theta_2} + (e^{V_D})^{1/\theta_2}\right]^{\theta_2/\theta_1}\}^{\theta_1/\theta_o} \\
&= (\alpha)^{1/\theta_o} \cdot G(e^{V_A}, \ldots, e^{V_D});
\end{aligned}$$

that is, $G(.)$ is homogeneous of degree $1/\theta_o$, which is positive if $\theta_o > 0$.

(3) $\lim_{e^{V_i} \to \infty} G(e^{V_A}, \ldots, e^{V_D}) = \infty$, for $i = A, B, C, D; \alpha$.

(4) First-order partial derivatives can be expressed as

$$\partial G/\partial e^{V_A} = 1/\theta_o \cdot e^{V_A(1/\theta_o - 1)}$$

$$\partial G/\partial e^{V_B} = \theta_1/\theta_o \cdot (e^{V_B/\theta_1} + e^{\delta_2 Y_2})^{\delta_1 - 1} \cdot 1/\theta_1 \cdot e^{V_B(1/\theta_1 - 1)}$$
$$\partial G/\partial e^{V_C} = \theta_1/\theta_o \cdot (e^{V_B/\theta_1} + e^{\delta_2 Y_2})^{\delta_1 - 1} \cdot \theta_2/\theta_1 \cdot (e^{V_C/\theta_2} + e^{V_D/\theta_2})^{\delta_2 - 1} \cdot 1/\theta_2 \cdot e^{V_C(1/\theta_2 - 1)}$$

Note that in this case there is no structural symmetry, and the different derivatives differ from each other. First-order derivatives are nonnegative if:

$$\theta_o \geq 0 \tag{3.A.11}$$

Other restrictions on the parameters θ can be deduced from the second-order mixed derivatives. In particular, it is sufficient to use only the following two mixed derivatives.

$$\begin{aligned}
\partial^2 G/\partial e^{V_B} \partial e^{V_C} &= \frac{1}{\theta_o} \cdot e^{V_B(\frac{1}{\theta_1} - 1)} \cdot \frac{\theta_1 - \theta_o}{\theta_o} \cdot (e^{V_B/\theta_1} + e^{\delta_2 Y_2})^{\delta_1 - 2} \\
&\quad \cdot \frac{1}{\theta_1} \cdot (e^{V_C/\theta_2} + e^{V_D/\theta_2})^{\delta_2 - 1} \cdot e^{V_C(\frac{1}{\theta_2} - 1)} \\
\partial^2 G/\partial e^{V_C} \partial e^{V_D} &= \frac{1}{\theta_o} \cdot e^{V_C(\frac{1}{\theta_2} - 1)} \cdot \frac{\theta_1 - \theta_o}{\theta_o} \cdot (e^{V_B/\theta_1} + e^{\delta_2 Y_2})^{\delta_1 - 2} \\
&\quad \cdot \frac{\theta_2}{\theta_1} \cdot \left[(e^{V_C/\theta_2} + e^{V_D/\theta_2})^{\delta_2 - 1}\right]^2 \cdot \frac{1}{\theta_2} \cdot e^{V_D(\frac{1}{\theta_2} - 1)} \\
&\quad + \frac{1}{\theta_o} \cdot e^{V_C(\frac{1}{\theta_2} - 1)} \cdot \frac{\theta_2 - \theta_1}{\theta_1} \cdot (e^{V_C/\theta_2} + e^{V_D/\theta_2})^{\delta_2 - 2} \\
&\quad \cdot \frac{1}{\theta_2} \cdot e^{V_D(\frac{1}{\theta_2} - 1)} \cdot (e^{V_B/\theta_1} + e^{\delta_2 Y_2})^{\delta_1 - 1}
\end{aligned} \tag{3.A.12}$$

Invoking inequality (3.A.11), it can be seen that the first one is nonpositive if:

$$0 \leq \theta_1 \leq \theta_o \tag{3.A.13}$$

Invoking (3.A.13) in the second one, it follows that the first term is always nonpositive and the second term is nonpositive if:

$$0 \leq \theta_2 \leq \theta_1 \tag{3.A.14}$$

Combining expressions (3.A.13) and (3.A.14), it follows that

$$0 \leq \theta_2 \leq \theta_1 \leq \theta_o \tag{3.A.15}$$

It can be shown that if inequality (3.A.15) holds, Condition (4) is always verified for the other second-order mixed derivatives not included in (3.A.12), as well as for higher-order mixed derivatives.

Choice probabilities for the multilevel hierarchical logit model described here can be obtained by substituting expression (3.A.10) in (3.A.1), yielding:

$$p[A] = \frac{e^{V_A}}{1/\theta_o} \cdot \frac{1/\theta_o \cdot e^{V_A(1/\theta_o - 1)}}{e^{Y_o}} = \frac{e^{V_A/\theta_o}}{e^{Y_o}}$$

$$p[B] = \frac{e^{V_B}}{1/\theta_o} \cdot \frac{\theta_1/\theta_o \cdot (e^{V_B/\theta_1} + e^{\delta_2 Y_2})^{\delta_1 - 1} \cdot 1/\theta_1 \cdot e^{V_B(1/\theta_1 - 1)}}{e^{Y_o}} = \frac{e^{V_B/\theta_1}}{e^{Y_o}} \cdot e^{(\delta_1 - 1)Y_1}$$

$$p[C] = \frac{e^{V_C}}{1/\theta_o}$$

$$\cdot \frac{\theta_1/\theta_o \cdot (e^{V_B/\theta_1} + e^{\delta_2 Y_2})^{\delta_1 - 1} \cdot \theta_2/\theta_1 \cdot (e^{V_C/\theta_2} + e^{V_D/\theta_2})^{\delta_2 - 1} \cdot 1/\theta_2 \cdot e^{V_C(1/\theta_2 - 1)}}{e^{Y_o}}$$

$$= \frac{e^{V_C/\theta_2}}{e^{Y_o}} \cdot e^{(\delta_1 - 1)Y_1} \cdot e^{(\delta_2 - 1)Y_2}$$

equal to the expressions (3.A.9)

The conditions on parameters θ obtained for the three models described so far are both necessary and sufficient; if they are not satisfied the function $G(\cdot)$ does not have the properties (1) through (4) and the models are not compatible with random utility theory.

3.A.4 Derivation of the Cross-nested Logit Model

The cross-nested logit model has a choice graph shown in Fig. 3.11 and can be obtained as a GEV model by specifying the function $G(.)$ as

$$G(.) = \sum_k \left(\sum_{i \in I_k} \alpha_{ik}^{1/\delta_k} e^{V_i/\theta_k} \right)^{\delta_k} \tag{3.A.16}$$

with $\delta_k = \theta_k/\theta_o$ and the membership parameters α_{ik} in the interval [0, 1]. In this case as well, it can be verified that $G(.)$ satisfies the four properties, given some restrictions on parameters θ_k.

In fact:

(1) $G \geq 0$ for any value of θ_k, V_i $(i = 1, \ldots, m)$, $a_{im}[0, 1]$.

(2)

$$G(\beta e^{V_1}, \ldots, \beta e^{V_m}) = \sum_k \left(\sum_{i \in I_k} \alpha_{ik}^{1/\delta_k} (\beta e^{V_i})^{1/\theta_k} \right)^{\delta_k}$$

$$= \sum_k \left(\beta^{1/\theta_k} \sum_{i \in I_k} \alpha_{ik}^{1/\delta_k} (e^{V_i})^{1/\theta_k} \right)^{\delta_k}$$

$$= \beta^{1/\theta_o} \sum_k \left(\sum_{i \in I_k} \alpha_{ik}^{1/\delta_k} (e^{V_i})^{1/\theta_k} \right)^{\delta_k}$$

$$= \beta^{1/\theta_o} \cdot G(e^{V_1}, \ldots, e^{V_m});$$

that is, $G(.)$ is homogeneous of degree $1/\theta_o$, which is positive if $\theta_o \geq 0$.

(3) $\lim_{e^{V_k} \to \infty} G(e^{V_1}, \ldots, e^{V_m}) = \infty$, for $k = 1, 2, \ldots, m$.

(4) The first-order partial derivative of $G(\cdot)$ with respect to any e^{V_j} is equal to:

$$G_j = \partial G(.)/\partial e^{V_j} = \sum_k \left[\delta_k \cdot \left(\sum_{i \in I_k} \alpha_{ik}^{1/\delta_k} e^{V_i/\theta_k} \right)^{\delta_k - 1} \cdot \frac{\alpha_{jk}^{1/\delta_k}}{\theta_k} \cdot (e^{V_j})^{\frac{1}{\theta_k} - 1} \right]$$

and is nonnegative if

$$\theta_o \geq 0 \tag{3.A.17}$$

Inequality (3.A.17) is implied by Condition (2) on the positivity of the homogeneity coefficient.

Moreover, second-order mixed derivatives are equal to:

$$\partial^2 G(.)/\partial e^{V_j} \partial e^{V_h} = \sum_k \left[\alpha_{hk}^{1/\delta_k} \cdot \frac{1}{\theta_k} (e^{V_h})^{\frac{1}{\theta_k} - 1} \cdot (\delta_k - 1) \cdot \left(\sum_{i \in I_k} \alpha_{ik}^{1/\delta_k} e^{V_i/\theta_k} \right)^{\delta_k - 2} \cdot \frac{\alpha_{jk}^{1/\delta_k}}{\theta_o} \cdot (e^{V_j})^{\frac{1}{\theta_k} - 1} \right]$$

If inequality (3.A.17) is satisfied, all terms of the summation are nonpositive if:

$$0 \leq \theta_k \leq \theta_o \quad \forall k \tag{3.A.18}$$

Thus the condition of nonpositivity is always satisfied (for any value of V_i, a_{ik}) if (3.A.18) is true.

It can be easily shown that Condition (4) for higher-order mixed derivatives is always verified if (3.A.18) holds.

Choice probabilities can be obtained by substituting the function $G(.)$ expressed by (3.A.16) in (3.A.1):

$$p[j] = \frac{e^{V_j}}{1/\theta_o} \cdot \frac{\sum_k [\frac{\alpha_{jk}^{1/\delta_k}}{\theta_o} \cdot (\sum_{i \in I_k} \alpha_{ik}^{1/\delta_k} e^{V_i/\theta_k})^{\delta_k - 1} \cdot (e^{V_j})^{\frac{1}{\theta_k} - 1}]}{\sum_k (\sum_{i \in I_k} \alpha_{ik}^{1/\delta_k} e^{V_i/\theta_k})^{\delta_k}}$$

$$= \frac{\sum_k [\alpha_{jk}^{1/\delta_k} e^{V_j/\theta_k} \cdot (\sum_{i \in I_k} \alpha_{ik}^{1/\delta_k} e^{V_i/\theta_k})^{\delta_k - 1}]}{\sum_k (\sum_{i \in I_k} \alpha_{ik}^{1/\delta_k} e^{V_i/\theta_k})^{\delta_k}} \tag{3.A.19}$$

which is the expression for the cross-nested logit model (3.3.49).

3.B. Random Variables Relevant for Random Utility Models

3.B.1 The Gumbel Random Variable

The Gumbel random variable is a continuous variable that plays a very important role in building logit-form random utility models. Below we describe the probability functions of this variable and illustrate some of its important properties. To facilitate the immediate application of the results to random utility models, the Gumbel variable is indicated by U (instead of X_G) and its expected value by V (instead of $E[X_G]$).

The probability density function of a Gumbel r.v. U with mean V and scale parameter θ is given by:

$$f_U(u) = 1/\theta \cdot \exp\left[-(u-V)/\theta - \Phi\right]\exp\left\{-\exp\left[-(u-V)/\theta - \Phi\right]\right\} \quad (3.B.1)$$

and its distribution function is:

$$F_U(u) = \exp\left\{-\exp\left[-(u-V)/\theta - \Phi\right]\right\} \quad (3.B.2)$$

where Φ is Euler's constant, approximately equal to 0.577.

The mean and the variance of the Gumbel variable are:

$$E[U] = V$$
$$\mathrm{Var}[U] = \sigma_U^2 = \frac{\pi^2\theta^2}{6} \quad (3.B.3)$$

From expressions (3.B.3) it can be deduced that the standard deviation of the Gumbel r.v. is directly proportional to the parameter θ. Figure 3.B.1 shows some probability density functions of the zero mean Gumbel r.v. for different values of parameter θ.

It can easily be demonstrated, by substitution in expression (3.B.2), that if U is a Gumbel variable with parameters (V,θ), any r.v. obtained from it by a linear transformation

$$Y = aU + b$$

is also a Gumbel r.v. with mean

$$E[Y] = aV + b$$

and the same parameter θ (same variance). From this result, it follows immediately that the residual of a random utility model $\varepsilon = U - V$ $(a = 1,\ b = -V)$ is a Gumbel r.v. with zero mean and parameter θ.

The Gumbel r.v. has the important property of stability with respect to maximization. In other words, if U_j, $j = 1,\ldots,N$, are independent Gumbel r.v. with different means V_j but the same parameter θ, the maximum of these variables:

$$U_M = \max_{j=1,\ldots,N} [U_j] \quad (3.B.4)$$

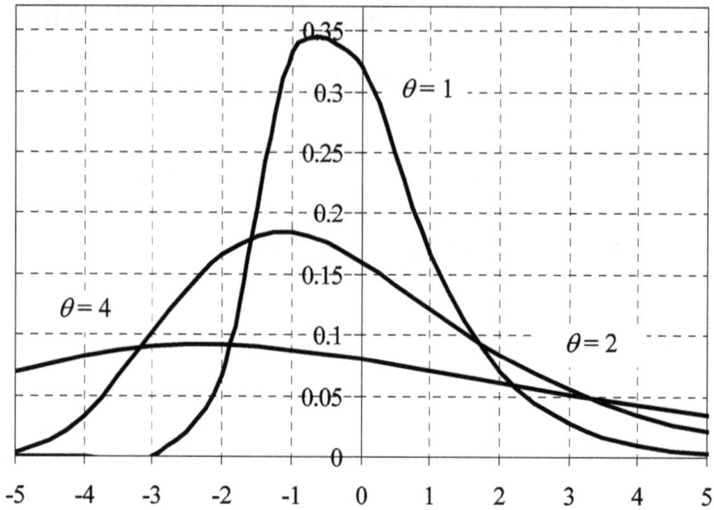

Fig. 3.B.1 Probability density functions of a Gumbel r.v.

is also a Gumbel r.v. with parameter θ.

In fact, the probability distribution function of U_M can be obtained as

$$F_{U_M}(u) = Pr(U_M < u) = Pr\left[\max_{j=1,\ldots,N}\{U_j\} \le u\right]$$

and from the independence of the U_j, it follows that:

$$Pr\left[\max_{j=1,\ldots,N}\{U_j\} \le u\right] = \prod_{j=1,\ldots,N} Pr[U_j < u] = \prod_{j=1,\ldots,N} F_{U_j}(u)$$

Substituting expression (3.B.2) for the Gumbel probability distribution function into the previous expression, it follows that

$$F_{U_M}(u) = \prod_{j=1,\ldots,N} \exp\{-\exp[-(u - V_j)/\theta - \Phi]\}$$

which yields:

$$F_{U_M}(u) = \exp\left[-\exp(-\Phi)\cdot\exp(-u/\theta)\cdot\sum_j \exp(V_j/\theta)\right] \tag{3.B.5}$$

If the EMPU variable described in Chap. 3 is denoted by V_M then:

$$V_M = \theta \ln \sum_j \exp(V_j/\theta) \tag{3.B.6}$$

and, when this is substituted in expression (3.B.5), the result is

$$F_{U_M}(u) = \exp\{-\exp[-(u - V_M)/\theta - \Phi]\}$$

which is still the probability distribution function of a Gumbel random variable with mean V_M and parameter θ, as can be immediately seen by comparison with (3.B.2).

The multinomial logit model can be obtained by using the definition of a random utility model (3.2.1) and the property of stability with respect to maximization of the Gumbel r.v. described above.

In fact, from (3.2.1) it follows that

$$p[j] = Pr(U_j > U_{M'})$$

with

$$U_{M'} = \max_{k \neq j}\{U_k\}$$

This probability can therefore be expressed as the product of the probability that the perceived utility U_j has a value within an infinitesimal neighborhood of x and the probability that $U_{M'}$ has a value less than x. The resulting probability element must obviously be integrated with respect to all possible values of x:

$$p[j] = Pr(U_j > U_{M'}) = \int_{-\infty}^{+\infty} F_{U_{M'}}(x) \cdot f_{U_j}(x)\, dx \tag{3.B.7}$$

where $F_{U_{M'}}$ and f_{U_j} are the probability distribution function and the probability density function of the random variables $U_{M'}$ and U_j, respectively. If the U_k are i.i.d. Gumbel variables with parameter θ and mean V_k, then $U_{M'}$, as shown above, is also a Gumbel variable with the same parameter θ and mean equal to:

$$V_{M'} = \theta \ln \sum_{k \neq j} \exp(V_k/\theta) \tag{3.B.8}$$

Expression (3.B.7) then becomes:

$$\begin{aligned}
p[j] &= \int_{-\infty}^{+\infty} \exp\{-\exp[-(x - V_{M'})/\theta - \Phi]\} \cdot \exp\{-\exp[-(x - V_j)/\theta - \Phi]\} \\
&\quad \times \exp[-(x - V_j)/\theta - \Phi] \cdot (1/\theta)\, dx \\
&= \int_{-\infty}^{+\infty} \exp\{-\exp[-(x - V_j)/\theta - \Phi] - \exp[-(x - V_{M'})/\theta - \Phi]\} \\
&\quad \times \exp[-(x - V_j)/\theta - \Phi] \cdot (1/\theta)\, dx \\
&= \exp(V_j/\theta - \Phi) \cdot \int_{-\infty}^{+\infty} \exp\{-\exp(-x/\theta) \cdot [\exp(V_j/\theta - \Phi) \\
&\quad + \exp[V_{M'}/\theta - \Phi]\} \exp(-x/\theta) \cdot (1/\theta)\, dx
\end{aligned}$$

$$
\begin{aligned}
&= \exp(V_j/\theta - \Phi) \\
&\quad \times \int_{-\infty}^{+\infty} \exp[-\exp(-x/\theta)]^{[\exp(V_j/\theta-\Phi)+\exp[V_{M'}/\theta-\Phi]} \exp(-x/\theta)\cdot(1/\theta)\,dx \\
&= \frac{\exp(V_j/\theta - \Phi)}{\exp(V_j/\theta - \Phi) + \exp(V_{M'}/\theta - \Phi)} \\
&\quad \times \left|\exp[-\exp(-x/\theta)]^{[\exp(V_j/\theta-\Phi)+\exp[V_{M'}/\theta-\Phi]}\right|_{-\infty}^{+\infty} \\
&= \frac{\exp(V_j/\theta)}{\exp(V_j/\theta) + \exp(V_{M'}/\theta)}
\end{aligned}
$$

and, substituting expression (3.B.8) for $V_{M'}$, it follows that

$$
p[j] = \frac{\exp(V_j/\theta)}{\exp(V_j/\theta) + \sum_{k\neq j}\exp(V_k/\theta)} = \frac{\exp(V_j/\theta)}{\sum_k \exp(V_k/\theta)}
$$

which is the multinomial logit model described in Sect. 3.3.1.

3.B.2 The Multivariate Normal Random Variable

The multivariate normal r.v., $\boldsymbol{X}_{MVN}$, is the generalization of the normal r.v. to n dimensions. Its probability density function is given by

$$
f_{X_{MVN}}(\boldsymbol{x}) = \left[(2\pi)^n \det(\boldsymbol{\Sigma}_X)\right]^{-1/2} \exp\left[-1/2(\boldsymbol{x} - \boldsymbol{\mu}_X)^{\mathrm{T}} \boldsymbol{\Sigma}_X^{-1}(\boldsymbol{x} - \boldsymbol{\mu}_X)\right] \quad (3.B.9)
$$

where $\det(\boldsymbol{\Sigma})$ denotes the determinant of the matrix $\boldsymbol{\Sigma}$.

The parameters of a multivariate normal r.v. are the vector $\boldsymbol{\mu}_X$ of the means, with components μ_{X_i}, and the positive semidefinite variance–covariance (or dispersion) matrix $\boldsymbol{\Sigma}_X$. In other words:

$$
E[\boldsymbol{X}_{MVN}] = \boldsymbol{\mu}_X, \qquad \boldsymbol{\Sigma}_{X_{MVN}} = \boldsymbol{\Sigma}_X
$$

The equiprobability surfaces of the multivariate normal variable, or the loci of points in the n-dimensional Euclidean space for which the density function is constant, have the equation:

$$
(\boldsymbol{x} - \boldsymbol{\mu}_X)^{\mathrm{T}} \boldsymbol{\Sigma}_X^{-1}(\boldsymbol{x} - \boldsymbol{\mu}_X) = C^2 \quad (3.B.10)
$$

where C is a constant. Expression (3.B.10) is the equation of an ellipsoid with $\boldsymbol{\mu}_X$ as its center (see Fig. 3.B.2).

Recall that the sum of two univariate normal random variables is again a normal random variable, a property known as invariance with respect to summation. Specifically, if X is distributed as $N(\mu_X, \sigma_X^2)$ and Y is distributed as $N(\mu_Y, \sigma_Y^2)$,

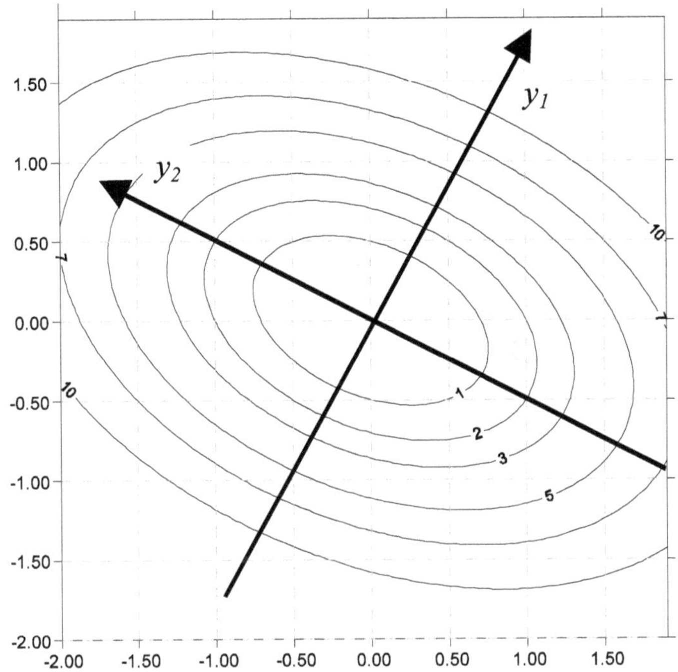

Fig. 3.B.2 Equiprobable surfaces of the multivariate normal r.v.

then $X+Y$ is distributed as $N(\mu_X+\mu_Y, \sigma_X^2+\sigma_Y^2+2\,\mathrm{cov}(X,Y))$; similarly, $X-Y$ is distributed as $N(\mu_X-\mu_Y, \sigma_X^2+\sigma_Y^2-2\,\mathrm{cov}(X,Y))$.

The multivariate normal r.v. has the property of invariance with respect to linear transformations, which can be considered an extension of the property of invariance with respect to summation of the univariate normal r.v. In other words, if $\boldsymbol{X}$ is a random vector with probability multivariate normal density function (3.B.9) and $\boldsymbol{A}$ is a matrix of dimensions $(m \times n)$, the vector $\boldsymbol{Y} = \boldsymbol{A}\boldsymbol{X}$ is also multivariate normal with mean vector and dispersion matrix given by

$$E[\boldsymbol{Y}] = \boldsymbol{A}E[\boldsymbol{X}] = \boldsymbol{A}\boldsymbol{\mu}_X,$$
$$\boldsymbol{\Sigma}_Y = E\big[\boldsymbol{A}(\boldsymbol{X}-\boldsymbol{\mu}_X)(\boldsymbol{X}-\boldsymbol{\mu}_X)^T\boldsymbol{A}^T\big] = \boldsymbol{A}\boldsymbol{\Sigma}_X\boldsymbol{A}^T$$

Furthermore, from (3.B.9) it can be easily deduced that if the n components of X_{MVN} are noncorrelated (i.e., the matrix $\boldsymbol{\Sigma}$ is diagonal), then they are also independent; that is, the probability density function (3.B.9) is the product of n density functions of univariate normal random variables with means μ_{X_i} and variances $\sigma_{X_i}^2$. It is worth recalling that two independent random variables are noncorrelated in any case.

Reference Notes

Random utility theory has stimulated, both in theory and in applications, the understanding and modeling of the mechanisms underlying travel demand. One of the first systematic accounts of its foundation can be found in the book by Domencich and McFadden (1975). The book formalizes the theoretical work carried out in the early 1970s on random utility models and on multinomial logit models in particular.

Theoretical analyses of random utility models can be found in Williams (1977), Manski (1977), and the book by Manski and McFadden (1981). The book by Ben-Akiva and Lerman (1985) gives a very comprehensive account of random utility theory, of logit family models, and of many applied issues dealt with in this chapter and in Chap. 8. A recent contribution covering advanced topics in random utility theory is represented by Train (2003).

Williams and Ortùzar (1982) analyze the limitations of random utility (or "compensatory") models and compare them with other behavioral discrete choice models. The paper also contains a comprehensive, albeit dated, bibliography on noncompensatory models. Detailed analysis of the state of the art in the mid-1980s on the use of random utility models in modeling travel demand can be found in the note by Horowitz (1985). More recent systematic reviews of random utility models can be found in Bath (1997) and in Ben-Akiva and Bierlaire (1999).

As for specific random utility models, references to the single-level hierarchical logit model can be found in Williams (1977) and Daly and Zachary (1978), and Daganzo and Kusnic (1993) discuss the multilevel hierarchical logit model in its most general form. The cross-nested logit model is implicitly encompassed in McFadden (1978); the first explicit formulation called "ordered GEV" can be traced back to Small (1987). Vovsha (1997), Vovsha and Bekhor (1998), Wen and Koppelman (2001), Papola (2004), and Abbe et al. (2007) provide further theoretical formulations and developments. The paired combinatorial logit model was first proposed by Chu (1989), and was subsequently elaborated by Koppelman and Wen (2000). The formulation reported in Sect. 3.3.4 is from Papola (2004).

Theoretical analysis of the covariances underlying the cross-nested logit model is provided by Marzano and Papola (2008). The GEV model was proposed by McFadden (1978) and subsequently generalized by Ben-Akiva and Francois (1983). The demonstration that GEV models are random utility models and the derivation of hierarchical logit models as GEV models is from Papola (1996) and the derivation of the cross-nested logit model as a GEV model is from Papola (2004).

Detailed analysis of the probit model is contained in the book by Daganzo (1979); for the calculation of probit choice probabilities reference can be made to Horowitz et al. (1982) and Langdon (1984). Reference to the factor analytic probit can be found in Ben-Akiva and Bierlaire (1999) and reference to the random coefficients (tastes) approach can be found in Ben-Akiva and Lerman (1985) and in Ortuzar and Willumsen (2001). The GHK method derives the name from its authors: Geweke (1991), Hajivassiliou and McFadden (1998), and Keane (1994); a different formulation can be found in Bolduc (1999).

The mixed logit model is also a rather recent development of random utility models. One of the first papers dealing with its theoretical and computational aspects was by Ben-Akiva and Bolduc (1996). Other references to this model may be found in Bolduc et al. (1996) and in Ben-Akiva and Bierlaire (1999); more recent developments and detailed analysis of model properties and applications can be found in Train (2003) and in the doctoral dissertation by Walker (2001).

The general approach to modeling choice set alternatives is contained in Manski (1977). A state-of-the-art review of explicit models of choice set generation and a number of specifications may be found in Ben-Akiva and Boccara (1995). The implicit availability perception approach is described in Cascetta and Papola (2001).

The expected maximum perceived utility function and its mathematical properties are dealt with in Daganzo's volume (1979). Reference can also be made to the work of Cantarella (1997), which draws on and generalizes Daganzo's results.

The definition of elasticity associated with random utility models and the expressions for the multinomial logit model are given in various texts; particular reference can be made to Domencich and McFadden (1975) and to Ben-Akiva and Lerman (1985). The results on elasticities of the single-level hierarchical logit model are from Koppelman (1989).

Chapter 4
Travel-Demand Models

4.1 Introduction

As stated in Chap. 1, travel demand derives from the need to carry out activities in multiple locations. Thus, the level and characteristics of travel demand are influenced by the activity system and the transportation opportunities in the area.

In order to analyze and design transportation systems, it is necessary to estimate the existing demand and to predict the changes in it that will result from the projects being studied and/or from changes in external factors. Mathematical demand models can be used for all these purposes.

A *travel-demand model* can be defined as a mathematical relationship between travel-demand flows and their characteristics on the one hand, and given activity and transportation supply systems and their characteristics.

A demand flow is an aggregation of individual trips, and each trip is the result of multiple choices made by the transportation system users, that is, an individual traveler in the case of passenger transportation or an operator (manufacturer, shipper, and carrier) for freight transportation. For a traveler, these choices range from long-term decisions, such as residence and employment location and vehicle ownership, to shorter-term decisions such as trip frequency, timing, destination, mode, and path. In freight transportation, long-term decisions influencing transportation demand include the location of production plants and purchasing/selling markets, ownership of a fleet of freight vehicles, storage facilities, and the like. Short-term decisions include such factors as shipment frequency, choice of mode, intermodal operator, and path. The choices underlying a journey are made with respect to different *choice dimensions*; these are defined by a set of available alternatives and by the values of their relevant attributes. For example, the mode choice dimension is defined by the alternative transportation modes available for a given origin–destination pair together with their attributes. In a given trip, the user may also make choices involving other dimensions, such as path and destination.

A large number of mathematical models have been developed to forecast travel demand[1]; the different models are based on different assumptions and have different specifications. Before describing some of these model families in detail, some classification criteria are introduced (see Fig. 4.1).

The first classification factor is related to the type of choice (i.e., choice dimension) that is implicitly or explicitly represented by the model. Decisions in some

---

[1]For now the discussion is in terms of passenger travel demand, even though many of the concepts introduced can be extended to freight transportation demand models. Section 4.7 deals specifically with freight models.

E. Cascetta, *Transportation Systems Analysis*,
Springer Optimization and Its Applications 29,
DOI 10.1007/978-0-387-75857-2_4,

TYPE OF CHOICE	Mobility or context models Travel models
SEQUENCE OF CHOICES	Trip-based demand models Trip chaining models Activity-based models
LEVEL OF DETAIL	Disaggregate models Aggregate models
BASIC ASSUMPTIONS	Behavioral models Descriptive models

Fig. 4.1 Classification of travel-demand models

choice dimensions influence individual trips indirectly, by defining the trip context or conditions. Decisions about residence and workplace locations, possession of a driver's license, and the number of cars owned by the household are examples of this type of dimension. Residence and workplace locations determine the origin and destination of work trips, having a driver's license makes the car available as a transportation mode, and so on. These choice dimensions and the models that represent them are known as *mobility choices* and *models*. Usually, mobility choices are relatively stable over time because there is a high cost associated with changing them; they can be assumed invariant in the short term.

Travel choices and *models* refer to the dimensions that characterize journeys (sequences of trips) and/or the individual trips that comprise journeys. Decisions about frequency, destination, transportation mode, and path are examples of this type of choice dimension.

The second classification factor relates to the approach taken for modeling travel demand, that is, for predicting the outcome of the travel choice decisions and representing the mutual effects of the different decisions on each other. *Trip-based travel-demand models* implicitly assume that the choices relating to each origin–destination trip are made independently of the choices for other trips within the same and other journeys. This approximation is made to simplify the analysis, and is reasonable when most of the journeys in the modeling period consist of round trips (origin–destination–origin).

Trip-chaining travel-demand models, on the other hand, assume that the choices concerning the entire journey influence each other. In this case, the choice of an intermediate destination, if any, takes into account the preceding or following destinations on the trip chain, the choice of transportation modes takes into account the whole sequence of trips in the chain, and so on. Models of this type have been studied for several years and have been applied to real situations, although less frequently than trip-based demand models. Examples of models of this type are presented in Sect. 4.4.

Finally, *activity-based demand models* predict travel demand as the outcome of the need to participate in different activities in different places and at different times. They therefore take into account the relationships among different journeys made by the same person during a day and, in the most general case, between journeys made by the various members of the same household. They are often implemented as microsimulation models, in which the decisions, activities, and trip-making of a large number of individual households and their members are explicitly represented. Models of this type are obviously more complex than those described previously and are aimed at understanding relationships between the demand for travel and the organization of the different activities of a person and his or her household. These models are presently at the research stage and are only discussed briefly in Sect. 4.5.

Models of all types can also be classified as either *aggregate* or *disaggregate*, depending on the level of detail of the representation of demand and/or the factors that influence it. In aggregate models, the variables (attributes) included in the model apply to a group of users (e.g., the average times or costs of all the trips between two traffic zones, or the average number of cars owned by families of a certain category). In disaggregate models, the variables refer to the individual user (e.g., the times or costs of travel between the actual origin and destination points of a trip, or the number of cars in a specific traveler's household). The appropriate level of aggregation of model variables depends on the purpose of demand modeling. The prevailing use considered in this book is modeling of the entire transportation system, as represented by a network model. This implies an aggregation level that is at least zonal because, as explained in Chaps. 1 and 2, the level-of-service variables obtained from network models relate to pairs of centroid nodes that represent traffic zones.[2]

The last classification factor considered here relates to the basic model assumptions. Models are called *behavioral* if they derive from explicit assumptions about users' choice behavior and *descriptive* if they capture the relationships between travel demand and activity and transportation supply-system variables without making specific assumptions about decision-makers' behavior. There are also mixed model systems in which some submodels are behavioral and others are descriptive.[3]

Finally, it should be noted that transportation demand models, as are all models used in engineering and econometrics, are schematic and simplified representations of complex real phenomena. They are intended to quantify certain relationships between the variables relevant to the problem under study. They should not be expected to reproduce reality perfectly, especially when the reality being modeled is

---

[2]It should also be noted that the appropriate level of aggregation might be different in a model's calibration and application phases. In other words, it is possible, and even advisable in some cases, to use disaggregate data for model specification and calibration, as shown in Chap. 8, while using aggregate (e.g., average) values of zone, user, and transportation system characteristics in model applications. This corresponds to the application of the aggregation techniques "by representative user" or "by category" described in Sect. 3.7.

[3]Differences between behavioral and descriptive models are becoming less important. Indeed, functional forms such as logit and hierarchical logit, which can be derived from random utility theory, are increasingly being used to predict aspects of demand that have no direct behavioral interpretation in terms of a decision-maker's choice. From this point of view, it would be more appropriate to classify the models based on their functional form, distinguishing between models that can or cannot be derived from random utility theory.

largely dependent on individual behavior, as is the case with transportation demand. Furthermore, as shown later, different models with different levels of accuracy and complexity can describe the same situation. However, more sophisticated models require more resources (data, specification and calibration effort, computing time, etc.), which must be justified by the application requirements.

The sections in this chapter present the characteristics of different types of transportation demand models, with an emphasis on passenger travel demand. Section 4.2 presents the partial share systems of trip-demand models. Individual submodels, including trip production (or frequency), distribution, mode choice, and path choice, as well as an example of an overall model system for interurban travel, are presented in Sect. 4.3. Sections 4.4 and 4.5 present trip-chaining and activity-based demand models, respectively. Section 4.6 discusses the interpretation of results obtained with demand models and the application of these models for different purposes. Finally, Sect. 4.7 describes some models used to predict freight transportation demand.

4.2 Trip-based Demand Model Systems

As previously stated, trip-based demand models[4] predict the average number of trips that have given characteristics and that are undertaken in a specific reference period (average trip flows). In formal terms, this can be expressed as follows.

$$d[K_1, K_2, \ldots] = d(\boldsymbol{SE}, \boldsymbol{T}; \boldsymbol{\beta})$$

where the average travel-demand flow between two zones having characteristics $K_1, K_2, \ldots, K_n$ is expressed as a function of a vector $\boldsymbol{SE}$ of socioeconomic variables related to the activity system and/or the decision-makers; and of a vector $\boldsymbol{T}$ of level-of-service attributes of the transportation supply system, typically obtained from the models described in Chap. 2.[5] Demand functions also involve a vector $\boldsymbol{\beta}$ of coefficients or parameters.[6]

Trip characteristics that are often considered relevant in trip-based demand modeling include:

i The user's class (category of socioeconomic characteristics)
o, d The zones of trip origin and destination

[4]Travel-demand models typically result from the integration of a number of submodels. In this respect it would be more appropriate to speak of a system of demand models. The definition of demand model used here corresponds to the microeconomic concept of an aggregate demand function for transportation services.

[5]Note that the vector $\boldsymbol{T}$ may include individual level of service or performance attributes as well as generalized costs, which are combinations of level-of-service attributes. The coefficients used to combine individual attributes into a generalized cost are among the model parameters.

[6]All the models presented in this chapter depend on coefficients or parameters that, for the time being, are assumed known. Model calibration, that is, the estimation of model parameter values, is discussed in detail in Chap. 8.

s The trip purpose, or more properly the pair of purposes[7]
h The time period, that is, the time band in which trips are undertaken
m The mode, or sequence of modes, used during the trip
k The trip path, that is, the series of links connecting centroids o and d over the network and representing the transportation service provided by mode(s) m

Therefore, with demand flow denoted by $d^i_{od}[s,h,m,k]$, the demand model can be formally expressed as

$$d^i_{od}[s,h,m,k] = d(\boldsymbol{SE}, \boldsymbol{T}) \tag{4.2.1}$$

Although different travel choices are generally dependent on each other, it is usually preferable, for reasons of analytical and statistical convenience,[8] to "decompose" the global demand function into a product of submodels, each of which relates to one or more choice dimensions.

The sequence most often used is the following.

$$d^i_{od}[s,h,m,k] = d^i_o \cdot [sh](\boldsymbol{SE}, \boldsymbol{T}).p^i[d/osh](\boldsymbol{SE}, \boldsymbol{T}) \cdot p^i[m/oshd](\boldsymbol{SE}, \boldsymbol{T}) \cdot p^i[k/oshdm](\boldsymbol{SE}, \boldsymbol{T}) \tag{4.2.2}$$

where

$d^i_o \cdot [sh](\boldsymbol{SE}, \boldsymbol{T})$ Is the trip production or frequency model, which gives the number of users in class i who, from origin zone o, undertake a trip for purpose s in time period h

$p^i[d/osh](\boldsymbol{SE}, \boldsymbol{T})$ Is the distribution model, which gives the fraction of users in class i who, undertaking a trip from origin zone o for purpose s in period h, travel to destination zone d

$p^i[m/oshd](\boldsymbol{SE}, \boldsymbol{T})$ Is the mode choice or mode split model, which gives the fraction of users in class i who, traveling between zones o and d for purpose s in period h, use mode m

$p^i[k/oshdm](\boldsymbol{SE}, \boldsymbol{T})$ Is the path choice model, which gives the fraction of users in class i who, traveling between zones o and d for purpose s in period h by mode m, use path k

---

[7] A trip is sometimes described as having a single purpose (e.g., work, study, etc.). This practice may cause confusion. It would be more precise to define the purpose s of a trip by a pair of purposes, that is, the activities carried out at the origin and at the destination. For example, work trips should be differentiated into home-to-work (H-W) and work-to-work (W-W) purposes, which are different. Trips for which the purpose home appears in the origin or destination are often indicated as home-based, and others as nonhome-based. The characterization of a trip by a pair of purposes also allows a more precise identification of the most relevant activity system variables.

[8] The use of a single model would require the definition of a choice set whose elementary alternatives are all feasible combinations of destinations, modes, and paths. This would lead to practical and econometric difficulties.

Superscript i designates a class of decision-makers having the same attributes, parameters, and model functional form. The system of models described above predicts the average trip-demand flow with its relevant characteristics by initially estimating the total number of trips (*trip productions*) from each zone o in the reference period $d_o[sh]$ and then splitting these trips between the possible destinations, modes, and paths. For this reason, the model is known as a *partial share model* (or system of models). Note that the first two models predict the demand's spatial and temporal characteristics, and therefore provide the elements of the origin–destination matrix.

The sequence of submodels in (4.2.2) reflects an assumption about the order in which decisions involving different choice dimensions are made, and therefore about how these decisions influence each other. The specification used in (4.2.2), corresponding to the model structure shown in Fig. 4.2, implies, for example, that destination choice depends only on trip production or frequency choice, whereas mode choice depends on destination and frequency choices. In other words, the decision-maker first chooses the trip destination from among all the available destination zones, and then the travel mode from among all the modes available for the chosen *od* pair.

Different submodel sequences are clearly possible; for example, some specifications proposed in the literature reverse the order of destination and mode choice in the sequence (4.2.2). Any sequence should be carefully reviewed in the calibration phase (see Chap. 8) and compared with reasonable alternatives, in order to determine the best.

Importantly, the user explicitly chooses each trip's mode and path, but other travel dimensions such as trip frequency and destination might depend on higher-level user choices such as residence and work locations (e.g., for regularly made trips such as home–work and home–study[9]). In these cases, the sequence (4.2.2) can be applied first estimating trip frequency and destination using descriptive models, and then mode and path choice using behavioral models.

As clarified later, upper-level choices (e.g., destination) are actually made taking into account the alternatives available at lower levels, such as the modes and paths available to reach the various possible destinations (see also Fig. 4.2).

Equation (4.2.2), because of its structure, is known as the *four-step model.* However, a greater or smaller number of levels can be used, and the fractions included in the models may differ from those shown. For example, it is possible to specify a six-level urban demand model that explicitly includes a trip production model $d^i_{o\cdot}[s]$ to represent the average number of class i users who travel from zone o over the entire day; a choice model for the time period h in which to make a trip of purpose s, $p^i[h/osx](\boldsymbol{SE}, \boldsymbol{T})$; and a model of parking location (d_p) and type (t_p) choice for auto trips (a) between origin o and final destination d, $p^i[d_p t_p/oshda](\boldsymbol{SE}, \boldsymbol{T})$:

[9]If period h is the whole day, it is also possible for these purposes to choose the number of trips to make (i.e., to return home for lunch or not).

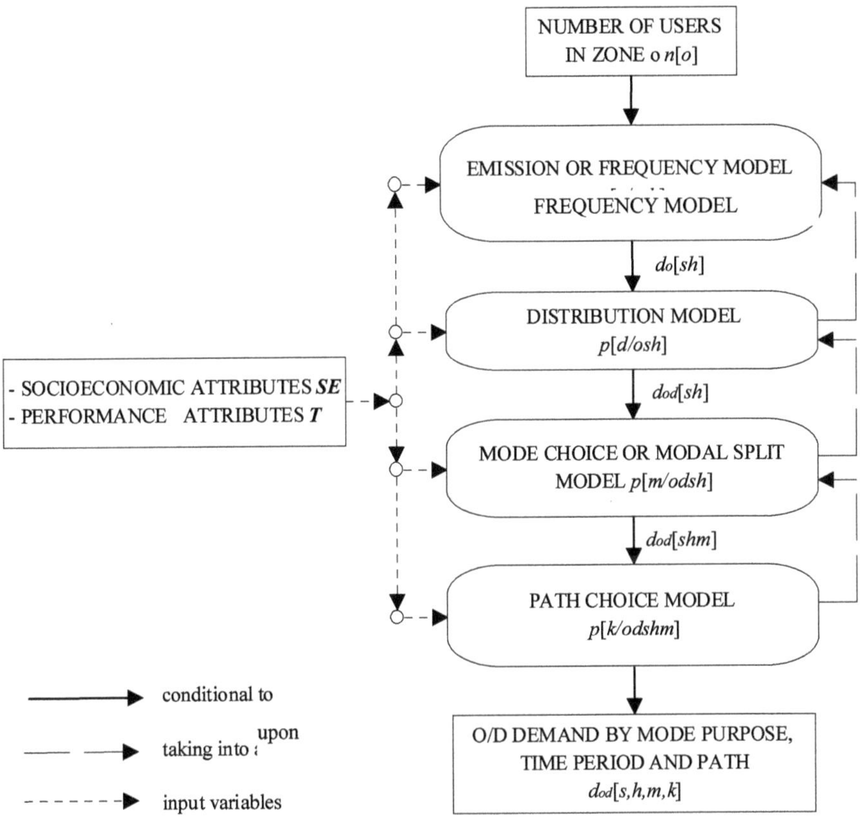

Fig. 4.2 Four-step trip-based travel-demand model system

$$d_{od}^{i}[s,h,a,t_p,d_p,k] = d_o^i \cdot [s](\boldsymbol{SE},\boldsymbol{T}) \cdot p^i[h/os](\boldsymbol{SE},\boldsymbol{T}) \cdot p^i[d/osh](\boldsymbol{SE},\boldsymbol{T}) \cdot p^i[a/oshd](\boldsymbol{SE},\boldsymbol{T}) \cdot p^i[t_p d_p/oshda](\boldsymbol{SE},\boldsymbol{T}) \cdot p^i[k/oshdat_p d_p](\boldsymbol{SE},\boldsymbol{T})$$

The model structures described here represent trip-based demand over all choice dimensions. This is common practice if the project being considered and/or the evaluation time horizon are such that existing values of performance and/or activity variables are likely to be modified significantly. In some short-term applications, a "reduced" version of the model can be used, for example, taking as given existing origin–destination matrices by purpose and user class $d_{od}^i[sh]$, and predicting only the mode and path choice decisions:

$$d_{od}^{i}[s,h,m,k] = d_{od}^i[sh] \cdot p^i[m/oshd](\boldsymbol{SE},\boldsymbol{T}) \cdot p^i[k/oshdm](\boldsymbol{SE},\boldsymbol{T})$$

Estimates of existing O-D matrices $d_{od}^i[sh]$ can be obtained using a variety methods, as shown in detail in Chap. 8.

4.2.1 Random Utility Models for Trip Demand

Regardless of the particular functional form used, each partial share in the previous structure can be modeled following a descriptive or a behavioral approach.

However, it is worthwhile to derive partial share model systems that are consistent with the general results of random utility theory presented in Chap. 3, where random utility models were introduced as a tool for representing choices from among a discrete set of alternatives $(1, \ldots, j, \ldots, m)$. Recall that, in the preceding section, a trip was viewed as the result of choices over multiple dimensions. In the most general case, therefore, random utility models for travel demand consider alternatives that represent sequences of choices in all the trip dimensions considered. In a four-step model, for example, an alternative might consist of making a particular number x of trips, for purpose s, in time period h, in order to reach destination d, by mode m, and path k. In this case the symbol j, which denoted a generic alternative in Chap. 3, is equivalent to a sequence $[x, d, m, k]$.

This section proposes two methods for defining a partial share system of models consistent with the hypotheses underlying random utility models. The first method factors a random utility model over the whole sequence of travel choice dimensions into a product of multiple random utility models, each having the same functional form as the original model but involving only a subset of the choice dimensions. The results presented in Chap. 3 on the multinomial logit and hierarchical logit models can be applied for this approach: such models, as was seen, are particularly suited to this purpose.

By contrast, the second method directly specifies the system of partial shares using random utility models, and then imposes conditions that ensure a consistent behavioral interpretation.

The factoring procedure is first described for a situation involving choice in only two dimensions, destination d and mode m; the more general case is considered subsequently. To simplify notation, the user class i, origin zone o, trip purpose s, and time period h are taken as understood here and in the rest of this section. Let us assume that the systematic utility associated with a particular choice alternative pair dm, V_{dm},[10] may be broken down into a part V_d that depends on destination d, and a part $V_{m/d}$ that, given the destination choice d, depends on mode m. This assumption is consistent with the hypothesis stated above that choice dimensions are considered in sequence: destination choice is affected by mode choice, but the latter, for a given destination, depends only on the attributes of alternative modes and not on those of the destination. The term V_d could be a function of the attributes of the destination, regardless of the mode used to reach it. For shopping trips, for example, attributes might include the number of shops or area of display space; an elementary specification might be:

$$V_d = \beta_1 SHOPS_d$$

[10]As noted, variables (systematic utility, EMPU, random residuals, etc.) are understood to depend on the origin zone o, trip purpose s, and time period h; thus notations such as V_{dm}, $V_{m/d}$, and $p[dm]$ are used instead of $V_{dm/osh}$, $V_{m/oshd}$, and $p[dm/osh]$, respectively.

The term $V_{m/d}$ is instead a function of attributes of both the mode and the destination, such as travel time and the monetary cost incurred in reaching d by mode m from o:

$$V_{m/d} = \beta_2 T_{m/d} + \beta_3 C_{m/d}$$

In conclusion, the perceived utility of alternative dm may be expressed:

$$U_{dm} = V_d + V_{m/d} + \varepsilon_{dm} \tag{4.2.3}$$

Assuming that the residuals ε_{dm} are i.i.d. Gumbel with parameter θ, the previous chapter showed that the probability of choosing alternative dm is given by the multinomial logit model:

$$p[dm] = \frac{\exp[(V_d + V_{m/d})/\theta]}{\sum_{d'} \sum_{m'/d'} \exp[(V_{d'} + V_{m'/d'})/\theta]} \tag{4.2.4}$$

where d' and m' are generic indexes and the sums are extended to all destinations and to all modes available for each destination for the user class in question.

Factoring (4.2.4) requires finding expressions for the probability of the mode choice given the destination $p[m/d]$, and of the destination choice $p[d]$.

The probability $p[m/d]$ may be obtained directly by applying the definition of the random utility model to (4.2.3):

$$\begin{aligned} p[m/d] &= Pr[V_d + V_{m/d} + \varepsilon_{dm} > V_d + V_{m'/d} + \varepsilon_{dm'}] \\ &= Pr[V_{m/d} + \varepsilon_{dm} > V_{m'/d} + \varepsilon_{dm'}] \quad \forall m' \neq m \end{aligned}$$

and from the assumptions made about the distribution of residuals, we again obtain the multinomial logit model:

$$p[m/d] = \frac{\exp[V_{m/d}/\theta]}{\sum_{m'} \exp[V_{m'/d}/\theta]} \tag{4.2.5}$$

The probability $p[d]$ of choosing destination d regardless of mode may be derived from the stability properties of Gumbel variables with respect to maximization. Indeed, if U_d^* stands for the utility associated with destination d by the most suitable mode, then:

$$U_d^* = V_d + \max_{m'}(V_{m'/d} + \varepsilon_{dm'}) \tag{4.2.6}$$

and, by the stability property, U_d^* is again Gumbel distributed with expected value

$$\begin{aligned} E\big[U_d^*\big] = E\big[V_d + \max_{m'}(V_{m'/d} + \varepsilon_{dm'})\big] &= V_d + \theta \ln \sum_{m'} \exp[V_{dm'}/\theta] \\ &= V_d + \theta Y_d \end{aligned} \tag{4.2.7}$$

where θ is, once again, the parameter associated with random variable U_d^* and Y_d is the logsum variable introduced in Sect. 3.3.1. This allows (4.2.6) to be expressed as

$$U_d^* = V_d + \theta Y_d + \varepsilon_d^* \tag{4.2.8}$$

where ε_d^* is still a Gumbel random variable $G(0, \theta)$ with zero mean and parameter θ.

Using the random utility model definition (3.3.6), the probability of choosing destination d may be calculated by replacing U_j with U_d^*, and a logit model is once again obtained:

$$p[d] = \frac{\exp[(V_d/\theta) + Y_d]}{\sum_{d'} \exp[(V_{d'}/\theta) + Y_{d'}]} \tag{4.2.9}$$

Finally, it is easy to verify that the product of $p[m/d]$ and $p[d]$, expressed respectively by (4.2.5) and (4.2.9), again gives $p[dm]$, expressed by (4.2.4).

A different partial share model may be obtained by using a hierarchical logit model. In this case, the elementary alternatives (dm) are grouped by destination: group I_d thus contains pairs (d, m') for all the available mode alternatives m' that serve destination d. In this case (see Sect. 3.3.2), it is assumed that the random residual ε_{dm} follows a Gumbel distribution with parameter θ_o and that can be broken down into the sum of two random variables η_d and $\tau_{m/d}$:

$$U_{dm} = V_{dm} + \varepsilon_{dm} = V_d + V_{m/d} + \eta_d + \tau_{m/d} \tag{4.2.10}$$

As shown in Sect. 3.3.2, the decomposition of ε_{dm} into the two components introduces a covariance between the residuals of alternatives dm and dm':

$$\text{Cov}(\varepsilon_{dm}, \varepsilon_{dm'}) = \text{Var}(\eta_d) = (\pi^2/6).(\theta_o^2 - \theta_d^2) \tag{4.2.11}$$

where θ_o and θ_d are the parameters of Gumbel distributions associated, respectively, with the root node and with all the intermediate decision nodes.

The behavioral interpretation of (4.2.11) is that the decision-maker perceives in a similar fashion the destination/mode alternatives that have the same destination but not those that have the same mode. Figure 4.3 shows schematically the two utility function structures corresponding to (4.2.3) and (4.2.10).

By applying the results of Sect. 3.3.2, the probability of choosing mode m conditional on destination d is again provided by a multinomial logit model, the expression for which may be obtained by substituting $j = m, k = d$, and $\theta = \theta_d$ in expression (3.3.12):

$$p[m/d] = \frac{\exp[V_{m/d}/\theta_d]}{\sum_{m'} \exp[V_{m'/d}/\theta_d]} \tag{4.2.12}$$

which is the same as (4.2.5) except for parameter θ.

By the same token, the destination choice probability may be obtained by (3.3.17):

$$p[d] = \frac{\exp[V_d/\theta_o + \delta Y_d]}{\sum_{d'} \exp[V_{d'}/\theta_o + \delta Y_{d'}]} \tag{4.2.13}$$

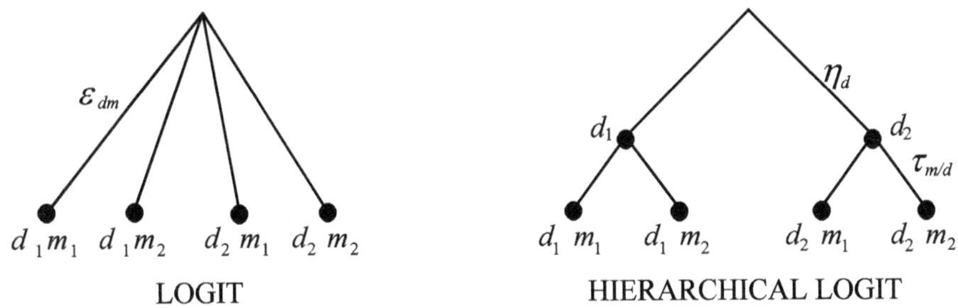

Fig. 4.3 Example of alternative utility function structures corresponding to a logit and hierarchical logit specification of a model for two destinations and two modes

where

$$\delta = \theta_d/\theta_o \tag{4.2.14}$$

The probability of choosing pair dm may thus be obtained from (4.2.12) and (4.2.13) as

$$p[dm] = p[d] \cdot p[m/d] = \frac{\exp[V_d/\theta_o + \delta Y_d]}{\sum_{d'} \exp[V_{d'}/\theta_o + \delta Y_{d'}]} \cdot \frac{\exp[V_{m/d}/\theta_d]}{\sum_{m'} \exp[V_{m'/d}/\theta_d]} \tag{4.2.15}$$

Note that the difference between the multinomial logit (4.2.4) and hierarchical logit models (4.2.15) lies in the value of the parameter δ defined in (4.2.14). As stated in Sect. 3.3.2, this parameter may take values between 0 and 1; for $\delta = 1$ the hierarchical logit model coincides with the logit.

Extension of the results to choices involving more than two dimensions is immediate. For example, the factored multinomial logit model for the sequence of choices $[d, m, k]$ becomes:

$$p[dmk] = \frac{\exp[V_d/\theta + Y_d]}{\sum_{d'} \exp[V_{d'}/\theta + Y_{d'}]} \cdot \frac{\exp[V_{m/d}/\theta + Y_{m/d}]}{\sum_{m'} \exp[V_{m'/d}/\theta + Y_{m'/d}]} \cdot \frac{\exp[V_{k/dm}/\theta]}{\sum_{k'} \exp[V_{k'/dm}/\theta]} \tag{4.2.16}$$

where the logsum variables are defined as

$$\begin{aligned} Y_d &= \ln \sum_{m'} \exp[V_{m'/d}/\theta + Y_{m'/d}] = \ln \sum_{m'} \sum_{k'} \exp\big[(V_{m'/d} + V_{k'/dm'})/\theta\big] \\ Y_{m/d} &= \ln \sum_{k'} \exp[V_{k'/dm}/\theta] \end{aligned} \tag{4.2.17}$$

The hierarchical logit model for these three choice dimensions takes the form:

$$p[dmk] = \frac{\exp[V_d/\theta_d + \delta_d Y_d]}{\sum_{d'} \exp[V_{d'}/\theta_d + \delta_d Y_{d'}]} \cdot \frac{\exp[V_{m/d}/\theta_m + \delta_m Y_{m/d}]}{\sum_{m'} \exp[V_{m'/d}/\theta_m + \delta_m Y_{m'/d}]} \cdot \frac{\exp[V_{k/dm}/\theta_k]}{\sum_{k'} \exp[V_{k'/dm}/\theta_k]} \tag{4.2.18}$$

where

$$\delta_d = \frac{\theta_m}{\theta_d}; \qquad \delta_m = \frac{\theta_k}{\theta_m} \quad \text{with} \begin{cases} \theta_d > \theta_m > \theta_k \\ \delta_d, \delta_m < 1 \end{cases}$$

and the inclusive variables Y have the expressions:

$$Y_d = \ln \sum_{m'} \exp[V_{m'/d}/\theta_m + \delta_m Y_{m'/d}] \tag{4.2.19}$$

$$Y_{m/d} = \ln \sum_{k'} \exp[V_{k'/dm}/\theta_k] \tag{4.2.20}$$

It is possible to define a form of factoring that is “weaker” than the one discussed here for logit and hierarchical logit models. In this second approach, the models that express the different steps of a partial step structure such as (4.2.2) are random utility models having different functional forms, such as logit for mode choice and probit for path choice. Therefore, the models corresponding to the sequence of partial choices cannot be obtained by factoring a single model that represents the choice of a compound alternative $[d, m, k]$. In this case, to maintain an interpretation consistent with the behavioral assumptions of random utility models, it is necessary for the model of each choice dimension to include an Expected Maximum Perceived Utility (EMPU) variable that reflects choice dimensions that are lower in the decision hierarchy.

For example, if in (4.2.10) we suppose that $\tau_{m/d}$ is distributed jointly as multivariate normal, the probability of choosing mode m in (4.2.12) will be given by a probit model, and the utility of destination choice is:

$$U_d^* = V_d + \max_{m'}(V_{m'/d} + \tau_{m'/d}) + \eta_d = U_d^* = V_d + s_d(V_{m'/d}) + \tau_d^* + \eta_d$$

where s_d is the EMPU that reflects mode choice. Moreover, if we assume that the sum of random variables τ_d^* and η_d is a Gumbel random variable $G(0, \theta)$ with zero mean and parameter θ, the destination choice model is a multinomial logit:

$$p[d] = \frac{\exp[(V_d + s_d(V_{m'/d})/\theta)]}{\sum_{d'} \exp[(V_{d'} + s_{d'}(V_{m'/d'})/\theta)]}$$

This approach may be extended to all choice dimensions by deriving partial share models analogous to those given by (4.2.16) and (4.2.18)

$$p[d, m, k] = p[d](V_d, s_d) \cdot p[m/d](\mathbf{V}_{m/d}, \mathbf{s}_{m/d}) \cdot p[k/dm](\mathbf{V}_{k/dm})$$

where the EMPU are expressed as:

$$s_{m/d} = E\Big[\max_{k'}(V_{k'/dm} + \tau_{k'/dm})\Big]$$

$$s_d = E\Big[\max_{m'}(V_{m'/d} + s_{m'/d} + \tau_{m'/d})\Big]$$

and the models that represent the various steps may have any functional form provided that they can be obtained from the assumptions of random utility models.

4.3 Examples of Trip-based Demand Models

This section describes some of the models often applied within a four-step structure, and also introduces some possible extensions such as inclusion of parking type and location choice within mode choice models. An example of an entire model system for interurban travel demand is presented at the end of the section.

4.3.1 Models of Spatial and Temporal Characteristics

4.3.1.1 Trip Production or Trip Frequency Models

A trip production or trip frequency model estimates the average number of trips $d^i_{o.}[sh]$ undertaken in period h for purpose s by a user of class i with origin in zone o; this is called the *trip rate* $m^i[osh]$. The total production of trips by users of class i for purpose s in period h by zone o can therefore be expressed as follows.

$$d^i_o[sh] = n^i[o]m^i[osh] \tag{4.3.1}$$

where $n^i[o]$ is the number of users in zone o belonging to class i.

As explained above, the trip production models used in applications fall into two main categories: descriptive models and behavioral models (or more properly, random utility models).

Descriptive Models As discussed in Sect. 4.2, descriptive models are generally used to represent regularly made trips, such as home-based work and home-based school trips.

Classification tables are the simplest *descriptive trip production models*. For each user class i, assumed to be homogeneous with respect to a given trip purpose, the average number of trips $m^i[osh]$ for purpose s in period h is directly estimated, most commonly from travel survey data. Figure 4.4 is an example of a classification table showing the daily trip rates for home-based work, school, and other trip purposes, obtained as the average of the trip rates estimated in the mid-1980s in five medium-sized Italian towns. Note the different definitions of user class adopted for different trip purposes: workers in the various economic sectors for home-based work trips, students of different levels for home-based school trips, and the family for home-based other purpose trips. The main limitation of classification table models is that trip frequencies and demand levels are not expressed as functions of socioeconomic variables other than those used to define the classes. In addition, limitations in data availability and the difficulty of forecasting the future number of users for detailed user classes generally keep the number of classes relatively small, even when a more detailed breakdown might be appropriate.

Trip rate regression models are more sophisticated. These models express the trip rate $m^i[osh]$ for a user of class i and for purpose s as a function, typically linear, of variables corresponding to the user class and the zone of origin:

$$m^i[osh] = \sum_j \beta_j X^i_{jo} \tag{4.3.2}$$

The attributes X_{jo} are usually the mean values of socioeconomic variables such as income, number of cars owned, and so on, but they may also include level-of-service attributes such as zonal accessibility, defined by the inclusive variable Y_x in (4.3.5) or by some other variable. The name trip rate regression is derived from the statistical model, linear regression, which is used to specify the variables X_j and to estimate the coefficients β_j.

In early applications, model (4.3.2) was specified at the level of traffic zones. Thus, its explanatory variables represented attributes of an entire zone (e.g., population, employment, number of shops, etc.) More recently, these models have been applied at a more disaggregate level, typically households and individuals. The application of model (4.3.2) at a disaggregate level, however, can lead to problems because some combinations of variable values and coefficients may result in negative trip rates. Hence it is better to use logit or other random utility specifications for disaggregate trip rate models.

Random Utility Models Behavioral models are generally applied to represent trips that are not regularly made. In a random utility framework, the trip rate $m^i[osh]$ can be expressed as

$$m^i[osh] = \sum_x x p^i[x/osh](\boldsymbol{SE}, \boldsymbol{T}) \tag{4.3.3}$$

where $p^i[x/osh](\boldsymbol{SE}, \boldsymbol{T})$ represents the probability that a user in zone o undertakes x trips for purpose s in period h. Alternatively, the trip rate $m^i[osh]$ can be obtained as the product of the outputs of two models: a trip production model that covers a longer time period, for example, the whole day g, and a departure time choice model:

$$m^i[osh] = \sum_x x p^i[x/osg](\boldsymbol{SE}, \boldsymbol{T}) \cdot \sum_{yh} y_h p^i[y_h/osx](\boldsymbol{SE}, \boldsymbol{T})$$

Purpose	Type of user	Trip rate
H-W	Worker in the Industrial sector	1.024
	Worker in the Service sector	1.084
	Worker in the Private Services sector	1.245
	Worker in the Public Services sector	0.931
H-Sc	Primary school student	0.84
	Lower secondary school student	0.87
	Upper secondary school student	0.86
	Vocational secondary school student	0.88
H-Sndg	Family	0.25
H-Sdg	Family	0.11
H-Ps	Family	0.16
H-Sr	Family	0.27
H-Acc	Family	0.11
H-oth	Family	0.13

Trip purpose code	Trip purpose
H-W	Home–work
H-Sc	Home–school
H-Sndg	Shopping for nondurable goods
H-Sdg	Shopping for durable goods
H-Ps	Personal services
H-Sr	Social–recreational
H-Acc	Accompanying others
H-Oth	Other purposes

Fig. 4.4 Daily urban trip production rates

where y_h represents the number of trips undertaken in period h out of all trips x made over the whole period $g[y_h = 0, 1, \ldots, x]$.

Specification of the full model requires definition of the alternatives, of the choice set and of the model that predicts choices from this set.

Definition of Choice Alternatives As stated, the choice alternatives in this case consist of different numbers of trips undertaken in period h.

Definition of Choice Set The choice set depends on the reference period. If h is a short period (i.e., the peak hour), so that the probability of undertaking more than one trip can be ignored, the choice set generally consists of two alternatives: one trip and no trip ($x = 0, 1$). For the sake of simplicity, the choice set is intentionally bounded ($x = 0, 1, 2$ or more) for larger periods.

Functional Form The binary and multinomial logit are the random utility models most frequently used to predict the trip frequency choice $p^i[x/osh]$ in (4.3.3). If h is so short that the probability of making more than one trip during the period is negligible, a binary logit model can be applied to the alternatives of undertaking the trip or not. Otherwise, a multinomial logit model gives the probability $p^i[x/osh]$ of undertaking x trips, with x equal to $0, 1, 2, \ldots, n$ or more trips:

$$p^i[x/osh] = \frac{\exp(V_x^i/\theta_o)}{\sum_{j=0,\ldots,n} \exp(V_j^i/\theta_o)} \tag{4.3.4}$$

Systematic utility functions include variables that represent the need or the possibility of carrying out activities connected with the purpose being modeled. These variables may relate either to the household or the individual. Household-level variables include, for example, total income and household size, whereas individual-level variables include occupational status, gender, family role, age, and so on. Other variables often used in the systematic utility of trip frequency models relate to the origin area, and especially its *accessibility* with respect to the possible destinations for the trip purpose. Accessibility can be expressed by the EMPU corresponding to the destination choice model, for example, the logsum Y_x given by the following expression for a logit distribution model,

$$Y_x = \ln \sum_{d'} \exp[V_{d'}/\theta_d + \delta_d Y_{d'}] \tag{4.3.5}$$

Figure 4.5 gives an example of a trip frequency model for the morning peak period in an urban area. A model of this type should be considered a method for quantitative analysis of the determinants of urban mobility[11] rather than an operational tool. Applying it to predict travel demand in an entire urban area would require a considerable amount of information. However, the same is not true of all behavioral models: operational trip frequency models are sometimes used to develop forecasts for large study areas; the intercity trip frequency models described in Sect. 4.3.4 are examples of this type of model.

Clearly, random utility models (4.3.3), or family or individual regression models (4.3.2) require more information[12] than the trip rate model (4.3.1). The latter, however, has the shortcoming of not being sensitive to variables other than those that define the user classes.

---

[11] Analysis of the model coefficients may suggest factors that influence urban trip-making for purposes other than commuting and study. For example, the results shown in Fig. 4.5 suggest that the frequency of activities (and trips) increases with income level. Greater accessibility of the residence zone with respect to the location of commercial activities increases shopping trip frequency, but is not significant for business and personal service trips. There is a greater tendency for women and unemployed persons to undertake trips; young people tend to have less mobility, in the time period considered, especially for shopping; there is a substitution effect with other members of the family for shopping (positive coefficient for the TOF variable), whereas there is a complementarity effect for other purposes (negative TOF coefficient). Carrying out other activities (coefficient of the TOP variable) reduces the time available to engage in the activity (trip purpose) considered and so on. Note, also, that the accessibility coefficient, in accordance with the behavioral interpretation of the model, should turn out to be within the interval (0, 1).

[12] The sample enumeration aggregation technique, described in Sect. 3.7, should therefore be used for more sophisticated model specifications.

$$V_{\text{TRIP}} = \beta_1 CA + \beta_2 WRK + \beta_3 AGE + \beta_4 INL + \beta_5 WMN + \beta_6 ACC$$
$$V_{\text{NOTRIP}} = \beta_7 TOP + \beta_8 TOF + \beta_9 NT$$

Type of variable	Name of variable	
Socioeconomic	Car availability	*CA*
	Working status	*WRK*
	Age	*AGE*
	Income level	*INL*
	Woman	*WMN*
Location	Accessibility	*ACC*
Time availability	No. of other trips made by the person for other purposes	*TOP*
Individual–family relationships	No. of trips of made by other family members for the same purpose	*TOF*
Alternative specific attributes (ASA)	NOTRIP	*NT*
CA	Dummy variable: 0 = car not available; 1 car available	
WRK	Dummy variable: 0 = nonworker; 1 = worker	
AGE	Dummy variable: 0 = ≤35 years; 1 = ≥35	
INL	Income level in 6 points scale: 0 = low income; 5 = high income	
WMN	Dummy variable: 0 = man, 1 = woman	

	No trip			Trip					
	TOP	*TOF*	*NT*	*CA*	*WRK*	*AGE*	*INL*	*WMN*	*ACC*
Shopping	0.55	0.61	1.35	0.24	−2.69	−2.53	0.08	0.60	0.11
t	5.4	3.7	5.4	1.2	−9.7	−8.0	1.5	3.8	1.7
Other purposes	0.22	−1.18	2.66	–	−0.34	−0.34	0.20	0.53	–
t	2.2	−10.9	15.3		−2.0	−2.0	3.5	3.3	

	Goodness-of-fit statistics		
	ρ^2	*% right*	*LR*
Shopping	0.431	0.847	1904
Other purposes	0.689	0.933	3041

Fig. 4.5 Trip frequency model for the morning peak period

4.3.1.2 Distribution Models

Distribution models express the percentage (probability) $p^i[d/osh]$ of trips made by users of class i going to destination d, given the origin zone o, purpose s, and time period h. For simplicity of notation, the user class index is omitted here.

Distribution models can be divided into descriptive and behavioral models.

Descriptive Models One of the best-known descriptive distribution models is the gravity model, whose name derives from its resemblance to Newton's law of gravity. In its typical formulation, this model provides the actual demand flow $d_{od}[sh]$ rather than the destination shares $p[d/osh]$ for each od pair:

$$d_{od}[sh] = \alpha d_o \cdot [sh]d \cdot {}_d[sh] f(C_{od}) \tag{4.3.6a}$$

where α is a constant, $d_o \cdot [sh]$ and $d \cdot {}_d[sh]$ represent, respectively, the total trip production from o and total trip attraction to d for purpose s in period h,[13] C_{od} is a variable related to the generalized transportation cost, and $f(C_{od})$ is an impedance (sometimes called *friction*) function that decreases with C_{od}. Typical expressions for this function are:

$$f(C_{od}) = \exp(-\beta C_{od}) \tag{4.3.7a}$$

$$f(C_{od}) = C_{od}^{-\beta} \tag{4.3.7b}$$

$$f(C_{od}) = C_{od}^{-\beta} \exp(-\beta C_{od}) \tag{4.3.7c}$$

In order to satisfy (1.3.1) and (1.3.2) of Sect. 1.3.3, the constant α is usually replaced by two factors that depend on the origin and destination zones (a *doubly constrained* gravity model):

$$d_{od}[sh] = A_o B_d d_o \cdot [sh] d \cdot {}_d[sh] f(C_{od}) \tag{4.3.6b}$$

where

$$A_o = 1/\sum_{d'} B_{d'} d_{\cdot d'} f(C_{od'}) \qquad B_d = 1/\sum_{o'} A_{o'} d_{o'\cdot} f(C_{o'd})$$

The two equations above are mutually dependent and therefore constants A_0 and B_0 are unknown quantities of a nonlinear equation system that can be solved by an iterative procedure.

When only one of these two conditions is satisfied, that is, (1.3.1) (a *singly constrained* gravity model[14]) in (4.3.6b), then $B_d = 1$ and

$$d_{od}[sh] = \frac{d_{o\cdot}[sh] \cdot d_{\cdot d}[sh] f(C_{od})}{\sum_{d'} d_{\cdot d'}[sh] f(C_{od'})} = d_{o\cdot}[sh] \cdot p[d/osh] \tag{4.3.6c}$$

[13] See Sect. 1.3.3. Trip attractions $d \cdot {}_d[sh]$ can be computed as a function of the zonal characteristics using models similar to those used to calculate trip productions $d_o \cdot [sh]$: for example, a trip attraction classification table or linear regression model.

[14] Gravity models originally derived their name from their similarity with Newton's law of universal gravitation. Singly and doubly constrained gravity models were subsequently derived from entropy maximization principles. In this approach, the entropy measure of a given trip distribution is expressed as a function of the number of possible microstates (i.e., individual trips between each origin–destination pair) that satisfy the distribution. The entropy function is then maximized subject to constraints on the total number of trips produced by (and in some models attracted to) each zone, and to the total cost (distance) of transportation. Distribution models that maximize this entropy are referred to as singly (and doubly) constrained gravity models. Although these models are still commonly used, they do not provide the flexibility of random utility models (whether these are interpreted behaviorally or not), and also do not allow for the introduction of attributes that account for the perceived attractiveness of different destinations. It should be pointed out that more sophisticated destination choice models are still relatively unstudied. Indeed, because of the possibility of spatial autocorrelation, the multinomial logit model's assumption of i.i.d. disturbances is questionable for traffic zones near each other. In this case cross-nested logit or probit models should be used. Models should also take account of travelers' different degrees of familiarity with potential destinations through choice set modeling procedures.

with

$$p[d/osh] = \frac{d_{\cdot d}[sh]f(C_{od})}{\sum_{d'} d_{\cdot d'}[sh]f(C_{od'})} \tag{4.3.8}$$

It is easy to verify that model (4.3.8) is invariant with respect to the aggregation or disaggregation of traffic zones, given equal "distance" from the origin. In other words, with a specification such as (4.3.8) the probability $p[d]$ of choosing a zone d that is aggregated from two smaller zones d_1 and d_2 is equal to the sum of the probabilities $p[d_1]$ and $p[d_2]$. Indeed, if the cost is constant:

$$C_{od} = C_{od_1} = C_{od_2} \quad \Rightarrow \quad f(C_{od}) = f(C_{od_1}) = f(C_{od_2})$$

then because $d_{.d} = d_{.d1} + d_{.d2}$ it follows that

$$\begin{aligned} p[d] &= \frac{d_{\cdot d} f(C_{od})}{d_{\cdot d} f(C_{od}) + \sum_{d' \neq d} d_{\cdot d'} f(C_{od'})} \\ &= \frac{d_{\cdot d_1} f(C_{od_1})}{d_{\cdot d_1} f(C_{od_1}) + d_{\cdot d_2} f(C_{od_2}) + \sum_{d' \neq d} d_{\cdot d'} f(C_{od'})} \\ &\quad + \frac{d_{.d_2} f(C_{od_2})}{d_{\cdot d_1} f(C_{od_1}) + d_{\cdot d_2} f(C_{od_2}) + \sum_{d' \neq d} d_{\cdot d'} f(C_{od'})} \\ &= p[d_1] + p[d_2] \end{aligned}$$

The property of invariance with respect to zonal aggregation is very useful in application because it provides results that do not depend on the particular level of spatial disaggregation that is used.

Random Utility Models Random utility distribution models represent the probability $p^i[d/osh]$ that a user of class i chooses destination d, given the origin zone o, purpose s, and time period h.

Definition of Choice Alternatives It is generally assumed that the zones in the study area zone system represent elementary destination choice alternatives. In reality, the destination where one chooses to carry out an activity is not a traffic zone but rather a specific location or locations (i.e., an office or a shopping center) within a traffic zone, and it is these specific locations that are the elementary destination alternatives. Therefore, a traffic zone should be modeled as a compound alternative that results from the aggregation of its elementary destination alternatives.

Different model functional forms can be derived depending on whether the elementary alternatives are taken to be the traffic zone or the specific destination locations; therefore the two cases are discussed separately.

(1) Alternative: Traffic Zone

Definition of Choice Set In this case, the choice set generally consists of all the traffic zones in the study area. This hypothesis is unrealistic because it leads to excessively large choice sets. It is easy to verify that in reality the user knows and considers only a small set of alternatives when choosing destinations. Therefore, the user's formation of a choice set should be modeled using one of the approaches presented in Sect. 3.5.

Functional Form Multinomial logit models are commonly used for destination choice modeling:

$$p[d/osh](\boldsymbol{SE}, \boldsymbol{T}) = \frac{\exp(V_d/\theta_d)}{\sum_{d'} \exp(V_{d'}/\theta_d)} \tag{4.3.9}$$

where $V_{d/osh} = E[U_{d/osh}]$ is the systematic utility of destination zone d and θ_d represents the Gumbel distribution parameter of $U_{d/osh}$.

In general, the attributes of the systematic utility $V_{d/osh}$ can be grouped into attributes of the activity system in zone d, or attractiveness attributes; and attributes that quantify the accessibility or cost of travel between zones o and d.

Attractiveness attributes are variables that measure the attractiveness of a zone as a destination; they might be a function of the number of employees (i.e., the number of workers of a given category) for home–work trips, the number of students of a certain grade school for home–study trips, the number of retail employees for home–shopping trips, and so on. Attractiveness attributes can also be alternative specific attributes, for example, a dummy variable equal to one for zones in the urban center zone and zero for the others, reflecting the greater symbolic value of the center for social and cultural reasons.

Cost attributes, as for nonbehavioral models, are variables expressing the generalized cost of a trip from o to d; therefore, their utility function coefficients β_k are negative. A wide variety of cost attributes can be considered, from the straight-line distance between zone centroids to generalized cost variables that take account of different contributions (walk time, in vehicle time, monetary cost, and so on) for each of the modes available between o and d.

From (4.3.9) it follows that

$$p[d/osh] = \frac{\exp(\beta_1 A_d - \beta_2 C_{od})}{\sum_{d'} \exp(\beta_1 A_{d'} - \beta_2 C_{od'})} \tag{4.3.10}$$

where A_d is the attractiveness variable of zone d and C_{od} the cost variable for traveling from origin o to destination d. In applications a logarithmic transformation of the attractiveness attribute ($A'_d = \ln(A_d)$) is usually adopted, hence (4.3.10) becomes:

$$p[d/osh] = \frac{A_d'^{\beta_1} \exp(-\beta_2 C_{od})}{\sum_{d'} A_{d'}'^{\beta_1} \exp(-\beta_2 C_{od'})} \tag{4.3.11}$$

Purpose	A_d	β_1	β_2
H-W	Firm employees	1.10	0.70
	Service employees	0.93	0.70
	Private service employees	0.93	0.83
	Public service employees	0.93	0.58
H-Sc	Elementary school students	0.90	2.52
	Primary school students	0.95	2.24
	Secondary school students	1.00	0.35
H-Ps	Service employees	0.91	0.78
H-Acc	Primary and secondary school students	0.20	1.35
H-Sndg	Trade employees	1.61	2.54

Fig. 4.6 Coefficients of a nonbehavioral urban trip distribution model

By taking β_1 equal to one in (4.3.11) there results a behavioral distribution model that is formally analogous to the gravity model (4.3.8), in which the sum of the trips attracted by a zone is replaced by the zone attractiveness and the cost function is of a negative exponential type (4.3.7a). Consequently, model (4.3.11) also satisfies the property of invariance with respect to zonal aggregation if the attractiveness variable satisfies $A_d = A_{d1} + A_{d2}$.

In this case, the difference between descriptive and behavioral models is merely a matter of interpretation (see footnote 3). For instance, model (4.3.11) can be used to represent the probability of shopping in destination zone d as a function of its utility, which is assumed to increase with zonal attractiveness and to decrease with trip cost; or alternatively it can be used to predict the fraction of individuals who travel to work in zone d, where this fraction tends to be greater for zones with a larger number of employees and which are easier to reach from the origin zone. This tendency exists because users tend to make mobility choices (choice of home and work location) so as to minimize the cost of home–work trips.

If in model (4.3.11) the logarithmic transformation of the cost attribute is also applied ($C_{od} = \ln(C_{od})$) it follows that

$$p[d/osh] = \frac{A'^{\beta_1}_{d} C'^{-\beta_2}_{od}}{\sum_{d'} A'^{\beta_1}_{d'} C'^{-\beta_2}_{od'}} \tag{4.3.12}$$

which is analogous to a gravity model (4.3.8) with cost function (4.3.7b).

As an example, Fig. 4.6 presents coefficients β_1 and β_2 of model (4.3.12) for selected user classes and for daily home–work, home–study, and home–other trips (personal services, accompanying others, and shopping for nondurable goods). The cost variable is the straight-line distance between zone centroids. The coefficients presented are typical values for average-size representative cities.

(2) Alternative: Elementary Destination

Definition of Choice Set In this case the choice dimension is represented by the choice of a specific destination location within a traffic zone. Because the real interest of the analyst is to reproduce the distribution of trips between traffic zones and not between elementary destinations, a procedure to aggregate elementary destinations into traffic zones is needed, in order to obtain a choice set analogous to the previous case.

Functional Form As previously noted, traffic zone d is a compound alternative composed of the aggregation of M_d elementary destination alternatives; a nested logit model is therefore usually used to predict $p^i[d/osh]$. In this case, (4.3.9) becomes:

$$p[d/osh] = \frac{\exp(V_{d/osh}/\theta_d + \delta_d Y_d)}{\sum_{d'} \exp(V_{d'/osh}/\theta_d + \delta_d Y_{d'})} = \frac{\exp[(V_{d/osh} + s_d)/\theta_d]}{\sum_{d'} \exp[(V_{d'/osh} + s_{d'})/\theta_d]} \quad (4.3.13)$$

where

$V_{d/osh} = E[U_{d/osh}]$ systematic utility of the traffic zone d, common to all elementary destinations in d

θ_d parameter of the Gumbel distribution of $U_{d/osh}$

$$s_d = \theta_r Y_d = \theta_r \ln \sum_{r'=1,\ldots,M_d} \exp(V_{r'/d}/\theta_r) \quad (4.3.14a)$$

(EMPU relative to the elementary destination choice)

$V_{r/d} = E[U_{r/d}]$ systematic utility of the elementary destination r conditional upon traffic zone d

θ_d parameter of the Gumbel distribution of $U_{r/d}$

$$\delta_d = \theta_r/\theta_d \le 1$$

As previously stated, the attributes in a distribution model include attractiveness attributes of the destination zone d and cost attributes associated with travel between the *od* pair. Inasmuch as most network supply models represent travel between zone centroids but not within zones, cost attributes change if the traffic zone changes, but not if the elementary destination changes, and therefore the transportation cost C_{od} is generally included in $V_{d/osh}$. Conversely, attractiveness attributes such as the number of employees in a certain category can be related to a single elementary destination, and so are usually part of $V_{r/d}$.

With some simple steps, another equivalent specification of s_d can be derived from (4.3.14a):

$$s_d = \overline{V}_d + \theta_r \ln M_d + \theta_r \ln\left[\frac{1}{M_d} \sum_{r'=1,\ldots,M_d} \exp\big[(V_{r'/d} - \overline{V}_d)/\theta_r\big]\right] \quad (4.3.14b)$$

where

$$\overline{V}_d = \frac{1}{M_d} \sum_{r'=1,\ldots,M_d} V_{r'/d}$$

This expression is particularly advantageous when the attractiveness of individual elementary destinations cannot be determined. Indeed (4.3.14b), except for the last term on the right side (which represents a heterogeneity term) can be easily calculated if the number M_d of the elementary destinations is known. Indeed, by setting:

$$V_{r/d} = A_r$$

Hence:

$$\sum_{r'=1,\ldots,M_d} V_{r/d} = \sum_{r'=1,\ldots,M_d} A_r = A_d$$

where A_r is the number of employees of the elementary destination r and A_d represents the number of employees within traffic zone d (generally known from statistical sources).

To understand the sense of (4.3.14b), the situation in which all the elementary destinations within d have the same systematic utility (i.e., an equal value of attractiveness) the heterogeneity term is equal to zero and the systematic utility of traffic zone d is given by the sum of the common term $V_{d/osh}$, of the utility of any elementary destination and of a positive "size" variable $\ln M_d$. The larger the elementary alternatives in d, the greater is M_d.

For some trip types the number of elementary destinations M_d in zone d can be calculated (e.g., the number of stores for shopping purposes). More frequently, the level of definition of trip purposes does not allow an accurate identification of the type of elementary destination and therefore of the number of elements in the choice set (e.g., for "other" purpose trips the actual elementary destination is unknown). In this case the size variable $\ln M_d$ can be replaced by a size function that estimates the unknown number of elementary alternatives in terms of other variables of the same zone (e.g., employees per sector, number of shops, etc.):

$$M_d = \sum_k \beta_k Z_{kd}$$

In this case it can be demonstrated that all the size function coefficients β_k but one can be identified; the unidentified coefficient can be arbitrarily set to one (see Chap. 5) and therefore (4.3.14b) becomes:

$$s_{d/osh} = \overline{V}_d + \theta_r \ln\left(Z_{1d} + \sum_{k=2}^{K} \beta_k Z_{kd} \right) \qquad (4.3.14c)$$

Examples of models with size functions are presented in Sect. 4.3.4.

4.3.2 Mode Choice Models

Mode choice models predict the fraction (or probability) $p^i[m/oshd]$ that users of class i select mode m to travel from zone o to zone d for trip purpose s in time period h. Mode choice is an example of a travel decision that can be easily modified for different journeys, and so for which performance or level-of-service attributes have considerable influence. It was no accident that the first random utility models were formulated to analyze transportation mode choice.

Definition of Choice Alternatives

In very simple cases the alternatives of a mode choice model are the individual transportation modes. In some cases "mixed" modes, that is, combinations of different modes such as car + train and car + bus, or different services of the same transportation mode (e.g., intercity, regional and night for the railway mode), are considered as choice alternatives. In interurban contexts, because of the high regularity and low frequency of transit services, the user is generally well informed about schedules and costs and tends to associate with each mode the utility of the most convenient service. In accordance with random utility theory, the logsum of lower choice dimensions (services) should be associated with the modes that offer them. To simplify the problem, some joint models of mode and service choice have been proposed.

Definition of Choice Set

Identification of the *relevant alternatives* depends on the transportation system under study. For example, modes such as walking or bicycle are typically considered to be choice alternatives in an urban system but, for obvious reasons, not for interurban systems. The definition of the choice set of each decision-maker is particularly important for mode choice models: not all transportation modes are available for all trips, either because of an objective impossibility (e.g., the personal car is not available to a user without a driving license) or because it is not perceived as an alternative for a particular trip (e.g., motorized modes may not be considered for very short trips).

Mode availability has been handled in mode choice models using the different approaches described in Sect. 3.5, usually via a combination of several heuristic methods. Objective nonavailability is usually dealt with by excluding the alternatives from the choice set of the decision-maker or user class; whereas contingent nonavailability or nonperception is generally accounted for by including availability/perception variables in the systematic utility specification. The attributes of car, bicycle, and motorcycle availability in the specification described in Fig. 4.7 should be interpreted in this way. Recently, IAP models that implicitly represent the probability of an alternative being available/perceived (as described in Sect. 3.5) have been applied to mode choice.

Functional Form

The systematic utility functions of mode choice models usually include level-of-service and socioeconomic attributes. As discussed in Chap. 2, *level-of-service* or *performance attributes* describe the characteristics of the service offered by the specific mode. Examples are travel time (possibly broken into access/egress time, waiting time, on-board time, etc.), monetary cost, service regularity, number of transfers, and so on. These attributes have negative coefficients because they usually represent disutilities for the user. In addition to level-of-service attributes, utility functions may include Alternative Specific Constants (ASCs) or *modal preference attributes*, variables that account for each mode's qualitative characteristics (e.g., the privacy of the car) or for attributes that are not otherwise included (e.g., service regularity for metro systems). In Chap. 3 it was shown that ASCs can be included in the systematic utility of all alternatives but one. Thus, after the effects of the other attributes in the utility function are accounted for, an ASC represents the remaining preference of users for a mode compared to a reference alternative. It follows that the coefficient of the ASC might have a positive or negative sign.

The ratios of level-of-service attribute coefficients in a linear utility function, also called the *marginal rates of substitution*, often have a meaningful interpretation. Among these, the rates of substitution between level-of-service attributes and monetary cost are particularly relevant, as these express the equivalent monetary value of the level-of-service attributes. If β_t and β_c are, respectively, the coefficients of travel time and monetary cost, the perceived Value of Time (VOT) implicit in mode choice behavior will be:

$$\text{VOT} = \frac{\beta_t}{\beta_c} \frac{[h^{-1}]}{[mon.unit^{-1}]} = [mon.unit/h] \tag{4.3.15}$$

Level-of-service attributes, and in particular times, monetary costs, and the like, should take into account alternatives in the "lower" choice dimension, in this case path choice. Thus, level-of-service attributes should refer to the different paths that the user can take on the network of each mode. This is done by using the EMPU of path choice which, in multinomial logit or hierarchical logit models, is the logsum variable $Y_{m/d}$. Sometimes, for the sake of simplicity, attributes are calculated only for the "minimum" cost path, although this introduces a theoretical inconsistency if path choice is not predicted with the deterministic utility (minimum cost) model described in the next section.

Socioeconomic attributes include characteristics of the decision-maker or her household. Typical examples are gender, age, family income, and car ownership and availability (number of cars owned by the household or the ratio between the cars owned and number of driving licenses).

Finally, in more sophisticated specifications some attributes may depend jointly on service and user characteristics. For example, monetary cost can be divided by user income, or differentiated by income level with different coefficients. In both cases the value of time varies by income, and is usually higher for users with higher income.

WALKING		
T_{walking}	Time (h)	−6.8237
BICYCLE		
T_{bk}	Time (h)	−8.2718
Nbcl/Nad	Number of bicycles owned in family per adult	0.6646
Bcl	Alternative specific attribute	−1.5818
MOTORCYCLE		
T_{mbk}	Time (h)	−8.2718
Age	Age variable (1 if ≤35 years, 0 otherwise)	0.6863
Nmbk/Nad	Number of scooters and motorbikes owned in family per adult	1.8572
Mbk	Alternative specific attribute	−2.3789
CAR		
T_{car}	Time (h)	−1.6142
Mc_{car}	Monetary cost (€)	−0.3338
Park	Parking (1 for priced parking destinations, 0 otherwise)	−1.1469
Hfam	Position in the family (1 if head of family, 0 otherwise)	0.4931
Ncar/Nad	Number of cars owned in family per adult	0.4014
Car	Alternative specific attribute	−1.7103
BUS		
T_{bus}	Total travel time (h)	−1.6142
Mc_{bus}	Monetary cost (€)	−0.3338
Ntrn	Number of transfers	−0.1772
Bus	Alternative specific attribute	−1.7827
$\ln L(\boldsymbol{\beta}_{\text{ML}})$		−475
$\ln L(\mathbf{0})$		−697
ρ^2		0.317
% right		0.651

Fig. 4.7 Alternatives, attributes, and coefficients of an MNL mode choice model for urban commuting trips

With respect to functional form, multinomial logit mode choice models are often used:

$$p^i[m/oshd] = \frac{\exp(V^i_{m/oshd})}{\sum_{m'} \exp(V^i_{m'/oshd})} \tag{4.3.16}$$

Figure 4.7 shows the alternatives, attributes, and coefficients of a logit mode choice model for commuting trips in a medium-sized Italian city. Other examples of MNL mode choice models are presented in Sect. 4.3.4, and in Chap. 8 on transportation demand estimation.

Hierarchical logit specifications are also being increasingly used. These models assume different levels of correlation between the perceived utilities of different mode groups, for example, private and public modes, and/or between different services of the same mode. A hierarchical logit mode choice model could also be used to predict the joint choice of mode and parking in urban areas.

In some applications to urban areas, specification of the systematic utility of the car mode includes level-of-service attributes related to parking, such as the time spent looking for a free parking space, parking cost, and walking distance to and from the parking space. In the most general case where several locations and types of parking are available, private modes such as auto are represented as groups of alternatives, each alternative corresponding to a specific parking location (d_p) and parking type (t_p) together with the given mode.

The lower-level multinomial logit model for parking choice can be specified as follows.

$$p^i[d_p t_p / oshda] = \frac{\exp(V^i_{d_p t_p})}{\sum_{d'_p t'_p} \exp(V^i_{d'_p t'_p})}$$

with

$$V^i_{d_p t_p} = \beta_{ts} Tsr_{d_p t_p} + \beta_c Mc^i_{d_p t_p} + \beta_{tw} Twl_{d_p/d}$$

where the variables are:

d_p, t_p Parking location (zone) and type (free on street, paid on-street, paid off-street, illegal etc.)

$Tsr_{d_p t_p}$ Average search time to find a parking space of type t_p in zone d_p

$Mc^i_{d_p t_p}$ Monetary cost (price or expected fine) of the alternative depending on the user class i (e.g., related to parking duration)

$Twl_{d_p/d}$ Time on foot needed to reach final destination d from location d_p

In this case, the logsum inclusive variable Y^i_p can be expressed as

$$Y^i_p = \ln \sum_{d'_p t'_p} \exp\left(V^i_{d'_p t'_p}\right)$$

and included in the systematic utility of the car alternative in the mode choice MNL model.

An example of a hierarchical logit mode and parking choice model in an urban area is given in Fig. 4.8.

4.3.3 Path Choice Models

Path choice models predict the fraction (or probability) $p^i[k/oshdm]$ of trips by users of class i on path k of mode m from o to d for trip purpose s in time period h. The path choice models used in practice are all behavioral, and the relevant attributes are, for the most part, performance or level-of-service variables obtained from the network supply models described in Chap. 2.

Path choice behavior and the models representing it depend on the type of service offered by the different transportation modes. In particular, the case where the

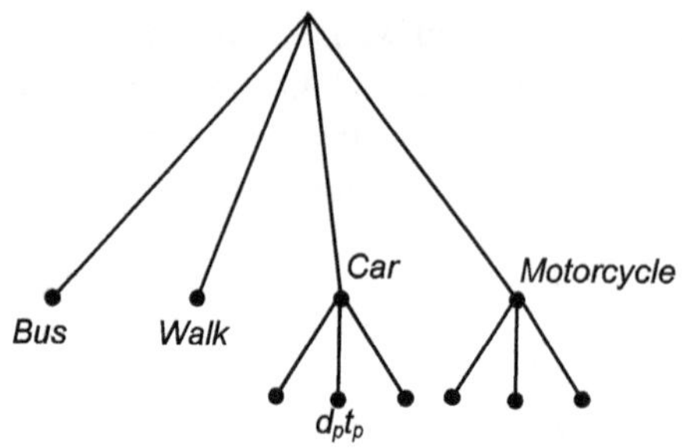

$$
\begin{aligned}
V_{\text{car}} &= \beta_{tb} \cdot T_{\text{car}} + \delta_p \cdot Y_p + \beta_c \cdot Mc_{\text{car}} + \beta_{\text{Car}} \cdot Car \\
V_{mbk} &= \beta_{tb} \cdot T_{mbk} + \delta_p \cdot Y_p + \beta_c \cdot Mc_{mbk} + \beta_{\text{Age}} \cdot Age + \beta_{Mbk} Mbk \\
V_{\text{bus}} &= \beta_{tb} \cdot T_b + \beta_{tw} \cdot tw_b + \beta_c \cdot Mc_b \\
V_{\text{walk}} &= \beta_{\text{twalk}} \cdot T_{\text{walk}} + \beta_{\text{Walk}} \cdot Walk
\end{aligned}
$$

with

T_{car}	= Car travel time [h]	Mc_{car}	= Monetary cost Car [€]
T_{mbk}	= Motorbike travel time [h]	Mc_{mbk}	= Monetary cost Motorbike
Tb_b	= Bus in vehicle time [h]	Mc_b	= Monetary cost Bus [€]
Tw_b	= Bus waiting time [h]		
Twl	= Walking time [h]		
Age	= Dummy variable of value 1 if age is <35 years, 0 otherwise		
Car, Mbk, Walk	= Mode Specific Attributes		

d_p = parking destination zone

$Yp = \ln \sum_{d'_p t'_p} \exp(V^i_{d'_p t'_p})$

t_p = type of parking: free limited duration; paid on street; paid off street; illegal

$V_{d_p t_p} = \beta_{ts} Tsr_{d_p t_p} + \beta_c Mc^i_{d_p t_p} + \beta_{tw} Twl_{d_p/d}$

With

Tsr = Average time spent finding a parking space [h]

Mc = Parking monetary cost [€]

Twl = Walking time from parking location to destination [h]

Model of parking choice

Tsr	*Mc*	*Twl*
−18.168	−3.358	−19.386

Model of modal choice

T	*Tw*	*Twl*	*Mc*	Y_p	*Age*	*Car*	*Mbk*	*Walk*
−1.961	−4.902	−4.314	−0.550	0.199	2.331	0.921	−1.631	3.127

Fig. 4.8 A hierarchical logit model of mode and parking choice in an urban area

whole path is chosen before starting the trip (*pre-trip choice*) can be distinguished from that where the path is chosen in two phases and is completely defined only during the trip itself (*pre-trip/en-route mixed choice*). Pre-trip choice behavior is usually assumed when representing path choice for continuous service systems; typical examples are road networks for private modes such as car, motorcycle, and the like. Pre-trip choice behavior is also assumed for scheduled transportation services with low frequency and high regularity, under the assumption that users know the service timetable and makes their decisions before beginning the trip (see Sect. 7.6.1). On the other hand, pre-trip/en-route mixed behavior is usually assumed for scheduled transportation systems with high frequency and/or low regularity, for example, urban transit systems (see Sect. 6.2).

As for all behavioral models, specification of a path choice model involves three phases: definition of the alternatives, identification of the set of possible alternatives (the choice set), and definition of the choice model. The first two phases are particularly important for path choice.

In the discussion below, behavioral assumptions and choice models are described separately for pre-trip and mixed path choice behavior, taking as examples road networks and transit networks with high frequency/low regularity. Path choice models for low-frequency/high-regularity scheduled services are covered in Chap. 7.

4.3.3.1 Path Choice Models for Road Networks

Definition of Choice Alternatives The assumption usually made for road networks is that, before making a trip, the user chooses a sequence of road segments to follow to the destination. This can be represented as a path,[15] a sequence of nodes and links on the graph that represents the road system, as described in Chap. 2. Only elementary (loopless) paths are considered, and thus their number is finite.

Definition of Choice Set Definition of the paths considered as choice alternatives, that is, definition of the choice set, is particularly important because the topological complexity of the network could generate an unrealistically large number of paths between a single O-D pair. The set of feasible paths K_{odm} connecting the centroids od over the mode m network should in principle be defined through an explicit choice set model, as described in Sect. 3.5. In practice, however, two types of heuristic approaches are used.

[15]Pure pre-trip choice behavior assumes that users do not ever modify the route that they choose at the beginning of their journey. In reality even for continuous service modes there are situations where users modify their routes by adapting to conditions encountered during the journey (e.g., accidents and unexpected traffic jams). This type of behavior is even more prevalent when real-time information technologies (variable message signs, radio traffic news, in-vehicle navigation systems) provide information on the current or predicted state of the network or suggest a route to take. Route choice models that take account of "mixed" behavior in continuous service networks are, however, still at the research stage and are not dealt with here. Furthermore, static assignment models are intended to simulate recurrent congestion: the route choice models used for static assignment can be assumed to reflect normal conditions and thus rule out accidents or other nonrecurrent events.

Selection criteria	Specification
Topological	A path is feasible (Dial efficient) if each link moves away from the origin and/or moves towards the destination; see Sect. 5.3.3 $K \in K_{od}$ if $Z_{o,i} < Z_{o,j}$ $\forall (i,j) \in k$
Comparison of costs	Paths with a generalized cost not exceeding the minimum cost by more than a factor of α $K \in K_{od}$ if $g_k \leq (1+\alpha)g_{\min}$
Progressive	The n paths with the lowest generalized costs
Multi-attribute	Minimum paths with respect to various attributes (usually performance variables such as travel time, monetary cost, motorway distance, etc.)
Behavioral	Paths excluding behaviorally unrealistic link sequences (e.g., repeated entrances and exits for the same motorway)
Distinctive	Paths overlapping for no more than a given percentage of their length

Fig. 4.9 Criteria for path feasibility on road networks

The *exhaustive approach* considers all elementary paths on the network. This approach may generate a large number of paths that share many links and so are correlated in their perceived (dis)utilities. Furthermore, given the computational complexity of explicitly enumerating all the paths in a network, this operation is usually carried out implicitly (*implicit path enumeration*) by algorithms that simultaneously calculate path choice probabilities and assign flows, as described in Chap. 5.

The *selective approach*, on the other hand, applies heuristic behavioral rules to identify only a subset of the elementary paths. For example, a path may not include more than one entrance and one exit from the same motorway, may not go farther away from the destination, may not have a generalized cost exceeding the minimum path cost by more than a given amount, and so on. Various criteria for the selection of feasible paths have been proposed in the literature. They relate to different application contexts (urban/interurban networks) and to different algorithms for generating paths and calculating choice probabilities and link flows. Examples of selection criteria are given in Fig. 4.9. In this table, $Z_{o,i}$ and $Z_{o,j}$ represent the minimum cost to reach node i and node j from origin o (see Chap. 5).

In general the selective approach requires *explicit path enumeration* between each O-D pair, and usually applies a combination of criteria. Chapter 5 describes some algorithms for path enumeration, and Fig. 4.10 depicts an example of the complete set of elementary paths and a selective set for an origin–destination pair. For more sophisticated feasible path generation models, the criteria to be used must be "calibrated" as are other parameters in the model. Calibration can be carried out by comparing the paths generated by the model with the paths perceived (or chosen) by a sample of users, adjusting the model to maximize the coverage of the latter by the former.

Some experimental results suggest that a good level of coverage of the paths used by users, at least for interurban networks, can be achieved by generating the

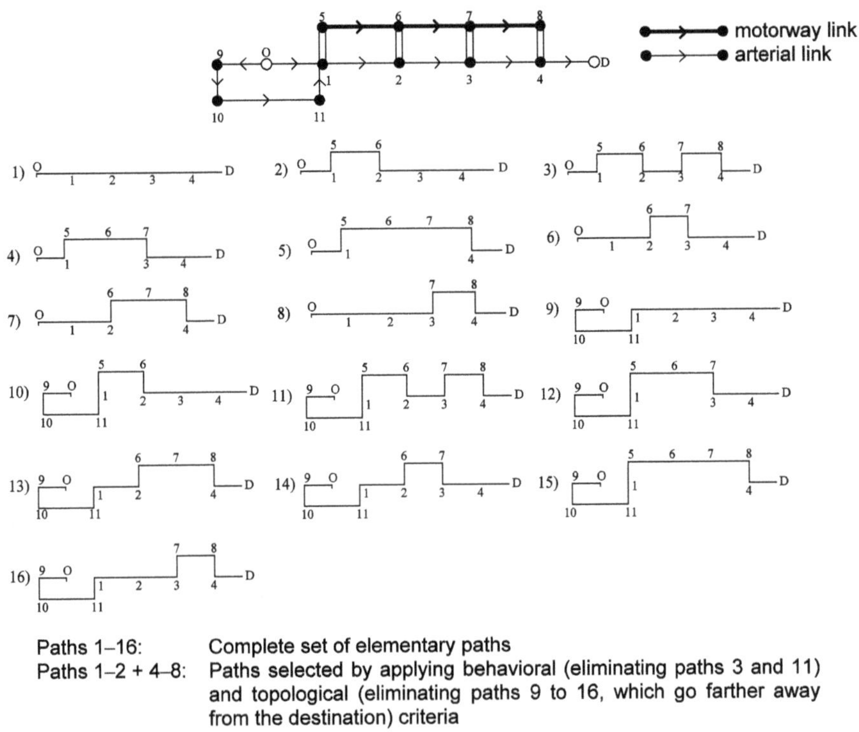

Fig. 4.10 Examples of exhaustive and selective set of feasible paths

first n paths for some intuitively reasonable criterion (e.g., minimum time, minimum monetary cost, maximum motorway use, etc.).

The selective approach guarantees better control of the feasibility of the generated paths while allowing the use of performance attributes that are not additive over links, as shown later. These advantages are obtained at the expense of greater computational complexity.

Conversely, implicit path enumeration methods are computationally more efficient and are typically used in the assignment models implemented in commercial software. However, it should be emphasized that there has been no systematic analysis of the computational complexity and memory requirements of the two methods. Recent literature suggests a growing tendency towards explicit path enumeration models in applications (see Chap. 5), perhaps because of the increasing power of the computing resources that are routinely available.

Functional Form Specification of a path choice model requires, as usual, definition of the attributes in the systematic utility function and of the joint probability distribution of random residuals, that is, the choice probability functional form. It is usually assumed that the variables influencing path choice are disutility attributes that enter in the utility function specification with negative coefficients,[16] for example, travel times, monetary cost, distance, and the like. Thus it follows that

$$U_k = V_k + \varepsilon_k \quad \forall k \in K_{odm} \tag{4.3.17a}$$

$$V_k = -g_k \tag{4.3.17b}$$

where g_k is the average generalized cost of path k expressed in utility units and K_{odm} is the set of paths connecting the pair od via mode m. Systematic utility and average cost should be differentiated by user class, V_k^i and g_k^i, although in what follows the superscript i is omitted for simplicity.

In Sect. 2.3.3 it was stated that the average path cost is usually a linear combination[17] of performance attributes with coefficients estimated from a path choice model:

$$g_k = \sum_n \beta_n z_{nk} \tag{4.3.18a}$$

If each path attribute z_{nk} is the sum of the corresponding link variables r_{nl}, the path cost g_k will be purely additive:

$$c_l = \sum_n \beta_n r_{nl} \qquad \mathbf{g}^{\mathrm{ADD}} = \sum_n \beta_n \mathbf{z}_n = \sum_n \beta_n \boldsymbol{\Delta}^{\mathrm{T}} \mathbf{r}_n = \boldsymbol{\Delta}^{\mathrm{T}} \mathbf{c} \tag{4.3.18b}$$

where δ_{lk} are the (0/1) elements of the link–path incidence matrix $\boldsymbol{\Delta}$ and c_l is the average cost of link l introduced in Chap. 2.

In some cases, however, the average cost might include some variables that cannot be obtained as the sum of link variables (*nonadditive cost* g_k^{NA}). This occurs, for example, if the monetary cost depends nonlinearly on path length, or if there is a dummy variable for minimum travel time or maximum motorway length paths. In the most general case, the expression (4.3.17b) therefore becomes:

$$V_k = -g_k^{\mathrm{ADD}} - g_k^{\mathrm{NA}}$$

Generally, nonadditive path cost variables require explicit path enumeration.

Figure 4.11 shows some example systematic utility specifications for path choice models in urban and interurban road networks.

The probability of choosing path k can be obtained with any random utility model and depends on the distribution of random residuals ε_k in (4.3.17a).

---

[16] More sophisticated specifications may also include socioeconomic attributes of the driver such as gender, income, and so on.

[17] When the generalized cost depends on a single attribute (such as travel time in urban networks), this is multiplied by a marginal utility coefficient β to convert it into utility.

PATH CHOICE MODEL FOR URBAN ROAD NETWORKS

$V_k = \beta_1 TTP_k + \beta_2 TTS_k + \beta_3 L_k + \beta_4 NTS_k + \beta_5 NLT_k + \beta_6 MTW_k$
TTP = Travel time on primary roads [h]
TTS = Travel time on secondary roads [h]
L = Total length [km]
NTS = Number of traffic-signal intersections on the path
NLT = Number of left turns
MTW = Dummy variable for the maximum motorway path

	TTP	*TTS*	*L*	*NTS*	*NLT*	*MTW*	ρ^2	*% right*	*L ratio*
	−16.462	−61.257	−9.601	−0.209	−2.296	3.158	0.403	0.532	844.344
t	−7.514	−16.445	−1.224	−1.143	−3.978	2.678			

PATH CHOICE MODEL FOR HEAVY VEHICLES IN INTERURBAN ROAD NETWORKS

$V_k = \beta_1 TT_k + \beta_2 Mc_k + \beta_3 ML_k + \beta_4 \text{Min}\, T_k + \beta_5 \text{Max}\, M_k + \beta_6 HVP_k + \beta_7 CF_k$
TT = Travel time [h]
Mc = Monetary cost [€]
ML = Total motorway length [km]
Min T = Dummy variable for minimum time path (0/1)
Max M = Dummy variable for maximum motorway use path (0/1)
HVP = Dummy variable for minimum time path for perishable and/or high value goods (0/1)
CF = Path commonality factor
VOT = Value of time [€/h]

	TT	*Mc*	*ML*	Min *T*	Max *M*	*HVP*	*CF*	*VOT*	ρ^2	*LR ratio*
	−4.525	−0.0165	0.013				−0.9524	68.5605	0.176	−2440
t	−19.3	−6.7	12.3				−12.9			
	−3.110	−0.0155	0.012			1.785	−0.839	50.1515	0.250	−2222
t	−14.2	−6.1	11			20.8	−11.6			
	−5.440	−0.018	0.012	2.292	2.585		−1.296	75.5555	0.306	−2055
t	−20.5	−6.9	10.5	19.9	20.1		−15.7			
	−3.650	−0.015	0.009	3.370	3.702	3.788	−1.205	60.8335	0.450	−1630
t	−14.1	−5.6	7.5	21.6	22	22.1	−14			

Fig. 4.11 Examples of multinomial logit path choice models in urban and interurban road networks

The simplest path choice model is the *deterministic utility* model, which is a special case of a random utility model in which the variance of the residuals ε_k is assumed to be equal to zero:

$$U_k = V_k = -g_k$$

In this case, a path k can be used only if, from among the set of alternative paths, its cost g_k is the least:

$$p[k/osdm] > 0 \quad \Rightarrow \quad g_k \leq g_h \quad \forall h \neq k,\ h, k \in K_{odm} \tag{4.3.19}$$

As already noted in Sect. 3.4, the deterministic utility model does not provide a unique path choice probability vector, except when there is a unique minimum cost

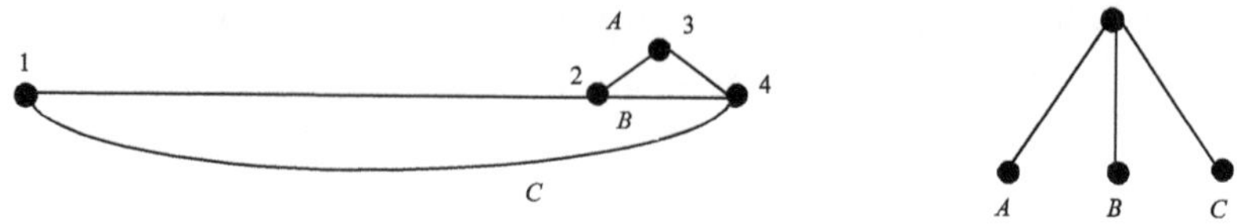

Path	Links	
A	(1.2) (2.3) (3.4)	$g_A=g_B=g_C=g$
B	(1.2) (2.4)	$p[A]=p[B]=p[C]=0.333$
C	(1.4)	$p[A]+p[B]=0.66$

Fig. 4.12 Application of a logit model to highly overlapping paths

path. In this latter case:

$$p[k/osdm] = \begin{cases} 1 & \text{if } g_k < g_h \quad \forall h \neq k,\ h,k \in K_{odm} \\ 0 & \text{otherwise} \end{cases} \tag{4.3.20}$$

Although deterministic choice models are arguably less realistic than general probabilistic models, for computational reasons they are often applied with implicit path enumeration to very congested networks. In such cases, they give results that are largely comparable with those obtained from probabilistic models, as shown in Sect. 5.4.5.

The probabilistic choice models generally used to calculate path choice probability are logit and probit. In this application, the multinomial logit model takes the form:

$$p[k/oshdm] = \frac{\exp(-g_k/\theta)}{\sum_{h \in K_{odm}} \exp(-g_h/\theta)} \tag{4.3.21}$$

The multinomial logit model results when the random residuals ε_k are assumed to be i.i.d. Gumbel variables with parameter θ, where θ is proportional to the residuals' standard deviation. As shown in Chap. 8, the parameter θ cannot be estimated separately for linear utility functions of the type (4.3.18a), so is assimilated in the coefficients β_h. The urban and interurban path choice models described in Fig. 4.11 have a multinomial logit specification.

The assumption of i.i.d. residuals that underlies the logit model and implies its independence of irrelevant alternatives property (see Sect. 3.3.1) is unrealistic when the paths in the choice set overlap (share links). In this case, it may be conjectured that the perceived costs of heavily overlapping paths are highly correlated, giving rise to choice probabilities that are smaller than those of other paths that have the same average costs but overlap less or not at all. In the extreme case of two paths that overlap almost completely, the MNL model gives them unrealistically large choice probabilities, as shown in Fig. 4.12. To reduce the effects of the IIA property, the multinomial logit model should be used with an explicit path enumeration method that eliminates highly overlapping paths.

Alternatively, if it is assumed that the residuals ε_k follow a multivariate normal distribution, the choice model has the probit form. The most widely used specification assumes that the variance of the random residuals is proportional to an additive path cost attribute z_k, and that the covariance of the residuals of two paths is proportional to the cumulative value of the cost attribute over the links that are shared by the two paths (z_{kh}):

$$\text{var}[\varepsilon_k] = \xi z_k, \quad k \in K_{odm} \tag{4.3.22a}$$

$$\text{cov}[\varepsilon_k, \varepsilon_h] = \xi z_{kh}, \quad h, k \in K_{odm} \tag{4.3.22b}$$

Usually, the variables z_k used to define the distribution differ from the actual path cost g_k (e.g., length or uncongested cost). These specifications satisfy the random utility model's property of additivity described in Sect. 3.4 and are useful in the analysis of the theoretical properties of equilibrium assignment models, as discussed in Chap. 5.

Note that the specification (4.3.22a), (4.3.22b) of the random residual variance–covariance matrix depends on a single calibration parameter ξ, and can be derived by applying the factor-analytic probit model described in Sect. 3.3.6 to the path choice context. To see this, assume that a perceived disutility u_l is associated to each link l, with:

$$u_l = E[u_l] + \eta_l = -c_l + \eta_l$$

The link random residuals, η_l $(l = 1, 2, \ldots, L)$, are independent normal variables $\eta_l \sim N(0, \boldsymbol{\sigma}_l)$ with:

$$\text{Var}[\eta_l] = \sigma_l = \xi r_l$$

$$\text{Cov}[\eta_l, \eta_j] = 0$$

$$\eta \sim \text{MVN}(\mathbf{0}, \boldsymbol{\Sigma}_\eta) \qquad \boldsymbol{\Sigma}_\eta = \xi \boldsymbol{DIAG}(\boldsymbol{r})$$

where r_l is the link-related performance variable corresponding to path attribute z and $\boldsymbol{DIAG}(\boldsymbol{r})$ is the $(n_L \times n_L)$ diagonal matrix containing these link variables. Assuming further that the path utility is the sum of its link utilities, it follows that

$$U_k = \sum_l \delta_{lk} u_l = E[U_k] + \varepsilon_k$$

$$E[U_k] = \sum_l \delta_{lk} E[u_l] = -\sum_l \delta_{lk} c_l = -g_k$$

$$\varepsilon_k = U_k - E[U_k] = \sum_l \delta_{lk}(u_l + c_l) = \sum_l \delta_{lk} \eta_l$$

$$\text{Var}[\varepsilon_k] = \sum_l \delta_{lk} \cdot \text{var}[\eta_l] = \sum_l \delta_{lk} \cdot \xi r_l = \xi z_k$$

$$\mathrm{Cov}[\varepsilon_k, \varepsilon_h] = E[\varepsilon_k, \varepsilon_h] = E\left[\sum_l \delta_{lk}\eta_l \cdot \sum_l \delta_{lh}\eta_l\right] = E\left[\sum_{l\in hk} \eta_l^2\right]$$

$$= \sum_{l\in hk} \mathrm{var}\left[\eta_l^2\right] = \xi z_{kh}$$

that is, the relationships (4.3.22a), (4.3.22b). Because the sum of normal variables is again a normal variable, then:

$$\boldsymbol{\varepsilon} \sim \mathrm{MVN}(\mathbf{0}, \boldsymbol{\Sigma})$$

where $\boldsymbol{\Sigma}$ is the variance–covariance matrix with elements given by (4.3.22a), (4.3.22b).

In other words, specification (4.3.22a), (4.3.22b) of the probit model can be obtained by applying the factor analytic probit to the path choice context with:

$$\boldsymbol{\varepsilon} = \boldsymbol{\Delta}^T \boldsymbol{\eta} = \boldsymbol{\Delta}^T \boldsymbol{\Sigma}_\eta^{1/2} \boldsymbol{\zeta} = \boldsymbol{\Delta}^T \left[\xi \cdot \boldsymbol{DIAG}(\boldsymbol{r})\right]^{1/2} \boldsymbol{\zeta} = \boldsymbol{F}\boldsymbol{\zeta}$$

where

- $\boldsymbol{\varepsilon}$ is the $(n_p \times 1)$ vector of multivariate normal distributed path random residuals, $\boldsymbol{\varepsilon} \sim \mathrm{MVN}(\mathbf{0}, \boldsymbol{\Sigma})$
- $\boldsymbol{\Delta}$ is the $(n_l \times n_p)$ link–path incidence matrix
- $\boldsymbol{\eta}$ is the $(n_l \times 1)$ vector of independent normal distributed link random residuals, $\boldsymbol{\eta} \sim \mathrm{MVN}(\mathbf{0}, \boldsymbol{\Sigma}_\eta)$
- $\boldsymbol{\zeta}$ is the $(n_l \times 1)$ vector of i.i.d. standard normal random variables, $\boldsymbol{\zeta} \sim \mathrm{MVN}(\mathbf{0}, \boldsymbol{I})$
- $\boldsymbol{F}$ $\boldsymbol{\Delta}^T \boldsymbol{\Sigma}_\eta^{1/2} = \boldsymbol{\Delta}^T [\xi \boldsymbol{DIAG}(\boldsymbol{r})]^{1/2}$ is the $(n_p \times n_l)$ matrix that maps the random vector $\boldsymbol{\zeta}$ into path choice random residuals $\boldsymbol{\varepsilon}$
- n_p is the total number of paths
- n_l is the total number of links, usually $n_l \ll n_p$

It can be easily shown that matrix $\boldsymbol{F}$ specified above, introduced in (3.3.62) and (3.3.63), yields (4.3.22a) and (4.3.22b) respectively.

This representation of the probit path choice model is also used in Sect. 5.3.3 for the specification of an algorithm for network assignment to uncongested networks.

The ability of the probit model to handle path overlapping, or perceived cost correlation, makes it particularly suitable for applications with exhaustive path generation (implicit enumeration). Furthermore, the difficulty of explicitly calculating probit choice probabilities can be overcome with algorithms that are based on Monte Carlo simulation, as mentioned in Sect. 3.3.6. These algorithms are discussed in detail in Chap. 5.

A modification to the logit path choice model was recently proposed to overcome the problems deriving from the logit IIA property while at the same time retaining a convenient analytical form. This modification is called the *C-logit* model and has the following specification.

$$p[k/oshdm] = \frac{\exp[(-g_k)/\theta - CF_k]}{\sum_{h\in K_{odm}} \exp[(-g_h)/\theta - CF_h]} \tag{4.3.23}$$

The term CF_k, known as the *commonality factor*, reduces the systematic utility of a path according to its degree of overlap with other paths. The commonality factor can be specified in various ways, for example, as

$$CF_k = \ln\left(1 + \sum_{h \neq k} \frac{z_{hk}}{(z_h z_k)^{1/2}}\right) \tag{4.3.24a}$$

where the attributes z_h, z_k, and z_{hk} are analogous to those described for the probit model. Expression (4.3.24a) shows immediately that the attribute CF_k is inversely proportional to path k's degree of independence from other paths, and is equal to zero if no other path shares links with path k. In this case:

$$z_{hk} = 0 \quad \forall h \neq k \rightarrow CF_k = \ln(1) = 0$$

Conversely the attribute CF_k is larger the more other paths share links with path k. For given path costs, the C-logit model (4.3.23) reduces the probability of choosing heavily overlapping paths and increases the probability of choosing nonoverlapping paths. Furthermore, in the limiting case of N completely overlapping paths, the C-logit choice probabilities tend to $1/N$ of the probability that a multinomial logit model would calculate if the N coincident paths were considered as one. These results are illustrated in Fig. 4.13, which presents logit, C-logit, and probit choice probabilities for a network similar to that in Fig. 4.12 and for different values of the coefficient of variation (cv). As can be seen, C-logit and probit probabilities are very similar and are lower than those obtained from the logit model for heavily overlapping paths. Some calibrations of interurban truck path choice models confirm the significance of the CF_k attribute (see Fig. 4.11).

Expression (4.3.24a) allows computation of the path commonality factor by adding up the values for the links making up the path; consequently, it lends itself to use in implicit path enumeration algorithms similar to Dial's (see Chap. 5).

Other specifications of CF have been proposed, including:

$$CF_k = \sum_{l \in k} w_{lk} \ln N_l \tag{4.3.24b}$$

where the summation is extended to all links l belonging to path k, w_{lk} is equal to the weight of link l in path k:

$$w_{lk} = \frac{r_l}{z_k}$$

and N_l is the number of paths between the same O-D pair using link l.

Expression (4.3.24b) takes into account the relative weight of shared links in the overall path cost; for example, if two paths h and k share the same link l:

$$w_{lh} > w_{lk} \rightarrow CF_k > CF_h$$

The attribute CF is larger for a path whose shared links contribute a larger fraction to its total length or cost.

Link	Paths A	B	C	Link costs
1	0	1	1	14
2	1	0	0	K
3	0	1	0	2
4	0	0	1	2

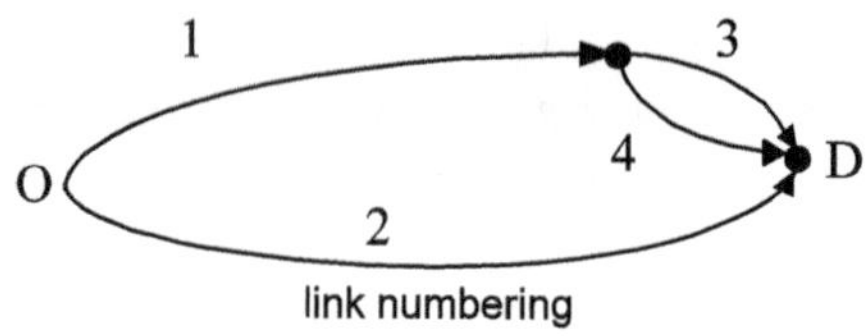

$K = 16$

Path	Cost	Logit ($\forall\theta$)	CLogit ($\forall\theta$)	Probit $\xi = 1$
A	16	0.333	0.478	0.450
B	16	0.333	0.261	0.275
C	16	0.333	0.261	0.275

$K = 17$

Path	Cost	Logit cv = 0.1	Logit cv = 0.3	Logit cv = 1.1	CLogit cv = 0.1	CLogit cv = 0.3	CLogit cv = 1.1	Probit cv = 0.1	Probit cv = 0.3	Probit cv = 1.1
A	17	0.091	0.227	0.302	0.156	0.350	0.442	0.162	0.342	0.421
B	16	0.454	0.387	0.349	0.422	0.325	0.279	0.419	0.329	0.289
C	16	0.454	0.387	0.349	0.422	0.325	0.279	0.419	0.329	0.289

Fig. 4.13 Comparison among path choice probabilities with logit, C-logit, and probit models

Another useful expression for CF is the following.

$$CF_k = \ln\left[1 + \sum_{h \neq k}\left(\frac{z_{hk}}{(z_h z_k)^2} \cdot \frac{z_k - z_{hk}}{z_h - z_{hk}}\right)\right] \tag{4.3.24c}$$

As (4.3.24c) shows, the CF of a path also depends on the cost of its nonshared links. In this way, the ratio CF_A/CF_B between the commonality factors of two paths increases as the overlap between them (the percentage of common cost with respect to the total one) increases, as $z_A > z_B$.

The C-logit model has a behavioral interpretation as an Implicit Availability Perception (IAP) model (discussed in Sect. 3.4) that simultaneously represents both the perception of paths as alternatives as well as the choice among the perceived alternatives. The commonality factor CF_k can in fact be interpreted as an attribute of the model, giving the degree of membership $\mu_{I_{odm}}(k)$ of path k in the set of perceived paths I_{odm}:

$$\mu_{I_{odm}}(k) \propto \exp(-CF_k) \tag{4.3.25}$$

that is, it is assumed that the perception of path k as an elementary alternative is larger if its overlap with other paths is smaller, and vice versa. On the other hand, the first-order IAP logit model described in Sect. 3.4 can be formally expressed as

$$p[k/odm] = \frac{\mu_{I_{odm}}(k) \cdot \exp(-g_k/\theta)}{\sum_{h \in K_{odm}} \mu_{I_{odm}}(h) \cdot \exp(-g_h/\theta)} \tag{4.3.26}$$

Substituting expression (4.3.25) into (4.3.26) gives expression (4.3.23).

4.3.3.2 Path Choice Models for Transit Systems

As stated in Chap. 2, public transportation systems offer services that are both non-continuous in space (i.e., only provided between discrete points such as stations or stops) and in time (i.e., available only at times corresponding to departures and arrivals). The supply models (transportation networks) representing such systems follow two main approaches: line-based and run-based. The choice between these depends on service frequency and regularity, and on the resulting assumptions about users' behavior. The discussion below refers to path choice models for scheduled services with frequencies high enough to justify a line-based[18] representation as described in Sect. 2.4.2.1 and restated in Fig. 4.14 for the reader's convenience.

This assumption is consistent with the assumption of within-day stationarity that underlies this chapter. In this representation, a path corresponds to a complete trip.

As is the case for modeling other choice contexts, complete specification of a path choice model for scheduled service networks involves three phases: definition of choice alternatives, identification of the set of alternatives, and specification of the model that predicts the choice among alternatives. This in turn implies selection of the attributes and systematic utility of the alternatives as well as the functional form of the choice model.

Definition of Choice Alternatives For high-frequency transportation services, it is unrealistic to assume that the only things that the user considers as pre-trip choice alternatives are the elementary paths on the graph that represents the service lines. If this were the case, a user would consider the paths defined by each of the lines connecting a given pair of stops to be different and mutually exclusive, even when these lines provide equivalent service. Consider a user traveling in the network represented by the graph in Fig. 4.15. If the user chose path b shown in Fig. 4.16 and line 5 belonging to it, he would, on arrival at stop F, refuse to board a vehicle of line 6 that happened to arrive at the stop earlier than a vehicle of line 5, despite the fact that the two lines are completely equivalent. This is not realistic.

To overcome these potential problems, a path choice model should allow for the possibility that users' pre-trip choice alternatives include multiple equivalent lines or, put differently, multiple paths on the graph that represents the public transportation services. The basic assumption in the definition of choice alternatives is that, prior to their trips, users of high-frequency transit systems do not have complete information on the service options that will be available. For example, users may

[18] Route choice for regular, low-frequency scheduled services with explicit run representation is usually assumed to be completely pre-trip and the models that represent it are analogous to those described for road networks. In this case, however, the choice alternatives are the single runs or sequences of runs that can be represented as paths on the diachronic network. This point is dealt with extensively in Chap. 7.

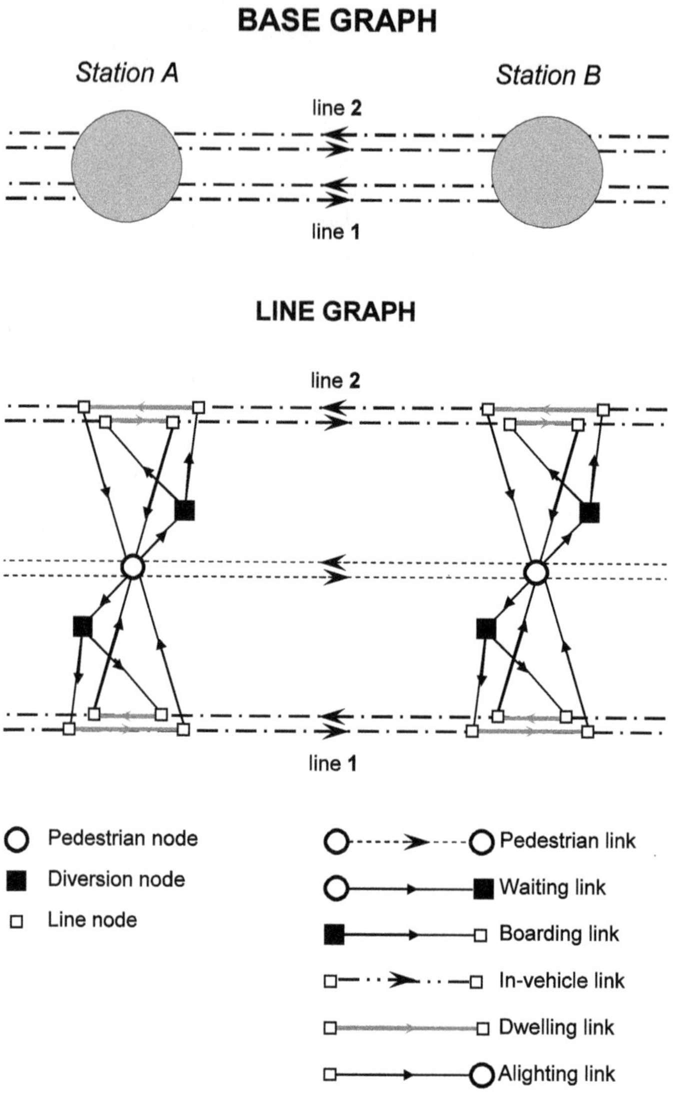

Fig. 4.14 Line-based representation of a scheduled transportation system

be unable to predict their arrival times at stops or the arrival times of the vehicles (trains, buses, etc.) on the different lines that call at each stop.

Under this hypothesis, it is assumed that the departing user chooses a *travel strategy* rather than a predetermined path. A strategy is a set of pre-defined travel alternatives together with decision rules that the user applies to select one of them in response to random or unknown events that may arise during the trip. In the example given in Fig. 4.15, one strategy could be to go to stop F and board the first

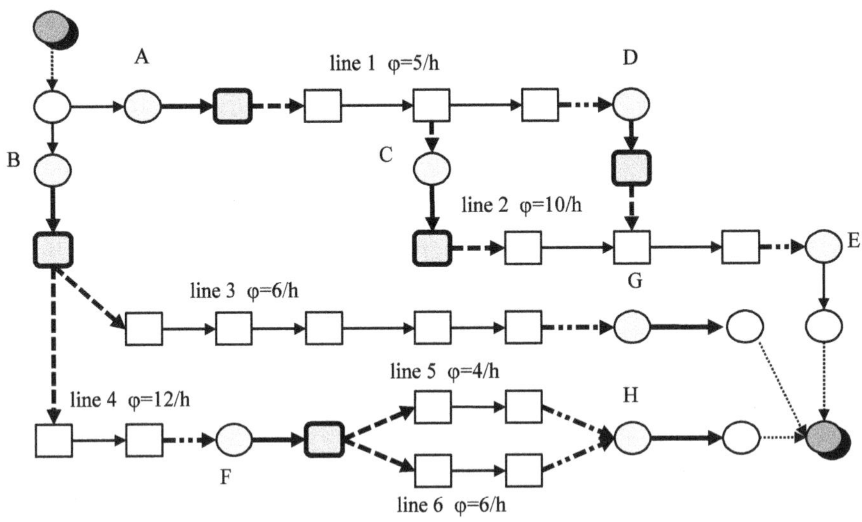

Fig. 4.15 Example of a transit line-based network

vehicle belonging to line 5 or 6; another possible strategy could be to go to stop F and board vehicles of line 5 only.

Two types of choice behavior are involved in choosing a path under the above assumptions.

En-route choice behavior underlies user choices made during the trip. This behavior describes how users respond to unknown or unpredictable events. The type of adaptive choice behavior and the set of alternatives to which it is applied define a strategy.

Pre-trip choice behavior underlies user choices made before departure. It includes the comparison of possible alternative strategies and the choice of one of them based on its characteristics or attributes, for example, its perceived average trip cost. Pre-trip choices are analogous to those made for path choice in continuous service networks and, in general, for choices in other dimensions.

It follows that the definition of choice alternatives (strategies) for high-frequency transit systems requires assumptions about en-route behavior. Usually it is assumed that en-route choices take place at diversion nodes (stops) m, and that the en-route decision rule is to board the first vehicle arriving on any of a given set of lines AL_m, called the set of *attractive lines*.[19] The choice of boarding and alighting stops, on the

[19]More complex rules of en-route behavior have been proposed. For example, the user may be assumed to decide between boarding the first arriving vehicle, or waiting for a vehicle of another line, based on a comparison of the expected cost of the different options, and given the information available at that moment. Such models of en-route behavior require a great deal of information and get close to the microsimulation of network journeys; for these reasons, they are not typically used for demand assignment to large scale networks. Models of this type are described in Chap. 7 for irregular scheduled services.

other hand, is assumed to be made pre-trip. To continue the example of Fig. 4.15, a strategy cannot include the option of alighting at stop C or at stop D of line 1, because it is assumed that these are pre-trip choices. This means that there are no events unknown to the user that would require a decision between either stop. Analogously, a strategy cannot include moving to stop A to take line 1 or to stop B to take line 4.

If it is assumed that user and vehicle arrivals at stops can be modeled as Poisson random processes with constant probability of arrival at any time, the probability of a boarding line l belonging to the set AL_m of attractive lines at stop m can be expressed as

$$Pr[l/m, AL_m] = \varphi_l \Big/ \sum_{n \in AL_m} \varphi_n \tag{4.3.27}$$

where φ_l represents the frequency (number of arrivals/time unit) of line l. Expression (4.3.27) also holds under an assumption of Poisson user arrivals and of deterministic equally spaced arrivals of vehicles on the lines belonging to AL_m.

In terms of the line-based graph, a travel strategy (i.e., a pre-trip adaptive choice alternative) can be represented by a subgraph known as a *hyperpath*. Elementary paths are possible strategies: they are strategies that do not include adaptive choices and are considered *simple hyperpaths*. Strategies that include one or more en-route stops with adaptive choices made there can be represented as the union of simple hyperpaths, having the property that multiple links emanate only from diversion nodes.[20] These subgraphs are known as *composite hyperpaths*. Figure 4.16 enumerates all the hyperpaths of the line network in Fig. 4.15.

Each diversion node m of hyperpath j corresponds to a set AL_{mj} of attractive lines belonging to that hyperpath. A *diversion probability* η_{lj} can be associated with the boarding links $l \equiv (m, n)$ that connect the diversion node m to the nodes n of the lines in AL_{mj}. This is the probability, expressed by (4.3.27), of using the line corresponding to link l of hyperpath j as a result of the random events that affect enroute choices:

$$\begin{aligned} \eta_{l,j} &= pr\big[l = (m,n)/m, AL_{mj}\big] \\ &= \varphi_l \Big/ \sum_{n \in AL_{m,j}} \varphi_n \quad \text{if } l \in AL_{mj} \text{ boarding link} \end{aligned} \tag{4.3.28}$$

Typically a diversion probability of one is assigned to all nonboarding links belonging to the hyperpath:

$$\eta_{lj} = 1 \quad \text{if } l \in j, l \text{ nonboarding link}$$

[20] A link emanating from a diversion node represents boarding a line serving the corresponding stop, as defined in Chap. 2 and represented graphically in Fig. 4.14. For a formal definition of a hyperpath in terms of graph variables, see Sect. 6.2.

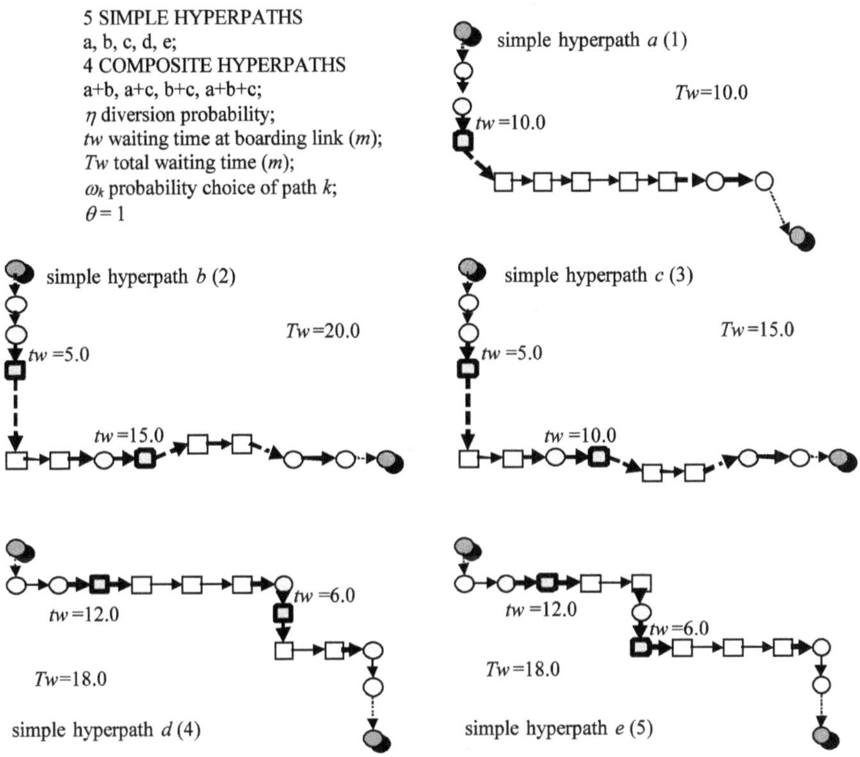

Fig. 4.16 **a** Enumeration of simple hyperpaths for the transit network of Fig. 4.15. **b** Enumeration of composite hyperpaths for the transit network of Fig. 4.15

and a zero probability is assigned to the links not belonging to hyperpath j:

$$\eta_{lj} = 0 \quad \text{if } l \notin j$$

For example, the diversion set AL_{m6} corresponding to diversion node m in composite hyperpath 6 in Fig. 4.16b consists of lines 3 and 4: $AL_{m6} = \{3, 4\}$ and the diversion probability of boarding link l on line 3 can be calculated as

$$\eta_{l6} = \varphi_3/(\varphi_3 + \varphi_4) = 6/18 = 0.33$$

Using the diversion probabilities η_{lj}, the probability ω_{kj} of following path k of hyperpath j during a given trip can be determined. Assuming statistical independence of the random events underlying en-route choices, the probability of following path k within hyperpath j is equal to the product of the diversion probabilities for all links l belonging to path k; that is,

$$\omega_{kj} = \prod_{l \in k} \eta_{l,j} \tag{4.3.29}$$

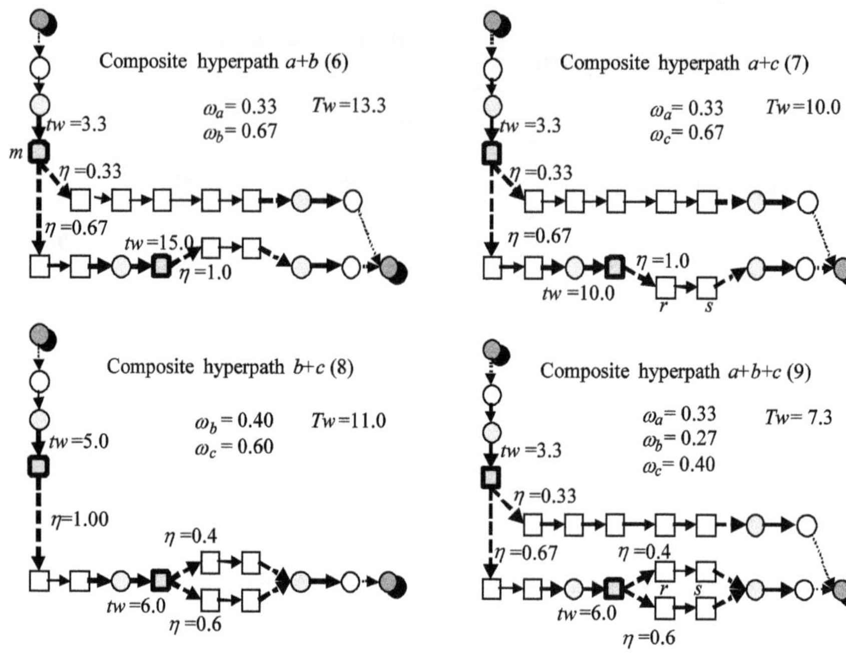

Fig. 4.16 (continued)

which yields:

$$\omega_{kj} = 0 \quad k \notin j$$

This probability is obviously equal to one if path k coincides with (simple) hyperpath j. Continuing with the previous example, the probability ω_{a6} of following path a within hyperpath 6 is equal to 0.33; the probability of following the same path within another hyperpath is different, for example $\omega_{a1} = 1$, $\omega_{a2} = 0$, and so on. Note that a path may belong to more than one hyperpath.

The probability λ_{lj} of traversing a link l of hyperpath j can also be calculated as the sum of the probabilities of following any of the paths k on hyperpath j that includes link l:

$$\lambda_{lj} = \sum_{k:l \in k} \omega_{kj} = \sum_{k} \delta_{lk} \omega_{kj} \tag{4.3.30}$$

where δ_{lk} is an element of the link–path incidence matrix. This yields:

$$\lambda_{lj} = 0 \quad \text{if } l \notin j$$

Continuing with the example in Fig. 4.16b, the probability of traversing all the links belonging to path b in hyperpath 2 is equal to one; the probability of traversing link (r, s) is equal to 0.67 in hyperpath 7 and to 0.40 in hyperpath 9. A user choosing a given strategy (or a hyperpath representing it) does not know before starting the trip which path and therefore which lines and links she will travel on because these depend on random events such as the sequence of vehicle arrivals at each stop. On different trips, the same user following the same strategy might use different lines, paths and links with probabilities given by (4.3.28), (4.3.29), and (4.3.30), respectively. Furthermore, on each trip she will experience different travel times and, in general, different costs. However, the expected value of these times and costs can be expressed as a function of the probabilities ω_{kj}, as shown shortly.

Definition of Choice Set Once choice alternatives (strategies and hyperpaths) have been defined, the issue of choice set definition can be considered. As was discussed for path choice on road networks, two general approaches can be followed to identify the set of feasible choice alternatives. In the *exhaustive approach*, all strategies (or the hyperpaths that represent them) are feasible. This approach is typically associated with implicit enumeration of the hyperpaths. In the *selective approach*, only the hyperpaths that satisfy certain conditions are feasible. For example, hyperpaths including paths with more than one transfer may be excluded from the choice set if there are direct paths and hyperpaths. In applications, the most commonly used approach is the exhaustive one, given the computational complexity associated with the explicit enumeration of hyperpaths.

Functional Form Specification of the choice model requires selection of the attributes and of the functional form of the random utility model. Let $J_{od,m}$ be the set of hyperpaths connecting the pair o, d on the network of the scheduled service transit mode (or modes) m. It is assumed that the perceived utility U_j of each hyperpath j belonging to $J_{od,m}$ has a negative systematic utility V_j equal to the mean cost x_j of the hyperpath:

$$U_j = V_j + \varepsilon_j = -x_j + \varepsilon_j \quad \forall j \in J_{odm} \tag{4.3.31}$$

The average cost of hyperpath x_j can be expressed as the sum of an additive part x_j^{ADD} and a nonadditive part x_j^{NA} that, in this case (and unlike that of path costs on continuous service networks), is always present:

$$x_j = x_j^{\mathrm{ADD}} + x_j^{\mathrm{NA}} \tag{4.3.32}$$

The additive cost x_j^{ADD} is a linear combination of the attributes (typically in-vehicle, boarding, alighting, dwelling, and access/egress times) associated with the nonwaiting links belonging to the hyperpath:

$$x_j^{\mathrm{ADD}} = \beta_b Tb_j + \beta_{br} Tbr_j + \beta_{al} Tal_j + \beta_d Td_j + \beta_a Ta_j \tag{4.3.33}$$

where the βs are the respective coefficients.

This cost can be obtained from the generalized costs of the individual links c_l and the probabilities of traversing the single links (λ_{lj}), or equivalently from the additive path costs g_k^{ADD} and the probabilities of following these paths ω_{kj}:

$$x_j^{\mathrm{ADD}} = \sum_k \omega_{kj} g_k^{\mathrm{ADD}} = \sum_k \omega_{kj} \left(\sum_{l \in k} c_l \right) = \sum_l \lambda_{lj} c_l \tag{4.3.34}$$

The nonadditive cost can be expressed as the sum of the waiting times (costs) Tw_j, as well as any further nonadditive costs, that is, costs that cannot be associated with single links such as fixed fares or transfer costs N_j.

$$x_j^{\text{NA}} = \beta_w Tw_j + \beta_N N_j$$

where β_w and β_N are the equivalent costs of the different nonadditive cost items (computed from their marginal rates of substitution with respect to cost).

The average waiting time (cost) Tw_j connected with hyperpath j can be calculated starting from waiting times tw_{lj} associated with each waiting link l that enters diversion node m; as discussed in Sect. 2.4.2.2, this can be expressed as

$$tw_{lj} = \begin{cases} \theta / \sum_{n \in AL_{m,j}} \varphi_n & \text{if } l \text{ is a diversion link} \\ 0 & \text{otherwise} \end{cases} \tag{4.3.35}$$

where θ is a parameter taking values from the interval [0.5–1], depending on the probability laws of user and vehicle arrivals (see Sect. 2.4.2.2).

The average total waiting cost Tw_j associated with hyperpath j can be expressed as

$$Tw_j = \sum_{k \in j} \omega_{kj} \left[\sum_{l \in k} tw_{lj} \right] = \sum_{l} \lambda_{lj} tw_{lj} \tag{4.3.36}$$

From (4.3.35) it follows that the waiting time tw_{lj} for diversion link l depends on the hyperpath, and therefore that the total waiting time Tw_j cannot be expressed as a linear combination of link attributes independently of the hyperpath; it is therefore a nonadditive hyperpath attribute.

The model of choice among alternative hyperpaths can be expressed formally as the probability q_j that hyperpath j has maximum perceived utility:

$$q_j = Pr[-x_j + \varepsilon_j \geq -x_{j'} + \varepsilon_{j'}] \quad \forall j', j, j' \in J_{od} \tag{4.3.37}$$

For hyperpath choice models there are again two possible approaches. The deterministic choice approach ($\text{Var}[\varepsilon_j] = 0$) assigns all the demand to the minimum generalized cost hyperpath(s). In contrast, the random utility approach, typically based on logit and probit forms, assigns a positive choice probability to all hyperpaths in the choice set. When applying the MNL model to hyperpath choice, however, the problems resulting from the IIA property are even more significant than in its applications to path choice because hyperpaths typically include a large number of overlapping lines. Alternatively, it is possible to use a probit model with a variance–covariance matrix structure similar to that for paths on road networks.

There are currently no examples in the literature of hyperpath choice models calibrated and validated from observed behavior; this can be explained at least in part by the difficulty of obtaining information on the alternatives (hyperpaths) chosen by users.

Finally, once the hyperpath choice probabilities have been calculated, it is possible to obtain path probabilities:

$$p[k/osdm] = \sum_j \omega_{kj} q_j \tag{4.3.38}$$

4.3.4 A System of Demand Models

This section presents the system of interurban passenger trip-demand models developed and used in the Information System for Transportation Monitoring and Planning in Italy (SIMPT). The system, presented schematically in Fig. 4.17, includes models for mobility choices (individual holding of driver's license and household automobile ownership) and partial share trip-demand models.

All of the models have a logit specification and the sequence of frequency/distribution/mode choice models has a three-level hierarchical logit structure with EMPUs that take into account the influence of "lower" choice dimensions on "upper" levels, as described in Sect. 4.2. The individual submodels and their variables are briefly described below.

The *driver's license holding model* (Fig. 4.18) is a binomial logit with license possession or nonpossession alternatives for each individual in a household. Its systematic utility attributes include the socioeconomic characteristics of the individual (age, gender, and professional status) and the household (income). The urbanization level of the residence zone is also significant. Densely urbanized zones usually have a more efficient public transportation system and provide better accessibility to various urban functions, reducing the need to use a car. The coefficients indicate that factors such as gender, age, professional status, and family income have a significant effect on license possession. Furthermore, it can be observed that the coefficients of socioeconomic variables that describe gender and age (women 18–48 and women > 48) are positive and increasing in the systematic utility of not holding a license. This result can be interpreted as an indicator of the delay with which the female population has gained access to car use, even though this gap is closing for younger generations.

The *car ownership model* (Fig. 4.19) predicts the choice of the number of cars owned in a household. The model is a trinomial logit, with alternatives 0, 1, 2, or more cars. The significant attributes are again household socioeconomic variables such as income, number of license holders, number of workers, and of students. The urbanization level of the residence zone reduces the utility (and the probability) of owning two or more cars, confirming the interpretation given for this variable in the license possession model.

The *trip-demand model system* estimates the average number $d^i_{od}[s,h,m,k]$ of interprovincial round trips undertaken by an individual i between the zones of residence o and destination d, for purpose s, in time period h, with mode m and

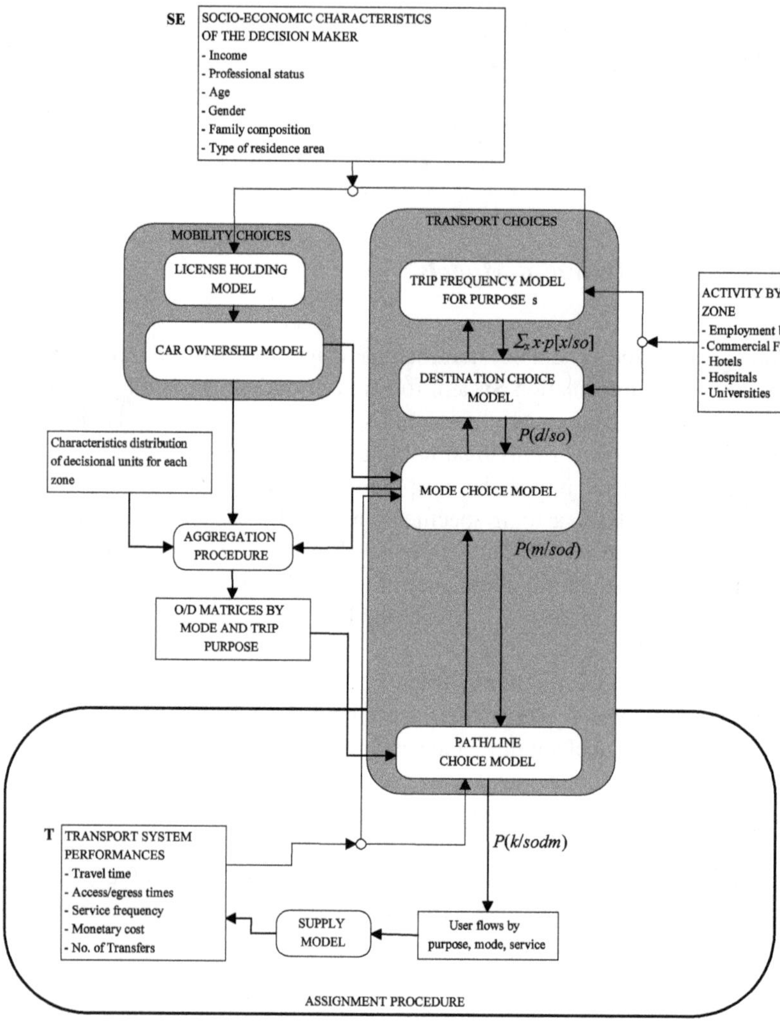

Fig. 4.17 Structure of a model system for interurban trip demand

path k:

$$d^i_{od}[s,h,m,k] = \sum_x x p^i[x/osh](\boldsymbol{SE},\boldsymbol{T}) \cdot p^i[d/osh](\boldsymbol{SE},\boldsymbol{T})$$

$$\cdot\, p^i[m/oshd](\boldsymbol{SE},\boldsymbol{T}) \cdot p^i[k/oshdm](\boldsymbol{SE},\boldsymbol{T}) \quad (4.3.39)$$

where

$p^i[x/osh]$ is the probability that individual i undertakes x interprovincial trips for purpose s in period h, obtained with the trip frequency model

$\rho^2 = 0.437$	Age 18–24 (0/1)	Age 25–56 (0/1)	Employed (0/1)	Average income 40–80 ml	High income > 80 ml	Dense urban zone (0/1)	Woman 18–48 (0/1)	Woman > 48 (0/1)	Asa
License	0.173	1.146	1.279	0.716	1.229				
t	2.1	16.4	19.5	10.2	5.9				
No license						0.262	1.197	2.384	−1.022
t						5.1	17.1	34.9	−16.2

Fig. 4.18 License holding model

$\rho^2 = 0.376$	ASA	No of workers	No of univ. stud.	Family head (0/1)	Average income 40–80 ml	High income > 80 ml	Dense urban zone (0/1)	No of licenses
0 cars	−1.33	−1.44	−0.99	−0.73				
t	−13.5	−17.2	−4.3	−7.2				
1 car	−0.48							1.06
t	−24.6							27.3
2 or more cars					1.01	1.53	−0.56	
t					12.4	6.1	−6.9	

Fig. 4.19 Car ownership model

$p^i[d/osh]$ is the probability of choosing destination d, obtained with the distribution model

$p^i[m/oshd]$ is the probability of choosing mode m, obtained with the mode choice model

$p^i[k/oshdm]$ is the probability of choosing path k in the mode m network, obtained with the path choice model

Five travel purposes are considered: commuting, professional business, study, recreation, and tourism, and other purposes.

The *trip-frequency model* $p^i[x/osh]$ has a logit structure with three alternatives: no trips, one trip, and more than one trip in the reference time period h (two winter weeks). The average number of trips undertaken by each individual is therefore obtained as a weighted average of the number of trips corresponding to each frequency class (respectively, zero, one, and the average number estimated by the sample) with weights given by the probability of choosing each frequency class (see (4.3.3)). The attributes in the systematic utility functions are the socioeconomic characteristics of the household (income level, number of members and cars in the household) and of the traveler (age group, professional status, license possession) and the inclusive utility associated with destination choice $[Y_o^i = \ln \sum_d \exp(V_{od}^i)]$. Because the model expresses the probability of undertaking journeys outside the province of residence, it includes a "self-attractivity" variable (e.g., total employment in the province) in the systematic utility of the no-trip alternative. This variable reflects the relatively small need to carry out activities outside the province for individuals who, other things equal, live in areas with more opportunities satisfying their needs. The accessibility variable in the utility of making one or more round-trips has a positive

$\rho^2 = 0.7061$		Total employment ($\times 10^6$) in zone O	Accessibility Y_o^i	Average income (40–80 ml)	High income > 80 ml	Male (0/1)	Manager (0/1)	ASA
0 journeys		0.11						
	t	4.8						
1 journey			0.14	0.61	1.53	0.96	0.33	−4.80
	t		2.3	5.3	7.2	4.9	10.2	−13.5
2 or more journeys			0.14	0.61	1.53	2.34	1.47	−5.592
	t		2.3			4.9	11.3	−14.5

Fig. 4.20 Travel frequency model for professional business purpose

$\rho^2 = 0.3129$	Y_{od}^i	Service employment X_{1d} ($\times 10^3$)	Size	Same region (0/1)
	0.334	1.000	0.913	1.787
t	61.3	–	13.8	42.3

Fig. 4.21 Destination choice model for professional business purpose

coefficient between zero and one, consistent with the behavioral interpretation of the hierarchical logit model. Figure 4.20 shows as an example the attributes and the coefficients calibrated for the professional business purpose trip-frequency model.

The *distribution model* $p^i[d/osh]$ has a multinomial logit specification. Its systematic utility includes the mode choice logsum variable Y_{od}^i to capture the (inverse) separation between two zones. In order to account for the unknown number of elementary destinations in each zone, size functions are used as zone attractiveness attributes (see Sect. 4.3.1.2). In summary, the utility function of the distribution model can be expressed as

$$V_{od}^i = \beta_1 Y_{od}^i + \beta_2 \ln\left(X_{1d}^i + \sum_{k=2}^{Ks} \beta_k X_{kd}^i \right) + \sum_{k=Ks+1}^{K} \beta_k X_{kd}^i$$

$$\text{with } Y_{od}^i = \ln \sum_m \exp\left(V_{odm}^i\right)$$

where the third term includes all the attributes common to the elementary destinations included in d, for example, a "same region" dummy variable introduced to represent the greater attractiveness, other attributes equal, of zones belonging to the same region. The size functions differ by trip purpose and include variables such as service and commerce employment, number of tourist facilities, and the like.

In the example presented in Fig. 4.21 for professional business trips, service employment is used in the size function as an indicator of the number of elementary destinations included in each zone. The coefficient of the logsum variable Y_{od}^i lies in the interval [0, 1].

The *mode choice model* $p^i[m/oshd]$ is a multinomial logit with six mode or service alternatives: car, bus, air, slow train (interregional, express), fast train (intercity), and night train. The (generic) attributes considered for each mode are total travel time and monetary cost. There are two different coefficients for monetary cost, one for low-income users and the other for medium- to high-income users. This accounts for the different willingness to pay and value of time of users with different incomes, as described in Sect. 4.3.2. The values of time (VOT) perceived by low-income and medium- to high-income users were found to be significantly different. In the example presented in Fig. 4.22 for the professional business purpose, the VOT is approximately 5.5 Euros per hour for low-income travelers and 12.5 Euros per hour for medium- to high-income travelers. For recreation and tourism and for other trip purposes, the VOT differences are less dramatic: for medium- to high-income individuals, the value of time is approximately 50% higher than for low-income travelers.

Other level-of-service attributes are also included in the model, such as the number of transfers and the average headway for scheduled modes/services. These modes also include a dummy variable equal to one if the destination zone is not a medium or large city. The negative coefficient of this variable can be interpreted as an (aggregated) measure of the difficulty of reaching the final destination from the service terminal (e.g., station) in low-density zones, due to less extensive local public transportation services. Finally, the model specification includes car availability (number of cars divided by the number of licensed drivers in the household) as a socioeconomic variable linked to the availability of that alternative.

The *path choice model* for the road network $p^i[k/oshdm]$ is also a multinomial logit model; the choice alternatives are obtained through an explicit path enumeration technique that eliminates heavily overlapping paths. The variables used measure level-of-service exclusively. Path choice predictions for scheduled service networks (slow train, fast train, bus, and air) are made by applying a logit model to a choice set of hyperpaths that are explicitly enumerated on the line-based network with heuristic feasibility rules. Path choice models are applied to origin–destination matrices by mode and trip purpose; these are obtained with the aggregation technique described below.

The *aggregation procedure* estimates aggregate origin–destination demand flows starting from individual representative trips. Because the models described involve multiple socioeconomic variables at the individual and household level, it would not be feasible to identify user classes characterized by equal values of these attributes. The aggregation procedure is based on the sample enumeration technique described in Sect. 3.7 with the identification of a representative sample of individuals and households and the application of zonal expansion factors calculated to match zonal values of aggregate target variables.

4.4 Trip-Chaining Demand Models*

As was stated in Sect. 4.1, traditional travel-demand models represent the trips that comprise a journey (a sequence of trips starting and ending at home) assuming that the decisions (choices) made for each trip are independent of those made for other trips in the same journey. It was also noted that these assumptions are reasonable

$\rho^2 = 0.758$	Time [h]	Mon. cost low inc. [€]	Mon. cost med-high [€]	Car avail.	Nonurban destin. (0/1)	No. of transf.	Time headway [h]	Asa Train IR	IC	Nite	Air	Bus
Car	−1.23	−0.22	−0.098	3.81								
Interregional	−1.23	−0.22	−0.098		−3.72	−0.97	−0.60	0.95				
Interurban	−1.23	−0.22	−0.098		−3.72	−0.97	−0.60		−0.54			
Night	−1.23	−0.22	−0.098		−3.72	−0.97	−0.60			9.96		
Air	−1.23	−0.22	−0.098		−3.72	−0.97	−0.60				−1.62	
Bus	−1.23	−0.22	−0.098		−3.72	−0.97	−0.60					−2.31
t	−26.2	−5.4	−15.7	30.3	−18.0	−5.4	−24.0	−0.6	−4.4	3.6	−12.7	−14.4

Fig. 4.22 Mode and service choice model: for professional business

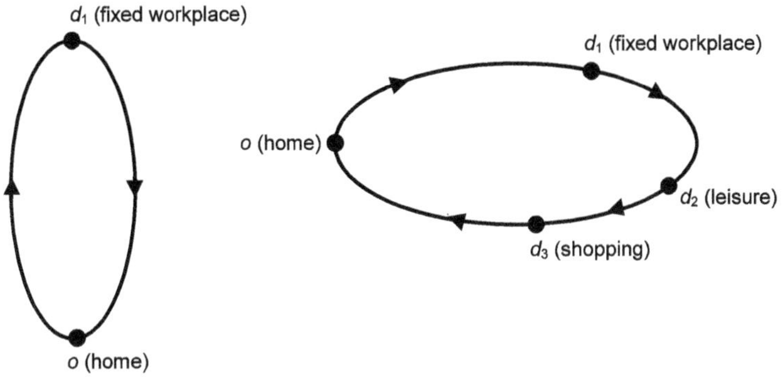

Fig. 4.23 Examples of round-trip and chain journeys

when the journey is a "round-trip" with a single destination and two symmetric trips.

However, human activities have become increasingly complex, especially in urban areas. One reflection of this in the domain of transportation is an increasing number of journeys that connect multiple and disparate activities in different locations, that is, journeys consisting of sequences of trips that influence each other in complex ways (Fig. 4.23). For example, if a personal car is not used for the first trip in a journey, it will not be available for subsequent trips either. A number of demand models have been proposed in the literature to address the sequence, or *chain*, of trips making up a journey. Some of these models represent the activities carried out (i.e., the different purposes of the journey) together with the trips that link them.

The mathematical models proposed to represent trip or activity chains do not have a standard structure as, for example, trip demand models do. This is due both to the relatively recent interest in these models (so there are fewer examples of them), and to the greater complexity of the phenomenon to be represented.

However, the most commonly used modeling structure, and the one closest to the structure described in the previous sections for single trips, is based on the concept of a *primary activity* (*destination*) for a particular journey. In other words, it is assumed that each journey is associated with a primary activity (or purpose), and that this activity is conducted in a particular place, known as the primary destination. Experimental studies suggest that the activity that the user perceives as primary for a particular journey is determined by relatively few criteria. These include:

- Hierarchical level of purpose (in decreasing order, workplace or study, services and professional business, other purposes)
- Duration of the activity (the primary activity is that which, within the highest hierarchical level, takes the most time)
- Distance from zone of residence (the primary activity, given the same hierarchical level and duration, is that which is carried out in the place farthest from the residence)

Adopting this definition, a system of demand models for trip sequences (journeys) can be specified with a partial share structure analogous to the standard four-step model described in Sect. 4.2. To avoid excessively complicated notation, it is assumed here that the journeys have at most two destinations (see Fig. 4.24). One of the possible partial share structures for trip chaining is the following.

$$\begin{aligned} d^i_{od_1d_2o}[s_1h_1m_1s_2h_2m_2h_3m_3] = n^i[o]p^i[x=1/os_1h_1](\boldsymbol{SE},\boldsymbol{T}) \\ \cdot\, p^i[d_1/os_1h_1](\boldsymbol{SE},\boldsymbol{T}) \\ \cdot\, p^i[s_2h_2/osh_1d_1](\boldsymbol{SE},\boldsymbol{T}) \\ \cdot\, p^i[d_2/os_1h_1d_1s_2h_2](\boldsymbol{SE},\boldsymbol{T}) \\ \cdot\, p^i[h_3/os_1h_1d_1s_2h_2d_2](\boldsymbol{SE},\boldsymbol{T}) \\ \cdot\, p^i[m_1m_2m_3/os_1h_1d_1s_2h_2d_2h_3](\boldsymbol{SE},\boldsymbol{T}) \end{aligned} \quad (4.4.1)$$

where

$d^i_{od_1d_2o}[s_1m_1h_1s_2m_2h_2m_3h_3]$ is the average number of journeys with origin in zone o undertaken by users of class i and composed of trips for primary activity s_1, carried out in zone d_1 in time period h_1, and secondary activity s_2, carried out in zone d_2 in time period h_2, and returning home in the time period h_3; these trips are undertaken with modes m_1, m_2, and m_3, respectively. Round-trips are a special case in which s_2 is the return trip home, d_2 coincides with the origin, and m_3 and h_3 are not meaningful

$p^i[x=1/os_1h_1](\boldsymbol{SE},\boldsymbol{T})$ is the frequency model expressing the probability that an individual of class i living in zone o undertakes a journey[21] for primary purpose s_1 in time period h_1

$p^i[d_1/os_1h_1](\boldsymbol{SE},\boldsymbol{T})$ is the primary destination choice model; it gives the probability that the journey for primary purpose s_1 undertaken in time period h_1 by individuals of class i in zone o has its primary destination in zone d_1

$p^i[s_2h_2/os_1h_1d_1](\boldsymbol{SE},\boldsymbol{T})$ is the journey type model; it gives the probability of undertaking a trip for a secondary purpose s_2 (which may or may not involve a secondary activity) in time period h_2 for a user of class i who has decided to undertake a primary journey in d_1 in time period h_1. Note that the time period h_2 may be before or after h_1; that is, the secondary destination may be reached before or after the primary one, as indicated in Fig. 4.24. Furthermore, if a trip is not undertaken for a secondary purpose, the journey is a round-trip and s_2 is the "return home" purpose

$p^i[d_2/os_1h_1d_1s_2h_2](\boldsymbol{SE},\boldsymbol{T})$ is the secondary destination choice model, expressing the probability of choosing zone d_2 to carry out activity s_2, if this is not the

[21] It is assumed that h_1 is defined such that the probability of undertaking more than one journey for the same purpose in the same time period is negligible.

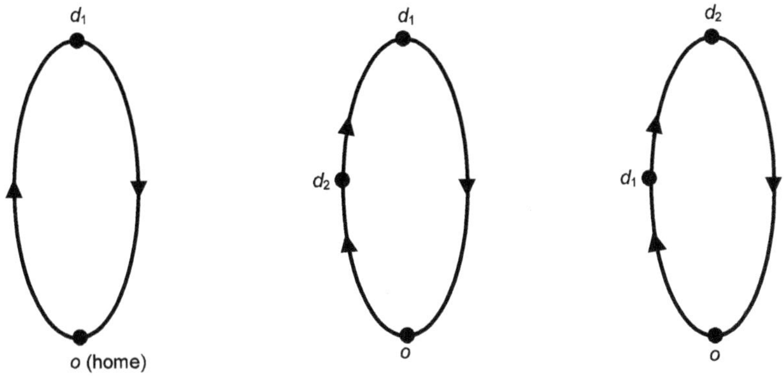

Fig. 4.24 Types of journey simulated by the model (4.4.1)

return trip home, in time period h_2 for a user who is undertaking a journey for primary purpose (activity) s_1 in zone d_1 in time period h_1. This model is obviously meaningless if the journey is a round-trip

$p^i[h_3/os_1h_1d_1s_2h_2d_2](\boldsymbol{SE}, \boldsymbol{T})$ is the return home time period distribution model; it gives the probability of returning home in time period h_3, conditional on all the elements that define the chain $(os_1h_1d_1s_2d_2h_2)$ or round-trip (os_1d_1) journey

$p^i[m_1m_2m_3/os_1h_1d_1s_2h_2d_2h_3](\boldsymbol{SE}, \boldsymbol{T})$ is the mode sequence choice model for the entire sequence of trips conditional on the elements defining it. Note that all mode choices are modeled simultaneously to take into account consistency constraints between successive trips. Some modes (in particular private modes) are available for later trips only if they have been used in the first trip

In all of the above, the parameters $\boldsymbol{SE}$ and $\boldsymbol{T}$ denote, as usual, the vectors of socioeconomic and level-of-service attributes included in the models.

Path choice models are equivalent to those described in Sect. 4.3.3. It is usually assumed that the probability of choosing a certain path depends exclusively on the origin–destination pair, the mode, and the time period of each single trip, and is not influenced by other trips within the same journey. For this reason, they are not presented here in order to simplify the analytical formulation.

Figure 4.25 is the graphical representation of the structure of the model systems described here. It can be observed that, just as in trip-demand model systems, some choice dimensions are conditional on others; for example, the journey type depends on the primary destination, and the secondary destination depends on the journey type and primary destination. Upper choice dimensions take into account the lower ones through EMPU variables that are represented by dotted arrows in Fig. 4.25. In the figure, some models in expression (4.4.1) have been further factored into the product of two models. In particular, the trip frequency models (primary, secondary, and return home) in a certain time period have been factored into the product of the probability of undertaking the trip and the probability of choosing a certain time

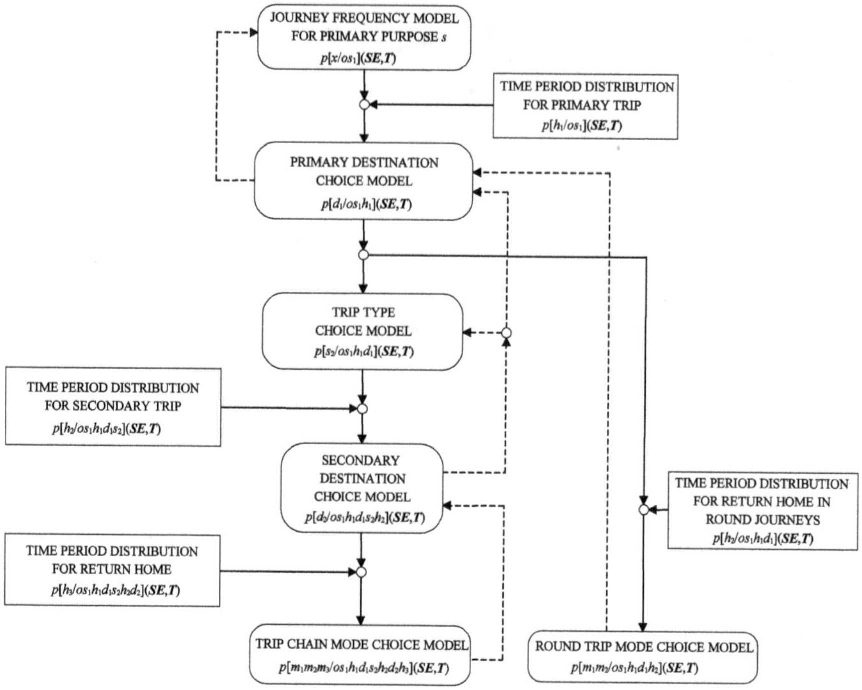

Fig. 4.25 Structure of a trip-chaining model system

period. The probability of returning home is assumed to be equal to one and is therefore not modeled.

Different specifications of the whole sequence as well as of individual models can be adopted within the partial share structure. A simplified model system that represents trip-chaining travel demand in urban areas is given here as an example.

The overall model is a hierarchical logit, with inclusive logsum variables linking the different choice dimensions; however, the distribution of trips (activities) in time periods h_1, h_2, and h_3 is assumed to be given. The system considers four possible primary purposes: work, study, other purposes constrained by destination (professional business, personal services, medical treatment, etc.), and other purposes not constrained by destination (shopping, recreational, other purposes).

The main models for primary purpose "other nonconstrained" are given below. The mode choice model is not included inasmuch as it is analogous to those described in previous sections, the only significant difference being that the choice alternatives are not single modes or services but rather feasible combinations of them, where feasibility is determined by the journey structure. For round-trip journeys, it is assumed that the return mode is the same as the outward mode; for chain journeys, it is assumed that if a car or motorcycle is used for the first trip, it must be used for the next two; but all combinations of walking and public transportation modes are possible.

	Y_{os_1}	*EMP*	*HSWF*	*STU*	*RETIRED*	*NOJOURNEY*
	0.1904	−0.5879	0.06948	0.5017	0.3607	0.2795
t	14.6	−26.2	3.10	12.7	18.6	8.30

Fig. 4.26 Parameters of the journey frequency model for nonconstrained other purposes

Journey frequency model $p^i[x/os_1h_1](\boldsymbol{SE},\boldsymbol{T})$. The journey frequency model is a binomial logit with systematic utilities of the two alternatives (to undertake or not a journey for the primary purpose) given by:

$$\begin{aligned} V^i_{\text{journey } s_1} &= \beta_1 Y^i_{os_1} + \beta_2 EMP + \beta_3 HSWF + \beta_4 STU + \beta_5 OTHER \\ V^i_{\text{Nojourney } s_1} &= \beta_6 Nojourney \end{aligned} \tag{4.4.2}$$

where

$Y^i_{os_1}$ $\ln\sum_{d_1}\exp(V^i_{os_1d_1})$ is the logsum variable corresponding to the primary destination choice for purpose s_1; it represents the accessibility of the residence zone with respect to all the possible destinations where the primary activity can be conducted

EMP is a dummy variable, equal to one if the individual is employed, zero otherwise

HSWF is a dummy variable, equal to one if the individual is a housewife, zero otherwise

STU is a dummy variable, equal to one if the individual is a high school or university student, zero otherwise

RETIRED is a dummy variable, equal to one if the individual is retired, zero otherwise

NOJOURNEY is the alternative specific attribute (ASA) of not undertaking a journey for primary purpose s_1

Figure 4.26 presents the parameters calibrated for model (4.4.2) for an average weekday. Accessibility of the residence zone increases the probability of undertaking the journey and the logsum inclusive variable has a coefficient in the interval (0, 1). The occupational status (category) of the individual considerably influences the probability of undertaking journeys for nonconstrained other purposes; employed individuals in particular show less utility for these trips compared with other purposes, other things equal, probably because of their reduced time available.

Primary destination choice model $p^i[d_1/os_1h_1](\boldsymbol{SE},\boldsymbol{T})$. The primary destination choice model is a multinomial logit with a systematic utility function of the form:

$$V^i_{od_1s_1h_1} = \beta_1 Y^i_{od_1h_1} + \beta_2 SZ_{d_1/o} + \beta_3 \ln(EMPret_{d_1} + \beta_4 EMPserv_{d_1}) \tag{4.4.3}$$

where

	$Y_{od_1h_1}$	$SZ_{d_1/o}$	$Size$	$EMPret_{d_1}(10^3)$	$EMPserv_{d_1}(10^3)$
	1.428	1.003	0.7725	1.000	0.065
t	19.1	9.70	19.4	–	2.73

Fig. 4.27 Parameters of the primary destination choice model for other unconstrained purposes

$Y^i_{od_1h_1}$ $\ln\sum_m \exp(V^i_{od_1mh_1})$ is the mode choice logsum variable, which accounts for the (dis)utility for user class i of moving from o to d_1 in departure interval h_1 using the available transportation modes

$SZ_{d_1/o}$ is a dummy variable equal to one if the zone d_1 is the residence zone o, zero otherwise

$EMPret_{d_1}$, $EMPserv_{d_1}$ are the total employment in the retail and service sectors, respectively, representing the attractiveness of each primary destination. Because the number of actual elementary destinations in each zone is unknown, this is approximated by means of a size function as described in Sect. 4.3.1.2

The coefficients shown in Fig. 4.27 indicate an increase in a zone's systematic utility as its attractiveness grows. Furthermore, the systematic utility increases as the logsum associated with mode choice increases or decreases the perceived mean cost. Also, the residence zone has an extra utility, probably due to the approximations in computing intrazonal level-of-service attributes.

Journey-type choice model $p^i[s_2/os_1h_1d_1h_2](\boldsymbol{SE},\boldsymbol{T})$. This model represents the choice between two alternatives: either undertaking a further trip on the journey for a secondary purpose (trip-chain journey) or returning home (round-trip journey). The model is therefore binary logit with the following systematic utility functions.

$$\begin{aligned} V_{\text{chain}} &= \beta_1 ML + \beta_2 EMP + \beta_3 STU + \beta_4 OTHER + \beta_5 MRNG \\ &\quad + \beta_6 AFTN + \beta_7 EVNG \\ V_{\text{round}} &= \beta_8 ROUND + \beta_9 DACC_{od_1} \end{aligned} \tag{4.4.4}$$

where

ML is a dummy variable, equal to one if the individual is male, zero otherwise

EMP is a dummy variable, equal to one if the person is employed, zero otherwise

STU is a dummy variable, equal to one if the person is a high school or university student, zero otherwise

OTHER is a dummy variable, equal to one if the person is a housewife, retired, or unemployed, zero otherwise

MRNG is a dummy variable, equal to one if the trip starts before 12:00 ($h_1 < 12$), zero otherwise

AFTN is a dummy variable, equal to one if the trip starts between 12:00 and 16:00 ($12 < h_1 < 16$), zero otherwise

EVNG is a dummy variable, equal to one if the trip starts between 16:00 and 20:00 ($16 < h_1 < 20$), zero otherwise

ROUND is the alternative specific attribute for the round-trip alternative

	ML	*EMP*	*STU*	*OTHER*	*MRNG*	*AFTN*	*EVNG*	*ROUND*	$DACC_{od_1}$
	1.708	0.4185	1.107	−0.3559	0.5295	−1.311	0.1835	4.4640	0.3934
t	24.8	7.30	11.10	−4.80	8.60	−11.9	3.10	61.9	7.8

Fig. 4.28 Parameters of the journey type choice model for "other unconstrained" purposes

$DACC_{od_1}$ is the accessibility differential of the residence zone o and primary destination zone d_1; accessibilities are calculated as logsum variables with respect to destination choice for the purpose being considered

The coefficients obtained from the calibrations are shown in Fig. 4.28. As can be seen, employees and students have, all else equal, a greater utility for chained trips, most likely because of their limited time budget. The systematic utility (and therefore probability) of undertaking chain trips is higher during the morning than the evening and, to an even greater extent, than the afternoon.

The role of the accessibility attribute $DACC_{od}$ deserves further comment. If a residence zone has a larger accessibility with respect to the possible destinations for "other unconstrained purposes" than the primary destination, the return home probability increases. On the other hand, if the residence zone has a lower accessibility, the probability of undertaking a chain trip increases. All else equal, a person who lives in the suburbs and undertakes a primary trip to the city center is more likely to undertake a trip chain than a person in the opposite situation who, once home, can undertake another journey for other purposes.

Secondary destination choice model $p^i[d_2/os_1h_1s_2h_2](\boldsymbol{SE},\boldsymbol{T})$. The secondary destination choice model is a multinomial logit with systematic utility functions similar to those described above for the primary destination choice model:

$$V^i_{od_2} = \beta_1 Y^i_{d_1d_2oh_2} + \beta_2 ZN_o + \beta_3 \ln(EMPret_{d_2} + \beta_4 EMPserv_{d_2}) \qquad (4.4.5)$$

where

$Y^i_{d_1d_2oh_2}$ mode choice model logsum inclusive variable, accounting for the (dis)utility of all modes from primary destination d_1 to secondary potential destination d_2 and to residence zone o

$SZ_{d_2/o}$ dummy variable equal to one if zone d_2 is the residence zone o, zero otherwise

$EMPret_{d_2}, EMPserv_{d_2}$ total employment in the retail and service sectors, respectively; these are included in the size function, which expresses the attractiveness of zone d_2 as a potential secondary destination

The coefficient estimates, shown in Fig. 4.29, are in line with expectations. All else equal, they give a larger utility to secondary destinations with lower generalized transportation cost and larger attraction capacity (greater number of elementary destinations).

	$Y^i_{d_1d_2oh_2}$	$SZ_{d_2/o}$	*Size*	$EMPret_{d_2}(10^3)$	$EMPser_{d_2}(10^3)$
	0.417	1.865	0.684	1.0000	0.618
t	2.90	5.00	3.80	–	1.0

Fig. 4.29 Parameters of the secondary destination choice model for "other unconstrained" purposes

4.5 Activity-Based Demand Models

Trip-chaining demand models provide the ability to represent relationships between the different trips that constitute an individual's travel chain, and so generalize considerably conventional trip-based models. However, they do not address the fundamental factors that determine the actual formation and choice of particular trip-chains and round-trips. To address such questions, it is necessary to consider explicitly the activities that individuals and households undertake, and that give rise to travel demand. Models that derive travel patterns from a representation of these more basic activities are called *activity-based demand models*. They are the subject of active research, and operational models have been implemented in a few urban areas.

This section provides a very brief overview of activity-based models and indicates some of the challenges that development and application of these models must confront. In view of the rapid pace of development in this area, specific current models are not described; the interested reader may refer to the literature for such information.

A number of factors account for the high level of interest in activity-based models. As noted above, the complexity of urban living has resulted in correspondingly complex tripmaking behavior. Trip-chaining is an important component of this behavior, but conventional demand models ignore this and trip-chaining models only predict choices from among pre-determined chains. Activity-based models, on the other hand, offer the possibility of understanding and predicting both the formation of trip-chains as well as the choice among them.

This improved understanding has very practical applications. Many of the activity and transportation system interventions that have been proposed to manage congestion can best be analyzed in terms of their effects on the activities from which travel demand is derived. For example, telecommuting and more flexible work schedules may affect the location and timing of work activities, and so also the demand for and time periods of home–work trips. Similarly, road user charges that vary by time of day or level of congestion affect the generalized travel cost that users associate with travel at different times, but the impacts on travel demand in different time periods depend in part on users' ability to rearrange or reschedule the activities that underlie their tripmaking. In both cases, the overall effect of the intervention may be to cause users to reorganize their entire schedule of activities and the resulting tripmaking. For example, greater work time flexibility may allow users to shop on their way to work, and so eliminate a separate shopping trip formerly made at a different time. Conversely, peak period road user charges may dissuade users from combining shopping or other trips with their primary return trip home from work, and cause these purposes to be accomplished through separate trips in off-peak periods.

As stated, activity-based models derive travel demand and its characteristics from users' involvement in other activities, for which the location and scheduling (timing and duration) are explicitly considered. The activities considered may include those undertaken at home, as well as those that require travel. Most frequently, activity-based models take residence and work locations as given, although some researchers have proposed incorporating these longer-term decisions into the modeling framework. Other distinguishing features of these models are their disaggregate focus, generally considering households and the individuals within them to be the basic decision-making units. In this context, the interrelationships between the activity and travel decisions made by different members of a household must be accounted for. Similarly, the identification of households and individuals in terms of user classes reflects their activity needs, commitments, and constraints, in addition to more conventional user class definition criteria such as income. This typically entails a much more detailed description of household characteristics than is common in conventional models. Finally, the activity patterns predicted by the models are translated into trip-chains, with the corresponding starting and ending locations, time periods, modes, and other attributes of the individual trips in the chain.

Broadly speaking, activity-based models follow one of two alternative approaches. Econometric activity-based models represent the various activity pattern and travel decisions using mathematical expressions that are susceptible to estimation using econometric methods. These models are frequently of random utility type, so the mathematical expressions specify the systematic utility functions and the associated distributions of the random residuals in a utility maximization framework, and can be estimated using methods discussed elsewhere in this book.

Alternatively, activity-based models may be implemented as computer simulations (generally probabilistic) of the activity and travel decision processes of individual households. These simulations may invoke random utility models to represent some components of the decision processes, but they typically also apply additional logic and rules to reflect aspects of the household's decision protocols that may not be convenient to express in purely mathematical form. A simulation model can reflect essentially any decision process that a household and its individual members apply to decide about the nature, location, and timing of their activities and of the trips between them. Of course, this generality brings with it considerable challenges in specifying, estimating, and validating the model and its components.

Regardless of the model type, the development and application of activity-based models must confront a number of difficult problems. To begin with, the data required to estimate and validate these models includes both conventional transportation survey data (describing the origins, destinations, purposes, times, modes, etc. for a sample of trips or trip-chains) as well as surveys of household activities to obtain details on household characteristics, in-home and outside activities, and constraints on decisions affecting activity participation and tripmaking. Particular models may require the collection of specific additional data such as stated preference surveys for estimation.

The number of possible activity organization and tripmaking choice alternatives – combinations of specific activities, their ordering in time, their scheduling and location, together with the travel mode and route taken to access them – is extremely large. Thus, activity-based models must implement a choice set generation step that prunes the set of possible alternatives to a smaller and more manageable size. In econometric models, this is frequently done by application of simple heuristics that generate a choice set that is considered reasonable. Simulation models may apply more complex search and selection rules to identify the pertinent choice set.

In practice, activity-based models are generally applied at the level of individual households (or of very detailed household classes), and their results are then aggregated. Thus, application of these models requires quite detailed information on the characteristics of households and individuals in the study area at the geographic level of model zones or finer. Typical sources of current and forecast household and population data do not generally provide breakdowns of the characteristics and location of the population at the required levels of detail. Consequently, it is often necessary to generate a synthetic population whose aggregate attributes match household and population characteristics known from available sources, and whose detailed attributes represent a reasonable joint distribution of characteristics subject to the aggregate constraints. (This can be thought of as filling in a multidimensional table given constraints on sums or averages of its rows, columns, or other sets of elements.)

For example, an activity-based model might require data on the occupational status of each member of a household, whereas available statistics might provide separate data on the distributions of household sizes, population ages, and occupational status of the working-age population by zone. A population generator would then develop a set of households with complete specification of the occupational status of their members, in such a way that the aggregate distributions are respected. A number of methods have been proposed to generate synthetic populations from standard data sources.

As mentioned, simulation-based models are generally probabilistic: repeated model executions with identical data give different outputs. Thus, simulation models must typically be run multiple times to generate a set of realizations sufficient to compute sample distributions, mean values, or other statistics of the output variables. Econometric models, which are typically based on random utility theory, may provide probabilities directly; however, because a complete activity-based model may comprise a number of separate econometric models, or include models for which the output probabilities cannot be computed analytically, determining the distribution or statistics of econometric model outputs may again require sampling multiple times from the model.

Most applications of activity-based models determine only the mean values of the model output variables: the average number of trip-chains having certain characteristics. There is increasing interest, however, in applying these models in an integrated supply–demand framework, where the model's output trip-chains are loaded on a network model and the resulting levels of service are fed back to the activity-based model, iterating until consistency is achieved. In this case, use of the mean values of activity-based model output variables as inputs to the network model may not give correct results because network levels of service vary nonlinearly with demand. It would be more correct to include the network model with the activity-based model in the sampling process, generating multiple joint realizations of both activity-based and network model outputs, and deriving the desired output statistics from the joint sample.

It can be seen that application of activity-based models involves a very considerable amount of computation: sampling multiple times from a large number of detailed household classes or even from a synthetic representation of every household in a study area. Because of the long model run-times on conventional computing hardware, there is increasing interest in running these models in high-performance computing environments such as on supercomputers or on a computer cluster.

In conclusion, activity-based models are at the frontier of travel-demand model development and application. They offer the prospect of representing very complex aspects of travel behavior, but present a number of challenges that researchers and advanced practitioners are working actively to overcome. Significant advances in this area of travel-demand modeling can be expected in the future.

4.5.1 A Theoretical Reference Framework

In this section a possible theoretical formulation for the specification of a system of models in activity-based-style is presented. The overall structure of the proposed framework is shared by several models proposed in the literature and is shown in Fig. 4.30.

This particular architecture aims to explicitly model all travel phenomena related to activity pattern and travel choices: from household weekly activities to individual single trips. It is composed of five submodels:

- *Weekly household activity model*, which reproduces the number and types of activities carried out by households within a week
- *Daily household activity model*, which reproduces the distribution over days of the week of all household activities
- *Daily individual activity list model*, which distributes daily activities among the household components
- *Daily individual activity pattern model*, which combines the individual daily activities leading to actual activity patterns and related trip-chain sequences
- *Trip-chain model*, which reproduces the organization of all trips provided within an activity pattern

Figure 4.30 shows that each choice level is related to the previous and subsequent levels. The three upper levels refer to longer-term decisions, because they reproduce the activity organization among household members in a fixed period of time, and the latter two levels represent shorter-term travel decisions. A possible approach to deal with the reciprocal relationships among the different submodels is to frame the

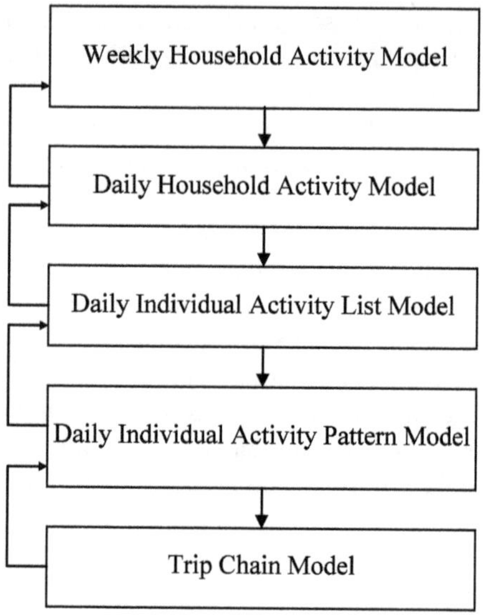

Fig. 4.30 Modeling architecture

overall modeling architecture within a hierarchical model approach where each submodel is conditional upon previous ones and takes into account subsequent models as in Fig. 4.2.

In the following subsections some more details are given for several submodels mainly regarding a possible definition and formalization of the choice alternatives.

4.5.1.1 Weekly Household Activity Model

The model aims to reproduce the whole set of activities carried out by a household within a week. Given a list of possible activities (work, study, shopping, sport, etc.), the generic alternative w^i is given by the set of activities carried out by a household of type i within a week. Formally we may write:

$$w^i = \left(x^i_{w;1}, x^i_{w;2}, \ldots, x^i_{w;a}, \ldots, x^i_{w;n_a}\right)$$
$$\forall i \in \{1, 2, \ldots, n_h\},\ \forall w^i \in \{1, 2, \ldots, C^i_w\} \quad (4.5.1)$$

where

- $x^i_{w;a}$ is the number of times that an activity of type a is performed by household i within a week in alternative w
- n_a is the number of possible activities
- n_h is the number of different household types
- C^i_w is the choice set, that is, the set of all possible weekly sets of activities for household i

Just as an example, alternative w^i could be composed by: $x^i_{w,1} = 12$ work activities, $x^i_{w,2} = 8$ study activities, and so on (assuming, for instance, that 1 stands for *Work* and 2 for *Study*).

Relevant attributes are the household's characteristics and may include the number and age of employed adults, the number and age of nonadults, the dwelling-place, income, number of driving licenses, number of cars, and so on, as well as a logsum variable related to the lower choice dimensions.

4.5.1.2 Daily Household Activity Model

In this case the model aims to reproduce how the set of weekly activities identified by the previous model is split into daily activity sets. The generic alternatives $d^i_{g/w}$, are given by any set of daily activities consistent with the weekly set of activities w^i. Formally we may write:

$$d^i_{g/w} = \left(x^i_{g/w;1}, x^i_{g/w;2}, \ldots, x^i_{g/w;a}, \ldots, x^i_{g/w;n_a}\right) \quad \forall g \in \{1, 2, \ldots, n_g = 7\} \tag{4.5.2}$$

where

$n_g(=7)$ is the number of days in a week

$x^i_{g/w;a}$ is the number of times that an activity of type a is carried out during day g by the household of type i given the weekly household set of activities w

The following constraints have to be satisfied.

$$\sum_{g=1}^{n_g} x^i_{g/w,a} = x^i_{w,a} \quad \forall a \in \{1, 2, \ldots, n_a\},\ \forall w^i \in C^i_w,\ \forall i \in \{1, 2, \ldots, n_h\} \tag{4.5.3}$$

For example if, as in the previous example, $x^i_{w,1} = 12$ work activities, $x^i_{w,2} = 8$ study activities, the following conditions have to be satisfied.

$$\sum_{g=1}^{7} x^i_{g/w,1} = 12, \qquad \sum_{g=1}^{7} x^i_{g/w,2} = 8 \quad 1 = \text{Work},\ 2 = \text{School}$$

Constraints (4.5.3) implicitly define the choice set $C^i_{g/w}$ of this choice dimension. However, it is useful in practical implementation to reduce the combinatorial complexity of the problem by dropping alternatives which are manifestly unfeasible or unlikely to occur. Relevant attributes are in principle similar to those of the previous models.

4.5.1.3 Daily Individual Activity List Model

This submodel reproduces the distribution of daily activities among the components of a household. This leads to daily individual activity lists that are the starting points for reproducing the daily travel choices of each individual. In this case the generic alternative $k^i_{r/g,w}$ is given by the daily activity list of each component r of household i, that is, types and numbers of activities he carries out during a day g given the daily set of household activities $d_{g/w}$:

$$k^i_{r/g,d_{g/w}} = \left(x^i_{r/g,d_{g/w};1}, \ldots, x^i_{r/g,d_{g/w};a}, \ldots, x^i_{r/g,d_{g/w};n_a}\right)$$
$$\forall r \in \left\{1, 2, \ldots, n^i_r\right\} \tag{4.5.4}$$

where

$x^i_{r/g,d_{g/w};a}$ is the number of times that an activity of type a is carried out by component r of household i in day g, given the daily set of household activities $d_{g/w}$

n^i_r is the number of components of the type i household

The following constraints have to be satisfied.

$$\sum_{r=1}^{n^i_r} x^i_{r/g,d_{g/w};a} = x^i_{g/w;a} \quad \forall a \in \{1, 2, \ldots, n_a\},\ \forall g \in \{1, 2, \ldots, n_g = 7\},$$
$$\forall w^i \in C^i_w,\ \forall i \in \{1, 2, \ldots, n_h\} \tag{4.5.5}$$

Once again, constraints (4.5.5) implicitly define the choice set of this submodel ($C^i_{r/g,w}$) but in order to reduce the combinatorial complexity of the problem, this can be reduced by dropping unlikely activity lists.

Relevant attributes are also in this case similar to those of the previous models but concern the specific individual and obviously include gender and occupational status.

4.5.1.4 Activity Pattern and Trip-Chain Models

This model reproduces how different activity patterns can be generated from a given daily individual activity list. Figure 4.31 exemplifies some possible activity patterns (right) that can be generated from a given daily individual activity list (left).

It is worth noting that the daily individual activity list provides the number of times each activity is carried out within the day (one in this case), except for *home* which can be repeated several times. The number of times (minus one) activity *home* is repeated in a given activity pattern implicitly determines the number of trip-chains related to that activity pattern. For instance, three chains are associated with the second activity pattern in Fig. 4.31 (H-P/D-O-H-W-H-L-H) because activity *home* is replicated four times.

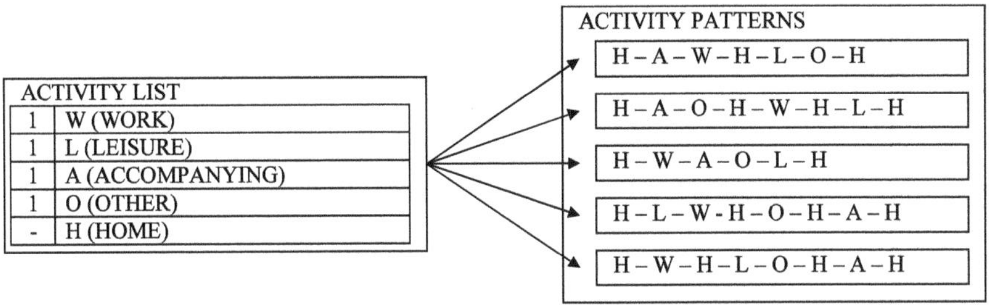

Fig. 4.31 Activity pattern production from a given activity list

Also in this case the number of possible activity patterns that can be associated with each activity list can be reduced by considering only those that are significant in the observed sample.

Relevant attributes are also in this case socioeconomic characteristics of the individual. The logsum variable related to the subsequent trip-chain model includes the generalized costs of the different chains. Therefore the choice of the activity pattern is influenced by the network congestion at different times of the day.

Given an activity pattern (i.e., a given succession of trip-chains), the role of the trip chain model (which can be similar to that described in Sect. 4.4) is to reproduce when and how these trip-chains are carried out within the day, introducing not only consistency within the generic trip-chain (as described in Sect. 4.4) but also among the different chains of the day, mainly in terms of activity duration and departure time.

4.6 Applications of Demand Models

To conclude the discussion of passenger travel-demand models, it is useful to comment on the nature, domains, and modalities of their application.

The "true" values of demand flows (present and predicted) are generally unknown to the analyst and as such must be represented as random variables. Demand models provide possibly unbiased estimates of the mean values of demand flows having particular characteristics (user class, purpose, time period, origin, destination, mode, path, etc.). In some cases the variances and covariances of the estimates can also be computed. For example, in a four-level demand model and a single trip for each purpose s in time period h, the demand flow $d_{od}[s,h,m,k]$ can be modeled as a multinomial random variable. In other words, the demand estimates obtained with a partial share model are the mean (expected) values of random variables that, assuming statistical independence of individual decisions, can be assumed to follow a multinomial distribution. It is therefore possible to express the variances and covariances of demand flows obtained from the models:

$$E\big[d_{od}[shmk]\big] = n[osh]p[xdmk/osh]$$

$$\mathrm{Var}\big[d_{od}[shmk]\big] = n[osh]p[xdmk/osh]\big[1 - p[xdmk/osh]\big] \tag{4.6.1}$$
$$\mathrm{Cov}\big[d_{od}[shmk]d_{od'}[shm'k']\big] = n[osh]p[xdmk/osh]p[xd'm'k'/osh]$$

The actual deviations of model estimates from "true" demand flows are certainly larger than what the variance (4.6.1) would suggest. After all, models, however sophisticated they may be, are only simplified representations of the complex phenomena underlying mobility. The probabilities $p[xdmk/osh]$ are therefore only estimates of real percentages whose deviation (variance) can only be determined empirically.

The practical uses of demand models fall into three categories: to estimate existing demand and its changes, for quantitative analysis of the characteristics of mobility, and as components of the system of demand–supply interaction (assignment) models. These three model application domains have a number of implications that are briefly discussed below.

Estimation of existing demand and changes in it. This is the classic use of demand models. Once specified and calibrated, models represent transportation demand and can be applied to existing activity and transportation supply systems to estimate unknown demand flows. Alternatively, the models can be used to forecast the changes in travel demand brought about by changes in the activity and/or transportation supply systems. For both these applications, a variety of techniques is available depending on the application context, and the models and their outputs can be integrated with other information available.

When models are applied to estimate existing demand and/or to predict changes in it, model results must be aggregated in order to obtain estimates of total demand flows between different origin–destination pairs. The different *aggregation techniques* described in Sect. 3.7 for aggregate and disaggregate random utility models can be used for this purpose. Aggregate models require aggregation by user class, implicitly assumed in expression (4.2.2), whereas disaggregate models can be aggregated using sample enumeration techniques with variables that correspond to the present situation or that are predicted for a future scenario. These topics are dealt with in more detail in Chap. 8.

Tools for quantitative mobility analysis. Demand models can also be used as *statistical tools for quantitative analysis* of mobility phenomena. In this case the models are seen as relationships that allow the influence on mobility of both socioeconomic and level-of-service variables to be evaluated. The emphasis here is not on model application to obtain aggregate demand estimates (present or future) but rather on specification and estimation of the coefficients of the model itself.

Some of the models described in this chapter could be used, for example, for a quantitative analysis of the effects of factors such as age, sex, income, and occupational status on different aspects of mobility. For this use, the model variables might be very detailed because neither their current values over the whole population of travelers nor their future values are required.

Demand models for assignment to transportation networks. The outputs of demand models are often used as inputs to assignment models, which predict the flows and service levels of various elements of the supply system in response to the demand. For this type of application, the models are considered to be demand functions. They express origin–destination flows by different modes during a time period as a function of socioeconomic variables $\boldsymbol{SE}$ and of generalized path costs $\boldsymbol{g}$. Path choice models are typically incorporated in the assignment models themselves.

In formulating assignment models, demand models are represented with a notation that is slightly different from the one used thus far. Because assignment models incorporate path choice, the path choice model is separated from those on other levels (choice dimensions). In this case, the generic partial share model becomes

$$d^i_{od}[hmk](\boldsymbol{SE},\boldsymbol{T}) = d^i_{od}[hm](\boldsymbol{SE},\boldsymbol{T})\, p^i_{od,k}\left(\boldsymbol{g}^i_{od,m}\right)$$

where, as shown in more detail in Chap. 5, $\boldsymbol{g}^i_{od,m}$ is the vector of generalized path costs corresponding to the O-D pair *od* on the mode m network and to user class i, and path choice attributes other than those that contribute to the generalized transportation cost are implicit. As stated in Sect. 4.3.3, path cost is the negative of systematic utility, $V_k = -g_k$. Generalized path costs $\boldsymbol{g}^i_{od,m}$ convert to common (cost) units the different components of the vector $\boldsymbol{T}$. It should also be noted that trip purpose s does not appear explicitly in the previous expression because, in an assignment context, the index i will denote the user group defined by the pair (user class, trip purpose).[22]

In assignment models the aggregate O-D flow for user class i is denoted by d^i_{odm} if the demand is considered inelastic (not affected by variations in generalized costs due to network congestion), and by $d^i_{odm}(\boldsymbol{s}(\boldsymbol{g}))$ if the demand is considered elastic in some or all dimensions. In elastic demand models, changes in other choice dimensions resulting from path cost changes are predicted using the EMPU variable $s_{m/od}$, corresponding to path choice on the mode network m in time period h for users of class i. The EMPU variables for all O-D pairs can be arranged in a column vector $\boldsymbol{s}_m$. The different notation for demand flow d^i_{odm} and demand functions $d^i_{odm}(\boldsymbol{s}(\boldsymbol{g}))$ does not mean that the latter cannot be obtained with the demand models described in this chapter. Rather, it underlines the dependence of demand on congestion-related costs in the analysis of interactions between elastic demand and supply (elastic demand assignment models). This notation is taken up in more detail in Chap. 5.

---

[22] Because a user class consists of individuals who can be represented by the same demand models (alternatives, parameters, and attributes), its definition depends on the models themselves, including the travel choice dimensions that they address. In Chap. 5 the classes are defined in terms of path choice models. Given the reduced number of attributes in these models, fewer classes might be used for assignment than for other choice dimensions. The assignment model classes can often be obtained by aggregating the more detailed classes. This is particularly true for individual-level models where, for the assignment model, individual trips can be aggregated to obtain O-D trip matrices for a given user group.

4.7 Freight Transportation Demand Models*

The demand for freight transportation is closely connected to the production and distribution of goods, that is, to the study area's economic system and its interactions with the external economic system. Many of the definitions presented for passenger travel demand can be extended to freight transportation demand, although their interpretation is in general very different. A system of freight demand models can be formally expressed as

$$d_{od}[K_1, K_2, \ldots] = d(\boldsymbol{SE}, \boldsymbol{T}, \boldsymbol{\beta}) \tag{4.7.1}$$

Here, demand flows represent movements of freight quantities (usually expressed in tons); the relevant characteristics, $K_1, K_2, \ldots$, are normally associated with *commodity type* (raw materials, semifinished products, finished products, etc.), with *sectors of economic activity*, with *characteristics of firms* (e.g., firm size, logistic organization), *transportation characteristics* (e.g., shipping frequency, size, and value) as well as with *transportation modes*.[23] The $\boldsymbol{SE}$ variables reflect the economics of production (value of production by sector, number, and size of production units, etc.) and consumption (household consumption, imports, etc.); and the transportation system variables $\boldsymbol{T}$ are related as before to the attributes of the different transportation modes and services (times, costs, service reliability, etc.). Vector $\boldsymbol{\beta}$ denotes as usual the model coefficients; it is not explicitly included in the discussion below.

These considerations suggest that the mechanisms underlying the formation of freight transportation demand and its fulfillment by transportation services are considerably more complex and interrelated than those for passenger demand. There is no single decision-maker for freight, but rather a complex and connected set of decision-makers responsible for production, logistics (storage and shipping), distribution, and marketing.

Schematically, the decision-makers who influence the level and composition of freight transportation demand can be grouped into three categories. *Producers* of goods and services decide how much and how to produce, and where and at what prices to sell; *consumers*, either intermediate (production companies) or final (households, businesses, public agencies), decide how much and what to consume; and *transportation companies*[24] decide how to provide transportation services.

Some of the classifications of passenger demand models can be extended to freight models. Models can be *disaggregate* or *aggregate* depending on whether their variables refer to disaggregate units such as individual companies or individual shipments, or to aggregate units such as all the companies of a given category

[23]The concept of *mode* is quite different in freight and passenger transportation. In freight transportation, it encompasses both physical (the sequence of transportation modes used for a consignment) and organizational (the sequence of entities that are responsible for the transportation) aspects of the movement. As a consequence, some authors consider freight mode choice as a choice of transportation *service* rather than of transportation *mode*.

[24]In practice, the entities involved in freight transportation supply are often classified as *shippers*, who organize the whole shipment, and *carriers*, who provide the actual transportation service(s).

and/or economic sector. Furthermore, freight demand models can be *behavioral* or *descriptive* depending on whether they are based on explicit assumptions regarding the behavior of market agents, or on empirical relationships between freight transportation demand and causal variables corresponding to the economic and/or transportation system.

Freight transportation demand models have been studied and applied to a lesser extent than passenger models, mainly because of the complexity of the underlying phenomena that influence freight transportation. There is no universal paradigm but rather only individual examples, which depend on the type of application and the data available. Just as with four-step passenger demand models, described in Sect. 4.2, the most recent and sophisticated freight demand model systems result from the integration of macroeconomic models, which represent the level (quantity) and spatial distribution of goods exchanged among different economic zones (leading to origin–destination matrices); and of transportation models, which simulate mode and path choice. Moreover, models that explicitly disaggregate macroeconomic demand to lower-level geographic units are often required in order to guarantee consistency among the different geographic units (e.g., zones) used at each modeling stage.

There is a broad and well-established body of literature on macroeconomic models; some suggestions for in-depth reading are presented in the bibliographical notes. In general, macroeconomic models can be classified based on their geographic level (international, national, regional/urban) and their adopted approach (generation, distribution, and joint generation–distribution models). Generation and distribution models have the same structure as the corresponding passenger models: the former are usually regression models or, rarely, random utility models, and the latter are singly constrained entropy/gravity or linear programming models. Joint generation–distribution models directly determine the origin–destination matrix through explicit representation of the pattern of economic exchanges among study area subareas and from/to external areas. Models in this category include *Spatial Price Equilibrium* (*SPE*), *Computable General Equilibrium* (*CGE*), and *input–output* models; in some applications doubly constrained gravity models are also applied.

SPE models represent the production and consumption of each zone and each economic sector through supply and demand curves that depend on prices. The determination of equilibrium prices, volumes of exchanged goods and transportation costs can be formulated, under certain assumptions, as a nonlinear programming problem subject to linear constraints. Although *SPE* models can be extended and generalized in a variety of ways, they have been criticized as lacking realism because of the deterministic assumptions embodied in some model formulations, according to which goods are traded between two zones only if the sale price in the origin (production) zone plus the transportation cost is equal to the sale price in the destination (consumption) zone. This leads to positive demand flows for a few origin–destination pairs and zero for the others (contrary to empirical evidence). Even if a modified model formulation, called *dispersed SPE (DSPE)*, partially overcomes this problem, other limits remain; many of these are due to the use of zonal demand and supply functions that do not take into account the relationships between economic sectors, and to the lack of data for calibrating these functions.

CGE models explicitly represent the economic behavior of households, government, and businesses, and represent the whole pattern of economic exchanges as the solution of an equilibrium problem. This formulation can take into account the spatial dimension of the problem and the effects of transportation level-of-service attributes (*spatial CGE* or *SCGE* models). Although the results to date are very encouraging, there are few examples of *SCGE* models of large-scale problems, mainly due to the lack of data needed for model calibration and application.

Input–output models start with an explicit representation of the interdependencies among the different sectors of the economy to predict the quantity of goods produced by and exchanged among different zones. This group includes a variety of models that differ from each other with respect to the elements of the economic system that they consider as fixed or variable, and with respect to their implicit or explicit representation of the price system. These models, when formulated at a regional level, have proved to be very flexible and practical tools. In this application they are called MultiRegional Input–Output (MRIO) models. They are described in the next section.

As previously stated, macroeconomic models are usually coupled with transportation models (typically mode and path choice models), which in most cases are identical to those already described for passenger transportation. Innovative models that use the tour-based approaches described in Sect. 4.4 to explicitly represent the choice of a transportation service rather than a transportation mode have also been described in the literature.

In the following, the general structure of multiregional input–output models is described (Sect. 4.7.1), and some models for freight mode choice are described in Sect. 4.7.2. Examples of both model types are drawn from an integrated model system that was used to predict freight demand in Italy, and whose structure is represented in Fig. 4.32.

4.7.1 Multiregional Input–Output (MRIO) models

The application of macroeconomic models to freight demand prediction usually involves two phases, as illustrated in Fig. 4.32. The first phase predicts the exchange (or trade) between economic sectors and regions in monetary terms; the second phase transforms these monetary exchanges into quantity exchanges (tons). This results in O-D matrices that are inputs to mode/service and path choice models.

A multiregional input–output model can be used for the first phase.

Such models assume that the study area is divided into n_z zones in accordance with the zoning principles described in Chap. 1. It should be noted that applications of macroeconomic models tend to use relatively large zones; this is due to the geographic level at which the statistical information required by the model is typically reported. Indeed, zones frequently coincide with entire geographic regions, hence the name MRIO. The transition to a finer zoning system, which is necessary for the representation of mode choice and network assignment, can be conducted in

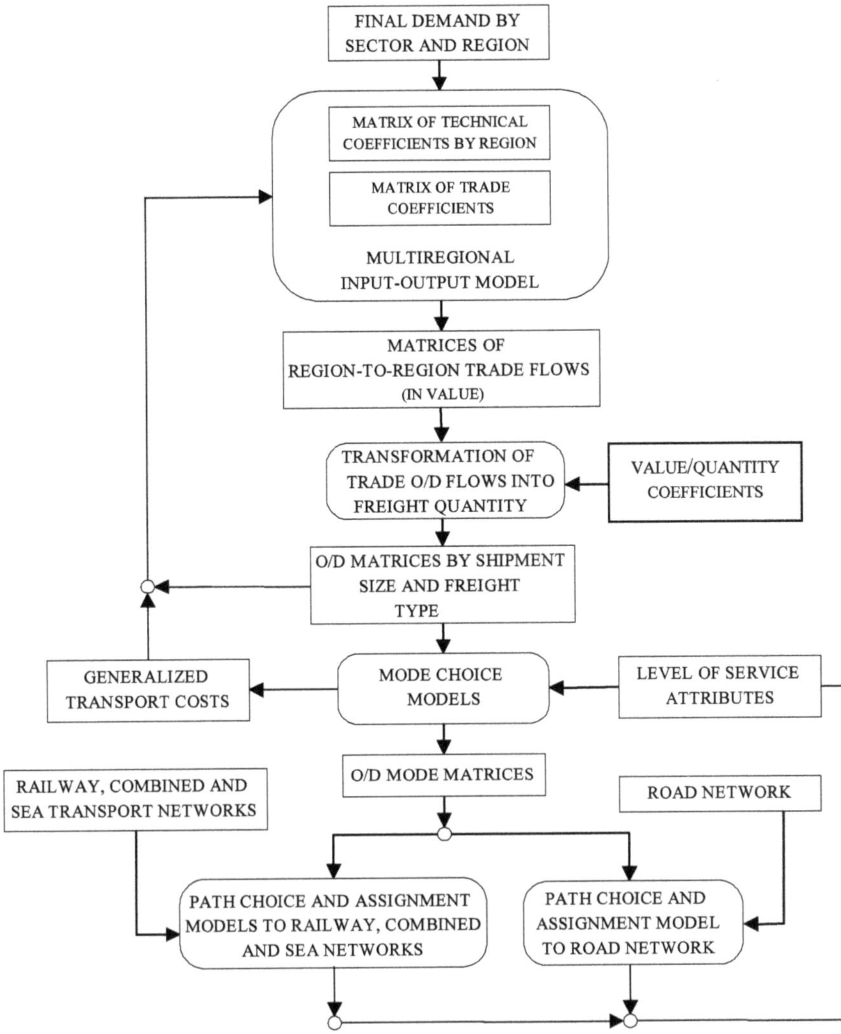

Fig. 4.32 Model system structure for freight transportation demand

the second phase, where monetary values are transformed into physical quantities, for example, using descriptive demand models. (Except where otherwise noted, the terms *zone* and *region* have equivalent meanings in the discussion below.)

Economic activities are divided into n_s sectors that represent the production and consumption of goods (e.g., agriculture and industrial sectors) or services (e.g., banking and commerce). The various actors within each sector are assumed to be homogeneous with respect to their economic behavior. A large number of small sectors would tend to ensure a more accurate description of significant economic phenomena and greater plausibility of the assumption of behavioral homogeneity;

		Sectors
Goods manufacturing	1	Agriculture, forestry, and fisheries
	2	Energy products
	3	Ferrous and nonferrous minerals and metals
	4	Nonmetallic minerals and products
	5	Chemical and pharmaceutical products
	6	Metal products and machinery
	7	Means of transportation
	8	Foods, drinks, and tobacco
	9	Textile products, clothing, leather goods, and footwear
	10	Paper, paper products, printing and publishing, other industrial products
	11	Wood, rubber
Service sectors	12	Buildings and civil engineering
	13	Retail, hotels, and public utilities
	14	Transportation and communication
	15	Banking and insurance
	16	Other services for sale
	17	Services not for sale

Final demand components
Household consumption
Public consumption
Investments
Changes in stock levels

a

Region i		Sectors of production				Final demand	Regional export	International export
		S_1	$\dots$	S_m	$\dots$			
Sectors of production	S_1	$\dots$	$\dots$	$\dots$	$\dots$	$\dots$	$\dots$	$\dots$
	$\dots$	$\dots$	$\dots$	$\dots$	$\dots$	$\dots$	$\dots$	$\dots$
	S_n	$\dots$	$\dots$	K_i^{nm}	$\dots$	Y_i^n	$Y_{\text{REG}i}^n$	$Y_{\text{EST}i}^n$
	$\dots$	$\dots$	$\dots$	$\dots$	$\dots$	$\dots$	$\dots$	$\dots$
Added value		$\dots$	$\dots$	$\dots$	$\dots$			
Value of production		$\dots$	$\dots$	X_i^m	$\dots$			
Regional import		$\dots$	$\dots$	$J_{\text{REG}i}^m$	$\dots$			
International import		$\dots$	$\dots$	$J_{\text{EST}i}^m$	$\dots$			

b

Fig. 4.33 **a** Sectors of the economy and components of final demand for the national model. **b** Simplified structure of a regional input–output table

on the other hand, in practice it is necessary to take into account the aggregation levels of available data. Figure 4.33a shows the 17 macrosectors used to represent the Italian economy for the above-mentioned system of national models.

As stated, input–output models use a table of sectoral interdependencies to represent the pattern of economic exchanges among sectors in a region. This fundamental instrument of economic analysis, known as a regional input–output table, is schematized in Fig. 4.33b; all variables are measured in monetary units, usually with respect to a given year.

To give a formal description of the MRIO model, it is necessary to introduce some new variables. Consider a zone (region) i and let:

K_i^{mn} be the value of the production in sector m (intermediate demand) used for the production of sector n in zone I

Y_i^m be the value of final demand of sector m in zone i. Figure 4.33a illustrates the final demand elements taken into account in the Italian national model

$\boldsymbol{Y}$ be the vector of final demand, with dimensions $(n_z \cdot n_s \times 1)$, obtained by ordering the elements Y_i^m for each sector m and each zone I[25]

$Y_{\mathrm{REG}i}^m$ be the value of exports in sector m from zone i to all other zones of the study area

$\boldsymbol{Y}_{\mathrm{REG}}$ be the vector of zonal exports, with dimensions $(n_z \cdot n_s \times 1)$

$Y_{\mathrm{EST}i}^m$ be the value of exports in sector m from zone i to outside the study area

$\boldsymbol{Y}_{\mathrm{EST}}$ be the vector of exports from the study area, with dimensions $(n_z \cdot n_s \times 1)$

X_i^m be the value of total production of sector m in zone i

$\boldsymbol{X}$ be the vector of total production, with dimensions $(n_z \cdot n_s \times 1)$

$J_{\mathrm{REG}i}^m$ be the value of imports in sector m to zone i from all other zones of the study area

$\boldsymbol{J}_{\mathrm{REG}}$ be the vector of zonal imports of dimensions $(n_z \cdot n_s \times 1)$

$J_{\mathrm{EST}i}^m$ be the value of imports in sector m to zone i from outside the study area

$\boldsymbol{J}_{\mathrm{EST}}$ be the vector of imports from outside the study area of dimensions $(n_z \cdot n_s \times 1)$

In detail, variables K_i^{mn} define a block of dimension $n_s \cdot n_s$ in the input–output table: a row m describes the value of the goods and services of sector m used for production by each other sector n of zone i. For instance, part of the output of the engineering industry (industrial machinery) may be used to produce goods within the same sector or used in other industrial sectors (e.g., the textile industry) or used for the production of services (such as office equipment). On the contrary a column n identifies the value of goods and services of each sector m needed for production of n in zone i. For instance, the production of goods in the chemical industry requires goods and services from all the other sectors (e.g., industrial machinery and metal products).

The sum of variables K_j^{mn} of row m, of the final demand and of the exports (to outside the study area) of goods and services of sector m from zone i provides the total demand for goods of sector m in zone i. Similarly, the sum of the variables K_j^{mn} of column n and of the *value added* provides the production value X_i^n of goods and services of sector n in zone i; the sum of production X_i^n and imports (from other zones and from outside the study area) of sector n to zone i defines the total supply (availability) of goods and services of sector n in i.

In the context of a multiregional study, the input–output table in question refers to a single zone, having exchanges both with other zones in the study area as well

[25]The structure presented here for vector $\boldsymbol{Y}$ is the same for all the vectors presented below, and is not repeated.

as with the external world. This approach can also be applied in a national input–output table, but in this case would take into account only trade with the rest of the world.

In light of the above, an equilibrium condition between the supply (column sum) and demand (row sum) of goods and services of sector m in zone i can be written as follows.

$$X_i^m + J_{\text{REG}i}^m + J_{\text{EST}i}^m = \sum_n K_i^{mn} + Y_i^m + Y_{\text{REG}i}^m + Y_{\text{EST}i}^m \tag{4.7.2}$$

All input–output models can be derived starting from the system of $n_z \cdot n_s$ (4.7.2). In applications, interest generally lies in assessing the production changes that result from changes in final demand and/or transportation supply; input–output models that consider the $n_z \cdot n_s$ production values X_i^m as unknowns to be solved for are called *demand-driven* models.

In order to derive a demand-driven MRIO model from equilibrium condition (4.7.2), it is first necessary to express the relationship between the (monetary) values of production and intermediate demand by defining *technical coefficients* a_i^{mn}:

$$a_i^{mn} = \frac{K_i^{mn}}{X_i^n}$$

which represent the value of the product of sector m (input) required to produce a unit of value of sector n (output) in zone i. These coefficients depend on the production technologies available in zone i; in general, the lower the coefficient a_i^{mn}, the more efficient is production in i because a lower input value is required to produce an output unit. The elements a_i^{mn} corresponding to a given zone i can be ordered in a square matrix $\boldsymbol{A}_i\,(n_s \times n_s)$, known as the *matrix of technical coefficients of zone i*. Different zones may have different production technologies and technical coefficient matrices. The matrices $\boldsymbol{A}_i$ can be arranged in a block diagonal matrix $\boldsymbol{A}$ of dimensions $(n_z \cdot n_s \times n_z \cdot n_s)$, in which each block relates to a zone. Figure 4.34 presents an example of some of the variables introduced, for a 3-region, 2-sector system (market).

Moreover, the specific other zones in the study area that are associated with the economic quantities of a given zone i (intermediate demand for production, import/export values, final demand) in (4.7.2) must be explicitly represented, through *trade coefficients*, in order to calculate interregional freight flows. Apart from some exceptions (discussed below), trade coefficients cannot generally be derived from the input–output table and therefore have to be estimated from surveys. Unlike technical coefficients, trade coefficients have been defined in a variety of different ways in the literature, each definition leading to a somewhat different formulation of the MRIO model. According to the first formulation proposed, a trade coefficient t_{ji}^{mn} expresses the percentage of goods and services of sector m in zone j that is used for producing goods and services of sector n in zone i. Because it is difficult to obtain the percentages t_{ji}^{mn} directly, trade coefficients can be hypothesized independent

Vector of Production by Sector $\boldsymbol{X}$ $(3 \cdot 2 \times 1)$

Region A	Sector 1	X_A^1
	Sector 2	X_A^2
Region B	Sector 1	X_B^1
	Sector 2	X_B^2
Region C	Sector 1	X_C^1
	Sector 2	X_C^2

Matrix of Technical Coefficients $\boldsymbol{A}$ $(3 \cdot 2 \times 3 \cdot 2)$

		Region A		Region B		Region C	
		Sector 1	Sector 2	Sector 1	Sector 2	Sector 1	Sector 2
Region A	Sector 1	a_A^{11}	a_A^{12}	0	0	0	0
	Sector 2	a_A^{21}	a_A^{22}	0	0	0	0
Region B	Sector 1	0	0	a_B^{11}	a_B^{12}	0	0
	Sector 2	0	0	a_B^{21}	a_B^{22}	0	0
Region C	Sector 1	0	0	0	0	a_C^{11}	a_C^{12}
	Sector 2	0	0	0	0	a_C^{21}	a_C^{22}

Matrix of Exchange or Trade Coefficients $\boldsymbol{T}$ $(3 \cdot 2 \times 3 \cdot 2)$

		Region A		Region B		Region C	
		Sector 1	Sector 2	Sector 1	Sector 2	Sector 1	Sector 2
Region A	Sector 1	t_{AA}^1	0	t_{AB}^1	0	t_{AC}^1	0
	Sector 2	0	t_{AA}^2	0	t_{AB}^2	0	t_{AC}^2
Region B	Sector 1	t_{BA}^1	0	t_{BB}^1	0	t_{BC}^1	0
	Sector 2	0	t_{BA}^2	0	t_{BB}^2	0	t_{BC}^2
Region C	Sector 1	t_{CA}^1	0	t_{CB}^1	0	t_{CC}^1	0
	Sector 2	0	t_{CA}^2	0	t_{CB}^2	0	t_{CC}^2

O/D Matrix of Value Exchanges $\boldsymbol{N}$ $(3 \cdot 2 \times 3 \cdot 2)$

		Region A		Region B		Region C	
		Sector 1	Sector 2	Sector 1	Sector 2	Sector 1	Sector 2
Region A	Sector 1	N_{AA}^{11}	N_{AA}^{12}	N_{AB}^{11}	N_{AB}^{12}	N_{AC}^{11}	N_{AC}^{12}
	Sector 2	N_{AA}^{21}	N_{AA}^{22}	N_{AB}^{21}	N_{AB}^{22}	N_{AC}^{21}	N_{AC}^{22}
Region B	Sector 1	N_{BA}^{11}	N_{BA}^{12}	N_{BB}^{11}	N_{BB}^{12}	N_{BC}^{11}	N_{BC}^{12}
	Sector 2	N_{BA}^{21}	N_{BA}^{22}	N_{BB}^{21}	N_{BB}^{22}	N_{BC}^{21}	N_{BC}^{22}
Region C	Sector 1	N_{CA}^{11}	N_{CA}^{12}	N_{CB}^{11}	N_{CB}^{12}	N_{CC}^{11}	N_{CC}^{12}
	Sector 2	N_{CA}^{21}	N_{CA}^{22}	N_{CB}^{21}	N_{CB}^{22}	N_{CC}^{21}	N_{CC}^{22}

Fig. 4.34 Variables for a 3-region, 2-sector MRIO model

of the usage sector in the destination zone, that is, $t_{ji}^{mn} = t_{ji}^{m} \forall n$, thereby making it possible to estimate them. This assumption yields the constraint:

$$t_{ei}^{m} + \sum_{j} t_{ji}^{m} = 1 \tag{4.7.3}$$

where t_{ei}^{m} represents the external (outside the study area) trade coefficients and the sum extends over all study area zones. Trade coefficients can be arranged in a matrix $\boldsymbol{T}$, known as the exchange or trade matrix, of dimensions $(n_Z \cdot n_s \times n_Z \cdot n_s)$, in which, for each pair of zones, there is a square diagonal submatrix. Each diagonal element of this submatrix provides, for the sector corresponding to the diagonal's row and column, the trade coefficient between the two zones. Note that the external trade coefficients t_{ei}^{m} do not appear explicitly in matrix $\boldsymbol{T}$, but instead influence the matrix values through constraint (4.7.3). Figure 4.34 presents an example of matrix $\boldsymbol{T}$ for a 3-region, 2-sector system.

Below, a MRIO model is derived, starting from (4.7.2) and assuming technical coefficients. Some extensions of the MRIO model, which allow for elasticity in the trade and technical coefficients, are then presented.

MRIO model with constant coefficients. By introducing the trade coefficients defined above, equilibrium condition (4.7.2) can be rewritten as follows.

$$\begin{aligned} J_{\mathrm{REG}i}^{m} &= \sum_{n} \sum_{j \neq i} t_{ji}^{m} K_{i}^{mn} + \sum_{j \neq i} t_{ji}^{m} Y_{i}^{m} \\ &= \left(\sum_{j \neq i} t_{ji}^{m} \right) \left(\sum_{n} K_{i}^{mn} + Y_{i}^{m} \right) \end{aligned} \tag{4.7.4a}$$

$$J_{\mathrm{EST}i}^{m} = \sum_{n} t_{ei}^{m} K_{i}^{mn} + t_{ei}^{m} Y_{i}^{m} = t_{ei}^{m} \left(\sum_{n} K_{i}^{mn} + Y_{i}^{m} \right) \tag{4.7.4b}$$

$$\sum_{n} K_{i}^{mn} = \sum_{n} t_{ii}^{m} K_{i}^{mn} + \sum_{n} \sum_{j \neq i} t_{ji}^{m} K_{i}^{mn} + \sum_{n} t_{ei}^{m} K_{i}^{mn} \tag{4.7.4c}$$

$$Y_{i}^{m} = t_{ii}^{m} Y_{i}^{m} + \sum_{j \neq i} t_{ji}^{m} Y_{i}^{m} + t_{ei}^{m} Y_{i}^{m} \tag{4.7.4d}$$

$$\begin{aligned} Y_{\mathrm{REG}i}^{m} &= \sum_{n} \sum_{j \neq i} t_{ij}^{m} K_{j}^{mn} + \sum_{j \neq i} t_{ij}^{m} Y_{j}^{m} \\ &= \sum_{j \neq i} \left[t_{ij}^{m} \left(\sum_{n} K_{j}^{mn} + Y_{j}^{m} \right) \right] \end{aligned} \tag{4.7.4e}$$

Equations (4.7.4a) and (4.7.4b) indicate that imports, from other zones in the study area and from the external world, respectively, of goods and services of sector m in zone i are used both for production reuse and to satisfy final demand. Similarly, (4.7.4c) and (4.7.4d) express that production reuse and final demand are satisfied through both internal production and study area and external imports.

The relations above also show that some of the trade coefficients, that is, t_{ii}^m and t_{ei}^m, can be directly calculated from input–output tables. From (4.7.4b) it follows that

$$t_{ei}^m = \frac{J_{\mathrm{EST}i}^m}{\sum_n K_i^{mn} + Y_i^m} \tag{4.7.5}$$

and combining (4.7.3) with (4.7.4a) it follows that

$$t_{ii}^m = 1 - \frac{J_{\mathrm{EST}i}^m + J_{\mathrm{REG}i}^m}{\sum_n K_i^{mn} + Y_i^m}$$

Substituting this expression for the technical coefficient into (4.7.4a)–(4.7.4e) and the latter into equilibrium condition (4.7.2) yields:

$$X_i^m = \sum_n \sum_j t_{ij}^m a_j^{mn} X_j^n + \sum_j t_{ij}^m Y_j^m + Y_{\mathrm{EST}i}^m$$

which can be expressed in vector terms as

$$\boldsymbol{X} = \boldsymbol{T}\boldsymbol{A}\boldsymbol{X} + \boldsymbol{T}\boldsymbol{Y} + \boldsymbol{Y}_{\mathrm{EST}} \tag{4.7.6}$$

Model (4.7.6) is usually applied to predict regional production by sector, that is, to calculate vector $\boldsymbol{X}$, starting from scenarios (assumptions) about the vector of study area final demand $\boldsymbol{Y}$ and external exports $\boldsymbol{Y}_{\mathrm{EST}}$. Once the vector $\boldsymbol{X}$ has been calculated, the matrix of O-D freight demands can be estimated, as shown later. The MRIO model with constant coefficients assumes that the elements of matrices $\boldsymbol{A}$ and $\boldsymbol{T}$ are constant and known (equal, e.g., to their current values). In this case, the solution of the linear equation system (4.7.6) can be expressed in closed form as

$$\boldsymbol{X} = (\boldsymbol{I} - \boldsymbol{T}\boldsymbol{A})^{-1} \cdot (\boldsymbol{T}\boldsymbol{Y} + \boldsymbol{Y}_{\mathrm{EST}}) \tag{4.7.7}$$

where $\boldsymbol{I}$ is the identity matrix of dimensions $(n_z \cdot n_s \times n_z \cdot n_s)$.

The MRIO model (4.7.7) is known as a model with *endogenous imports*, because it represents a case where an increase in final demand is met by an increase in both internal production as well as imports from outside the study area. In other words, an increase in final demand Y_i^m yields an increase in production X_i^m and hence also in intermediate inputs K_i^{mn}, under the assumption of constant technical coefficients. As a consequence, the denominator of (4.7.5) increases as Y_i^m increases and therefore, in order to keep the external trade coefficients constant, the numerator represented by foreign imports also has to increase. Consequently, in a MRIO model with endogenous imports the level of external imports increases consistently with changes in final demand, whereas the ratio between external imports and total availability, which can be seen as a measure of the dependence of the study area economy on the external world, remains constant.

Figure 4.35 gives a numerical example of the application of model (4.7.7) to a situation with three regions and two sectors. Analysis of the results provides some general indications about the performance of MRIO models. If a zone's final demand increases, the production in other zones also increases. The example presents two scenarios. The second assumes an increase in the final demand of region A, which causes an increase in production of the different sectors in the same region and in other regions. Furthermore, the production increase in region B is greater than that in region C because the former has higher exchange coefficients with region A than the latter, for example, because of lower transportation costs. It can also be observed that because the increase in final demand in region A is greater for sector 2 (+300) than for sector 1 (+200) and the production technology of sector 2 makes greater use of intermediate products of the same sector, the production increase in sector 2 is greater than that of sector 1 in all regions.

MRIO models with variable coefficients. Application of the MRIO model with constant coefficients assumes that the exchange and technical coefficients are independent of variables such as production level, relative prices, and generalized transportation costs. This hypothesis is only reasonable for short-term forecasts. To overcome this limitation, various extensions of model (4.7.7) have been proposed in which the exchange coefficients (matrix $\boldsymbol{T}$) and/or the technical production coefficients (matrix $\boldsymbol{A}$) are expressed as functions of other transportation and economic variables. In this sense, these extensions can be referred to as variable coefficient models.

In an initial specification, known as a *MRIO model with elastic trade coefficients*, the coefficients t_{ij}^m are obtained from an explicit descriptive or random utility model that simulates the choice of supply zone. It is usually assumed for a number of reasons (product heterogeneity within sectors, market mechanisms differing from pure competition, omitted attributes, etc.) that a zone's imports come from multiple zones (probabilistic model), rather than exclusively from the zone(s) that has (have) minimum acquisition cost (deterministic model).

The systematic utility of acquiring from zone i the product m used in zone j, V_{ij}^m, is usually a function of several variables among which are the total production of sector m in zone i, X_i^m, and the average unit acquisition cost q_{ij}^m:

$$V_{ij}^m = V\left(X_i^m, q_{ij}^m\right) \tag{4.7.8}$$

In applications, acquisition source percentages are determined with a multinomial logit model:

$$t_{ij}^m = \frac{\exp(V_{ij}^m)}{\sum_k \exp(V_{kj}^m)} \tag{4.7.9}$$

In general, then, the overall trade matrix is a function of the vector $\boldsymbol{X}$ and of the acquisition cost matrix $\boldsymbol{q}$.

$$\boldsymbol{T} = \boldsymbol{T}\left(\boldsymbol{V}(\boldsymbol{X}, \boldsymbol{q})\right) \tag{4.7.10}$$

Interpretation of the attributes included in the specification of acquisition source percentages requires further comment. The value of total production of sector m in zone i, X_i^m, can be considered a proxy of supply diversity. This attribute should

Technical Coefficient Matrix $\mathbf{A}$ $(3 \cdot 2 \times 3 \cdot 2)$

		Region A		Region B		Region C	
		Sector 1	Sector 2	Sector 1	Sector 2	Sector 1	Sector 2
Region A	Sector 1	0.30	0.10	0.00	0.00	0.00	0.00
	Sector 2	0.20	0.40	0.00	0.00	0.00	0.00
Region B	Sector 1	0.00	0.00	0.40	0.20	0.00	0.00
	Sector 2	0.00	0.00	0.30	0.70	0.00	0.00
Region C	Sector 1	0.00	0.00	0.00	0.00	0.35	0.20
	Sector 2	0.00	0.00	0.00	0.00	0.25	0.40

Matrix of Exchange or Trade Coefficients $\mathbf{T}$ $(3 \cdot 2 \times 3 \cdot 2)$

		Region A		Region B		Region C	
		Sector 1	Sector 2	Sector 1	Sector 2	Sector 1	Sector 2
Region A	Sector 1	0.50	0.00	0.30	0.00	0.10	0.00
	Sector 2	0.00	0.40	0.00	0.35	0.00	0.15
Region B	Sector 1	0.30	0.00	0.60	0.00	0.20	0.00
	Sector 2	0.00	0.35	0.00	0.50	0.00	0.25
Region C	Sector 1	0.20	0.00	0.10	0.00	0.70	0.00
	Sector 2	0.00	0.25	0.00	0.15	0.00	0.60

Vectors of final demand $\boldsymbol{Y}$ (2 hypotheses)

	Sector	Y_1	Y_2
Region A	1	100	300
	2	200	500
Region B	1	400	400
	2	200	200
Region C	1	300	300
	2	300	300

Vector of production by sector $\boldsymbol{X}$ $(3 \cdot 2 \times 1)$ for the 2 hypotheses $\boldsymbol{Y}$

	Results		
	Sector	X_1	X_2
Region A	1	498	697
	2	734	1045
Region B	1	770	964
	2	945	1290
Region C	1	625	766
	2	771	1011

Fig. 4.35 Numerical example of a 3-region, 2-sector MRIO model, assuming no external trade flows

actually be used through its logarithm ($\ln X_i^m$) and considered as a size function (see Sect. 4.3.1.2) expressing the unknown number of elementary choice alternatives. If there were other attributes M_{kj}^m correlated to the number of production units, the size function would have the more general expression:

$$\ln\left(X_i^m + \sum_k \gamma_k M_{ki}^m\right)$$

A nonbehavioral interpretation of (4.7.8) and (4.7.9) simply reflects the observation zones with lower acquisition cost and larger productions are associated with higher acquisition percentages. This may be due to agglomeration behavior in which production units tend to be set up near their supply and/or distribution markets.

In the most general case, the average unit acquisition cost q_{ij}^m can be expressed as a function of the average unit price (price index) of products m in i, p_i^m, and of

obtained by solving the input–output model or one of its generalizations such as

$$N = TADg(X) + TDg(Y) \tag{4.7.15}$$

where the matrices $Dg(X)$ and $Dg(Y)$ are obtained by arranging the elements of the vectors X and Y, respectively, along the main diagonal of a square matrix with $(n_z \cdot n_s)$ rows and columns.

Finally, matrix N provides the total flows N_{ij}^n of goods produced in sector n in zone i and consumed in zone j. These flows are expressed in monetary units and can be computed by adding up the values corresponding to all consumption sectors:

$$N_{ij}^n = \sum_m N_{ij}^{nm}$$

The last step is the transformation of the O-D matrices N_{ij}^n from values into physical quantities (tons) by freight class (market segments). This transformation is normally done using value/quantity coefficients estimated for the current situation, and then modified exogenously according to the forecasting scenarios.[26] Freight classes, identified on the basis of shipment size and/or of the manufacturing company, are closely linked to the structure and attributes of the mode choice models that are discussed in the next section.

In conclusion, a number of models having different levels of complexity and different input data requirements are available for the prediction of freight transportation demand. The most highly structured formulations of such models aim to represent the entire economy and then derive from that the demand for trade in goods. However, this wider approach requires a considerable amount of data, much of which might not be necessary if the aim of the modeling is limited to the prediction of freight transportation demand.

A further consideration concerns the interaction between macroeconomic and transportation models. The formulations described above assume that generalized transportation costs c_{ij}^m are known. However, these depend on the production costs of carriers such as road and railway haulage companies, which depend in turn on a variety of factors including the level-of-service variables for the various modes (travel times, congestion levels, etc.) as well as the carriers' production structure (production functions). It is therefore possible, at least in principle, to introduce additional feedback cycles and related equilibrium problems between generalized transportation costs and goods (and passenger) flows on the various modal networks through mode and path choice models.

[26] Value/quantity transformation coefficients can differ significantly from unit market prices because they capture the differences between physical goods movements and commercial transactions. For example, a single commercial transaction may correspond to several freight movements due to intermediate storage locations and so on. Given the increasing relevance of freight logistics on transportation demand, value/quantity transformation coefficients should be explicitly modeled as functions of relevant variables of the logistic cycle of the industrial sector to which they apply.

4.7.2 Freight Mode Choice Models

A number of formulations have been proposed for models that represent the choice by freight shippers and carriers from among available transportation modes and services. These models are derived using a variety of approaches (descriptive, microeconomic, inventory, random utility). The following section discusses freight mode choice models that are based on the random utility paradigm, because they are consistent with the general approach to demand modeling adopted in this book, and many of the models proposed using other approaches can be considered generalizations of random utility models.

Random utility models applied to represent freight mode choice can be characterized as *aggregate* or *disaggregate* according to the data used for their specification, calibration, and application. Aggregate models are based on data and attributes corresponding to aggregate freight flows between different zones with available transportation modes. These models mainly use level-of-service attributes (e.g., average consignment times, average prices, etc.) Although they are simple to apply, aggregate models have proved to have only limited analysis capabilities because many important decision factors cannot be taken into account without a greater level of disaggregation.

For these reasons, disaggregate mode choice models have been the more actively studied in recent years. These typically follow the random utility paradigm and can be divided into two types: *consignment models*, which represent mode choice for individual consignments; and *logistic models*, which represent a sequence of logistic choices including consignment size and frequency, as well as the transportation mode.

Consignment mode choice models are more frequently used in applications. They usually have a functional form that belongs to the logit family, most often of the multinomial logit type although hierarchical logit models have also been proposed in some applications. Choice alternatives typically correspond to the transportation modes available for a given consignment (truck, train, ship, air) and different services are also frequently distinguished (e.g., conventional railway or intermodal road/railway). The level-of-service attributes normally used include travel time, cost, and reliability. Other attributes frequently included in specifications correspond to characteristics of the consignment (e.g., size, goods class, frequency) and of the firm (e.g., annual invoicing, availability of own trucks, or availability of railway sidings). Figure 4.36 shows an example of a consignment mode choice model calibrated for the Italian national model.

Logistic mode choice models are newer and so far have had few applications despite their theoretical interest and their usefulness for evaluating innovative supply combinations (logistic + transportation services). These models represent mode choice in the context of the logistic decisions made by the firm that chooses the transportation mode; depending on the particular situation, this firm might be the vendor or purchaser of the items being transported. It is assumed that the choice of a transportation mode depends on the *logistic cost* of its use, which in turn is made up of different components such as

Alternatives: Train, Road, Combined Rail + Road

$V_{\text{train}} = \beta_{T_t} T_t + \beta_{\text{Mc}} Mc_t + \beta_{p>30} \cdot p > 30 + \beta_{\text{HVG}} \cdot HVG + \beta_{\text{train}} \cdot TRAIN$
$V_{\text{road}} = \beta_{T_r} T_r + \beta_{\text{Mc}} Mc_r + \beta_{\text{PSH}} \cdot PSH$
$V_{\text{combined}} = \beta_{T_c} T_c + \beta_{\text{Mc}} Mc_c + \beta_{\text{COMB}} \cdot COMB$
T_t = Train travel time
T_r = Road travel time
T_c = Combined travel time
Mc_t = Train monetary cost
Mc_r = Road monetary cost
Mc_c = Combined monetary cost
$p > 30$ = Dummy variable: 1 if the shipment weights more than 30 t, 0 otherwise
PSH = Dummy variable: 1 if goods are perishable, 0 otherwise
HVG = Dummy variable: 1 for of high value goods, 0 otherwise
$TRAIN$ = Alternative Specific Attributes (ASA)
$COMB$

	T_t	T_r	T_c	Mc	$p > 30$	PSH	HVG	$TRAIN$	$COMB$
	−0.06	−0.15	−0.12	−1.47	1.20	0.86	−0.64	0.29	−3.34
t	−1.7	−2.2	−2.0	−3.2	0.6	1.1	−1.2	0.5	−2.5

Fig. 4.36 Example of freight consignment mode choice model

- Costs associated with order management
- Costs of transportation (transportation service rates)
- Costs of loss and damage
- Costs of capital immobilized during transportation
- Costs of carrying inventory
- Costs of stockout (inadequate inventory to meet demand)
- Costs connected with the nonavailability or delayed arrival of equipment for transportation
- Costs of unreliability (early or delayed arrival and related costs of longer storage or stocking larger inventories)

Logistic costs depend on a number of factors such as the total (annual) quantity of shipments during a given commercial relation, the average frequency and size of the shipments, and the value of the goods. Furthermore, they depend on the characteristics of the service offered by the different modes such as price, reliability of shipment times, and the possibility of theft and damage. Direct information on all the components of the logistic cost is very difficult to obtain, so it is assumed that the systematic utility function for each mode j is a combination of variables X^i_{jk} linked to the logistic cost items of a certain commercial relation i and that the coefficients β_k are the unknown cost factors. As things stand, considerable information is required to specify and calibrate such models, and their current use is mostly limited to analysis of the factors that influence mode choice rather than to large-scale applications.

Reference Notes

The literature on transportation demand models is very broad and covers a period of more than 40 years.

The first partial share demand model systems were formulated in the 1950s and 1960s although with time they have undergone a number of developments, both formal and interpretive. A descriptive treatment of the traditional system can be found in the books by Wilson (1974) and Hutchinson (1974).

Since the mid-1970s, a number of travel forecasting model systems based on random utility theory have been proposed. Examples can be found in the books by Domencich and McFadden (1975), Richards and Ben-Akiva (1975), Manheim (1979), Ben-Akiva and Lerman (1985), and Ortuzar and Willumsen (2001). The systems of random utility models proposed in the literature are mainly based on factoring logit and hierarchical logit models. The general formulation of systems of partial share models based on different random utility models integrated through EMPU variables, as proposed in Sect. 4.2.1, is original.

Among the first examples of trip generation models based on cross-classification tables, the work of Oi and Shuldiner (1962) should be mentioned. An example of behavioral models of trip frequency at the urban level is contained in Biggiero (1991), and at the interurban level in Cascetta et al. (1995).

Distribution models with size functions were proposed by Richards and Ben-Akiva (1975), Koppelman and Hauser (1978), and Kitamura et al. (1979); a summary can be found in Ben-Akiva and Lerman (1985). References to descriptive or gravity distribution models can be found in Wilson (1974). An example of an urban behavioral destination choice model with explicit choice set simulation is contained in Cascetta and Papola (2009).

Descriptions of mode split models of the logit or nested logit type are extremely numerous in the literature; the books by Ben-Akiva and Lerman (1985) and Ortuzar and Willumsen (2001) give many examples.

Cascetta (1995) contains a systematic analysis of the different hypotheses underlying path choice models. Relatively few path choice models for road networks have been calibrated from empirical data. Models of this type include those by Ben-Akiva et al. (1984), Cascetta et al. (1995), and Russo and Vitetta (1995). Specification of the probit path choice model is described in Sheffi (1985). The C-Logit model is described in Cascetta et al. (1996). Vovsha and Bekhor (1998) proposed the first cross-nested logit formulation for path choice, which they called link-nested logit. Marzano and Papola (2004a) provide a comparison of the theoretical and operational properties of different random utility path choice models, and propose a new cross-nested logit formulation for path choice, called path multilevel logit.

A systematic analysis of path choice models for schedule-based transit networks is provided by Nuzzolo and Russo (1997) and by Nuzzolo et al. (2003). The interpretation of pre-trip/en-route behavior is described in Cascetta and Nuzzolo (1986), the concept of travel strategy is formulated in Spiess and Florian (1989), and the representation of a travel strategy as a network hyperpath is proposed by Nguyen and Pallottino (1988).

Theoretical contributions and applications of trip frequency, distribution, mode, and path choice models have been made for both urban and interurban contexts. The most exhaustive Italian application in this field is the demand model system for medium-size urban networks implemented within the context of the *Progetto Finalizzato Trasporti* sponsored by the CNR (National Council for Research), whose results are summarized in Cascetta and Nuzzolo (1988). The system of interurban travel-demand forecasting models described in Sect. 4.3.4 was calibrated for the Italian National Modal System SIMPT and is described in Cascetta et al. (1995).

Several trip-chaining (journey) demand models are described in the literature; an analysis and bibliographical commentary can be found in Ben-Akiva et al. (1996). Trip-chaining models based on the concept of primary destination (activity) are described in Antonisse et al. (1986) and Algers et al. (1993). The model system described in Sect. 4.4 is based on the work of Cascetta et al. (1995).

One of the first contributions on the activity-based approach is given by Adler and Ben-Akiva (1979) who proposed a model explicitly considering the daily activity program. An interesting review on the subject is provided by Jones et al. (1990). Golob and McNally (1997) sought to explicitly model all the interactions within the family and among the different activities of the day from both a spatial and a temporal perspective. Explicit activity participation models can be found in Ben-Akiva and Bowman (1998) and in McNally (2000). Interesting recent contributions have been made by Bhat et al. (2004), Olaru and Smith (2005), and Lee and McNally (2006), whereas application to real cases can be found in Ben-Akiva and Bowman (1998), Bowman and Ben-Akiva (2001) and in Bifulco et al. (2003) who also propose the modeling architecture presented in this book.

The literature on freight demand models follows different classification criteria. The contributions by Harker (1985), Picard and Nguyen (1987), Zlatoper and Austrian (1989), Mazzarino (1997), and Regan and Garrido (2001), among others, are noteworthy. SPE models are described in Frietz et al. (1983) and in the books by Harker (1985, 1987). An introduction to CGE and SCGE models can be found in Bergman (1990). A CGE model for predicting freight demand at a national level has been proposed by Roson (1993).

Chenery (1953), Izard (1951), Moses (1955), and Leontief and Strout (1963) contributed to the development of the MRIO model with constant coefficients. Miller and Blair (1985) provide a systematic overview of input–output techniques. Leontief and Costa (1987) and Costa and Roson (1988) propose some applications of the MRIO model to freight transportation demand prediction. Application of the MRIO model with elastic trade coefficients to Italian freight demand prediction is described in Cascetta et al. (1996). Its generalization to include price equilibrium and production constraints was introduced by de la Barra (1989) and, more recently, by Zhao and Kockelman (2003), whose literature review provides further references. The derivation of the MRIO model from the input–output table row–column balance constraints is taken from Marzano and Papola (2004b), who also provide a taxonomy of different input–output models proposed in the literature.

The literature on freight mode split models is quite substantial. An analysis of factors influencing the behavior of operators can be found in the volume by Bayliss

(1988); analysis and classification of the different mode split models is provided in Winston (1983); and examples of disaggregate consignment models calibrated in Italy are in Nuzzolo and Russo (1995). Modenese Vieira (1992) and Russo and Cartenì (2005) provide a description of the theoretical assumptions of logistic random utility models together with some empirical results.

Chapter 5
Basic Static Assignment to Transportation Networks

5.1 Introduction

Traffic assignment models simulate the interaction of demand and supply on a transportation network. These models allow calculation of performance measures and user flows for each supply element (network link), resulting from origin–destination (O-D) demand flows, path choice behavior, and the mutual interactions between supply and demand.

Assignment models combine the supply and demand models described in the previous chapters; for this reason they are also referred to as demand–supply interaction models. More specifically, as seen in Chap. 4, path choices and flows depend on generalized path costs; moreover, demand flows themselves are generally influenced by path costs in choice dimensions such as mode and destination. Furthermore, as seen in Chap. 2, link and path performance measures and costs generally depend on flows as a result of congestion. There is therefore a circular dependency among demand, flows, and costs; assignment models represent this dependency. Figure 5.1 illustrates the general modeling framework.

Assignment models play a central role in comprehensive transportation system models because their outputs describe the state of the system, or rather, the mean state and its variation. Assignment model outputs, in turn, are inputs required for design and/or evaluation of transportation projects.

5.1.1 Classification of Assignment Models

The system state simulated through assignment models depends on assumptions about user behavior (demand functions, path choice, available information) and the approach used for representing supply–demand interactions. Several classification criteria may be applied.

First, the fundamental classification factor for assignment models is the approach used for studying supply and demand interactions. One approach, *user equilibrium*

Giulio Erberto Cantarella is co-author of this chapter.

E. Cascetta, *Transportation Systems Analysis*, Springer Optimization and Its Applications 29,
DOI 10.1007/978-0-387-75857-2_5,

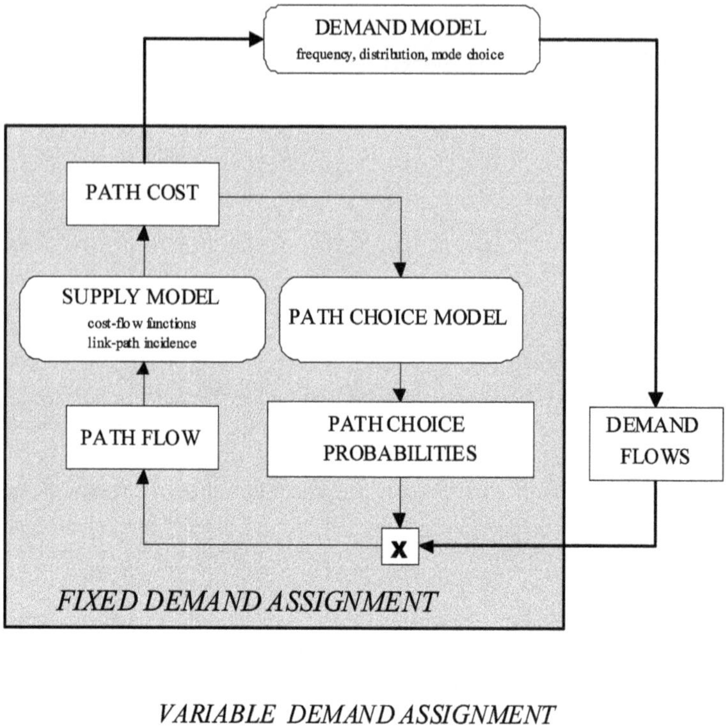

Fig. 5.1 Schematic representation of assignment models

assignment,[1] represents equilibrium configurations of the system, that is, configurations in which demand, path, and link flows are consistent with the costs that they produce in the network. From a mathematical point of view, equilibrium assignment can be defined as the problem of finding a flow vector that reproduces itself based on the correspondence defined by the supply and demand models. This problem can be easily formulated with fixed point models, or else with variational inequality or optimization models, as shown in the following sections.

The alternative approach for representing supply–demand interactions leads to *between-period* (or *day-to-day*) *dynamic process* assignment models. In this case it

[1] The concept of equilibrium in transportation systems can be compared with supply–demand equilibrium in classical economics. The analogy, however, is more formal than substantial. As seen in Chap. 2, transportation network "supply" (travel cost) functions express the average cost of using a facility as a function of the number of its users. Economic supply functions, on the other hand, relate the service quantity to be produced to the production cost and the sale price of the service. In a given transportation system, and therefore for a given service supply, the equilibrium condition defines the congruence between the demand and the functioning of the supply system, whereas equilibrium in a market defines the congruence between the behavior of two "groups": consumers and producers. Furthermore, some special aspects of the transportation system, such as the network structure of the supply, make the mathematical treatment of the problem more complex.

is assumed that the system evolves over time (i.e., in successive reference periods), through possibly different feasible states, as a result of changes in the number of users undertaking trips, path choices, supply performance, and so on. One of the mechanisms that drives the changes from one state to another is the dependency between flows and costs. In a given reference period the system state – defined by the demand, path, and link flows and the corresponding costs – may be internally inconsistent, and this may cause a change towards a different state in the following reference periods. Dynamic process assignment models explicitly simulate the evolution of the system state based on the mechanisms underlying path choice and information acquisition, which in turn determine user choices in successive reference periods. By analogy, equilibrium assignment could be termed within-day static assignment. *Dynamic process* models can be further categorized as *deterministic* or *stochastic*, depending on whether the system state is modeled using deterministic or stochastic (random) variables.

The dependence of link performance variables on flows is the other main supply-based classification factor. When link costs are independent of flows (i.e., congestion effects are negligible), *UNcongested network* (*UN*) assignment models result. On the other hand, if link costs depend on flows, *congested network* assignment models are obtained.

Assignment models can be classified based on assumptions regarding *supply characteristics*. The first classification factor is the nature of the transportation service being represented; service can be classified as either *continuous* or *scheduled*, as introduced in Chaps. 1 and 2.

Assignment models can be distinguished based on their hypotheses regarding path choice behavior presented in Sect. 4.3.3. In general, the particular path followed for a trip may result from a sequence of decisions made before and during the trip; these are referred to as *pre-trip* and *en-route* choices, respectively. Pre-trip choice, which takes place at the origin before a journey is begun, considers as alternatives either single paths to be followed without deviation from origin to destination, or decision strategies for en-route choice among paths. En-route choices involve a strategy for determining the path to follow as a result of decisions made during the journey in response to information received while traveling. Many models consider only *fully pre-trip* behavior, where the pre-trip choice is between alternative O-D paths, and the chosen path is followed unswervingly to the destination. In all cases, user choice takes into account the cost attributes of the choices offered by the network. For example, the pre-trip path choice model represents the choice of single paths or hyperpaths as a function of the corresponding cost attributes. Models based on random utility theory are typically used to simulate these choices. In particular, *deterministic* choice models assume that the perceived utility of a path is deterministic, and that users will only choose the alternative(s) having maximum average utility (minimum average cost). On the other hand, *probabilistic* or *stochastic* choice models assume that the perceived utility of a path is a random variable, and express the probability that users will choose each of the available alternatives, as described in Sect. 4.3.

With respect to demand segmentation, assignment models are called *multiuser class* models if users are subdivided into several classes. Users in different classes

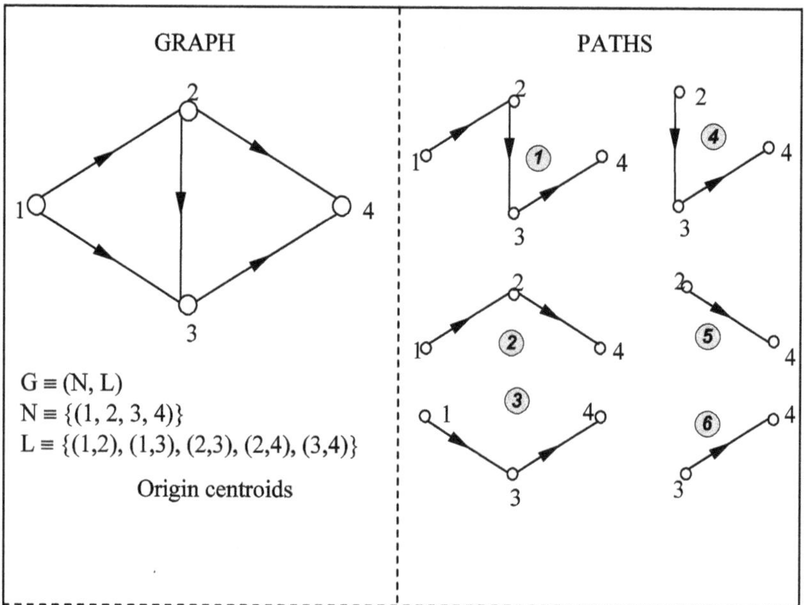

LINK-PATH INCIDENCE MATRIX

O-D Pairs	1-4			2-4		3-4
Links \ Paths	1	2	3	4	5	6
1	1	1	0	0	0	0
2	0	0	1	0	0	0
3	1	0	0	1	0	0
4	0	1	0	0	1	0
5	1	0	1	1	0	1

Fig. 5.3 Example of a graph and its link–path incidence matrix

5.2.1 *Supply Model*

Transportation supply is simulated with a (congested) network model, as described in Chap. 2. A (generalized) cost c_a is associated with each link a; if travel time

$$\boldsymbol{g} = \boldsymbol{\Delta}^T \cdot \boldsymbol{c} + \boldsymbol{g}^{NA}$$

$$\begin{bmatrix} 6 \\ 4 \\ 2 \\ \hline 4 \\ 2 \\ \hline 1 \end{bmatrix} = \begin{bmatrix} 1 & 0 & 1 & 0 & 1 \\ 1 & 0 & 0 & 1 & 0 \\ 0 & 1 & 0 & 0 & 1 \\ \hline 0 & 0 & 1 & 0 & 1 \\ 0 & 0 & 0 & 1 & 0 \\ \hline 0 & 0 & 0 & 0 & 1 \end{bmatrix} \cdot \begin{bmatrix} 2 \\ 1 \\ 3 \\ 2 \\ 1 \end{bmatrix} + \begin{bmatrix} 0 \\ 0 \\ 0 \\ \hline 0 \\ 0 \\ \hline 0 \end{bmatrix}$$

$$\boldsymbol{g} = \boldsymbol{g}^{ADD} + \boldsymbol{g}^{NA} = \boldsymbol{\Delta}^{T}\boldsymbol{c} + \boldsymbol{g}^{NA}$$

Fig. 5.4 Example of the relationship between link costs and path costs (nonadditive costs are zero for the sake of simplicity)

t_a is the only component of cost, it yields: $c_a = \beta t_a$. Furthermore, each path k is associated with a *path cost*[2] g_k, consisting of two types of cost attribute:

Linkwise additive (or generic) path costs that are obtained by adding up the corresponding costs of the links on the path, regardless of the particular O-D pair and/or path (for instance travel time); these costs may depend on link flows in the case of congested networks; *Linkwise nonadditive (or specific) path costs* that are specific to the path and/or O-D pair, in the sense that they cannot be determined by adding up the generic costs of the links on the path (for instance, some types of tolls or fees). In the following analysis, these costs are assumed to be independent of congestion. Therefore, we do not consider path costs that are simultaneously nonadditive and dependent on congestion. Let:

- $\boldsymbol{c}$ be the link cost vector, with entries c_a
- $\boldsymbol{g}_{od}^{ADD}$ be the vector of additive path costs for users of O-D pair *od*, consisting of elements $g_k^{ADD}, k \in K_{od}$
- $\boldsymbol{g}_{od}^{NA}$ be the vector of nonadditive costs for users of O-D pair *od*, consisting of elements $g_k^{NA}, k \in K_{od}$
- $\boldsymbol{g}_{od}$ be the vector of total path costs for users of O-D pair *od*, consisting of elements $g_k, k \in K_{od}$

The relationship between link costs and path costs is given for each O-D pair *od* by the following equations (see Figs. 5.3 and 5.4):

$$\begin{aligned} \boldsymbol{g}_{od}^{ADD} &= \boldsymbol{\Delta}_{od}^{T}\boldsymbol{c} \quad \forall od \\ \boldsymbol{g}_{od} &= \boldsymbol{g}_{od}^{ADD} + \boldsymbol{g}_{od}^{NA} = \boldsymbol{\Delta}_{od}^{T}\boldsymbol{c} + \boldsymbol{g}_{od}^{NA} \quad \forall od \end{aligned} \tag{5.2.1}$$

The above relation can be expressed using matrix notation. Let:

- $\boldsymbol{g}^{ADD} = [\boldsymbol{g}_{od}^{ADD}]_{od}$ be the overall vector of additive path costs, consisting of the vectors of additive path costs $\boldsymbol{g}_{od}^{ADD}$ for all O-D pairs
- $\boldsymbol{g}^{NA} = [\boldsymbol{g}_{od}^{NA}]_{od}$ be the overall vector of nonadditive path costs, consisting of the vectors of nonadditive path costs $\boldsymbol{g}_{od}^{NA}$ for all O-D pairs

---

[2]In the following sections, the indices that designate the specific origin and destination served by path k are usually omitted, because each path is uniquely associated with an O-D pair. On occasion, however, the O-D and path indices are both specified for emphasis.

Fig. 5.5 Example of the relationship between link flows and path flows

$$f = \boldsymbol{\Delta} \boldsymbol{h}$$

$$\boldsymbol{f} \quad = \quad \boldsymbol{\Delta} \quad \cdot \quad \boldsymbol{h}$$

$$\begin{bmatrix} 335 \\ 665 \\ 494 \\ 1341 \\ 1959 \end{bmatrix} = \left[\begin{array}{ccc|cc|c} 1 & 1 & 0 & 0 & 0 & 0 \\ 0 & 0 & 1 & 0 & 0 & 0 \\ 1 & 0 & 0 & 1 & 0 & 0 \\ 0 & 1 & 0 & 0 & 1 & 0 \\ 1 & 0 & 1 & 1 & 0 & 1 \end{array}\right] \cdot \left[\begin{array}{c} 90 \\ 245 \\ 665 \\ \hline 404 \\ 1096 \\ \hline 800 \end{array}\right]$$

$\boldsymbol{g} = [\boldsymbol{g}_{od}]_{od}$ be the overall vector of the total path costs, consisting of the vectors of total path costs $\boldsymbol{g}_{od}$ for all O-D pairs

A flow f_a is associated with each link a. Link flows are measured in units commensurate with demand flows. Let:

$\boldsymbol{f}$ be the link flow vector, with entries f_a.

In congested networks, as described in Chap. 2, link costs depend on link flows through the cost functions:

$$\boldsymbol{c} = \boldsymbol{c}(\boldsymbol{f}) \tag{5.2.2}$$

In turn, link flows depend, through the *network flow propagation* model, on the flow associated with each path. In particular, for a given O-D pair, the path flows induce the corresponding O-D specific link flows through the link-path incidence matrix. Furthermore, the total flow on a link is the sum of the flows induced by all paths and all O-D pairs. (Demand, path, and link flows are assumed to be expressed in consistent units.) Let:

$\boldsymbol{h}_{od}$ be the path flow vector for users of O-D pair *od*, the elements of which are the flows h_k for all $k \in K_{od}$

$\boldsymbol{f}^{od}$ be the vector of O-D specific link flows f_a^{od}, resulting from the trips for O-D pair *od* over available paths

The relationship between link flows and path flows is expressed by the following equations (Fig. 5.5):

$$\boldsymbol{f}^{od} = \boldsymbol{\Delta}_{od} \boldsymbol{h}_{od} \quad \forall od$$

from which

$$\boldsymbol{f} = \sum_{od} \boldsymbol{f}^{od} = \sum_{od} \boldsymbol{\Delta}_{od} \boldsymbol{h}_{od} \tag{5.2.3}$$

All the above relations can be expressed using matrix notation. Let:

$\boldsymbol{h} = [\boldsymbol{h}_{od}]_{od}$ be the overall vector of path flows, consisting of the vectors of path flows

$\boldsymbol{h}_{od}$ for all O-D pairs

The whole supply model is defined by (5.2.1) to (5.2.3) which combine to express the relationship between path costs and path flows that was introduced in Chap. 2:

$$\boldsymbol{g}_{od} = \boldsymbol{\Delta}_{od}^T \boldsymbol{c}\left(\sum_{od} \boldsymbol{\Delta}_{od} \boldsymbol{h}_d\right) + \boldsymbol{g}_{od}^{\text{NA}} \quad \forall od \tag{5.2.4}$$

The above relations can be expressed using matrix notation.

$$\boldsymbol{g} = \boldsymbol{\Delta}^T \boldsymbol{c}(\boldsymbol{\Delta h}) + \boldsymbol{g}^{\text{NA}}$$

If the cost functions (are continuous and) have continuous first derivatives with respect to link flows, the supply model (is also continuous and) has continuous first derivatives with respect to path flows. The presence of nonadditive path costs guarantees that any linear transformation does not modify the results of the model.

5.2.2 Demand Model

As stated earlier, it is assumed here that O-D demand flows are known and independent of cost variations; thus path choice – the way that paths flow themselves through the network – is the only choice dimension explicitly simulated. It is also assumed that the demand flows for different O-D pairs are expressed in consistent units. For private passenger modes such as a car, for example, they are typically measured in vehicles or drivers per unit of time, whereas for public (scheduled) transport modes they are usually expressed in terms of passengers per unit of time. Let:

$d_{od} \geq 0$ be the demand flow for O-D pair *od*, defined by the elements of the O-D matrix corresponding to the purpose, mode, and time band being analyzed

$\boldsymbol{d}$ the demand vector, whose components are the demand values d_{od} for each O-D pair *od*

Path choice behavior is simulated with random utility models, assuming that the relevant component of the systematic utility is equal to the negative of the generalized path cost (utility function; Sect. 4.3.3):

$$\boldsymbol{V}_{od} = -\beta \boldsymbol{g}_{od} + \boldsymbol{V}_{od}^{\circ} \quad \forall od \tag{5.2.5}$$

where

β is a utility parameter[3] (see Chap. 3), which is omitted in the following because it is assumed included in the scale parameter within the choice function, introduced below (see Sects. 4.2 and 5.5)

$\boldsymbol{V}_{od}$ is a vector whose elements consist of the systematic path utilities $V_k, k \in K_{od}$, for users of O-D pair *od*

[3]Note that this parameter is measured in units inverse with the utility. Therefore a change in the measurement units of the cost-related attributes does not affect the systematic utility value.

$\boldsymbol{V}^{\circ}_{od}$ is a vector whose elements are the parts of the systematic utility that depend on attributes other than path costs (such as users' socioeconomic attributes); with no loss of generality, from a mathematical point of view attributes in vector $\boldsymbol{V}^{\circ}_{od}$ may be considered within nonadditive path cost vector or vice versa, hence for simplicity this term is generally omitted in the following sections (clearly any change of the reference utility value does not modify the results of the model)

Thereafter, (path or link) costs are assumed measured in units commensurate with the utility by using appropriate coefficients (with the same meaning of β coefficients introduced in Chap. 3).

Path choice probabilities depend on the systematic utilities of the available paths through the path choice function. Let:

$p_{od,k} = p[k/od] \geq 0$ be the probability that a user on a trip from origin o to destination d will use path k, $k \in K_{od}$, with $\sum_{k \in K_{od}} p_{od,k} = 1$

$\boldsymbol{p}_{od} \geq \boldsymbol{0}$ be the vector of path choice probabilities for users of O-D pair od, whose elements are the probabilities $p_{od,k}$, $k \in K_{od}$, with $\boldsymbol{1}^{\mathrm{T}} \boldsymbol{p}_{od} = 1$

As seen in Sect. 4.2, a random utility model used to simulate path choice is given by

$$p_{od,k} = p[k/od] = Prob[V_k - V_j \geq \varepsilon_j - \varepsilon_k \; \forall j \in K_{od}] \quad \forall od, k$$

$$\boldsymbol{p}_{od} = \boldsymbol{p}_{od}(\boldsymbol{V}_{od}) \quad \forall od$$

where ε_j denotes the random residual corresponding to the perceived utility of path j. If the random residuals are equal to zero ($\varepsilon_j = 0$), then the variance–covariance matrix of the random residuals is null ($\boldsymbol{\Sigma} = 0$), and the resulting choice model is deterministic. On the other hand, if the variance–covariance matrix of the random residuals is nonnull and nonsingular, $|\boldsymbol{\Sigma}| \neq 0$, then the model is probabilistic (see Sect. 3.2).

A relation between path choice probabilities and path costs for O-D pair od, known as the *path choice map*, is obtained by combining the path choice function with the systematic utility function:

$$p_{od,k} = p_{od,k}(\boldsymbol{V}_{od}) = p_{od,k}(-\boldsymbol{g}_{od}) \quad \forall od, k$$

$$\boldsymbol{p}_{od} = \boldsymbol{p}_{od}(\boldsymbol{V}_{od}) = \boldsymbol{p}_{od}(-\boldsymbol{g}_{od}) \quad \forall od$$

The flow h_k on path k connecting O-D pair od, $k \in K_{od}$, is simply given by the product of the demand flow d_{od} and the probability of choosing path k:

$$h_k = d_{od}\, p_{od,k}$$

and is measured in demand units. Thus, for each O-D pair, the relationship between path flows, path choice probabilities and demand flows is given by:

$$\boldsymbol{h}_{od} = d_{od}\, \boldsymbol{p}_{od}(\boldsymbol{V}_{od}) \quad \forall od \tag{5.2.6}$$

The whole demand model is defined by the relations (5.2.5) and (5.2.6) which, combined, describe the relationship between path flows and path costs:

$$\boldsymbol{h}_{od} = d_{od}\,\boldsymbol{p}_{od}(-\boldsymbol{g}_{od}) \quad \forall od \tag{5.2.7}$$

The above equation (5.2.7) is a particular specification of (4.2.2) consistent with the assumptions introduced at the beginning of this section. It should be noted that the choice function $\boldsymbol{p}_{od}()$ may vary with O-D pair.

All the above relations can be expressed using matrix notation (Fig. 5.6). Let:

$\boldsymbol{P}$ be the path choice probability matrix, with a column for each O-D pair od, a row for each path k, and element (k, od) given by $p[k/od]$ if path k connects the O-D pair, otherwise zero ($\boldsymbol{P}$ is a block diagonal matrix with blocks given by the vectors $\boldsymbol{p}_{od}$)

The previous equations become:

$$\boldsymbol{P} = \boldsymbol{P}(\boldsymbol{V}) = \boldsymbol{P}(-\boldsymbol{g})$$

$$\boldsymbol{h} = \boldsymbol{P}(\boldsymbol{V})\boldsymbol{d}$$

$$\boldsymbol{h} = \boldsymbol{P}(-\boldsymbol{g})\boldsymbol{d}$$

Different probabilistic path choice models ($|\boldsymbol{\Sigma}| \neq 0$; see Sect. 4.3.3) can be specified according to different assumptions on the joint probability density function of perceived utilities or random residuals. In any case a (one-to-one) function $\boldsymbol{p}_{od}()$ is obtained. An example is provided in Fig. 5.6a. Some useful general requirements for stochastic assignment are discussed below.

Continuity of the path choice model, $\boldsymbol{p}_i = \boldsymbol{p}_i(\boldsymbol{g}_i)$, assures that small changes in path costs induce small changes in choice probabilities. If it is also continuously differentiable it has a continuous Jacobian, $\mathbf{Jac}[\boldsymbol{p}_i(\boldsymbol{g}_i)]$. This feature, assured by commonly used joint probability density functions, guarantees continuity of the resulting SNL function. Thus it is useful to state existence of stochastic user equilibrium.

Monotonicity of the path choice model, $\boldsymbol{p}_i = \boldsymbol{p}_i(\boldsymbol{g}_i)$, ensures that an increase in the cost of a path k induces a decrease in the corresponding choice probability. More generally, the path choice model, $\boldsymbol{p}_i = \boldsymbol{p}_i(\boldsymbol{g}_i)$, should be nonincreasing monotone with respect to path costs. This feature guarantees monotonicity of the resulting SNL function. Hence it is useful to state uniqueness of solutions of stochastic user equilibrium. It is ensured if no other parameter of the perceived utility joint probability density functions depends on the mean, say the systematic utility. The resulting choice function is called *invariant* (see Sect. 3.4).

Independence of linear transformations of utility ensures that no change in the scale of the utility affects the model results (as guaranteed by commonly used random residual joint probability density functions, such as Gumbel, or Normal distributions). For instance, it is not relevant whether travel time is measured in hours or minutes.

In addition to the above mathematical requirements, some modeling requirements presented below are useful to effectively simulate path choice behavior.

Fig. 5.6a Example of demand model with probabilistic path choice

$$g^T = [6 \quad 4 \quad 2 \quad 4 \quad 2 \quad 1]$$

$$p_{od,k} = \frac{\exp(-g_{od,k}/\theta)}{\sum_{j \in K_{od}} \exp(-g_{od,j}/\theta)};$$

$$\theta = 2$$

$$\boldsymbol{p}_{14} = \begin{bmatrix} 0.090 \\ 0.245 \\ 0.665 \end{bmatrix}; \qquad \boldsymbol{p}_{24} = \begin{bmatrix} 0.269 \\ 0.731 \end{bmatrix}; \qquad \boldsymbol{p}_{34} = [1.000]$$

P Matrix

Path \ O-D pair	1–4	2–4	3–4
1	0.090	0	0
2	0.245	0	0
3	0.665	0	0
4	0	0.269	0
5	0	0.731	0
6	0	0	1.000

$$\boldsymbol{h} = \boldsymbol{P} \cdot \boldsymbol{d}$$

$$\begin{bmatrix} 90 \\ 245 \\ 665 \\ 404 \\ 1097 \\ 800 \end{bmatrix} = \begin{bmatrix} 0.090 & 0 & 0 \\ 0.245 & 0 & 0 \\ 0.665 & 0 & 0 \\ 0 & 0.269 & 0 \\ 0 & 0.731 & 0 \\ 0 & 0 & 1.000 \end{bmatrix} \cdot \begin{bmatrix} 1000 \\ 1500 \\ 800 \end{bmatrix}$$

Similarity of perception of partially overlapping paths rules out counterintuitive results. Indeed two partially overlapping paths are likely not perceived as two totally separate paths. Introducing a positive covariance between any two overlapping paths can simulate similarity, as in the probit choice model, or a communality factor as in the C-logit choice model (see Sect. 4.3.3).

Independence of link segmentation (within the network model) ensures that if a link is further divided into sublinks and link costs redefined such that path costs are not affected, path perceived utility distribution is not affected either, nor are choice probabilities. This feature is clearly guaranteed for path explicit formulations of the distribution of perceived utility (e.g., logit model). If the distribution of perceived utility is formulated from link distributions (e.g., some probit specifications) this feature is only guaranteed for distributions stable w.r.t. summation (e.g., Normal distribution).

Negativity of perceived utility ensures that no user perceives a positive utility to travel along any path. This feature is ensured by assuming lower bounded distributions (for instance, log-normal, or Gamma). According to this feature a nonelementary path is always a worse choice than the elementary path within it, thus supporting the assumption of considering elementary paths alone. On the other hand, if this feature is not presented, a nonelementary path may be a better choice than the el-

Fig. 5.6b Example of a demand model with deterministic path choice

$$g^T = [6 \quad 4 \quad 2 \quad 4 \quad 2 \quad 1]$$

$$p_{od,k} \begin{cases} \in [0,1] & \text{if } g_{od,k} = \min_{j \in k_{od}} g_{od,j} \\ = 0 & \text{if } g_{od,k} > \min_{j \in k_{od}} g_{od,j} \end{cases}$$

$$\sum_{k \in k_{od}} p_{od,k} = 1$$

$$p_{14} = \begin{bmatrix} 0 \\ 0 \\ 1 \end{bmatrix}; \qquad p_{24} = \begin{bmatrix} 0 \\ 1 \end{bmatrix}; \qquad p_{34} = [1]$$

P Matrix

Path \ O-D pair	1–4	2–4	3–4
1	0	0	0
2	0	0	0
3	1	0	0
4	0	0	0
5	0	1	0
6	0	0	1

$$h_{\text{DET}} = P \cdot d$$

$$\begin{bmatrix} 0 \\ 0 \\ 1000 \\ 0 \\ 1500 \\ 800 \end{bmatrix} = \begin{bmatrix} 0 & 0 & 0 \\ 0 & 0 & 0 \\ 1 & 0 & 0 \\ 0 & 0 & 0 \\ 0 & 1 & 0 \\ 0 & 0 & 1 \end{bmatrix} \cdot \begin{bmatrix} 1000 \\ 1500 \\ 800 \end{bmatrix}$$

ementary path within it; hence, nonelementary paths should be included within the path choice set (which may no longer be finite), possibly leading to unrealistic situations (some algorithmic drawbacks may also arise). Several adopted distributions (Gumbel, MVN) fail to satisfy this requirement, even though this condition is not relevant in practice.

Deterministic path choice models ($\boldsymbol{\Sigma} = \mathbf{0}$; see Sect. 3.2) usually result in a one-to-many map because, if there are several minimum cost paths between an O-D pair *od*, the choice probability vector $\boldsymbol{p}_{\text{DET},od}$, and therefore the path flow vector $\boldsymbol{h}_{\text{DET},od}$, are not uniquely defined. An example is given in Fig. 5.6b. General requirements discussed above can be quite easily extended to a deterministic choice model.

It can be useful to reformulate the deterministic demand model (5.2.7) as a system of inequalities. This system is obtained by applying to each O-D pair condition (3.4.11a) on deterministic choice probabilities $\boldsymbol{p}_{\text{DET},od}$; it is repeated here for the convenience of the reader:

$$(\boldsymbol{V}_{od})^T(\boldsymbol{p}_{od} - \boldsymbol{p}_{\text{DET},od}) \leq 0 \quad \forall \boldsymbol{p}_{od}: \boldsymbol{p}_{od} \geq \mathbf{0},\ \mathbf{1}^T \boldsymbol{p}_{od} = 1\ \forall od$$

The corresponding (weighted) average cost AC_{od} is obtained by dividing by the demand flow:

$$AC_{od} = TC_{od}/d_{od} = (\boldsymbol{g}_{od})^T \boldsymbol{h}_{od}/d_{od} \quad \forall od$$

The total network cost TC is given by the sum of the total O-D costs over all O-D pairs:

$$TC = \sum_{od} TC_{od} = \sum_{od} \sum_{k \in K_{od}} h_k g_k = \sum_k h_k g_k = \boldsymbol{g}^T \boldsymbol{h}$$

The network-level average cost AC is obtained by weighting the average costs of all the O-D pairs by the corresponding demand flows, that is, by weighting the path costs by the path flows:

$$\begin{aligned} AC &= \left(\sum_{od} AC_{od} d_{od}\right) \Big/ \left(\sum_{od} d_{od}\right) = \left(\sum_{od} \sum_{k \in K_{od}} h_k g_k\right) \Big/ \left(\sum_{od} \sum_{k \in K_{od}} h_k\right) \\ &= \left(\sum_{od} TC_{od}\right) \Big/ \left(\sum_{od} d_{od}\right) = TC/d_{..} = \boldsymbol{g}^T \boldsymbol{h}/\boldsymbol{1}^T \boldsymbol{h} = \boldsymbol{g}^T \boldsymbol{h}/\boldsymbol{1}^T \boldsymbol{d} \end{aligned}$$

where $d_{..} = \sum_{od} d_{od} = \sum_{od} \sum_{k \in K_{od}} h_k = \boldsymbol{1}^T \boldsymbol{h} = \boldsymbol{1}^T \boldsymbol{d}$ denotes the total demand flow.

With reference to additive and nonadditive path costs, the following also holds:

$$TC = (\boldsymbol{g}^{\mathrm{ADD}})^T \boldsymbol{h} + (\boldsymbol{g}^{\mathrm{NA}})^T \boldsymbol{h} = (\boldsymbol{\Delta}^T \boldsymbol{c})^T \boldsymbol{h} + (\boldsymbol{g}^{\mathrm{NA}})^T \boldsymbol{h} = \boldsymbol{c}^T \boldsymbol{f} + (\boldsymbol{g}^{\mathrm{NA}})^T \boldsymbol{h}$$

an expression that, when nonadditive path costs are zero ($\boldsymbol{g}^{\mathrm{NA}} = \boldsymbol{0}$), reduces to:

$$TC = \boldsymbol{c}^T \boldsymbol{f} = \sum_a f_a c_a \tag{5.2.10}$$

In other words, in the absence of nonadditive costs the sum of the link costs multiplied by the corresponding flows coincides with the total network cost.

An *Expected Maximum Perceived Utility* (or *EMPU*), s_{od}, can be associated with each O-D pair *od*; it depends on the path choice model (see Sect. 3.4). The EMPU is a function of the systematic utilities of the available paths (neglecting here the other attributes $\boldsymbol{V}^{\circ}_{od}$ for the sake of simplicity):

$$s_{od} = s_{od}(\boldsymbol{V}_{od}) = s_{od}(-\boldsymbol{g}_{od}) = s_{od}\left(-\boldsymbol{\Delta}_{od}^T \boldsymbol{c} - \boldsymbol{g}_{od}^{\mathrm{NA}}\right) \quad \forall od \tag{5.2.11}$$

Recall (see Sect. 3.4) that the EMPU is greater than or equal to the maximum systematic utility and therefore to the average systematic utility as well:

$$s_{od} \geq \max(\boldsymbol{V}_{od}) \geq (\boldsymbol{V}_{od})^T \boldsymbol{p}_{od} = (\boldsymbol{V}_{od})^T \boldsymbol{h}_{od}/d_{od} \quad \forall od$$

The EMPU is therefore greater than or equal to the negative of the minimum cost over all the paths, which in turn is greater than or equal to the negative of the average

O-D pair	*Path*	*Cost*	*Flow*	*Total cost*	*Average cost*	$-\min(g)$	$\exp(-C/\theta)$	*Average EMPU* $s = \theta \times \ln(\sum \exp(-C/\theta))$	*Total EMPU*
1–4	1	6	90	540			0.00248		
	2	4	245	980			0.01832		
	3	**2**	665	1330			0.13534		
	Total		1000	2850			0.15613		−1857
					2.85	2.00		−1.85	
2–4	4	4	404	1616			0.01832		
	5	**2**	1096	2192			0.13534		
	Total		1500	3808			0.15365		2810
					2.54	2.00		−1.87	
3–4	6	**1**	800	800			0.36788		
	Total		800	800			0.36788		800
					1.00	1.00		−1.00	
Total network values			3300	7458					5467
Average network values					2.26	1.75		−1.66	

Fig. 5.7 Performance indicators for the network in Fig. 5.3

cost:

$$s_{od} \geq -\min(\boldsymbol{g}_{od}) \geq -(\boldsymbol{g}_{od})^T \boldsymbol{h}_{od}/d_{od} = -AC_{od} \quad \forall od$$

The total EMPU, *TS*, is defined as the sum of each O-D pair's EMPU multiplied by the corresponding demand flow:

$$TS = \sum_{od} d_{od} s_{od}(\boldsymbol{V}_{od}) = \sum_{od} d_{od} s_{od}(-\boldsymbol{g}_{od}) = \sum_{od} d_{od} s_{od}\left(-\boldsymbol{\Delta}_{od}^T \boldsymbol{c} - \boldsymbol{g}_{od}^{\text{NA}}\right)$$

The corresponding average EMPU, *AS*, is obtained by dividing by the total demand flow:

$$AS = \sum_{od} d_{od} s_{od} \Big/ \sum_{od} d_{od} = \sum_{od} d_{od} s_{od}/d_{..} = TS/d_{..}$$

In conclusion, the total cost is an estimate, made without considering the effect of dispersion, of the disutility users receive when distributing themselves among paths according to path flows $\boldsymbol{h}$, whereas the EMPU is the disutility users perceive when making path choices leading to path flows $\boldsymbol{h}$ including the effect of dispersion. From the preceding considerations, the following relations hold between the total and average values of EMPU and cost.

$$TS \geq -TC \qquad AS \geq -AC$$

Numerical examples of network indicators are presented in Fig. 5.7.

As examples, the preceding relationships are applied to two different path choice models for which the EMPU can be calculated in closed form. The first example is a logit path choice model with parameter θ_{od}, which gives (see Sect. 3.4):

$$TS = \sum_{od} d_{od}\theta_{od} \ln\Big(\sum_{k\in K_{od}} \exp(V_k/\theta_{od})\Big) = \sum_{od} d_{od}\theta_{od} \ln\Big(\sum_{k\in K_{od}} \exp(-g_k/\theta_{od})\Big)$$
$$\geq -\sum_{od} d_{od} \min(\boldsymbol{g}_{od}) \geq -\sum_{od} d_{od} \sum_{k\in K_{od}} g_k(h_k/d_{od}) = -TC$$

The second example is a deterministic path choice model, for which the EMPU is equal to both the maximum systematic utility and the average systematic utility (Sect. 3.4); the total EMPU is thus equal to the negative of the total cost:

$$TS = \sum_{od} d_{od} \max(\boldsymbol{V}_{od}) = \sum_{od} d_{od} \boldsymbol{V}^T_{od} \boldsymbol{p}_{od} = \sum_{od} d_{od} \boldsymbol{V}^T_{od} (\boldsymbol{h}_{od}/d_{od}) = -TC$$

because, in this case, elements of the choice probability $\boldsymbol{p}_{od}$ vector and therefore the path flow vector $\boldsymbol{h}_{od}$ are nonzero only for minimum cost paths (Sect. 4.3.3).

5.3 Uncongested Networks

Assignment to uncongested networks is based on the assumptions that costs do not depend on flows.[5] In other words, path flows, and thus link flows, are obtained from path choice probabilities that are themselves computed from flow-independent link performance attributes and costs.

Uncongested assignment models are used for the analysis of relatively uncongested road transportation systems (generally, link cost functions are almost flat with respect to flows for flow-capacity ratios up to values around 0.50–0.70). They are also often used for analyzing public transport systems, for which costs may be assumed independent of link passenger flows if the available capacity is sufficient. Furthermore, uncongested network assignment models are a key component of congested network assignment models, which are described in the following sections.

UNcongested network (UN) assignment models are defined by the demand model (5.2.7), expressing path flows as a function of path costs and demand flows:

$$\begin{aligned} \boldsymbol{h}_{\mathrm{UN},od} &= \boldsymbol{h}_{\mathrm{UN},od}(\boldsymbol{g}_{od}; d_{od}) = d_{od}\boldsymbol{p}_{od}(-\boldsymbol{g}_{od}) \quad \forall od \\ \boldsymbol{h}_{\mathrm{UN}} &= \boldsymbol{h}_{\mathrm{UN}}(\boldsymbol{g}; \boldsymbol{d}) = \boldsymbol{P}(-\boldsymbol{g})\boldsymbol{d} \end{aligned} \tag{5.3.1}$$

[5] In the literature these are sometimes referred to as *network loading models*. In this book that term refers to a specific component of the supply model, and is an alternative expression for *network flow propagation models*. *Network loading* is intended to capture the effects of users moving over the network and inducing link loads, rather than the full range of demand–supply interactions implied by assignment models. This meaning of the term is also well established in the context of within-day dynamic supply models.

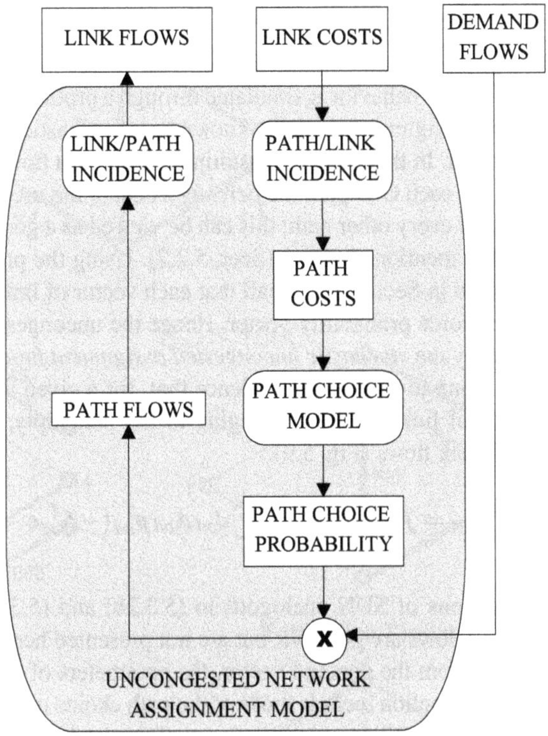

Fig. 5.8 Schematic representation of uncongested network assignment models

The path costs $\boldsymbol{g}$ can be obtained from the link costs $\boldsymbol{c}$ with (5.2.1), and the link flows $\boldsymbol{f}$ corresponding to the path flows $\boldsymbol{h}$ are given by (5.2.3). Figure 5.8 depicts these relationships graphically, applying the framework in Fig. 5.1 to the case of uncongested network assignment.

General uncongested network assignment models can also be expressed in terms of link variables by combining (5.3.1) with (5.2.1) and (5.2.3). The result is called the *uncongested network assignment map*, which associates a link flow vector with each demand flow vector and link cost vector, and can be expressed in an aggregate or disaggregate way as

$$\boldsymbol{f}_{\mathrm{UN}} = \boldsymbol{f}_{\mathrm{UN}}(\boldsymbol{c}; \boldsymbol{d}) = \sum_{od} d_{od} \boldsymbol{\Delta}_{od} \boldsymbol{p}_{od}\left(-\boldsymbol{\Delta}_{od}^T \boldsymbol{c} - \boldsymbol{g}_{od}^{\mathrm{NA}}\right) \quad \forall \boldsymbol{c}$$
$$\boldsymbol{f}_{\mathrm{UN}} = \boldsymbol{f}_{\mathrm{UN}}(\boldsymbol{c}; \boldsymbol{d}) = \boldsymbol{\Delta} \boldsymbol{P}(-\boldsymbol{\Delta}^T \boldsymbol{c} - \boldsymbol{g}^{\mathrm{NA}})\boldsymbol{d} \quad \forall \boldsymbol{c} \tag{5.3.2}$$

Note that link flows depend nonlinearly on the link costs, but linearly on the demand flows, so that the effect of each O-D pair can be evaluated separately.

In the next sections, probabilistic and deterministic path choice models, which lead respectively to stochastic and deterministic uncongested network assignment models and algorithms, are considered in turn.

lated when explicit path enumeration can be carried out as shown below; otherwise algorithms described in Sect. 5.3.3 can be used.

When paths are explicitly enumerated, path costs can be easily computed from link costs by applying the link–path incidence relationship (5.2.1). Nonadditive costs can be easily handled. Similarly, path flows can be obtained by applying the demand model (5.2.7) and its extensions, and link flows can be computed from path flows using the congruence relationship (5.2.3). Eventually, EMPU, given by $s_{od} = s_{od}(-\boldsymbol{\Delta}_{od}^T \boldsymbol{c} - \boldsymbol{g}_{od}^{\mathrm{NA}})$, which is related to the path choice alternatives available for O-D pair *od*, can also be readily calculated.

It should be recalled that, for probit path choice models, it is not possible analytically to calculate choice probabilities or to evaluate the demand model (5.2.7) and its extensions. Nonetheless, unbiased estimates of path choice probabilities and of the corresponding path flows can be obtained in the probit case by applying a Monte Carlo sampling technique.[6] The method generates a random vector realization, where each component of the vector is considered the perceived cost random residual of an O-D path. The corresponding path perceived cost is computed by adding the path systematic cost to the residual. The perceived costs of all O-D paths are computed in this way. For each O-D pair, the demand flow is assigned to the path with the minimum perceived cost. These steps are repeated for each of a sample of m random vector realizations, and the resulting path flows are averaged. These averages are unbiased estimates of the stochastic uncongested network path flows:

$$\bar{\boldsymbol{h}}^m = \sum_{j=1,m} \boldsymbol{h}^j / m$$

where

$\boldsymbol{h}^j = \boldsymbol{h}_{\mathrm{SPA}}(\boldsymbol{g} + \boldsymbol{\varepsilon}^j)$ is the vector of path flows obtained by assigning the demand flow of each O-D pair to the shortest path w.r.t. the perceived path costs $\boldsymbol{g} + \boldsymbol{\varepsilon}^j$

$\boldsymbol{g}$ is the vector of systematic path costs

$\boldsymbol{\varepsilon}^j \leftarrow MVN(\boldsymbol{0}, \boldsymbol{\Sigma})$ is the jth (in a sample of m) perceived path cost random residual vector; in probit path choice, $\boldsymbol{\varepsilon}^j$ is obtained as a realization of a multivariate normal random variable with zero mean and variance–covariance matrix $\boldsymbol{\Sigma}$

$\boldsymbol{h}^m$ is an unbiased estimate of the SUN assignment path flow vector, obtained from a sample of m perceived path cost vectors

Moreover, the average perceived shortest path cost, computed with respect to the paths that connect an O-D pair, is an unbiased estimate of EMPU associated with the O-D pair path choice alternatives.

In practice, the path flow estimate $\bar{\boldsymbol{h}}^m$ can be obtained by evaluating the following recursive equations up to $j = m$, starting with $j = 0$ and $\bar{\boldsymbol{h}}^o = 0$:

$$j = j + 1$$

[6] In fact, this approach can be adopted for any random residual distribution.

$$\boldsymbol{\varepsilon}^j \leftarrow MVN(\mathbf{0}, \boldsymbol{\Sigma})$$
$$\boldsymbol{h}^j = \boldsymbol{h}_{\mathrm{SPA}}(\boldsymbol{g} + \boldsymbol{\varepsilon}^j)$$
$$\bar{\boldsymbol{h}}^j = \left((j-1)\bar{\boldsymbol{h}}^{j-1} + \boldsymbol{h}^j\right)/j$$

In applications, direct use of this approach can be computationally burdensome because of the need to generate multiple realizations of a multivariate normal random variable with nonzero covariances, $\boldsymbol{\varepsilon}^j \leftarrow MVN(\mathbf{0}, \boldsymbol{\Sigma})$. On the other hand, the method allows arbitrary covariance structures (due, e.g., to positive or negative correlations between the perceived cost random residuals of different links). When this generality is not required, it is convenient to generate perceived path costs from link costs, adopting the same approach described in Sect. 5.3.3.

5.3.2 Models for Deterministic Assignment

Under the assumption of deterministic path choice behavior, the demand flow of each O-D pair is assigned to the minimum cost path(s) (i.e., paths with maximum systematic utility), whereas no flow is assigned to other paths. For this reason, Deterministic UNcongested network (DUN) assignment is also known as all-or-nothing assignment.[7] In general, as has already been noted, multiple path choice probability vectors may correspond to a single vector of link and path costs. It follows that the general uncongested network assignment relationship (5.3.2) must be specified as the *deterministic uncongested network assignment map* $\boldsymbol{h}_{\mathrm{DUN}} = \boldsymbol{h}_{\mathrm{DUN}}(\boldsymbol{g}; \boldsymbol{d}) \in S_h$, which is a one-to-many (or point-to-set) map between path costs and flows. In other words, because there may be several alternative minimum cost paths connecting an origin to a destination, a given path and link cost vector may correspond to multiple vectors of deterministic uncongested network path and link flows. Consequently, study of the properties of deterministic network loading frequently uses indirect formulations, equivalent to (5.3.2), based on the formulation of the deterministic demand model as a system of inequalities (5.2.7b). Summing the inequalities (5.2.7b) over all O-D pairs yields expression (5.3.4):

$$\boldsymbol{g}^T(\boldsymbol{h} - \boldsymbol{h}_{\mathrm{DUN}}) \geq 0 \quad \forall \boldsymbol{h} \in S_h \tag{5.3.4}$$

The resultant path (or link) flows satisfy Wardrop's first principle. Figure 5.10 presents an example of the deterministic uncongested network assignment model.

[7] When using a stochastic uncongested network assignment, a positive choice probability can be associated with a path whose systematic cost is greater than the minimum; it is equal to the probability that the path has maximum perceived utility (or minimum perceived cost) in the path choice set. Because of this, O-D flows are spread over multiple paths and stochastic uncongested network assignment is sometimes referred to as multipath assignment, as compared to all-or-nothing assignment that corresponds to the deterministic case.

Regardless of the model adopted, the link flow vector (or rather one of the vectors) resulting from deterministic uncongested network assignment can easily be calculated when using path choice models based on explicit path enumeration. When nonadditive path costs are equal to zero, a link flow vector can easily be obtained without explicit path enumeration using procedures based on algorithms for the calculation of minimum cost paths (see Sect. 5.3.3), or by directly solving optimization models (5.3.7) and (5.3.8).[8]

5.3.3 Algorithms Without Explicit Path Enumeration

Algorithms for assignment to noncongested networks and those for determining the minimum cost paths on which they are based, are exact algorithms, in the sense that convergence to the solution sought is guaranteed in a finite number of steps, which generally depends on the number of nodes.

Shortest Path Algorithms

Modeling of path choice behavior in assignment algorithms frequently involves identification of the shortest paths between pairs of nodes. In particular, assignment algorithms that incorporate deterministic path choice assumptions require the identification of the shortest path (or paths) between each pair of nodes, whereas stochastic uncongested network assignment algorithms that incorporate probabilistic path choice models sometimes compute shortest paths as a step in the processing. Furthermore, models that construct a relevant path set by applying a selective approach and explicitly enumerating paths (described in Sect. 4.3.3) generally involve the solution of a shortest path problem. For example, the relevant path set could be specified as the set of paths that minimize different link attributes such as distance, monetary cost, and travel time; alternatively, they might be identified as the first k shortest paths with respect to some link attribute.

If only *elementary paths* (those without loops) are relevant, there are a finite number of them and in principle they could be enumerated for each pair of origin and destination nodes. The shortest such path could then be identified by inspection. When explicit enumeration of all paths is not feasible due to their large number, as is often the case, algorithms that avoid explicit enumeration must be adopted. These are described here.

Applications in transportation network assignment typically do not require the determination of the shortest path between all possible pairs of nodes, but only between pairs of origin and destination nodes (O-D pair) relative to centroids (introduced in Chap. 1). It should be remembered that each centroid is represented in the

[8]Note that by using definition (5.2.9) in Sect. 5.2.3, for the feasible set of link flows S_f, the optimization problem known in the literature as the *linear minimum cost multicommodity flow problem* is obtained.

network model by two unconnected nodes: an origin node, with only exiting links, and a destination node, with only entering links (Chap. 2).

Nonetheless, rather than computing the shortest path for each individual O-D pair in turn, it is often easier to compute the set of shortest paths between an origin (or destination) node and all other network nodes (including the possible destination nodes), looping over the origins (or destinations) until shortest paths for all O-D pairs have been found. This approach is usually more computationally efficient than determining all O-D paths one at a time, and corresponds more closely to the typical processing logic of assignment algorithms (which generally treat all flows from an origin or to a destination in one step). This section therefore describes the basic structure of algorithms for computing shortest paths from an origin node o to all network nodes (*forward shortest paths*), or from all network nodes to a destination node d (*backward shortest paths*).[9]

For simplicity, the performance variable associated with each link is referred to as cost, inasmuch as in practice it often represents a generalized transportation cost. However, it could just as well be any other performance measure (distance, travel time, etc.). Only link-additive performance measures are considered unless otherwise noted. Moreover, the link performance variable is assumed to be nonnegative. Let:

$c_a = c_{ij} \geq 0$ be the cost on link $a = (i, j)$

$Z_{i,j} \geq 0$ be the cost of the shortest path between any pair of nodes i and j; note that in general it may happen that $Z_{i,j} \neq Z_{j,i}$ (due, e.g., to one-way streets, slopes, etc.)

Shortest path costs satisfy the *triangle inequality*:

$$Z_{i,j} + Z_{j,k} \geq Z_{i,k} \quad \forall i, j, k$$

This can be seen by noting that if, for a pair of nodes i and k, there were a node j for which $Z_{i,j} + Z_{j,k} < Z_{i,k}$, then the cost of the path from i to k through node j would be less than $Z_{i,k}$, contradicting the definition of $Z_{i,k}$ as the cost of the shortest path from i to k. Because i and k are arbitrary, this relationship holds in particular for origin and destination nodes, and shortest paths between them.

The triangle inequality implies that link costs and shortest path costs satisfy the *Bellman principle*, which states that a shortest path is itself made up of shortest paths:

If link (i, j) belongs to the shortest path between o and j

$$\text{then } Z_{o,i} + c_{ij} = Z_{o,j} \qquad \text{otherwise } Z_{o,i} + c_{ij} \geq Z_{o,j}$$

More generally:

If link (i, j) belongs to the shortest path between o and d

$$\text{then } Z_{o,i} + c_{ij} + Z_{j,d} = Z_{o,d} \qquad \text{otherwise } Z_{o,i} + c_{ij} + Z_{j,d} \geq Z_{o,d}$$

---

[9]The two problems are obviously equivalent because it is sufficient to change the directions of all the network links to convert one problem to the other.

If there is only one shortest path between each pair of nodes in a network, the second assertion of each of the above two formulations of the Bellman principle holds as a strict inequality. It can easily be seen that, for an uncongested network, the Bellman principle is equivalent to the first Wardrop principle discussed above.

It should be recognized that if there is only one shortest path between each pair of nodes (or, when there are several shortest paths, if only one is considered), the set of shortest paths from an origin node o to the other network nodes forms a *forward tree*[10] $T(o)$ rooted at node o. Any forward tree can be described by specifying, for each node j, the unique link that enters it (or equivalently by specifying the initial node of this entering link). Similarly, the set of shortest paths from all network nodes to a destination node d forms a *backward tree* $T(d)$ rooted at node d. Any backward tree can be described by specifying the unique link that exits from each node i (or equivalently by specifying the final node of this exiting link). The use of the same notation for forward trees from an origin o and for backward trees towards a destination d is not ambiguous, because we only consider trees rooted at the origin or destination nodes: in this case, the type of root (origin or destination) defines the type of tree (forward or backward).

Given any forward tree $T(o)$ from origin node o, let:

$X_{T(o),i} \geq 0$ be the cost along the unique path from node o to node i in tree $T(o)$

It follows that

$$X_{T(o),i} + c_{ij} = X_{T(o),j} \quad \forall (i, j) \in T(o)$$

A tree $T(o)$ from origin node o is the shortest path tree (or is one such tree when there are multiple shortest paths between some pairs of nodes) if and only if the following condition, deduced from the Bellman principle, is verified.

$$X_{T(o),i} + c_{ij} \geq X_{T(o),j} \quad \forall (i, j) \notin T(o) \tag{5.3.9}$$

In this case, the values $X_{T(o),i}$ are the shortest path costs $Z_{o,i}$.

Similarly, given a backward tree $T(d)$ towards destination node d, let:

$X_{i,T(d)} \geq 0$ be the cost along the unique path from node i to destination d in tree $T(d)$

It follows that

$$c_{ij} + X_{j,T(d)} = X_{i,T(d)} \quad \forall (i, j) \in T(d)$$

In this case, a tree $T(d)$ to destination node d is the shortest path tree (or is one such tree when there are multiple shortest paths between some pairs of nodes) if and only if the following condition is verified.

$$c_{ij} + X_{j,T(d)} \geq X_{i,T(d)} \quad \forall (i, j) \notin T(d) \tag{5.3.10}$$

---

[10] In a directed graph, a tree rooted at node n, $T(n)$, is a subgraph having the property that a single path connects node n and every other node in the graph. In a forward tree, the root has only exiting links and the paths are oriented from the root towards every other node. In a backward tree, the root has only entering links, and the paths are oriented from every other node towards the root.

In this case the values $X_{i,T(d)}$ are again the shortest path costs $Z_{i,d}$.

The algorithms commonly used to compute forward (resp., backward) shortest path trees are based on the iterative updating of the values $X_{T(o),i}$ (resp., $X_{i,T(d)}$), called the node *labels*. In each iteration a node is chosen, and the labels of immediately downstream (resp., upstream) nodes are examined and updated as required. Iterations continue until condition (5.3.9) (resp., (5.3.10)) holds everywhere, at which point the minimum path costs have been found. Bookkeeping operations carried out along with the label updates enable the specific minimum path to (resp., from) each node to be traced.

The number of steps that an algorithm requires to compute the minimum path tree depends on its strategy for choosing, in each iteration, the node at which to verify whether further updating steps are needed.

Under the assumption of nonnegative costs, a node label cannot be updated if it is examined from a node with a higher label. This observation is the basis of the class of label-setting algorithms which, in each iteration, set (make permanent) the label of the node with the lowest label among those that have not yet been set. The algorithm then updates the labels of adjacent nodes. Label-setting algorithms need to maintain information about the ordering of nodes according to their labels; different algorithms employ different data structures for this purpose, and their efficiency depends strongly on this. The algorithms require as many iterations as there are nodes, because each iteration sets one node label. Note that the nodes are set in order of increasing labels (shortest path cost).

Label-correcting algorithms do not examine nodes in order of their labels and so are generally simpler to implement. On the other hand, node labels become permanent only at the end of the algorithm. In these algorithms, the number of updating steps depends on the node choice strategy.

Examples of updates for a forward tree from origin o, and for a backward tree towards destination d, are shown in Figs. 5.11a and 5.11b.

When there are multiple shortest paths between a particular O-D pair, the set of shortest paths from an origin (or towards a destination) is no longer a tree. The algorithms presented above will determine only one of the shortest paths; the particular one identified depends on the order in which the nodes are examined. The algorithms can easily be modified to compute all possible shortest paths, although in practice this is rarely done.

Algorithms for Uncongested Network Deterministic Assignment

Under the assumption of deterministic path choice behavior, all users traveling from an origin to a destination choose the shortest path between them (Sect. 5.3.2); this leads to deterministic uncongested network assignment. Algorithms for DUN assignment are known as *all-or-nothing* assignment algorithms.

As observed above, if multiple shortest paths connect an O-D pair, then path flows, and therefore link flows, are not uniquely defined. However, shortest path algorithms usually compute a single path between each O-D pair. The specific path

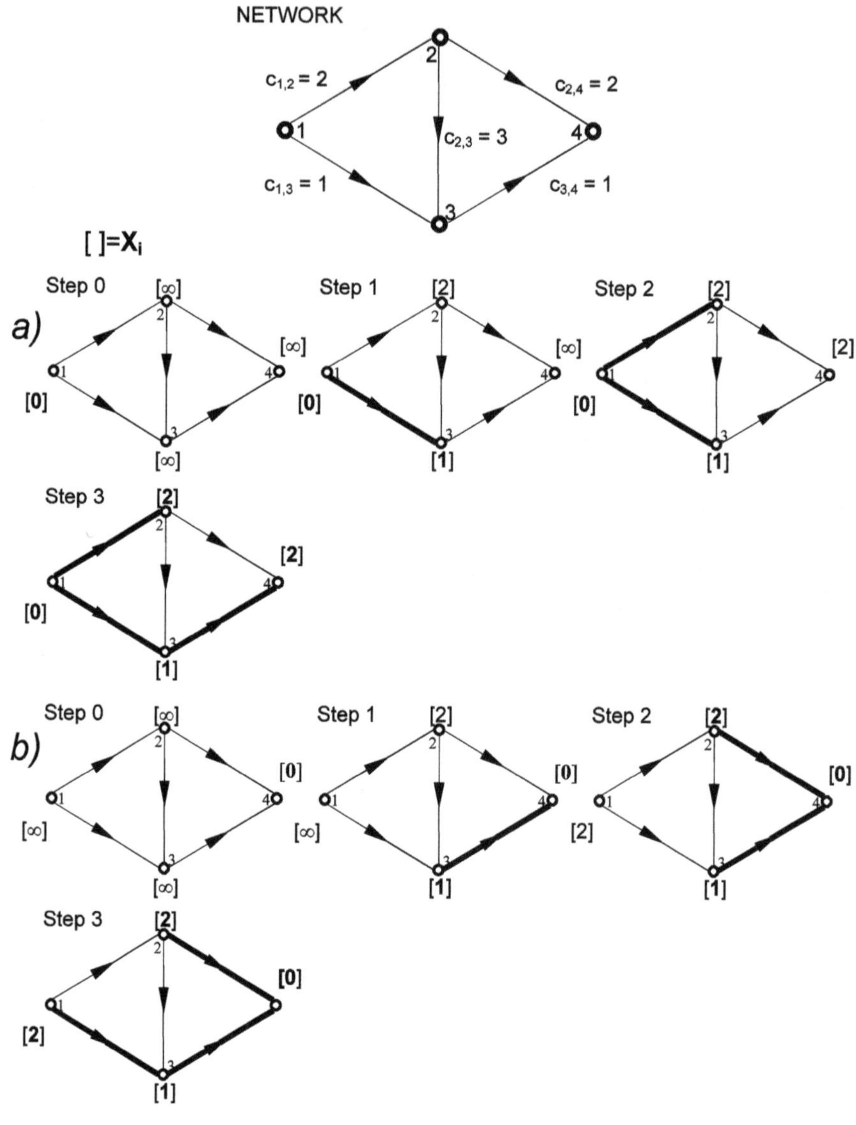

Fig. 5.11 Example of forward (**a**) and backward (**b**) label-setting shortest path algorithms

identified depends on the implementation details of the algorithm and in particular on the ordering of the nodes.

Link flows can therefore be calculated by assigning all the flow of each O-D pair to the links of a shortest O-D path, and nothing to the links of other paths. In practice, all-or-nothing algorithms generally process the entire tree of shortest paths from an origin or to a destination, rather than individual shortest O-D paths.

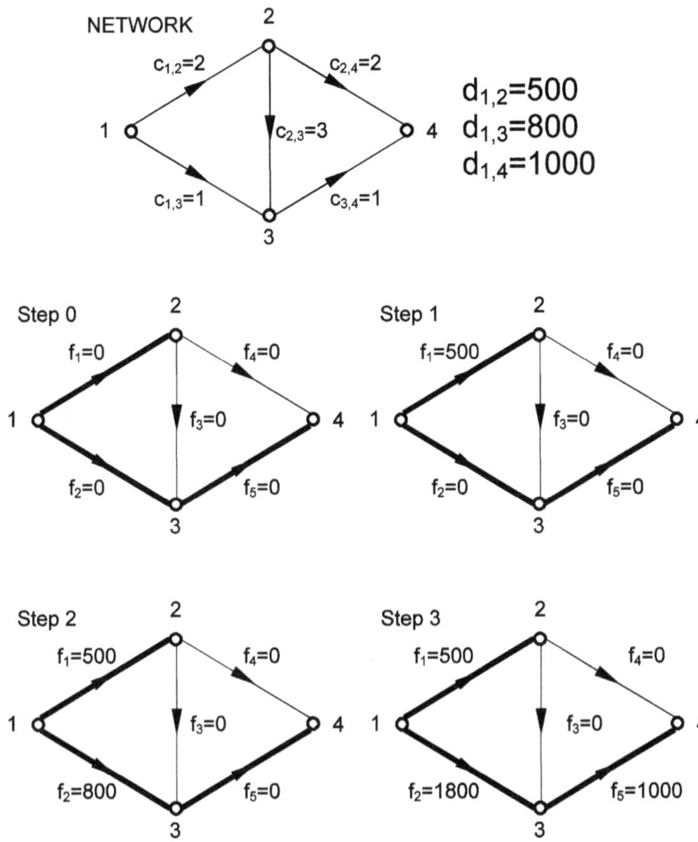

Fig. 5.12 Example of sequential forward algorithm for DUN assignment

They can be implemented with two different approaches. Both start with an empty network.

In the *sequential* approach, once a shortest path tree from an origin o has been calculated, the O-D demand d_{od} from the origin towards each destination d is added to the flows on all the links on the path from o to d. The DUN link flows result when all O-D specific flows have been accumulated on each link in this way. An example of the sequential algorithm is given in Fig. 5.12. The procedure is analogous if the shortest path tree towards each destination d is calculated.

In contrast to the sequential approach, other DUN assignment algorithms follow a *simultaneous* approach. Simultaneous algorithms are computationally more efficient and can be extended to DUN assignment models for transit networks (shortest hyperpaths) as described in Sect. 6.2. These algorithms are particularly efficient if each shortest path tree designates the nodes in order of increasing minimum cost from the origin (or to the destination). As discussed above, such an order is automatically obtained from label-setting shortest path algorithms.

Simultaneous algorithms from an origin are based on the calculation of the flow entering each node, defined as the sum of the flows on the links incident to the node. Considering one origin o at a time, each destination node d is initially assigned the corresponding demand flow d_{od} as its entering flow; all other nodes are tentatively assigned a zero entering flow. Once the tree of shortest paths from origin o has been calculated, the algorithm examines each node i in decreasing order of minimum cost, starting with the node farthest from origin o (i.e., the node i with the highest value Z_{oi}), working backward until o is reached. The flow entering a node i is assigned to the unique previous link in the shortest path tree, and added to the flow entering the initial node of this link. The order adopted is such that, when a node i is examined, all nodes farther from the origin have already been examined. Consequently there cannot be any node still to be examined from which the flow could contribute to the flow entering node i.[11]

For each O-D pair *od*, the EMPU associated with deterministic path choice is given by the cost on the shortest path, $s_{od} = Z_{od}$.

An example of the application of a simultaneous algorithm is given in Fig. 5.13. The procedure is analogous if shortest path trees towards each destination d are calculated.

Algorithms for Uncongested Network Stochastic Assignment

In stochastic assignment to noncongested networks it is assumed that each user associates with each path connecting its O-D pair a value of perceived utility represented with a random variable, whose expected value is given by the opposite of the path cost (see Sect. 4.3.3).[12]

Below we describe first an algorithm without explicit path enumeration in the case of a probit path choice model based on a Monte Carlo technique. The algorithm may also be applied to different choice models, assuming that random residuals relative to paths are distributed according to a multivariate variable which may be obtained from independent univariate variables relative to links. We then describe an algorithm without explicit path enumeration relative to a particular implementation of the logit model to represent path choice.

---

[11]Using a simultaneous algorithm, given an origin (or a destination), two additions are carried out for each link in the shortest path tree, regardless of the tree structure; that is, the algorithm requires $2(n-1)$ additions, where n is the number of nodes. Using a sequential algorithm, on the other hand, the number of additions depends on the structure of the shortest path tree. This number ranges between the number of links of the tree, $n-1$, when the paths within the tree do not overlap at all; and the value $n_d(n-n_d-1)+n_d = n_d(n-n_d)$ (assuming $n > n_d$ where n_d is the number of destinations) in the case of maximum overlap.

[12]According to these hypotheses, a positive choice may also be associated with a nonminimum (systematic) cost path, given by the probability of the path being the maximum perceived utility, that is, the minimum perceived cost. This is why stochastic network loading is sometimes termed multipath assignment, contrasting with all-or-nothing assignment to deterministic loading.

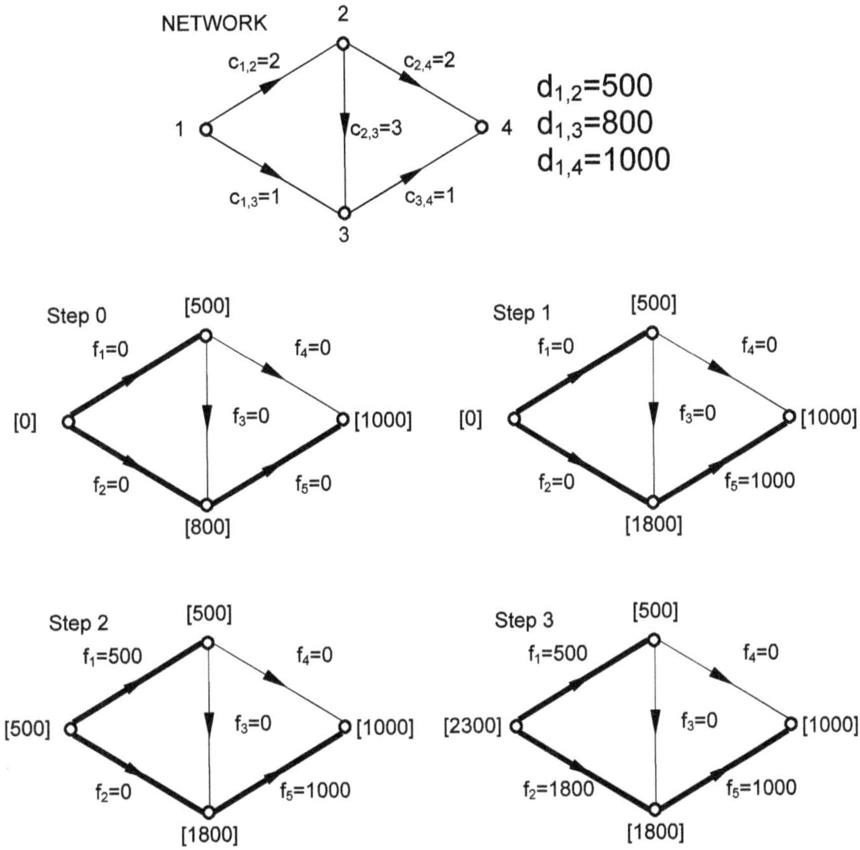

Fig. 5.13 Example of simultaneous forward algorithm for DUN assignment

Monte Carlo Algorithms

This algorithm is commonly used in the hypothesis of a probit path choice model. In this case it proves to be an exact algorithm with finite convergence, apart from the numerical estimation errors described below. The algorithm may also be applied to different choice models, assuming that random residuals relative to paths are distributed according to a multivariate variable which may be obtained from independent univariate variables relative to links.

The probit path choice model results from the assumption that the perceived path utility random residuals follow an $MVN(\mathbf{0}, \boldsymbol{\Sigma})$ multivariate Normal distribution, with zero mean and variance–covariance matrix $\boldsymbol{\Sigma}$. This model can account for overlapping paths by introducing a positive covariance between the perceived utilities of two paths sharing some links, but it does not allow calculation in closed form of path choice probabilities. However, unbiased estimates of path choice probabilities and their corresponding SUN path and link flows can be obtained using a Monte Carlo technique somewhat similar to the algorithm described in Sect. 3.3.6.

An assignment algorithm that does not require explicit path enumeration can be developed for any path choice model specification, probit or otherwise, assuming that users associate with each path a perceived utility that can be modeled by a random variable whose expected value is given by the negative path cost (see Sect. 4.3.3):

$$\boldsymbol{U} = \boldsymbol{V} + \boldsymbol{\varepsilon} = -\boldsymbol{g} + \boldsymbol{\varepsilon} \tag{5.3.11}$$

with

$$E[\boldsymbol{g}] = \boldsymbol{g} = -\boldsymbol{V} = -E[\boldsymbol{U}] \qquad \text{Var}[\boldsymbol{g}] = \boldsymbol{0}$$

$$E[\boldsymbol{\varepsilon}] = \boldsymbol{0} \qquad \text{Var}[\boldsymbol{\varepsilon}] = \text{Var}[\boldsymbol{U}] = \boldsymbol{\Sigma}$$

where

$\boldsymbol{U}$ is the vector of perceived path utilities, with expected value $\boldsymbol{V} = E[\boldsymbol{U}]$ and variance–covariance matrix $\text{Var}[\boldsymbol{U}] = \boldsymbol{\Sigma}$

$\boldsymbol{g} = -\boldsymbol{V}$ is the vector of path costs, given by the negative of the systematic path utility vector $\boldsymbol{V}$

$\boldsymbol{\varepsilon} = \boldsymbol{U} - E[\boldsymbol{U}]$ is the vector of path utility random residuals

Because of the assumption of additive path costs, the relationship between link and path costs expressed by the link–path incidence matrix $\boldsymbol{\Delta}$ allows us to express (5.3.11) in terms of link utilities, costs and random residuals. Let:

$\boldsymbol{u}$ be the vector of perceived link utilities, with expected value $\boldsymbol{v} = E[\boldsymbol{u}]$ and variance–covariance matrix $\text{Var}[\boldsymbol{u}] = \boldsymbol{\Sigma}_a$

$\boldsymbol{c} = -\boldsymbol{v}$ be the vector of link costs, given by the negative of the systematic link utility vector $\boldsymbol{v}$

$\boldsymbol{\eta} = \boldsymbol{u} - E[\boldsymbol{u}]$ be the vector of link utility random residuals

It follows that

$$\boldsymbol{U} = \boldsymbol{\Delta}^T \boldsymbol{u} \tag{5.3.12}$$

$$\boldsymbol{g} = \boldsymbol{\Delta}^T \boldsymbol{c} \tag{5.3.13}$$

$$\boldsymbol{\varepsilon} = \boldsymbol{\Delta}^T \boldsymbol{\eta} \tag{5.3.14}$$

thus

$$\boldsymbol{u} = -\boldsymbol{c} + \boldsymbol{\eta} \tag{5.3.15}$$

with

$$E[\boldsymbol{c}] = \boldsymbol{c} = -E[\boldsymbol{u}] \qquad \text{Var}[\boldsymbol{c}] = \boldsymbol{0}$$

$$E[\boldsymbol{\eta}] = \boldsymbol{0} \qquad \text{Var}[\boldsymbol{\eta}] = \text{Var}[\boldsymbol{u}] = \boldsymbol{\Sigma}_a$$

Because relationships (5.3.12) to (5.3.14) are linear, the variance–covariance matrix of path random residuals $\boldsymbol{\Sigma}$ depends on the variance–covariance matrix of link random residuals $\boldsymbol{\Sigma}_a$ through the relationship:

$$\boldsymbol{\Sigma} = \boldsymbol{\Delta}^T \boldsymbol{\Sigma}_a \boldsymbol{\Delta} \tag{5.3.16}$$

These results can be interpreted as a specification of a path choice model in which users perceive the costs of individual links, and the perceived cost of a path is equal to the sum of the perceived costs of its links.

A SUN algorithm that does not require explicit path enumeration can be developed from the above relationships, together with the assumptions that the choice set consists of all elementary paths, and that the variance–covariance matrix $\boldsymbol{\Sigma}$ has the structure described in Sect. 3.3. Let:

- g_k be the cost of path k
- g_{kj} be the cost on the links shared by paths k and j
- $\sigma_{kk} = \sigma_k^2$ be the variance of the random residual of path k, a main diagonal element of the variance–covariance matrix $\boldsymbol{\Sigma}$
- σ_{kj} be the covariance between the random residuals of paths k and j, an off-diagonal element of the variance–covariance matrix $\boldsymbol{\Sigma}$
- ξ be the proportionality coefficient between path costs and elements of the variance–covariance matrix (expressed in units that are consistent with costs and utilities)

Under these assumptions about the structure of the variance–covariance matrix, it follows that

$$\sigma_k^2 = \sigma_{kk} = \xi g_k$$
$$\sigma_{kj} = \xi g_{kj}$$

Referring to the relationship between link and path costs expressed by the link path incidence matrix $\boldsymbol{\Delta}$, then:

$$g_k = \sum_a \delta_{ak} c_a = \sum_a \delta_{ak}^2 c_a$$
$$g_{jk} = \sum_a \delta_{ak} \delta_{aj} c_a$$

and

$$\sigma_{kk} = \xi \sum_a \delta_{ak}^2 c_a$$
$$\sigma_{kj} = \xi \sum_a \delta_{ak} \delta_{aj} c_a$$

If we indicate by $\boldsymbol{DIAG}(\boldsymbol{c})$ the diagonal matrix whose main diagonal elements are given by link costs $\boldsymbol{c}$, we have:

$$\boldsymbol{\Sigma} = \xi \boldsymbol{\Delta}^T \boldsymbol{DIAG}(\boldsymbol{c}) \boldsymbol{\Delta} \tag{5.3.17}$$

Turning now to the particular case of SUN assignment with probit path choice, it may be assumed that each link random residual η_a is independently distributed as a univariate normal $N(0, \sigma_a^2)$ random variable with zero mean and variance $\sigma_a^2 = \xi c_a$. Therefore the vector $\boldsymbol{\eta}$ is a multivariate normal $MVN(\mathbf{0}, \boldsymbol{\Sigma}_a)$ random variable with zero mean and diagonal variance–covariance matrix defined by:

$$\boldsymbol{\Sigma}_a = \xi \boldsymbol{DIAG}(\boldsymbol{c}) \tag{5.3.18}$$

In this case the path random residuals deriving from the linear relationship (5.3.14), $\boldsymbol{\varepsilon} = \boldsymbol{\Delta}^T \boldsymbol{\eta}$, follow a multivariate normal $MVN(\mathbf{0}, \boldsymbol{\Sigma})$ distribution with variance–covariance matrix given by the relationship (5.3.16) which, combined with (5.3.18), provides the relationship (5.3.17).

Therefore, a realization of normally distributed path random residuals, $\boldsymbol{\varepsilon} \sim MVN(\mathbf{0}, \boldsymbol{\Sigma})$, can be obtained from a realization of link random residuals $\boldsymbol{\eta}$, obtained by independently drawing the residual η_a of each link a from a univariate normal $N(0, \sigma_a^2)$ distribution.

It should be stressed that the link attributes used to define the variance–covariance matrix through relation (5.3.17) may be different from the actual link costs $\boldsymbol{c}$ that express the systematic utility of links, $\boldsymbol{v} = -\boldsymbol{c}$, and therefore of paths, $\boldsymbol{V} = -\boldsymbol{g}$. For example, it might be assumed that the perceived degree of similarity of two overlapping paths, expressed by the covariance of their random residuals, is proportional to the length of the links that they share, but that the systematic link cost is a function of the travel time (dependent on flows for congested networks). These assumptions ensure that the SUN assignment function is nonincreasing monotone with respect to (congested) link costs and has symmetric Jacobian (Sect. 5.3.1). The (sufficient) condition for uniqueness of the resulting stochastic equilibrium is therefore ensured (as described in Sect. 5.4.1), as is the convergence of stochastic equilibrium algorithms described in Sect. 5.4.2.

From an algorithmic point of view, in order to calculate SUN assignment flows with a probit path choice model, a sample of normally distributed perceived link cost vector realizations must be generated. For each perceived link cost vector realization in the sample, the demand flow for each O-D pair is assigned to the perceived shortest path using a DUN (all-or-nothing) assignment algorithm, described in the next subsection. The average of the link flows obtained for the different link cost vectors of the sample is an unbiased estimate of the probit SUN link flows. The algorithm can be stated formally by introducing the following variables.

c_a the (systematic) cost on arc a

$\eta_a^j \leftarrow N(0, \sigma_a^2 = \xi c_a)$ the jth (in a sample of m) realization of the perceived cost random residual for link a, obtained by drawing from a normal distribution with zero mean and variance $\sigma_a^2 = \xi c_a$

$r_a^j = c_a + \eta_a^j$ the jth perceived cost for link a

$\boldsymbol{r}^j = [r_a^j]_a$ the jth vector of perceived link costs, with elements r_a^j

$\boldsymbol{f}_{\text{DNL}}^j = \boldsymbol{f}_{\text{DUN}}(\boldsymbol{r}^j)$ the deterministic uncongested network assignment link flow vector corresponding to link costs $\boldsymbol{r}^j$ (computed as described in the next subsection)

$\bar{f}^m$ an unbiased estimate of the vector of stochastic uncongested network assignment link flows

With m elements in the sample, we have:

$$\bar{f}^m = \sum_{j=1,m} f^j_{\text{DNL}}/m$$

The link flow estimate $\bar{f}^m$ can be obtained by evaluating the following recursive equations up to $j = m$, starting with $j = 0$ and $\bar{f}^0 = 0$.

$$j = j + 1$$
$$\eta^j_a \leftarrow N(0, \sigma^2_a = \xi c_a) \quad \forall a$$
$$r^j = [c_a + \eta^j_a]_a$$
$$\bar{f}^j = ((j-1)\bar{f}^{j-1} + f_{\text{DNL}}(r^j))/j$$

For each O-D pair *od*, the average of the various shortest path costs Z^j_{od} obtained from the different realizations of the link random residuals is an unbiased estimate of the negative path choice EMPU variable: $\bar{s}^m_{od} = -\sum_{j=1,m} Z^j_{od}/m$.

Unlike other algorithms in this chapter, this algorithm, which is an example of the class of Monte Carlo algorithms, does not yield exact link flow values, but only a sequence of unbiased estimates whose precision increases with the number of iterations.

In practice, the algorithm continues until a stop criterion is met: for example, a pre-assigned maximum number of iterations $j_{\max}$. The algorithm could also terminate when the relative difference between the link flow vector estimates in two successive iterations falls below a pre-assigned threshold δ using a suitable vector norm $|\bar{f}^j - \bar{f}^{j-1}|/|\bar{f}^{j-1}| < \delta$, such as $\max_a |\bar{f}^j_a - \bar{f}^{j-1}_a|/|\bar{f}^{j-1}_a| < \delta$. However, this criterion is not very effective because, as the number of iterations j increases, it tends to be verified in any case, so it is effectively the same as specifying a maximum number of iterations. More correctly, the algorithm should be stopped when the sample estimate of the precision of link flows falls below a given threshold, $\max_a[\text{var}(\bar{f}^m_a)^{(1/2)}/\bar{f}^m_a] \leq \delta$. Alternatively, a statistical test of equality between two successive averages can be used. It can easily be proved that, whatever convergence criterion is adopted, the calculation time is roughly equal to m times the time needed to carry out a deterministic uncongested network assignment (with any of the algorithms described in the next subsection).

An example of the Monte Carlo algorithm is given in Fig. 5.14.

Dial Algorithm

For logit path choice models, link flows can be computed without explicit path enumeration using an algorithm known in the literature as the Dial algorithm, after its

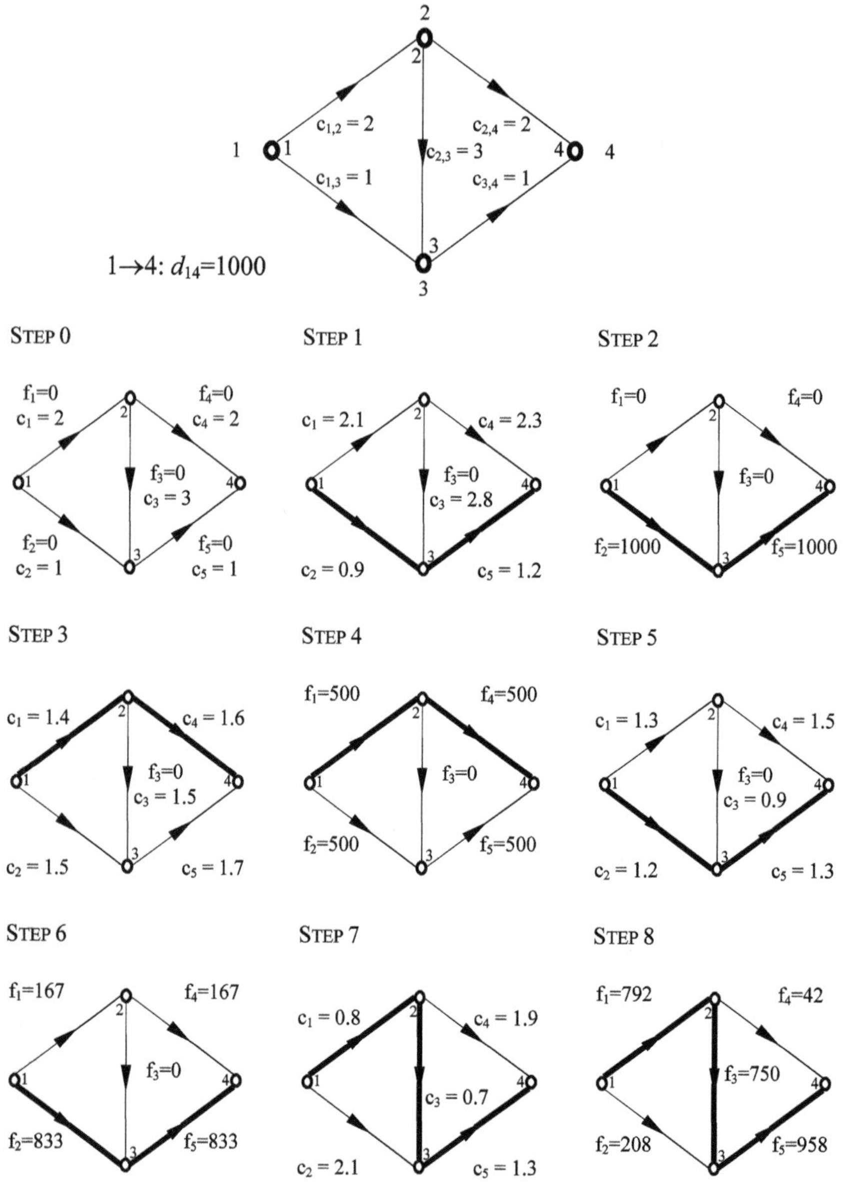

Fig. 5.14 Example of the Monte Carlo algorithm for probit SUN assignment

author. This algorithm is based on a particular specification of the logit path choice model, for which the set of relevant paths consists only of efficient paths with respect to the origins; these paths are made up of links $a = (i, j)$, termed *efficient links*, such that the cost of the shortest path from an origin o to the link initial node i is less than the cost of the shortest path from the origin to the final node j; that is,

$Z_{o,i} < Z_{o,j}$. Note that if link costs are strictly positive, the links of a shortest path tree are efficient by definition and therefore shortest paths are among the efficient paths. Thus the efficiency condition must be tested only for links that do not belong to the shortest path tree. Efficient paths with respect to the destinations can be defined analogously. It is also possible to define efficient paths with respect to both origins and destinations; in this case, each O-D pair must be analyzed separately, resulting in lower computational efficiency. Note that if link costs are strictly positive, the links of a shortest path tree are uniquely defined.

Even though the algorithm was originally proposed with reference to the relevant path sets described above, any other relevant path set resulting in an acyclic graph will also work, because in this case at least one complete ordering of nodes may be built up consistently with the acyclic graph. Let $\text{pos}(i)$ be the position of node i in such an order that origin o has the first position, $\text{pos}(0) = 1$. Relevant paths are made up of links $a = (i, j)$, termed *efficient links*, such that the position of the link initial node i is less than the position of the final node j; that is, $\text{pos}(i) < \text{pos}(j)$.

This path choice set definition is a topological selective approach, in the sense defined in Sect. 4.3.3. For brevity, the discussion here considers only the case of efficient paths with respect to the origins.

Figures 5.15a through 5.15c illustrate efficient paths from origin 1 to destination 4 for the same network topology but different link cost vectors. Notice that with configuration (a), only the shortest paths are efficient. This is no longer the case for the costs shown in (b) and (c). These examples show that efficiency does not depend only on topology.

Theoretical analysis of the Dial algorithm is based on an equivalent formulation of the logit path choice model that highlights the role of link costs in determining path costs. This formulation allows simultaneous analysis of all paths to all destinations from a given origin o. Recall from Sect. 4.3.3.2 that the logit probability $p_{od,k}$ that users traveling from origin o to destination d choose path k is given by

$$p_{od,k} = \exp(-g_k/\theta) \Big/ \sum_{j \in K_{od}} \exp(-g_j/\theta) \propto \exp(-g_k/\theta) \tag{5.3.19}$$

where

$\theta = (\sqrt{6}/\pi)\sigma$ is the scale parameter of the logit model, which is proportional to the standard deviation of the random residuals

g_k is the cost of path k

K_{od} is the set of (relevant) paths connecting the O-D pair od

If (additive) path costs g_k are expressed as the sum of link costs c_{ij} through the congruence relationship (5.2.1), expression (5.3.19) yields:

$$p_{od,k} \propto \exp\left(-\sum_{(i,j)\in k} c_{ij}/\theta\right) = \prod_{(i,j)\in k} \exp(-c_{ij}/\theta) \tag{5.3.20}$$

Alternatively, if each path is considered to be a sequence of nodes j and links (i, j), the probability $p_{od,k}$ of choosing a path k can be expressed as the prod-

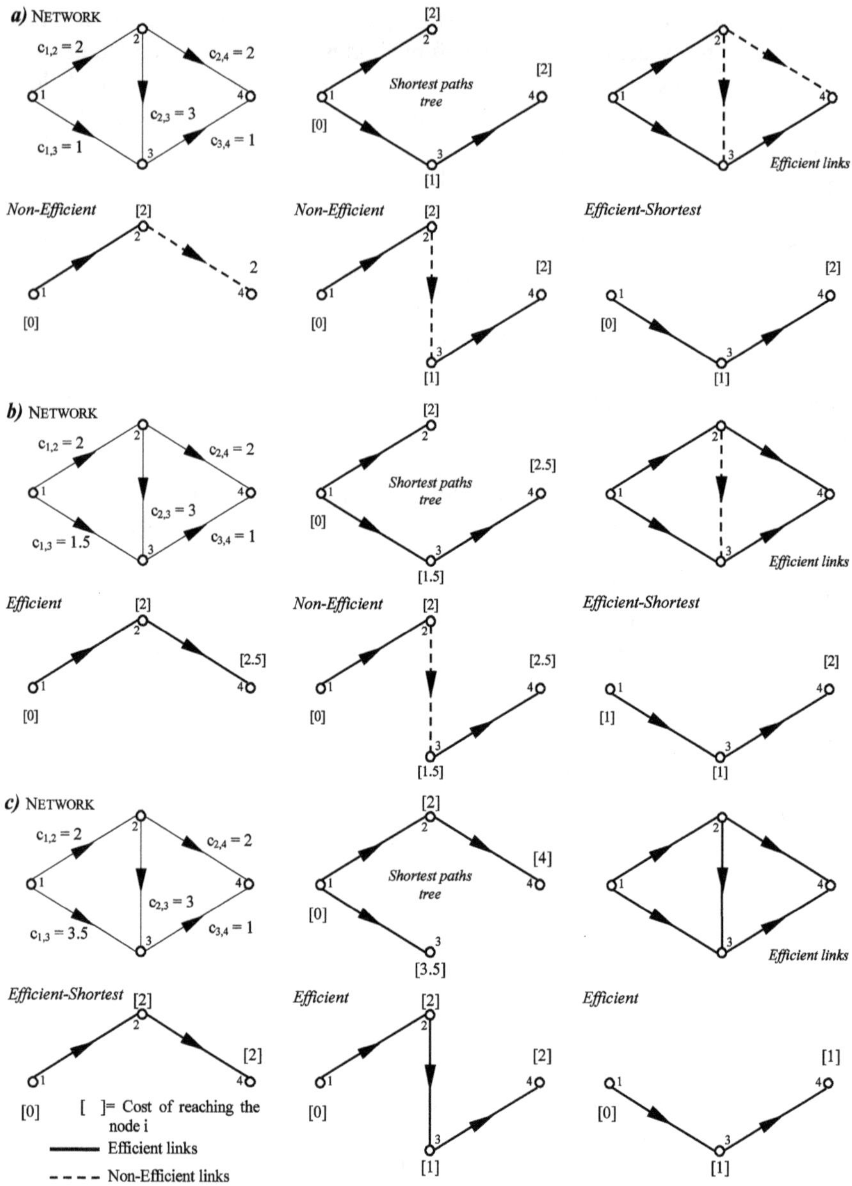

Fig. 5.15 Examples of efficient paths

uct of the conditional probabilities $Pr[(i, j)/j]$ of choosing each link (i, j) of path k, conditional on the link's downstream node j being on the path (Fig. 5.16):

$$p_{od,k} = \prod_{(i,j)\in k} Pr\big[(i, j)/j\big] \tag{5.3.21}$$

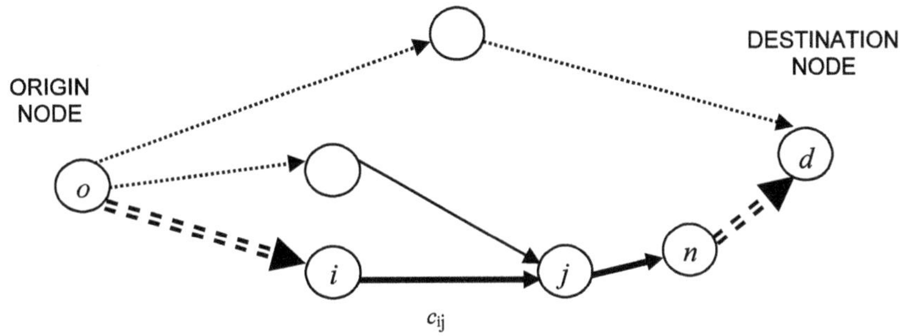

Fig. 5.16 Path k from origin o to destination d through link (i, j)

The two probabilities $p_{od,k}$ calculated with (5.3.21) and (5.3.20) will be equal if the probability $Pr[(i, j)/j]$ is given by a particular logit model. In this model, the alternatives in the choice set are the efficient links (i, j) incident to (entering) node j. The systematic utility $V_{ij/j}$ of each such alternative is the sum of the negative link cost c_{ij} and a logsum variable Y_i that takes into account the utilities of all the efficient paths from the origin o to the initial node i of the link. The model parameter is θ:

$$Pr\big[(i, j)/j\big] = \exp(V_{ij/j}/\theta) \Big/ \sum_{(m,j)\in BS(j)} \exp(V_{mj/j}/\theta) \tag{5.3.22}$$

$$V_{ij/j} = -c_{ij} + \theta Y_i \tag{5.3.23}$$

$$Y_i = \ln\left(\sum_{(n,i)\in BS(i)} \exp(V_{ni/i}/\theta) \right) \tag{5.3.24}$$

where

$BS(j)$ is the backward star of node j: the set of links (i, j) incident to node j
Y_i is the logsum variable of the utilities of the links incident to node i.

The relationships (5.3.22) to (5.4.24) yield:

$$Pr\big[(i, j)/j\big] = \exp\big((-c_{ij} + \theta Y_i)/\theta\big) \Big/ \sum_{(m,j)\in BS(j)} \exp\big((-c_{mj} + \theta Y_m)/\theta\big) = w_{ij}/W_j$$

with

$$\begin{aligned} w_{ij} &= \exp\big((-c_{ij} + \theta Y_i)/\theta\big) = \exp(-c_{ij}/\theta)\exp(Y_i) \\ &= \exp(-c_{ij}/\theta) \sum_{(n,i)\in BS(i)} \exp(V_{ni/i}/\theta) \end{aligned}$$

$$W_j = \sum_{(m,j)\in BS(j)} \exp\big((-c_{mj} + \theta Y_m)/\theta\big) = \sum_{(m,j)\in BS(j)} w_{ij}$$

The probability $Pr[(i, j)/j]$ of choosing link (i, j) conditional on the final node j can therefore be expressed as the ratio between a weight w_{ij} associated with link (i, j), and a weight W_j associated with node j. Note that the definition of the link weights yields:

$$w_{ij} = \exp(-c_{ij}/\theta) \sum_{(n,i)\in BS(i)} \exp(V_{ni/i}/\theta) = \exp(-c_{ij}/\theta) W_i$$

Furthermore, nonefficient links (links (i, j) with $Z_{o,i} \geq Z_{o,j}$) have weight $w_{ij} = 0$, consistent with the assumption that a link (i, j) belongs to an efficient path if and only if the shortest path from the origin to its initial node i is less than the shortest path to its final node j.

From the above, the link weights w_{ij}, node weights W_j, and probabilities $Pr[(i, j)/j]$ can all be determined using a system of recursive equations equivalent to relations (5.3.22)–(5.3.24). They are computed for each link, starting from the origin o with $W_o = 1$, and proceeding to other nodes i in order of increasing minimum cost $Z_{o,i}$:

$$w_{ij} = \begin{cases} \exp(-c_{ij}/\theta) W_i & \text{if } Z_{o,i} < Z_{o,j} \\ 0 & \text{if } Z_{o,i} \geq Z_{o,j} \end{cases} \tag{5.3.25}$$

$$W_j = \sum_{(m,j)\in BS(j)} w_{mj} \tag{5.3.26}$$

$$Pr\big[(i, j)/j\big] = w_{ij}/W_j \tag{5.3.27}$$

Substituting relationships (5.3.25) to (5.3.27) in (5.3.21) yields expression (5.3.20).

In fact, the weights of the path nodes, apart from the origin and the destination, are irrelevant. Because the final node of one link is the initial node of the next link along the path, these weights appear in both the numerator and the denominator of successive factors in the product, and cancel (see Fig. 5.17):

$$\begin{aligned} p_{od,k} &= \prod_{(i,j)\in k} W_i \exp(-c_{ij}/\theta)/W_j \\ &= \prod_{(i,j)\in k} \exp(-c_{ij}/\theta) W_o/W_d \propto \prod_{(i,j)\in k} \exp(-c_{ij}/\theta) \end{aligned}$$

The Dial algorithm for SUN assignment is based on the iterative calculation of the weights of the nodes and links, for each origin o, using relationships (5.3.25) and (5.3.26). The processing of nodes in order of increasing minimum cost from the origin ensures that the recursive relationships (5.3.25) and (5.3.26) can be applied, that is, that when the weight w_{ij} of a link (i, j) is to be computed, the weight W_i of its initial node i has already been determined.

Of course the same condition occurs if the relevant path set results in an acyclic graph, because in this case nodes may be completely ordered w.r.t. the acyclic graph.

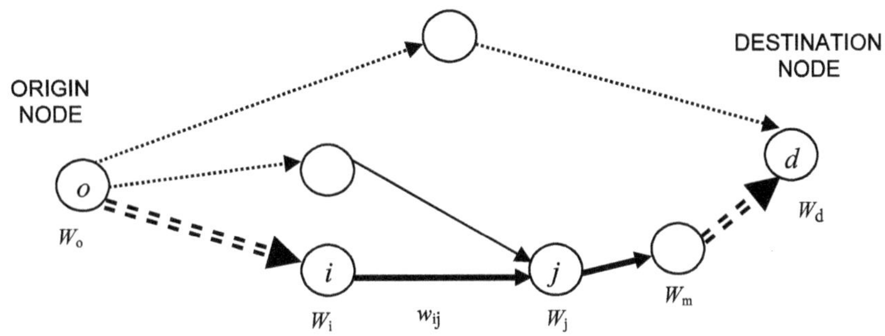

$$p_{od,k} = \frac{W_o \exp(-\alpha c_{oa})}{W_a} \cdots \times \cdots \frac{W_i \exp(-\alpha c_{ij})}{W_j} \times \frac{W_j \exp(-\alpha c_{jm})}{W_m} \cdots \times \cdots \frac{W_z \exp(-\alpha c_{zd})}{W_i}$$

Fig. 5.17 Node and link weights

Let $pos(i)$ be the position of node i as introduced above; processing nodes by increasing position from the origin ensures that the recursive relationships (5.3.25) and (5.3.26) can be applied, that is, that when the weight w_{ij} of a link (i, j) is to be computed, the weight W_i of its initial node i has already been determined.

When the weights of all the nodes and links are known, the demand flow d_{od} to each destination d is assigned to the network by starting at the destination and proceeding in reverse order of node cost, splitting each node's flow among its incident links according to the probabilities in expression (5.3.27). (This is somewhat similar to the simultaneous DUN assignment procedure described above.)

For an origin o, the path choice EMPU for a destination d is given by the destination's inclusive variable $s_{od} = Y_d$. Figure 5.18 provides an example application of the Dial algorithm. The computation time for the algorithm is two or three times greater than that needed for DUN assignment to the same network.

The algorithm can be extended to calculate SUN assignment link flows for C-logit path choice models (described in Sect. 4.3.3.1), provided that one O-D pair is examined at a time and an appropriate specification of the commonality factor is adopted.

Observe that the shortest paths used to define efficient paths can be calculated using link attributes that are different from the link costs $\boldsymbol{c}$ used to determine path choice probabilities. For example, efficient paths could be defined in terms of their physical length (or other attribute), while simulating users' choice among these paths using a cost proportional to their travel time. In this case, the shortest paths and the distances $Z_{o,i}$ would be calculated from the physical lengths of the links, and link weights w_{ij} and node weights W_i would be calculated using the costs (times) c_{ij}. With this approach, the set of efficient paths is independent of link costs. This property is important for stochastic equilibrium assignment because it is necessary to guarantee that the SUN function is increasing monotone in terms of the (congested) link costs and has a symmetric Jacobian (see Sect. 5.4.1). Therefore, the (sufficient) condition for uniqueness of stochastic equilibrium is ensured, as is the convergence of the stochastic equilibrium algorithms described in Sect. 5.4.2.

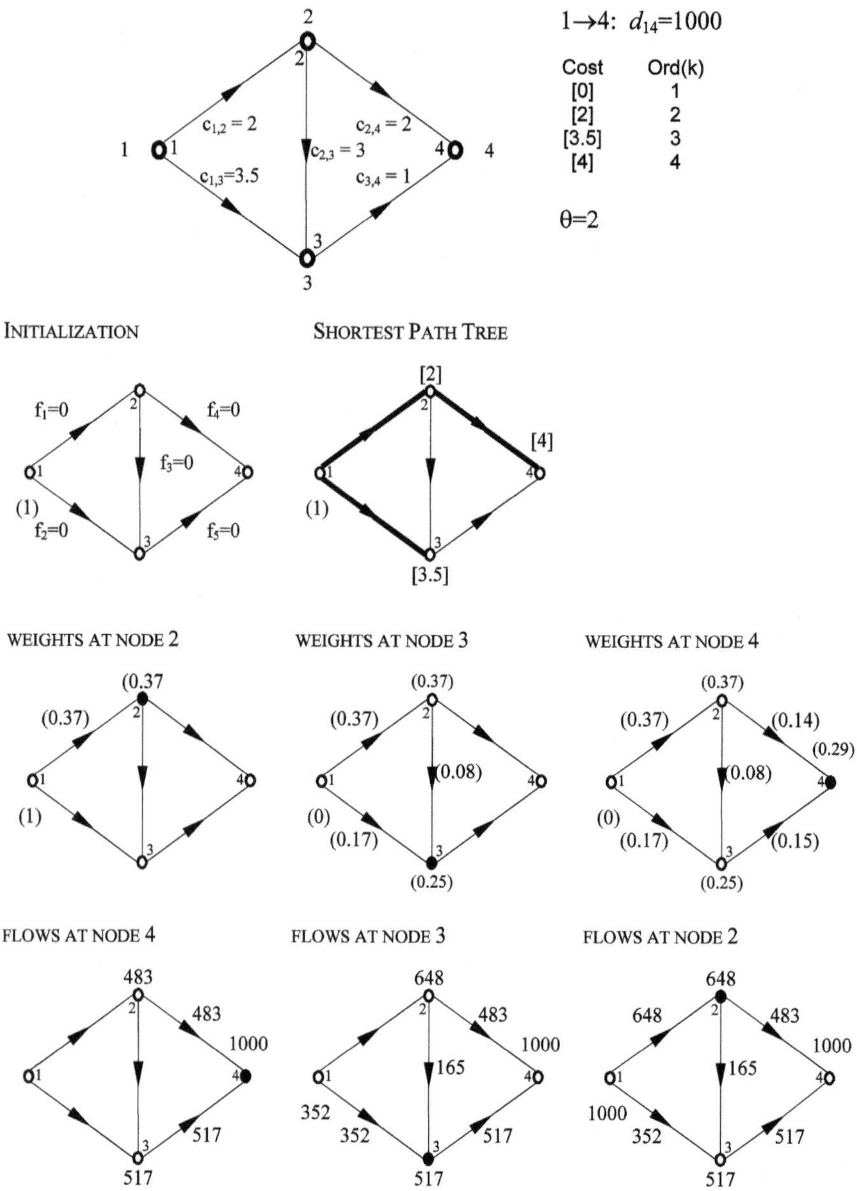

Fig. 5.18 Application of the Dial algorithm for logit SUN assignment

5.4 Congested Networks: Equilibrium Assignment

Equilibrium assignment is generally expressed by fixed point models, that is, systems of nonlinear equations, or by variational inequalities. Hence only asymptotically converging algorithms are available.

We consider here the situation where O-D demands are fixed, but link performance measures and costs depend on link flows through the performance and cost functions introduced in Chap. 2. Conversely, link flows depend on link costs through the path choice probabilities, as described by the uncongested network assignment map. The user equilibrium approach to the study of the supply–demand interactions assumes that the state of the real-world system can be represented by a configuration of path flows that is consistent with the corresponding path costs.[13] Equilibrium path flows and costs are defined by a system of nonlinear equations obtained by combining the supply model (5.2.4) with the demand model (5.2.5)–(5.2.6):

$$\begin{aligned}
\boldsymbol{g}^*_{od} &= \boldsymbol{\Delta}^T_{od}\boldsymbol{c}\Big(\sum_{od}\boldsymbol{\Delta}_{od}\boldsymbol{h}^*_{od}\Big) + \boldsymbol{g}^{\text{NA}*}_{od} \quad \forall od \\
\boldsymbol{V}^*_{od} &= -\boldsymbol{g}^*_{od} \quad \forall od \\
\boldsymbol{h}^*_{od} &= d_{od}\,\boldsymbol{p}_{od}\big(\boldsymbol{V}^*_{od}\big) \quad \forall od
\end{aligned}$$

or

$$\begin{aligned}
\boldsymbol{g}^*_{od} &= \boldsymbol{\Delta}^T_{od}\boldsymbol{c}\Big(\sum_{od}\boldsymbol{\Delta}_{od}\boldsymbol{h}^*_{od}\Big) + \boldsymbol{g}^{\text{NA}*}_{od} \quad \forall od \\
\boldsymbol{h}^*_{od} &= d_{od}\,\boldsymbol{p}_{od}\big(-\boldsymbol{g}^*_{od}\big) \quad \forall od
\end{aligned}$$

Equivalent equilibrium assignment models expressed in terms of link variables can be formulated by the system of nonlinear equations obtained by combining the uncongested network assignment map (5.3.2) with the flow-dependent cost functions (5.2.2):

$$\begin{aligned}
\boldsymbol{c}^* &= \boldsymbol{c}(\boldsymbol{f}^*) \\
\boldsymbol{f}^* &= \sum_{od} d_{od}\,\boldsymbol{\Delta}_{od}\,\boldsymbol{p}_{od}\big(-\boldsymbol{\Delta}^T_{od}\boldsymbol{c}^* - \boldsymbol{g}^{\text{NA}}_{od}\big)
\end{aligned}$$

or

$$\begin{aligned}
\boldsymbol{c}^* &= \boldsymbol{c}(\boldsymbol{f}^*) \\
\boldsymbol{f}^* &= \boldsymbol{f}_{\text{UN}}(\boldsymbol{c}^*; \boldsymbol{d})
\end{aligned}$$

The above system of equations shows that, in congested networks, link flows may depend nonlinearly on demand flows (unlike uncongested network assignment). Thus, in this case, the effect of each O-D pair cannot be evaluated separately.

---

[13] This assumption can be justified by considering the equilibrium configuration as a state towards which the system evolves (see Sect. 6.5). According to this interpretation, the equilibrium approach is valid for the analysis of the recurrent congestion conditions of the system, in other words, for those conditions systematically brought about by a sufficiently large sequence of reference periods to guarantee that the system will achieve the equilibrium state (and remain in it for a sufficient length of time).

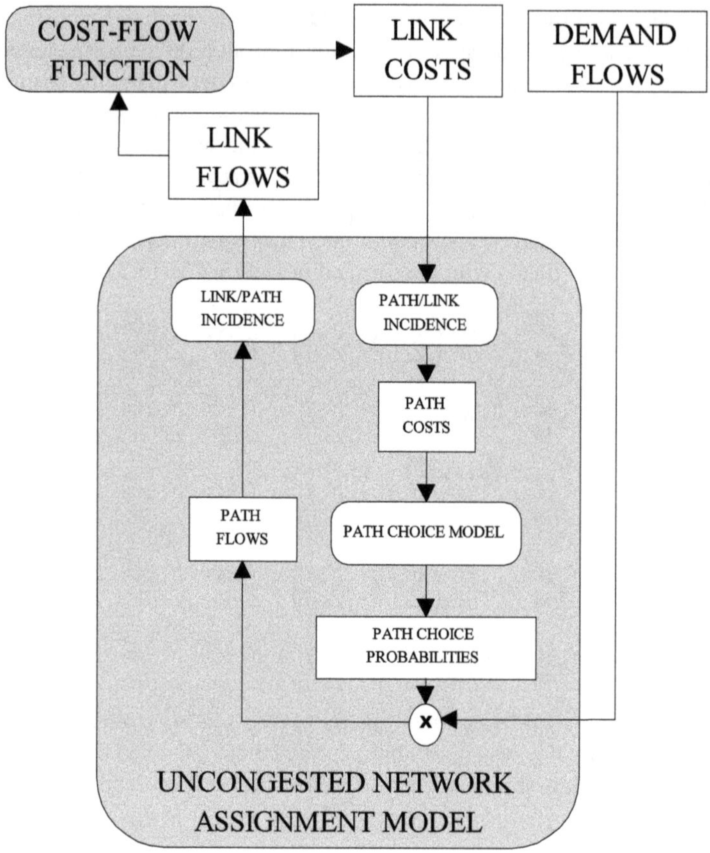

Fig. 5.19 Schematic representation of fixed demand equilibrium assignment models

The circular dependence between flows and costs expressed by the equilibrium approach is depicted in Fig. 5.19. This figure particularizes the general framework in Fig. 5.1 for the fixed demand assumption made in this section, and highlights the role of the uncongested network assignment model in the equilibrium framework.

The formulation and analysis of the theoretical properties (existence and uniqueness) of equilibrium flows (and costs) depend on the type of model adopted to simulate path choices: probabilistic or deterministic. This selection defines, respectively, stochastic and deterministic equilibrium assignment models and corresponding solution algorithms, which are the subjects of the following sections.

In general, algorithms for calculating equilibrium flows are based on recursive equations which, starting from an initial feasible link flow vector $\boldsymbol{f}^0 \in S_f$, generate a sequence of feasible link flow vectors:

$$\boldsymbol{f}^k = \varphi(\boldsymbol{f}^{k-1}) \in S_f$$

In each step, an assignment algorithm attempts to improve the solution estimate obtained in preceding steps, but an exact equilibrium solution will not generally be found in a finite number of steps. However, if at any step k the equilibrium flow vector is generated, all subsequent elements of the sequence will remain equal to the equilibrium vector:

$$\boldsymbol{f}^k = \boldsymbol{f}^* \quad \Rightarrow \quad \boldsymbol{f}^j = \boldsymbol{f}^* \quad j > k$$

Furthermore, if link flow vectors in two successive steps are equal, they are the equilibrium vector:

$$\boldsymbol{f}^k = \boldsymbol{f}^{k-1} \quad \Rightarrow \quad \boldsymbol{f}^k = \boldsymbol{f}^*$$

Under certain assumptions on the cost functions and the path choice model, it can be demonstrated that the sequence defined by the recursive equations converges to the equilibrium flow vector $\boldsymbol{f}^*$, provided that it is unique:

$$\lim_{k \to \infty} \boldsymbol{f}^k = \boldsymbol{f}^*$$

Below it is worth distinguishing the particular case of cost functions with a symmetric Jacobian from the general case. Remember that the algorithms described, reported for the sake of example, are only those more widely used and more simply implemented, and are essentially based on calculating link cost and flow functions of assignment to noncongested networks, using the algorithms described in the previous section.

5.4.1 Models for Stochastic User Equilibrium

Stochastic User Equilibrium (SUE) assignment is obtained by applying the equilibrium approach to congested networks under the assumption of probabilistic path choice behavior. The resulting path flows $\boldsymbol{h}^*$ correspond to the condition in which, for each O-D pair, the perceived cost of the paths used at equilibrium is less than or equal to the perceived cost of every other path. Equilibrium path flows can be expressed as the solution of a fixed-point model defined on the feasible path flow set S_h and obtained by combining the supply model (5.2.4) with the demand model (5.2.7):

$$\boldsymbol{h}^*_{od} = d_{od}\, \boldsymbol{p}_{od}\left(-\boldsymbol{\Delta}^T_{od}\boldsymbol{c}\left(\sum_{od} \boldsymbol{\Delta}_{od}\boldsymbol{h}^*_{od}\right) - \boldsymbol{g}^{\mathrm{NA}}_{od}\right) \quad \forall od \tag{5.4.1}$$

with

$$\boldsymbol{h}^* = \left[\boldsymbol{h}^*_{od}\right]_{od} \in S_h$$

An equivalent fixed-point model using link flow variables $\boldsymbol{f}^*$ (and therefore defined on the feasible link flow set S_f) can be obtained by combining the stochastic uncongested network assignment function (5.3.3) (disaggregated here by O-D pair to facilitate the analysis) with the flow-dependent cost functions (5.2.2):

$$\boldsymbol{f}^* = \boldsymbol{f}_{\text{SUN}}\big(\boldsymbol{c}(\boldsymbol{f}^*)\big) \quad \text{or} \quad \boldsymbol{f}^* = \sum_{od} d_{od} \boldsymbol{\Delta}_{od} \boldsymbol{p}_{od}\big(-\boldsymbol{\Delta}_{od}^T \boldsymbol{c}(\boldsymbol{f}^*) - \boldsymbol{g}_{od}^{\text{NA}}\big) \qquad (5.4.2)$$

with

$$\boldsymbol{f}^* \in S_f$$

The corresponding equilibrium costs can be obtained with the equations reported in Sect. 5.2. Fixed-point models expressed in terms of link or path cost variables are also possible to develop.

An example of stochastic equilibrium using a logit path choice model for a two-link/path network is given in Fig. 5.20. The stochastic equilibrium pattern is obtained at the intersection of the curves representing the supply and (inverse) demand equations. Note that the stochastic equilibrium configuration does not correspond to equal (systematic) costs on the two paths, which means that the intersection point of the two curves does not correspond to a zero value of the difference $g_1 - g_2$. In other words, at stochastic equilibrium, some travelers have higher (systematic) path costs than others. This result obviously depends on the assumptions made about path choice behavior. The perceived path cost is modeled as a random variable and therefore some users may choose higher (systematic) cost paths because they perceive them as least cost.

The existence and uniqueness of stochastic equilibrium flows and costs are guaranteed, respectively, by the continuity and the monotonicity of the cost functions, provided that the path choice model guarantees the continuity and monotonicity of the stochastic uncongested network assignment function (as described below). Note that these conditions for existence and uniqueness are only sufficient; that is, there can be noncontinuous and/or nonmonotone cost functions that also give rise to a unique equilibrium configuration. In the following, existence and uniqueness are explicitly analyzed for equilibrium link flow variables only; these conditions then ensure the existence and uniqueness of the corresponding link costs $\boldsymbol{c}^* = \boldsymbol{c}(\boldsymbol{f}^*)$, and of the path costs and flows $\boldsymbol{g}^*$ and $\boldsymbol{h}^*$, obtained through relations (5.2.1) and (5.2.7) respectively.

Continuity of the Stochastic Uncongested Network Assignment Function. If the path choice model is a continuous function having continuous first partial derivatives with respect to path costs, as is the case for typical probabilistic ($|\boldsymbol{\Sigma}| \neq 0$) models, then the stochastic uncongested network assignment function is also continuous and has continuous first partial derivatives with respect to link costs. In this case, in other words, a "small" variation in link costs induces a "small" variation in link flows.

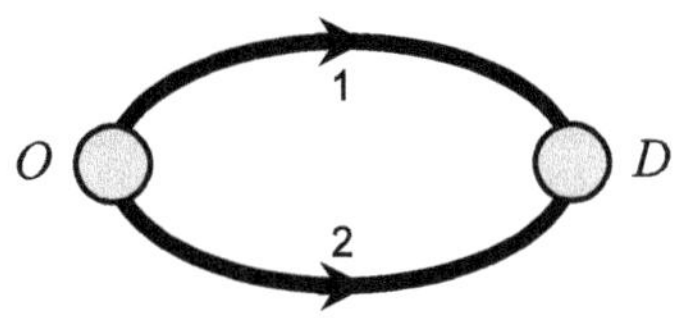

$$c_1 = c_0\left(1 + a_1\frac{f_1}{Cap_1}\right)^{\gamma} = 30\left(1 + 2\left(\frac{f_1}{500}\right)^4\right);$$

$$c_2 = c_0\left(1 + a_2\frac{f_2}{Cap_2}\right)^{\gamma} = 350\left(1 + 2\left(\frac{f_2}{500}\right)^4\right);$$

$$h_1 + h_2 = d_{od} = 1000; \quad h_1 = f_1; \quad h_2 = f_2;$$

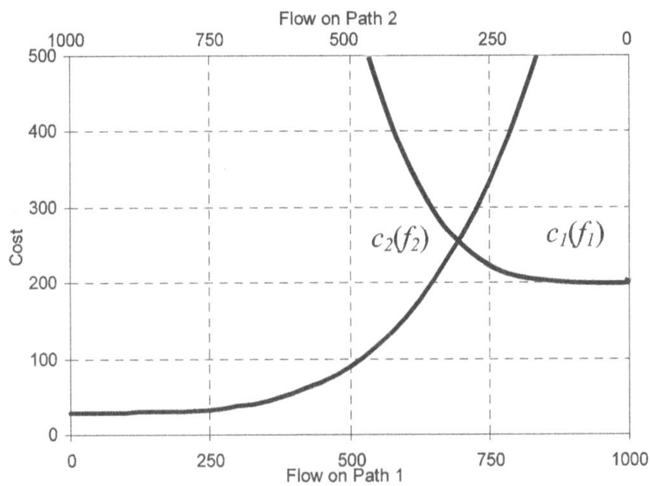

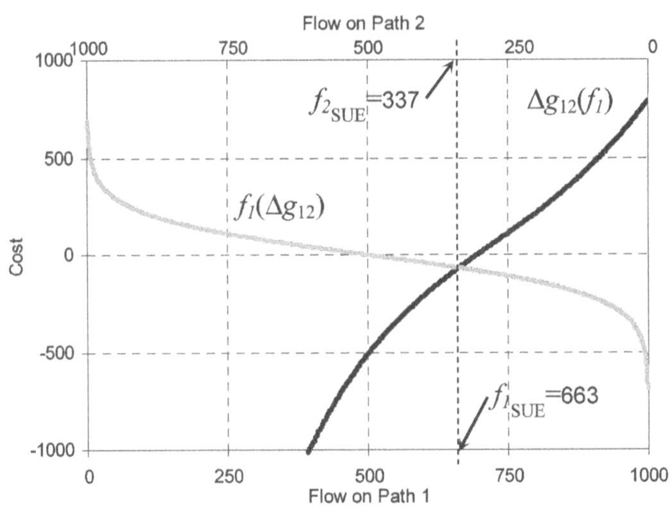

Supply equation $\quad \Delta g_{1,2}(f_1) = g_1(f_1) - g_2(f_2 = d - f_1)$

Demand equation $\quad f_1(\Delta g_{1,2}) = d_{o,d}\dfrac{1}{1 + \exp(\Delta g_{1,2}/\theta)}$

$\quad f_2 = d_{od} - f_1$

Fig. 5.20 Example of Stochastic User Equilibrium (SUE; $\theta = 100$)

Existence of Stochastic User Equilibrium Link Flows. The fixed-point model (5.4.2) has at least one solution if the cost function $\boldsymbol{c} = \boldsymbol{c}(\boldsymbol{f})$ and the path choice function $\boldsymbol{p}_{od} = \boldsymbol{p}_{od}(\boldsymbol{V}_{od})$ (which defines the stochastic uncongested network assignment function $\boldsymbol{f} = \boldsymbol{f}_{\text{SUN}}(\boldsymbol{c}; \boldsymbol{d})$) are both continuous.

The equilibrium solution $\boldsymbol{f}^*$ is a fixed point of the composite function $\boldsymbol{y} = \boldsymbol{f}_{\text{SUN}}(\boldsymbol{c}(\boldsymbol{x}))$ which, under the above assumptions (and for a connected network), is a continuous function defined over the nonempty, compact, and convex set S_f. Furthermore, the function $\boldsymbol{y} = \boldsymbol{f}_{\text{SUN}}(\boldsymbol{c}(\boldsymbol{x}))$ assumes values only in the feasible set S_f; thus all of the assumptions of Brouwer's theorem on the existence of fixed points are satisfied (see Appendix A).

The continuity of the cost functions over the feasible flow set (and therefore the existence of the equilibrium solution) requires that the cost functions be defined for any feasible link flow value, even if a particular link flow is greater than the physical capacity of that link (recall, however, that link flows are bounded above by the demand flows). Still, if explicit capacity constraints are added, the set of feasible flows might be empty: there may be no link flow vector that corresponds to the travel demand and simultaneously does not exceed the capacity of each network link. Such a limit case corresponds to an excess of demand compared to the available capacity of the system.

Monotonicity of the Stochastic Uncongested Network Assignment Function. If the path choice model is defined by a nondecreasing monotone function of the systematic utility, as in the case of additive probabilistic models (as demonstrated in Sect. 3.4), the stochastic uncongested network assignment function is monotone nonincreasing with respect to link costs. Thus, if the cost of one or more of the links increases, the flow (or flows) on these links decreases, and vice versa. This property is formally expressed as

$$\left(\boldsymbol{f}_{\text{SUN}}(\boldsymbol{c}') - \boldsymbol{f}_{\text{SUN}}(\boldsymbol{c}'')\right)^T (\boldsymbol{c}' - \boldsymbol{c}'') \le 0 \quad \forall \boldsymbol{c}', \boldsymbol{c}''$$

Given any two link cost vectors $\boldsymbol{c}'$ and $\boldsymbol{c}''$, consider the following notation.

$$\boldsymbol{g}'_{od} = \boldsymbol{\Delta}_{od}^T \boldsymbol{c}' + \boldsymbol{g}_{od}^{\text{NA}} \qquad \boldsymbol{V}'_{od} = -\boldsymbol{g}'_{od} \qquad \boldsymbol{p}'_{od} = \boldsymbol{p}_{od}(\boldsymbol{V}'_{od})$$

$$\boldsymbol{h}'_{od} = d_{od}\,\boldsymbol{p}'_{od} \qquad \boldsymbol{f}' = \sum_{od} \boldsymbol{\Delta}_{od} \boldsymbol{h}'_{od}$$

$$\boldsymbol{g}''_{od} = \boldsymbol{\Delta}_{od}^T \boldsymbol{c}'' + \boldsymbol{g}_{od}^{\text{NA}} \qquad \boldsymbol{V}''_{od} = -\boldsymbol{g}''_{od} \qquad \boldsymbol{p}''_{od} = \boldsymbol{p}_{od}(\boldsymbol{V}''_{od})$$

$$\boldsymbol{h}''_{od} = d_{od}\,\boldsymbol{p}''_{od} \qquad \boldsymbol{f}'' = \sum_{od} \boldsymbol{\Delta}_{od} \boldsymbol{h}''_{od}$$

Assuming that the path choice model is additive and the choice map is monotone nondecreasing (see Sect. 3.4) we obtain:

$$\left(\boldsymbol{p}_{od}(\boldsymbol{V}'_{od}) - \boldsymbol{p}_{od}(\boldsymbol{V}''_{od})\right)^T (\boldsymbol{V}'_{od} - \boldsymbol{V}''_{od}) \geq 0 \quad \forall od$$

and it follows from the nonnegativity of the demand flow $d_{od} \geq 0$ that

$$d_{od}\left(\boldsymbol{p}_{od}(\boldsymbol{V}'_{od}) - \boldsymbol{p}_{od}(\boldsymbol{V}''_{od})\right)^T (\boldsymbol{V}'_{od} - \boldsymbol{V}''_{od}) \geq 0 \quad \forall od$$

$$(\boldsymbol{h}'_{od} - \boldsymbol{h}''_{od})^T (\boldsymbol{V}'_{od} - \boldsymbol{V}''_{od}) \geq 0 \quad \forall od$$

$$\sum_{od} (\boldsymbol{h}'_{od} - \boldsymbol{h}''_{od})^T (\boldsymbol{V}'_{od} - \boldsymbol{V}''_{od}) \geq 0$$

Because $\boldsymbol{V}_{od} = -\boldsymbol{g}_{od} = -\boldsymbol{\Delta}_{od}^T \boldsymbol{c} - \boldsymbol{g}_{od}^{\mathrm{NA}}$, the above reduces to:

$$-\sum_{od} (\boldsymbol{h}'_{od} - \boldsymbol{h}''_{od})^T (\boldsymbol{g}'_{od} - \boldsymbol{g}''_{od}) \geq 0$$

$$\sum_{od} (\boldsymbol{h}'_{od} - \boldsymbol{h}''_{od})^T \left(\boldsymbol{\Delta}_{od}^T \boldsymbol{c}' + \boldsymbol{g}_{od}^{\mathrm{NA}} - \boldsymbol{\Delta}_{od}^T \boldsymbol{c}'' - \boldsymbol{g}_{od}^{\mathrm{NA}}\right) \leq 0$$

$$\sum_{od} (\boldsymbol{h}'_{od} - \boldsymbol{h}''_{od})^T \boldsymbol{\Delta}_{od}^T (\boldsymbol{c}' - \boldsymbol{c}'') \leq 0$$

from which $(\boldsymbol{f}' - \boldsymbol{f}'')^T (\boldsymbol{c}' - \boldsymbol{c}'') \leq 0$ follows.

Note that two different vectors of link costs $\boldsymbol{c}'$ and $\boldsymbol{c}''$ usually generate two different vectors of additive path costs $\boldsymbol{\Delta}_{od}^T \boldsymbol{c}'$ and $\boldsymbol{\Delta}_{od}^T \boldsymbol{c}''$ and therefore two vectors of systematic utility $\boldsymbol{V}'_{od}$ and $\boldsymbol{V}''_{od}$. Thus, the assumption that the path choice model is additive (see Sect. 3.4) with respect to the path systematic utility is equivalent to the assumption that, for each O-D pair, the distribution parameters of the path utility random residuals $\boldsymbol{\varepsilon}_{od}$ (such as the parameter θ in a logit model or the variance–covariance matrix $\boldsymbol{\Sigma}$ in a probit model) do not depend on the additive path costs, and therefore on the link costs relevant to congestion. However, they may depend on other reference variables (such as distance, free flow costs, etc.).[14] Note that, under this assumption, the Jacobian of the function $\boldsymbol{f}_{\mathrm{SUN}} = \boldsymbol{f}_{\mathrm{SUN}}(\boldsymbol{c}), \boldsymbol{Jac}[\boldsymbol{f}_{\mathrm{SUN}}(\boldsymbol{c})] = \sum_{od} d_{od} \boldsymbol{\Delta}_{od} \boldsymbol{Jac}[\boldsymbol{p}_{od}(-\boldsymbol{\Delta}_{od}^T \boldsymbol{c} - \boldsymbol{g}_{od}^{\mathrm{NA}})] \boldsymbol{\Delta}_{od}^T$ is symmetric and negative semidefinite, because the Jacobian $\boldsymbol{Jac}[\boldsymbol{p}_{od}(-\boldsymbol{\Delta}_{od}^T \boldsymbol{c} - \boldsymbol{g}_{od}^{\mathrm{NA}})]$ is symmetric and positive semidefinite (see Sect. 3.4).

Uniqueness of Stochastic User Equilibrium Link Flows. The fixed-point model (5.4.2) has at most one solution if the link cost functions $\boldsymbol{c} = \boldsymbol{c}(\boldsymbol{f})$ are strictly in-

[14] If the random residual variance of a path depended on the path cost, then, as the cost increased, the corresponding increase in variance might lead to an increase in the path choice probability itself.

creasing[15] over the set of feasible link flows:

$$\left[\boldsymbol{c}(\boldsymbol{f}') - \boldsymbol{c}(\boldsymbol{f}'')\right]^T (\boldsymbol{f}' - \boldsymbol{f}'') > 0 \quad \forall \boldsymbol{f}' \neq \boldsymbol{f}'' \in S_f$$

and the path choice models are additive (and are expressed by continuous functions $\boldsymbol{p}_{od} = \boldsymbol{p}_{od}(\boldsymbol{V}_{od})$ with continuous first partial derivatives).

As previously shown, under this assumption the stochastic uncongested network assignment function $\boldsymbol{f}_{\text{SUN}}(\boldsymbol{c})$ is monotone nonincreasing with respect to the link costs:

$$\left[\boldsymbol{f}_{\text{SUN}}(\boldsymbol{c}') - \boldsymbol{f}_{\text{SUN}}(\boldsymbol{c}'')\right]^T (\boldsymbol{c}' - \boldsymbol{c}'') \leq 0 \quad \forall \boldsymbol{c}', \boldsymbol{c}''$$

The proof is then completed by *reductio ad absurdum*. Suppose that two different equilibrium link flow vectors existed, $\boldsymbol{f}_1^* \neq \boldsymbol{f}_2^* \in S_f$. Then with $\boldsymbol{c}_1^* = \boldsymbol{c}(\boldsymbol{f}_1^*)$ and $\boldsymbol{c}_2^* = \boldsymbol{c}(\boldsymbol{f}_2^*)$, the equilibrium definition $\boldsymbol{f}_1^* = \boldsymbol{f}_{\text{SUN}}(\boldsymbol{c}_1^*)$ and $\boldsymbol{f}_2^* = \boldsymbol{f}_{\text{SUN}}(\boldsymbol{c}_2^*)$ and the monotonicity of the stochastic uncongested network assignment function with $\boldsymbol{c}' = \boldsymbol{c}_1^*$ and $\boldsymbol{c}'' = \boldsymbol{c}_2^*$ yield:

$$\left[\boldsymbol{f}_1^* - \boldsymbol{f}_2^*\right]^T (\boldsymbol{c}_1^* - \boldsymbol{c}_2^*) \leq 0$$

From the monotonicity of the cost functions, with $\boldsymbol{f}' = \boldsymbol{f}_1^* \neq \boldsymbol{f}'' = \boldsymbol{f}_2^*$, it also follows that

$$\left[\boldsymbol{c}_1^* - \boldsymbol{c}_2^*\right]^T (\boldsymbol{f}_1^* - \boldsymbol{f}_2^*) > 0$$

Thus, there is a contradiction between the monotonicity of the cost functions and that of the stochastic uncongested network assignment function.

A sufficient condition for the strict monotonicity of the cost functions is that the Jacobian matrix $\boldsymbol{Jac}[\boldsymbol{c}(\boldsymbol{f})]$ of the cost vector $\boldsymbol{c}(\boldsymbol{f})$ is positive definite over the set S_f (see Appendix A). In the case of separable cost functions, $c_a = c_a(f_a)$, the Jacobian matrix is diagonal and its elements are the derivatives of the cost functions of each link with respect to the corresponding link flow. In the usual case when cost functions are increasing with respect to flow,[16] the derivatives are positive, the Jacobian matrix is positive definite, and the equilibrium flow vector $\boldsymbol{f}^*$ is unique. However, there are real situations in which the cost functions are not monotone.

In applications, the nonuniqueness of equilibrium, or the difficulty of demonstrating it a priori, gives rise to problems in both computation and interpretation. Although it is only possible to demonstrate convergence of the solution algorithms if the solution is unique (see Sect. 5.4.2), nonunique equilibria leave open the possibility that a particular calculated equilibrium flow vector may not be the appropriate one with which to design or evaluate the transportation system under study. In other

---

[15]In the case of logit or probit path choice models, for which path choice probabilities are strictly greater than zero regardless of cost, it is possible to demonstrate the uniqueness of equilibrium flows even for cost functions that are monotone but not strictly so: $[\boldsymbol{c}(\boldsymbol{f}') - \boldsymbol{c}(\boldsymbol{f}'')]^T (\boldsymbol{f}' - \boldsymbol{f}'') \geq 0$ $\forall \boldsymbol{f}', \boldsymbol{f}'' \in S_f$.

[16]The link cost functions reported in Chap. 2 are all strictly increasing with respect to link flows.

words, the system may attain different equilibrium patterns, and each of these would need to be verified and analyzed.

Stochastic equilibrium link flows can be calculated with various algorithms, the simplest of which use the stochastic uncongested network assignment function as described in the next section.

Appendix 5.A at the end of this chapter presents some optimization models for fixed demand SUE with separable cost functions, which may be used for dealing with SUE under some limiting assumptions. It should be noted that it is hard, if not impossible, to extend them to deal with issues addressed in Chap. 6. At this point, they are presented mainly for the purpose of completeness.

5.4.2 Algorithms for Stochastic User Equilibrium

As mentioned above, the fixed-point problem (5.4.2) can be solved with an algorithm that generates a sequence of estimated link flow solution vectors $\boldsymbol{f}^k$, starting from an initial feasible solution $\boldsymbol{f}^0 \in S_f$. The flow estimate at step k is obtained by combining the solution estimate at step $k-1$ with an *auxiliary* flow vector $\boldsymbol{f}^k_{\text{SUN}}$, obtained from a SUN assignment based on link costs that correspond to the estimate at step $k-1$. The algorithm can be described by the following system of recursive equations, starting from $\boldsymbol{f}^0 \in S_f$ and $k = 0$.

$$k = k + 1 \tag{5.4.3}$$

$$\boldsymbol{c}^k = \boldsymbol{c}(\boldsymbol{f}^{k-1}) \tag{5.4.4}$$

$$\boldsymbol{f}^k_{\text{SUN}} = \boldsymbol{f}_{\text{SUN}}(\boldsymbol{c}^k) \tag{5.4.5}$$

$$\boldsymbol{f}^k = \boldsymbol{f}^{k-1} + 1/k\left(\boldsymbol{f}^k_{\text{SUN}} - \boldsymbol{f}^{k-1}\right) \tag{5.4.6}$$

In general, this procedure is known as the Method of Successive Averages (MSA). Because the solution estimate at iteration k, $\boldsymbol{f}^k$, is the average of flows from the first k SUN assignments, the algorithm is called the Flow Averaging (MSA-FA) algorithm. Note that the cost vector $\boldsymbol{c}^k$ is always feasible, in the sense that it represents the exact costs that result from a feasible flow pattern. An initial solution estimate $\boldsymbol{f}^0 \in S_f$ can easily be obtained from a SUN assignment using free flow costs $\boldsymbol{f}^0 = \boldsymbol{f}_{\text{SUN}}(\boldsymbol{c}(\boldsymbol{f} = \boldsymbol{0}))$. A fixed point is found if the auxiliary SUN flows are equal to the current solution estimate:

$$\boldsymbol{f}^k_{\text{SUN}} - \boldsymbol{f}^{k-1} = 0$$

In practice, the algorithm is stopped when the relative difference between the SUN link flows and the current solution estimate at iteration k falls below a pre-assigned threshold δ, as measured using a suitable vector norm $|\boldsymbol{f}^k_{\text{SUN}} - \boldsymbol{f}^{k-1}|/|\boldsymbol{f}^{k-1}| < \delta$; one such norm is $\max_a |f^k_{\text{SUN},a} - f^{k-1}_a|/|f^{k-1}_a|$.

The convergence speed of the MSA algorithm close to the solution may be rather slow because the step length $1/k$ gets increasingly smaller. Therefore, it might be best after a certain number of iterations to restart the algorithm, using the current solution estimate as the new initial solution. This approach leads to a multiphase algorithm, where each phase is characterized by a pre-determined maximum number of MSA iterations, increasing with each successive phase: for example, 5 iterations in the first phase, 10 in the second, 15 in the third, and so on. This approach is, however, a heuristic one whose convergence properties remain unknown.

If the cost functions $\boldsymbol{c} = \boldsymbol{c}(\boldsymbol{f})$ are continuous and strictly monotone increasing, and if the SUN assignment function $\boldsymbol{f} = \boldsymbol{f}_{\text{SUN}}(c)$ is continuous and monotone non-increasing, then the fixed-point problem (5.4.2) has a unique solution, as shown in Sect. 5.4.1. Under these assumptions, application of Blum's theorem (see Appendix A) guarantees that, if the Jacobian of the cost functions is symmetric, the sequence of link flow solution estimate vectors $\boldsymbol{f}^k$ generated by the MSA-FA algorithm almost certainly converges to the equilibrium link flow vector. An example application of the MSA-FA algorithm is given in Figs. 5.21a and 5.21b.

Monotonicity of the SUN assignment function is ensured if the distribution of path choice model random residuals does not depend on the congestion level. With a logit path choice model, this condition is met if the parameter θ and the definition of efficient paths are independent of the link costs $\boldsymbol{c}$ (they might depend, however, on free flow costs, or on other link attributes that do not vary with congestion). Analogously, with a probit path choice model, this condition is met if the random residual variance–covariance matrix $\boldsymbol{\Sigma}$ is independent of the link costs $\boldsymbol{c}$ (but again it might depend on free flow costs or on other attributes that do not vary with congestion).

In the case of a probit path choice model, the Monte Carlo assignment algorithm only provides an unbiased estimate of SUN flows, as was seen in Sect. 5.3.3. In this case, Blum's theorem guarantees almost definite convergence of the MSA-FA algorithm. The convergence threshold δ that can be achieved depends on the number of iterations carried out within the SUN assignment algorithm. (Because the SUN assignment is executed as one step of the MSA, its Monte Carlo iterations are called inner iterations, whereas the flow averaging iterations of the MSA are called outer iterations.)

To improve the overall efficiency of the SUE algorithm, a two-phase approach is sometimes used. In this approach, the Monte Carlo SUN algorithm is first run with a small number (say 1–3) of inner iterations until the MSA algorithm finds a flow vector close to the equilibrium solution. The previously discussed stop criterion cannot be applied in this phase because of the small number of inner iterations; thus termination of the first phase is usually based on a comparison of successive solution estimates $\boldsymbol{f}^k \cong \boldsymbol{f}^{k-1}$. In the second phase, a larger number (say 30–60) of inner iterations is run, depending on the convergence threshold, and the correct stop criterion can be used. Another approach is to let the maximum number of inner iterations of the SUN algorithm increase with the outer iteration index of the MSA algorithm: for example, two iterations within the SUN algorithm for the first ten iterations of the MSA algorithm, then four for the next ten, and so on.

The MSA algorithm could also be applied to SUE problems with nonseparable cost functions; in this case, however, convergence cannot be guaranteed. A different stochastic equilibrium algorithm for nonseparable cost functions (asymmetric Jacobian) can be obtained by applying the method of successive averages to link costs rather than flows. This results in the Cost Averaging (MSA-CA) algorithm, specified by the following system of recursive equations, starting with $\boldsymbol{f}^0 \in S_f$, $\boldsymbol{c}^0 = c(\boldsymbol{f}^0)$, and $k = 0$:

$$k = k + 1 \tag{5.4.7}$$

$$\boldsymbol{f}^k = \boldsymbol{f}_{\text{SUN}}(\boldsymbol{c}^{k-1}) \tag{5.4.8}$$

$$\boldsymbol{x}^k = \boldsymbol{c}(\boldsymbol{f}^k) \tag{5.4.9}$$

$$\boldsymbol{c}^k = \boldsymbol{c}^{k-1} + 1/k(\boldsymbol{x}^k - \boldsymbol{c}^{k-1}) \tag{5.4.10}$$

Note that the link flow vector $\boldsymbol{f}^k$ at each iteration k is feasible, in the sense that it represents the flows resulting from an SUN assignment based on feasible costs.

The algorithm terminates if the SUN flows calculated with costs $\boldsymbol{x}^k$ are equal to the flow vector $\boldsymbol{f}^k$:

$$\boldsymbol{f}_{\text{SUN}}\big(\boldsymbol{c}(\boldsymbol{f}^k)\big) - \boldsymbol{f}^k = 0$$

In practice, the algorithm terminates when the difference $\boldsymbol{f}_{\text{SUN}}(\boldsymbol{c}(\boldsymbol{f}^k)) - \boldsymbol{f}^k$ is below a pre-assigned threshold δ, as determined using a suitable norm, as discussed above. Note that the termination test is computationally demanding, because it requires a further SUN assignment.

The convergence of the MSA-CA algorithm is, in general, slower than that of the MSA-FA algorithm.[17] From a practical point of view, it may be convenient to perform some initial iterations using the MSA-FA algorithm in order to approach the equilibrium solution, and then to apply the MSA-CA algorithm using the current solution as the initial solution (two-phase algorithm). The considerations discussed for the MSA-FA algorithm with probit path choice model apply also in this case.

An application of Blum's theorem (see Appendix A) shows that convergence of the MSA-CA algorithm is ensured if the conditions for existence and uniqueness of the solutions hold and the Jacobian of the SUN function is symmetric. Existence and uniqueness require, respectively, continuous and strictly increasing monotone cost functions and a continuous and nondecreasing monotone SUN function. The last condition is met if the distribution of random residuals in the path choice model is independent of congestion. In this case, moreover, the Jacobian of the SUN function is symmetric (as noted in Sect. 5.3.1).

The stochastic equilibrium problem with nonseparable cost functions can also be solved through the inverse cost function algorithm mentioned in the bibliographic notes. It could also be solved by applying the diagonalization algorithm, as de-

---

[17] Some computational results suggest that the speed of convergence can be increased by reducing the step length by a factor $\beta \in]0, 1[$, $\boldsymbol{c}^k = \boldsymbol{c}^{k-1} + \beta/k(\boldsymbol{y}^k - \boldsymbol{c}^{k-1})$.

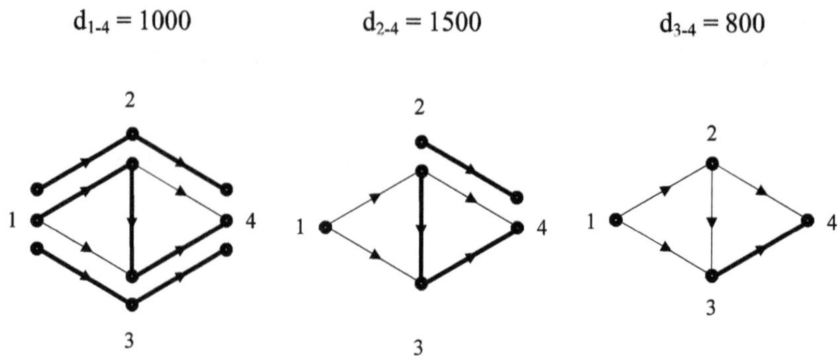

PATHS

Paths	Nodes
1	1-2-4
2	1-2-3-4
3	1-3-4
4	2-3-4
5	2-4
6	3-4

PARAMETERS OF COST FUNCTIONS

Arc	c_0	a	Cap	γ
1-2	10	2	1000	4
1-3	22	2	1000	4
2-3	13	2	2500	4
2-4	20	2	1000	4
3-4	11	2	3300	4

COST FUNCTION TYPE

$$c_1 = c_0\left(1 + a\frac{f}{Cap}\right)^{\gamma}$$

LOGIT PARAMETER

$$\theta = 30$$

ITERATIONS

Iteration		Link	$c(f=0)$	f^0_{SNL}	f^0
0		1-2	10	675	675
		1-3	22	325	325
		2-3	13	1060	1060
		2-4	20	1115	1115
		3-4	11	2185	2185
Iteration (k)	$\mu = 1/k$	Link	$c(f^{k-1})$	f^k_{SNL}	f^k
1	1.000	1-2	14	480	480
		1-3	22	520	520
		2-3	14	1403	1403
		2-4	92	577	577
		3-4	11	2723	2723
2	0.500	1-2	11	668	574
		1-3	25	332	426
		2-3	20	1009	1206
		2-4	20	1159	868
		3-4	11	2141	2432
3	0.333	1-2	12	617	589
		1-3	23	383	411
		2-3	15	1098	1170
		2-4	32	1019	919
		3-4	11	2281	2381
4	0.250	1-2	12	596	591
		1-3	23	404	409
		2-3	15	1133	1161
		2-4	37	963	930
		3-4	11	2337	2370

Fig. 5.21a Example of the MSA-FA algorithm for SUE assignment with link variables (first iterations only)

scribed for deterministic equilibrium in next subsection. SUE under some limiting assumptions can also be solved through optimization techniques based on models presented in Appendix 5.A at the end of this chapter.

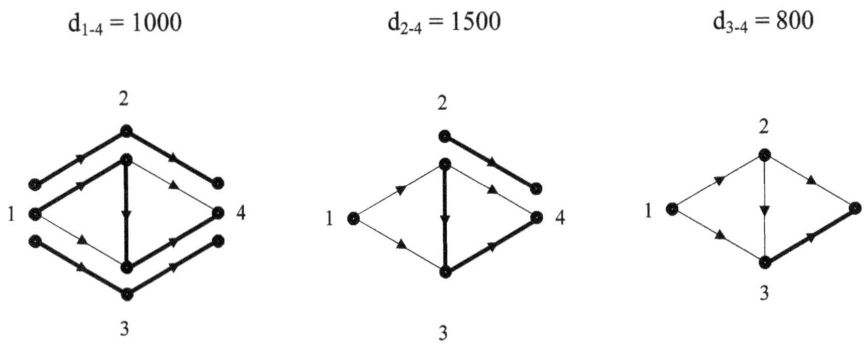

PATHS

Paths	Nodes
1	1-2-4
2	1-2-3-4
3	1-3-4
4	2-3-4
5	2-4
6	3-4

PARAMETERS OF COST FUNCTIONS

Arc	c_0	a	Cap	γ
1-2	10	2	1000	4
1-3	22	2	1000	4
2-3	13	2	2500	4
2-4	20	2	1000	4
3-4	11	2	3300	4

COST FUNCTION TYPE

$$c_1 = c_0\left(1 + a\frac{f}{Cap}\right)^{\gamma}$$

LOGIT PARAMETER

$$\theta = 30$$

FIRST ITERATIONS

Iteration		Path	$C(F=0)$	p^0	F^0_{SNL}	F^0
0		1	30	0.360	360	360
		2	34	0.315	315	315
		3	33	0.325	325	325
		4	24	0.467	700	700
		5	20	0.533	800	800
		6	11	1.000	800	800
Iteration (k)	**$\mu = 1/k$**	**Path**	**$C(F^{k-1})$**	**p^k**	**F^k_{SNL}**	**F^k**
1	1.000	1	106	0.046	46	46
		2	39	0.435	435	435
		3	34	0.520	520	520
		4	25	0.905	1357	1357
		5	92	0.095	143	143
		6	11	1.000	800	800
2	0.500	1	31	0.394	394	220
		2	42	0.274	274	354
		3	36	0.332	332	426
		4	31	0.410	615	986
		5	20	0.590	885	514
		6	11	1.000	800	800
3	0.333	1	44	0.281	281	240
		2	38	0.336	336	348
		3	35	0.383	383	411
		4	26	0.545	817	930
		5	32	0.455	683	570
		6	11	1.000	800	800
4	0.250	1	50	0.352	352	254
		2	38	0.148	148	366
		3	34	0.500	500	380
		4	26	0.296	444	890
		5	37	0.704	1056	610
		6	11	1.000	800	800

Fig. 5.21b Example of the MSA-FA algorithm for SUE assignment with path

5.4.3 Models for Deterministic User Equilibrium

Deterministic User Equilibrium (DUE) assignment is obtained by applying the equilibrium approach for congested networks under the assumption of deterministic path choice behavior. Deterministic equilibrium link flows $\boldsymbol{f}^*$, path flows $\boldsymbol{h}^*$, and the corresponding costs $\boldsymbol{c}^*$ and $\boldsymbol{g}^*$ can be determined with a fixed-point model obtained by simultaneously applying the supply model (5.2.4) and the demand model (5.2.7), as in the stochastic equilibrium case (an alternative is to utilize the deterministic uncongested network assignment map and flow-dependent cost functions). In this case, however, there are some mathematical complications arising from the fact that the deterministic demand model is expressed (such as the corresponding deterministic uncongested network assignment map[18]) by a one-to-many map, as was noted in Sect. 5.2.2 (and in 5.3.2).

For this reason, the properties of deterministic equilibrium are usually studied through indirect formulations. The most general is the variational inequality formulation based on the specification of the deterministic demand model as the system of inequalities (5.2.7b):

$$\boldsymbol{g}(\boldsymbol{h}^*)^T(\boldsymbol{h}-\boldsymbol{h}^*) \geq 0 \quad \forall \boldsymbol{h} \in S_h \tag{5.4.11}$$

By combining the demand model obtained by summing (5.2.7b) on all O-D pairs with the supply model (5.2.4), expression (5.4.11) is obtained. In the case of congested networks, therefore, the resulting path (or link) flows correspond to the condition expressed by Wardrop's first principle.

Equivalent variational inequality models expressed in terms of link flows are obtained by combining the link cost functions (5.2.2) with the inequality systems (5.3.5) or (5.3.6) that represent deterministic uncongested network assignment:

$$\boldsymbol{c}(\boldsymbol{f}^*)^T(\boldsymbol{f}-\boldsymbol{f}^*) \geq 0 \quad \forall \boldsymbol{f} \in S_f \tag{5.4.12}$$

$$\boldsymbol{c}(\boldsymbol{f}^*)^T(\boldsymbol{f}-\boldsymbol{f}^*) + (\boldsymbol{g}^{\mathrm{NA}})^T(\boldsymbol{h}-\boldsymbol{h}^*) \geq 0 \quad \forall \boldsymbol{f} = \boldsymbol{\Delta h},\ \forall \boldsymbol{h} \in S_h \tag{5.4.13}$$

Expressions (5.4.12) and (5.4.13) apply, respectively, to cases with zero and nonzero nonadditive path costs. Note that expressions (5.4.11)–(5.4.13) are different from those used for deterministic uncongested assignment in that the path and link costs depend on flows. In the presence of nonadditive path costs, the considerations presented in Sect. 5.3.2 hold, and (5.4.13) can be expressed in terms of link flows

[18]For the deterministic uncongested network assignment map, it is possible to demonstrate properties analogous to those of the stochastic uncongested network assignment function. In particular, the deterministic uncongested network assignment map is semicontinuous, and the set of flows associated with each link's cost is nonempty, compact, and convex. Furthermore, the map is monotone nonincreasing with respect to link costs. These properties permit analysis of the existence and uniqueness of the deterministic user equilibrium flow configurations analogously to the analysis carried out for stochastic user equilibrium flows in Sect. 5.4.1.

f^* and of the total nonadditive cost G^{NA*} at deterministic equilibrium:

$$\begin{aligned} &c(f^*)^T(f - f^*) + (G^{NA} - G^{NA*}) \geq 0 \\ &\forall f = \Delta h, \; G^{NA} = (g^{NA})^T h, \; \forall h \in S_h \end{aligned} \tag{5.4.14}$$

An example of deterministic user equilibrium assignment for a two-link/path network is shown in Fig. 5.22. Note that the deterministic equilibrium flows correspond to the intersection point of the supply and demand curves (in this case, step curves) and they result in costs that are equal for the two paths since both are used.

Conditions ensuring the existence and uniqueness of deterministic equilibrium link flows and costs are similar to those described for stochastic equilibrium. In particular, the continuity and monotonicity of the cost functions guarantee, respectively, the existence and uniqueness of the solution. It should be noted once again that these existence and uniqueness conditions are only sufficient; there may exist nonmonotone cost functions that give rise to a unique equilibrium vector.

Existence of Deterministic User Equilibrium Link Flows. The variational inequalities (5.4.11)–(5.4.13) have at least one solution if the cost functions are continuous functions defined on the nonempty, compact, and convex set of the feasible path flows S_h or link flows S_f.

This is a general property of variational inequalities, which can be proved using Brouwer's theorem (see Appendix A).

The considerations regarding the continuity of cost functions discussed for SUE models apply also for DUE models. The existence of equilibrium link flows ensures the existence of the corresponding link costs $c^* = c(f^*)$, and of path costs and flows g^* and h^*, given by the expressions reported in Sect. 5.2.

Uniqueness of Deterministic User Equilibrium Link Flows. The variational inequality (5.4.14), which expresses deterministic equilibrium in terms of link flows, has at most one solution if the link cost functions $c = c(f)$ are strictly increasing with respect to link flows:

$$\left[c(f') - c(f'')\right]^T (f' - f'') > 0 \quad \forall f' \neq f'' \in S_f$$

The same result holds for the variational inequality (5.4.12), which is a special case of (5.4.14) when nonadditive costs are zero.

The proof is by *reductio ad absurdum*. Assume that there exist two different equilibrium link flow vectors $f_1^* \neq f_2^* \in S_f$, corresponding to two different feasible path flow vectors, $h_1^* \neq h_2^* \in S_F$, and that $G_1^{NA*} = (g^{NA})^T h_1$ and $G_2^{NA*} = (g^{NA})^T h_2^*$ are the relative values of total nonadditive cost. Because f_1^* is

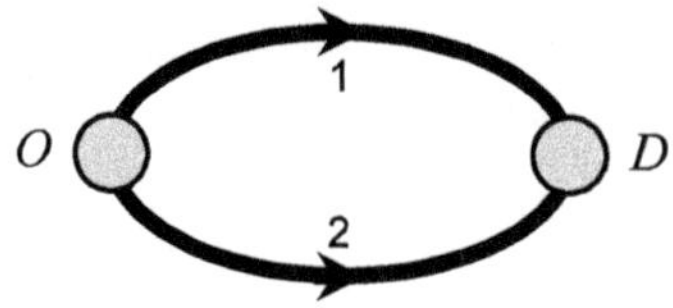

$$c_1 = c_{0,1}\left(1 + a_1 \frac{f_1}{Cap_1}\right)^{\gamma} = 30\left(1 + 2\left(\frac{f_1}{500}\right)^4\right);$$

$$c_2 = c_{0,1}\left(1 + a_2 \frac{f_2}{Cap_2}\right)^{\gamma} = 200\left(1 + 2\left(\frac{f_2}{500}\right)^4\right);$$

$$F_1 + F_2 = d_{od} = 1000; \quad F_1 = f_1; \quad F_2 = f_2;$$

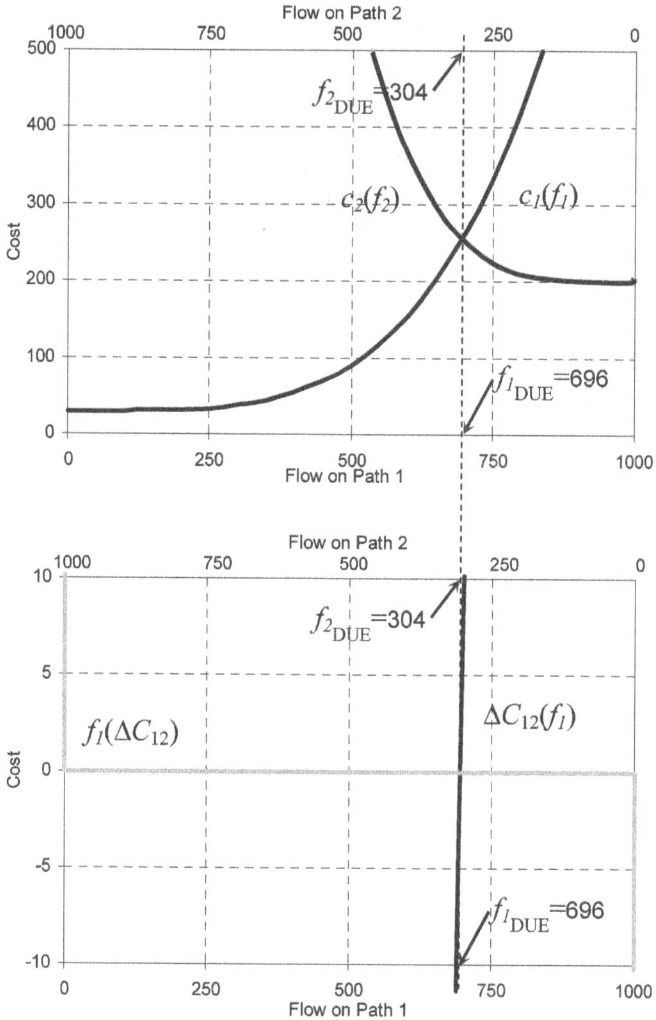

Supply equation $\quad \Delta C_{1,2}(f_1) = C_1(f_1) - C_2(f_2 = d - f_1)$

Demand equation $\quad f_1(\Delta C_{1,2}) = \begin{cases} 0 & \text{if } \Delta C_{1,2} > 0 \\ \in [0, d_{1,2}] & \text{if } \Delta C_{1,2} = 0 \\ d_{1,2} & \text{if } \Delta C_{1,2} < 0 \end{cases}$

$$f_2 = d_{od} - f_1$$

Fig. 5.22 Example of Deterministic User Equilibrium (DUE)

an equilibrium flow vector, $\boldsymbol{f}_1^*$ and $G^{\mathrm{NA}*}$ must satisfy (5.4.14); letting $\boldsymbol{f} = \boldsymbol{f}_2^* \in S_f$ and $G^{\mathrm{NA}} = G_2^{\mathrm{NA}*}$ then yields:

$$\boldsymbol{c}(\boldsymbol{f}_1^*)^T(\boldsymbol{f}_2^* - \boldsymbol{f}_1^*) + (G_2^{\mathrm{NA}*} - G_1^{\mathrm{NA}*}) \geq 0$$

Furthermore, $\boldsymbol{f}_2^*$ and $G_2^{\mathrm{NA}*}$ must also satisfy (5.4.14); again letting $\boldsymbol{f} = \boldsymbol{f}_1^* \in S_f$ and $G^{\mathrm{NA}} = G_1^{\mathrm{NA}*}$ yields:

$$\boldsymbol{c}(\boldsymbol{f}_2^*)^T(\boldsymbol{f}_1^* - \boldsymbol{f}_2^*) + (G_1^{\mathrm{NA}*} - G_2^{\mathrm{NA}*}) \geq 0$$

Adding the two above relationships gives:

$$\boldsymbol{c}(\boldsymbol{f}_1^*)^T(\boldsymbol{f}_2^* - \boldsymbol{f}_1^*) + \boldsymbol{c}(\boldsymbol{f}_2^*)^T(\boldsymbol{f}_1^* - \boldsymbol{f}_2^*) \geq 0$$

or

$$\left[\boldsymbol{c}(\boldsymbol{f}_1^*) - \boldsymbol{c}(\boldsymbol{f}_2^*)\right]^T(\boldsymbol{f}_1^* - \boldsymbol{f}_2^*) \leq 0$$

which contradicts the monotonicity of the cost functions.

The considerations regarding the monotonicity of the cost functions already expressed for stochastic equilibrium also hold for the deterministic model. Moreover, the uniqueness of equilibrium link flows ensures the uniqueness of the corresponding equilibrium link and path costs, $\boldsymbol{c}^* = \boldsymbol{c}(\boldsymbol{f}^*)$ and $\boldsymbol{g}^* = \boldsymbol{\Delta}^T\boldsymbol{c}^* + \boldsymbol{g}^{\mathrm{NA}}$. In general, however, uniqueness of link flows, and therefore of link and path costs, does not ensure the uniqueness of path flows, because there might exist different path flow vectors that induce the same link flow vector $\boldsymbol{f}^*$, and that correspond to the equilibrium costs $\boldsymbol{c}^*$ and $\boldsymbol{g}^*$.

The nonuniqueness of DUE path flows is not particularly relevant in practice if the main objective of equilibrium analysis is the modeling of link flows. However, knowledge of path flows is useful or necessary in some applications (such as the estimation of the O-D flows from traffic counts, described in Chap. 8); in such cases, this characteristic of deterministic equilibrium assignment may result in theoretical and/or algorithmic drawbacks.

Formulation with Optimization Models. Under certain assumptions on the cost functions, fixed demand deterministic equilibrium assignment problems can also be formulated as optimization models. These models allow the use of simple and efficient solution algorithms (described in the following). In particular, under the assumptions of separable cost functions and absence of nonadditive path costs, deterministic equilibrium is given by the solution to:

$$\boldsymbol{f}^* = \operatorname{argmin} \sum_a \int_0^{f_a} c_a(y_a)\, dy_a \quad \boldsymbol{f} \in S_f \tag{5.4.15}$$

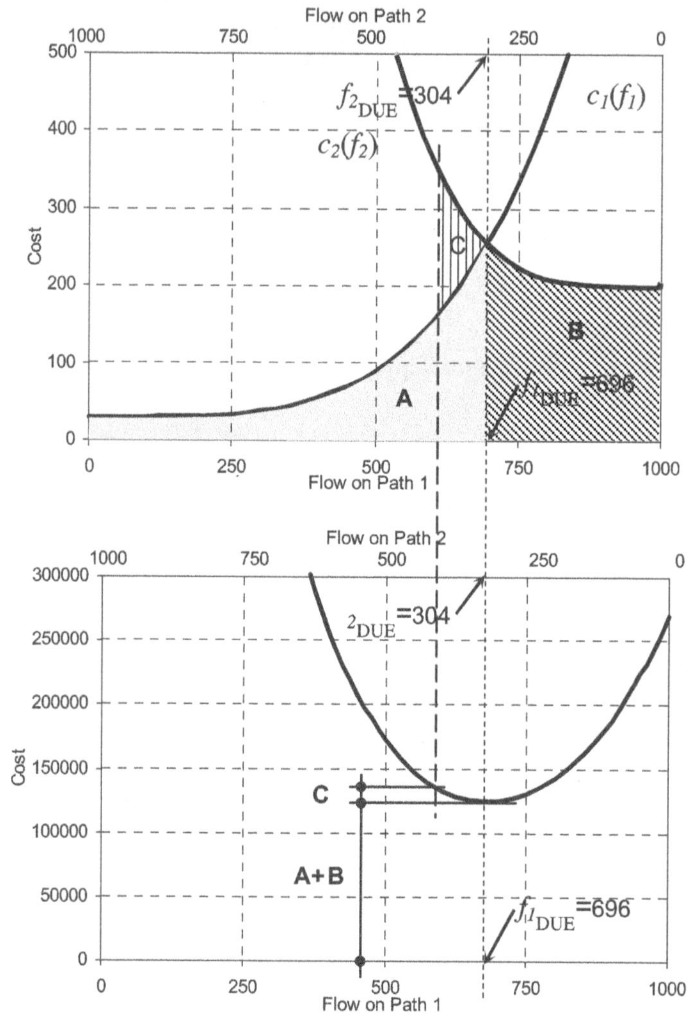

Fig. 5.23 Example of an optimization model for the DUE flows of Fig. 5.22

Figure 5.23 is a graphic illustration of the model (5.4.15) and a diagram of the function $z(\boldsymbol{f}) = \sum_a \int_0^{f_a} \boldsymbol{c}_a(y_a)\, dy_a$, known as the *integral cost*, for the two-link network introduced in Fig. 5.22. (The relation between the integral cost and the total cost $\boldsymbol{c}(\boldsymbol{f})^T \boldsymbol{f}$ is analyzed in Sect. 5.4.4.) Note that the point where the function $z(\boldsymbol{f})$ attains a minimum corresponds to the value of the flows for which the path costs are equal, which are the deterministic equilibrium flows (because both the paths are used).

The formulation (5.4.15) can be extended to nonseparable cost functions as long as they have a symmetric Jacobian (separable functions, with diagonal Jacobian, are clearly a special case of this):

$$f^* = \text{argmin} \int_0^f \boldsymbol{c}(\boldsymbol{y})^T \, d\boldsymbol{y} \quad \boldsymbol{f} \in S_f \tag{5.4.16}$$

The assumption that the cost functions have a symmetric Jacobian is critical for the formulation of the model (5.4.16) because, in general, the value of a line integral depends on the path of integration. However, when the Jacobian $\boldsymbol{Jac}[\boldsymbol{c}(\cdot)]$ of the integrand $\boldsymbol{c}(\cdot)$ is symmetric, Green's theorem ensures that the value of the integral does not depend on the path of integration (because the set is convex).[19] In this case, the integral depends only on the limits of integration. Indeed, because the lower limit is zero, the value of the integral depends only on the link flow. It is worth pointing out that in practice the Jacobian of nonseparable cost functions is rarely symmetric because the way in which the flow on a link i affects the cost of another link j is generally different from the way in which the flow on link j affects the cost on link i.

The relationship between solutions $\boldsymbol{f}^*$ of the constrained optimization model (5.4.15) and an equilibrium vector can be analyzed by verifying their relationship with solutions of the variational inequality (5.4.12), as shown below (the demonstrations refer to general features of optimization problems and variational inequalities; see Appendix A).[20]

Equivalence of Optimization Model for DUE. If the cost functions $\boldsymbol{c}(\boldsymbol{f})$ are continuous with continuous first partial derivatives and symmetric Jacobian, a vector $\boldsymbol{f}^*$ solving the optimization model (5.4.15) is an equilibrium flow vector (but not necessarily vice versa).

The function $z(\boldsymbol{f}) \int_0^f \boldsymbol{c}(\boldsymbol{y})^T \, d\boldsymbol{y}$ is differentiable with a continuous gradient because $\nabla z(\boldsymbol{f}) = \boldsymbol{c}(\boldsymbol{f})$, and therefore its minimum points satisfy the necessary condition for a minimum:

$$\nabla z(\boldsymbol{f}^*)^T (\boldsymbol{f} - \boldsymbol{f}^*) \geq 0 \quad \forall \boldsymbol{f} \in S_f$$

Because $\nabla z(\boldsymbol{f}^*) = \boldsymbol{c}(\boldsymbol{f}^*)$, (5.4.12) holds. Furthermore, the function $z(\boldsymbol{f})$ is differentiable, and therefore continuous, on a compact (and convex) set; it therefore has at least one minimum point, consistent with the existence conditions of the solutions of (5.4.12).

---

[19]If a function $\boldsymbol{c}(\boldsymbol{f})$ has a symmetric Jacobian $\boldsymbol{Jac}[\boldsymbol{c}(\boldsymbol{f})]$, it is the gradient of a function $z(\boldsymbol{f})$, $\nabla z(\boldsymbol{f}) = \boldsymbol{c}(\boldsymbol{f})$, and conversely. In this case, furthermore, the Jacobian of $\boldsymbol{c}(\boldsymbol{f})$, $\boldsymbol{Jac}[\boldsymbol{c}(\boldsymbol{f})]$ is the (symmetric) Hessian matrix of $z(\boldsymbol{f})$, $\boldsymbol{Hess}[z(\boldsymbol{f})]$.

[20]Under the same assumptions, a direct (although more complicated) demonstration that the equivalent optimization model solutions are equilibrium values is also possible; it is obtained by applying the theory of constrained optimization. Note that the equivalence conditions are stricter than those necessary to define the variational inequality models.

If the cost functions $\boldsymbol{c}(\boldsymbol{f})$ are continuous and with continuous first partial derivatives and symmetric positive semidefinite Jacobian $\boldsymbol{Jac}[\boldsymbol{c}(\boldsymbol{f})]$, a vector $\boldsymbol{f}^*$ solving the fixed optimization model (5.4.15) is an equilibrium flow vector, and vice versa.

Under the above assumptions, $z(\boldsymbol{f})$ is differentiable with continuous gradient and continuous positive semidefinite Hessian matrix, because $\nabla z(\boldsymbol{f}) = \boldsymbol{c}(\boldsymbol{f})$ and $\boldsymbol{Hess}[z(\boldsymbol{f})] = \nabla^2 z(\boldsymbol{f}) = \boldsymbol{Jac}[\boldsymbol{c}(\boldsymbol{f})]$. Therefore $z(\boldsymbol{f})$ is convex, and its minimum points $\boldsymbol{f}^*$ are defined by the necessary and sufficient condition:

$$\nabla z(\boldsymbol{f}^*)^T(\boldsymbol{f} - \boldsymbol{f}^*) \geq 0 \quad \forall \boldsymbol{f} \in S_f$$

Because $\nabla z(\boldsymbol{f}^*) = \boldsymbol{c}(\boldsymbol{f}^*)$, (5.4.12) holds. (Furthermore, $z(\boldsymbol{f})$ is convex on a convex set, and therefore has at least one minimum point, consistent with the existence conditions of the solutions of (5.4.12).)

If nonadditive path costs differ from zero, the optimization model becomes:

$$\boldsymbol{f}^*, G^{\text{NA}*} = \text{argmin} \int_0^{\boldsymbol{f}} \boldsymbol{c}(\boldsymbol{y})^T d\boldsymbol{y} + G^{\text{NA}} \tag{5.4.17}$$

$$\boldsymbol{f} = \boldsymbol{\Delta h}$$

$$G^{\text{NA}} = (\boldsymbol{g}^{\text{NA}})^T \boldsymbol{h}$$

$$\boldsymbol{h} \in S_h$$

The model (5.4.17) has properties analogous to those shown above for model (5.4.16).

When the cost function $\boldsymbol{c}(\boldsymbol{f})$ has a symmetric positive definite Jacobian, the objective functions of models (5.4.15), (5.4.16), and (5.4.17), respectively, have a single minimum point (i.e., they are unimodal). In particular, the objective function of model (5.4.15) is strictly convex and therefore has a single minimum point, consistent with the uniqueness conditions presented for variational inequality models, because under this assumption the cost functions are strictly increasing. On the other hand, the objective function of model (5.4.17) is convex with a single minimum point, inasmuch as it is the sum of a function that is strictly convex with respect to the variables $\boldsymbol{f}$ and a linear function with respect to the variable G^{NA}.

5.4.4 Algorithms for Deterministic User Equilibrium

Deterministic user equilibrium link flows can be calculated with various algorithms that directly solve the variational inequality or optimization models (in the case of cost functions with symmetric Jacobian).[21] Some simple algorithms that use deterministic network loading are described in the following.

---

[21] Note that, by using the definition (5.2.9) in Sect. 5.2.3, for the feasibility set of link flows S_f an optimization problem is obtained which is known in the literature as convex minimum cost multicommodity flow.

The optimization problem (5.4.16), having a nonlinear objective function and linear constraints, can be solved using an adaptation of the Frank–Wolfe algorithm (see Appendix A). Starting from an initial feasible solution $\boldsymbol{f}^0 \in S_f$, this algorithm generates a sequence of feasible link flow vectors $\boldsymbol{f}^k$ by solving a sequence of linear problems that approximate problem (5.4.17); each such problem is defined in terms of the current solution estimate $\boldsymbol{f}^{k-1}$. The solution of each linear problem identifies a direction along which the objective function is minimized to determine the new solution estimate $\boldsymbol{f}^k$.

Specifically, the objective function $z(\boldsymbol{f})$ is approximated around a point $\bar{\boldsymbol{f}} \in S_f$ by a linear function $\bar{z}(\boldsymbol{f})$, using a first-order Taylor's series approximation:

$$z(\boldsymbol{f}) \cong z(\bar{\boldsymbol{f}}) + \nabla z(\bar{\boldsymbol{f}})^T(\boldsymbol{f} - \bar{\boldsymbol{f}}) = \bar{z}(\boldsymbol{f})$$

The optimization problem (5.4.16) is thus approximated by a linear programming problem, that is, a problem with linear objective function $\bar{z}(\boldsymbol{f})$ and the same linear constraints $\boldsymbol{f} \in S_f$:

$$\underset{f \in Sf}{\operatorname{argmin}}\, z(\boldsymbol{f}) \cong \underset{f \in Sf}{\operatorname{argmin}}\, \bar{z}(\boldsymbol{f}) = \underset{f \in Sf}{\operatorname{argmin}}\, z(\bar{\boldsymbol{f}}) + \nabla z(\bar{\boldsymbol{f}})^T(\boldsymbol{f} - \bar{\boldsymbol{f}})$$

or

$$\underset{f \in Sf}{\operatorname{argmin}}\, z(\boldsymbol{f}) \cong \underset{f \in Sf}{\operatorname{argmin}}\, \nabla z(\bar{\boldsymbol{f}})^T \boldsymbol{f} \tag{5.4.18}$$

Note that the gradient of the objective function $\bar{z}(\boldsymbol{f})$ of problem (5.4.16) at a point $\bar{\boldsymbol{f}}$ is equal to the link cost vector evaluated at that point, $\nabla z(\boldsymbol{f}) = \boldsymbol{c}(\boldsymbol{f})$. Hence expression (5.4.18) becomes:

$$\underset{f \in Sf}{\operatorname{argmin}}\, z(\boldsymbol{f}) \cong \underset{f \in Sf}{\operatorname{argmin}}\, c(\bar{\boldsymbol{f}})^T \boldsymbol{f} \tag{5.4.19}$$

The linear optimization problem expressed by (5.4.19) consists of finding a feasible link flow vector that minimizes total travel costs in a network where link costs are given by the fixed vector $c(\boldsymbol{f})$. The solution to this problem is obtained by assigning all the flow of each O-D pair to the minimum cost path between them. Thus, this problem corresponds to the optimization model (5.3.7) described in Sect. 5.3.2 for deterministic unncongested network assignment, and it can be solved with one of the DUN algorithms described in Sect. 5.3.3. A DUN algorithm is formally denoted as

$f_{\text{DUN}}(c)$ DUN link flows corresponding to link cost vector $\boldsymbol{c}$

The Frank–Wolfe algorithm for the calculation of DUE link flows with fixed demand and with cost functions having symmetric Jacobian can be described by the following system of recursive equations, starting at $\boldsymbol{f}^0 \in S_f$ and $k = 0$.

$$k = k + 1$$

$$\boldsymbol{c}^k = \boldsymbol{c}(\boldsymbol{f}^{k-1}) \tag{5.4.20}$$

$$f^k_{\text{DUN}} = f_{\text{DUN}}(c^k) \tag{5.4.21}$$

$$\mu^k = \underset{\mu \in [0,1]}{\operatorname{argmin}}\, \psi(\mu) = z\big(f^{k-1} + \mu(f^k_{\text{DUN}} - f^{k-1})\big) \tag{5.4.22}$$

$$f^k = f^{k-1} + \mu^k\big(f^k_{\text{DUN}} - f^{k-1}\big) \tag{5.4.23}$$

The MSA-FA algorithm presented for stochastic equilibrium, (5.4.3) to (5.4.6), is quite similar to the Frank–Wolfe algorithm. The main difference is in the determination of the step size μ^k: in the MSA it is $1/k$ and so depends only on the iteration index, whereas in the Frank–Wolfe algorithm it results from an optimization problem (5.4.22). However, the MSA-FA algorithm may show a slower convergence.

Equation (5.4.22) defines a one-dimensional nonlinear optimization problem in the scalar variable μ that can be solved with a line search algorithm such as the bisection algorithm (see Appendix A). The bisection algorithm requires the derivative of the function $\psi(\mu) = z(f^{k-1} + \mu(f^k_{\text{DUN}} - f^{k-1}))$, which can easily be obtained from the link costs:

$$\begin{aligned} d\psi(\mu)/d\mu &= \nabla z\big(f^{k-1} + \mu(f^k_{\text{DUN}} - f^{k-1})\big)^T \big(f^k_{\text{DUN}} - f^{k-1}\big) \\ &= c\big(f^{k-1} + \mu(f^k_{\text{DUN}} - f^{k-1})\big)^T \big(f^k_{\text{DUN}} - f^{k-1}\big) \end{aligned}$$

Note that in order to apply the bisection algorithm it is not necessary to actually compute the value of the function $\psi(\mu)$.

From expression (5.4.23) it can be deduced that the solution estimate at iteration k, f^k, is a convex combination of the first k DUN assignments; it is thus a feasible solution, $f^k \in S_f$, because DUN assignment outputs are feasible and the set of feasible flows S_f is convex. An initial feasible solution $f^0 \in S_f$ can easily be obtained, for example, with a DUN algorithm using free flow costs, $f^0 = f_{\text{DUN}}(c(f = \mathbf{0}))$.

The algorithm stops when the product of the objective function gradient and the descent direction is greater than or equal to zero (see Appendix A):

$$\nabla z(f^{k-1})^T (f^k_{\text{DUN}} - f^{k-1}) = c(f^{k-1})^T (f^k_{\text{DUN}} - f^{k-1}) \geq 0$$

It can easily be deduced that if the algorithm stops, the current solution estimate f^k is the DUE flow vector. In practice, the algorithm terminates when the absolute value of the product $c(f^{k-1})^T (f^k_{\text{DUN}} - f^{k-1})$ is below a stop threshold δ, which is defined relative to the total cost to avoid dependence on the measurement units:

$$\left|(c^k)^T (f^k_{\text{DUN}} - f^{k-1})\right| / \left((c^k)^T f^{k-1}\right) < \delta$$

Convergence of this algorithm near the solution may be rather slow because it tends to zigzag; thus, a number of algorithms that modify the descent direction $f^k_{\text{DUN}} - f^{k-1}$ have been developed (some of which are referred to in Appendix A). An example application of the Frank–Wolfe algorithm is given in Fig. 5.24.

If the cost functions $c = c(f)$ are continuous with continuous first partial derivatives and symmetric positive definite Jacobian, the function $z(f)$ has only one

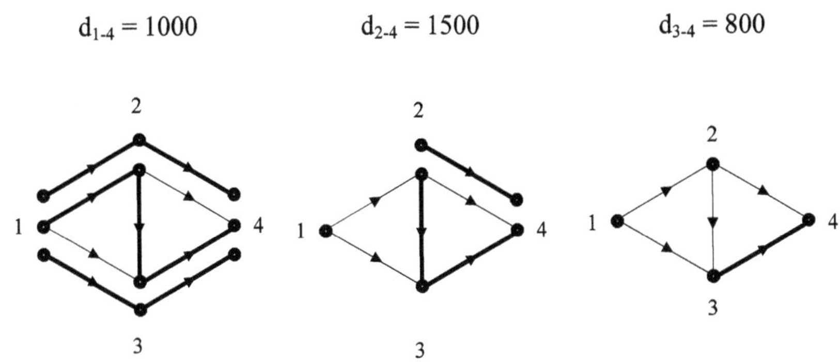

PATHS

Paths	Nodes
1	1-2-4
2	1-2-3-4
3	1-3-4
4	2-3-4
5	2-4
6	3-4

PARAMETERS OF COST FUNCTIONS

Arc	c_0	a	Cap	γ
1-2	10	2	1000	4
1-3	22	2	1000	4
2-3	13	2	2500	4
2-4	20	2	1000	4
3-4	11	2	3300	4

COST FUNCTION TYPE

$$c_1 = c_0\left(1 + a\frac{f}{Cap}\right)^{\gamma}$$

ITERATIONS

Iteration		Link	$c(f=0)$	f^0_{AoN}	f^0
		1-2	10	1000	1000
		1-3	22	0	0
0		2-3	13	0	0
		2-4	20	2500	2500
		3-4	11	800	800
Iteration (k)	**μ**	**Link**	**$c(f^{k-1})$**	**f^k_{AoN}**	**f^k**
		1-2	810	0	304
		1-3	22	1000	696
1	0.69592	2-3	13	1500	1044
		2-4	25920	0	760
		3-4	53	3300	2540
		1-2	67	1000	408
		1-3	720	0	592
2	0.14868	2-3	147	2500	1260
		2-4	807	0	647
		3-4	457	3300	2653
		1-2	153	1000	488
		1-3	369	0	512
22	0.00006	2-3	216	2500	1275
		2-4	694	0	714
		3-4	478	3300	2586
		1-2	153	1000	488
		1-3	369	0	512
		2-3	216	0	1275
23	0.00002	2-4	694	2500	714
		3-4	478	800	2586
		2-4	23	1500	679
		3-4	11	1800	2620

Fig. 5.24 Example of the Frank–Wolfe algorithm for DUE assignment

minimum point, $\boldsymbol{f}^*$, as stated in Sect. 5.4.3. In this case, the function $\psi(\mu)$ also has only one minimum point. Under these assumptions, it can be demonstrated by using results from optimization theory that the sequence of (feasible) link flow vectors $\boldsymbol{f}^k$ generated by the Frank–Wolfe algorithm converges to the vector of DUE link flows.

The calculation of fixed demand DUE link flows for *nonseparable cost functions* (including the case of asymmetric Jacobian) is based on algorithms that solve the variational inequality model (5.4.12).

Variational inequality can be solved using the diagonalization algorithm. This algorithm generates a sequence of feasible link flow vectors $\boldsymbol{f}^k$, starting from an initial feasible solution $\boldsymbol{f}^0 \in S_f$, by solving a sequence of separable cost function problems that approximate problem (5.4.12). In particular, at a solution estimate $\boldsymbol{f} \in S_f$ the cost function of link a, $c_a(\boldsymbol{f})$, is approximated by a separable cost function, $\bar{c}_a(f_a)$, obtained by diagonalizing the function's Jacobian:

$$c_a(f_1, \ldots, f_{a-1}, f_a, f_{a+1}, \ldots) \cong \bar{c}_a(\bar{f}_1, \ldots, \bar{f}_{a-1}, f_a, \bar{f}_{a+1}, \ldots) = \bar{c}_a(f_a) \quad \forall a$$

Therefore, variational inequality (5.4.12) is approximated by a variational inequality with separable cost functions $\bar{c}_a(f_a)$:

$$\boldsymbol{c}(\boldsymbol{f}^*)^T(\boldsymbol{f} - \boldsymbol{f}^*) \cong \sum_a \bar{c}_a(f_a)(f_a - f_a^*) \geq 0 \quad \forall \boldsymbol{f} \in S_f \tag{5.4.24}$$

which, in turn, is equivalent to problem (5.4.16) for fixed demand DUE assignment with symmetric Jacobian cost functions. Thus problem (5.4.24) can be solved as described previously. Let:

$\boldsymbol{f}_{\text{DUE}}[\boldsymbol{c}(\cdot)]$ be the DUE link flows resulting from link cost functions $\boldsymbol{c}(\cdot)$ with symmetric Jacobian; $\boldsymbol{f}_{\text{DUE}}$ can be calculated, for example, with the Frank–Wolfe algorithm

The diagonalization algorithm can be described by the following system of recursive equations, starting with $\boldsymbol{f}^0 \in S_f$ and $k = 0$:

$$\bar{c}_a^k(f_a) = c_a\left(f_a^{k-1}, \ldots, f_{a-1}^{k-1}, f_a, f_{a+1}^{k-1}, \ldots\right) \quad \forall a \tag{5.4.25}$$

$$\boldsymbol{f}^k = \boldsymbol{f}_{\text{DUE}}\left[\bar{c}_a^k(f_a)\right] \tag{5.4.26}$$

The algorithm is therefore equivalent to performing a sequence of DUE assignments with separable cost functions. These are obtained by defining a new cost function for each link. The only variable in the new link cost function is the corresponding link flow; flows on other links are set equal to the previous equilibrium solution estimate. The diagonalization algorithm can also be applied by averaging over the successive DUE vectors for separable cost functions, as described by the following system of recursive equations, starting with $\boldsymbol{f}^0 \in S_f$ and $k = 0$.

$$k = k + 1$$

$$\bar{c}_a^k(f_a) = c_a\left(f_1^{k-1}, \ldots, f_{a-1}^{k-1}, f_a, f_{a+1}^{k-1}, \ldots\right) \quad \forall a$$

$$\boldsymbol{f}_{\text{DUE}}^k = \boldsymbol{f}_{\text{DUE}}\left[\bar{c}_a^k(f_a)\right]$$

$$f^k = f^{k-1} + (1/k)\left(f^k_{\text{DUE}} - f^{k-1}\right)$$

It can easily be deduced that, in both cases, if the diagonalization algorithm converges to a solution, this is the equilibrium DUE assignment for the problem with nonseparable cost functions. Consistent with the results described in Sect. 5.4.3, if the cost functions are continuous and differentiable with positive definite Jacobian, variational inequality (5.4.12) has one and only one solution. Under this assumption, the sequence of link flow vectors f^k generated by the diagonalization algorithm converges to the equilibrium link flow vector under some technical conditions on the maximum value of an appropriate norm of the Jacobian matrix. In practice, to speed up the application of the algorithm, the convergence threshold of the Frank–Wolfe algorithm is decreased at each iteration of the diagonalization algorithm; alternatively a deterministic uncongested network assignment $f_{\text{DUN}}(\cdot)$ is heuristically substituted for the symmetric deterministic equilibrium $f_{\text{DUE}}[\cdot]$ assignment.

5.4.5 Relationship Between Stochastic and Deterministic Equilibrium

The deterministic path choice model underlying deterministic equilibrium models can be considered a special case of a random utility model in which the variance of the random residuals is null. For this reason, stochastic equilibrium flows are increasingly closer to deterministic equilibrium flows as the random residual variance goes to zero. Figure 5.25 shows the curves expressing the demand model for the example used in Figs. 5.20 and 5.22 for various values of the parameter θ (which is proportional to the standard deviation of random residuals of the path choice model). The figure clearly shows that, as the variance decreases, the probabilistic demand curve progressively approaches the curve corresponding to the deterministic model, and SUE flows approach DUE flows.

Deterministic and stochastic models give similar results in the case of very congested networks. If link flows are close to capacity, the derivatives of the cost functions, representing the cost variations introduced by an additional user, are most likely larger than the random residuals. In other words, a flow distribution very different from deterministic equilibrium would induce large cost differences between the different paths, and these are likely to be correctly perceived by almost all the users.

This effect is shown in Fig. 5.26, where the link cost functions vary in such a way that their derivatives increase but their flow values where they intersect remain fixed. In other words, the DUE flows remain unchanged as the system becomes more congested, and thus more sensitive to small flow variations. As the figure shows, as the cost curves vary, SUE flows change and approach DUE flows.

The closeness of deterministic and stochastic equilibrium flows means that for very congested networks it is possible to use DUE assignment as an approximation

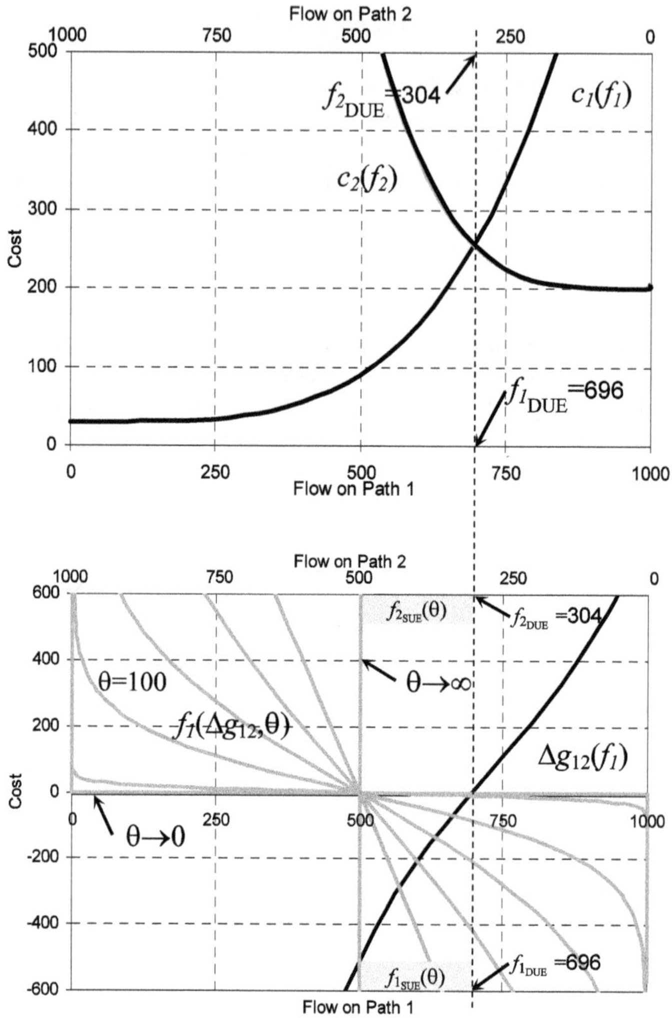

Fig. 5.25 Relationship between SUE and DUE flows for different random residual standard deviations (see Figs. 5.20 and 5.22)

to SUE assignment. This is good for practical problems, because DUE flows are easier to compute, as shown in Chap. 7. However, it should be noted that for other applications (assignment to lightly congested or nonuniformly congested networks, estimation of the O-D matrix from traffic counts, etc.) the deterministic model is not a good substitute for the stochastic one. Furthermore, as pointed out in the preceding section, it is not possible to guarantee the uniqueness of deterministic equilibrium path flows, nor (as shown in Sect. 6.4) of flows per user class in the case of multiclass assignment.

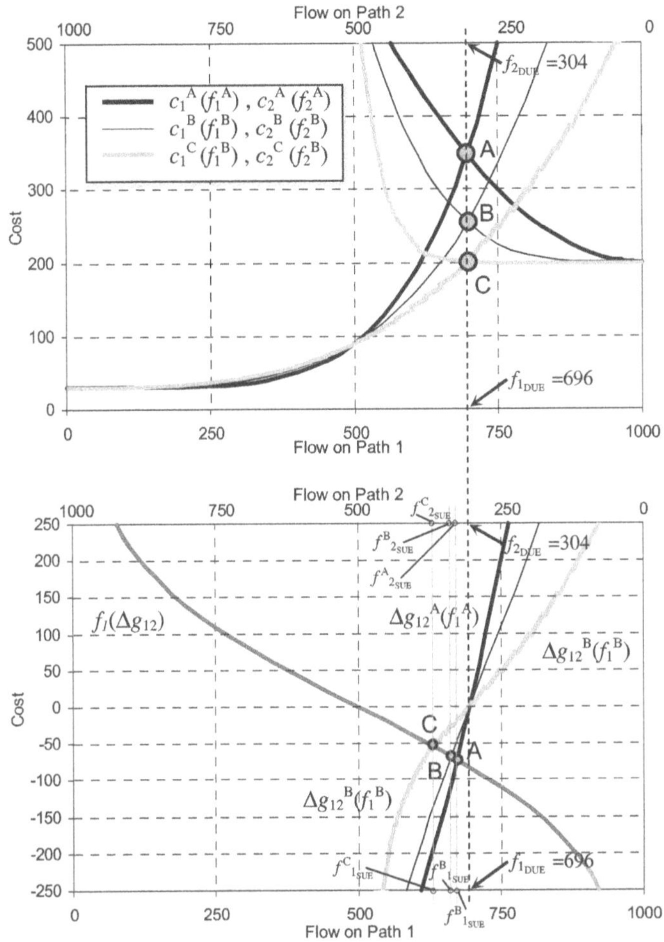

Cost functions

$$c_1 = c_{0,1}\left(1 + a_1 \frac{f_1}{Cap_1}\right)^{\gamma_1};$$

$$c_2 = c_{0,2}\left(1 + a_2 \frac{f_2}{Cap_2}\right)^{\gamma_2};$$

Parameters of the Cost Functions

Case	$c_{0,1}$	$c_{0,2}$	a_1	a_2	Cap_1	Cap_2	γ_1	γ_2
A	30	200	2	2	500	500	5.06	2
B	30	200	2	2	500	500	4	4
C	30	200	2	2	500	500	3.17	12

$\theta = 100$

Fig. 5.26 Relationship between SUE and DUE flows for varying cost functions

5.4.6 System Optimum Assignment[*]

System optimal assignment models derive from assumptions that are significantly different from those underlying user equilibrium models. Indeed, it is assumed that users cooperate to minimize total system cost, rather than try individually to minimize their own costs as in user equilibrium models (or they are indifferent to costs, e.g., freight transport). For congested networks, the two assignment problems are generally different; for uncongested networks, however, the problem is the same as the deterministic assignment problem. Note that under the assumptions of system optimal assignment, some users may follow a nonminimum (perceived or systematic) cost path. Wardrop's second principle expresses the assignment condition under which the total cost on the network is minimal; it is also known as System Optimum (SO) assignment.

Knowledge of system optimum flows can be a useful reference element in the analysis of congested networks. Although the behavioral assumptions underlying SO are not realistic for the modeling of individual tripmaker behavior, minimization of total costs corresponds to (one of) the typical system management objectives that network operators attempt to achieve through available control instruments (prices, traffic-light regulation, service frequency, etc.).[22] Furthermore, SO assignment can be applied for the assignment of flow units that lack autonomous decision capability, such as freight vehicles.

SO assignment is defined by an optimization model expressed in terms of link flows, with an objective function consisting of the total cost presented in Sect. 5.2 (ignoring nonadditive path costs for the sake of simplicity):

$$\begin{aligned} \boldsymbol{f}_{\mathrm{SO}} &= \operatorname{argmin} \boldsymbol{c}(\boldsymbol{f})^T \boldsymbol{f} \\ \boldsymbol{f} &\in S_f \end{aligned} \tag{5.4.27}$$

Note that it is unnecessary to introduce assumptions on the symmetry of the cost function Jacobian to formulate system optimum assignment through an optimization model (which in this case is the direct formulation, rather than an equivalent indirect one as with DUE). The existence and uniqueness of optimum system flows and costs are discussed below.

Existence. The optimization model (5.4.27) has at least one solution if the cost functions, $\boldsymbol{c} = \boldsymbol{c}(\boldsymbol{f})$, are continuous.

Under these assumptions, the objective function $z(\boldsymbol{f}) = \boldsymbol{c}(\boldsymbol{f})^T \boldsymbol{f}$, is continuous on the nonempty (under the assumption of a connected network), and compact (as well as convex) set S_f, and therefore has at least one minimum point (see Appendix A).

Existence and uniqueness. The optimization model (5.4.27) has one and only one solution if the cost functions $\boldsymbol{c}(\boldsymbol{f})$ have continuous first and second partial derivatives; their Jacobian $\boldsymbol{Jac}[\boldsymbol{c}(\boldsymbol{f})]$ is continuous and positive definite (cost functions

[22] Formal models for supply design are dealt with in Chap. 9.

are strictly increasing); and the Hessian matrix $\boldsymbol{Hess}[c_a(\boldsymbol{f})]$ of each cost function $c_a = c_a(\boldsymbol{f})$ is positive semidefinite (each cost function is convex).

Under these assumptions, the cost functions have continuous first derivatives and are therefore differentiable and continuous, a condition guaranteeing the existence of a solution. Furthermore, the gradient $\nabla z(\boldsymbol{f})$ of the function $z(\boldsymbol{f}) = \boldsymbol{c}(\boldsymbol{f})^T \boldsymbol{f}$ is given by

$$\nabla z(\boldsymbol{f}) = \boldsymbol{Jac}\big[\boldsymbol{c}(\boldsymbol{f})\big]\boldsymbol{f} + \boldsymbol{c}(\boldsymbol{f})$$

and its Hessian matrix $\boldsymbol{Hess}[z(\boldsymbol{f})]$ is given by

$$\boldsymbol{Hess}\big[z(\boldsymbol{f})\big] = \boldsymbol{Jac}\big[\nabla z(\boldsymbol{f})\big] = \boldsymbol{Jac}\big[\boldsymbol{c}(\boldsymbol{f})\big]^T + \sum_a f_a \boldsymbol{Hess}\big[c_a(\boldsymbol{f})\big] + \boldsymbol{Jac}\big[\boldsymbol{c}(\boldsymbol{f})\big]$$

Both $z(\boldsymbol{f})$ and $\boldsymbol{Hess}[z(\boldsymbol{f})]$ are continuous, so that the function $z(\boldsymbol{f}) = \boldsymbol{c}(\boldsymbol{f})^T \boldsymbol{f}$ is twice differentiable. Finally, the Hessian matrix $\boldsymbol{Hess}[z(\boldsymbol{f})]$ is symmetric positive definite because it is the sum of symmetric positive semidefinite matrices and of symmetric positive definite matrices. Therefore, the function $z(\boldsymbol{f}) = \boldsymbol{c}(\boldsymbol{f})^T \boldsymbol{f}$, defined over the convex set S_f, is strictly convex and has one and only one minimum point.

System optimum flows do not generally coincide with DUE flows, as is shown by the example illustrated in Fig. 5.27. The figure shows that, with respect to the DUE flows, the shift of some users to a path that is slightly more expensive but less congested significantly reduces the total cost borne by all users. However, if link costs are independent of flows (i.e., if $\boldsymbol{Jac}[\boldsymbol{c}(\boldsymbol{f})] = 0$), the solutions to the two problems coincide.

If $\boldsymbol{f}^*$ is a minimum point of the function $z(\boldsymbol{f}) = \boldsymbol{c}(\boldsymbol{f})^T \boldsymbol{f}$ it follows that

$$\nabla z(\boldsymbol{f}^*)^T(\boldsymbol{f} - \boldsymbol{f}^*) \geq 0 \quad \forall \boldsymbol{f} \in S_f$$

and because

$$\nabla z(\boldsymbol{f}^*) = \boldsymbol{Jac}\big[\boldsymbol{c}(\boldsymbol{f})\big]\boldsymbol{f} + \boldsymbol{c}(\boldsymbol{f}):$$

$$\big(\boldsymbol{Jac}\big[\boldsymbol{c}(\boldsymbol{f})\big]\boldsymbol{f} + \boldsymbol{c}(\boldsymbol{f})\big)^T(\boldsymbol{f} - \boldsymbol{f}^*) \geq 0 \quad \forall \boldsymbol{f} \in S_f$$

a condition that is different in general from the variational inequality (5.4.15) that expresses deterministic equilibrium. However, if link costs are independent of the flows ($\boldsymbol{Jac}[\boldsymbol{c}(\boldsymbol{f})] = 0$) then the above inequality coincides with the variational inequality; the deterministic user equilibrium problem is reduced to the deterministic uncongested network assignment problem and can be expressed by the model (5.3.7), which is equivalent in this case to model (5.4.27) that expresses the system optimum assignment problem.

Of particular interest is the example in Figs. 5.28a and 5.28b, known in the literature as *Braess' paradox*. The paradox involves a network where the addition of a new link causes an increase in the total cost under deterministic equilibrium assignment, while leaving the system optimum total cost unchanged. In the first case,

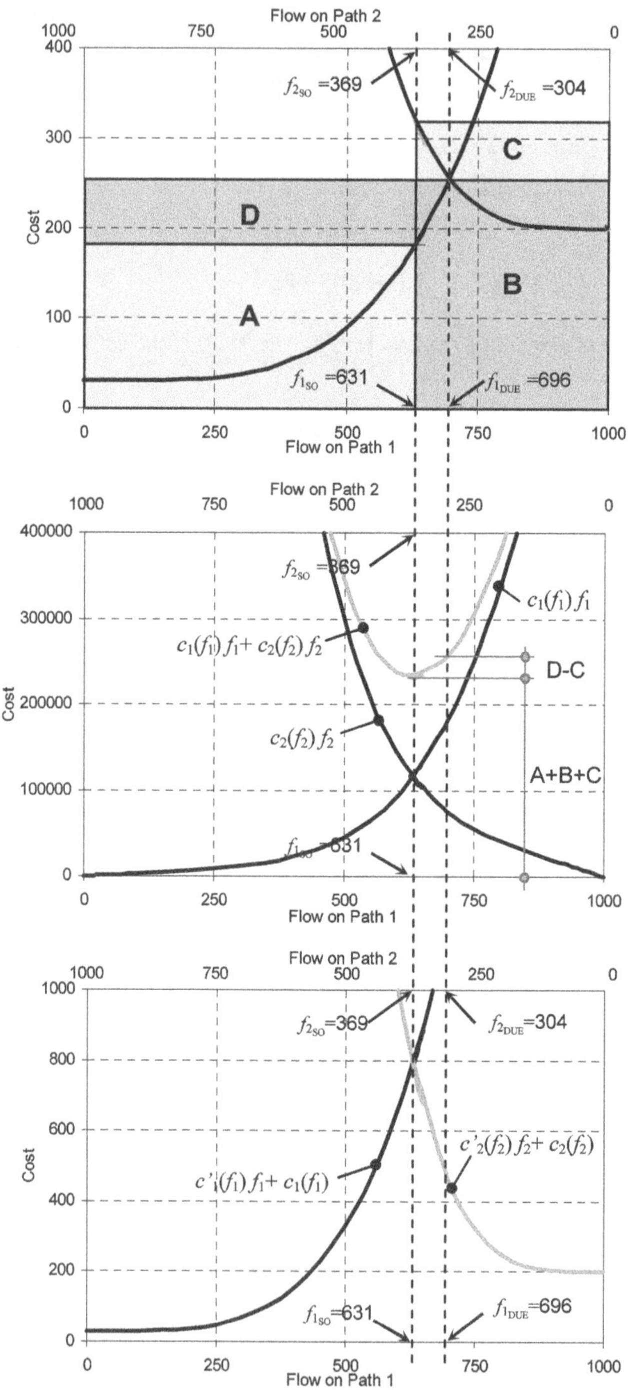

Fig. 5.27 System optimum (SO) flows on the test network of Fig. 5.22

the SO link flows minimize the total cost. For this reason, the addition of a link cannot increase the overall system cost because the SO link flow pattern corresponding to zero flow on the new link is a feasible solution of the new SO problem. In user equilibrium, on the other hand, the objective of each individual is to minimize her own transport cost and the equilibrium link flow pattern corresponding to the introduction of a new link may cause an increase in total cost. It should be pointed out, however, that conditions analogous to the Braess paradox are not often found in real systems.[23]

The system optimum model (5.4.27) can be reformulated to be formally analogous to the DUE optimization model (5.4.15). To this end, consider the marginal cost function $\boldsymbol{b}(\boldsymbol{f})$. It is the gradient $\nabla z(\boldsymbol{f})$ of the function $z(\boldsymbol{f}) = \boldsymbol{c}(\boldsymbol{f})^T \boldsymbol{f}$:

$$\boldsymbol{b}(\boldsymbol{f}) = \nabla z(\boldsymbol{f}) = \boldsymbol{Jac}\left[\boldsymbol{c}(\boldsymbol{f})\right]^T \boldsymbol{f} + \boldsymbol{c}(\boldsymbol{f}) \tag{5.4.28}$$

The interpretation of the function (5.4.28) is more straightforward in the case of separable cost functions $\boldsymbol{c}(\boldsymbol{f})$, where the functions $\boldsymbol{b}(\boldsymbol{f})$ are also separable. Under this assumption, if the first derivative of the link a cost function $c_a(f_a)$ is denoted by $c'_a(f_a)$, it follows that

$$\boldsymbol{b}_a(f_a) = c'_a(f_a) f_a + c_a(f_a)$$

In the general (nonseparable) case, if the cost functions have continuous first and second derivatives, the Jacobian $\boldsymbol{Jac}[\boldsymbol{b}(\boldsymbol{f})]$ of $\boldsymbol{b}(\boldsymbol{f})$ is symmetric. In this case the line integral of $\boldsymbol{b}(\boldsymbol{f})$ between the limits $\boldsymbol{0}$ and $\boldsymbol{f}$ does not depend on the path of integration, and the integral's value coincides with the total cost:

$$\int_o^{\boldsymbol{f}} \boldsymbol{b}(\boldsymbol{y})^T d\boldsymbol{y} = \boldsymbol{c}(\boldsymbol{f})^T \boldsymbol{f}$$

If the cost functions $\boldsymbol{c}(\boldsymbol{f})$ have continuous first and second derivatives, the marginal costs $\boldsymbol{b}(\boldsymbol{f})$ have continuous cost derivatives; thus they are differentiable and continuous. Furthermore, the Jacobian $\boldsymbol{Jac}[\boldsymbol{b}(\boldsymbol{f})]$ of the gradient function $\boldsymbol{b}(\boldsymbol{f}) = \nabla z(\boldsymbol{f})$, coinciding with the Hessian matrix of the function $z(\boldsymbol{f})$ is symmetric.

System optimum assignment can therefore be formulated as an optimization model using the marginal cost function $\boldsymbol{b}(\boldsymbol{y})$ defined in (5.4.28):

$$\boldsymbol{f}_{\mathrm{SO}} = \mathrm{argmin}\, z(\boldsymbol{f}) = \int_0^{\boldsymbol{f}} \boldsymbol{b}(\boldsymbol{y})\, d\boldsymbol{y} \quad \boldsymbol{f} \in S_f \tag{5.4.29}$$

The optimization model (5.4.29) is formally analogous to the optimization model (5.4.15) for the (symmetric) DUE and can be solved with the same algorithms described (see Fig. 5.26).

---

[23] Results reported in the literature indicate that cost functions characterized by a stronger form of monotonicity preclude the occurrence of Braess' paradox.

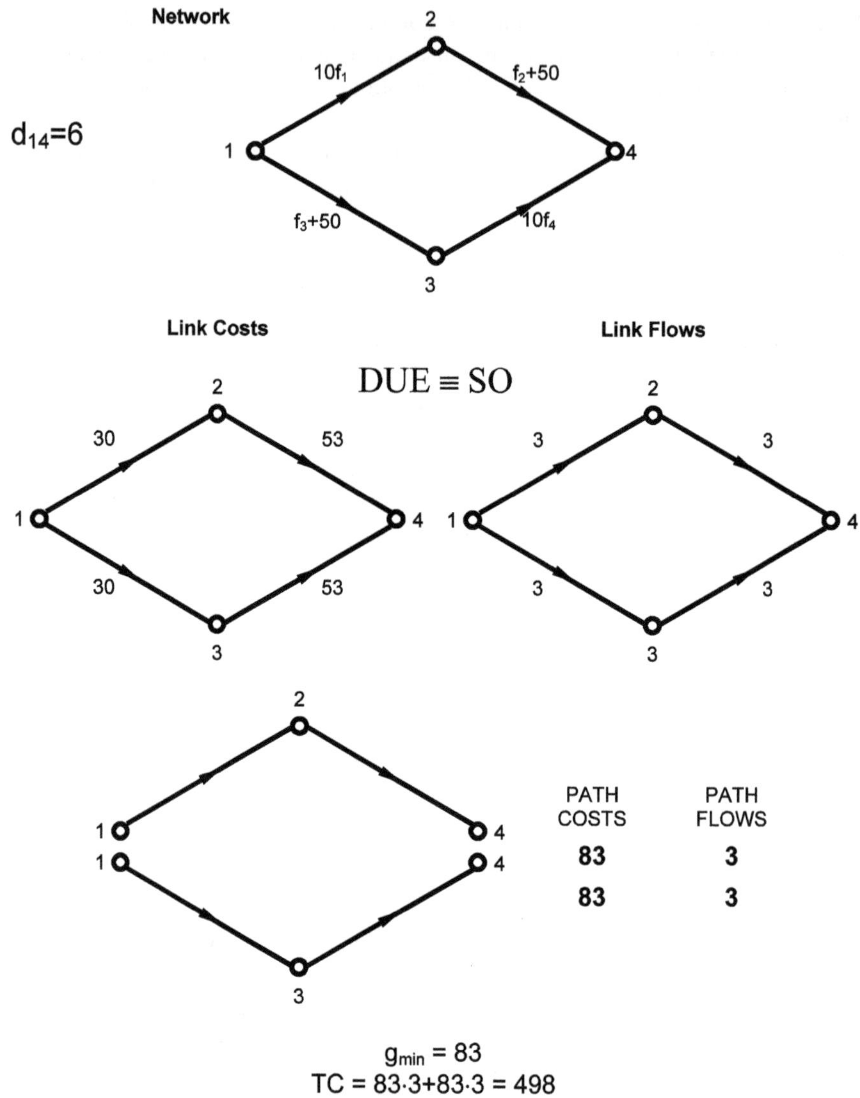

Fig. 5.28a Example of Braess' paradox

An operational interpretation of the model (5.4.29) is that if link costs were modified so as to make the costs that users perceive coincide with the marginal costs $\boldsymbol{b}(\boldsymbol{f})$, then individual deterministic path choice based on such costs would lead to a flow pattern that minimized the total cost $\boldsymbol{c}(\boldsymbol{f})^T\boldsymbol{f}$. One way (but not the only one) of enforcing these costs is by introducing flow-dependent link tolls equal to $\boldsymbol{b}(\boldsymbol{f}) - \boldsymbol{c}(\boldsymbol{f}) = \boldsymbol{Jac}[\boldsymbol{c}(\boldsymbol{f})]\boldsymbol{f}$. If the link cost functions are separable, this expression for the toll value reduces to $c'_a(f_a)f_a$. Chapter 9 returns to this point in the discussion of supply design models.

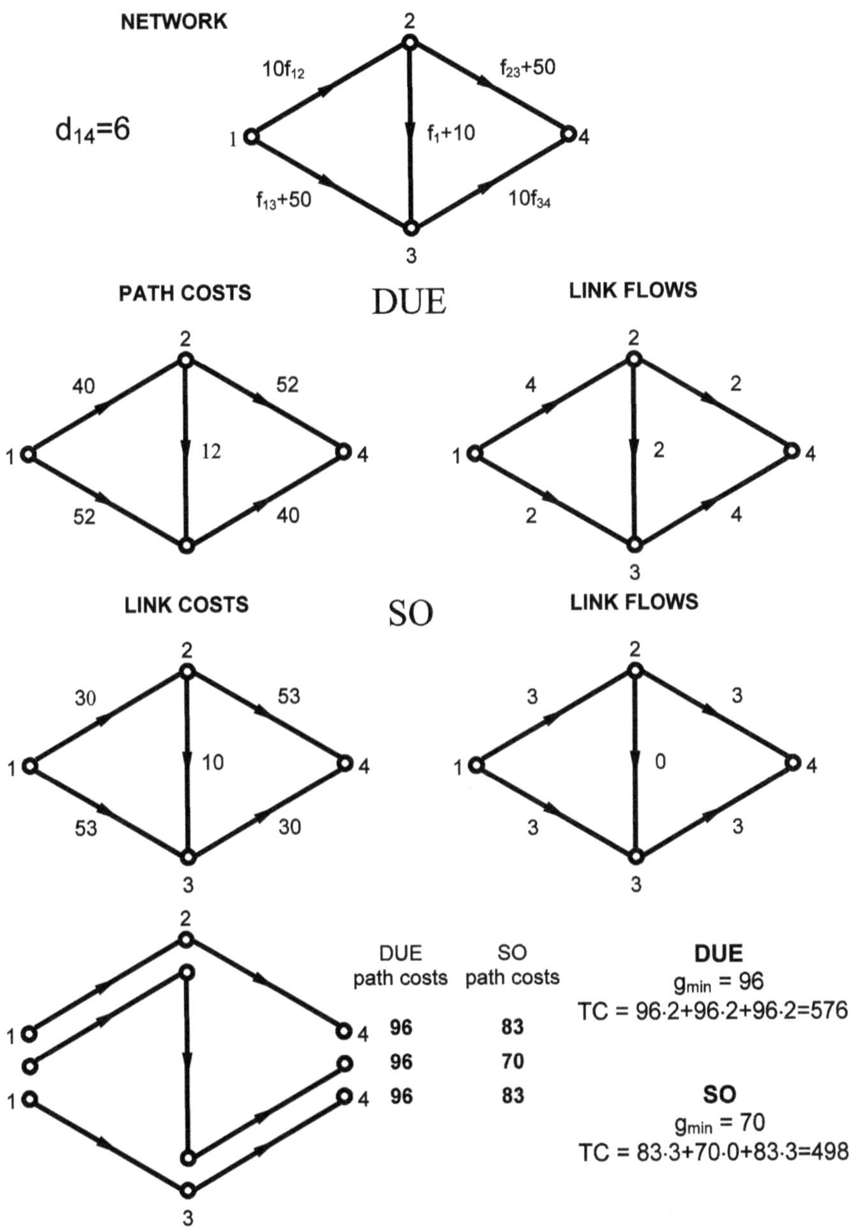

Fig. 5.28b Example of Braess' paradox

Finally, it may be deduced through similar arguments that system optimum assignment does not generally coincide with stochastic equilibrium. In this case as well, it is possible to derive conditions sufficient to ensure the equivalence of the two problems; however, these lead to rather unrealistic cost functions.

The Frank–Wolfe algorithm, presented as a solution method for symmetric deterministic user equilibrium (Sect. 5.4.4), can also be applied to solve the SO, as formulated through the optimization model (5.4.27). The algorithm is described by the following system of recursive equations, starting from $\boldsymbol{f}^0 \in S_f$:

$$\boldsymbol{b}^k = \nabla z(\boldsymbol{f}^{k-1}) = \boldsymbol{Jac}\big[\boldsymbol{c}(\boldsymbol{f}^{k-1})\big]\boldsymbol{f}^{k-1} + \boldsymbol{c}(\boldsymbol{f}^{k-1}) \tag{5.4.30}$$

$$\boldsymbol{f}^k_{\mathrm{DUN}} = \boldsymbol{f}_{\mathrm{DUN}}(\boldsymbol{g}^k) \tag{5.4.31}$$

$$\mu^k = \operatorname*{argmin}_{\mu \in [0,1]} \psi(\mu) = z\big(\boldsymbol{f}^{k-1} + \mu(\boldsymbol{f}^k_{\mathrm{DUN}} - \boldsymbol{f}^{k-1})\big) \tag{5.4.32}$$

$$\boldsymbol{f}^k = \boldsymbol{f}^{k-1} + \mu^k\big(\boldsymbol{f}^k_{\mathrm{DUN}} - \boldsymbol{f}^{k-1}\big) \tag{5.4.33}$$

Note that, unlike deterministic user equilibrium, calculation of the gradient of function $z(\boldsymbol{f})$ in (5.4.30) requires calculation of the cost function Jacobian, a task that is easy only for separable functions. Equation (5.4.32) defines the step size μ^k as a solution to the one-dimensional nonlinear optimization problem in the scalar variable μ. This model can be solved with any of a variety of line search algorithms, such as the golden section algorithm (see Appendix A), which avoid the use of the derivative of the function $\psi(\mu)$ (because this derivative depends on the gradient of the function $z(\boldsymbol{f})$ and therefore on the Jacobian of the cost functions).

The algorithm stops if the scalar product of the gradient of the objective function and the descent direction is greater than or equal to zero (see Appendix A):

$$\begin{aligned}&\nabla z(\boldsymbol{f}^{k-1})^T\big(\boldsymbol{f}^k_{\mathrm{DUN}} - \boldsymbol{f}^{k-1}\big)\\ &\quad = \big(\boldsymbol{f}^k_{\mathrm{DUN}} - \boldsymbol{f}^{k-1}\big)^T\big(\boldsymbol{Jac}\big(\boldsymbol{c}(\boldsymbol{f}^{k-1})\big)\boldsymbol{f}^{k-1} + \boldsymbol{c}(\boldsymbol{f}^{k-1})\big) \geq 0\end{aligned}$$

In order to avoid calculating the gradient of the function $z(\boldsymbol{f})$, the algorithm can terminate when the relative difference between the values of the function $z(\boldsymbol{f})$ in two successive iterations is below a stopping threshold δ:

$$\big|z(\boldsymbol{f}^k) - z(\boldsymbol{f}^{k-1})\big| / z(\boldsymbol{f}^{k-1}) < \delta$$

The function $z(\boldsymbol{f})$ is strictly convex, and has a unique minimum point, if the Jacobian $\boldsymbol{Jac}[\boldsymbol{c}(\boldsymbol{f})]$ of the cost functions $\boldsymbol{c}(\boldsymbol{f})$ is continuous and positive definite (the cost functions are strictly increasing) and each link cost function $c_a = c_a(\boldsymbol{f})$ has a Hessian matrix $\boldsymbol{Hess}[c_a(\boldsymbol{f})]$ that is continuous and positive semidefinite (each cost function is convex). The function $\psi(\mu)$ is strictly convex if the function $z(\boldsymbol{f})$ is strictly convex. Under these conditions, it can be shown that the sequence of (feasible) link flow vectors $\boldsymbol{f}^k$ generated by the Frank–Wolfe algorithm converges to the SO link flow vector.

5.5 Result Interpretation and Parameter Calibration

It is worth recalling that none of the models formulated under the assumption of within-day stationarity allow modeling of queuing due to oversaturation phenomena, which cannot be analyzed in a static context. Thus, if the solution of such an assignment model yields a flow exceeding link capacity, the results of the model may be used as indicators of critical points on the network, but can no longer be interpreted rigorously as estimates of system steady state. Within-day dynamic assignment models, described in Chap. 7, should be adopted to analyze such situations more accurately.

Regardless of the application, assignment models should be seen as simplified representations of real complex phenomena. Thus, the link flows resulting from any assignment model[24] might more correctly be denoted as $\boldsymbol{f}^{\text{SIM}}$. Assuming that the flows occurring in the real transportation system are represented by a random variable, due to uncertainty about their values, with expected value $\boldsymbol{f}$, the link flows $\boldsymbol{f}^{\text{SIM}}$ may be considered only an estimate of the expected value $\boldsymbol{f}$. The relation between actual flows and the flows resulting from an assignment model can be formally expressed as

$$\boldsymbol{f} = \boldsymbol{f}^{\text{SIM}} + \boldsymbol{\varepsilon}^{\text{SIM}} = \boldsymbol{\Delta P}^{\text{SIM}}\boldsymbol{d} + \boldsymbol{\varepsilon}^{\text{SIM}} \tag{5.5.1}$$

The matrix $\boldsymbol{P}^{\text{SIM}}$ represents the path choice fractions resulting from the assignment model and it generally differs from the matrix $\boldsymbol{P}$ of actual fractions. The vector $\boldsymbol{\varepsilon}^{\text{SIM}}$ represents the deviations between (the expected value of) actual flows and the flows resulting from the assignment of demand $\boldsymbol{d}$. These errors derive from the simplifying assumptions adopted in the system definition (delimitation of study area and zoning); in the specification of supply, path choice, and supply–demand interaction models; and in the estimation of the average demand flow $\boldsymbol{d}$. Different assumptions will produce different flows $\boldsymbol{f}^{\text{SIM}}$ and errors $\boldsymbol{\varepsilon}^{\text{SIM}}$. This point is dealt with in greater detail in Sect. 8.5. For now, note that even if the actual average demand flows were assigned to the network, other error sources would produce assignment errors $\boldsymbol{\varepsilon}^{\text{SIM}}$.

With respect to the choice of the supply–demand interaction model, some experimental evidence indicates that the more realistic the underlying assumptions, the smaller the assignment errors. For example, for given network and demand flows, both stochastic and deterministic equilibrium models estimate link flows closer to the observed ones than those resulting from uncongested network assignment models; probabilistic models are more accurate than deterministic ones for lightly congested or nonuniformly congested networks; and hyperpath assignment models are more precise than path-based assignment models for high-frequency and low-regularity public transport systems.

Figure 5.29 reports some experimental curves showing the assignment errors obtained with different assignment models for an urban road network against the counted flows. The assignment errors are measured through the relative standard

[24]Completely analogous considerations can be expressed with regard to the other variables that result from the assignment model, such as link costs, path costs and flows, performances, and so on.

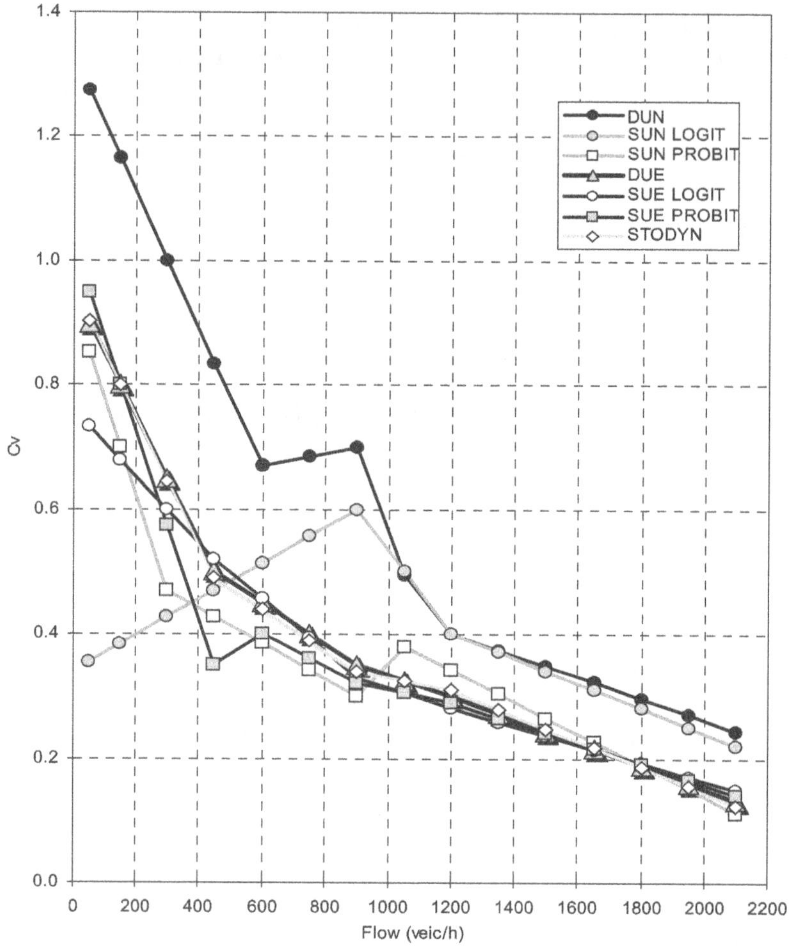

Fig. 5.29 Experimental relationships between the relative standard deviation of assignment errors and measured flow

deviation *Cv* of the assignment errors, that is, the ratio between the standard deviation of the errors between computed and assigned flows in a given range of measured flows, and the average flow in that range. In Fig. 5.29 DUN and SUN refer to assignments to noncongested networks (made by using zero-flow costs), DUE and SUE refer to equilibrium assignments to congested networks, and STODYN refers to results obtainable with a stochastic model such as those described in Sect. 6.5.3. Unfortunately, despite the very large number of applications to real transportation systems, the literature provides few systematic comparative analyses of different assignment models based on large databases, so that general conclusions on the relative merits of the different models in different application contexts cannot currently be drawn.

Specification and calibration of path choice models can be carried out using disaggregate and/or aggregate data. Disaggregate specification and calibration consists of the selection of the functional form and the attributes (specification) and the statistical estimation of the coefficients (calibration) on the basis of the paths chosen by a (random) sample of users. Methodologies for disaggregate specification and calibration of path choice models are completely analogous to those used for any random utility model and are described in Chap. 8. In the case of path choice models, however, disaggregate data are not easy to collect and analyze.[25] Thus aggregate specification and calibration techniques are often adopted. These techniques specify and calibrate path choice models by minimizing a measure of distance, usually the squared errors between simulated flows f^{SIM} and the flows f^{OBS} counted on some links. Aggregate calibration of path choice models requires the application of an assignment model, and are considered again more formally in Sect. 8.6.

5.5.1 Specification and Calibration of Assignment Models

Assignment models, as with all of the mathematical models described in this volume, should be calibrated. The specification of the model and its parameters should reproduce as closely as possible the available data on the system state. However, assignment models are affected by a wide variety of assumptions and parameters, because they incorporate all of the assumptions and the parameters of demand and supply models described in this volume. For this reason, a calibration procedure formally derived from the theory of statistical interference has not been proposed. Some partial procedures aimed at selecting assumptions and parameters specific to the assignment model have been applied in a limited number of cases. These usually assume that the supply model and demand functions or O-D flows have been calibrated separately, and focus on the choice of the supply–demand interaction model and the specification and calibration of path choice models.

5.A. Optimization Models for Stochastic Assignment

This appendix presents some optimization models for stochastic assignment that can be used for stochastic assignment under certain limiting assumptions. When they can be applied, they provide results equivalent to those from the more general fixed-point models discussed in the previous sections. However, it is hard or

[25]In reality, it is often a complex task to determine the path actually followed during a journey. Also, even when a path choice model has been specified and calibrated on disaggregated data, it is useful to carry out an aggregated recalibration. From a theoretical point of view, this can be seen as a correction of the parameters to compensate for the errors of the disaggregate model aggregation process.

even impossible to extend them to deal with the issues addressed. Hence they are presented mainly for the purpose of completeness.

These optimization models can also be compared with the deterministic assignment optimization models described earlier. Equivalent optimization models can be used to specify mathematical programming algorithms for the calculation of stochastic assignment link flows. In some special cases, these algorithms can be reduced to the fixed-point algorithms described earlier (e.g., the MSA-FA for stochastic equilibrium), but more generally they are still an open research area. Furthermore, equivalent optimization models for stochastic assignment can be included in bilevel optimization formulations of models for supply design models or O-D demand estimation using traffic counts.

In the following, nonadditive path costs are assumed equal to zero. Formal proofs are not reported, because they can be long and quite cumbersome.

5.A.1 Uncongested Network: Stochastic Assignment

For the logit path choice model with parameter θ independent of link costs, it can be demonstrated that SUN link and path flows are solutions of the following optimization model.

$$(\boldsymbol{f}_{\mathrm{SUN}}, \boldsymbol{h}_{\mathrm{SUN}}) = \operatorname{argmin} \sum_a c_a f_a + \theta \sum_k h_k (\ln h_k - 1) \quad \boldsymbol{f} = \boldsymbol{\Delta h}, \quad \boldsymbol{h} \in S_h \tag{5.A.1a}$$

Note that path flows appear explicitly as variables. In terms of the path flows alone, because $\sum_a c_a f_a = \sum_k h_k g_k$, it follows that

$$(\boldsymbol{h}_{\mathrm{SUN}}) = \operatorname{argmin} \sum_k h_k g_k + \theta \sum_k h_k (\ln h_k - 1) \tag{5.A.1b}$$

$$\boldsymbol{h} \in S_h$$

It can easily be seen that the objective functions in models (5.A.1a) and (5.A.1b) are convex if path flows are nonnegative.

In both models (5.A.1a) and (5.A.1b), the second term of the objective function goes to zero when parameter θ goes to zero, that is, when the variance of the path choice random residuals becomes small. In this case, the path choice model becomes deterministic and both models (5.A.1a) and (5.A.1b) coincide with the optimization model described in Sect. 5.3.2 for the DUN assignment.

5.A.2 Congested Network: Stochastic User Equilibrium

As with the previous model, for a logit path choice model with parameter θ independent of link costs, it may be demonstrated that, if cost functions have a symmetric

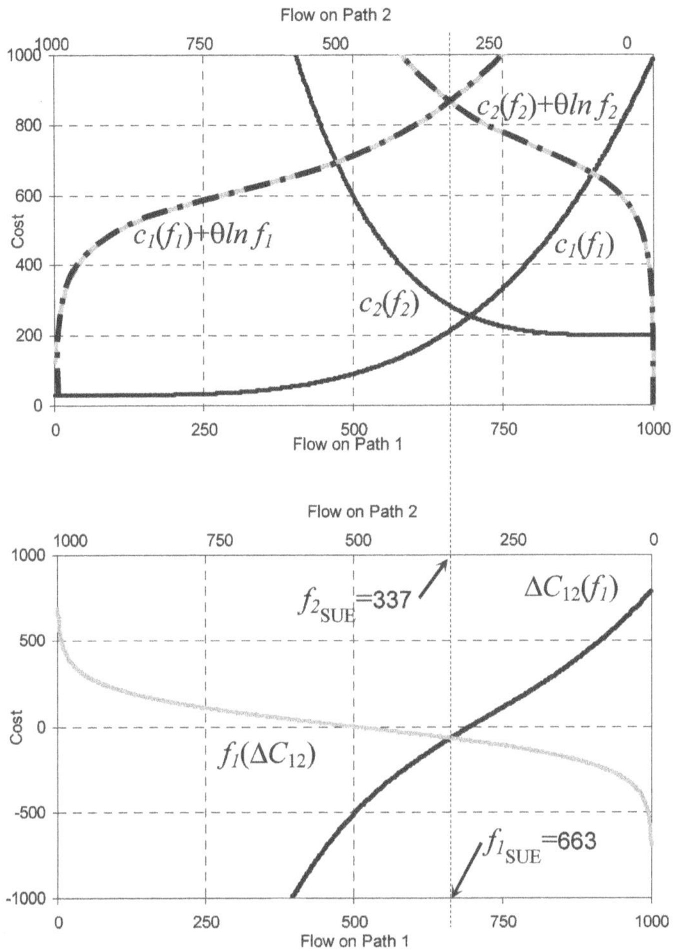

Fig. 5.A.1 Equivalent optimization model of SUE: logit assignment

Jacobian, then stochastic equilibrium (SUE) link and path flows are solutions of the following optimization model.

$$(\boldsymbol{f}^*, \boldsymbol{h}^*) = \text{argmin} \int_0^{\boldsymbol{f}} \boldsymbol{c}(\boldsymbol{y})^T d\boldsymbol{y} + \theta \sum_k h_k(\ln h_k - 1) \tag{5.A.2a}$$

$$\boldsymbol{f} = \boldsymbol{\Delta h}, \quad \boldsymbol{h} \in S_h$$

Note that path flows appear explicitly as variables. Because $\boldsymbol{f} = \boldsymbol{\Delta h}$, model (5.A.2a) can be expressed in terms of path flows alone:

$$\boldsymbol{h}^* = \text{argmin} \int_0^{\Delta h} \boldsymbol{c}(\boldsymbol{y})^T d\boldsymbol{y} + \theta \sum_k h_k(\ln h_k - 1) \tag{5.A.2b}$$

$$\boldsymbol{h} \in S_h$$

The objective functions of models (5.A.2a) and (5.A.2b) are (strictly) convex if the path flows are nonnegative and the cost functions are (strictly) increasing.

Considering the relationship with the corresponding DUE model, the second term of (5.A.2a) and (5.A.2b) goes to zero as the parameter θ goes to zero, that is, as the variance of the random residuals gets smaller. In this case, the path choice model becomes deterministic and model (5.A.2a) coincides with the optimization model described in Sect. 5.4.3 for DUE with symmetric Jacobian cost functions. Figure 5.A.1 illustrates the equivalent optimization model for SUE logit assignment in a simple two-link network.

In the case of a general additive path choice model and cost functions with a symmetric Jacobian, it can be shown that equilibrium link flows are a solution of the following unconstrained optimization model.

$$\boldsymbol{f}^* = \underset{\boldsymbol{f}}{\operatorname{argmin}} \sum_{od} d_{od} s_{od}\left(-\boldsymbol{\Delta}_{od}^T \boldsymbol{c}(\boldsymbol{f})\right) + \boldsymbol{c}(\boldsymbol{f})^T \boldsymbol{f} - \int_0^{\boldsymbol{f}} \boldsymbol{c}(\boldsymbol{y})^T d\boldsymbol{y} \qquad (5.A.3)$$

where $s_{od} = s_{od}()$ is the path choice EMPU for O-D pair *od*. Unlike the equivalent optimization model for DUE assignment network, constraints $\boldsymbol{f} \in S_f$ are not needed because they can be proven to be satisfied by all solutions of the model.

Reference Notes

Assignment Models

The traffic assignment problem has been the subject of extensive research for several decades. Exhaustive analyses of the models (and algorithms) for uncongested network and user equilibrium assignment are reported in the books by Sheffi (1985), Thomas (1991), Ortuzar and Willumsen (2001), and Patriksson (1994), the latter being mainly devoted to deterministic assignment models. For deterministic assignment models, the article by Florian and Hearn (1995) can also be referred to and the state of the art for stochastic assignment models is described in Cantarella and Cascetta (1998).

However, the approach proposed in this chapter to assignment models – or more properly to models of supply–demand interaction on transportation networks – is original. This approach allows, through a minimal set of hypotheses and equations, a consistent specification of uncongested network assignment models as well as fixed-point models and variational inequality models for user equilibrium on congested networks; elsewhere in the literature these are usually obtained independently of one another. This approach is based on the fixed-point approach proposed by Daganzo (1983) and further developed by Cantarella (1997).

The proposed classification of assignment models is another original contribution of this book. Deterministic user equilibrium models with separable cost functions and system optimum models were first formulated using optimization models in the pioneering work of Beckman et al. (1956), based on Wardrop's (1952) statement of traffic assignment principles. But it was not until the 1970s, with the increased power and availability of computing resources, that the assignment problem began to receive ongoing theoretical attention and to see a greater number and variety of applications. Recently, Maher et al. (2005) proposed a relation between system optimum and stochastic equilibrium.

The extension of the optimization model to symmetric deterministic equilibrium, the formulation of asymmetric deterministic equilibrium using variational inequality models, and the establishment of corresponding equilibrium existence and uniqueness conditions, are dealt with in the work of Dafermos (1971, 1972, 1980, 1982b) and Smith (1979). These articles also describe extensions of DUE models to variable demand and multiclass assignment. More complex optimization models proposed by various authors for asymmetric deterministic user equilibrium assignment are described and compared by Hearn et al. (1984). Bernstein and Smith (1994) analyzed deterministic equilibrium with lower semicontinuous link cost-flow functions.

Stochastic User Equilibrium (SUE) models were introduced by Daganzo and Sheffi (1977) (see also Daganzo (1979) for SUE probit models). Optimization models for symmetric SUE were proposed by Fisk (1980) in the case of the logit path choice, and by Daganzo and Sheffi (1982) and Sheffi and Powell (1982) in the general case.

Fixed-point models for SUE assignment were introduced by Daganzo (1983), who also analyzed variable demand assignment (with the hypernetwork approach referred to in Sect. 6.3) and multiclass assignment (Sect. 6.4). The compact notation and the related reformulation of the optimization problem for SUE models was first adopted by Cascetta (1987). Cantarella (1997) developed a general treatment using fixed point models of multimodal/multiclass variable demand equilibrium assignment, pre-trip/en-route path choice behavior, including stochastic as well as deterministic user equilibrium. In Cantarella and Cascetta (1998) the general problem of the stochastic equilibrium assignment models was discussed.

An analysis of stochastic assignment models with different formulations of random residuals was developed by Mirchandani and Soroush (1987). Nielsen (1997) analyzed the advantages and drawbacks of several distributions of link perceived costs. Cantarella and Binetti (2002) described and analyzed Gammit path choice models within stochastic equilibrium assignment. Watling (1999) proposed a generalization of SUE models by expressing moments of the distribution of multinomially distributed path flows. Bifulco (1993) proposed some extensions to simulate parking policies.

Some extension and application may be found in Nielsen et al. (1998), Nielsen (2000), and Nielsen et al. (2002). A national scale application was proposed by Russo and Vitetta (1995).

The introduction of hard link capacity constraints in deterministic or stochastic equilibrium models, studied by several authors in the context of a static approach, has been thoroughly analyzed by Ferrari (1997) for deterministic models. Bell (1995) proposed an application for a particular stochastic equilibrium model.

A further line of research relates to equilibrium models in which the values of an (uncongested) cost attribute are distributed among the users; monetary costs, for example, might follow the distribution of value of time. These models can be considered an extension of multiclass assignment models to an infinite number of classes when, for example, the Value of Time (VoT) is represented by a continuous random variable. Deterministic equilibrium has been specified in this case with extensions of variational inequality models by Leurent (1993, 1995, 1996) and by Marcotte and Zhu (1996), Marcotte et al. (1996), as well as by Dial (1996). The extension of stochastic equilibrium fixed point models has been dealt with by Cantarella and Binetti (1998).

Assignment Algorithms

General formulations of the assignment algorithms are reported in the books cited above. The literature proposes several algorithms for finding the shortest path tree, which are useful for deterministic uncongested network assignment. Comprehensive treatment of algorithms for transportation networks and a comparison of their performances can be found in Gallo and Pallottino (1988), Ahuja et al. (1993), Cherkassky et al. (1996), and Pallottino and Scutellà (1997).

Implementation of stochastic uncongested network assignment algorithms is discussed in Sheffi (1985). For the logit path choice model, the Dial algorithm described in Sect. 5.3.3 is an original generalization of the algorithm described in the original work by Dial (1971); see also Van Vliet (1981). An adaptation of Dial's algorithm to the C-logit path choice model is described in Russo and Vitetta (2003).

The Monte Carlo approach to stochastic uncongested network assignment was first proposed by Burrell (1968). Its application to probit SUN assignment is described in Sheffi and Powell (1982). Maher and Hughes (1997, 1998) have proposed an approach to probit SUN assignment based on Clark's approximation. Other applications of the SUE assignment models are in Maher (1997, 1998).

The adaptation of the Frank–Wolfe algorithm to the calculation of deterministic equilibrium flows is described in the original works of Le Blanc et al. (1975) and Nguyen (1976). As noted, many improvements to this algorithm have been proposed, such as the PARTAN (Florian and Spiess 1983), or other variations, namely in Fukushima (1984) and Lupi (1986). An interpretation of the Frank–Wolfe algorithm as a variational inequality algorithm is described in Van Vliet (1987). The diagonalization algorithm for nonseparable cost functions is analyzed in Florian and Spiess (1982); other algorithms for nonseparable cost functions are described in Nguyen and Dupuis (1984) and Hearn et al. (1984).

The MSA-FA algorithm for stochastic equilibrium is covered in Sheffi and Powell (1982), and its convergence is demonstrated in Powell and Sheffi (1982) as an optimization algorithm. Daganzo (1983) described the MSA-FA algorithm as a fixed-point algorithm, following Blum (1954), as well as the inverse cost function algorithm. The MSA-CA algorithm, and the internal cycle fixed-point algorithms for variable demand assignment are covered in Cantarella (1997). External cycle MSA algorithms described in Sect. 5.4.2 are an original contribution of this book. Other algorithms for the solution of logit SUE symmetric models, based on the minimization model proposed by Fisk (1980), are described in Bell et al. (1993), Chen and Alfa (1991a, 1991b), and Damberg et al. (1996).

For the logit path choice model, the optimization model presented for the SUE assignment problem derived from Fisk (1980). For the probit path choice model a broad presentation may be found in Sheffi (1985), who also reported proof of existence and uniqueness. For the probit path choice model a broad presentation may be found in Sheffi (1985), who also reported proof of existence and uniqueness (see also Daganzo, 1982). For the logit path choice model with parameter θ independent of link costs, the optimization model presented for the SUN assignment problem has been derived from the SUE assignment one for this book. Other algorithms for stochastic equilibrium (with separable cost functions) under some limiting assumptions may be developed by resolving optimization models presented in the chapter appendix.

Chapter 6
Advanced Models for Traffic Assignment to Transportation Networks

6.1 Introduction

Assignment to a transportation network has already been introduced in Chap. 5. Here we continue the analysis of assignment in the absence of within-day dynamics. In Sect. 6.2 we describe assignment with preventive-adaptive path choice, Sect. 6.3 covers the extension to the case of variable demand and/or multimodal assignment, and Sect. 6.4 deals with multiclass assignment models. In Sect. 6.5 we introduce assignment with day-to-day dynamics (dynamic process). As each of these extensions may be combined in banal fashion with each of the others, combined cases are not treated explicitly (e.g., multiclass assignment with variable demand); in Sect. 6.6 we present an overall scheme that allows straightforward comparison of these extensions. Also in this chapter, as in Chap. 5, the algorithms described are only those used more commonly or that are simpler to implement.

6.2 Assignment with Pre-trip/En-route Path Choice

Treatment of assignment with preventive-adaptive path choice uses the following scheme adopted in Chap. 5: definitions and assignment to a noncongested and equilibrium assignment, so as to make comparison easier.

6.2.1 Definitions, Assumptions, and Basic Equations

The previous chapter dealt with the situation in which users, before starting their trip, choose between alternative paths that they then follow without deviation to their destination. However, the analysis can quite easily be extended to include both pre-trip and en-route path choice behavior. This is relevant, for example, when modeling public transport systems with high frequency and/or low reliability. In this case (as was seen in Sect. 4.3.3.2) the appropriate pre-trip choice alternatives are en-route path choice strategies that can be represented by network hyperpaths (see Sect. 4.3.3.2), whereas en-route choices are made during the trip itself at each diversion (waiting) node where different lines are available (Fig. 6.1). The approach

---

Giulio Erberto Cantarella is co-author of this chapter.

E. Cascetta, *Transportation Systems Analysis*,
Springer Optimization and Its Applications 29,
DOI 10.1007/978-0-387-75857-2_6,

described in this section can be applied to other transportation systems once en-route diversion nodes and the related choice behavior have been specified.

The main modifications required to handle such cases concern the demand model defined in Sect. 5.2 by (5.2.7). In particular, the difference between path and hyperpath costs and flows is defined by the path choice probabilities within the hyperpaths. Referring to notation introduced in Sect. 4.3.3.2 (see Fig. 6.1), let:

$\omega_{od,kj}$ be the conditional probability of choosing path k within hyperpath j for a user of O-D pair od

$\boldsymbol{\Omega}_{od}$ be the matrix of conditional path choice probabilities $\omega_{od,kj}$ within the hyperpaths for O-D pair od

By analogy with the path definitions in Sect. 5.2, additive and nonadditive costs can be considered for each hyperpath. Let:

$\boldsymbol{x}_{od}^{\mathrm{ADD}}$ be the hyperpath additive cost vector for users of O-D pair od

$\boldsymbol{x}_{od}^{\mathrm{NA}}$ be the hyperpath nonadditive cost vector for users of O-D pair od

$\boldsymbol{x}_{od}$ be the vector of the total hyperpath costs for users of O-D pair od

As was seen in Sect. 4.3.3.2, the hyperpath additive costs $\boldsymbol{x}_{od}^{\mathrm{ADD}}$ are usually defined by a linear combination of on-board time $\boldsymbol{Tb}$, access/egress times $\boldsymbol{Ta}$, and boarding and alighting times $\boldsymbol{Tbr}$ and $\boldsymbol{Tal}$, all converted to utility units by suitable coefficients:

$$\boldsymbol{x}_{od}^{\mathrm{ADD}} = \beta_b \boldsymbol{Tb} + \beta_{br} \boldsymbol{Tbr} + \beta_{al} \boldsymbol{Tal} + \beta_d \boldsymbol{Td} + \beta_a \boldsymbol{Ta} \quad \forall od$$

Furthermore, nonadditive hyperpath costs $\boldsymbol{x}_{od}^{\mathrm{NA}}$ frequently include performance attributes that cannot be computed from generic link costs. Examples include the waiting time $\boldsymbol{Tw}_{od}$ and number of transfers $\boldsymbol{N}_{od}$, again converted to utility units by suitable coefficients:

$$\boldsymbol{x}_{od}^{\mathrm{NA}} = \beta_w \boldsymbol{Tw}_{od} + \beta_N \boldsymbol{N}_{od} \quad \forall od$$

The relationship between the hyperpath costs and the additive path costs is expressed by the following.

$$\boldsymbol{x}_{od}^{\mathrm{ADD}} = \boldsymbol{\Omega}_{od}^T \boldsymbol{g}_{od}^{\mathrm{ADD}} \quad \forall od$$

In the following, it is assumed for simplicity of notation that any nonadditive path costs $\boldsymbol{g}_{od}^{\mathrm{NA}}$ have been included in the nonadditive hyperpath costs $\boldsymbol{x}_{od}^{\mathrm{NA}}$, and therefore the path costs $\boldsymbol{g}_{od}$ coincide with the additive costs $\boldsymbol{g}_{od}^{\mathrm{ADD}}$ (see Fig. 6.2).

$$\boldsymbol{x}_{od} = \boldsymbol{\Omega}_{od}^T \boldsymbol{g}_{od} + \boldsymbol{x}_{od}^{\mathrm{NA}} \quad \forall od \tag{6.2.1}$$

The choice of strategy, that is, of the hyperpath representing its topology, is simulated by a random utility model in which the systematic utility of a hyperpath is the negative of its systematic cost, analogously to (5.2.5) (Sect. 4.3.3.2):

$$\boldsymbol{V}_{od} = -\boldsymbol{x}_{od} + \boldsymbol{V}_{od}^{\circ} = -\boldsymbol{\Omega}_{od}^T \boldsymbol{g}_{od}^{\mathrm{ADD}} - \boldsymbol{x}_{od}^{\mathrm{NA}} + \boldsymbol{V}_{od}^{\circ} \quad \forall od \tag{6.2.2}$$

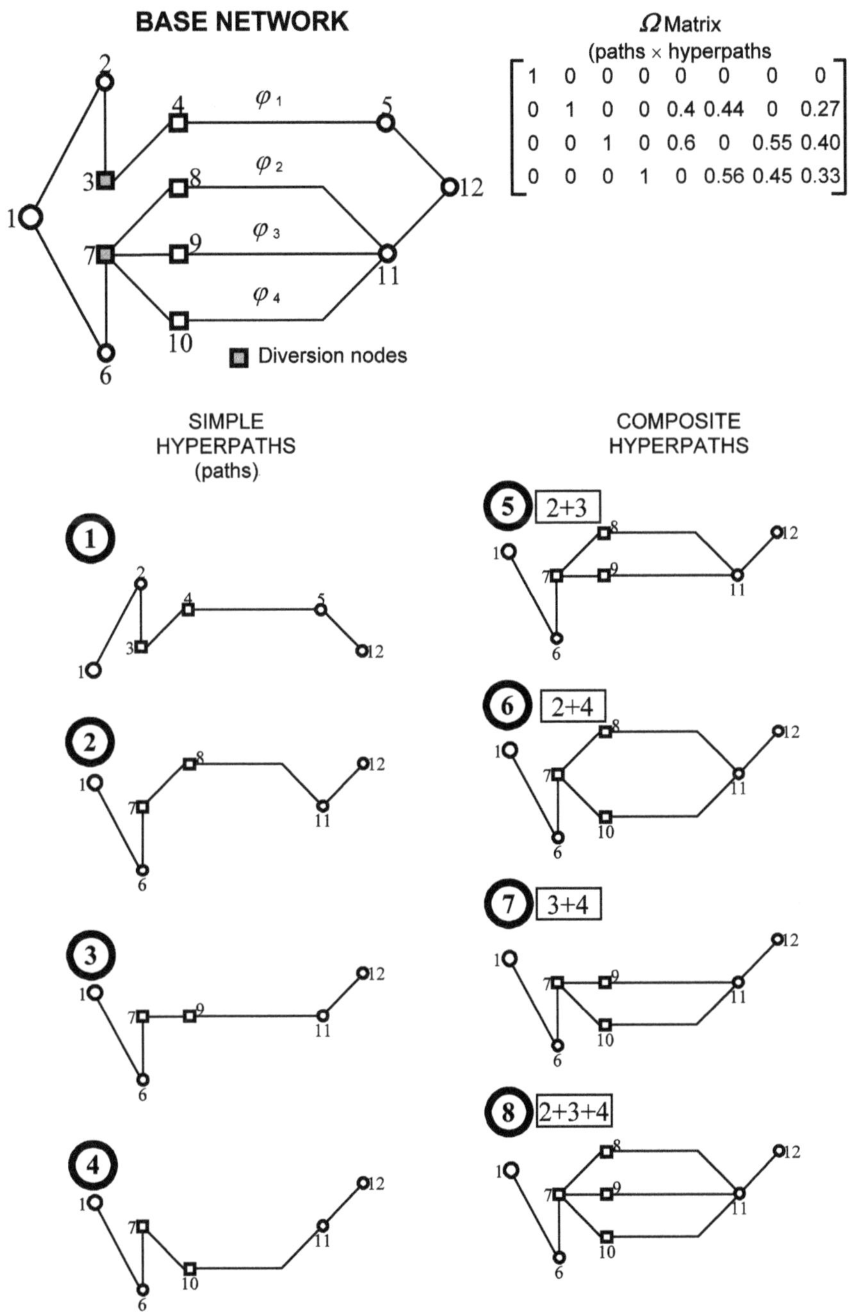

Fig. 6.1 Example of conditional path choice matrix

Fig. 6.2 Relationship between path and hyperpath costs for the network in Fig. 6.1

$$\boldsymbol{x} = \boldsymbol{\Omega}^T \boldsymbol{g} + \boldsymbol{x}^{\mathrm{NA}}$$

$$\underset{\boldsymbol{x}}{\begin{bmatrix} 471 \\ 461 \\ 481 \\ 816 \\ 474 \\ 667 \\ 630 \\ 594 \end{bmatrix}} = \underset{\boldsymbol{\Omega}^T}{\begin{bmatrix} 1 & 0 & 0 & 0 \\ 0 & 1 & 0 & 0 \\ 0 & 0 & 1 & 0 \\ 0 & 0 & 0 & 1 \\ 0 & 0.40 & 0.60 & 0 \\ 0 & 0.44 & 0 & 0.56 \\ 0 & 0 & 0.55 & 0.45 \\ 0 & 0.27 & 0.40 & 0.33 \end{bmatrix}} \bullet \underset{\boldsymbol{g}}{\begin{bmatrix} 421 \\ 421 \\ 451 \\ 771 \end{bmatrix}} + \underset{\boldsymbol{x}^{\mathrm{NA}}}{\begin{bmatrix} 50 \\ 40 \\ 30 \\ 45 \\ 35 \\ 50 \\ 35 \\ 45 \end{bmatrix}}$$

where

$\boldsymbol{V}_{od}$ is a vector with an element for each hyperpath j, given by the systematic utility V_j of the hyperpath j for users of O-D pair *od*

$\boldsymbol{V}^{\circ}_{od}$ is a vector with elements that capture the effects of attributes that cannot be assigned to hyperpath costs (such as user socioeconomic attributes); these are ignored for simplicity

The hyperpath choice probabilities depend on the systematic utilities of the hyperpaths, and therefore on the systematic costs. Let[1]:

$q[j/od]$ be the probability that a user chooses hyperpath j during a trip from origin o to destination d (with trip purpose, time band, and mode not explicitly indicated)

$\boldsymbol{q}_{od}$ be the vector of hyperpath choice probabilities for users of O-D pair *od*; its elements are the probabilities $q[j/od]$ for all available hyperpaths j; this set is assumed to be nonempty (each O-D pair is connected by at least one hyperpath) and finite (only elementary hyperpaths are considered)

As shown in Sect. 4.3.3.2, hyperpath choice probabilities can be expressed through random utility models as

$$q[j/od] = Prob[V_j - V_{mj} \geq \varepsilon_m - \varepsilon_j \ \forall m] \quad \forall od, j$$

$$\boldsymbol{q}_{od} = \boldsymbol{q}_{od}(\boldsymbol{V}_{od}) \quad \forall od$$

where ε_j is the random residual corresponding to the perceived utility of hyperpath j.

Combining the hyperpath choice model with the systematic utility specification gives a relation between hyperpath choice probabilities and costs for O-D pair *od*,

[1] As was done in previous sections, it is assumed that users belong to a single class, are traveling for the same purpose, and have the same hyperpath choice models. Generalization to multiclass assignment is dealt with in Sect. 5.4.

$$Q = Q(V) = Q(-x)$$

$$y = Q(V)d$$

$$h = \Omega y = \Omega Q d$$

$$\begin{array}{ccc} h = & \Omega & \bullet\; y \\ \begin{bmatrix} 138 \\ 244 \\ 283 \\ 214 \end{bmatrix} & \begin{bmatrix} 1 & 0 & 0 & 0 & 0 & 0 & 0 & 0 \\ 0 & 1 & 0 & 0 & 0.4 & 0.44 & 0 & 0.27 \\ 0 & 0 & 1 & 0 & 0.6 & 0 & 0.55 & 0.4 \\ 0 & 0 & 0 & 1 & 0 & 0.56 & 0.45 & 0.33 \end{bmatrix} & \begin{bmatrix} 138 \\ 139 \\ 136 \\ 97 \\ 137 \\ 113 \\ 117 \\ 0 \end{bmatrix} \end{array}$$

Fig. 6.3 Relationship between path and hyperpath flows for the network in Fig. 6.1

known as the *hyperpath choice map*:

$$\boldsymbol{q}_{od} = \boldsymbol{q}_{od}(\boldsymbol{V}_{od}) = \boldsymbol{q}_{od}(-\boldsymbol{x}_{od}) \quad \forall od$$

The (average) flow y_j on the hyperpath j connecting O-D pair *od* is given by the product of the corresponding demand flow d_{od} and hyperpath choice probability:

$$y_j = d_{od} q[j/od]$$

and is measured in demand flow units. Let:

$\boldsymbol{y}_{od}$ be the hyperpath flow vector for O-D pair *od*, whose elements are the flows y_j for all available hyperpaths j

For each O-D pair *od*, the relation between hyperpath choice probabilities and flows and demand flows, analogous to (5.2.6), is expressed by

$$\boldsymbol{y}_{od} = d_{od} \boldsymbol{q}_{od}(\boldsymbol{V}_{od}) \quad \forall od \tag{6.2.3}$$

Each path k that connects O-D pair *od* may belong to several hyperpaths, so that the flow h_k is given by the sum of the hyperpath flows y_j weighted by the probability $\omega_{od,kj}$ that path k is used within the hyperpath j (an example is reported in Fig. 6.3):

$$h_k = \sum_h \omega_{od,kj} y_j \quad \forall k$$

$$\boldsymbol{h}_{od} = \boldsymbol{\Omega}_{od} \boldsymbol{y}_{od} = d_{od} \boldsymbol{\Omega}_{od} \boldsymbol{q}_{,od} \quad \forall od \tag{6.2.4}$$

The complete demand model, for situations of combined pre-trip and en-route path choice behavior, is defined by relations (6.2.1)–(6.2.2) that specify the systematic utility, and by relations (6.2.3)–(6.2.4) that define the path flows. When combined, these lead to a relation between path flows and costs that generalizes expression (5.2.7):

$$\boldsymbol{h}_{od} = d_{od}\boldsymbol{\Omega}_{od}\boldsymbol{q}_{,od}\left(-\boldsymbol{\Omega}_{od}^{T}\boldsymbol{g}_{od} - \boldsymbol{x}_{od}^{\mathrm{NA}}\right) \quad \forall od \tag{6.2.5}$$

By combining the demand model (6.2.5) with the supply model (5.2.4), the assignment models described in Chap. 5 can be extended to handle combined pre-trip and en-route path choice behavior. In this case, it is useful to express the relation between link and hyperpath flows and costs. Let:

- $\lambda_{od,aj}$ be the probability that users of O-D pair *od* traverse link *a* within hyperpath *j*
- $\boldsymbol{\Lambda}_{od}$ be the matrix, for users of O-D pair *od*, of link-hyperpath traversing probabilities $\lambda_{od,aj}$ for each link *a* within each hyperpath *j*

The relationship between link-hyperpath traversing probabilities and path choice probabilities within a hyperpath (analogous to the link-path incidence relationships described in Chap. 2 and repeated in Sect. 5.2) is expressed by the following relations (Fig. 6.4).

$$\lambda_{od,aj} = \sum_{k} \delta_{od,ak}\omega_{od,kj} \quad \forall l\ \forall od$$
$$\boldsymbol{\Lambda}_{od} = \boldsymbol{\Delta}_{od}\boldsymbol{\Omega}_{od} \quad \forall od \tag{6.2.6}$$

The relationship among hyperpath costs, link costs, and additive path costs is expressed by the following equation, obtained by combining expressions (6.2.1), (6.2.6), and (5.2.1) (nonadditive path costs $\boldsymbol{g}_{od}^{\mathrm{NA}}$ have been included in the nonadditive hyperpath costs $\boldsymbol{x}_{od}^{\mathrm{NA}}$; thus path costs coincide with additive costs: $\boldsymbol{g}_{od} = \boldsymbol{g}_{od}^{\mathrm{ADD}}$).

$$\boldsymbol{x}_{od} = \boldsymbol{x}_{od}^{\mathrm{ADD}} + \boldsymbol{x}_{od}^{\mathrm{NA}} = \boldsymbol{\Omega}_{od}^{T}\boldsymbol{\Delta}_{od}^{T}\boldsymbol{c} + \boldsymbol{x}_{od}^{\mathrm{NA}} = \boldsymbol{\Lambda}_{od}^{T}\boldsymbol{c} + \boldsymbol{x}_{od}^{\mathrm{NA}} \quad \forall od \tag{6.2.7}$$

Similarly, the relationship between link and hyperpath flows is expressed by the following equation, obtained by combining expressions (6.2.4), (6.2.6), and (5.2.3) (Fig. 6.5).

$$\boldsymbol{f} = \sum_{od} \boldsymbol{\Delta}_{od}\boldsymbol{h}_{od} = \sum_{od} \boldsymbol{\Delta}_{od}\boldsymbol{\Omega}_{od}\boldsymbol{y}_{od} = \sum_{od} \boldsymbol{\Lambda}_{od}\boldsymbol{y}_{od} \tag{6.2.8}$$

Combined pre-trip/en-route behavior assignment can therefore be expressed by relations (6.2.7)–(6.2.8) and (6.2.3)–(6.2.4), together with the cost functions (5.2.2). It clearly follows from this formulation that the pre-trip assignment models expressed by relations (5.2.1) and (3.5.6) are special cases that can be obtained from combined pre-trip/en route assignment models by setting $\boldsymbol{\Omega}_{od} = \boldsymbol{I}$, for which it would follow that $\boldsymbol{\Delta}_{od} = \boldsymbol{\Lambda}_{od}$, $\boldsymbol{y}_{od} = \boldsymbol{h}_{od}$ and $\boldsymbol{x}_{od} = \boldsymbol{g}_{od}$. In fact, in pre-trip assignment each hyperpath corresponds to a single path (simple hyperpath), and en-route choices are not considered.

The set of feasible hyperpaths and link flows S_y and S_f are defined, as in Sect. 5.2.3, by

$$S_y = \left\{\boldsymbol{y} = [\boldsymbol{y}_{od}]_{od} : \boldsymbol{y}_{od} \geq \boldsymbol{0}, \boldsymbol{1}^{T}\boldsymbol{y}_{od} = d_{od} \quad \forall od\right\} \tag{6.2.9}$$

$$\boldsymbol{h} = \boldsymbol{\Omega}\, \boldsymbol{Q}(-\boldsymbol{\Omega}^T \boldsymbol{g} - \boldsymbol{x}^{\text{NA}})\boldsymbol{d}$$
$$\boldsymbol{\Lambda} = \boldsymbol{\Delta\Omega}$$

Λ (Links × Hyperpaths) =

	1	**2**	**3**	**4**	**5**	**6**	**7**	**8**
1-2	1	0	0	0	0	0	0	0
1-6	0	1	1	1	1	1	1	1
2-3	1	0	0	0	0	0	0	0
3-4	1	0	0	0	0	0	0	0
4-5	1	0	0	0	0	0	0	0
5-12	0	1	1	1	1	1	1	1
6-7	0	1	1	1	1	1	1	1
7-8	0	1	0	0	0.4	0.44	0	0.27
7-9	0	0	1	0	0.6	0	0.55	0.40
7-10	0	0	1	0	0.6	0	0.55	0.40
8-11	0	0	1	0	0.6	0	0.55	0.40
9-11	0	0	0	1	0	0.56	0.45	0.33
10-11	0	0	0	1	0	0.56	0.45	0.33
11-12	0	1	1	1	1	1	1	1

= **Δ** (Links × Paths) • **Ω** (Paths × Hyperpaths)

	1	**2**	**3**	**4**
1-2	1	0	0	0
1-6	0	1	1	1
2-3	1	0	0	0
3-4	1	0	0	0
4-5	1	0	0	0
5-12	1	0	0	0
6-7	0	1	1	1
7-8	0	1	0	0
7-9	0	0	1	0
7-10	0	0	0	1
8-11	0	1	0	0
9-11	0	0	1	0
10-11	0	0	0	1
11-12	0	1	1	1

	1	**2**	**3**	**4**	**5**	**6**	**7**	**8**
1	1	0	0	0	0	0	0	0
2	0	1	0	0	0.4	0.44	0	0.27
3	0	0	1	0	0.6	0	0.55	0.40
4	0	0	0	1	0	0.56	0.45	0.33

Fig. 6.4 Incidence and traversing probability matrices for the network of Fig. 6.1

$$S_f = \left\{ \boldsymbol{f} = \sum_{od} \boldsymbol{\Lambda}_{od}\, \boldsymbol{y}_{od},\, [\boldsymbol{y}_{od}]_{od} \in S_y \right\} \tag{6.2.10}$$

As in Sects. 5.2.1 and 5.2.2, all the above relationships can be expressed in matrix terms. Let:

$\boldsymbol{\Omega}$ be the overall matrix of conditional path choice probabilities for all paths, all hyperpaths, and all O-D pairs, obtained by placing side by side the blocks $\boldsymbol{\Omega}_{od}$ corresponding to each O-D pair od

$\boldsymbol{x}^{\text{ADD}}$ be the overall vector of hyperpath additive costs, consisting of the hyperpath additive cost vectors $\boldsymbol{x}_{od}^{\text{ADD}}$ for each O-D pair od

$$x = x^{\text{ADD}} + x^{\text{NA}} = \boldsymbol{\Omega}^T \boldsymbol{\Delta}^T c + G^{\text{NA}} = \boldsymbol{\Lambda}^T c + x^{\text{NA}}$$

$$f = \boldsymbol{\Delta} h = \boldsymbol{\Delta}\boldsymbol{\Omega} y = \boldsymbol{\Lambda} y$$

$$f = \boldsymbol{\Delta} \bullet h$$

$$\begin{bmatrix} 126 \\ 874 \\ 126 \\ 126 \\ 126 \\ 126 \\ 874 \\ 254 \\ 356 \\ 264 \\ 254 \\ 356 \\ 264 \\ 874 \end{bmatrix} \begin{bmatrix} 1 & 0 & 0 & 0 \\ 0 & 1 & 1 & 1 \\ 1 & 0 & 0 & 0 \\ 1 & 0 & 0 & 0 \\ 1 & 0 & 0 & 0 \\ 1 & 0 & 0 & 0 \\ 0 & 1 & 1 & 1 \\ 0 & 1 & 0 & 0 \\ 0 & 0 & 1 & 0 \\ 0 & 0 & 0 & 1 \\ 0 & 1 & 0 & 0 \\ 0 & 0 & 1 & 0 \\ 0 & 0 & 0 & 1 \\ 0 & 1 & 1 & 1 \end{bmatrix} \begin{bmatrix} 126 \\ 254 \\ 356 \\ 264 \end{bmatrix}$$

$$f = \boldsymbol{\Lambda} \bullet y$$

$$\begin{bmatrix} 126 \\ 874 \\ 126 \\ 126 \\ 126 \\ 874 \\ 874 \\ 254 \\ 356 \\ 356 \\ 356 \\ 264 \\ 264 \\ 874 \end{bmatrix} \begin{bmatrix} 1 & 0 & 0 & 0 & 0 & 0 & 0 & 0 \\ 0 & 1 & 1 & 1 & 1 & 1 & 1 & 1 \\ 1 & 0 & 0 & 0 & 0 & 0 & 0 & 0 \\ 1 & 0 & 0 & 0 & 0 & 0 & 0 & 0 \\ 1 & 0 & 0 & 0 & 0 & 0 & 0 & 0 \\ 0 & 1 & 1 & 1 & 1 & 1 & 1 & 1 \\ 0 & 1 & 1 & 1 & 1 & 1 & 1 & 1 \\ 0 & 1 & 0 & 0 & 0.4 & 0.44 & 0 & 0.27 \\ 0 & 0 & 1 & 0 & 0.6 & 0 & 0.55 & 0.40 \\ 0 & 0 & 1 & 0 & 0.6 & 0 & 0.55 & 0.40 \\ 0 & 0 & 1 & 0 & 0.6 & 0 & 0.55 & 0.40 \\ 0 & 0 & 0 & 1 & 0 & 0.56 & 0.45 & 0.33 \\ 0 & 0 & 0 & 1 & 0 & 0.56 & 0.45 & 0.33 \\ 0 & 1 & 1 & 1 & 1 & 1 & 1 & 1 \end{bmatrix} \begin{bmatrix} 126 \\ 93 \\ 122 \\ 79 \\ 157 \\ 126 \\ 138 \\ 159 \end{bmatrix}$$

Fig. 6.5 Relationship among link, path, and hyperpath flows for the network of Fig. 6.1

x^{NA} be the overall vector of hyperpath nonadditive costs, consisting of the hyperpath nonadditive cost vectors x_{od}^{NA} for each O-D pair *od*

x be the overall vector of total hyperpath costs, consisting of the vectors of the total hyperpath cost vectors x_{od} for each O-D pair *od*

y be the overall vector of hyperpath flows, consisting of the vectors of the hyperpath flows y_{od} for each O-D pair *od*

$\boldsymbol{Q}$ be the hyperpath choice probability matrix, with a column for each O-D pair *od* and a row for each hyperpath j, with entries given by $q[j/od]$ if hyperpath j connects the O-D pair, and zero otherwise (in other words, the matrix $\boldsymbol{Q}$ is block diagonal with blocks given by the vectors q_{od})

$\boldsymbol{\Lambda}$ be the overall matrix of link-hyperpath traversing probabilities, consisting of the blocks Λ_{od} for each O-D pair *od*

6.2.2 Uncongested Networks

The *uncongested network assignment* models described in Sect. 5.3 can therefore easily be extended to the case of combined pre-trip/en-route choice. In particular, the uncongested network assignment model can be expressed in terms of link flows by an equation similar to (5.3.2):

$$\boldsymbol{f}_{\mathrm{UN}} = \boldsymbol{f}_{\mathrm{UN}}(\boldsymbol{c};\boldsymbol{d}) = \sum_{od} d_{od}\boldsymbol{\Lambda}_{od}\boldsymbol{q}_{,od}\left(-\boldsymbol{\Lambda}_{od}^{T}\boldsymbol{c} - \boldsymbol{x}_{od}^{\mathrm{NA}}\right) \tag{6.2.11}$$

Uncongested network assignment models with combined pre-trip/en-route behavior can be probabilistic or deterministic depending on the hyperpath choice model adopted. In the case of probabilistic choice behavior, the Stochastic UNcongested network (SUN) assignment models can be expressed by a function similar to (5.3.3):

$$\boldsymbol{f}_{\mathrm{SUN}} = \boldsymbol{f}_{\mathrm{SUN}}(\boldsymbol{c};\boldsymbol{d}) = \sum_{od} d_{od}\boldsymbol{\Lambda}_{od}\boldsymbol{q}_{od}\left(-\boldsymbol{\Lambda}_{od}^{T}\boldsymbol{c} - \boldsymbol{x}_{od}^{\mathrm{NA}}\right) \tag{6.2.12}$$

which retains the properties of continuity and monotonicity discussed in Sect. 5.4.1.

In the case of deterministic choice behavior, the relationship between hyperpath flow and costs can be expressed with a system of inequalities similar to (5.3.4):

$$\boldsymbol{x}^{T}(\boldsymbol{y} - \boldsymbol{y}_{\mathrm{DUN}}) \geq 0 \quad \forall \boldsymbol{y} \in S_y \tag{6.2.13}$$

If nonadditive path costs are not explicitly considered, then by substituting (6.2.7) and (6.2.8) in (6.2.13) it follows that

$$\boldsymbol{c}^{T}(\boldsymbol{f} - \boldsymbol{f}_{\mathrm{DUN}}) + (\boldsymbol{x}^{\mathrm{NA}})^{T}(\boldsymbol{y} - \boldsymbol{y}_{\mathrm{DUN}}) \geq 0 \quad \forall \boldsymbol{f} = \boldsymbol{\Lambda}\boldsymbol{y},\ \forall \boldsymbol{y} \in S_y \tag{6.2.14}$$

Because of the presence of nonadditive costs, the considerations discussed in Sect. 5.3.3 hold, and (6.2.14) can be expressed in terms of link flows $\boldsymbol{f}_{\mathrm{DUN}}$ and total nonadditive costs $X_{\mathrm{DUN}} = (\boldsymbol{x}^{\mathrm{NA}})^{T}\boldsymbol{y}_{\mathrm{DUN}}$ corresponding to deterministic assignment:

$$\boldsymbol{c}(\boldsymbol{f}_{\mathrm{DUN}})^{T}(\boldsymbol{f} - \boldsymbol{f}_{\mathrm{DUN}}) + (X - X_{\mathrm{DUN}}) \geq 0 \quad \forall \boldsymbol{f} = \boldsymbol{\Lambda}\boldsymbol{y},\ \forall X = (\boldsymbol{x}^{\mathrm{NA}})^{T}\boldsymbol{y},\ \forall \boldsymbol{y} \in S_y$$

Below we describe shortest hyperpath algorithms as an extension of shortest path algorithms. Then their application to uncongested network assignment is presented.

Shortest Hyperpath Algorithms

The algorithms described in Sect. 5.3.3 for the computation of shortest paths can be extended to identify shortest hyperpaths, such as those relevant to modeling the pre-trip/en-route path choice behavior described in Sect. 6.2. In the following, for the sake of simplicity, it is assumed that all origins and destinations are connected;

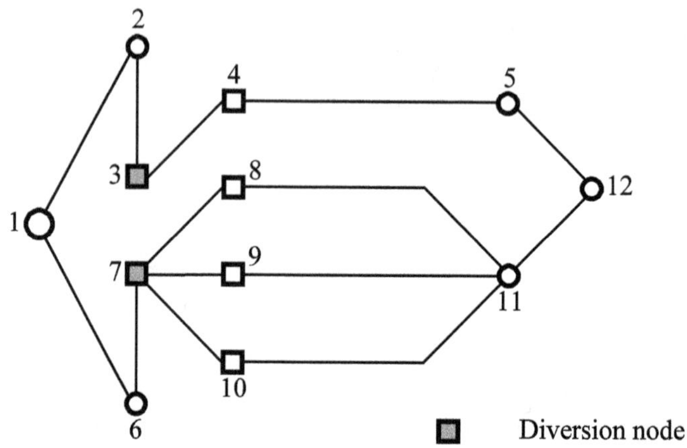

Fig. 6.6 Diversion nodes and adjacent elements

that is, there is at least one hyperpath from each origin to each node, and from each node to each destination. For a transit system, it is assumed that the only costs due to en-route choices are waiting times at stops; this leads to nonadditive hyperpath costs, as described. By analogy to shortest path problems, let:

$c_a = t_{mn} \geq 0$ be the cost of link $a = (m, n)$, corresponding to travel time components such as boarding, on board travel, alighting, and access/egress times. These attributes are associated to the corresponding types of links. (If needed, the different time components can be multiplied by appropriate conversion coefficients that, for the sake of simplicity, are not explicitly indicated here.) As explained above, the waiting time associated with a waiting link is not an intrinsic network attribute because it depends on the hyperpath under consideration. It can be derived from the frequency of the lines as described below. Let:

$Z_{o,d} \geq 0$ be the cost of the shortest hyperpath between nodes o and d

Diversion nodes, where en-route choices are made, correspond here to stops where users decide which transit line to board (see Sects. 4.3.3.2 and 6.2). Consistent with the transit network model described in Sect. 2.3, from each diversion node m there are boarding links $a = (m, n)$ connecting to the different lines available at the stop, and a waiting link connects the stop node to the diversion node m (Fig. 6.6). Let:

DN be the set of diversion nodes

$pr(m)$ be the stop node preceding diversion node m, connected to it by the waiting link $(pr(m), m)$

$\varphi_{m,n} > 0$ be the frequency of the line accessed through boarding link $a = (m, n)$. This value and the boarding time $t^s_{m,n}$ are associated with each boarding link. For the sake of simplicity, all boarding times are assumed constant and equal in the following; $t^s_{m,n} = t^s$

The topology of a hyperpath j is defined by a sequence of nodes with the property that at most one link may exit from a nondiversion node $n \notin DN$, and multiple (boarding) links $a = (m, n)$ may exit from a diversion node $m \in DN$. (Examples of hyperpaths are given in the figures in Sect. 4.3.3.2.) When the topology of a hyperpath is known, a waiting time can be defined for each waiting link as a function of the frequencies of the lines that belong to that hyperpath. Because of randomness in the arrivals of users and, possibly, vehicles at a stop, the waiting time is a random variable. In what follows, we are concerned only with the average (expected value) of this and related random variables. For a hyperpath j, let:

$X^j_{m,d}$ be the cost or travel time from node m to node d along hyperpath j
$AL_{m,j}$ be the set of boarding links from diversion node m in hyperpath j
Φ^j_m be the sum of the frequencies of the lines that belong to hyperpath j and are available at diversion node m
$t^{w,j}_m$ be the waiting time on the (unique) waiting link $(pr(m), m)$ that enters diversion node m on hyperpath j

Assuming random user arrivals, the (average) waiting time is inversely proportional to the sum of the frequencies of the lines in the hyperpath. The proportionality parameter $\theta \in [0.5, 1.0]$ depends on the service regularity (see Sect. 2.3):

$$\Phi^j_m = \left(\sum_{(m,n)\in AL_{m,j}} \varphi_{mn} \right) \tag{6.2.15}$$

$$t^m_{w,j} = \theta \Big/ \left(\sum_{(m,n)\in AL_{m,j}} \varphi_{mn} \right) = \theta / \Phi^j_m \tag{6.2.16}$$

The average travel time from diversion node m to destination d is the frequency-weighted average of the travel times on the lines accessible from node m in hyperpath j (as noted in Sect. 6.2):

$$X^j_{m,d} = \sum_{(m,n)\in AL_{m,j}} \left(t^s + X^j_{n,d}\right)\left(\varphi_{mn}/\Phi^j_m\right) \tag{6.2.17}$$

The average travel time $X^j_{pr(m),d}$ to reach destination d from the stop node $pr(m)$ connected to diversion node m can be defined as the sum of the average time from the diversion node $X^j_{m,d}$ and the average waiting time $t^{w,j}_m$:

$$X^j_{pr(m),d} = X^j_{m,d} + t^{w,j}_m \tag{6.2.18}$$

Relation (6.2.17) allows us to express $Z_{m,d}$, the average minimum travel time from a diversion node m to the destination d, as the frequency-weighted average of the minimum times along the lines from node m that belong to the shortest hyperpath j^*:

$$Z_{m,d} = \sum_{(mn)\in AL_{m,j*}} \left(t^s + Z_{n,d}\right)\left(\varphi_{m,n}/\Phi^{j^*}_m\right)$$

The shortest travel time $Z_{pr(m),d}$ from the stop node $pr(m)$ connected to diversion node m can be obtained by summing the shortest travel time from the diversion node $Z_{m,d}$ and the waiting time t_m^{w,j^*}:

$$Z_{pr(m),d} = Z_{m,d} + t_m^{w,j^*}$$

(to be compared with the shortest travel time along the access network). All the above relations can be used to extend the Bellmann principle to the shortest hyperpath problem.

It should be noted that if the forward shortest hyperpath tree from an origin o to all the other nodes were searched, it would be necessary, at each stop, to distinguish users by destination, to take account of the different lines available. For this reason, it is useful to adopt algorithms based on an extension of the *backward updating step*, previously defined for shortest paths, that allows the determination of the shortest hyperpath tree from all nodes towards the destination.

Now consider a hyperpath j (not necessarily the shortest one); the backward updating step from node n is similar to the step already described for shortest paths (Sect. 5.3.3) unless node n is the end of a boarding link (m,n) (see Fig. 6.6). In this case, the updating step must be extended to check whether including the boarding link $a = (m,n)$ in hyperpath j will reduce the average travel time from node $pr(m)$. Let:

$\Phi_m^j, t_m^{w,j}$ be the values at node m of the cumulative frequency and the average waiting time of this hyperpath, as defined by (6.2.15) and (6.2.16)

The average travel times from waiting node m and stop node $pr(m)$ on hyperpath j are given by (6.2.17) and (6.2.18), respectively. Note that the node $pr(m)$ might be connected to the destination d through other paths using the access links.

In what follows it is assumed that a *label-setting* algorithm will be adopted, for reasons that become clear below. Thus, let:

$Z_{n,d}$ be the minimum cost or travel time between line node n and destination d, already known when node n is examined

If boarding link (m,n) is added to hyperpath j, a further line with frequency $\varphi_{m,n}$ is available at stop node m. Therefore, there is an additional path available to reach destination d. The new hyperpath j' that includes this path reduces the average travel time from node $pr(m)$ to destination d if:

$$X_{pr(m),d}^{j'} \leq X_{pr(m),d}^{j} \tag{6.2.19}$$

To analyze the implications of (6.2.19), note that at node m the hyperpath j' has a larger cumulative frequency and a smaller waiting time than hyperpath j. This can be seen by applying the relationships (6.2.15) and (6.2.16) (Fig. 6.7):

$$\Phi_m^{j'} = \Phi_m^j + \varphi_{mn} \tag{6.2.20}$$

$$t_m^{w,j'} = \theta / \Phi_m^{j'} = t_m^{w,j}\left[\Phi_m^j / \left(\Phi_m^j + \varphi_{mn}\right)\right] \tag{6.2.21}$$

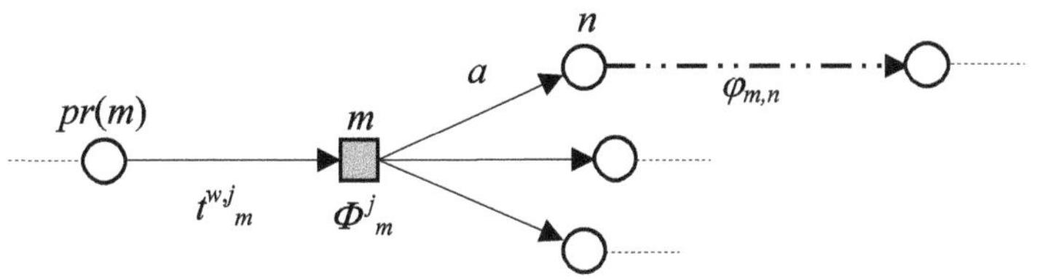

Fig. 6.7 Diversion node, waiting link, boarding links

The inclusion of the additional line causes a change in the average travel time from node m to destination d. From (6.2.17):

$$\begin{aligned} X^{j'}_{m,d} &= X^{j}_{m,d}\left[\Phi^j_m/(\Phi^j_m+\varphi_{mn})\right]+\left(Z_{n,d}+t^s_{mn}\right)\left[\varphi_{mn}/(\Phi^j_m+\varphi_{mn})\right] \\ &= X^{j}_{m,d}+\left(Z_{n,d}+t^s_{mn}-X^{j}_{m,d}\right)\varphi_{mn}/\left(\Phi^j_m+\varphi_{mn}\right) \end{aligned} \quad (6.2.22)$$

because $[\Phi^j_m/(\Phi^j_m+\varphi_{mn})] = 1-[\varphi_{mn}/(\Phi^j_m+\varphi_{mn})]$. Thus, after the introduction of boarding link (m,n), the average travel time from waiting node $pr(m)$ to destination d through diversion node m becomes:

$$X^{j'}_{pr(m),d} = X^{j'}_{m,d} + t^{w,j}_m \quad (6.2.23)$$

or

$$X^{j'}_{pr(m),d} = X^{j}_{m,d}+\left(Z_{n,d}+t^s_{mn}-X^{j}_{m,d}\right)\varphi_{mn}/\left(\Phi^j_m+\varphi_{mn}\right)+t^{w,j}_m\left[\Phi^j_m/\left(\Phi^j_m+\varphi_{mn}\right)\right]$$

Combining the above relationship with condition (6.2.19), we obtain:

$$\begin{aligned} &X^{j}_{m,d}+\left(Z_{n,d}+t^s_{mn}-X^{j}_{m,d}\right)\varphi_{mn}/\left(\Phi^j_m+\varphi_{mn}\right)+t^{w,j}_m\left[\Phi^j_m/\left(\Phi^j_m+\varphi_{mn}\right)\right] \\ &\quad\le X^{j}_{m,d}+t^{w,j}_m \end{aligned} \quad (6.2.24)$$

$$\left(Z_{n,d}+t^s_{mn}-X^{j}_{m,d}\right)\varphi_{mn}/\left(\Phi^j_m+\varphi_{mn}\right) \le t^{w,j}_m\,\varphi_{mn}/\left(\Phi^j_m+\varphi_{mn}\right)$$

$$Z_{n,d}+t^s_{mn} \le X^{j}_{m,d}+t^{w,j}_m$$

because $[\varphi_{mn}/(\Phi^j_m+\varphi_{mn})] > 0$. Therefore link $l = (m,n)$ is worth including if:

$$Z_{n,d}+t^s_{mn} \le X^{j}_{pr(m),d} \quad (6.2.25)$$

On the other hand, given a hyperpath j' that contains boarding link $a = (m,n)$, it is not possible to reduce the total travel time by excluding link a from the hyperpath if condition (6.2.25) is verified (and conversely if the condition is not verified). Therefore condition (6.2.25) is both necessary and sufficient. Condition (6.2.25) shows that, to reduce the average travel time and find the shortest hyperpath, it is worth including a new line if the travel time with the new line, including boarding time, is less than the travel time, including waiting time, without the line. If this is so, inclusion of the new line reduces the waiting time so that even if the average travel time from the diversion node increases, the average travel time from the stop node decreases.

The shortest hyperpath for a pair (o,d) might not include any waiting links (and therefore boarding, line, and alighting links). In this case it consists only of access links, meaning that the shortest path on the access network has a lower cost than any paths using a transit line.

The algorithms for calculating the tree of shortest hyperpaths towards a destination d are similar to those described in Sect. 5.3.3 for shortest paths. The main difference is the updating step that, for hyperpaths, also includes operations to update the tentative diversion node label, using condition (6.2.25) and relations (6.2.22) and (6.2.23) to update the labels (average travel times). In this way, a stop node might be connected to destination d by other paths through the access links adjacent to it. The tree of shortest hyperpaths towards a destination node can be described by the unique link that exits from each nondiversion node, and the set of boarding links that exit from each diversion node; these boarding links identify the lines included in the shortest hyperpath.

Note that the node made permanent at each iteration should be the one with the least value of label among nonpermanent nodes, and the updating step should be performed from this node. Therefore, to identify a shortest hyperpath tree towards a destination, *label-setting algorithms* should be adopted.

In addition, consider a further boarding link (m,r) not included in hyperpath j such that $Z_{r,d} \leq Z_{n,d}$, or $Z_{r,d} + t^s \leq Z_{n,d} + t^s$. If condition (6.2.25) is verified for link (m,r), and therefore it is worth including link (m,n) to reduce the average cost, it is also verified for link (m,n), and including also link (m,r) is even more appropriate. Observation further supports the adoption of label-setting algorithms, in which the updating of the line nodes n connected to a diversion node m through boarding links (m,n) is carried out by increasing values of $Z_{n,d}$. Otherwise it would be necessary to check, at each new inclusion, whether some of the boarding links already included should be removed.

Label-setting algorithms terminate after as many updating steps as there are nodes, because at each step a node label is made permanent. Node labels are made permanent in order of increasing minimum costs or travel times to the destination. At the end of the algorithm, the waiting times, specific to the shortest hyperpaths, and the set of boarding links for each diversion node, are also determined. The shortest hyperpath tree $T(d)$, towards destination d, can be described by the one link exiting from each node n, but several boarding links may exit from a diversion node.

It is worth noting that a hyperpath connecting an O-D pair may well not contain any waiting link (likewise any boarding link, or line, or alighting), and only contain pedestrian links; that is, the cost of the shortest pedestrian path cannot be improved by riding transit.

Algorithms for Uncongested Network Assignment with Hyperpaths

Uncongested network assignment models, which hold for noncongested networks, are often adopted to analyze public transportation systems in which it may be roughly assumed that costs do not depend on user flows. Moreover, noncongested network assignment algorithms constitute an element of equilibrium assignment algorithms (for congested networks) described in the sections below.

If hyperpaths are explicitly enumerated, then calculation of link flows is straightforward using the sequence of relations given in Sect. 6.2. In general, however, as already noted, explicit enumeration of hyperpaths is extremely burdensome, and UN assignment algorithms that avoid explicit enumeration are adopted, making use of the shortest hyperpath algorithms previously described.

Deterministic hyperpath-based UN assignment algorithms assume that users choose the shortest hyperpath between each O-D pair (Sect. 6.2). In this case, the shortest hyperpath tree algorithm identifies a shortest hyperpath between each O-D pair. Link flows can be calculated by assigning the demand flow for each O-D pair to the links of the shortest hyperpath, and summing over all O-D pairs. If there are multiple shortest hyperpaths for some O-D pairs, then hyperpath flows, and therefore link flows, are not uniquely defined.

The backward simultaneous algorithm, discussed for DUN assignment (all-or-nothing) in Sect. 5.3.3, can be extended to handle shortest hyperpaths. In this case as well, the algorithm requires that we know the order of node labels on shortest hyperpath trees to each destination. The operations performed at a diversion node must be modified. In this case the exit flow must be divided among all boarding links included in the hyperpath tree, proportionally to their probabilities (depending on line frequencies). The application of DUN algorithms to shortest hyperpaths yields the link flows f_{DUN} as a function of both the costs $\boldsymbol{c}$ of nonwaiting links and the line frequencies $\boldsymbol{\varphi}$. It is also possible to calculate the hyperpath total nonadditive cost $X^{\text{NA}}_{\text{DUN}}$, given by the total waiting time, which can be determined with shortest hyperpath algorithms without explicit enumeration.

Stochastic uncongested network assignment algorithms with probit choice models can easily be extended to transit networks by extending all-or-nothing algorithms as described below. The extension essentially requires multiple sampling of perceived link costs (and possibly frequencies), as in the Monte Carlo algorithm described in Sect. 5.3.3. However, very few examples of this approach have appeared in the literature. The generalization to logit hyperpath choice models without explicit hyperpath enumeration is still at the research stage (see bibliographical notes).

6.2.3 Congested Networks: Equilibrium Assignment

The *fixed demand equilibrium assignment* models described in Sect. 5.4 can easily be extended to situations of combined pre-trip/en-route path choice. It is usually assumed, as is done for nonadditive path costs, that nonadditive hyperpath costs (such as waiting times at transit stops) are not affected by congestion; that is, they do not depend on the link flows.[2] Under this hypothesis, a system of equations in terms of equilibrium *path variables*, namely costs $\boldsymbol{g}^*$ and flows $\boldsymbol{h}^*$, is obtained by combining the supply model (5.2.4) with the demand model (6.2.5):

$$\boldsymbol{g}^*_{od} = \boldsymbol{\Delta}^T_{od}\boldsymbol{c}\left(\sum_{od} \boldsymbol{\Delta}_{od}\boldsymbol{h}^*_{od}\right) \quad \forall od$$

$$\boldsymbol{h}^*_{od} = d_{od}\boldsymbol{\Omega}_{od}\boldsymbol{q}_{od}\left(-\boldsymbol{\Omega}^T_{od}\boldsymbol{g}^*_{od} - \boldsymbol{x}^{\mathrm{NA}}_{od}\right) \quad \forall od$$

An analogous formulation in terms of equilibrium *hyperpath variables*, again costs and flows, is also possible:

$$\boldsymbol{x}^*_{od} = \boldsymbol{\Lambda}^T_{od}\boldsymbol{c}\left(\sum_{od} \boldsymbol{\Lambda}_{od}\boldsymbol{y}^*_{od}\right) + \boldsymbol{x}^{\mathrm{NA}}_{od} \quad \forall od$$

$$\boldsymbol{y}^*_{od} = d_{od}\boldsymbol{q}_{od}(-\boldsymbol{x}_{od}) \quad \forall od$$

As in the case of assignment with fully pre-trip path choice behavior, an equivalent formulation in terms of link variables can be expressed by the system of equations obtained by combining the uncongested network assignment map (6.2.11) with the cost functions (6.2.2):

$$\boldsymbol{c}^* = \boldsymbol{c}(\boldsymbol{f}^*)$$

$$\boldsymbol{f}^* = \boldsymbol{f}_{\mathrm{UN}}(\boldsymbol{c}^*; \boldsymbol{d}) = \sum_{od} d_{od}\boldsymbol{\Lambda}_{od}\boldsymbol{q}_{od}\left(-\boldsymbol{\Lambda}^T_{od}\boldsymbol{c}^* + \boldsymbol{x}^{\mathrm{NA}}_{od}\right)$$

In the case of Stochastic User Equilibrium (SUE), a fixed-point model similar to model (5.4.2) in link flows is obtained:

$$\boldsymbol{f}^* = \boldsymbol{f}_{\mathrm{SUN}}\left(\boldsymbol{c}(\boldsymbol{f}^*); \boldsymbol{d}\right) = \sum_{od} d_{od}\boldsymbol{\Lambda}_{od}\boldsymbol{q}_{od}\left(-\left(\boldsymbol{\Lambda}^T_{od}\boldsymbol{c}(\boldsymbol{f}^*) + \boldsymbol{x}^{\mathrm{NA}}_{od}\right)\right) \qquad (6.2.26)$$

with

$$\boldsymbol{f}^* \in S_f$$

Stochastic user equilibrium can also be formulated with fixed-point models in terms of path or hyperpath flow variables, or link, path, or hyperpath cost variables; these formulations are not reported here for the sake of brevity. Under the assumption of flow-independent nonadditive costs, the conditions for existence and uniqueness analyzed in Sect. 5.4.1 still hold; in particular, the cost-flow functions for on-board, access, boarding, and alighting links must be, respectively, continuous

[2]In other words, it is assumed that service congestion affects the perceived cost of on-board time, but not waiting time. This precludes modeling a situation in which congestion causes some users to wait longer because the vehicles are too crowded to board.

and/or strictly increasing.[3] Extension of the results described for the case of flow-dependent nonadditive costs (such as waiting costs) is not straightforward and is not pursued here.

Deterministic User Equilibrium (DUE) assignment can be analyzed with variational inequality models. In particular, expressing the hyperpath cost functions as $\boldsymbol{x}(\boldsymbol{y}) = \boldsymbol{\Lambda}^T \boldsymbol{c}(\boldsymbol{\Lambda y}) + \boldsymbol{x}^{\mathrm{NA}}$, models similar to the variational inequality (5.4.11)–(5.4.13) are obtained:

$$\boldsymbol{x}(\boldsymbol{y}^*)^T(\boldsymbol{y} - \boldsymbol{y}^*) \geq 0 \quad \forall \boldsymbol{y} \in S_y$$

$$\boldsymbol{c}(\boldsymbol{f}^*)^T(\boldsymbol{f} - \boldsymbol{f}^*) + (\boldsymbol{x}^{\mathrm{NA}})^T(\boldsymbol{y} - \boldsymbol{y}^*) \geq 0 \quad \forall \boldsymbol{f} = \boldsymbol{\Lambda y},\ \forall \boldsymbol{y} \in S_y$$

Nonadditive hyperpath costs can be handled as described in Sect. 5.4.3. The above expression can be formulated in terms of link flows $\boldsymbol{f}^*$ and total nonadditive cost X^*:

$$\boldsymbol{c}(\boldsymbol{f}^*)^T(\boldsymbol{f} - \boldsymbol{f}^*) + (X - X^*) \geq 0 \quad \forall \boldsymbol{f} = \boldsymbol{\Lambda y},\ \forall X = (\boldsymbol{x}^{\mathrm{NA}})^T \boldsymbol{y},\ \forall \boldsymbol{y} \in S_y$$

The optimization models described in the previous sections and in the appendix for deterministic or stochastic assignment can also be easily applied in this case, within the limits of the assumptions.

Algorithms for Fixed-Demand Equilibrium Assignment with Hyperpaths

The algorithms described in Sects. 5.4.2 and 5.4.4 for fixed demand stochastic or deterministic equilibrium assignment can be extended to situations with pre-trip and en-route path choice. The main modification occurs in the calculation of the UN flows with the procedure described in the previous section. Furthermore, it is necessary to consider explicitly the nonadditive waiting time component of hyperpath costs.

In the case of stochastic equilibrium, the fixed-point problem (6.2.26) can be solved with the MSA-FA and MSA-CA algorithms already described, where each iteration involves a stochastic uncongested network assignment to the hyperpaths.

In the case of symmetric deterministic equilibrium, the optimization model (5.4.16) becomes:

$$\begin{aligned} &(\boldsymbol{f}^*, X^{\mathrm{NA}*}) = \operatorname{argmin} z(\boldsymbol{f}, X) = \int_0^{\boldsymbol{f}} \boldsymbol{c}(\boldsymbol{x})^T\, d\boldsymbol{x} + X^{\mathrm{NA}} \\ &\boldsymbol{f} = \boldsymbol{\Lambda y},\ X^{\mathrm{NA}} = (\boldsymbol{x}^{\mathrm{NA}})^T \boldsymbol{y}, \quad \boldsymbol{y} \in S_y \end{aligned} \tag{6.2.27}$$

where

[3] In the case of logit or probit path choice models, for which a path has a choice probability strictly greater than zero independent of cost, it can be demonstrated that the uniqueness of the equilibrium flows is also ensured in the case of cost functions which are not strictly monotone: $(\boldsymbol{c}(\boldsymbol{f}') - \boldsymbol{c}(\boldsymbol{f}''))^T(\boldsymbol{f}' - \boldsymbol{f}'') \geq 0,\ \forall \boldsymbol{f}', \boldsymbol{f}'' \in S_f$.

$\boldsymbol{x}^{\mathrm{NA}}$ is the vector of nonadditive hyperpath costs (i.e., waiting times), consisting of the vectors of nonadditive hyperpath costs $\boldsymbol{x}_{od}^{\mathrm{NA}}$ for each O-D pair od (these costs are assumed to be independent of congestion)

$X^{\mathrm{NA}} = (\boldsymbol{x}^{\mathrm{NA}})^T \boldsymbol{y}$ is the total nonadditive cost corresponding to the hyperpath flow vector $\boldsymbol{y}$

This model can be solved with the Frank–Wolfe algorithm described in Sect. 5.4.4, considering as problem variables the link flow vector $\boldsymbol{f}$ and the nonadditive hyperpath total cost (total waiting time) X^{NA}.

The following variables are needed to describe the algorithm.

$\nabla z(\boldsymbol{f}, X^{\mathrm{NA}}) = [\boldsymbol{c}(\boldsymbol{x}), 1]$ gradient of the function $z(\boldsymbol{f}, X^{\mathrm{NA}})$

$\boldsymbol{f}_{\mathrm{DUN}} \in S_f$ link flows resulting from DUN assignment to hyperpaths as a function of the total costs $\boldsymbol{c}$ on nonwaiting links, and the line frequencies $\boldsymbol{\varphi}$

$X_{\mathrm{DUN}}^{\mathrm{NA}}$ total nonadditive hyperpath cost resulting from the nonadditive hyperpath assignment as a function of nonwaiting link costs $\boldsymbol{c}$ and line frequencies $\boldsymbol{\varphi}$

$(\boldsymbol{f}_{\mathrm{DUN}}, X_{\mathrm{DUN}}^{\mathrm{NA}}) = \mathrm{DUN}(\boldsymbol{c}, \boldsymbol{\varphi})$ a function giving $\boldsymbol{f}_{\mathrm{DUN}}$ and $X_{\mathrm{DUN}}^{\mathrm{NA}}$ in terms of $\boldsymbol{c}$ and $\boldsymbol{\varphi}$

Given an initial solution, $(\boldsymbol{f}^0, X^{\mathrm{NA}0})$, that can easily be found with a DUN assignment algorithm using zero flow costs, $(\boldsymbol{f}^0, X^{\mathrm{NA}0}) = \mathrm{DUN}(\boldsymbol{c}(\boldsymbol{f} = \boldsymbol{0}), \boldsymbol{\varphi})$, the Frank–Wolfe algorithm for the solution of the model (6.2.27) can be described by the following system of recursive equations.

$$\boldsymbol{c}^k = \boldsymbol{c}(\boldsymbol{f}^{k-1}) \tag{6.2.28}$$

$$(\boldsymbol{f}_{\mathrm{DUN}}^k, X_{\mathrm{DUN}}^{\mathrm{NA}k}) = \mathrm{DUN}(\boldsymbol{c}^k, \boldsymbol{\varphi}) \tag{6.2.29}$$

$$\mu^k = \operatorname*{argmin}_{\mu \in [0,1]} \psi(\mu) = z\big((\boldsymbol{f}^{k-1} + \mu(\boldsymbol{f}_{\mathrm{DUN}}^k - \boldsymbol{f}^{k-1})), (X^{\mathrm{NA}k-1} + \mu(X_{\mathrm{DUN}}^{\mathrm{NA}k} - X^{\mathrm{NA}k-1}))\big) \tag{6.2.30}$$

$$\boldsymbol{f}^k = \boldsymbol{f}^{k-1} + \mu^k(\boldsymbol{f}_{\mathrm{DUN}}^k - \boldsymbol{f}^{k-1}) \tag{6.2.31}$$

$$X^{\mathrm{NA}^k} = X^{\mathrm{NA}k-1} + \mu^k(X_{\mathrm{DUN}}^{\mathrm{NA}k} - X^{\mathrm{NA}k-1}) \tag{6.2.32}$$

Equation (6.2.30) defines a one-dimensional nonlinear optimization problem in the scalar variable μ that can be solved with any of a number of algorithms, such as the bisection algorithm (see Appendix A). This algorithm uses the derivative of the objective function $\psi(\mu)$, which can be easily computed from link costs:

$$\begin{aligned} d\psi(\mu)/d\mu &= \nabla z[(\boldsymbol{f}^{k-1} + \mu(\boldsymbol{f}_{\mathrm{DUN}}^k - \boldsymbol{f}^{k-1})), (X^{\mathrm{NA}k-1} \\ &\quad + \mu(X_{\mathrm{DUN}}^{\mathrm{NA}k} - X^{\mathrm{NA}k-1}))]^T \cdot [(\boldsymbol{f}_{\mathrm{DUN}}^k - \boldsymbol{f}^{k-1}), (X_{\mathrm{DUN}}^{\mathrm{NA}k} - X^{\mathrm{NA}k-1})] \\ &= [\boldsymbol{c}(\boldsymbol{f}^{k-1} + \mu(\boldsymbol{f}_{\mathrm{DUN}}^k - \boldsymbol{f}^{k-1})), 1]^T \\ &\quad [(\boldsymbol{f}_{\mathrm{DUN}}^k - \boldsymbol{f}^{k-1}), (X_{\mathrm{DUN}}^{\mathrm{NA}k} - X^{\mathrm{NA}k-1})] \end{aligned}$$

$$= \boldsymbol{c}\big(\boldsymbol{f}^{k-1} + \mu(\boldsymbol{f}^{k}_{\mathrm{DUN}} - \boldsymbol{f}^{k-1})\big)^{T}\big(\boldsymbol{f}^{k}_{\mathrm{DUN}} - \boldsymbol{f}^{k-1}\big) + \big(X^{\mathrm{NA}\,k}_{\mathrm{DUN}} - X^{\mathrm{NA}\,k-1}\big)$$

Note that the algorithm does not require calculation of the function $\psi(\mu)$.

If the cost functions $\boldsymbol{c} = \boldsymbol{c}(\boldsymbol{f})$ are continuous with continuous first partial derivatives and with positive definite symmetric Jacobian, the term $\int_0^f \boldsymbol{c}(\boldsymbol{v})^T d\boldsymbol{v}$ is a strictly convex function of $\boldsymbol{f}$. In this case the function $z(\boldsymbol{f}, X^{\mathrm{NA}})$ has one and only one minimum point $(\boldsymbol{f}^*, X^{\mathrm{NA}*})$, as already seen above. Furthermore, the function $\psi(\mu)$ has one and only one minimum point. In this case, results of optimization theory can be invoked to demonstrate that $\boldsymbol{f}^k$, the sequence of (feasible) link flow vectors generated by the Frank–Wolfe algorithm, converges to the deterministic equilibrium link flow vector, as do the values $X^{\mathrm{NA},k}$.

Deterministic equilibria with nonseparable cost functions can be analyzed using variational inequality models that are expressed in terms of link flows $\boldsymbol{f}^*$ and total nonadditive cost $X^{\mathrm{NA}*}$. This problem can be solved with the diagonalization algorithm.

6.3 Equilibrium Assignment with Variable Demand

In variable demand assignment models, the O-D demand flows are assumed to depend on transportation costs. These models simulate supply–demand interactions when path cost variations due to variations in congested link costs[4] influence user behavior other than path choice (such as the decision to travel, to what destination, by what mode, etc.). The dependence of demand on cost is expressed by the demand models described in Chap. 4.

If demand models are based on random utility theory, the demand flow for each O-D pair generally depends on the values of the (systematic) utilities associated with the paths available for the various O-D pairs, through the EMPU of path choice. This can be seen as an "average" over the systematic utilities (i.e., costs) of the available paths. This is described in Sect. 3.4 and in Sect. 4.2 on the general structure of demand models.

For uncongested networks, variable demand assignment is not meaningful, because path costs, EMPUs, and thus demand flows are independent of link flows. Link and path flows can then be obtained using the uncongested network assignment models described in Sect. 5.3. For congested networks, by contrast, costs depend on flows, and a further mutual dependence between flows and costs is introduced through the demand function.

For variable demand equilibrium assignment, it is useful to distinguish between single- and multimode problems. In *single-mode assignment*, dealt with in

[4]Fixed-demand assignment models also occur when demand flows are assumed to depend on flow-independent path cost attributes, such as free-flow times or generalized costs.

Sect. 6.3.1, there is one mode for which link costs depend on flows, and either the demand elasticity does not depend on mode split at all, or link costs for all other modes are not congestion-dependent. In the latter case, level-of-service attributes of the uncongested modes are known before the solution of the assignment model and play a role similar to fixed parameters of the congested mode's demand model. Once the congested mode equilibrium assignment has been solved and the cost attributes of this mode have been determined, demand for the other (uncongested) modes can be obtained and assigned using uncongested network assignment models, one mode at a time.

In *multimode assignment*, dealt with in Sect. 6.3.2, there is more than one mode with link costs that depend on flows (congested modes). In this case, the cost attributes of congested modes cannot be known before the solution of the assignment model, and the equilibrium assignment problem must be solved simultaneously (at least for the congested modes). Note that the various congested modes may have separate supply (network) and path choice models.

To clarify the difference between the two types of variable demand assignment, consider a situation involving the choice between two modes, car and bus. If bus travel times are independent of the link flows, its level-of-service attributes are independent of congestion. They can be calculated through the network model and then used as fixed parameters of the mode choice model that computes demand flows for the car mode. The known costs of the bus mode and the cost functions for the car mode allow the specification of a single-mode congested assignment problem with variable demand to determine car mode flows and costs. When this model is solved and the car mode equilibrium attributes are found, the bus mode demand flows will be determined and an uncongested network assignment can be performed for the buses. On the other hand, if the costs of both modes depend on the network flows, it is necessary to assign the demand of both modes at the same time to find the congested cost pattern for each of them. These costs have to be consistent with the mode choice, path choices, and the network flows of both modes.

6.3.1 Single-Mode Assignment

The *demand function* for travel between O-D pair *od* by mode *m* during time band *h* (not explicitly indicated in the following) can be expressed as

$$d_{od} = d_{od}(\boldsymbol{s}) \quad \forall od$$

or in matrix terms:

$$\boldsymbol{d} = \boldsymbol{d}(\boldsymbol{s})$$

where

$\boldsymbol{d}$ is the demand flow vector, with element d_{od} for each O-D pair *od*
$\boldsymbol{s}$ is the path choice EMPU vector, with element s_{od} for each O-D pair *od*

In general, the demand function simulates the dependence between demand flows and EMPU, and will vary depending on the particular choice dimensions that are considered variable with respect to congestion costs. For example, if demand is variable with respect to destination choice, the demand flow d_{od} depends only on the elements of the vector $\boldsymbol{s}$ for O-D pairs having the same origin zone o, $d_{od} = d_{od}(s_{od1}, \ldots, s_{odn}, \ldots)$. If the demand flow d_{od} of O-D pair od depends only on the EMPU of the same O-D pair, we have the special case of separable demand functions $d_{od} = d_{od}(s_{od})$; this may arise in the case of variable trip frequency or trip production models.

The EMPU depends in turn on the values of the path systematic utility through relation (5.2.8) given in Sect. 5.2:

$$s_{od} = s_{od}(\boldsymbol{V}_{od}) \quad \forall od$$

Note that the EMPU is defined as a utility and consistently measured. Thus, the EMPU of path choice models is negative, because the systematic utility of each path is generally negative, being the additive inverse of the corresponding systematic cost.

From the systematic utility expression (5.2.5) it follows that:

$$d_{od} = d_{od}\big(\boldsymbol{s}(\boldsymbol{V})\big) = d_{od}\big(\boldsymbol{s}(-\boldsymbol{g})\big) \quad \forall od \tag{6.3.1}$$

or in matrix notation:

$$\boldsymbol{d} = \boldsymbol{d}\big(\boldsymbol{s}(\boldsymbol{V})\big) = \boldsymbol{d}\big(\boldsymbol{s}(-\boldsymbol{g})\big)$$

If destination choice, for example, is simulated with a logit model having parameter θ_1, and path choice is simulated with a logit model having parameter θ_2, an elementary specification of the previous expression could be:

$$d_{od} = d_{o.} \exp\big((\beta_1 A_d + \beta_2 s_{od})/\theta_1\big) \Big/ \sum_j \exp\big((\beta_1 A_j + \beta_2 s_{oj})/\theta_1\big) \quad \forall od$$

$$s_{od} = \theta_2 \ln\Bigg(\sum_{k \in K_{od}} \exp(-g_k/\theta_2)\Bigg) \quad \forall od$$

where

$d_{o.}$ is the total flow leaving from zone o, assumed constant
A_d is the attraction attribute of the destination zone d
β_1, β_2 are conversion coefficients in the systematic utility function

In the variable demand assignment models described below, it is assumed that the demand flow d_{od} for each O-D pair od is nonnegative and bounded above by a positive value; that is, $d_{od} \in [0, d_{od,\max}]$. It is therefore possible to define the set of feasible demand flow vectors as

$$S_d = \big\{\boldsymbol{d} : d_{od} \in [0, d_{od,\max}]\ \forall od\big\}$$

Under this hypothesis, the sets of feasible path and link flows, S_h and S_f, respectively, described in Sect. 5.2, are compact and convex (and nonempty if the network is connected), as in the case of fixed demand.

Three approaches can be followed, as described below.

Internal Approach The general variable demand model can be written:

$$\boldsymbol{h}_{od} = d_{od}\big(s(-\boldsymbol{g})\big)\boldsymbol{p}_{od}(-\boldsymbol{g}_{od}) \quad \forall od \tag{6.3.2}$$

in matrix notation:

$$\boldsymbol{h} = \boldsymbol{P}(-\boldsymbol{g})\boldsymbol{d}\big(s(-\boldsymbol{g})\big)$$

Note that expression (6.3.2) is the equivalent of expression (5.2.7) that was derived for the case of fixed demand. On the other hand, the supply model remains unchanged, as expressed by the relation (5.2.4).

The variable demand single-mode equilibrium approach assumes that the state of the system can be represented by a path flow configuration $\boldsymbol{h}^*$ that is mutually consistent with the corresponding path costs $\boldsymbol{g}^*$, as defined by the supply model (5.2.4) and the demand model (6.3.2):

$$\boldsymbol{g}^* = \boldsymbol{\Delta}^T \boldsymbol{c}(\boldsymbol{\Delta h}^*) + \boldsymbol{g}^{\text{NA}}$$
$$\boldsymbol{h}^* = \boldsymbol{P}(-\boldsymbol{g}^*)\boldsymbol{d}\big(s(-\boldsymbol{g}^*)\big)$$

The corresponding equilibrium demand flows $\boldsymbol{d}^*$ are given by (6.3.1). An equivalent formulation of the variable demand single-mode equilibrium assignment model can be developed in terms of link variables. In this case, the system of equations in terms of equilibrium link flows $\boldsymbol{f}^*$ is obtained by combining the cost functions (5.2.2) with the equation obtained from the combination of the uncongested network assignment map (5.3.2), the demand function (6.3.1), and the path cost expression (5.2.1):

$$\boldsymbol{c}^* = \boldsymbol{c}(\boldsymbol{f}^*)$$
$$\boldsymbol{f}^* = \boldsymbol{f}_{\text{UN-EL}}(\boldsymbol{c}^*; \boldsymbol{d}) = \boldsymbol{\Delta P}(-\boldsymbol{\Delta}^T\boldsymbol{c}^* - \boldsymbol{g}^{\text{NA}})\boldsymbol{d}\big(s(-\boldsymbol{\Delta}^T\boldsymbol{c}^* - \boldsymbol{g}^{\text{NA}})\big)$$

The analysis of variable demand equilibrium assignment can easily be carried out for the internal approach through direct extension of the fixed demand equilibrium assignment models described in Sect. 5.4, distinguishing the cases of stochastic and deterministic equilibrium.

External Approach The circular dependence between demand flows and costs can also be expressed externally to the equilibrium between (link and path) flows and costs. At the inner level, for a given vector of demand flows, (fixed demand) equilibrium link flows and costs are defined by the path choice model and by the cost functions. At the outer level, the equilibrium between the costs resulting from the (fixed demand) equilibrium assignment and the demand flows defined by the demand functions is defined. Let:

$\boldsymbol{f}_{\text{UE-FIX}} = \boldsymbol{f}_{\text{UE-FIX}}(\boldsymbol{d})$ be the implicit correspondence between the fixed demand equilibrium link flows $\boldsymbol{f}_{\text{UE-FIX}}$ and the demand flows $\boldsymbol{d}$. This correspondence is defined by the solution of one of the models described in Sect. 5.4. It is a function (one-to-one correspondence) if equilibrium flows are unique.

External variable demand equilibrium assignment can therefore be formulated with a system of nonlinear equations:

$$\boldsymbol{d}^* = \boldsymbol{d}\left(\boldsymbol{s}\left(-\boldsymbol{\Delta}^T \boldsymbol{c}(\boldsymbol{f}^*)\right)\right)$$
$$\boldsymbol{f}^* = \boldsymbol{f}_{\text{UE-FIX}}(\boldsymbol{d}^*)$$

Combining the two previous equations results in a fixed-point problem (with an implicitly defined function) with respect to the demand flows $\boldsymbol{d}^*$ or link flows $\boldsymbol{f}^*$. Formulations with respect to link cost or EMPU are also possible. The external approach can be adopted to define solution procedures, but it is difficult to analyze theoretically.

Hypernetwork Approach It is also possible to adapt fixed demand assignment models to deal with variable demand by expanding the network model with appropriately defined links into so-called hypernetworks. (This hypernetwork approach is not related to the hyperpath approaches discussed above.) Behavior in nonpath choice dimensions can thus be simulated as can path choice in a modified network. This approach is difficult to generalize, and can be used in some cases only. The expanded network model also contains fictitious links that simulate frequency, destination, mode, or other travel choice behavior in the same way that path choice behavior is simulated in conventional networks.

This approach can be applied only to some demand functions, and is briefly described below with reference to deterministic equilibrium. For the sake of simplicity, frequency is assumed to be the only variable demand dimension. The hypernetwork approach to model elasticity of other demand components is similar. For each O-D pair *od*, a fictitious path consisting of a single link is added to the network. To satisfy the demand conservation constraint, a flow equal to the excess demand flow, $h_k^0 = d_{od,\max} - d_{od}$, is assigned to this path; this flow represents the potential demand flow that is not traveling (Fig. 6.8). Let:

$\boldsymbol{d}_{\max}$ be the maximum demand flow vector
$\boldsymbol{h}^0 = \boldsymbol{d}_{\max} - \boldsymbol{d}$ be the vector of excess path flows
$\boldsymbol{f}^0 = \boldsymbol{f}^0$ be the vector of excess link flows

A fictitious cost function $c_{od}^0 = c_{od}^0(\boldsymbol{f}^0)$ can be associated with each such new link. This function is obtained from the inverse demand function that relates minimum cost to demand flows as discussed in Sect. 6.3.1.2:

$$Z_{od}(d) = Z_{od}(d_{\max} - d_{\max} + d) = Z_{od}(d_{\max} - h^0) = g_{od}^0(\boldsymbol{h}^0) = c_{od}^0(\boldsymbol{f}^0)$$

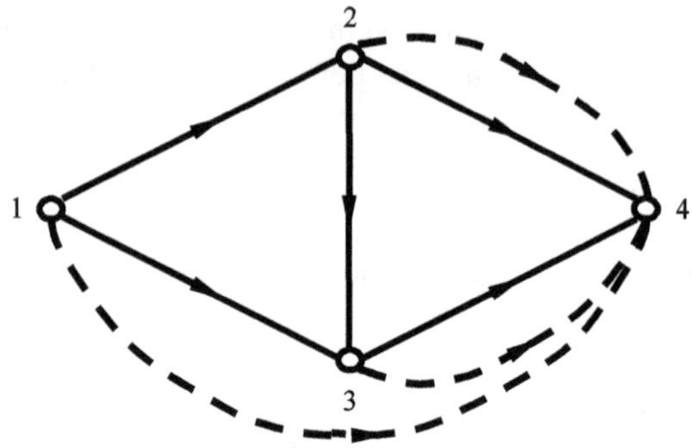

Fig. 6.8 Hypernetwork approach

It can easily be verified that the variational inequality model (5.4.11) for fixed demand deterministic equilibrium applied to this network is equivalent to the variational inequality model (6.3.6) for variable demand deterministic equilibrium applied to the original network. Thus, the variable demand DUE problem can be solved by applying a fixed demand DUE algorithm to the expanded network.

6.3.1.1 Models for Stochastic User Equilibrium

The fixed demand SUE path flow fixed-point model (5.4.1) can easily be extended to variable demand situations by combining the supply model (5.2.4) and the demand model (6.3.2):

$$\boldsymbol{h}^* = \boldsymbol{P}\big(-\boldsymbol{\Delta}^T \boldsymbol{c}(\boldsymbol{\Delta h}^*) - \boldsymbol{g}^{\mathrm{NA}}\big)\boldsymbol{d}\big(\boldsymbol{s}\big(-\boldsymbol{\Delta}^T \boldsymbol{c}(\boldsymbol{\Delta h}^*) - \boldsymbol{g}^{\mathrm{NA}}\big)\big) \tag{6.3.3}$$

with

$$\boldsymbol{h}^* \in S_h$$

The equivalent fixed demand SUE link flow fixed-point model (5.4.2) for the fixed demand SUE problem can also be easily extended to variable demand:

$$\boldsymbol{f}^* = \sum_{od} d_{od}\big(\boldsymbol{s}\big(-\boldsymbol{\Delta}^T \boldsymbol{c}(\boldsymbol{f}^*) - \boldsymbol{g}^{\mathrm{NA}}\big)\big)\boldsymbol{\Delta}_{od}\boldsymbol{p}_{od}\big(-\boldsymbol{\Delta}_{od}^T \boldsymbol{c}(\boldsymbol{f}^*) - \boldsymbol{g}_{od}^{\mathrm{NA}}\big) \tag{6.3.4}$$

with

$$\boldsymbol{f}^* \in S_f$$

Equilibrium link costs are given by $\boldsymbol{c}^* = \boldsymbol{c}(\boldsymbol{f}^*)$, and therefore the corresponding demand flows are given by $d_{od}^* = d_{od}(\boldsymbol{s}(-\boldsymbol{\Delta}^T \boldsymbol{c}^* - \boldsymbol{g}^{\mathrm{NA}}))$.

The analysis of the existence and uniqueness of solutions is a straightforward extension of the results given in Sect. 5.4.1. It requires explicit assumptions on the demand functions that are sufficient to ensure continuity and monotonicity of the following stochastic uncongested network assignment function (with variable demand).

$$
\begin{aligned}
f_{\text{SUN-EL}}(c) &= f_{\text{SUN}}\big(c^*; d\big(s\big(-\Delta^T c - g^{\text{NA}}\big)\big)\big) \\
&= \sum_{od} d_{od}\big(s\big(-\Delta^T c - g^{\text{NA}}\big)\big)\Delta_{od}\, p_{od}\big(-\Delta_{od}^T c - g_{od}^{\text{NA}}\big)
\end{aligned}
$$

Below, existence and uniqueness are analyzed explicitly only for equilibrium link flows. These properties of the equilibrium link flows also ensure the existence and uniqueness of the corresponding equilibrium link costs $c^* = c(f^*)$, path costs and flows g^* and h^*, and demand flows d_{od}^*.

Existence of Variable Demand Stochastic User Equilibrium Assuming that each O-D pair is connected and that demand flows are bounded, the fixed-point model formulated in terms of link flows (6.3.4) has at least one solution if the cost functions $c = c(f)$ and the component functions of $f_{\text{SUN-EL}}$ are all continuous. These component functions are the path choice probability functions $p_{od} = p_{od}(V_{od})$, EMPU functions $s_{od} = s_{od}(V_{od})$, and demand functions $d_{od} = d_{od}(s)$.

The proof is similar to that in Sect. 5.4.1 for fixed demand.

Monotonicity of the Variable Demand Stochastic Uncongested Network Assignment Function If path choice models are defined by functions that are monotone nondecreasing with respect to the systematic utilities, as is the case of probabilistic additive models (with $|\Sigma| \neq 0$; see Sect. 3.4), and if demand functions are nonnegative, bounded, and nondecreasing with respect to the EMPU:

$$
\big[d(s') - d(s'')\big]^T (s' - s'') \geq 0 \quad \forall s', s''
$$

then the variable demand stochastic uncongested network assignment function is monotone nonincreasing with respect to link costs. Thus if the cost of a set of links increases, the corresponding flows do not increase. This property is expressed formally as

$$
\big(f_{\text{SUN-EL}}(c') - f_{\text{SUN-EL}}(c'')\big)^T (c' - c'') \leq 0 \quad \forall c', c''
$$

Under these assumptions, given the two systematic utility vectors V'_{od} and V''_{od}, corresponding to the paths that connect O-D pair *od*, the following relations involving the corresponding path choice probabilities and the EMPU hold (see Sect. 3.4).

$$
\begin{aligned}
p_{od}(V'_{od})^T (V'_{od} - V''_{od}) &\geq s_{od}(V'_{od}) - s_{od}(V''_{od}) \\
s_{od}(V'_{od}) - s_{od}(V''_{od}) &\geq p_{od}(V''_{od})^T (V'_{od} - V''_{od})
\end{aligned}
$$

Letting $s'_{od} = s_{od}(V'_{od})$ and $s''_{od} = s_{od}(V''_{od})$, multiplying the first relation by $d_{od}(s') \geq 0$ and the second by $d_{od}(s'') \geq 0$ gives:

$$
d_{od}(s')\, p_{od}(V'_{od})^T (V'_{od} - V''_{od}) \geq d_{od}(s')\big(s_{od}(V'_{od}) - s_{od}(V''_{od})\big)
$$

$$d_{od}(s'')\big(s_{od}(V'_{od}) - s_{od}(V''_{od})\big) \geq d_{od}(s'')p_{od}(V''_{od})^T(V'_{od} - V''_{od})$$

Hence, summing over all O-D pairs:

$$\sum_{od} d_{od}(s')p_{od}(V'_{od})^T(V'_{od} - V''_{od}) \geq \sum_{od} d_{od}(s')\big(s_{od}(V'_{od}) - s_{od}(V''_{od})\big)$$

$$\sum_{od} d_{od}(s'')\big(s_{od}(V'_{od}) - s_{od}(V''_{od})\big) \geq \sum_{od} d_{od}(s'')p_{od}(V''_{od})^T(V'_{od} - V''_{od})$$

Furthermore, from the monotonicity of the demand functions, it follows that

$$\sum_{od} d_{od}(s')\big(s_{od}(V'_{od}) - s_{od}(V''_{od})\big) \geq \sum_{od} d_{od}(s'')\big(s_{od}(V'_{od}) - s_{od}(V''_{od})\big)$$

Therefore, the following expression is obtained.

$$\sum_{od} d_{od}(s')p_{od}(V'_{od})^T(V'_{od} - V''_{od}) \geq \sum_{od} d_{od}(s'')p_{od}(V''_{od})^T(V'_{od} - V''_{od})$$

from which, letting $h'_{od} = d_{od}(s')p'_{od}$ and $h''_{od} = d_{od}(s'')p''_{od}$, we deduce:

$$\sum_{od} (h'_{od} - h''_{od})^T(V'_{od} - V''_{od}) \geq 0$$

Given two different link cost vectors c' and c'', let

$$g'_{od} = \Delta^T_{od}c' + g^{NA}_{od} \qquad V'_{od} = -g'_{od}$$

$$g''_{od} = \Delta^T_{od}c'' + g^{NA}_{od} \qquad V''_{od} = -g''_{od}$$

Therefore, analogous to the exposition in Sect. 5.3.1, with

$$f' = \sum_{od} \Delta_{od}h'_{od} \qquad f'' = \sum_{od} \Delta_{od}h''_{od}$$

we finally obtain: $(f' - f'')^T(c' - c'') \leq 0$.

Note that, under these assumptions, the Jacobian $Jac[f_{\text{SUN-EL}}(c)]$ is symmetric negative semidefinite because the Jacobian $Jac[p_{od}(V_{od})]$ is symmetric positive semidefinite (see Sect. 3.4).

Uniqueness of Variable Demand Stochastic User Equilibrium The link flow fixed-point model (6.3.4) has at most one solution if the link cost functions $c = c(f)$ are strictly increasing with respect to the feasible link flows:

$$\big[c(f') - c(f'')\big]^T(f' - f'') > 0 \quad \forall f' \neq f'' \in S_f;$$

if demand functions are nonnegative, bounded, and nondecreasing with respect to the EMPU:

$$\big[d(s') - d(s'')\big]^T(s' - s'') \geq 0 \quad \forall s', s''$$

and if path choice models are additive, in the sense defined in Sect. 5.3.1, and expressed by continuous functions $\boldsymbol{p}_{od} = \boldsymbol{p}_{od}(\boldsymbol{V}_{od})$ with continuous first partial derivatives.

The proof is similar to that provided in Sect. 5.4.1 for fixed demand. Under the above assumptions, the variable demand SUN function $\boldsymbol{f}_{\text{SUN-EL}}(\boldsymbol{c})$ is monotone nonincreasing with respect to the link costs.

The considerations expressed in Sect. 5.4.1 on the existence and uniqueness of the solutions, and on the continuity and monotonicity of the cost functions, can be directly extended to variable demand models. As for the demand functions, their monotonicity implies that variations in path cost induce opposite variations in EMPUs and therefore in demand flows. In other words, the increase in a link cost, and therefore in the cost of the paths including it, cannot induce an increase in the demand flows between the O-D pairs connected by these paths. This property is always guaranteed if the demand functions are defined through probabilistic choice models, which are invariant with respect to the EMPU of path choice (deterministic demand models also satisfy the monotonicity requirement).

6.3.1.2 Models for Deterministic User Equilibrium

When path choice behavior is simulated with a deterministic model, the EMPU is given, as stated in Sect. 3.4, by the maximum systematic utility, or the negative of the minimum path cost:

$$s_{od} = s_{od}(\boldsymbol{V}_{od}) = \max_{k \in K_{od}} (V_{od,k}) = -\min_{k \in K_{od}} (g_{od,k}) = -Z_{od} \quad \forall od$$

where

$Z_{od} = -s_{od}$ is the minimum cost of the paths connecting O-D pair od
$\boldsymbol{Z} = -\boldsymbol{s}$ is the vector of the minimum path costs between all O-D pairs

The demand functions $d(\cdot)$ are, in the case of deterministic assignment, usually expressed in terms of minimum cost, that is, the negative of the EMPU:

$$d_{od} = d_{od}(-\boldsymbol{s}) = d_{od}(\boldsymbol{Z}) \quad \forall \boldsymbol{Z}\ \forall od$$

or equivalently

$$\boldsymbol{d} = \boldsymbol{d}(\boldsymbol{Z}) \quad \forall \boldsymbol{Z} \tag{6.3.5a}$$

As an example, consider the case of a logit model that simulates destination choice, analogous to that described previously, whereas path choice is simulated with a deterministic model. Expression (6.3.5a) becomes:

$$d_{od} = d_o \exp\big((\beta_1 A_d - \beta_2 Z_{od})/\theta_1\big) \Big/ \sum_j \exp\big((\beta_1 A_j - \beta_2 Z_{oj})/\theta_1\big) \quad \forall od$$

where

d_o is the total flow leaving zone o, assumed constant
A_d is the attraction attribute of the destination zone d
θ_1 is the destination choice model logit parameter
β_1, β_2 are the systematic utility conversion coefficients

The indirect formulation of fixed demand deterministic equilibrium through variational inequality models, described in Sect. 5.4.3, can be extended to variable demand. For this purpose, it is necessary to assume that the demand functions (6.3.5a) are invertible[5]; that is, it is possible to define the *inverse demand function*[6] giving, for each demand flow vector $\boldsymbol{d}$, the corresponding vector of minimum path costs $\boldsymbol{Z}$. This is the vector of minimum path costs that, through the demand function, would generate the demand vector $\boldsymbol{d}$:

$$\boldsymbol{Z} = \boldsymbol{Z}(\boldsymbol{d}) \quad \forall \boldsymbol{d} \in S_d \tag{6.3.5b}$$

The inverse demand function (6.3.5b) has the same properties of continuity and monotonicity as the demand function (6.3.5a). In particular, it is strictly decreasing if (and only if) the demand function (6.3.5a) is strictly decreasing. Thus, for an increase in demand flows, the inverse demand function associates a decrease in costs. This property is guaranteed if the demand function is defined by additive probabilistic choice models specified in terms of minimum path costs (or by deterministic models).

The variational inequality formulation of variable demand deterministic equilibrium assignment can be achieved by extending the path flow model (5.4.11) described in Sect. 5.4.3 for fixed demand. In the case of variable demand, this model becomes (excluding nonadditive path costs for simplicity of notation):

$$\boldsymbol{g}(\boldsymbol{h}^*)^T(\boldsymbol{h} - \boldsymbol{h}^*) - \boldsymbol{Z}(\boldsymbol{d}^*)^T(\boldsymbol{d} - \boldsymbol{d}^*) \geq 0 \quad \forall \boldsymbol{h} \in S_h \ \forall \boldsymbol{d} \in S_d \tag{6.3.6}$$

In fact, applying condition (3.4.11a) on the deterministic choice probabilities $\boldsymbol{p}_{\text{DET},od}$ (as introduced in Sect. 3.4) to each O-D pair od yields:

$$\boldsymbol{V}_{od}^T \boldsymbol{p}_{\text{DET},od} = \max(\boldsymbol{V}_{od}) \quad \forall od$$

Given path costs $\boldsymbol{g}^*_{od}$, let $Z^*_{od} = \min(\boldsymbol{g}^*_{od})$ be the minimum path cost for each O-D pair od. Assuming $\boldsymbol{V}^*_{od} = -\boldsymbol{g}^*_{od}$ yields $\max(\boldsymbol{V}^*_{od}) = -Z^*_{od}$. Furthermore, let d^*_{od} be the demand flow corresponding to minimum cost Z^*_{od}; that is, $\boldsymbol{Z}^* = \boldsymbol{Z}(\boldsymbol{d}^*)$ is consistent with the inverse demand function. Multiplying the above equation by the

[5]A strictly monotone continuous function is always invertible, and an invertible and continuous function is strictly monotone (see Appendix A).

[6]It should be noted that it is usually very difficult to get closed form expressions for the inverse demand functions $\boldsymbol{Z} = \boldsymbol{Z}(\boldsymbol{d})$, even in the case of simple demand models. This characteristic considerably limits the application to variable demand deterministic equilibrium of variational inequality models (but not of fixed-point models). In the case of logit-type demand models, an equivalent optimization model can be adopted, as shown below.

nonnegative demand flow $d^*_{od} \geq 0\ \forall od$ yields:

$$\left(\boldsymbol{g}^*_{od}\right)^T \boldsymbol{h}_{\mathrm{DET},od} = Z^*_{od} d^*_{od} \quad \forall od \tag{a}$$

because

$$\boldsymbol{h}_{\mathrm{DET},od} = d^*_{od} \boldsymbol{p}_{\mathrm{DET},od} \quad \forall od$$

Generally, the following condition also holds (see Sect. 3.4).

$$\boldsymbol{V}^T_{od} \boldsymbol{p}_{od} \leq \max(\boldsymbol{V}_{od}) \quad \forall \boldsymbol{p}_{od} : \boldsymbol{p}_{od} \geq \boldsymbol{0}, \qquad \boldsymbol{1}^T \boldsymbol{p}_{od} = 1 \quad \forall od$$

Given the path costs $\boldsymbol{g}^*_{od}$, with $\boldsymbol{V}^*_{od} = -\boldsymbol{g}^*_{od}$ and $\max(\boldsymbol{V}^*_{od}) = -Z^*_{od}$, multiplying the above equation by any feasible demand flow $d_{od} \geq 0\ \forall od$ yields:

$$\left(\boldsymbol{g}^*_{od}\right)^T \boldsymbol{h}_{od} \leq Z^*_{od} d_{od} \quad \forall \boldsymbol{h}_{od} : \boldsymbol{h}_{od} \geq \boldsymbol{0}, \qquad \boldsymbol{1}^T \boldsymbol{h}_{od} = d_{od} \quad \forall d_{od} \geq 0\ \forall od$$

thus

$$\left(\boldsymbol{g}^*_{od}\right)^T \boldsymbol{g}_{od} \leq Z^*_{od} d_{od} \quad \forall \boldsymbol{h}_{od} : \boldsymbol{h} \in S_h,\ \forall d_{od} : \boldsymbol{d} \in S_d\ \forall od \tag{b}$$

because $\boldsymbol{h}_{od} = d_{od} \boldsymbol{p}_{od}\ \forall od$.

Subtracting (a) from (b) yields:

$$\left(\boldsymbol{g}^*_{od}\right)^T (\boldsymbol{g}_{od} - \boldsymbol{g}_{\mathrm{DET},od}) \leq Z^*_{od}\left(d_{od} - d^*_{od}\right) \quad \forall \boldsymbol{h}_{od} : \boldsymbol{h} \in S_h,\ \forall d_{od} : \boldsymbol{d} \in S_d\ \forall od$$

Summing up the above equation for all O-D pairs, and letting $\boldsymbol{Z}^* = \boldsymbol{Z}(\boldsymbol{d}^*)$, a deterministic demand model with variable demand is obtained:

$$(\boldsymbol{g}^*)^T (\boldsymbol{h} - \boldsymbol{h}_{\mathrm{DET}}) \leq \boldsymbol{Z}(\boldsymbol{d}^*)^T (\boldsymbol{d} - \boldsymbol{d}^*) \quad \forall \boldsymbol{h} \in S_h,\ \forall \boldsymbol{d} \in S_d$$

Combining the above demand model (b) with the supply model (5.2.4), say $\boldsymbol{g}(\boldsymbol{h}^*) = \boldsymbol{\Delta}^T \boldsymbol{c}(\boldsymbol{\Delta h}^*) + \boldsymbol{g}^{\mathrm{NA}}$, relation (6.3.6) is obtained.

Expression (6.3.6) can easily be reformulated in terms of link flows, extending the model (5.4.12) described in Sect. 5.4.3. Expressing equilibrium path costs in terms of link costs according to the supply model, it follows, as in (5.2.4), that

$$\boldsymbol{c}(\boldsymbol{f}^*)^T (\boldsymbol{f} - \boldsymbol{f}^*) - \boldsymbol{Z}(\boldsymbol{d}^*)^T (\boldsymbol{d} - \boldsymbol{d}^*) \geq 0 \quad \forall \boldsymbol{f} \in S_f\ \forall \boldsymbol{d} \in S_d \tag{6.3.7}$$

The existence of (link or path) flows and costs and the uniqueness of link flows and costs as well as of the demand flows for variable demand deterministic user equilibrium are guaranteed respectively by the continuity and monotonicity of the cost functions and of the (inverse) demand functions.[7]

---

[7]To this end, note that both models (6.3.6) and (6.3.7) can be expressed as a variational inequality defined for a suitable function $\varphi(\boldsymbol{x})$, with vector $\boldsymbol{x}$ drawn from a suitable set $S : \varphi(\boldsymbol{x}^*)^T (\boldsymbol{x} - \boldsymbol{x}^*) \geq 0,\ \forall \boldsymbol{x} \in S$. In particular, in the model (6.3.6), the vector $\boldsymbol{x}$ is defined by the path and demand flow vectors $\boldsymbol{h}$ and $\boldsymbol{d}$; the set S is defined by the product of the sets of feasible path and demand flows S_h and S_d; and the function $\varphi(\boldsymbol{x})$ is defined by the path cost functions and the negative of the inverse demand function $\boldsymbol{g}(\boldsymbol{h})$ and $\boldsymbol{Z}(\boldsymbol{d})$. The same holds for the model (5.4.7) expressed in terms of link flows and demand flows.

Existence of Variable Demand Deterministic User Equilibrium Variational inequalities (6.2.6) and (6.2.7) have at least one solution if the cost functions, defined over the nonempty, compact, and convex set of feasible path or link flows, and the inverse demand functions, defined over the nonempty, closed, and bounded interval of demand values, are both continuous.

The proof is similar to that described for fixed demand in Sect. 5.4.1.

Uniqueness of Variable Demand Deterministic User Equilibrium Link Flows The variational inequality (6.3.7) expressed in terms of link flows has at most one solution if the link cost functions $\boldsymbol{c} = \boldsymbol{c}(\boldsymbol{f})$ are strictly increasing with respect to link flows:

$$\left[\boldsymbol{c}(\boldsymbol{f}') - \boldsymbol{c}(\boldsymbol{f}'')\right]^T (\boldsymbol{f}' - \boldsymbol{f}'') > 0 \quad \forall \boldsymbol{f}' \neq \boldsymbol{f}'' \in S_f$$

and the inverse demand functions, $\boldsymbol{Z} = \boldsymbol{Z}(\boldsymbol{d})$, are strictly decreasing[8] with respect to the demand flows (i.e., the demand functions are strictly decreasing with respect to the minimum cost):

$$\left[\boldsymbol{Z}(\boldsymbol{d}') - \boldsymbol{Z}(\boldsymbol{d}'')\right](\boldsymbol{d}' - \boldsymbol{d}'') < 0 \quad \forall \boldsymbol{d}' \neq \boldsymbol{d}'' \in S_d$$

The proof, parallel to that described for fixed demand in Sect. 5.4.3, is performed by a *reductio ad absurdum*. If there existed two different equilibrium link flow vectors $\boldsymbol{f}_1^* \neq \boldsymbol{f}_2^* \in S_f$, corresponding to two feasible demand flow vectors $\boldsymbol{d}_1^*, \boldsymbol{d}_2^* \in S_d$ (not necessarily different), they both would satisfy (6.3.7) and therefore, with $\boldsymbol{f} = \boldsymbol{f}_2^* \in S_f$ and $\boldsymbol{d} = \boldsymbol{d}_2^* \in S_d$, we would have:

$$\boldsymbol{c}(\boldsymbol{f}_1^*)^T (\boldsymbol{f}_2^* - \boldsymbol{f}_1^*) - \boldsymbol{Z}(\boldsymbol{d}_1^*)^T (\boldsymbol{d}_2^* - \boldsymbol{d}_1^*) \geq 0$$

Furthermore, $\boldsymbol{f}_2^*$ and $\boldsymbol{d}_2^*$ would also respect (6.3.7) and therefore, with $\boldsymbol{f} = \boldsymbol{f}_1^* \in S_f$ and $\boldsymbol{d} = \boldsymbol{d}_1^* \in S_d$, we would have:

$$\boldsymbol{c}(\boldsymbol{f}_2^*)^T (\boldsymbol{f}_1^* - \boldsymbol{f}_2^*) - \boldsymbol{Z}(\boldsymbol{d}_2^*)^T (\boldsymbol{d}_1^* - \boldsymbol{d}_2^*) \geq 0$$

Adding the above two relations gives:

$$\begin{aligned} &\boldsymbol{c}(\boldsymbol{f}_1^*)^T (\boldsymbol{f}_2^* - \boldsymbol{f}_1^*) - \boldsymbol{Z}(\boldsymbol{d}_1^*)^T (\boldsymbol{d}_2^* - \boldsymbol{d}_1^*) + \boldsymbol{c}(\boldsymbol{f}_2^*)^T (\boldsymbol{f}_1^* - \boldsymbol{f}_2^*) \\ &\quad - \boldsymbol{Z}(\boldsymbol{d}_2^*)^T (\boldsymbol{d}_1^* - \boldsymbol{d}_2^*) \geq 0 \end{aligned}$$

or

$$\left[\boldsymbol{c}(\boldsymbol{f}_1^*) - \boldsymbol{c}(\boldsymbol{f}_2^*)\right]^T (\boldsymbol{f}_1^* - \boldsymbol{f}_2^*) - \left[\boldsymbol{Z}(\boldsymbol{d}_1^*) - \boldsymbol{Z}(\boldsymbol{d}_2^*)\right]^T (\boldsymbol{d}_1^* - \boldsymbol{d}_2^*) \leq 0$$

which, if $\boldsymbol{d}_1^* \neq \boldsymbol{d}_2^*$, contradicts the assumption of the monotonicity of the cost functions and the inverse demand functions. Analogously, if there existed two different

[8]Note that strict monotonicity is needed here, in contrast to stochastic user equilibrium.

vectors of feasible demand flows $\boldsymbol{d}_1^* \neq \boldsymbol{d}_2^* \in S_d$ and corresponding vectors of equilibrium link flows $\boldsymbol{f}_1^*, \boldsymbol{f}_2^* \in S_f$, this would again result in a contradiction.

Note that, as in the case of fixed demand, the uniqueness of link flows and equilibrium demand does not imply the uniqueness of equilibrium path flows.

Formulation with Optimization Models Variable demand deterministic user equilibrium can also be formulated with optimization models. These allow simple solution algorithms to be used (see Sect. 6.3.1.3). Equivalent optimization models require that cost functions and inverse demand functions have symmetric Jacobians. In particular, assuming for the sake of simplicity the absence of nonadditive path costs, the model (5.4.14) can be extended in the following form.

$$(\boldsymbol{f}^*, \boldsymbol{d}^*) = \operatorname{argmin} \int_0^{\boldsymbol{f}} \boldsymbol{c}(\boldsymbol{x})^T d\boldsymbol{x} - \int_0^{\boldsymbol{d}} \boldsymbol{Z}(\boldsymbol{y})^T d\boldsymbol{y} \tag{6.3.8}$$

$$\boldsymbol{f} \in S_f$$

$$\boldsymbol{d} \in S_d$$

In general, formulation (6.3.8) is of limited use in practice because it is difficult to express the inverse demand function $\boldsymbol{Z} = \boldsymbol{Z}(\boldsymbol{d})$ in closed form, and therefore to prove the symmetry of its Jacobian.

This condition holds, however, if the demand model is of the logit type, like that described at the beginning of this section. In this case, the following holds.

$$\int_0^{\boldsymbol{d}} \boldsymbol{Z}(\boldsymbol{y})\, d\boldsymbol{y} = (\theta_1/\beta_2) \sum_{od} (d_{od} \ln d_{od} - d_{od}) + (\beta_1/\beta_2) \sum_{od} (A_d d_{od}) \tag{6.3.9}$$

with

$$\sum_{od} d_{od} = d_o \quad \forall o$$

Analogously, the integral (6.3.9) can be explicitly computed for logit mode choice model demand with attributes independent of congestion for the other transportation modes.

6.3.1.3 Algorithms

This section briefly describes extensions of the fixed demand equilibrium assignment algorithms to variable demand equilibrium assignment problems. The algorithms described can also be adapted to solve multimode equilibrium assignment problems, but this is not discussed in detail.

As seen in Sect. 6.3, variable demand assignment models assume that the levels of O-D demand flow depend on congested transportation costs. This assumption implies that users' behavior on choice dimensions other than path choice (e.g., mode, destination) is influenced by variations in path costs resulting from variations in congestion levels. In single-mode assignment, it is assumed that only one mode's costs depend on congestion. In this case, the dependence of demand flows on path costs can be expressed by demand functions that depend on the EMPU function for the path choice model (see Sect. 6.3):

$$\boldsymbol{s} = \boldsymbol{s}(\boldsymbol{V} = -\boldsymbol{g})$$

$$\boldsymbol{d} = \boldsymbol{d}(\boldsymbol{s})$$

Calculation of link and demand flows for variable demand (single-mode) equilibrium assignment can be carried out applying any of three different approaches described below.

Internal cycle algorithms are based on an extension of the algorithms that solve fixed demand equilibrium assignment problems described in Sect. 5.4. It is straightforward to extend to variable demand problems the MSA-FA or MSA-CA algorithms presented therein for fixed demand stochastic equilibrium. In each iteration, these algorithms compute the EMPUs and therefore the demand flows corresponding to costs in the previous iteration, before proceeding to UN assignment of that demand. This approach is simple to apply with or without explicit path enumeration.

In the case of deterministic assignment for noncongested networks (without explicit path enumeration) the algorithms described in Sect. 5.3.3 may be extended. In particular, for each origin o, using an algorithm to determine the minimal path tree, one also calculates the minimum costs Z_{od} to each destination d; this value, except for its sign, constitutes the value of satisfaction $s_{od} = -Z_{od}$, between the od pair from which one can determine the demand flow d_{od} to be loaded onto the minimum path from o to d, and hence determine the link flows.

In the case of logit SUN (without explicit path enumeration), the Dial algorithm described in Sect. 5.3.2 can easily be extended. In particular, for each origin o, after the calculation of the node weights W_i and the link weights w_{ij}, in the first phase of the algorithm, the inclusive variable Y_d is obtained for each destination d. This variable is the EMPU s_{od} between the O-D pair od. The demand flow d_{od} can thus be computed and loaded on the network with the Dial algorithm.

In the case of probit SUN (without explicit path enumeration), the Monte Carlo algorithm described in Sect. 5.3.2 can also be quite easily extended: for each O-D pair od, the average of the shortest path costs corresponding to the sampled perceived costs is an unbiased estimate of the negative EMPU $\bar{s}_{od}$. From these estimates, the demand flows d_{od} can be estimated and, from them, link flows can in turn be determined:

$$\bar{\boldsymbol{s}}^m = \bar{\boldsymbol{s}}^m(\boldsymbol{c})$$

$$\bar{\boldsymbol{d}} = \boldsymbol{d}(\bar{\boldsymbol{s}})$$

$$\bar{\boldsymbol{f}}^m = \bar{\boldsymbol{f}}^m(\boldsymbol{c}, \boldsymbol{d})$$

where

$\bar{\boldsymbol{s}}^m = \bar{\boldsymbol{s}}^m(\boldsymbol{c})$ is a vector of unbiased estimates of the EMPUs for all O-D pairs od, obtained with a sample of m perceived link cost realizations with mean $\boldsymbol{c}$

$\bar{f}^m = \bar{f}^m(\boldsymbol{c}, \boldsymbol{d})$ is an unbiased estimate of SUN link flows resulting from demand flows $\boldsymbol{d}$ and a sample of m vectors of perceived link costs with mean $\boldsymbol{c}$

Note that direct application of this approach, given a vector $\boldsymbol{c}$, requires two repetitions of the estimation process, first for the EMPUs and then for link flows. Thus other approaches are usually adopted when SUN is embedded within an algorithm for stochastic equilibrium.

In the case of deterministic UN assignment (without explicit path enumeration), the algorithms described in Sect. 5.3.3 can again be easily extended. In particular, the algorithm for determining the shortest path tree from each origin o gives the minimum cost Z_{od} between o and all destinations d. The negatives of these values are the EMPUs, $s_{od} = -Z_{od}$, from which demand flows d_{od} can be computed and assigned to the links of the shortest path between o and d.

Whatever procedure is adopted for UN assignment – stochastic or deterministic, with or without explicit path enumeration – the MSA-FA algorithm for internal cycle variable demand equilibrium can be defined by the following system of recursive equations, given $\boldsymbol{f}^0 \in S_f$ and $\boldsymbol{d}^0 \in S_d$ at $k = 0$.

$$k = k + 1 \tag{6.3.10}$$

$$\boldsymbol{c}^k = \boldsymbol{c}(\boldsymbol{f}^{k-1}) \tag{6.3.11}$$

$$\boldsymbol{f}^k_{\mathrm{UN}} = \boldsymbol{f}_{\mathrm{UN}}\left(\boldsymbol{c}^k, \boldsymbol{d}\left(\boldsymbol{s}\left(-\boldsymbol{\Delta}^T \boldsymbol{c}^k\right)\right)\right) \tag{6.3.12}$$

$$\boldsymbol{f}^k = \boldsymbol{f}^{k-1} + 1/k\left(\boldsymbol{f}^k_{\mathrm{UN}} - \boldsymbol{f}^{k-1}\right) \tag{6.3.13}$$

where

$\boldsymbol{f}_{\mathrm{UN}}(\boldsymbol{c}, \boldsymbol{d})$ are the link flows resulting from a UN assignment with costs $\boldsymbol{c}$ and demand flows $\boldsymbol{d}$

$\boldsymbol{d} = \boldsymbol{d}(\boldsymbol{s}(-\boldsymbol{\Delta}^T \boldsymbol{c}))$ are the demand flows corresponding to the EMPUs that result from link costs $\boldsymbol{c}$

The internal cycle MSA-FA algorithm can be further extended by averaging both EMPU values and link flows, as described by the following system of recursive equations, given $\boldsymbol{f}^0 \in S_f$, $\boldsymbol{s}^0 = \boldsymbol{s}(-\boldsymbol{\Delta}^T \boldsymbol{c}(\boldsymbol{f}^0))$ and $k = 0$.

$$k = k + 1 \tag{6.3.14}$$

$$\boldsymbol{c}^k = \boldsymbol{c}(\boldsymbol{f}^{k-1}) \tag{6.3.15}$$

$$\boldsymbol{d}^k = \boldsymbol{d}(\boldsymbol{s}^{k-1}) \tag{6.3.16}$$

$$\left(\boldsymbol{s}^k_{\mathrm{UN}}, \boldsymbol{f}^k_{\mathrm{UN}}\right) = \mathrm{UN}(\boldsymbol{c}^k, \boldsymbol{d}^k) \tag{6.3.17}$$

$$\boldsymbol{s}^k = \boldsymbol{s}^{k-1} + 1/k\left(\boldsymbol{s}^k_{\mathrm{UN}} - \boldsymbol{s}^{k-1}\right) \tag{6.3.18}$$

$$\boldsymbol{f}^k = \boldsymbol{f}^{k-1} + 1/k\left(\boldsymbol{f}^k_{\mathrm{UN}} - \boldsymbol{f}^{k-1}\right) \tag{6.3.19}$$

where

$(s_{\mathrm{UN}}, \boldsymbol{f}_{\mathrm{UN}}) = \mathrm{UN}(\boldsymbol{c}, \boldsymbol{d})$ are the EMPU and flows resulting from a UN assignment with link costs $\boldsymbol{c}$ and demand flows $\boldsymbol{d}$; they can be computed simultaneously using one of the procedures described in Sect. 5.3.3.

This algorithm, called MSA-FSA, is particularly useful with probit path choice models because it avoids the double Monte Carlo application at each iteration.

Convergence of the MSA-FA and MSA-FSA algorithms for equilibrium problems with nonseparable cost functions (asymmetric Jacobian) has not been proved. In this case, it is possible to adopt an immediate extension of the MSA-CA algorithm.[9] In particular, the MSA-CA algorithm can be described by the following system of recursive equations, given $\boldsymbol{f}^0 \in S_f, \boldsymbol{c}^0 = \boldsymbol{c}(\boldsymbol{f}^0)$ and $k = 0$:

$$k = k + 1 \tag{6.3.20}$$

$$\boldsymbol{f}^k_{\mathrm{UN}} = \boldsymbol{f}_{\mathrm{UN}}\big(\boldsymbol{c}^{k-1}, \boldsymbol{d}\big(\boldsymbol{s}\big(-\boldsymbol{\Delta}^T \boldsymbol{c}^{k-1}\big)\big)\big) \tag{6.3.21}$$

$$\bar{\boldsymbol{c}}^k = c(\boldsymbol{f}^k) \tag{6.3.22}$$

$$\boldsymbol{c}^k = \boldsymbol{c}^{k-1} + 1/k\big(\bar{\boldsymbol{c}}^k - \boldsymbol{c}^{k-1}\big) \tag{6.3.23}$$

Note that the link flow vector $\boldsymbol{f}^k = \boldsymbol{f}_{\mathrm{UN}}(c^{k-1})$ at iteration k is feasible.

In general, it is possible to average both demand flows and link costs, with an algorithm called MSA-CDA. The algorithm is described by the following system of recursive equations, given $\boldsymbol{f}^0 \in S_f, \boldsymbol{d}^0 \in S_d, \boldsymbol{c}^0 = c(\boldsymbol{f}^0)$ and $k = 0$.

$$k = k + 1 \tag{6.3.24}$$

$$\boldsymbol{f}^k = \boldsymbol{f}_{\mathrm{UN}}(\boldsymbol{c}^{k-1}, \boldsymbol{d}^{k-1}) \tag{6.3.25}$$

$$\bar{\boldsymbol{d}}^k = \boldsymbol{d}\big(\boldsymbol{s}(-\boldsymbol{A}^T \boldsymbol{c}^k)\big) \tag{6.3.26}$$

$$\bar{\boldsymbol{c}}^k = c(\boldsymbol{f}^k) \tag{6.3.27}$$

$$\boldsymbol{d}^k = \boldsymbol{d}^{k-1} + 1/k(\bar{\boldsymbol{d}}^k - \boldsymbol{d}^{k-1}) \tag{6.3.28}$$

$$\boldsymbol{c}^k = \boldsymbol{c}^{k-1} + 1/k(\bar{\boldsymbol{c}}^k - \boldsymbol{c}^{k-1}) \tag{6.3.29}$$

The convergence of the internal cycle algorithms described above has been analyzed only for separable demand functions $d_i = d_i(s_i)$. In this case, the fixed demand equilibrium conditions already discussed for the MSA-FA and MSA-CA algorithms continue to hold, with the further requirement that the demand functions $d_i = d_i(s_i)$ be continuous, differentiable, nondecreasing monotone, and bounded.

Among the internal cycle algorithms, the equivalent optimization problem (6.3.8) could be solved with the Frank–Wolfe algorithm for variable demand symmetric deterministic equilibrium. However, this approach would require explicit formulation of the inverse demand function $\boldsymbol{Z}(\boldsymbol{d})$, expressing the minimum costs $\boldsymbol{Z}$ in terms of

[9] In the case of asymmetric Jacobian it is also possible to adopt the diagonalization algorithm (described in Sect. 5.4 for fixed demand equilibrium), but no convergence proof has been provided.

demand flows $\boldsymbol{d}$; moreover the inverse demand function would need to have a symmetric Jacobian. Both these conditions are difficult to meet in practice. In any case, the resulting algorithm would require modifications of the DUN algorithm.

External cycle algorithms solve a formulation of variable demand equilibrium assignment models in which the circular dependence between demand flows and costs is expressed externally to the flow-cost equilibrium. As stated in Sect. 6.3, this defines a two-level problem. Equilibrium between flows and costs is computed at the inner level for a given set of demand flows. The outer level computes equilibrium between the costs resulting from the inner-level equilibrium assignment and demand flows obtained from demand functions. Let:

$\boldsymbol{f}_{\text{UE-FIX}} = \boldsymbol{f}_{\text{UE-FIX}}(\boldsymbol{d})$ be the implicit correspondence between fixed demand equilibrium link flows $\boldsymbol{f}_{\text{UE-FIX}}$ and demand flows $\boldsymbol{d}$. This correspondence expresses the solution of one of the models described in Sect. 5.4. If the equilibrium link flow vector is unique for a given demand vector, the above correspondence is a one-to-one function. Its value can be calculated with one of the algorithms described in Sect. 5.4.

Variable demand equilibrium assignment can be formulated with a system of nonlinear equations:

$$\boldsymbol{d}^* = \boldsymbol{d}\left(\boldsymbol{s}\left(-\boldsymbol{\Delta}^T \boldsymbol{c}(\boldsymbol{f}^*)\right)\right) \tag{6.3.30}$$

$$\boldsymbol{f}^* = \boldsymbol{f}_{\text{UE-FIX}}(\boldsymbol{d}^*) \tag{6.3.31}$$

Combining the two equations (6.3.30) and (6.3.31), we obtain a combined fixed-point problem (with an implicitly defined function) in either the demand flows $\boldsymbol{d}^*$ or the link flows $\boldsymbol{f}^*$:

$$\boldsymbol{d}^* = \boldsymbol{d}\left(\boldsymbol{s}\left(-\boldsymbol{\Delta}^T \boldsymbol{c}\left(\boldsymbol{f}_{\text{UE}}(\boldsymbol{d}^*)\right)\right)\right) \tag{6.3.32}$$

$$\boldsymbol{f}^* = \boldsymbol{f}_{\text{UE-FIX}}\left(\boldsymbol{d}\left(\boldsymbol{s}\left(-\boldsymbol{\Delta}^T \boldsymbol{c}(\boldsymbol{f}^*)\right)\right)\right) \tag{6.3.33}$$

The fixed-point problem can also be formulated in terms of link costs or EMPU values.

The simplest external cycle algorithms are based not only on the iterative application of a fixed demand equilibrium assignment algorithm for calculating link flows and costs with given demand flows but also on the demand function for calculating demand flows with given costs and EMPUs. In particular, an external cycle algorithm of this type can be specified by the following system of recursive equations, given an initial feasible value of the demand flows $\boldsymbol{d}^0 \in S_d$ at $k = 0$.

$$k = k + 1$$

$$\boldsymbol{f}^k = \boldsymbol{f}_{\text{UE-FIX}}(\boldsymbol{d}^{k-1}) \tag{6.3.34}$$

$$\boldsymbol{c}^k = \boldsymbol{c}(\boldsymbol{f}^k) \tag{6.3.35}$$

$$\boldsymbol{s}^k = \boldsymbol{s}(-\boldsymbol{\Delta}^T \boldsymbol{c}^k) \tag{6.3.36}$$

$$\boldsymbol{d}^k = \boldsymbol{d}(\boldsymbol{s}^k) \tag{6.3.37}$$

The initial value of the demand flows $\boldsymbol{d}^0$ can be obtained, for example, with EMPUs corresponding to zero flow link costs: $\boldsymbol{c}^0 = \boldsymbol{c}(\boldsymbol{f} = \boldsymbol{0}), \boldsymbol{s}^0 = \boldsymbol{s}(-\boldsymbol{\Delta}^T \boldsymbol{c}^0), \boldsymbol{d}^0 = \boldsymbol{d}(\boldsymbol{s}^0)$.

A more sophisticated external cycle algorithm is obtained by applying the MSA to the fixed-point problem (6.3.32) in demand flows $\boldsymbol{d}^*$. The resulting algorithm is described by the following system of recursive equations, given $\boldsymbol{d}^0 \in S_d$ and $k = 0$.

$$k = k + 1 \tag{6.3.38}$$

$$\boldsymbol{f}^k = \boldsymbol{f}_{\text{UE-FIX}}(\boldsymbol{d}^{k-1}) \tag{6.3.39}$$

$$\boldsymbol{c}^k = \boldsymbol{c}(\boldsymbol{f}^k) \tag{6.3.40}$$

$$\boldsymbol{s}^k = \boldsymbol{s}(-\boldsymbol{\Delta}^T \boldsymbol{c}^k) \tag{6.3.41}$$

$$\boldsymbol{d}^k = \boldsymbol{d}^{k-1} + (1/k)\big(\boldsymbol{d}(\boldsymbol{s}^k) - \boldsymbol{d}^{k-1}\big) \tag{6.3.42}$$

Similarly, an external cycle algorithm can be specified by applying the MSA method to the fixed-point problem (6.3.33) in link flows. This produces an algorithm described by the following system of recursive equations, given $\boldsymbol{f}^0 \in S_f$ and $k = 0$.

$$k = k + 1 \tag{6.3.43}$$

$$\boldsymbol{c}^k = \boldsymbol{c}(\boldsymbol{f}^{k-1}) \tag{6.3.44}$$

$$\boldsymbol{s}^k = \boldsymbol{s}(-\boldsymbol{\Delta}^T \boldsymbol{c}^k) \tag{6.3.45}$$

$$\boldsymbol{d}^k = \boldsymbol{d}(\boldsymbol{s}^k) \tag{6.3.46}$$

$$\boldsymbol{f}^k = \boldsymbol{f}^{k-1} + (1/k)\big(\boldsymbol{f}_{\text{UE-FIX}}(\boldsymbol{d}^k) - \boldsymbol{f}^{k-1}\big) \tag{6.3.47}$$

In both cases, termination tests should compare the value computed in the previous iteration, ($\boldsymbol{d}^{k-1}$ or $\boldsymbol{f}^{k-1}$) with the auxiliary value obtained within the iteration ($\boldsymbol{d}(\boldsymbol{s}^k)$ or $\boldsymbol{f}_{\text{UE-FIX}}(\boldsymbol{d}^*)$).

Other algorithms can be specified by applying the MSA method to fixed-point problems expressed in terms of EMPUs, link costs, or pairs of variables. In any of these cases, it is easily deduced that, if an external cycle algorithm converges to a solution, then this is the equilibrium solution sought. The convergence of external algorithms has not yet been completely analyzed nor have convergence conditions on assignment models and demand functions been established. External algorithms are easily implemented starting from existing fixed demand assignment implementations, and can accommodate a wide variety of demand functions.

Note the difference between the external ((6.3.38) to (6.3.42)) and internal ((6.3.10) to (6.3.13)) cycle algorithms. In the former, a fixed demand equilibrium assignment, requiring several UN assignments, is performed in each iteration and the resulting link flows are averaged. Conversely, in the internal cycle algorithm only one UN assignment is performed in each iteration and the resulting link flows are averaged. No systematic comparisons of the two approaches have been published. From the purely computational point of view, the relative efficiency is certainly related to the relative complexity of computing UN flows and demand flows.

6.3.2 Multimode Equilibrium Assignment

The previous models can be extended to multimode assignment in which mode attributes, useful for simulating mode choice behavior, depend on congested costs for more than one mode. Obviously, in addition to mode and path choice, demand models can be variable with respect to other choice dimensions, such as frequency and destination. To specify these models it is useful to modify the notation used in Sect. 5.2 by introducing a further subscript to designate the mode m. Let:

$\boldsymbol{\Delta}_{od,m}$	be the link-path incidence matrix for the O-D pair *od* and mode *m*
$\boldsymbol{\Delta}$	be the overall link-path incidence matrix, obtained by arranging the blocks $\boldsymbol{\Delta}_{od,m}$ side by side, corresponding to each O-D pair *od* and each mode *m*
$\boldsymbol{c}$	be the link cost vector c_a
$\boldsymbol{g}^{\text{ADD}}_{od,m}$	be the additive path cost vector for the O-D pair *od* and mode *m*
$\boldsymbol{g}^{\text{ADD}}$	be the overall additive path cost vector, composed of the vectors $\boldsymbol{g}^{\text{ADD}}_{od,m}$ corresponding to each O-D pair *od* and each mode *m*
$\boldsymbol{g}^{\text{NA}}_{od,m}$	be the additive path cost vector for the O-D pair *od* and the mode *m*
$\boldsymbol{g}^{\text{NA}}$	be the overall nonadditive path cost vector, composed of vectors $\boldsymbol{g}^{\text{NA}}_{od,m}$ corresponding to each O-D pair *od* and each mode *m*
$\boldsymbol{g}_{od,m}$	be the total path cost vector for the O-D pair *od* and the mode *m*
$\boldsymbol{g}$	be the overall total path cost vector, composed of the vectors $\boldsymbol{g}_{od,m}$ corresponding to each O-D pair *od* and each mode *m*
$\boldsymbol{h}_{od,m}$	be the path flow vector for of the O-D pair *od* and the mode *m*
$\boldsymbol{h}$	be the overall path flow vector, composed of the path flow vectors $\boldsymbol{h}_{od,m}$ corresponding to each O-D pair *od* and each mode *m*

In general, in the case of multimode assignment it is appropriate to consider explicitly both user flows per mode (e.g., car passengers and motorcycles, individual transportation modes, and passengers on buses, trolley-buses, trams, etc., collective transportation modes) and the corresponding vehicle flows. Hence it would be necessary to introduce one variable of vehicle flow and one for passengers for each mode with reference to each link. It is thus possible to analyze both vehicle onboard congestion due to the number of users (flow) present, and congestion due to the possible mix of vehicle flows. This circumstance does not obviously occur if the vehicle flows of the various modes are physically separate, for example, cars, buses with dedicated lanes, or underground trains.

However, under some simple assumptions adopted in applicative practice, it is not necessary to introduce two types of flow variables, but is sufficient to consider only passenger flows: vehicle flows of individual transportation modes are assumed to be linearly related to the relative passenger flows through the crowding coefficient, and to be measured by equivalent vehicles. In addition, with reference to public transportation modes, vehicle flows are assumed to be predetermined (resulting from service scheduling) and also expressed by equivalent vehicles. Hence this flow is considered a constant flow present on the links. Both these types of flows contribute to determining the cost on shared links.

For instance, consider the presence of two modes, car (A) and bus (B) without overlapping lines, with reference to a link a relative to the road network (see Chap. 2). In general, it would be necessary to introduce the following flow variables.

$f_a[P,A]$ flow of passengers (P) in cars (A) on link a
$f_a[V,A]$ flow of vehicles (V) of cars (A) on link a
$f_a[V,B]$ flow of vehicles (V) of buses (B) on link a

Note that there is no point introducing bus passenger flows on link a. As regards mode A, if ω is the crowding coefficient, then: $f_a[P,A] = \omega f_a[V,A]$. Because it is not necessary to consider on-board car congestion, it is sufficient to use only the flow variable $f_a[P,A]$, assuming that the path flows and demand flows that combine to determine this variable are measured in passengers. The flow $f_a[P,A]$ may be considered the characteristic flow of mode A for link a, f_a^A. Assuming that the cost on link a, c_a, is a function of the overall flow crossing it, then: $c_a = c_a(f_a[P,A]/\omega + f_a[V,B])$, that is, the crowding coefficient, is used to adjust the capacities in the cost functions, whereas flow $f_a[V,B]$ may be considered a parameter of the function, insofar as it is not variable. The term $f_a[V,B]$ is expressed in equivalent cars using an appropriate coefficient greater than 1. For link l of the bus service network (see Chap. 2) let:

$f_a[P,B]$ be the passenger flow (P) in buses (B) on link a
$f_a[V,B]$ be the vehicle flow (V) of buses (B) on link a

Note that there is no point introducing car flows on link a. The cost on link a, c_a is made up by the result of two congestion effects. On-board time on link a, tb_a, depends on vehicle flows on the corresponding road link a, already introduced according to a nonseparable function: $tb_a = tb_a(f_a[P,A]/\omega + f_a[V,B])$. The disutility due to the bus crowding coefficient on link a, dr_a, depends on flow $f_a[P,B] : dr_a = dr_a(f_a[P,B])$. Thus flow $f_a[P,B]$ may be considered the characteristic flow of mode B for link a, f_a^B. Hence:

$$c_a = w tb_a\big(f_a[P,A]/\omega + f_a[V,B]\big) + dr_a\big(f_a[P,B]\big) = c_a\big(f_a^A, f_a^B\big)$$

where w is a suitable coefficient of homogenization. Flow $f_a[V,B]$ combines with other similar flows to form flow $f_a[V,B]$. Such considerations may be easily extended to the case of more than one mode with partly overlapping public transportation lines. Thus it is sufficient to have one flow variable per link.

To conclude the analysis of the supply model, it is generally assumed that the cost functions are nonseparable. It is also assumed that a link may be used by more than one mode, for example, in the case of pedestrian links crossed by users of the "foot" mode and public transport mode.[10] Let:

[10]In the special case, not relevant for the analysis below, where each link is used by one mode only, and the cost on a link depends only on the flows of the corresponding mode, the entire network is separable into independent modal networks that share only the centroid nodes.

$\boldsymbol{f}^{od,m}$ be the vector of mode- and O-D-specific link flows, with entries given by the flow on link a, $f_a^{od,m}$, corresponding to the pair od and mode m

$\boldsymbol{f}$ be the overall link flow vector

$\boldsymbol{c}$ be the link cost vector

In analogy with the results presented in Sect. 5.2, and assuming that link flows for each pair od and each mode m are measured in commensurate units, the following holds.

$$f_a = \sum_m \sum_{od} f_a^{od,m}$$

The following relationships (analogous to (5.2.1)–(5.2.3)) relate the variables introduced,

$$\boldsymbol{g}_{od,m} = \boldsymbol{g}_{od,m}^{\text{ADD}} + \boldsymbol{g}_{od,m}^{\text{NA}} = \boldsymbol{\Delta}_{od,m}^T \boldsymbol{c} + \boldsymbol{g}_{od,m}^{\text{NA}} \quad \forall od, m \tag{6.3.48}$$

$$\boldsymbol{c} = \boldsymbol{c}(\boldsymbol{f}) \tag{6.3.49}$$

$$\boldsymbol{f} = \sum_m \sum_{od} \boldsymbol{f}^{od,m} = \sum_m \sum_{od} \boldsymbol{\Delta}_{od,m} \boldsymbol{h}_{od,m} \tag{6.3.50}$$

The multimodal supply model is expressed by the following relationship (analogous to (5.2.4)).

$$\boldsymbol{g}_{od,m} = \boldsymbol{\Delta}_{od,m}^T \boldsymbol{c}\left(\sum_m \sum_{od} \boldsymbol{\Delta}_{od,m} \boldsymbol{h}_{od,m}\right) + \boldsymbol{g}_{od,m}^{\text{NA}} \quad \forall od, m \tag{6.3.51}$$

Path choice behavior can be simulated with a random utility model, possibly different for each mode. For example, a deterministic model might be used for public transport modes, whereas probit models might be specified for car and truck modes. Assuming for simplicity completely pre-trip choice behavior, let:

$\boldsymbol{V}_{od,m}$ be the vector of systematic utilities for paths related to the O-D pair od and the mode m

$p[k/odm]$ be the probability of using path k for a trip from origin o to destination d by mode m (with purpose and time band not explicitly indicated)

$\boldsymbol{p}_{od,m}$ be the vector of path choice probabilities for the O-D pair od and mode m

$d_{od,m}$ be the demand flow of the users between the O-D pair od with mode m, element of the O-D matrix for mode m

The following relationships (analogous to (5.2.5) and (5.2.6)) hold between the variables introduced.

$$\boldsymbol{V}_{od,m} = -\boldsymbol{g}_{od,m} + \boldsymbol{V}_{od,m}^{\circ} \quad \forall od, m \tag{6.3.52}$$

$$\boldsymbol{h}_{od,m} = d_{od,m} \boldsymbol{p}_{od,m}(\boldsymbol{V}_{od,m}) \quad \forall od, m \tag{6.3.53}$$

where

$\boldsymbol{V}^{\circ}_{od,m}$ is a vector with elements consisting of the systematic utility components that depend on attributes other than path costs (such as the socioeconomic characteristics of the users). It is omitted in the following for simplicity of notation

The demand flow $d_{od,m}$ for the pair od on mode m is generally defined by a system of demand models that includes a mode choice submodel, and is therefore a function of the path choice EMPU for the various modes (analogous to (6.3.1)):

$$d_{od,m} = d_{od,m}(\boldsymbol{s}) \quad \forall od, m \tag{6.3.54}$$

where

$\boldsymbol{s}$ is the vector of the path choice EMPU, with a component $s_{od,m}$ for each O-D pair od and each mode m

Finally, the EMPU depends on the vector of systematic utilities (analogous to (5.2.8)):

$$\boldsymbol{s} = \boldsymbol{s}(\boldsymbol{V}) \tag{6.3.55}$$

Thus, the whole multimode demand model is expressed by the equation (analogous to (6.3.2)):

$$\boldsymbol{h}_{od,m} = d_{od,m}\big(\boldsymbol{s}(-\boldsymbol{g})\big)\,\boldsymbol{p}_{od,m}(-\boldsymbol{g}_{od,m}) \quad \forall od, m \tag{6.3.56}$$

Note that the demand model (6.3.56) is an extension of model (5.2.7) derived in the case of fixed demand. It is also a particular specification of the general partial share demand model (4.2.2) introduced in Chap. 4.

By combining supply and demand models we may formulate models for multi-mode equilibrium assignment analogous to the variable demand single-mode user equilibrium assignment described in the previous subsection. The fixed-point models are more flexible and easy to formulate, while retaining the properties described, if the mode choice model within the demand model is specified as a random utility model:

$$\boldsymbol{f}^* = \sum_{od,m} d_{od,m}\big(\boldsymbol{s}(-(\boldsymbol{\Delta}^T \boldsymbol{c}(\boldsymbol{f}^*) + \boldsymbol{g}^{\mathrm{NA}}))\big)\,\boldsymbol{\Delta}_{od,m}\,\boldsymbol{p}_{od,m}\big(-(\boldsymbol{\Delta}^T_{od,m}\boldsymbol{c}(\boldsymbol{f}^*) + \boldsymbol{g}^{\mathrm{NA}}_{od,m})\big)$$

The analysis of existence and uniqueness of the solutions is a simple extension of that developed in Sect. 6.3.1 for single-mode user equilibrium. In particular, to prove existence the mode choice model needs to be specified by continuous functions, whereas to prove uniqueness it needs to be specified by monotone functions, in the sense defined in Sect. 5.3.1. These conditions hold for invariant probabilistic models expressed by continuous functions with continuous first partial derivatives.

6.4 Multiclass Assignment

The assignment models described in the previous sections were developed under the assumption that users are homogeneous with respect to relevant behavioral models and parameters. In the following, these models are extended to deal with the case of *multiclass assignment*, that is, under the assumption that users fall into a number of distinct classes. Users of a given class share all the behavioral characteristics such as specification, parameters, and attributes of the relevant demand models, including path choice. All these features may be different from those of other classes. Users of a given class share the same category and trip purpose as defined in Chap. 4.[11] The definition of the user classes depends on the type of application. For example, in urban systems, classes may be identified on the basis of trip purpose, socioeconomic category and activity duration (influencing parking duration) because different travel costs (parking tolls) and different time values may be associated with these characteristics. In intercity systems, classes may be defined by vehicle type (auto, light and heavy commercial vehicles), trip purpose, and socioeconomic characteristics, because motorway tolls, time values, and path choice models may be different.

In what follows, for the sake of simplicity, reference is made to fixed demand single-mode assignment with fully pre-trip path choice behavior. The results can easily be extended to models with pre-trip/en-route choice behavior and/or with variable demand.

The notation presented in Sect. 5.2 remains valid, but a further subscript i, indicating the user class, is added. Some straightforward changes in notation are described below. Let:

$\boldsymbol{\Delta}_{od,i}$ be the link-path incidence matrix for the O-D pair *od* and class I[12]

$\boldsymbol{\Delta}$ be the overall link-path incidence matrix obtained by arranging side by side the blocks $\boldsymbol{\Delta}_{od,i}$ corresponding to each O-D pair *od* and class i

$d_{od,i}$ be the demand flow for the O-D pair *od* and class i (for a given mode and time band)

$\mathbf{d}$ be the demand vector, with elements consisting of the demand flows $d_{od,i}$

It is assumed that demand flows of each user class are measured in common units, using conversion coefficients as required for users with different effects on congestion (see Sect. 2.3). For individual modes, such as car, demand flows are typically expressed in vehicles per unit time, whereas for public modes they are typically expressed in passengers per unit time.

Transport supply is simulated with a network model analogous to those described in Chap. 2. However, the cost of traversing link a may be different for users of different classes. A cost and flow is therefore associated with each link a and each class i. Let:

[11] In the limit, each segment can consist of a single user, and in this way disaggregated assignment models are obtained. Models of this type are at present only in the research stage.

[12] Different classes corresponding to the same O-D pair may have different incidence matrices if they have different available path sets.

f_a^i be the flow of user class i on link a
$\boldsymbol{f}^i$ be the link flow vector for class i, with entries f_a^i
$f_a = \sum_i f_a^i$ be the total flow on the link a, the sum of the flows of the various classes and measured in units commensurate with the demand flows
$\boldsymbol{f} = \sum_i \boldsymbol{f}^i$ be the vector of the total link flow, with entries f_a^i
c_a^i be the cost on link a for class i
$\boldsymbol{c}^i$ be the link cost vector for class i, with entries c_a^i

The average cost of a path for users of class i can be expressed as the sum of two terms: *additive path costs* with respect to class i link costs, possibly dependent on congestion; and *nonadditive path costs*, which include all the specific path and/or class costs, and are assumed to be independent of congestion. Let:

$\boldsymbol{g}_{od,i}^{\text{ADD}}$ be the additive path cost vector for O-D pair od and class i
$\boldsymbol{g}_{od,i}^{\text{NA}}$ be the nonadditive path cost vector for O-D pair od and class i
$\boldsymbol{g}_{od,i}$ be the total path cost vector for O-D pair od and class i

Consistency between link and path costs for each O-D pair od and each class i, as in Chap. 2, is expressed by the following relation (analogous to (5.2.1)).

$$\begin{aligned} \boldsymbol{g}_{od,i}^{\text{ADD}} &= \boldsymbol{\Delta}_{od,i}^T \boldsymbol{c}^i \quad \forall od\ \forall i \\ \boldsymbol{g}_{od,i} &= \boldsymbol{g}_{od,i}^{\text{ADD}} + \boldsymbol{g}_{od,i}^{\text{NA}} = \boldsymbol{\Delta}_{od,i}^T \boldsymbol{c}^i + \boldsymbol{g}_{od,i}^{\text{NA}} \quad \forall od\ \forall i \end{aligned} \tag{6.4.1}$$

Congestion phenomena are simulated by assuming that the cost c_a^i is a function of the class flows on the same link a, and possibly on other links. Thus, we consider cost functions that are nonseparable with respect to class flows as well as link flows. This effect is usually represented using cost functions similar to those described in Chap. 2, in which the congested link performance attributes for each class depend on the total link flows[13]:

$$\boldsymbol{c}^i = \boldsymbol{c}^i(\boldsymbol{f}^1, \ldots, \boldsymbol{f}^i, \ldots) = \boldsymbol{c}^i(\boldsymbol{f}) = \boldsymbol{c}^i\left(\sum_i \boldsymbol{f}^i\right) \quad \forall i \tag{6.4.2}$$

For example, the road link travel time for car users can depend on the total flow of the other vehicle types (motorcycles, trucks, etc.), converted into commensurate units. The cost functions of different classes, for example, cars and trucks, may be different, but it is assumed that they all depend on the overall link flow.

Consistency between link and path flows is expressed by the following relation (analogous to (5.2.3)).

$$\boldsymbol{f}^i = \sum_{od} \boldsymbol{\Delta}_{od,i} \boldsymbol{h}_{od,i} \quad \forall i \tag{6.4.3}$$

[13]It is also possible to specify cost functions for class i depending only on the flow $\boldsymbol{f}^i$; these models, however, are seldom adopted as they do not correspond to known congestion phenomena.

The multiclass supply model is thus described by the following equation (analogous to (5.2.4)), obtained by combining (6.4.1) to (6.4.3).[14]

$$\boldsymbol{g}_{od,i} = \boldsymbol{\Delta}^T_{od,i}\boldsymbol{c}^i\left(\sum_i\sum_{od}\boldsymbol{\Delta}_{od,i}\boldsymbol{h}_{od,i}\right) + \boldsymbol{g}^{\mathrm{NA}}_{od,i} \quad \forall od\ \forall i \tag{6.4.4}$$

Path choice behavior for each class i can be simulated through a random utility model having systematic utility equal to the negative of the systematic path cost:

$$\boldsymbol{V}_{od,i} = -\boldsymbol{g}_{od,i,} + \boldsymbol{V}^{\circ}_{od,i} \quad \forall od\ \forall i \tag{6.4.5}$$

where

$\boldsymbol{V}_{od,i}$ is a vector with elements consisting of the systematic utility $V_{od,i,k}$ of path k connecting the pair od for the class i

$\boldsymbol{V}^{\circ}_{od,i}$ is a vector of systematic utility attributes other than those included in path costs, for simplicity of notation taken as understood in the following

Path choice probabilities depend on the systematic utilities of alternative paths through the path choice model. Let:

$\boldsymbol{p}_{od,i} = \boldsymbol{p}_{od,i}(\boldsymbol{V}_{od,i})$ be the path choice probabilities vector for O-D pair od and class i

$\boldsymbol{h}_{od,I}$ be the path flow vector for O-D pair od and class i

The path choice model is expressed (analogously to (5.2.6)) by

$$\boldsymbol{h}_{od,i} = d_{od,i}\,\boldsymbol{p}_{od,i}(\boldsymbol{V}_{od,i}) \quad \forall od\ \forall i \tag{6.4.6}$$

The complete demand model is obtained by combining (5.2.5) and (5.2.6):

$$\boldsymbol{h}_{od,i} = d_{od,i}\,\boldsymbol{p}_{od,i}(-\boldsymbol{g}_{od,i}) \quad \forall od\ \forall i \tag{6.4.7}$$

If behavior in other dimensions, such as mode and destination choice, also depends on path costs, then variable demand multiuser assignment models, such as those discussed in Sect. 6.3, are obtained. Extensions of the models to combined pre-trip/en-route path choice behavior are analogous to those presented in Sect. 6.2.

Multiclass assignment models can be specified by combining the supply model (6.4.4) with the demand model (6.4.7). In the following sections, multiclass assignment models are analyzed separately for the special case where the congestion function of each class is a linear transformation of a common congestion function (*undifferentiated congestion*), and for the case of congestion functions that differ between classes (*differentiated congestion*).

[14]The supply model (6.4.4) can also be interpreted as an instance of the general network supply model (5.2.4) in which each physical link is represented by several network links, one for each class.

6.4.1 Undifferentiated Congestion Multiclass Assignment

In *undifferentiated congestion multiclass assignment*, it is assumed that the cost function of each class can be expressed as a linear transformation of a cost function that is common to all the classes and that depends on total link flows. These costs are called *reference costs*. Therefore, multiclass equilibrium assignment can be formulated in terms of total flows and reference link costs. Under these assumptions, expression (6.4.2) for the link cost function becomes:

$$c_a^i = c_a^i(\boldsymbol{f}) = \gamma_i \bar{c}_a(\boldsymbol{f}) + c_{0,a}^i \quad \forall i \tag{6.4.8}$$

where

$\bar{c}_a = \bar{c}_a(\boldsymbol{f})$ is the reference cost function of link a
$\gamma_i \geq 0$ is the ratio (assumed independent of the link) between the link cost for class i and the reference cost; if $\gamma_i = 0$ the class i costs are uncongested
$c_{0,a}^i$ is the cost of link a specific to class i, assumed independent of congestion

All costs are assumed to be expressed (through conversion coefficients) in units commensurate with the utility. The reference cost function $\bar{c}_a(\boldsymbol{f})$ may represent disutility related to the average travel time, and $c_{0,a}^i$ may represent the disutility connected to monetary costs, possibly different for different classes and/or with different substitution coefficients. The coefficients γ_i can express the ratios between class-specific and average travel times.

Using expression (6.4.8), the consistency between link and path costs is expressed for each O-D pair od and class i by the following relation.

$$\boldsymbol{g}_{od,i} = \boldsymbol{\Delta}_{od,i}^T (\gamma_i \bar{\boldsymbol{c}} + \boldsymbol{c}_0^i) + \boldsymbol{g}_{od,i}^{\mathrm{NA}} \quad \forall i\ \forall od$$

$$\boldsymbol{g}_{od,i} = \gamma_i \boldsymbol{\Delta}_{od,i}^T \bar{\boldsymbol{c}} + \boldsymbol{\Delta}_{od,i}^T \boldsymbol{c}_0^i + \boldsymbol{g}_{od,i}^{\mathrm{NA}} \quad \forall i\ \forall od$$

where

$\bar{\boldsymbol{c}}$ is the vector of reference link costs
$\boldsymbol{c}_0^i$ is the vector of class i specific link costs
$\boldsymbol{g}_{od,i}^{\mathrm{NA}}$ is the vector of nonadditive path costs for O-D pair od and class i
$\boldsymbol{g}_{od,i}$ is the total path cost vector for O-D pair od and class i

The average cost of a path between O-D pair od for a user of class i therefore consists of two components:

- *Additive (and generic) costs*, the sum of reference link costs, possibly dependent on congestion, given by $\gamma_i \boldsymbol{\Delta}_{od,i}^T \bar{\boldsymbol{c}}$;
- Congestion-independent path costs consisting of:
  - *(Additive and) Class-specific costs*, the sum of class-specific link costs, given by $\boldsymbol{\Delta}_{od,i}^T \boldsymbol{c}_{i,0}$;
  - *Nonadditive costs*, which cannot be expressed as the sum of link costs, given by $\boldsymbol{g}_{od,i}^{\mathrm{NA}}$.

Let:

$\boldsymbol{g}_{od,i}^{\text{SPNA}} = \gamma_i \boldsymbol{\Delta}_{od,i}^T \boldsymbol{c}_0^i + \boldsymbol{g}_{od,}{}^{i\text{NA}}$ be the vector of specific and/or nonadditive path costs for O-D pair od and class i

A relationship between link and path costs analogous to (6.4.1) can be formulated:

$$\boldsymbol{g}_{od,i} = \gamma_i \boldsymbol{\Delta}_{od,i}^T \bar{\boldsymbol{c}} + \boldsymbol{g}_{od,i}^{\text{SPNA}} \quad \forall_{od}\ \forall i \tag{6.4.9}$$

The undifferentiated congestion multiclass supply model is thus described by the following relation obtained by combining (6.4.3) with (6.4.8) and (6.4.9) and the reference cost functions given by (5.2.2):

$$\boldsymbol{g}_{od,i} = \gamma_i \boldsymbol{\Delta}_{od,i}^T \bar{\boldsymbol{c}}\left(\sum_i \sum_{od} \boldsymbol{\Delta}_{od,i} \boldsymbol{h}_{od,i}\right) + \boldsymbol{g}_{od,i}^{\text{SPNA}} \quad \forall od\ \forall i \tag{6.4.10}$$

Path choice behavior is simulated by a random utility model, expressed by (6.4.6), in which the systematic utility of a path is equal to the negative of the path average cost for class i, as expressed in the relation (6.4.5). In the case of a logit path choice model, parameter γ_i cannot be identified separately from parameter θ. In the case of a deterministic path choice model, γ_i is not relevant because it does not change the maximum systematic utility alternative, that is, the minimum cost path.[15]

Under the given assumptions, undifferentiated congestion multiclass assignment models can therefore be defined with respect to total path or link flows, consistent with reference link costs and the interaction between classes. The considerations expressed in the previous sections are still valid. In particular, the sets of feasible path S_F and link S_f flows are defined as in Sect. 5.2.

Undifferented congestion uncongested network multiclass assignment models are expressed by

$$\boldsymbol{f}_{\text{UN}}(\bar{\boldsymbol{c}}; \boldsymbol{d}, \boldsymbol{\gamma}) = \sum_{od,i} d_{od,i} \boldsymbol{\Delta}_{od,i} \boldsymbol{p}_{od,i}\left(-\gamma_i \boldsymbol{\Delta}_{od,i}^T \bar{\boldsymbol{c}} - \boldsymbol{g}_{od,i}^{\text{SPNA}}\right) \tag{6.4.11}$$

The stochastic uncongested network assignment function retains the properties of continuity and monotonicity discussed in Sect. 5.3.3 if the coefficients γ_i are non-negative. In the case of deterministic assignment, systems of inequalities analogous to those presented in Sect. 5.3.3 can be developed.

Undifferentiated congestion equilibrium multiclass assignment models are defined by the system of equations obtained by combining the supply model (6.4.10) and the demand model (6.4.7). An equivalent formulation, in terms of total link flows $\boldsymbol{f}$ and reference link costs $\bar{\boldsymbol{c}}$, can be expressed by the system of equations obtained by combining the UN assignment map (6.4.11) with the reference cost functions given by (5.2.2). Stochastic or deterministic user equilibrium assignment can

---

[15]More generally, note that the results of deterministic path choice models are not modified even by a nonlinear relationship between systematic utilities and path cost, as long as this relationship is strictly increasing.

be formulated with fixed-point or variational inequality models, respectively, analogous to the models presented in the previous sections. Continuity and monotonicity of the link reference cost functions are required for the existence and uniqueness of the equilibrium solution.

It can easily be deduced that parameters γ_i (assumed to be nonnegative) do not alter the existence and uniqueness conditions of equilibrium solutions. They do influence the value of the SUE solution and, as noted earlier, do not influence DUE assignment.

Finally, it must be noted that in stochastic equilibrium, once the equilibrium total link flows $\boldsymbol{f}^*$ are known, it is possible to compute equilibrium reference costs $\bar{\boldsymbol{c}}^*$ and therefore class-specific link and path costs, $\boldsymbol{c}^i$ and $\boldsymbol{g}_i$, respectively. From these costs, class-specific path flows $\boldsymbol{h}_i$ and hence link flows $\boldsymbol{f}^i$ can be obtained:

$$\boldsymbol{f}^i = \sum_{od} \boldsymbol{\Delta}_{od,i} \boldsymbol{p}_{od,i}\left(-\gamma_i \boldsymbol{\Delta}^T_{od,i} \bar{\boldsymbol{c}}(\boldsymbol{f}) - \boldsymbol{g}^{\mathrm{SPNA}}_{od,i}\right) \quad \forall i$$

The existence and uniqueness of stochastic equilibrium total link flows ensure the existence and uniqueness of class-specific flows. On the other hand, in the case of deterministic models, multiple class-specific link flows could be associated with the same link cost vector if there were several minimum cost paths. Thus, in the case of deterministic multiclass equilibrium, the existence of total equilibrium link flows ensures the existence of class flows, but the uniqueness of total link flows does not guarantee the uniqueness of class-specific link flows. To guarantee the uniqueness of class link flows in this case, an explicit formulation in terms of class flows is necessary, as in the case of differentiated congestion assignment.

6.4.2 Differentiated Congestion Multiclass Assignment

Differentiated congestion multiclass assignment models can be formulated with respect to the path or link flows of each class. These must be consistent with the corresponding costs experienced by each class. In the case of congested network assignment, cost functions generally differ for each class, and depend on the total flow of all classes (6.4.2). The single-class assignment models described in previous sections can easily be extended by considering link flows and costs per class, defining for each class i the sets of feasible path and link flow vectors S^i_h and S^i_f, respectively.

Differentiated congestion multiclass uncongested network assignment models can be expressed in terms of class link flows by combining (6.4.13) with the demand model (6.4.7):

$$\boldsymbol{f}^i_{\mathrm{UN}}(\boldsymbol{c}^i; \boldsymbol{d}_i) = \sum_{od} d_{od,i} \boldsymbol{\Delta}_{od,i} \boldsymbol{p}_{od,i}\left(-\left(\boldsymbol{\Delta}^T_{od,i} \boldsymbol{c}^i + \boldsymbol{g}^{\mathrm{NA}}_{od,i}\right)\right) \quad \forall \boldsymbol{c}_i \ \forall i \qquad (6.4.12)$$

The stochastic uncongested network assignment function retains the properties of continuity and monotonicity discussed in Sect. 5.4.3, which are useful to prove existence and uniqueness of equilibrium flows as discussed below. In the case of deterministic uncongested network assignment, systems of inequalities analogous to those presented in Sect. 5.3.3 can be specified.

Differentiated congestion multiclass equilibrium assignment models are defined by combining the supply model (6.4.4) and the demand model (6.4.7). An equivalent formulation in terms of link variables can be expressed by combining the UN assignment map (6.4.12) with the cost functions (6.4.2). Extension to variable or multimodal demand assignment (Sect. 6.3) or to combined pre-trip/en-route path choice behavior (Sect. 6.2) is relatively straightforward.

Stochastic multiclass equilibrium can be formulated with fixed-point models analogous to those described in the previous sections, and deterministic multiclass user equilibrium can also be formulated with variational inequality models.

Existence conditions for multiclass equilibrium require continuity of the cost functions $\boldsymbol{c}^i()$ for each class i with respect to the flows of the various classes, $\boldsymbol{f}^1, \ldots, \boldsymbol{f}^i, \ldots$. Note that continuity with respect to the total flows $\boldsymbol{f}$ also ensures the continuity with respect to the individual class flows $\boldsymbol{f}^i$, and therefore the existence of an equilibrium.

Uniqueness conditions for multiclass equilibrium require, for each class i, the monotonicity of the cost functions $\boldsymbol{c}^i = \boldsymbol{c}^i()$ with respect to the flows of the various classes, $\boldsymbol{f}^1, \ldots, \boldsymbol{f}^i, \ldots$, as defined by the following condition,

$$\sum_i [\boldsymbol{c}^i(\boldsymbol{f}^1, \ldots, \boldsymbol{f}^i, \ldots) - \boldsymbol{c}^i(\boldsymbol{y}^1, \ldots, \boldsymbol{y}^i, \ldots)]^T (\boldsymbol{f}^i - \boldsymbol{y}^i) > 0$$
$$\forall (\boldsymbol{f}^1, \ldots, \boldsymbol{f}^i, \ldots) \neq (\boldsymbol{y}^1, \ldots, \boldsymbol{y}^i, \ldots) : \boldsymbol{f}^i, \quad \boldsymbol{y}^i \in S_f^i \ \forall i$$

or

$$\sum_i \left[\boldsymbol{c}^i\left(\sum_j \boldsymbol{f}^j\right) - \boldsymbol{c}^i\left(\sum_j \boldsymbol{y}^j\right)\right]^T (\boldsymbol{f}^i - \boldsymbol{y}^i) > 0$$
$$\forall (\boldsymbol{f}^1, \ldots, \boldsymbol{f}^i, \ldots) \neq (\boldsymbol{y}^1, \ldots, \boldsymbol{y}^i, \ldots) : \boldsymbol{f}^i, \quad \boldsymbol{y}^i \in S_f^i \ \forall i \qquad (6.4.13)$$

It should be noted that strict monotonicity of the class cost functions with respect to total link flows, as defined by the following condition

$$[\boldsymbol{c}^i(\boldsymbol{f}) - \boldsymbol{c}^i(\boldsymbol{x})]^T (\boldsymbol{f} - \boldsymbol{x}) > 0 \quad \forall i$$
$$\forall \boldsymbol{f} = \sum_j \boldsymbol{f}^j \neq \boldsymbol{x} = \sum_j \boldsymbol{x}^j : \boldsymbol{f}^i, \quad \boldsymbol{x}^i \in S_f^i \ \forall i$$

or

$$\left[\boldsymbol{c}^i\left(\sum_j \boldsymbol{f}^j\right) - \boldsymbol{c}^i\left(\sum_j \boldsymbol{x}^j\right)\right]^T \sum_j (\boldsymbol{f}^j - \boldsymbol{x}^j) > 0 \quad \forall i \qquad (6.4.14)$$
$$\forall \sum_j \boldsymbol{f}^j \neq \sum_j \boldsymbol{x}^j : \boldsymbol{f}^i, \quad \boldsymbol{x}^i \in S_f^i \ \forall i$$

does not imply strict monotonicity of the class cost functions with respect to class flows, as defined by (6.4.13). (The sum over index i of inequalities (6.4.14) does not necessarily imply condition (6.4.13).[16]) Therefore, equilibrium uniqueness cannot be concluded under these conditions.

It should also be noted that, in a multiclass assignment model, the symmetry of the cost function Jacobian that is necessary for an optimization model formulation of DUE, relates not only to the effect of link flow on the costs of different links but also of class flow on the cost of other classes on the same link. Similarly, separability of cost functions requires that c_a^i, the cost of class i on link a,$_a$ depends only on f_a^i, the flow $_a$ of the same class on the same link. This second condition is almost never satisfied in applications.

In general, the problem of differentiated congestion multiclass equilibrium assignment can be formulated by extending the corresponding single-class models. However, the (sufficient) uniqueness conditions are seldom satisfied.

6.5 Interperiod Dynamic Process Assignment**

User equilibrium models define a priori the relevant state of the system as that in which average demand and costs are mutually consistent. In contrast, dynamic process assignment models simulate the evolution of the system over a sequence of similar periods (days or portions of days[17]), and the possible convergence of the system over time to a stable condition. For this reason, dynamic process models are also known as *nonequilibrium models*. As was noted in Chap. 1, this type of dynamic is known as *interperiod* or *day-to-day dynamics*. Dynamic process models are based on (nonlinear) time-discrete dynamic systems theory or on stochastic process theory, according to whether the state of the system is described by deterministic or random variables.

Dynamic process models, which are a sector of growing research interest, can be seen as a generalization of equilibrium models because they simulate the convergence of the supply–demand system towards possibly different equilibrium states, and the transient states visited due to modifications in supply and/or demand. Furthermore, under some rather mild assumptions, equilibrium configurations of the system described in previous sections can be modeled as attractors of the system, that is, states in which the system stops evolving. Finally, the dynamic approach al-

[16] Note that the two conditions (6.4.13) and (6.4.14) coincide if two flow vectors are considered that differ only in terms of class flows. The same circumstance obviously occurs in the case of a single-user class.

[17] For the sake of simplicity, the generic reference period is identified as a "day". Note that the periods need not be successive. For example, if the aim is not to explicitly simulate the development of the system but only to study its convergence properties, reference can be made only to weekdays or to periods of fictitious behavior updating.

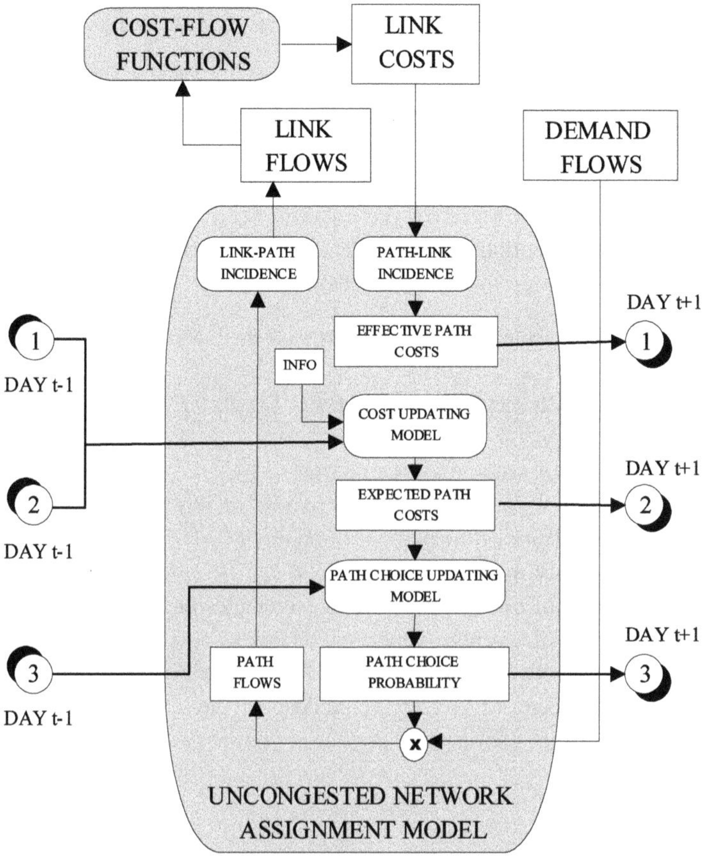

Fig. 6.9 Schematic representation of Dynamic Process (DP) assignment models

lows analysis of the stability of equilibrium configurations and provides a complete statistical description of the system's evolution.

In general, the specification of a dynamic process model requires a more detailed representation of users' behavior than does the specification of an equilibrium model. It requires in particular the explicit modeling of two phenomena (Fig. 6.9) that are not relevant in the equilibrium approach:

- The users' choice updating behavior, that is, how present choices are influenced by the choices made on previous days, including phenomena such as habit (*choice updating model*);
- The users' learning and forecasting mechanisms, that is, how experience and information on previous transport costs influence present choices, including phenomena such as memory and information diffusion (*utility updating model*).

6.5.1 Definitions, Assumptions, and Basic Equations

This section presents the basic relationships that define a dynamic process assignment model. For the sake of clarity, fixed demand single-mode single-class[18] assignment is considered. It is also assumed that path choice behavior is probabilistic and fully pre-trip. Some of the variables presented in Sect. 5.2 need to be redefined in order to associate them with the evolution of the system over a sequence of reference periods (interperiod or day-to-day dynamics). Let:

t	be the generic reference period, assumed for the sake of simplicity to be a day
$\boldsymbol{\Delta}_{od}$	be the link path incidence matrix for O-D pair *od*, assumed to be independent of the day
$\boldsymbol{\Delta}$	be the total link path incidence matrix
$\boldsymbol{h}^t_{od}$	be the vector of path flows for O-D pair *od* on day t
$\boldsymbol{h}^t$	be the total vector of the path flows on day t
$\boldsymbol{f}^t$	be the vector of the link flows on day t
$\boldsymbol{r}^t_n$	be the vector of nth link performance attributes on day t
$\boldsymbol{c}^t$	be the vector of (average) link costs on day t
$\boldsymbol{g}^t_{od}$	be the vector of (average) path costs for O-D pair *od* on day t
$\boldsymbol{g}$	be the total vector of (average) path costs on day t

6.5.1.1 Supply Model

Supply is simulated by applying the relations (5.2.1)–(5.2.3) to the costs and flows on day t. Ignoring for simplicity any nonadditive path costs ($\boldsymbol{g}^{\mathrm{NA}}_{od} = \boldsymbol{0}$), it follows that

$$g^t_{od} = \boldsymbol{\Delta}^T_{od}\boldsymbol{c}^t$$
$$\boldsymbol{g}^t = \boldsymbol{\Delta}^T\boldsymbol{c}^t \tag{6.5.1}$$

$$\boldsymbol{c}^t = \boldsymbol{c}(\boldsymbol{f}^t) \tag{6.5.2}$$

$$\boldsymbol{f}^t = \sum_{od} \boldsymbol{\Delta}_{od}\boldsymbol{h}^t_{od}$$
$$\boldsymbol{f}^t = \boldsymbol{\Delta}\boldsymbol{h}^t \tag{6.5.3}$$

---

[18] A dynamic process assignment model can also be multiclass and applied to different levels of aggregation by considering, for each O-D pair, homogeneous classes of users, each consisting, in the extreme case, of a single user (completely disaggregated assignment).

Combining (6.5.1)–(6.5.3), we obtain the following relation between path costs $\boldsymbol{g}^t$ and path flows $\boldsymbol{h}^t$ on day t.

$$\begin{aligned} \boldsymbol{g}^t_{od} &= \boldsymbol{\Delta}^T_{od}\boldsymbol{c}\Big(\sum_{od}\boldsymbol{\Delta}_{od}\boldsymbol{h}^t_{od}\Big) \quad \forall od \\ \boldsymbol{g}^t &= \boldsymbol{\Delta}^{Tc}(\boldsymbol{\Delta}\boldsymbol{h}^t) \end{aligned} \tag{6.5.4}$$

Equations (6.5.4) define the *supply model* corresponding to day t. It is readily apparent that the relation (6.5.4) is analogous to (5.2.4) that defines the supply model in the static case.

6.5.1.2 Demand Model

The modeling of day-to-day dynamic path choice behavior requires extending the static demand model relations (5.2.5)–(5.2.7). In particular, the relationships between the costs on different days and the attributes influencing user choices, as well as the choice updating mechanisms on subsequent days, must be made explicit. Let:

- $d_{od} \geq 0$ be the demand flow for the users of O-D pair *od*, assumed to be independent of the day for the sake of simplicity (consistent with the fixed demand hypothesis)
- $\boldsymbol{d}$ be the demand vector, whose components are the demand values d_{od} for each O-D pair
- $\boldsymbol{V}^t_{od}$ be the vector of systematic path utilities forecast on day t by the users of O-D pair *od*
- $\boldsymbol{V}^t$ be the total vector of systematic path utilities forecast on day t

The *utility updating model* simulates the way in which perceived utilities on day t are influenced by utilities and costs on previous days (and possibly by other sources of information). In principle, a disaggregate assignment model could model the updating of the individual utility of user i by expressing the dependence of $U^{i,t}_k$, the perceived utilities for all paths k on day t, on the perceived utilities on previous days and on the corresponding actual costs. This can be expressed symbolically as

$$U^{i,t}_{od} = U\big(U^{i,t-1}_{od} U^{i,t-2}_{od}, \ldots, g^{t-1}_{od}, g^{t-2}_{od}, \ldots\big)$$

This model, however, is not applicable to aggregate assignment. Furthermore, it would be complex to specify choice models based on random utility theory given the serial correlation of the day t random residuals with those of previous days. The models proposed in the literature are special cases; they assume that utility updating is applied to average (systematic) utilities through a function $\boldsymbol{V}()$, known as a *filter*. The filter is a generalization of the systematic utility function that is defined in the static case by relation (5.2.5):

$$\boldsymbol{V}_{od}^{t} = \boldsymbol{V}_{od}\left(\boldsymbol{V}_{od}^{t-1}, \boldsymbol{g}_{od}^{t-1}, \boldsymbol{V}_{od}^{t-2}, \boldsymbol{g}_{od}^{t-2}, \ldots\right) \quad \forall od$$
$$\boldsymbol{V}^{t} = \boldsymbol{V}(\boldsymbol{V}^{t-1}, \boldsymbol{g}^{t-1}, \boldsymbol{V}^{t-2}, \boldsymbol{g}^{t-2}, \ldots)$$

For the sake of simplicity, it is assumed in the following that the expected (or predicted) average utilities on day t depend only on the actual costs $\boldsymbol{g}^{t-1}$ and the expected utilities $\boldsymbol{V}^{t-1}$ on the previous day.

$$\begin{aligned} \boldsymbol{V}_{od}^{t} &= \boldsymbol{V}_{od}\left(\boldsymbol{V}_{od}^{t-1}, \boldsymbol{g}_{od}^{t-1}\right) \quad \forall od \\ \boldsymbol{V}^{t} &= \boldsymbol{V}(\boldsymbol{V}^{t-1}, \boldsymbol{g}^{t-1}) \end{aligned} \tag{6.5.5}$$

Note that, under this assumption, the actual costs on days prior to $t-1$ still influence the choice behavior on day $\boldsymbol{t}$, because they influence the expected utility $\boldsymbol{V}^{t-1}$ on the previous day.

A simple example of a utility updating model is defined by an *exponential filter* in which the expected utility on day $\boldsymbol{t}$ is expressed by a convex combination of the previous day's expected utility $\boldsymbol{V}^{t-1}$, and the (negative of the) actual path costs $-\boldsymbol{g}^{t-1}$, as defined by the supply model (6.5.4). Relation (6.5.5) then becomes:

$$\boldsymbol{V}^{t} = -\beta\boldsymbol{g}^{t-1} + (1-\beta)\boldsymbol{V}^{t-1} \quad \forall od \tag{6.5.6}$$

where

$\beta \in]0, 1]$ is the average weight attributed by the users to the actual costs on day $t-1$; if $\beta = 1$, the expected utility is equal to the negative actual cost on day $t-1$, and the costs on previous days do not influence user behavior. This parameter is usually assumed to be independent of the day and may differ according to user class

Given the linear relationship between link and (additive) path costs, the exponential filter can also be applied to link costs:

$$\boldsymbol{x}^{t} = \beta\boldsymbol{c}^{t-1} + (1-\beta)\boldsymbol{x}^{t-1} \tag{6.5.7}$$

where $\boldsymbol{x}^{t}$ is the vector of expected link costs on day $\boldsymbol{t}$. In this case, the expected path utilities on day $\boldsymbol{t}$ are given by[19]

$$\boldsymbol{V}_{od}^{t} = -\boldsymbol{\Delta}_{od}^{T}\boldsymbol{x}^{t}$$
$$\boldsymbol{V}^{t} = -\boldsymbol{\Delta}^{T}\boldsymbol{x}^{t}$$

[19]Note that the two cost updating models, or systematic utility models, correspond to two assumptions that differ in terms of their underlying behavioral mechanism. In the case of model (6.5.5), it is assumed that the user remembers and averages path costs on successive days; whereas in the case of model (6.5.5), it is assumed that the user remembers the costs of individual links, which are put together later to obtain the path values. The two formulations are equivalent for the assumptions made here, but they might not be for other cost updating models and/or in the presence of nonadditive path costs.

The *choice updating model* simulates the way in which choices on day t are influenced by choices made on previous days. The most general approach can be expressed by a square matrix $\boldsymbol{R}^t$, known as a *conditional choice matrix*, which has a number of rows and columns equal to the number of paths. The elements $r^t_{k,j} \in [0,1]$ are the conditional path choice fractions, that is, the fraction of users choosing path k on day t given that path j was chosen on day $t-1$. Because $r^t_{k,j} = 0$ if paths k and j do not connect the same O-D pair, the following holds, $\sum_{k \in Kod} r^t_{k,j} = 1$ $\forall j \in K_{od}$.

The path flow vector on day t, $\boldsymbol{h}^t$, can be expressed as the product of the conditional choice matrix $\boldsymbol{R}^t$ and the path flow vector on the previous day, $\boldsymbol{h}^{t-1}$:

$$h^t_k = \sum_{j \in Kod} r^t_{k,j} h^{t-1}_j \quad \forall k \in K_{od} \; \forall od$$

$$\boldsymbol{h}^t_{od} = \boldsymbol{R}^t_{od} \boldsymbol{h}^{t-1}_{od} \quad \forall od$$

$$\boldsymbol{h}^t = \boldsymbol{R}^t \boldsymbol{h}^{t-1}$$

Note that the path flow vector on day t is feasible, $\boldsymbol{h}^t \in S_h$, if the path flow vector on the previous day is feasible, $\boldsymbol{h}^{t-1} \in S_h$ (i.e., if it is nonnegative and satisfies the demand conservation constraint).

The elements of the conditional choice matrix (or rather their average values) $\boldsymbol{R}^t$ can be simulated with a random utility model involving the expected utilities on day t (and possibly other days and/or other attributes not expressed here). In this way, we obtain a generalization of the path choice models used in the static case:

$$\boldsymbol{R}^t_{od} = \boldsymbol{R}_{od}\left(\boldsymbol{V}^t_{od}\right)$$

$$\boldsymbol{R}^t = \boldsymbol{R}(\boldsymbol{V}^t)$$

Combining the two previous relationships, a generalization of the static model relation (5.2.6) is obtained:

$$\begin{aligned} \boldsymbol{h}^t_{od} &= \boldsymbol{R}_{od}\left(\boldsymbol{V}^t_{od}\right)\boldsymbol{h}^{t-1}_{od} \quad \forall od \\ \boldsymbol{h}^t &= \boldsymbol{R}(\boldsymbol{V}^t)\boldsymbol{h}^{t-1} \end{aligned} \tag{6.5.8}$$

A simple example of a choice updating model for the modeling of the conditional choice matrix is the exponential filter model. This model assumes that each day some users repeat the choices made the previous day, and others reconsider (although do not necessarily change) their choices with a probability independent of the choice made on the previous day:

$$r^t_{kk} = \alpha p^t_k + (1-\alpha) \quad \forall k \in K_{od} \; \forall od$$

$$r^t_{kj} = \alpha p^t_j \quad \forall j \neq k, \; j \in K_{od} \; \forall k \in K_{od} \; \forall od$$

where

$p_k^t \in]0, 1]$ is the probability that on day t a user reconsidering the choice made the previous day chooses the path $k \in K_{od}$

$\alpha \in]0, 1]$ is the probability that a user reconsiders the choice made the previous day. Therefore $(1-\alpha)$ is the probability that the previous day's choice is repeated; if $\alpha = 1$ all the users reconsider their previous day choices; this parameter is usually assumed to be independent of the day[20] but may differ by user class

Under this model, it follows that

$$h_k^t = \sum_{j \in Kod} \alpha p_k^t h_j^{t-1} + (1-\alpha) h_k^{t-1}$$
$$= \alpha p_k^t \sum_{j \in Kod} h_j^{t-1} + (1-\alpha) h_k^{t-1} \quad \forall k \in K_{od}\ \forall od$$

Because $d_{od} = \sum_{j \in Kod} h_k$, we obtain:

$$\boldsymbol{h}_{od}^t = \alpha d_{od} \boldsymbol{p}_{od}^t + (1-\alpha) \boldsymbol{h}_{od}^{t-1}$$

The path choice probability p_k^t is usually obtained with one of the path choice models described in Sect. 4.3.3, $\boldsymbol{p}_{od}^t = \boldsymbol{p}_{od}(\boldsymbol{V}_{od}^t)$. The relation (6.5.8) therefore becomes (cf. (5.2.6)):

$$\boldsymbol{h}_{od}^t = \alpha d_{od} \boldsymbol{p}_{od}\left(\boldsymbol{V}_{od}^t\right) + (1-\alpha) \boldsymbol{h}_{od}^{t-1} \tag{6.5.9}$$

By combining the two recursive equations (6.5.5) and (6.5.8), we get a relationship between the path flows $\boldsymbol{h}^t$ on day t and path costs $\boldsymbol{g}^{t-1}$ on day $t-1$ which defines the demand model corresponding to day t:

$$\begin{aligned} \boldsymbol{h}_{od}^t &= \boldsymbol{R}_{od}\left(\boldsymbol{V}_{od}\left(\boldsymbol{V}_{od}^{t-1}, \boldsymbol{g}_{od}^{t-1}\right)\right) \boldsymbol{h}_{od}^{t-1} \quad \forall od \\ \boldsymbol{h}^t &= \boldsymbol{R}\left(\boldsymbol{V}(\boldsymbol{V}^{t-1}, \boldsymbol{g}^{t-1})\right) \boldsymbol{h}^{t-1} \end{aligned} \tag{6.5.10}$$

This relation is a generalization of the static case (5.2.7). If exponential filters are adopted to formulate utility and choice updating models, expression (6.5.6) becomes:

$$\boldsymbol{h}_{od}^t = \alpha d_{od} \boldsymbol{p}_{od}\left(-\beta \boldsymbol{g}_{od}^{t-1} + (1-\beta) \boldsymbol{V}_{od}^{t-1}\right) + (1-\alpha) \boldsymbol{h}_{od}^{t-1} \tag{6.5.11}$$

6.5.1.3 Approaches to Dynamic Process Modeling

A dynamic process model is identified by the combination of the recursive equations (6.5.10) that define the choice model, and the recursive equations (6.5.5) that specify

[20]In some more sophisticated choice updating model formulations, the parameter is replaced by a model that expresses the probability of reconsidering the choices as a function of socioeconomic attributes and service-level type (difference between expected values and actual values, information, etc.).

how the expected utilities and the supply model (6.5.4) are updated. The state of the system on day t is defined by the vectors of predicted systematic utilities $\boldsymbol{V}^t$ and by the path flows $\boldsymbol{h}^t$; these variables capture the net results of the utility and choice updating models as a function of the state on the previous day[21]:

$$\boldsymbol{V}^t = \boldsymbol{V}\big(\boldsymbol{V}^{t-1}, \boldsymbol{\Delta}'\boldsymbol{c}(\boldsymbol{\Delta h}^{t-1})\big) \tag{6.5.12}$$

$$\boldsymbol{h}^t = \boldsymbol{R}(\boldsymbol{V}^t)\boldsymbol{h}^{t-1} \tag{6.5.13}$$

The set of feasible states S, known as the *state space*, is defined by the vectors of expected path utilities $\boldsymbol{V}^t \in R^n$, and the feasible path flows $\boldsymbol{h}^t \in S_h : S = S_h \times R^n$.

Given an initial state, the recursive equations (6.5.12) and (6.5.13) define a dynamic process model (Fig. 6.9). If the vectors of path flows $\boldsymbol{h}^t$ and predicted utilities $\boldsymbol{V}^t$ are modeled as deterministic variables, a *deterministic process* model results; whereas if they are modeled as random variables, a *stochastic process* model is obtained (Fig. 6.10). A deterministic process model can also be interpreted as a process that approximates the expected values of the corresponding stochastic process.

Note that the terms stochastic and deterministic have different meanings when they refer to dynamic process formulations versus path choice models in assignment. In the former case, they relate to the actual representation of the system, that is, to assumptions made by the analyst about the deterministic or probabilistic nature of the state variables. In the latter, they relate to assumptions made in modeling path choices, that is, the absence or presence of a random residual in the utility functions, and therefore the form of path choice models. Equilibrium models, whether deterministic or stochastic, imply a deterministic system representation.

Below we briefly analyze the implications and some theoretical results with regard to the two types of dynamic process. Note that a model of a deterministic process may also be interpreted as an average process that approximates the expected value of the corresponding stochastic process. To make the text clear for readers unfamiliar with dynamic processes some brief theoretical comments are also included.

6.5.2 Deterministic Process Models

Deterministic process models derive from the assumption that the path flows and utilities predicted on day t are represented by deterministic variables, that is, that the actual flows and utilities coincide with their average values. System evolution over time, in terms of path flows and utilities, is defined by the recursive equations (6.5.12) and (6.5.13). A model of this type allows analysis of the evolution of the

[21]The adoption of different formulations for the cost and choice updating models can lead to different definitions of the system state. For example, if a moving average filter of k previous days is specified for the cost updating model, the state of the system on day t is defined by the path flows and costs on those k days.

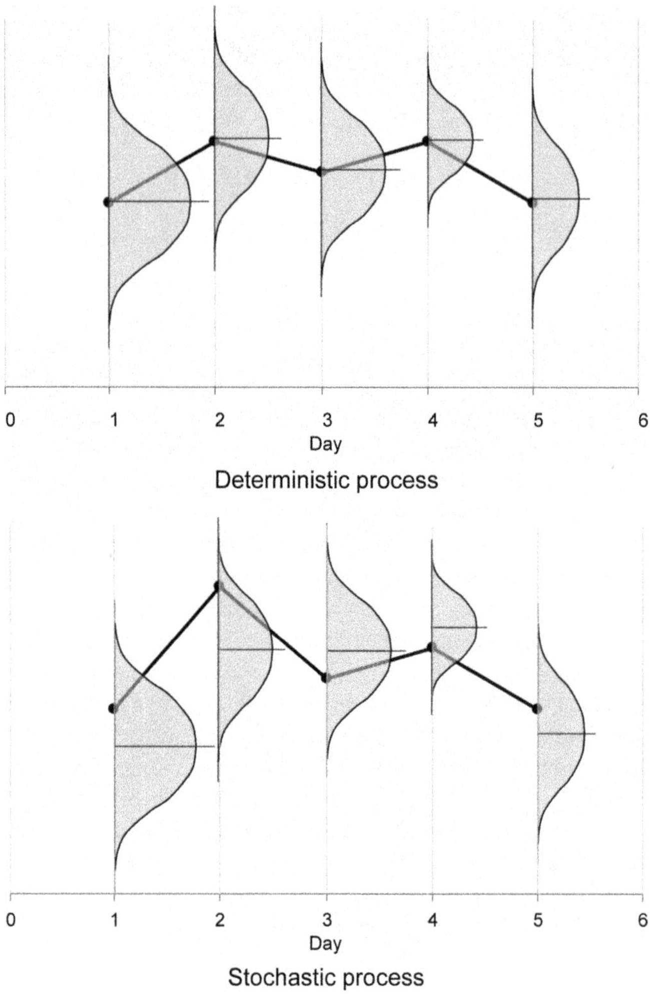

Fig. 6.10 Graphic representation of deterministic and stochastic process models

system over time, including in particular whether it converges and, if so, towards which subset of the state space.

In the theory of (nonlinear) time-discrete dynamic systems, given a transition function $\boldsymbol{x}^t = \psi(\boldsymbol{x}^{t-1})$ relating the state on day t to the state on the previous day $t-1$, any proper subset $A \subset S$ of the state space $S = \{\boldsymbol{x}^t\} \subseteq R^N$, having a dimension strictly smaller than the dimension N of S,[22] is called an *attractor* if:

[22]In other words, N is the number of the components of the vector that describes the state of the system.

	Types of attractor A	Number of points in A	Dimension of A ($<N$)
Nonchaotic	Fixed-point The system always occupies the same point	1	0
Trajectories starting from near states remain close	k-Periodic The system periodically occupies k points	k	0
	Quasi-periodic The system moves on a torus (or a set of tori)	∞	Integer
Chaotic	A-periodic The system moves in a fractal set	∞	Noninteger

Fig. 6.11 Attractors of a deterministic dynamic process

- The system cannot evolve towards a state outside the attractor starting from a state inside it;
- The attractor is properly contained in another subset $B \subseteq S$ (called the *basin* of A), such that if the initial state is contained in B the final state tends to be contained in A;
- A is minimal in the sense that it does not properly contain other attractors.

In other words, if the initial state is sufficiently close to the attractor, the system evolves towards it and, once reached, does not leave. Note that a system may have several attractors, each with its own basin.[23] A classification of attractors is given in Fig. 6.11 (examples are given in Fig. 6.12). If a fixed-point state, say $\boldsymbol{x}^* = \psi(\boldsymbol{x}^*)$ is reached, the system stops evolving, even though it may be not an attractor.

Recursive equations (6.5.12) and (6.5.13) are an instance of the transition function ψ relating the state on day t to the state on the previous day $t-1$:

$$(\boldsymbol{h}^t, \boldsymbol{V}^t) = \psi(\boldsymbol{h}^{t-1}, \boldsymbol{V}^{t-1}) \tag{6.5.14}$$

In the case of model (6.5.14) the dimension of the state space is $N = 2n_p$, where n_p is the number of paths.

If a fixed-point state $(\boldsymbol{h}^*, \boldsymbol{V}^*)$ (not necessarily an attractor) is reached, the evolution of the system stops:

$$(\boldsymbol{h}^t, \boldsymbol{V}^t) = (\boldsymbol{h}^{t-1}, \boldsymbol{V}^{t-1}) = (\boldsymbol{h}^*, \boldsymbol{V}^*)$$

that is,

$$(\boldsymbol{h}^*, \boldsymbol{V}^*) = \psi(\boldsymbol{h}^*, \boldsymbol{V}^*) \tag{6.5.15}$$

[23]The boundary points between different attractor basins are singular points of behavior (saddle points, e.g.) that can be ignored in a first analysis: small variations from such initial states move the development of the system towards the basin of an attractor.

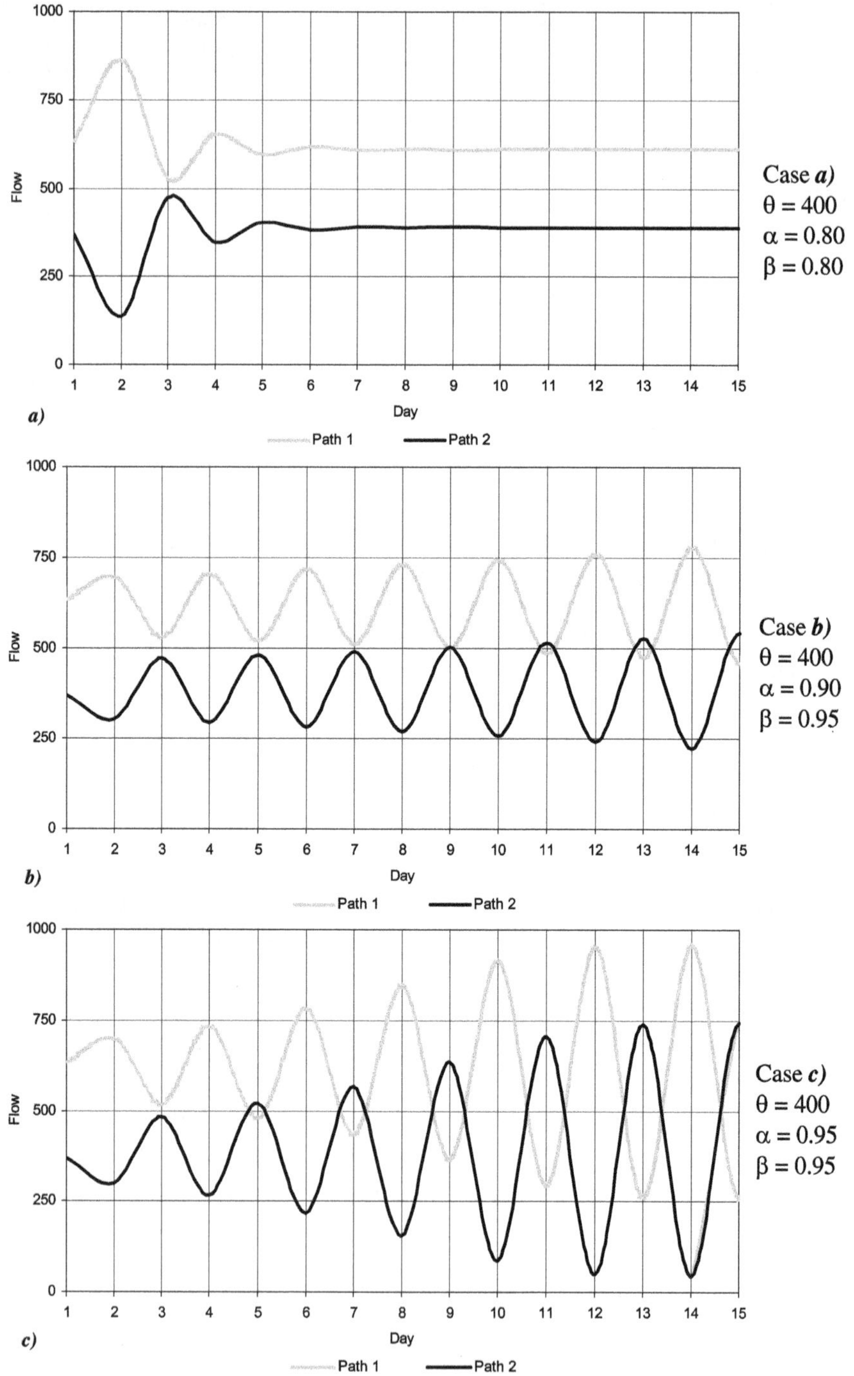

Fig. 6.12 Evolution of a deterministic process model for the system in Fig. 5.20. (The parameter θ has a different value here in order to highlight the evolution over time)

This condition, combined with (6.5.12) and (6.5.13), leads to:

$$V^* = V\big(V^*, \boldsymbol{\Delta}' c(\boldsymbol{\Delta} h^*)\big)$$
$$h^* = R(V^*)h^*$$

In general, fixed-point states depend on the utility and choice updating models (and are different from equilibrium states).

An example of a deterministic process is obtained by adopting exponential filter specifications for the utility and choice updating models presented in Sect. 6.5.1.2. In this case, (6.5.6) can be reformulated as

$$V^t = -\beta \boldsymbol{\Delta}^T c(\boldsymbol{\Delta} h^{t-1}) + (1-\beta)V^{t-1} \tag{6.5.16}$$
$$h^t = \alpha P(V^t)d + (1-\alpha)h^{t-1} \tag{6.5.17}$$

Similarly, the model can be expressed in terms of link flows and expected costs:

$$x^t = \beta c(f^{t-1}) + (1-\beta)x^{t-1} \tag{6.5.18}$$
$$f^t = \alpha f_{\mathrm{SUN}}(x^t) + (1-\alpha)f^{t-1} \tag{6.5.19}$$

Fixed-point states of the process defined by (6.5.16) and (6.5.17) are given by

$$g^* = \boldsymbol{\Delta}^T c(\boldsymbol{\Delta} h^*) \tag{6.5.20}$$
$$h^* = P(-g^*)d \tag{6.5.21}$$

and for the process defined by (6.5.18)–(6.5.20) in terms of link flows and costs by:

$$c^* = c(f^*) \tag{6.5.22}$$
$$f^* = f_{\mathrm{SUN}}(c^*) \tag{6.5.23}$$

In this case, it can be immediately verified that the formulations in terms of path and link variables are equivalent. Furthermore, the fixed-point states coincide with the stochastic user equilibrium states defined in Sect. 5.4, and the conditions of existence and uniqueness discussed still hold. Note also that the definition, existence, and uniqueness of fixed-points do not depend on the parameters α and β, which specify the choice and utility updating filters, respectively.[24]

Examples of the evolution of the transportation system described in Fig. 5.20 (Chap. 5) for different values of the parameters are given in Fig. 6.12. It should be noted that for some values of the parameters, link flows converge to a fixed-point state that coincides with the SUE configuration.

By applying the theory of nonlinear dynamic systems, it is possible to identify conditions ensuring that a fixed-point state is (*locally*) *stable*; that is, it has an attraction basin that is (a subset of) the state space S. In particular, if the transition

[24]This condition, which is not generally valid, can be extended to a larger class of cost (but not of choice) updating models.

function $\boldsymbol{x}^t = \psi(\boldsymbol{x}^{t-1})$ is continuous and differentiable with continuous Jacobian $\boldsymbol{Jac}[\psi(\boldsymbol{x}^{t-1})]$, a fixed-point $\boldsymbol{x}^*$ is stable if all the eigenvalues[25] λ_j^* of the Jacobian at the fixed-point $\boldsymbol{Jac}[\psi(\boldsymbol{x}^*)]$ have absolute values less than one: $|\lambda_j^*| < 1$. This condition constrains the eigenvalues to lie in the interior of a circle of unit radius on the complex plane (Argand plane).

To facilitate the comparison with equilibrium, the following analysis considers the model formulated in terms of link flows and costs (6.5.18), (6.5.19). Assume also that the transition function $(\boldsymbol{f}^t, \boldsymbol{c}^t) = \psi(\boldsymbol{f}^{t-1}, \boldsymbol{c}^{t-1})$ is continuous and differentiable with a continuous Jacobian $\boldsymbol{Jac}[\psi(\boldsymbol{f}^{t-1}, \boldsymbol{c}^{t-1})]$. It is worth noting that the transition function Jacobian, and therefore its eigenvalues, depend on the utility and choice updating models, which therefore influence the stability of a fixed-point state. In this case, the dynamic system is defined by $2n_L$ variables where n_L is the number of links and the Jacobian has $2n_L$ eigenvalues, two for each link a, denoted by λ_a and λ_{nL+a}. Under these assumptions, a fixed-point state defined by (6.5.22) and (6.5.23) is stable if all the eigenvalues of the Jacobian calculated at the fixed-point $\boldsymbol{Jac}[\psi(\boldsymbol{f}^*, \boldsymbol{c}^*)]$ have absolute values less than one:

$$\left|\lambda_a^*\right| < 1 \quad \forall a$$

$$\left|\lambda_{nL+a}^*\right| < 1 \quad \forall a$$

The Jacobian $\boldsymbol{Jac}[\psi(\boldsymbol{f}, \boldsymbol{c})]$ of the transition function $(\boldsymbol{x}, \boldsymbol{y}) = \psi(\boldsymbol{f}, \boldsymbol{c})$ for the model (6.5.18) and (6.5.19) at the point $(\boldsymbol{f}, \boldsymbol{c})$ is given by

$$\boldsymbol{Jac}\left[\psi(\boldsymbol{f}, \boldsymbol{c})\right] = \begin{array}{|c|c|} \hline (1-\beta)\boldsymbol{I} & \beta \boldsymbol{J}_c \\ \hline \alpha(1-\beta)\boldsymbol{J}_f & (1-\alpha)\boldsymbol{I} + \alpha\beta \boldsymbol{J}_f \boldsymbol{J}_c \\ \hline \end{array}$$

where

$\boldsymbol{J}_c = \boldsymbol{Jac}[\boldsymbol{c}(\boldsymbol{f})]$ is the Jacobian of the cost functions at point $\boldsymbol{f}$: if it is positive definite, the cost functions are strictly increasing

$\boldsymbol{J}_f = \boldsymbol{Jac}[\boldsymbol{f}_{\text{SUN}}(\boldsymbol{c})] = \sum_i d_i \boldsymbol{\Delta}_i \boldsymbol{Jac}[\boldsymbol{p}_i(-\boldsymbol{\Delta}_i^T \boldsymbol{c})]\boldsymbol{\Delta}_i^T$ is the Jacobian of the stochastic uncongested network assignment function; under the assumptions that guarantee the monotonicity of the SUN assignment function (see Sect. 5.3.1), it is symmetric and negative semidefinite

The elements of the Jacobian $\boldsymbol{Jac}[\psi(\boldsymbol{f}, \boldsymbol{c})]$ depend on the parameters α and β, which specify the choice and utility updating filters, respectively. Therefore the values of these parameters affect the stability of a fixed-point.

[25] An eigenvalue of a square matrix $\boldsymbol{J}$ is a number λ such that: $\boldsymbol{J\omega} = \lambda\boldsymbol{\omega}$, with $\boldsymbol{\omega} \neq \boldsymbol{0}$. The vector $\boldsymbol{\omega}$ is called an eigenvector of the matrix $\boldsymbol{J}$ corresponding to the eigenvalue λ. Eigenvalues are solutions of the algebraic equation $|\boldsymbol{J} - \lambda\boldsymbol{I}| = 0$ (where $|\cdot|$ denotes the determinant function), and are equal in number to the dimension of the matrix $\boldsymbol{J}$ (possibly with repetitions). A general real matrix may have real or complex eigenvalues (and eigenvectors) that occur in conjugate pairs; a symmetric matrix can only have real eigenvalues (and eigenvectors).

In the special case in which $\alpha = \beta = 1$, the Jacobian becomes

$$\boldsymbol{Jac}[\psi(\boldsymbol{f},\boldsymbol{c})] = \begin{array}{|c|c|}\hline \mathbf{0} & \boldsymbol{J}_C \\ \hline \mathbf{0} & \boldsymbol{J}_f\boldsymbol{J}_C \\ \hline \end{array}$$

and the eigenvalues are given by

$$\lambda_a = \gamma_j \quad \forall a$$

$$\lambda_{n+a} = 0 \quad \forall a$$

where γ_j is one of the n_L eigenvalues of the matrix $\boldsymbol{J}_f\boldsymbol{J}_C$.

The elements[26] of the matrix $\boldsymbol{J}_f\boldsymbol{J}_C$, and therefore its eigenvalues, depend on the parameters of the system such as the demand flows, the link capacities, the random residuals variance, and so on.

In the more general case, if $\alpha \in]0, 1]$ and/or $\beta \in]+0, 1]$, for each of the n_L eigenvalues of the matrix $\boldsymbol{J}_f\boldsymbol{J}_C$, two eigenvalues λ_a and λ_{nL+a} of the Jacobian $\boldsymbol{Jac}[\psi(\boldsymbol{f},\boldsymbol{c})]$ can be defined as a function of the parameters α and β. The stability condition can be rewritten as a function of the n_L eigenvalues γ_a; it is now represented by an ellipse on the complex plane that must contain the eigenvalues γ_a. In other words, if the system parameters are such that the points representing the n_L eigenvalues γ_a are contained in the ellipse, the fixed-point, or the system's equilibrium state, is stable. This ellipse, whose semiaxes depend only on the parameters α and β, is symmetrical with respect to the real axis and intersects it at two points (see Fig. 6.13).

In general, an increase in demand flows and/or a decrease in link capacities and/or a reduction in the variance of the random residuals tends to move the eigenvalues γ_a outside the stability region, whereas an increase in the parameters α and β tends to reduce the area of the region. Note that whatever the values of α and β, the ellipse is to the left of the point on the real axis with coordinates $\gamma_R = 1, \gamma_I = 0$. Therefore, if all the eigenvalues γ_a have a real part less than one, $\gamma_{R,a} < 1$, it is always possible to find sufficiently small values of the parameters α and β defining an ellipse that includes all the eigenvalues γ_a. In this case, the stability of the fixed-point would be ensured. If the parameters α and β do not satisfy this condition, the fixed-point is not stable even if it is unique and, according to results of the theory of nonlinear dynamic systems, the system may converge towards quasi-periodic or aperiodic attractors. Conversely, if some eigenvalues have a real part greater than one, $\gamma_{R,k} \geq 1$, the fixed-point is not stable for any values of α and β; yet there may be other (stable) fixed-points.

In the system described in Fig. 6.12, for example (for a given path choice and the supply models), as the parameters α and β increase, the system evolves towards attractors other than the fixed-point state, which in turn becomes unstable. This effect is shown in Fig. 6.13.

---

[26]The elements of the matrix $\boldsymbol{J}_f\boldsymbol{J}_C$, and therefore its eigenvalues, are dimensionless; the stability of a fixed-point is therefore not influenced by the unit of measurement adopted.

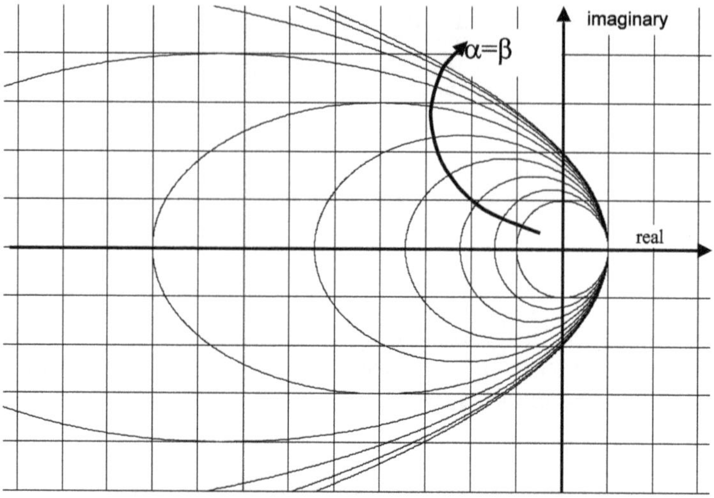

Fig. 6.13 Stability regions of a fixed-point state for $\alpha = \beta$

It is interesting to analyze the relationship between the above considerations and the stochastic equilibrium uniqueness conditions described in Sect. 5.4.1. In particular, it was shown that, under the assumption that the SUN assignment function is monotone with symmetric negative semidefinite Jacobian as described in Sect. 5.3.1, if the cost functions have a positive definite Jacobian $\boldsymbol{J}_c$ (strictly increasing) the stochastic equilibrium is unique. In this case, it can be shown that the eigenvalues γ_a of the matrix $\boldsymbol{J}_f \boldsymbol{J}_c$ always have a nonpositive real part, $\gamma_{R,a} \leq 0$. In accordance with the previous considerations, this excludes the possibility of multiple fixed-point states, and therefore of multiple equilibria. Also, it is interesting to note that if the Jacobian of the cost functions is symmetric, each of the eigenvalues γ_a of the matrix $\boldsymbol{J}_f \boldsymbol{J}_c$ is real (and nonpositive).

A deterministic process model can also be used as an algorithm to find fixed-point attractors, that is, stochastic equilibrium states. In this case, the model can be defined as a *dynamic process algorithm*; the parameters α and β have no behavioral interpretation and are simply chosen to guarantee the convergence of the algorithm (i.e., the stability of the fixed-point).[27]

6.5.3 Stochastic Process Models

Stochastic process models derive from the assumption that $(\boldsymbol{V}^t, \boldsymbol{h}^t)$, the predicted utilities and path flows on day t, are random vectors. These models allow one to ob-

[27] For this purpose, compare the algorithm MSA-FA, described in Sect. 5.4.2, with model (6.5.18)–(6.5.19), assuming: $\alpha = 1/t$, $\beta = 1$.

tain a statistical description of the system states and to model explicitly phenomena such as the randomness of link and path performance. The state of the system on day t, $(\boldsymbol{V}^t, \boldsymbol{h}^t)$, can therefore be interpreted as a particular realization of random vectors $(\boldsymbol{W}^t, \boldsymbol{X}^t)$. Expressions (6.5.12) and (6.5.13) define the expected values of $\boldsymbol{W}^t$ and $\boldsymbol{X}^t$ as a function of the state $(\boldsymbol{V}^{t-1}, \boldsymbol{h}^{t-1})$ on the previous day and the vector of actual path costs, expressed by the random vector $\boldsymbol{G}^{t-1}$:

$$\begin{aligned} &\boldsymbol{V}^t \leftarrow \boldsymbol{W}^t \\ &\text{with} \quad E[\boldsymbol{W}^t] = \boldsymbol{V}(\boldsymbol{V}^{t-1}, \boldsymbol{G}^{t-1}) \end{aligned} \tag{6.5.24}$$

$$\begin{aligned} &\boldsymbol{h}^t \leftarrow \boldsymbol{H}^t \\ &\text{with} \quad E[\boldsymbol{H}^t] = \boldsymbol{R}(\boldsymbol{V}^t)\boldsymbol{h}^{t-1} \end{aligned} \tag{6.5.25}$$

(where $x \leftarrow X$ means that x is a realization of the random variable X).

Equation (6.5.24) expresses the randomness of $\boldsymbol{V}^t$, the vector of average perceived utilities across users on day t. The expected value $E[\boldsymbol{W}^t]$ depends on the actual value of the average path cost on day $t-1$. The randomness of path costs $\boldsymbol{G}^{t-1}$ might simulate various factors. One of the most important is the uncertainty about link costs which, for a given value of h^{t-1}, might take on values $\boldsymbol{c}^{t-1}$ different from the average values $\boldsymbol{c}(\boldsymbol{\Delta h}^{t-1})$. In this case, the link costs $\boldsymbol{c}^{t-1}$ can be modeled as the realization of a random vector $\boldsymbol{C}^t$, and path costs are a linear transformation of $\boldsymbol{C}^t$:

$$\boldsymbol{c}^{t-1} \leftarrow \boldsymbol{C}^{t-1} \qquad \boldsymbol{g}^{t-1} \leftarrow \boldsymbol{G}^{t-1} = \boldsymbol{\Delta}^T \boldsymbol{C}^{t-1}$$

with

$$E[\boldsymbol{C}^{t-1}] = \boldsymbol{c}(\boldsymbol{\Delta h}^{t-1}) \qquad E[\boldsymbol{G}^{t-1}] = \boldsymbol{\Delta}^T \boldsymbol{c}(\boldsymbol{\Delta h}^{t-1})$$

The randomness of the path flow vector $\boldsymbol{h}^t$ derives from the unpredictability of user path choices. It is often assumed that the choices made by different individual users are independent of each other, but are all made with probabilities given by the demand model $\boldsymbol{R}(\boldsymbol{V}^t)$, that depends on the average perceived utilities across users. Under these assumptions $\boldsymbol{H}^t_{od}$, the path flow vector for O-D pair *od* on day t, is a multinomial random variable.

The stochastic process (6.5.24) and (6.5.25) is called a *discrete time, homogeneous Markov process*. It is Markovian because the state on the day t depends only on the state on the previous day $t-1$. It is homogeneous because the cost and network assignment functions and the cost and choice adjustment parameters are independent of the day. It is discrete time because the evolution over time is described by the integer day index t.

Given an initial state $(\boldsymbol{h}^o, \boldsymbol{V}^o)$, a model of this kind theoretically allows the determination, for each subsequent day t, of the probability that the system is in a particular state $(\boldsymbol{h}^1, \boldsymbol{V}^1)$ of the state space. The probability function $\phi^t(\boldsymbol{h}, \boldsymbol{V})$ is recursively defined as the probability that the system is in state $(\boldsymbol{h}^t, \boldsymbol{V}^t)$ on day t, conditional on its being in state $(\boldsymbol{h}^{t-1}, \boldsymbol{V}^{t-1})$ on the previous day $t-1$:

$$\phi^t(\boldsymbol{h}, \boldsymbol{V}) = Pr[\boldsymbol{h}^t = \boldsymbol{h}, \boldsymbol{V}^t = \boldsymbol{V} / \boldsymbol{h}^{t-1}, \boldsymbol{V}^{t-1}]$$

Under these assumptions, the path flows $\boldsymbol{h}^t$ are a realization of a discrete random vector, and the predicted utilities $\boldsymbol{V}^t$ are generally a realization of a continuous random vector. Thus the function $\phi^t(\boldsymbol{h}, \boldsymbol{V})$ must be considered a joint probability mass function with respect to $\boldsymbol{h}$, and a joint probability density function with respect to $\boldsymbol{V}$. In applications to transportation systems, it is often interesting to know the probabilities of path flows (and therefore of link flows). The marginal probability function $\pi^t(\boldsymbol{h})$ of the path flows $\boldsymbol{h}$ on day t is given by:

$$\pi^t(\boldsymbol{h}) = Pr[\boldsymbol{h}^t = \boldsymbol{h}/\boldsymbol{h}^{t-1}, \boldsymbol{V}^{t-1}] = \int_V \phi^t(\boldsymbol{h}, \boldsymbol{V})\,dV$$

According to the theory of stochastic processes, an *ergodic set* is a minimal subset of the state space such that there is a zero probability that the system transitions to a state outside it starting from a state inside it. An ergodic set is minimal in the sense that it does not properly contain any other ergodic subsets. With each ergodic set is associated a probability function, known as the *stationary probability distribution*, expressing the probability that, as $t \to \infty$, the system is in a state that belongs to the set:

$$\pi^*(\boldsymbol{h}) = \lim_{t\to\infty} \pi^t(\boldsymbol{h})$$

Only states belonging to the ergodic set have a nonzero stationary probability. A stochastic process is called *stationary* or *ergodic* if it has, respectively, at least one or only one stationary probability distribution $\pi^*(\boldsymbol{h})$. For the specific case discussed here, this stationary probability distribution is $\pi^*(\boldsymbol{h}, V)$.[28] A stochastic ergodic process is said to be *regular* if its probability distribution converges towards the unique stationary probability distribution regardless of the initial state (or its distribution).

In this case, a unique (stationary) probability distribution of the system state can be associated with each system specification independently of the initial state. The stationary probabilities $\pi^*(\boldsymbol{h})$, one for each vector $\boldsymbol{h}$ belonging to the ergodic set, can be interpreted as the probabilities of observing the system in the state corresponding to the path flow vector $\boldsymbol{h}$ during any period of observation t sufficiently far from the initial one. Relevant statistics (average, variances, etc.) can be calculated with a single (pseudo-)realization of the process, simulated with Monte Carlo techniques. The transient states visited from a given initial state toward a new stationary distribution following modifications in supply and/or demand can also be analyzed. The probability distribution of each day can be estimated by averaging several (pseudo-) realizations of the process for the same transient day t.

---

[28] A stochastic process can be interpreted as a deterministic process in the (infinite-dimensional) space of density functions $\pi(\boldsymbol{h})$, whose state on day t is given by $\pi^t(\boldsymbol{h})$. In this interpretation, an ergodic set is a fixed-point state of the deterministic process. The properties of stationarity, ergodicity, and regularity correspond to the existence, uniqueness, and (global) stability of this fixed-point state, which is a deterministic process attractor in that it is (globally) stable.

A special case, often adopted in applications,[29] is obtained if the randomness of the vector of average predicted utilities is ignored, that is, if it is assumed that $\boldsymbol{V}^t = \boldsymbol{W}^t$. This is equivalent to assuming that the costs realized on day t coincide with the average values given by the cost functions:

$$\boldsymbol{V}^t = \boldsymbol{V}\big(\boldsymbol{V}^{t-1}, \boldsymbol{\Delta}^T \boldsymbol{c}(\boldsymbol{\Delta} \boldsymbol{h}^{t-1})\big) \tag{6.5.26}$$

$$\begin{aligned} &\boldsymbol{h}^t \leftarrow \boldsymbol{H}^t \\ &\text{with} \quad E[\boldsymbol{H}^t] = \boldsymbol{R}(\boldsymbol{V}^t)\boldsymbol{h}^{t-1} \end{aligned} \tag{6.5.27}$$

In this case, the marginal probability $\pi^t(\boldsymbol{h})$ (i.e., the probability on day t of path flows $\boldsymbol{h}$) is given by

$$\begin{aligned} \pi^t(\boldsymbol{h}) &= Pr[\boldsymbol{h}^t = \boldsymbol{h}/\boldsymbol{h}^{t-1}, \boldsymbol{V}^{t-1}] \\ &= Pr\big[\boldsymbol{h}^t = \boldsymbol{h}, \boldsymbol{V}^t = \boldsymbol{V}\big(\boldsymbol{\Delta}^T \boldsymbol{c}(\boldsymbol{\Delta}\boldsymbol{h}^{t-1}), \boldsymbol{V}^{t-1}\big)/\boldsymbol{h}^{t-1}, \boldsymbol{V}^{t-1}\big] \end{aligned}$$

(An example of the more complicated case of a stochastic process with random costs and thus random expected utilities is presented in Sect. 7.6.2.)

It can be demonstrated that the regularity of stochastic processes defined by (6.5.26) and (6.5.27) is ensured given the rather general assumptions that the network is connected and that the cost functions and the SUN assignment function are continuous. In this case, therefore, a unique probability distribution of path and link flows can be associated with each demand and supply specification, independently of the initial state, and all relevant statistics can be calculated with a single (pseudo-)realization of the process, simulated by Monte Carlo techniques.

According to the law of large numbers, as demand flows increase, the evolution of the system described by a stochastic process better approximates the evolution of the corresponding deterministic process model. In this case, the expected values of the path and link flows resulting from a stochastic process model can be well approximated by a corresponding deterministic process, simulating the evolution of average values (the process of averages). From this point of view, stochastic process models seem more suitable for disaggregate detailed analyses, whereas deterministic processes are best suited for the modeling of the evolution of system averages at an aggregate level, and for equilibrium stability analyses.

An example of a stochastic process can be obtained by applying exponential filters for the specification of utility and choice updating models:

$$\boldsymbol{V}^t = -\beta \boldsymbol{\Delta}^T \boldsymbol{c}(\boldsymbol{\Delta}\boldsymbol{h}^{t-1}) + (1-\beta)\boldsymbol{V}^{t-1} \tag{6.5.28}$$

$$\begin{aligned} &\boldsymbol{h}^t \leftarrow \boldsymbol{H}^t \\ &\text{with} \quad E[\boldsymbol{H}^t] = \alpha \boldsymbol{P}(\boldsymbol{V}^t)\boldsymbol{d} + (1-\alpha)\boldsymbol{h}^{t-1} \end{aligned} \tag{6.5.29}$$

---

[29]The proposed formulation could easily be extended to consider the costs as random variables as well. Note, however, that by adopting a probabilistic path choice model, a perceived utility of randomness is introduced; this can also be attributed implicitly to the randomness of the attributes that appear in the systematic utility (in this case, the path costs).

Similarly, in terms of predicted link flows and costs, we have:

$$x^t = \beta c(f^{t-1}) + (1-\beta)x^{t-1} \tag{6.5.30}$$

$$f^t \leftarrow F^t \quad \text{with} \quad E[F^t] = \alpha f_{\text{SUN}}(x^t) + (1-\alpha)f^{t-1} \tag{6.5.31}$$

Another particular stochastic process model is a *renewal process*. These processes are such that the state on day t is a realization of a probability distribution independent from previous days. In this case, the Markovian property of the system expressed by (6.5.24) and (6.5.25) does not hold. This condition can be formally expressed as

$$\phi^t(h, V) = Pr[h^t = h, V^t = V/h^{t-1}, V^{t-1}] = Pr[h^t = h, V^t = V]$$

If the joint probability function $\phi^t(h, V)$ is constant for each t, the renewal process is stationary. Under these assumptions, renewal process models can simulate systems for which the expected (predicted) utilities of users are independent of the actual costs incurred on previous days (e.g., they may be based on long-term averages or on uncongested values) and there are no habit effects (e.g., $\alpha = 1$ in models (6.5.29) and (6.5.31)). An example of a renewal process model in the case of a stochastic supply model with random costs can be expressed by the following equations.

$$V^t = W^t \text{ (MVN variable)}$$

with

$$E[W^t] = -\Delta^T c_o = -g_o$$

$$h^t \leftarrow H^t \text{ (multinomial variable)}$$

with

$$E[H^t] = P(V^t)d$$

A renewal process model is specified for simulating within-day dynamic irregular transit systems in Sect. 7.6.2.

Finally, it should be noted that regularity of a stochastic process is a weaker property than fixed-point existence, uniqueness, and stability of the corresponding deterministic process. In other words, the stability of a system, in the engineering sense, requires not only that there exist a unique stationary distribution towards which the system state distribution converges, but also that the stationary distribution be unimodal.

Comparison between deterministic or stochastic models and equilibrium models (with fixed demand) is illustrated in Figs. 6.14 and 6.15 which report the various formulations. Further examples of dynamic process models are reported in Sect. 7.6.2.

	Single-class single-mode assignment	Assignment with pre-trip en-route	Assignment with variable demand	Multimode assignment	Differentiated congestion multiclass	Nondifferentiated congestion multiclass
Section	5.2–5.3–5.4	6.2	6.3.1	6.3.2	6.4.2	6.4.1
Supply model	$\boldsymbol{g}_{od} = \Delta_{od}^T \boldsymbol{c} + \boldsymbol{g}_{og}^{\text{NA}}$	$\boldsymbol{g}_{od}^{\text{NA}}$ in $\boldsymbol{x}_{od}^{\text{NA}}$	$\boldsymbol{g}_{od} = \boldsymbol{\Delta}_{od}^T \boldsymbol{c} + \boldsymbol{g}_{od}^{\text{NA}}$	$\boldsymbol{\Delta}_{od,m}^T \boldsymbol{c} + \boldsymbol{g}_{od\,m}^{\text{NA}}$	$\boldsymbol{g}_{od,i} = \Delta_{od,i}^T \boldsymbol{c}^i + \boldsymbol{g}_{og,i}^{\text{NA}}$	$\boldsymbol{g}_{od,i} = \gamma_i \boldsymbol{\Delta}_{od,i}^T \boldsymbol{c} + \boldsymbol{g}_{og,i}^{\text{spNA}}$
	$\boldsymbol{c} = \boldsymbol{c}(\boldsymbol{f})$	$\boldsymbol{c} = \boldsymbol{c}(\boldsymbol{f})$	$\boldsymbol{c} = \boldsymbol{c}(\boldsymbol{f})$	$\boldsymbol{c} = \boldsymbol{c}(\boldsymbol{f})$	$\boldsymbol{c}^i = \boldsymbol{c}^i(\sum_i \boldsymbol{f}^i)$	$\boldsymbol{c} = \boldsymbol{c}(\boldsymbol{f})$
	$\boldsymbol{f} = \sum_{od} \boldsymbol{\Delta}_{od} \boldsymbol{h}_{od}$	$\boldsymbol{f} = \sum_{od} \boldsymbol{\Delta}_{od} \boldsymbol{h}_{od}$	$\boldsymbol{f} = \sum_{od} \boldsymbol{\Delta}_{od} \boldsymbol{h}_{od}$	$\sum_m \sum_{od} \boldsymbol{\Delta}_{od,m} \boldsymbol{h}_{od,m}$	$\boldsymbol{f}^i = \sum_{od} \boldsymbol{\Delta}_{od,i} \boldsymbol{h}_{od,i}$	$\boldsymbol{f} = \sum_i \sum_{od} \boldsymbol{\Delta}_{od,i} \boldsymbol{h}_{od,i}$
Demand model	$\boldsymbol{V}_{od} = -\boldsymbol{g}_{od} + \boldsymbol{V}_{od}^\circ$	$\boldsymbol{V}_{od} = -\Omega_{od}^T \boldsymbol{g}_{od} - \boldsymbol{x}_{od}^{\text{NA}} + V_{od}^\circ$	$\boldsymbol{V}_{od} = -\boldsymbol{g}_{od} + \boldsymbol{V}_{od}^\circ$	$\boldsymbol{V}_{od,m} = -\boldsymbol{g}_{od,m} + V_{od,m}^\circ$	$\boldsymbol{V}_{od,i} = -\boldsymbol{g}_{od,i} + V_{od,I}^\circ$	$\boldsymbol{V}_{od,i} = -\boldsymbol{g}_{od,i} + V_{od,I}^\circ$
	$\boldsymbol{h}_{od} = d_{od} \boldsymbol{p}_{od}(\boldsymbol{V}_{od})$	$\boldsymbol{h}_{od} = d_{od} \boldsymbol{\Omega}_{od} \boldsymbol{q}_{od}(\boldsymbol{V}_{od})$	$\boldsymbol{h}_{od} = d_{od}(\boldsymbol{s}(\boldsymbol{V})) \boldsymbol{p}_{od}(\boldsymbol{V}_{od})$	$\boldsymbol{h}_{od,m} = d_{od,m} \boldsymbol{p}_{od,m}(\boldsymbol{V}_{od,m})$	$\boldsymbol{h}_{od,i} = d_{od,i} \boldsymbol{p}_{od,i}(\boldsymbol{V}_{od,i})$	$\boldsymbol{h}_{od,i} = d_{od,i} \boldsymbol{p}_{od,i}(\boldsymbol{V}_{od,i})$

Fig. 6.14a Synopsis of user equilibrium assignment models

Section	Single-class single-mode assignment 5.2–5.3–5.4	Assignment with pre-trip en-route 6.2	Assignment with variable demand 6.3.1	Multimode assignment 6.3.2	Differentiated congestion multiclass 6.4.2	Nondifferentiated congestion multiclass 6.4.1
Uncongest. network assignment MAP	$\boldsymbol{f}_{\text{UN}}(\boldsymbol{c};\boldsymbol{d}) = \sum_{od} d_{od}\boldsymbol{\Delta}_{od} \times \boldsymbol{p}_{od}(-\boldsymbol{\Delta}_{od}^T\boldsymbol{c} - \boldsymbol{g}_{od}^{\text{NA}})$	$\boldsymbol{f}_{\text{UN}}(\boldsymbol{c};\boldsymbol{d}) = \sum_{od} d_{od}\boldsymbol{\Lambda}_{od} \times \boldsymbol{q}_{od}(-\boldsymbol{\Lambda}_{od}^T\boldsymbol{c} - \boldsymbol{x}_{od}^{\text{NA}})$	$\boldsymbol{f}_{\text{UN}}(\boldsymbol{c}) = \sum_{od} d_{od}(\boldsymbol{s}(-\boldsymbol{\Delta}^T\boldsymbol{c} - \boldsymbol{g}^{\text{NA}}))\boldsymbol{\Delta}_{od} \times \boldsymbol{p}_{od}(-\boldsymbol{\Delta}_{od}^T\boldsymbol{c} - \boldsymbol{g}_{od}^{\text{NA}})$	$\boldsymbol{f}_{\text{UN}}(\boldsymbol{c}) = \sum_{m,od} d_{od,m} \times (\boldsymbol{s}(-\boldsymbol{\Delta}^T\boldsymbol{c} - \boldsymbol{g}^{\text{NA}})) \times \boldsymbol{\Delta}_{od,m}\boldsymbol{p}_{od,m} \times (-\boldsymbol{\Delta}_{od,m}^T\boldsymbol{c} - \boldsymbol{g}_{od,m}^{\text{NA}})$	$\boldsymbol{f}_{\text{UN}}^i(\boldsymbol{c}^i;\boldsymbol{d}_i) = \sum_{od} d_{od,i}\boldsymbol{\Delta}_{od,i} \times \boldsymbol{p}_{od,i}(-\boldsymbol{\Delta}_{od,i}^T\boldsymbol{c}^i - \boldsymbol{g}_{od,i}^{\text{NA}})$	$\boldsymbol{f}_{\text{UN}}(\boldsymbol{c};\boldsymbol{d},\boldsymbol{\gamma}) = \sum_{i,od} d_{od,i}\boldsymbol{\Delta}_{od,i} \times \boldsymbol{p}_{od,i}(-\gamma_i\boldsymbol{\Delta}_{od,i}^T\boldsymbol{c}^i - \boldsymbol{g}_{od,i}^{\text{SPNA}})$
Link-based equations for equilibrium	$\boldsymbol{c}^* = \boldsymbol{c}(\boldsymbol{f}^*)$ $\boldsymbol{f}^* = \boldsymbol{f}_{\text{UN}}(\boldsymbol{c}^*;\boldsymbol{d})$	$\boldsymbol{c}^* = \boldsymbol{c}(\boldsymbol{f}^*)$ $\boldsymbol{f}^* = \boldsymbol{f}_{\text{UN}}(\boldsymbol{c}^*;\boldsymbol{d})$	$\boldsymbol{c}^* = \boldsymbol{c}(\boldsymbol{f}^*)$ $\boldsymbol{f}^* = \boldsymbol{f}_{\text{UN}}(\boldsymbol{c}^*)$	$\boldsymbol{c}^* = \boldsymbol{c}(\boldsymbol{f}^*)$ $\boldsymbol{f}^* = \boldsymbol{f}_{\text{UN}}(\boldsymbol{c}^*)$	$\boldsymbol{c}^{i*} = \boldsymbol{c}^i(\sum_i \boldsymbol{f}^{i*})$ $\boldsymbol{f}^{i*} = \boldsymbol{f}_{\text{UN}}^i(\boldsymbol{c}^{i*};\boldsymbol{d})$	$\boldsymbol{c}^* = \boldsymbol{c}(\boldsymbol{f}^*)$ $\boldsymbol{f}^* = \boldsymbol{f}_{\text{UN}}(\boldsymbol{c}^*;\boldsymbol{d},\boldsymbol{\gamma})$
Fixed-point model for equilibrium	$\boldsymbol{f}^* = \boldsymbol{f}_{\text{UN}}(\boldsymbol{c}(\boldsymbol{f}^*);\boldsymbol{d})$	$\boldsymbol{f}^* = \boldsymbol{f}_{\text{UN}}(\boldsymbol{c}(\boldsymbol{f}^*);\boldsymbol{d})$	$\boldsymbol{f}^* = \boldsymbol{f}_{\text{UN}}(\boldsymbol{c}(\boldsymbol{f}^*))$	$\boldsymbol{f}^* = \boldsymbol{f}_{\text{UN}}(\boldsymbol{c}(\boldsymbol{f}^*))$	$\boldsymbol{f}^{i*} = \boldsymbol{f}_{\text{UN}}^i(\boldsymbol{c}^i(\sum_i \boldsymbol{f}^{i*});\boldsymbol{d})$	$\boldsymbol{f}^* = \boldsymbol{f}_{\text{UN}}(\boldsymbol{c}(\boldsymbol{f}^*);\boldsymbol{d},\boldsymbol{\gamma})$

Fig. 6.14b Synopsis of user equilibrium assignment models

	Single-class single-mode assignment with fixed demand	Deterministic process (general example)	Deterministic process (simple example)	Stochastic process (general example)	Stochastic process (simple example)
Section	5.2–5.3–5.4	6.5.2	6.5.2	6.5.3	6.5.3
Supply model	$\boldsymbol{g}_{od} = \boldsymbol{\Delta}_{od}^T \boldsymbol{c} + \boldsymbol{g}_{od}^{\mathrm{NA}}$	$\boldsymbol{g}_{od}^t = \boldsymbol{\Delta}_{od}^T \boldsymbol{c}^t + \boldsymbol{g}_{od}^{\mathrm{NA}}$	$\boldsymbol{g}_{od}^t = \boldsymbol{\Delta}_{od}^T \boldsymbol{c}^t + \boldsymbol{g}_{od}^{\mathrm{NA}}$	$\boldsymbol{g}_{od}^t = \boldsymbol{\Delta}_{od}^T \boldsymbol{c}^t + \boldsymbol{g}_{od}^{\mathrm{NA}}$	$\boldsymbol{g}_{od}^t = \boldsymbol{\Delta}_{od}^T \boldsymbol{c}^t + \boldsymbol{g}_{od}^{\mathrm{NA}}$
	$\boldsymbol{c} = \boldsymbol{c}(\boldsymbol{f})$	$\boldsymbol{c}^t = \boldsymbol{c}(\boldsymbol{f}^t)$	$\boldsymbol{c}^t = \boldsymbol{c}(\boldsymbol{f}^t)$	$\boldsymbol{c}^t = \boldsymbol{c}(\boldsymbol{f}^t)$	$\boldsymbol{c}^t = \boldsymbol{c}(\boldsymbol{f}^t)$
	$\boldsymbol{f} = \sum_{od} \boldsymbol{\Delta}_{od} \boldsymbol{h}_{od}$	$\boldsymbol{f}^t = \sum_{od} \boldsymbol{\Delta}_{od} \boldsymbol{h}_{od}^t$	$\boldsymbol{f}^t = \sum_{od} \boldsymbol{\Delta}_{od} \boldsymbol{h}_{od}^t$	$\boldsymbol{f}^t = \sum_{od} \boldsymbol{\Delta}_{od} \boldsymbol{h}_{od}^t$	$\boldsymbol{f}^t = \sum_{od} \boldsymbol{\Delta}_{od} \boldsymbol{h}_{od}^t$
Demand model	$\boldsymbol{V}_{od} = -\boldsymbol{g}_{od} + \boldsymbol{V}_{od}^{\circ}$	$\boldsymbol{V}_{od}^t = \boldsymbol{V}_{od}(\boldsymbol{g}_{od}^{t-1}, \boldsymbol{V}_{od}^{t-1})$	$\boldsymbol{V}_{od}^t = -\beta \boldsymbol{g}_{od}^{t-1} + (1-\beta)\boldsymbol{V}_{od}^{t-1}$	$\boldsymbol{V}_{od}^t = \boldsymbol{V}_{od}(\boldsymbol{g}_{od}^{t-1}, \boldsymbol{V}_{od}^{t-1})$	$\boldsymbol{V}_{od}^t = -\beta \boldsymbol{g}_{od}^{t-1} + (1-\beta)\boldsymbol{V}_{od}^{t-1}$
	$\boldsymbol{h}_{od} = d_{od} \boldsymbol{p}_{od}(\boldsymbol{V}_{od})$	$\boldsymbol{h}_{od}^t = \boldsymbol{R}_{od}(\boldsymbol{V}_{od}^t)\boldsymbol{h}_{od}^{t-1}$	$\boldsymbol{h}_{od}^t = \alpha d_{od} \boldsymbol{p}_{od}(\boldsymbol{V}_{od}^t) + (1-\alpha)\boldsymbol{h}_{od}^{t-1}$	with $E[\boldsymbol{H}_{od}^t] = \boldsymbol{R}_{od}(\boldsymbol{V}_{od}^t)\boldsymbol{h}_{od}^{t-1}$	with $E[\boldsymbol{H}_{od}^t] = \alpha d_{od} \boldsymbol{p}_{od}(\boldsymbol{V}_{od}^t) + (1-\alpha)\boldsymbol{h}_{od}^{t-1}$

Fig. 6.15a Synopsis of user equilibrium and dynamic process assignment models (for simplicity, systematic utility is not considered to be a random variable)

	Single-class single-mode assignment with fixed demand	Deterministic process (general example)	Deterministic process (simple example)	Stochastic process (general example)	Stochastic process (simple example)
Section	5.2–5.3–5.4	6.5.2	6.5.2	6.5.3	6.5.3
Uncongest. network assignment MAP	$\boldsymbol{f}_{\mathrm{UN}}(\boldsymbol{c};\boldsymbol{d}) = \sum_{od} d_{od}\boldsymbol{\Delta}_{od} \times \boldsymbol{p}_{od}(-\boldsymbol{\Delta}_{od}^T\boldsymbol{c} - \boldsymbol{g}_{od}^{\mathrm{NA}})$	Not available	$\boldsymbol{f}_{\mathrm{UN}}(\boldsymbol{c};\boldsymbol{d}) = \sum_{od} d_{od}\boldsymbol{\Delta}_{od} \times \boldsymbol{p}_{od}(-\boldsymbol{\Delta}_{od}^T\boldsymbol{c} - \boldsymbol{g}_{od}^{\mathrm{NA}})$	Not available	$\boldsymbol{f}_{\mathrm{UN}}(\boldsymbol{c};\boldsymbol{d}) = \sum_{od} d_{od}\boldsymbol{\Delta}_{od} \times \boldsymbol{p}_{od}(-\boldsymbol{\Delta}_{od}^T\boldsymbol{c} - \boldsymbol{g}_{od}^{\mathrm{NA}})$
System of link-based equations	$\boldsymbol{c}^* = \boldsymbol{c}(\boldsymbol{f}^*)$ $\boldsymbol{f}^* = \boldsymbol{f}_{\mathrm{UN}}(\boldsymbol{c}^*;\boldsymbol{d})$	Not available	$\boldsymbol{c}^t = \beta\boldsymbol{c}(\boldsymbol{f}^{t-1}) + (1-\beta)\boldsymbol{c}^{t-1}$ $\boldsymbol{f}^t = \alpha\boldsymbol{f}_{\mathrm{UN}}(\boldsymbol{x}^t) + (1-\alpha)\boldsymbol{f}^{t-1}$	Not available	$\boldsymbol{x}^t = \beta\boldsymbol{c}(\boldsymbol{f}^{t-1}) + (1-\beta)\boldsymbol{x}^{t-1}$ $\boldsymbol{f}^t \leftarrow \boldsymbol{F}^t$ with $E[\boldsymbol{h}^t] = \alpha\boldsymbol{f}_{\mathrm{UN}}(\boldsymbol{x}^t) + (1-\alpha)\boldsymbol{f}^{t-1}$

Fig. 6.15b Synopsis of user equilibrium and dynamic process assignment models (for simplicity, systematic utility is not considered to be a random variable)

6.6 Synthesis and Application Issues

The assignment models described in the previous sections are summarized, using the notation introduced in this chapter, in Fig. 6.14 where different models for user equilibrium assignment are compared, and in Fig. 6.15 where "basic" equilibrium models are compared with dynamic process assignment models.

In general, in the case of uncongested networks and fixed demand, the assignment model defines a relationship between output link flows and costs, and input demand flows on the other hand. This relationship is expressed by the UN assignment map. In the case of congested networks and/or variable demand, the assignment relationship includes link cost functions and/or demand functions as inputs; this relationship is implicitly defined by equilibrium assignment or dynamic process models.

Reference Notes

Extension of deterministic assignment to pre-trip/en-route path choice behavior for transit networks was proposed by Nguyen and Pallottino (1988) and Spiess and Florian (1989). Extensions of DUE assignment models to transit networks and the analysis of its theoretical properties can be found in Nguyen and Pallottino (1988) and Wu et al. (1994). Recently, Bouzaiene-Ayari et al. (1995, 1997) analyzed several approaches to represent user behavior at a bus stop within assignment models, including congested waiting times. The algorithm for computing shortest hyperpaths and its applications extension to DUE appear in Nguyen and Pallottino (1988), as well as Spiess and Florian (1989), and Wu et al. (1994). A comprehensive review of hyperpaths and related topics may be found in Gallo et al. (1993). Algorithms for stochastic assignment with logit hyperpath choice are described by Nguyen et al. (1993). Extension to stochastic assignment was analyzed by Cantarella (1997) and Cantarella and Vitetta (2000).

As already stated in the bibliographical note in Chap. 5, the use of fixed-point models for stochastic (or deterministic) equilibrium was introduced by Daganzo (1983), who also analyzed multiclass assignment with variable demand (following the hypernetwork approach cited). Cantarella (1997) developed a general treatment using fixed-point models of multimodal/multiclass variable demand equilibrium assignment, pre-trip/en-route path choice behavior, including stochastic as well as deterministic user equilibrium.

Internal cycle algorithms for assignment with variable demand are treated by Cantarella (1997). External cycle algorithms constitute an original contribution in this book. A general presentation may be found in Cantarella and Cascetta (1998). See also the bibliographical survey reported therein.

A further line of development consists of equilibrium models in which a (noncongested) cost attribute such as monetary cost is distributed among users, following, for example, the distribution of the value of time (for application to stochastic assignment and an analysis of the literature, see Cantarella and Binetti (1998)). Such models may be considered an extension of multiclass assignment models to an infinite number of classes should the value of time be represented by a continuous random variable.

In recent years, dynamic process (nonequilibrium) models for representing supply–demand interaction have received increasing attention from transportation system analysis researchers (Cantarella and Cascetta 1995). Initial contributions (Daganzo and Sheffi 1977; and Horowitz 1984) analyzed particular models for the study of equilibrium stability; Cascetta (1987, 1989) proposed stochastic process models to represent supply–demand interactions rather than to analyze equilibrium configurations. Since then, stochastic and deterministic process models have been proposed by various authors including Davis and Nihan (1993) and Watling (1996, 1999).

Chapter 7
Intraperiod (Within-Day) Dynamic Models*

7.1 Introduction

The models presented in the previous chapters describe the steady-state behavior of the transport system. Invariance in time of the variables concerned means it can be represented by a single snapshot; a representation that we could therefore call *static*. The result of an equilibrium assignment, for example, photographs that particular condition of system behavior in which path (or link) flows and costs are mutually consistent and, as stated above, stationary. This configuration could be observed in the real world only if demand, path choices, and supply system remained constant for a sufficiently long period of time that the system could reach a steady-state condition. Thus, although difficult to observe in reality, as described in Chap. 1, it may be assumed as being representative of average system conditions in the simulation period adopted.

However, the simplifications induced by this assumption of stationarity do not allow the system's internal behavior to be represented. Indeed, with this approach we only know the model response – path and link costs and flows – to the input stimulus: O-D demand flows. This correspondence is then independent of the previous system history (which is why such models are also known as *models without memory*). In other words, the models that follow this approach do not allow us to explicitly represent the inner dynamics of the system and describe its causal evolution in time. For these reasons, in systems theory one refers to them as *static models.*[1]

Clearly, given the impossibility of representing the internal behavior of the system we cannot effectively simulate transport systems such as highly congested urban road systems. Indeed, in this case it is impossible to represent the dynamics of some important phenomena, especially the creation, propagation, and dissipation of queues (in other words, congestion), a phenomenon that occurs, for example, when a link is temporarily oversaturated. Just as it is not possible to reproduce effects caused by nonstationary demand or supply (see, e.g., demand peaks in rush hours or

---

[1]A simple example of a static model for a physical system may be taken from standard circuit theory. Consider a circuit consisting of two resistances in series, r and R, linked around a source of voltage V, and of an intermediate socket between the two resistances (voltage divider system). Every time the voltage V is varied at the source (input), we obtain a different voltage v at the intermediate socket (output). This adaptation is not instantaneous despite the very high velocity at which it occurs. A complete description of the phenomenon would require assessment of the transient phase. For practical purposes it is sufficient to know only the final voltage v and we can thus adopt a simple input–output functional relationship that can be obtained from standard circuit theory: $v = V \cdot R/(R + r)$. As this formula is algebraic, each value of V has only one corresponding value of v. Note that the model is unable to describe the transient that follows the variation in voltage in input V, which is why it is called *static*.

E. Cascetta, *Transportation Systems Analysis*,
Springer Optimization and Its Applications 29,
DOI 10.1007/978-0-387-75857-2_7,

temporary reductions in link capacity), that is, the effects of all real-time control or user information strategies (i.e., strategies based on knowledge of the current system state; see, e.g., traffic responsive systems). Moreover, for low-frequency public transport services (e.g., two flights a day), it is worth questioning the assumptions introduced in previous chapters of supply stationarity (and continuity) and of mixed preventive–adaptive user choice behavior. In summary, it may be stated that where transient system behavior needs to be evaluated, or furthermore, the system itself functions only transiently, a modeling approach based on the stationarity hypothesis clearly proves ineffective.

For the above reasons various dynamic models were recently developed, usually termed *Dynamic Traffic Assignment* models (*DTA*) or also models with within-day dynamics, given their capacity to reproduce dynamics within the reference period of the system in question. The need to explicitly simulate the system's inner workings and reproduce its evolution in time leads to major reformulation of the models described in the previous chapters, as regards the demand models and, to an even greater extent, supply models. Especially in the latter, simple algebraic relations that, under the assumption of within-day stationarity, correlate the variables involved, are in many cases substituted by differential equations that describe the evolution in time of the same variables.

It is worth noting that the flow configuration resulting from a static equilibrium assignment may be viewed within the more general framework of dynamic assignment. Indeed, the response of a dynamic model in which both the input variables and the supply model are stationary may verge towards the configuration of flows and costs identified by static equilibrium assignment.[2]

This chapter therefore covers supply, demand, and within-day assignment models. Their formulation and level of complexity depend on the type of supply system concerned. As we saw in Chap. 2, transport services and corresponding supply models may be divided into two main classes: continuous and discrete. The former refers to services available at any moment and accessible from arbitrary locations, such as services supplied by individual road modes (car, bicycle, etc.). The latter concerns services available only at certain times that can be accessed only from specific locations (stops, terminals, stations, airports, etc.).

Hence in Sect. 7.2 we introduce within-day dynamic supply models for continuous services that, as stated above, are significantly more complex than static ones. For the same services, the following section presents demand models, which are a direct extension of static ones. Section 7.4 covers assignment models, obtained by combining supply models and demand models. Section 7.5 presents a specific formulation that allows solving the assignment problem on large road networks. For discrete services, given their intrinsically discrete structure in time and space, static models of supply, demand, and assignment may be directly extended with within-day dynamics (see Sect. 7.6). Finally Appendix 7.A provides some details about supply models applied in Sect. 7.5.

---

[2] At least within the limits in which dynamic supply models are consistent with static ones.

7.2 Supply Models for Transport Systems with Continuous Service

The first dynamic models of traffic flow proposed in the 1950s represented traffic flow basing on the analogy with the lines of water flow in rivers. Using this approach, individual vehicles are treated as a continuous (one-dimensional) fluid, for which variables such as flow, density, and velocity can be defined at each point in space and time. Evolution in time of these state variables is modeled using a partial derivative equation that comprises both the conservation of mass (vehicles) and an experimental relation between flow and density. In accordance with a classification based on the level of detail in the models, this approach to modeling vehicle flow is called *macroscopic*.

Macroscopic models may be further divided according to their representation of space, assuming that time is always treated continuously. In *space-continuous* models the state variables are defined at each point in space, even if their resolution generally requires discretization both of space and time. By contrast, *space-discrete* models (known also as *link-based* models) are closer to static models: the basic variables affecting link performance, such as density or speed, do not vary along the link. Their resolution requires, however, at least a discretization of time.

Mesoscopic models represent road flow at the level of detail of the single vehicle (or group of vehicles generally called a *packet*). In this case, although representation of the traffic is discrete, the movement of each individual vehicle depends on laws that describe relations between aggregate flow variables (e.g., mean speed as a function of density), or on probabilistic functions (see models based on the analogy with gas kinetics, which describe speed distribution dynamics).

Microscopic models describe the movements of individual vehicles as the result of individual disaggregate choices and interactions with other vehicles and with the road environment. Path choice, decisions to accelerate or change lanes, behavior at intersections, and so on, of each individual vehicle, are generally explicitly modeled. Moreover, each flow entity has its own characteristics that may include: vehicle characteristics, such as type or access to trip information; vehicle performance, such as maximum acceleration or maximum speed; and driver characteristics, such as reaction time or desired speed.

Microscopic models consider the driver–vehicle system as a single entity and do not simulate closed-cycle interactions between the driver and vehicle. However, the control process of the vehicle effected by the driver is captured by *submicroscopic* models that explicitly represent activities such as steering, gear-changing, or brake- and accelerator-pedal control. Use of such models is limited to detailed analysis of the vehicle–driver–environment system and is not extended to flow propagation problems on the network, which is why it is not included in the classification introduced below.

Figure 7.1 presents a classification of nonstationary network flow models for continuous services. The classification proposed is based both on the representation of user flow (continuous or discrete) and on the performance functions adopted (aggregate or disaggregate).

<table>
<tr><th rowspan="2">Flow Representation</th><th colspan="3">Performance functions</th></tr>
<tr><th colspan="2">AGGREGATE
(explicit capacity)</th><th>DISAGGREGATE</th></tr>
<tr><td rowspan="2">CONTINUOUS</td><td colspan="2">MACRO-SIMULATION</td><td rowspan="2">•</td></tr>
<tr><td>space discrete</td><td>space continuous</td></tr>
<tr><td>DISCRETE</td><td colspan="2">MESO-SIMULATION</td><td>MICRO-SIMULATION</td></tr>
</table>

Fig. 7.1 Classification of dynamic supply models for continuous services

The field of application of the various approaches varies appreciably. For example, macro- and mesoscopic models are particularly suitable for simulating the road system in designing control strategies: they permit explicit state–space modeling, which may be easily included in optimization schemes. Although dependent on network dimensions and processing capacity, the calculation time for such models is generally appreciably lower than real-time, allowing their use in online applications. By contrast, in microscopic simulation models the calculation times increase considerably with the increase in road system congestion, that is, with the increase in vehicle numbers to be simulated, generally restricting their use to offline test applications.

Aside from the above considerations, the choice of model to be used is generally influenced by the level of detail required by the application. For example, when the individual behavior of system users has to be simulated, disaggregate models such as microscopic ones have to be used.

However, it is worth noting that a more detailed simulation model does not necessarily supply more accurate results. This is why great importance is attached to the availability of real data on which to calibrate such models. The quantity and complexity of data to be recorded increase with the level of detail of the approach adopted, and might thus be a significant reason for choosing the approach itself.

From the modeling point of view, macroscopic and mesoscopic dynamic supply models, which adopt aggregate performance functions, express system flows and performance according to the path flows and physical characteristics of the system as in the static case. Figure 7.2 shows the general structure of such a dynamic supply model.

As may be noted, this structure is very similar to that of static models, introduced in Chap. 2. The only difference lies in the dependence of the flow propagation on link performance: the number of users on a link at a given moment depends on travel times required by users to reach that link which, in turn, depends on the number of users encountered on the network links in the previous instants (i.e., on the congestion level encountered). This circular dependence between the network flow propagation model and the link performance model does not allow the resolution

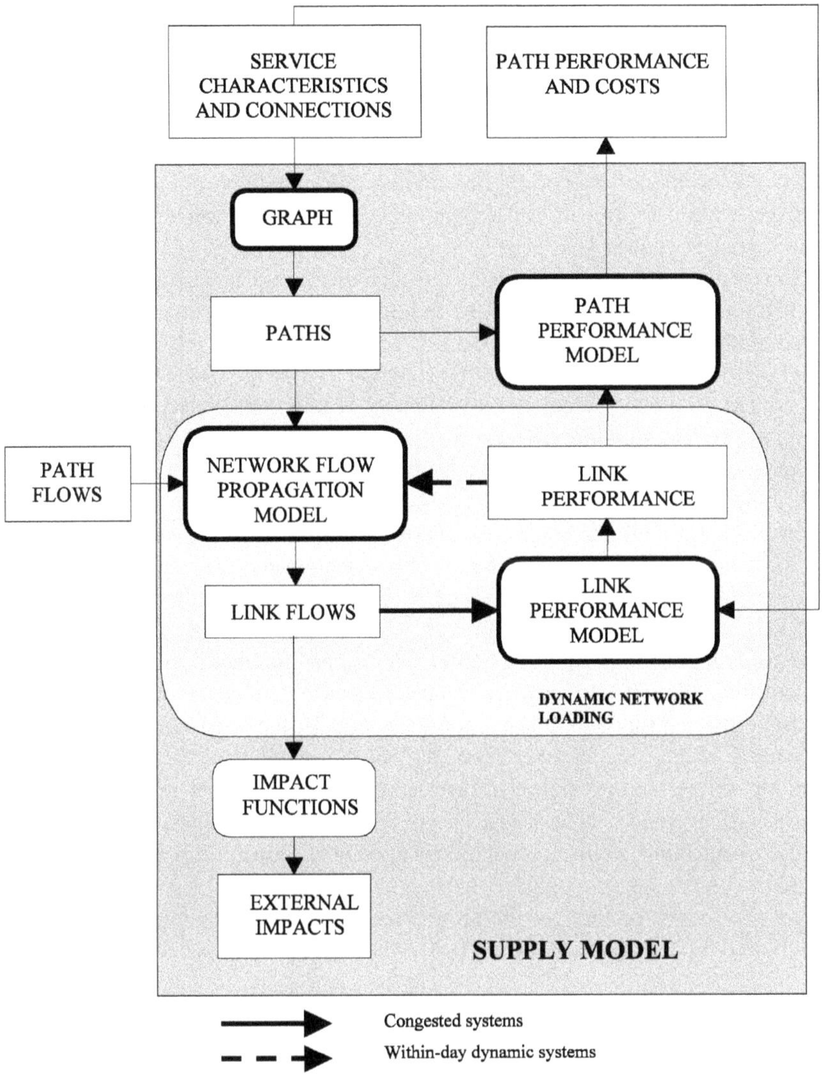

Fig. 7.2 Diagram of supply models with within-day dynamics

of the two models in sequential fashion, as in the static case, and gives rise to a problem known in the literature as *Dynamic Network Loading* (DNL). Although the structure of the dynamic supply model, as represented in Fig. 7.2, is formally analogous to the static case, the relations comprising the models and generally the set of relations that allow us to express path costs as a function of path flows (overall supply model) are no longer linear nor algebraic, but often differential (e.g., in the time variable). In this regard, the introduction of time dependence of the variables ensures that travel times assume a twofold role: as in the static case, they represent part of the generalized cost perceived by the user, that is, link and path costs; on the other hand, they enter as an independent variable in the various components of the supply model, ensuring its temporal consistency.

As regards the individual components of the supply model, for macroscopic and mesoscopic models all the concepts and notation concerning the graph models introduced for continuous transport services in the static case (centroid nodes, paths, etc.) are extended directly to within-day dynamics. The other components are covered in the following sections.

Microscopic supply models deserve special mention. For such models, no structure like that represented in Fig. 7.2 can be identified. As stated above, the performance functions of these models are disaggregate; that is, they refer to the individual vehicle. Thus, link and path performance (such as average travel time on a link in a time interval) are not calculated explicitly, but can be obtained downstream of the simulation by aggregating the performance observed for individual vehicles (e.g., by averaging the travel times of vehicles that crossed the link in that interval).

The same topological and functional representation of supply departs considerably from that used in the static case and in macro- and mesoscopic dynamic models. Some link characteristics are no longer exogenous to the supply model (and, as such, input data of the model), but can be extrapolated from the results of one or more simulations (see, e.g., link capacity, which is no longer calculated a priori and supplied as input to the model together with other link characteristics, but is also a result of simulations, as it can be calculated as the maximum number of vehicles that can cross the link). In general, for such models, the level of detail of the supply is much higher than in other cases, and all the elements of the road systems are explicitly represented (vehicles, infrastructure, organization, etc.). Due to its complexity and the considerable differences over the modeling structure represented in Fig. 7.2, description of this approach lies beyond the scope of this chapter and is not covered below.

In the following section, we therefore present space-discrete macroscopic supply models, and Sect. 7.2.2 deals with mesoscopic models.

7.2.1 Space-Discrete Macroscopic Models

7.2.1.1 Variables and Consistency Conditions

In this section we introduce (i) the variables of space-discrete continuous-flow dynamic models, following a classification into three groups: topological variables, flow/concentration variables, and time/cost variables; (ii) temporal consistency conditions; and (iii) network consistency conditions of the variables.

Topological Variables The topological characteristics of a trip are represented by using a graph. Let:

a be the index of a link of length L_a

k be the index of a path, consisting of a sequence of adjacent links $a_1^k, a_2^k, \ldots, a_{nk}^k$ where a_i^k, with $i = 1, 2, \ldots, n_k$, is the ith link of path k and n_k the number of links along k

a_{i+1}^k be the link that follows a_i^k on path k

a_{i-1}^k be the link that precedes a_i^k on path k

Flow and Concentration Variables For the sake of analytical convenience we assume that all the flow and concentration variables are (continuous and) differentiable functions of time $\tau \geq 0$. The terms temporal profile or temporal trajectory are used to underline the dependence of the dynamic variables upon time. Let:

$h^k(\tau) \geq 0$ be the *instantaneous flow* of users who begin their trip at time τ and follow path k. Assuming flow and time are continuous, the users are considered as infinitesimal particles of fluid that leave the origin of path k with an instantaneous flow $h^k(\tau)$. The number of users leaving on path k in the infinitesimal interval of width $d\tau$ around τ is thus given by $h^k(\tau) \cdot d\tau$ (see Fig. 7.3)

$\boldsymbol{h}_{od}(\tau) \geq \boldsymbol{0}$ be the vector of instantaneous path flows that leave at instant τ, made up by the instantaneous flows on all paths connecting origin–destination pair $od(h_k(\tau) : k \in K_{od})$

$\boldsymbol{h}(\tau) \geq \boldsymbol{0}$ be the vector of path flows leaving at time τ, for all O-D pairs

$f_{a_i^k,s}^k(\tau) \geq 0$ be the instantaneous flow of users who follow path k and cross-section s of link a_i^k at time τ. Unlike the static case, link flow at time τ generally varies from section to section (see Fig. 7.4). Of all the sections of a link, the entry ($s = 0$) and exit ($s = L_{a_i^k}$) sections are of special interest

$u_{a_i^k}^k(\tau) = f_{a_i^k,0}^k(\tau) \geq 0$ be the instantaneous flow traveling along path k and entering link a_i^k at time τ *(entry flow)*

$w_{a_i^k}^k(\tau) = f_{a_i^k,L_{a_i}}^k(\tau) \geq 0$ be the instantaneous flow traveling along path k and exiting link a_i^k at time τ *(exit flow)*

$n_{a_i^k}^k(\tau) \geq 0$ be the load of users traveling along k, on link a_i^k at time τ

$f_{a,s}(\tau), u_a(\tau), w_a(\tau), n_a(\tau)$ be, respectively, the total flow crossing section s from link a, the total flows entering and exiting from a, and the total load on a, all at time τ

The relations between specific path link variables and total link variables defined above are as follows.

$$f_{a,s}(\tau) = \sum_k \delta_{ak} f_{a,s}^k(\tau) \quad \forall a \tag{7.2.1a}$$

$$u_a(\tau) = \sum_k \delta_{ak} u_a^k(\tau) \tag{7.2.1b}$$

$$w_a(\tau) = \sum_k \delta_{ak} w_a^k(\tau) \tag{7.2.1c}$$

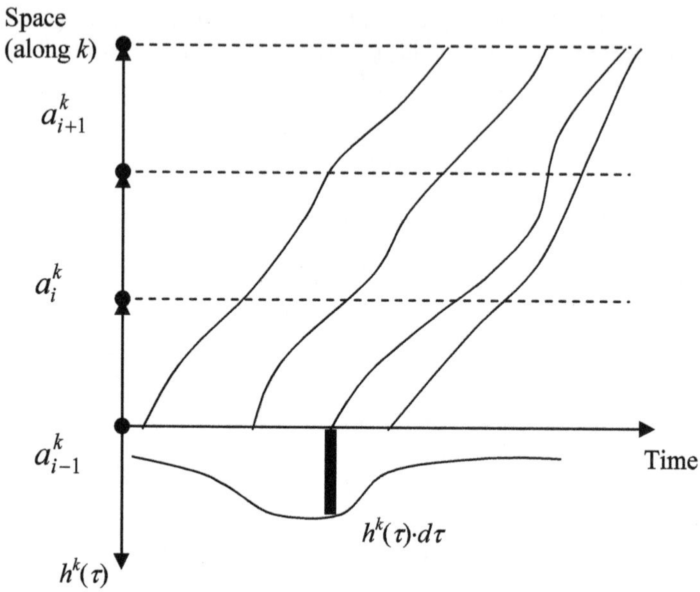

Fig. 7.3 Temporal profile of a path flow and its trajectories on the network

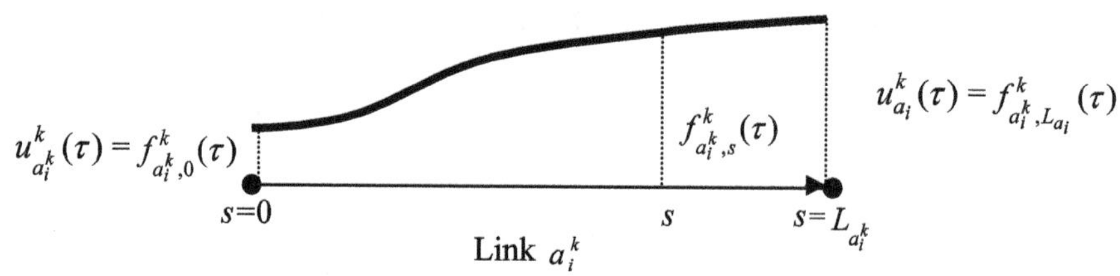

Fig. 7.4 Instantaneous flows through different cross-sections of a link

$$n_a(\tau) = \sum_k \delta_{ak} n^k_a(\tau) = L_a k_a(\tau) \tag{7.2.1d}$$

where δ_{ak} is the element on row a and column k of the link-path incidence matrix Δ, L_a the length of link a, and $k_a(\tau)$ the density of users on link a at time τ (see Chap. 2).

Travel Time and Cost Variables In supply models with within-day dynamics, as stated above, the travel time variable plays a twofold role. As in static models it represents a level of service and link performance variable, included as a component of the generalized cost perceived by users. However, the variable is required to ensure internal time congruence of relations defined between the dynamic model variables. Hence travel time is distinct from other performance variables.

Link and path travel times may also assume different values with variations in time, as both the transport supply and the level of network congestion may vary in time. We therefore need to introduce a series of new variables correlated to travel time. We assume that also travel times are (continuous and) differentiable functions of absolute time τ. Hence let:

$t^f_{a^k_i}(\tau) \geq 0$ be the *forward link travel time* function, which yields the quantity of time taken by the fluid particle entering link a^k_i at time τ, to cross the link

$t^b_{a^k_i}(\tau) \geq 0$ be the *backward link travel time* function, which yields the quantity of time taken by the fluid particle exiting from link a^k_i at time τ, to cross the link

$\tau^L_{a^k_i}(\tau) \geq \tau$ be the *exit time* function, which yields the time in which the fluid particle entering link a^k_i at time τ leaves the link

$\tau^E_{a^k_i}(\tau) \leq \tau$ be the *entry time* function, which yields the time in which a fluid particle leaving link a^k_i at time τ entered the link

$ec_{a^k_i}(\tau)$ be the *generalized extra-cost* required to cross link a^k_i entering at time τ. The generalized extra-cost expresses all the perceived disutility components for link a^k_i other than travel time. It includes performance variables such as tolls (variable in time), converted into disutility units

Below it is shown that exit and entry time functions are "well-defined" (in other words, a single exit or entry time corresponds to each time, τ) due to the congruence conditions introduced. Moreover, between the exit and entry time functions the following relation holds,

$$\tau'' = \tau^L_{a^k_i}(\tau') \quad \Leftrightarrow \quad \tau' = \tau^E_{a^k_i}(\tau'') \quad \forall \tau', \tau'' : \tau' \leq \tau'',$$

whence it derives that the entry time function is the *inverse of the exit time function* (used below in its place and indicated simply by $\tau^{-1}_{a^k_i}(\cdot)$) and vice versa. Hence:

$$\tau^E_{a^k_i}(\cdot) = \tau^{-1}_{a^k_i}(\cdot).$$

By definition, between the travel time functions and the exit time functions the following relations hold (see Fig. 7.5).

$$\tau^L_{a^k_i}(\tau) = \tau + t^f_{a^k_i}(\tau) \tag{7.2.2a}$$

$$\tau^{-1}_{a^k_i}(\tau) = \tau - t^b_{a^k_i}(\tau). \tag{7.2.2b}$$

The time variables may also be associated with path fractions (see Fig. 7.6). In particular:

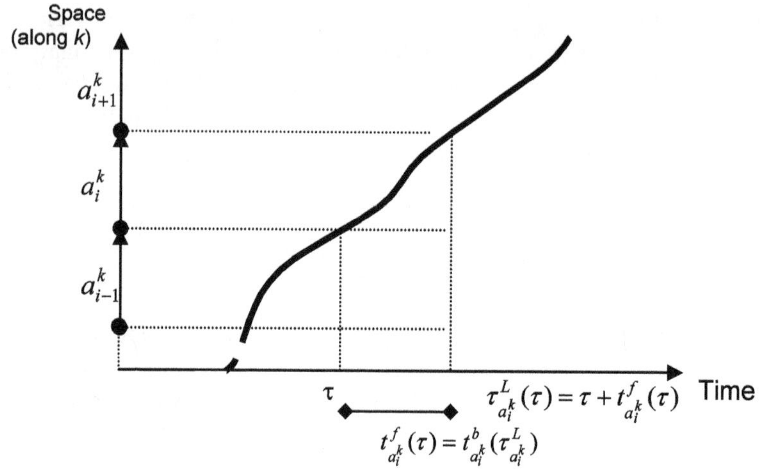

Fig. 7.5 Temporal consistency of forward and backward link travel times

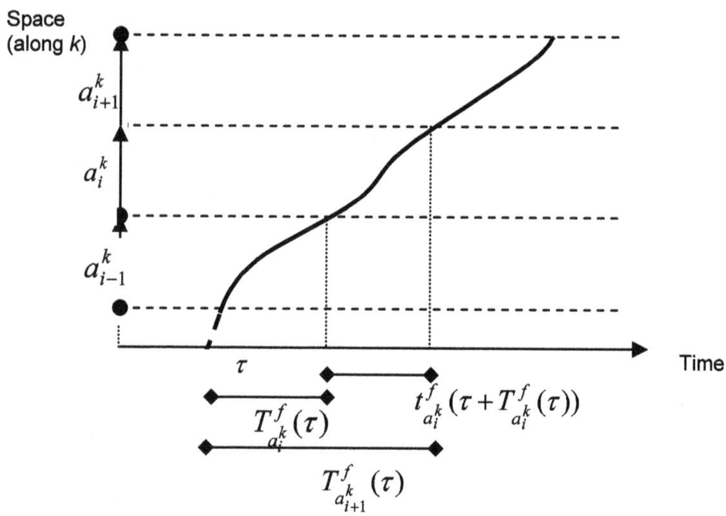

Fig. 7.6 Temporal consistency of forward and backward path fractions travel times

$T_{a_i^k}^f(\tau)$ is the *forward travel time of the path fraction up to* a_i^k, in other words, the amount of travel time required to reach the beginning of link a_i^k starting at time τ from the origin node (i.e., the first link of path k, a_1^k)

$T_{a_i^k}^b(\tau)$ be instead, is *the backward travel time of the path fraction up to* a_i^k, in other words, the amount of travel time required to reach the beginning of link a_i^k at time τ, having followed path k

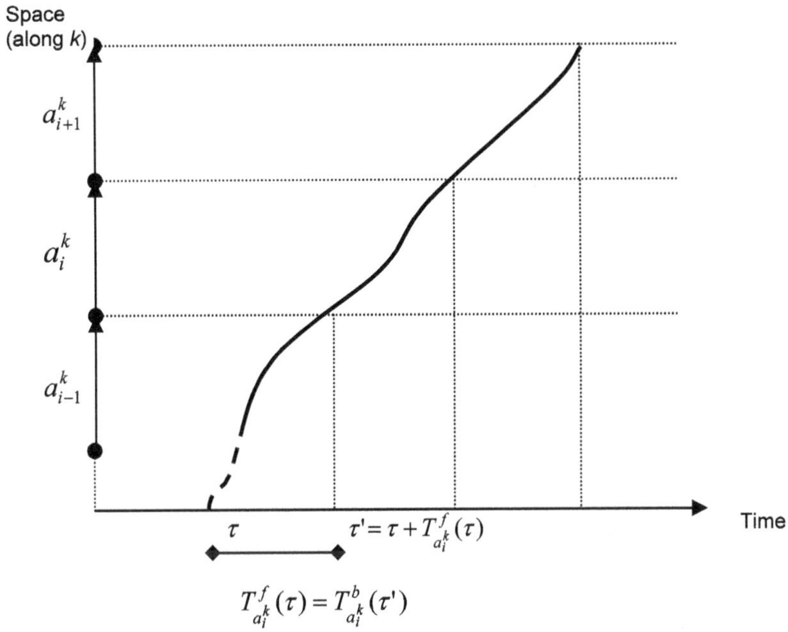

Fig. 7.7 Relation between link travel times and path fraction travel times

Temporal Consistency and the FIFO Condition Temporal consistency of the supply model requires that the following relations hold between forward link travel time functions and corresponding backward functions (see Fig. 7.5):

$$t_{a_i^k}^f(\tau) = t_{a_i^k}^b\left(\tau + t_{a_i^k}^f(\tau)\right)$$

$$t_{a_i^k}^b(\tau) = t_{a_i^k}^f\left(\tau - t_{a_i^k}^b(\tau)\right)$$

and analogous relations between forward travel time of a path fraction and the corresponding backward functions (see Fig. 7.6):

$$T_{a_i^k}^f(\tau) = T_{a_i^k}^b\left(\tau + T_{a_i^k}^f(\tau)\right)$$

$$T_{a_i^k}^b(\tau) = T_{a_i^k}^f\left(\tau - T_{a_i^k}^b(\tau)\right)$$

Moreover, travel times of a path fraction and link travel times are related by the following equations (see Fig. 7.7).

$$T_{a_{i+1}^k}^f(\tau) = T_{a_i^k}^f(\tau) + t_{a_i^k}^f\left(\tau + T_{a_i^k}^f(\tau)\right) \tag{7.2.3a}$$

$$T_{a_{i+1}^k}^b(\tau) = t_{a_i^k}^b(\tau) + T_{a_i^k}^b\left(\tau - t_{a_i^k}^b(\tau)\right) \tag{7.2.3b}$$

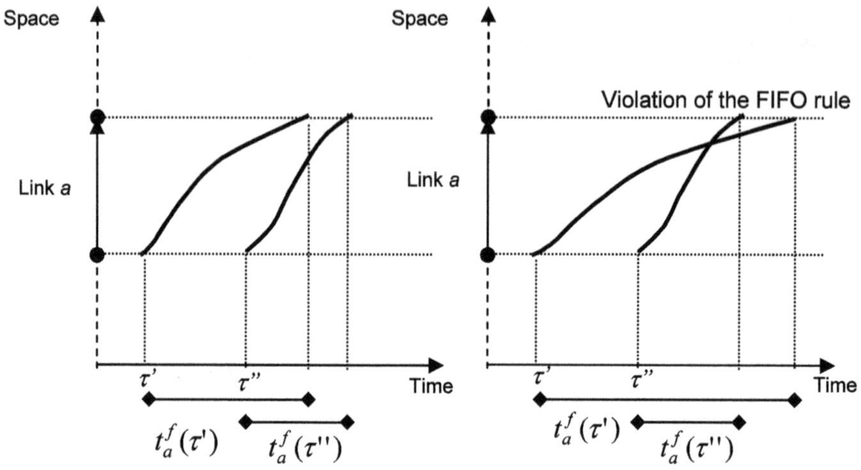

Fig. 7.8 Representation of the FIFO rule on a link

Furthermore, having assumed that user flows are represented with a partially compressible one-dimensional fluid, the travel time functions must be such that an infinitesimal fluid particle entering link a at time τ'' can never reach nor overtake another particle that has entered the same link at a previous time $\tau' < \tau''$. Otherwise, this would mean that the fluid entering between τ' and τ'' may be compressed into a null space (infinite density), in other words that the assumption of one-dimensional fluid (or absence of turbulence along the link) is violated. This condition is generally reported in the literature as a strict condition of First In–First Out (FIFO) and may be formulated, respectively, in terms of forward or backward link travel times, as (see Fig. 7.8):

$$\tau' + t^f_{a_i^k}(\tau') < \tau'' + t^f_{a_i^k}(\tau'') \tag{7.2.4a}$$

$$\tau' - t^b_{a_i^k}(\tau') < \tau'' - t^b_{a_i^k}(\tau'') \quad \forall \tau' < \tau'' \tag{7.2.4b}$$

or, in equivalent fashion, exploiting relations (7.2.2), as

$$\tau^L_{a_i^k}(\tau') < \tau^L_{a_i^k}(\tau'')$$

$$\tau^{-1}_{a_i^k}(\tau') < \tau^{-1}_{a_i^k}(\tau'') \quad \forall \tau' < \tau''.$$

It can be easily shown that, if all the link travel time functions observe the FIFO rule, path fraction travel times also respect it. Analogously with link time functions, we can then write:

$$\tau' + T^f_{a_i^k}(\tau') < \tau'' + T^f_{a_i^k}(\tau'') \tag{7.2.5a}$$

$$\tau' - T^b_{a_i^k}(\tau') < \tau'' - T^b_{a_i^k}(\tau'') \quad \forall \tau' < \tau''. \tag{7.2.5b}$$

When the strict inequality of the above relations is replaced by a weak inequality, we obtain the weak FIFO condition which, for brevity's sake, is not covered below.

Equations (7.2.4) and (7.2.5) imply that $(\tau + t_a^f(\tau))$ and $(\tau - t_a^b(\tau))$ (i.e., $\tau_a^L(\tau)$ and $\tau_a^{-1}(\tau)$), and $(\tau + T^f_{a_i^k}(\tau))$ and $(\tau - T^b_{a_i^k}(\tau))$, are strictly increasing functions of time τ, and hence well-defined and invertible (as stated above). Besides, if the FIFO condition were not imposed, two particles could cross the same section at the same time. Hence, for the same section and at a single instant in time, we would have different values of speed and acceleration, and the inverse function of the exit time, $\tau_a^{-1}(\cdot)$, would not be correctly defined.

Assuming that functions $t_a^f(\tau)$ and $t_a^b(\tau)$ can be differentiated, we can demonstrate that the following condition is equivalent to respecting the FIFO rule.

$$\frac{dt_a^f(\tau)}{d\tau} > -1 \quad \forall\tau.$$

Indeed, (7.2.4a) may be rewritten as

$$\frac{t_a^f(\tau'') - t_a^f(\tau')}{\tau'' - \tau'} > -1 \quad \forall\tau' < \tau'',$$

hence

$$\lim_{\tau''\to\tau'} \frac{t_a^f(\tau'') - t_a^f(\tau')}{\tau'' - \tau'} > -1.$$

The similar condition in terms of backward travel time is:

$$\frac{dt_a^b}{d\tau}(\tau) < 1 \quad \forall\tau.$$

This last relation is particularly suited to immediate physical interpretation: in order that the FIFO rule is not violated, link travel time must not decrease more rapidly than the advancing of absolute time.

Network Consistency Network consistency requires that the time profiles of path flow and link flow variables satisfy the conservation equations in each instant.[3] Hence, for the first link a_1^k along path k (i.e., which exits from origin centroid node):

$$u^k_{a_1^k}(\tau) = h^k(\tau) \quad \forall k. \tag{7.2.6a}$$

For each pair of adjacent links a_i^k, a_{i+1}^k, up to the last link along path k, *flow conservation at the nodes*, at time τ, requires that entry flows and exit flows satisfy

[3] As in telecommunications networks and unlike hydraulic or electrical networks, the flows that move on a transport network should be distinguished by origin and destination; we should also underline that they can also be distinguished by path.

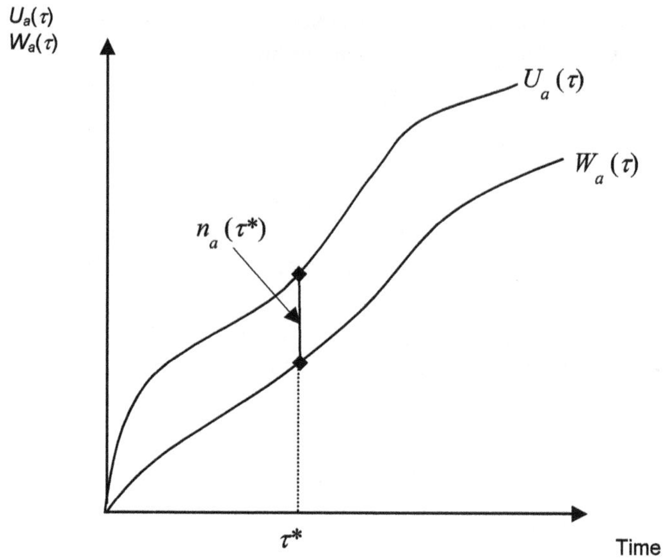

Fig. 7.9 Relation between users' load on a link at one time and cumulative entry and exit flows from the link up to the same time

the following equation,

$$w^k_{a^k_i}(\tau) = u^k_{a^k_{i+1}}(\tau) \quad \forall k, i. \tag{7.2.6b}$$

By contrast, *flow conservation on the link*, for each link a^k_i along path k, at time τ, is expressed by the following differential equation,

$$\frac{dn^k_{a^k_i}(\tau)}{d\tau} = u^k_{a^k_i}(\tau) - w^k_{a^k_i}(\tau) \quad \forall k, i. \tag{7.2.7}$$

This equation is equivalent to the finite-difference equation (2.2.2) introduced in Chap. 2 and referred to *observed variables* (entry flow and number of users present on a road segment). Once integrated, it yields the loading on the link at time τ:

$$\begin{aligned} n^k_{a^k_i}(\tau) &= \int_0^\tau u^k_{a^k_i}(t)\,dt - \int_0^\tau w^k_{a^k_i}(t)\,dt + n^k_{a^k_i}(0) \\ &= U^k_{a^k_i}(\tau) - W^k_{a^k_i}(\tau) + n^k_{a^k_i}(0) \end{aligned} \tag{7.2.8}$$

where:

$$U^k_{a^k_i}(\tau) = \int_0^\tau u^k_{a^k_i}(t)\,dt \quad \text{and} \quad W^k_{a^k_i}(\tau) = \int_0^\tau w^k_{a^k_i}(t)\,dt$$

are the cumulative flows up to time τ, respectively, for entry and exit from link a^k_i, of users following path k. Equation (7.2.8) thus expresses the relation between the

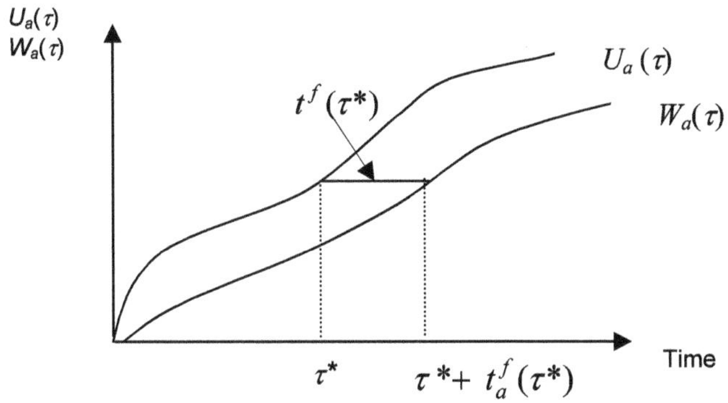

Fig. 7.10 Necessary condition for the FIFO rule (forward travel time)

number of users on a link (i.e., link load) and the cumulative entry and exit flows (see Fig. 7.9).

7.2.1.2 Network Flow Propagation Model

In this section we introduce the network flow propagation model, which expresses the relation between path flows and link flows.

Having defined in the previous section *flow conservation* at time τ for network nodes and links, we first need to formalize the problem of flow propagation in time, along a link. Helping us in this purpose is the hypothesis that the flow is partly compressible and one-dimensional, hence that the FIFO rule holds. If the latter holds, the total number of vehicles entering link a_i^k up to time τ equals the total number of vehicles exiting from the same link up to time $\tau + t_a^f(\tau)$ (see Fig. 7.10)[4]:

$$U^k_{a_i^k}(\tau) = W^k_{a_i^k}\left(\tau + t^f_{a_i^k}(\tau)\right) \quad \forall i, k \tag{7.2.9a}$$

and, similarly, proceeding backwards:

$$U^k_{a_i^k}\left(\tau - t^b_{a_i^k}(\tau)\right) = W^k_{a_i^k}(\tau) \quad \forall i, k. \tag{7.2.9b}$$

Differentiating the previous relations we obtain:

$$u^k_{a_i^k}(\tau) = w^k_{a_i^k}\left(\tau + t^f_{a_i^k}(\tau)\right) \cdot 1 + \frac{dt^f_{a_i^k}(\tau)}{d\tau} \quad \forall i, k \tag{7.2.10a}$$

[4]Equations (7.2.9) are not sufficient conditions for respect of the FIFO rule, as they also hold in the presence of overtaking maneuvers. In this case, they are satisfied provided, for each vehicle on the link, the difference between the number of vehicles overtaking and being overtaken is nil.

$$u^k_{a^k_i}\left(\tau - t^b_{a^k_i}(\tau)\right) \cdot 1 - \frac{dt^b_{a^k_i}(\tau)}{d\tau} = w^k_{a^k_i}(\tau) \quad \forall i, k. \tag{7.2.10b}$$

Assuming that the FIFO rule holds, each of (7.2.10) expresses *the dynamic propagation of the flow along a link*. Equation (7.2.10a) states that, if the flow on link a^k_i at time τ is decelerating $(dt/d\tau > 0)$, the flow exiting the link after the travel time required to cross it will be less than the flow entering at τ. Vice versa, if the flow on the link is accelerating $(dt/d\tau < 0)$, the flow exiting the link will exceed that entering. The same conclusions may be reached by (7.2.10b). However, when the link travel time at time τ is constant $(dt/d\tau = 0)$ (in noncongested networks with constant supply or in steady-state conditions in the system this occurs at every time and for every network link), the flow exiting from a link is simply translated in time compared with the entry flow. In Sect. 7.2.1.4 we show that, in the case of steady-state behavior of the system (see static models), the hypothesis of stationary flows entering the network also means that entry and exit flows from each link are the same and constant; that is: $u^k_{a^k_i} = w^k_{a^k_i}$.

Moreover, any of (7.2.10), under the hypothesis of positive entry flows $u_a(\tau) > 0$, and positive exit flows $w_a(\tau) > 0$, guarantees that the FIFO rule holds. The opposite does not hold true, because the assumption of FIFO rule validity and respect of any of (7.2.10) means only that entry and exit flows have the same sign.

Dynamic flow propagation along a path may therefore be derived from propagation along a link. Indeed, the flow traveling along path k and entering link a_{i+1} at time τ (i.e., the entry flow $u^k_{a_i}(\tau)$) may be expressed as a function of path flow h^k exiting from the centroid node at a previous instant in time (out of simplicity, we omit the superscript k in the link notation, meaning $a_i = a^k_i$). To this end (7.2.10) and (7.2.6) can be applied. In the case of backward travel times, for example, by substituting (7.2.6b) into (7.2.10b), we obtain:

$$u^k_{a_{i+1}}(\tau) = u^k_{a_i}\left(\tau - t^b_{a_i}(\tau)\right) \cdot 1 - \frac{dt^b_{a_i}(\tau)}{d\tau} \quad \forall i, k. \tag{7.2.11}$$

The entry flow $u^k_{a_i}(\tau - t^b_{a_i}(\tau))$ in the right-hand side of (7.2.11) may in turn be expressed by (7.2.10b) as a function of the entry flow on the previous link, a_{i-1}, in a previous moment:

$$u^k_{a_i}\left(\tau - t^b_{a_i}(\tau)\right) = u^k_{a_{i-1}}\left(\tau - t^b_{a_i}(\tau) - t^b_{a_{i-1}}\left(\tau - t^b_{a_i}(\tau)\right)\right)$$
$$\cdot \left(1 - \frac{dt^b_{a_{i-1}}(\tau - t^b_{a_i}(\tau))}{d\tau}\right). \tag{7.2.12}$$

Substituting (7.2.12) into (7.2.11) and iterating the previous steps until we obtain in the second member of (7.2.11) the entry flow on the first link of path k, we obtain the relation sought between the flow entering a link at a certain time τ (having followed path k) and the corresponding path flow; in other words, the flow on path

k which, having left at a previous time, reached the above link exactly in τ:

$$u^k_{a_{i+1}}(\tau) = h^k\big(\tau - T^b_{a_{i+1}}(\tau)\big) \cdot \prod_{j=1}^{i}\left(1 - \frac{dt^b_{a_j}(\tau^L_{a_j})}{d\tau}\right) \tag{7.2.13a}$$

where $T^b_{a_{i+1}}(\tau)$, backward travel time of the path fraction up to a^k_{i+1}, is given by the sum of the travel times on all the links that precede a_{i+1} (each calculated in the instant of exiting from the link itself):

$$T^b_{a_{i+1}}(\tau) = \sum_{j=1}^{i} t^b_{a_j}(\tau^L_{a_j}) \tag{7.2.14a}$$

in which $\tau^L_{a_j}$ is the exit time from link a_j, calculated from τ by subtracting link travel times up to a_j:

$$\tau^L_{a_j} = \tau - t^b_{a_i}(\tau) - t^b_{a_{i-1}}\big(\tau - t^b_{a_i}(\tau)\big) - \cdots - t^b_{a_{j+1}}\big(\tau - t^b_{a_i}(\tau) - \cdots\big).$$

The similar expression in terms of forward travel times is:

$$u^k_{a_i}\big(\tau + T^f_{a_i}(\tau)\big) = \frac{h^k(\tau)}{\prod_{j=1}^{i-1}\big(1 + \frac{dt^f_{a_j}(\tau + T^f_{a_j}(\tau))}{d\tau}\big)} \tag{7.2.13b}$$

where the time required to reach the beginning of link a_i following path k and leaving at τ, $T^f_{a_{i+1}}(\tau)$, may be expressed as

$$T^f_{a_i}(\tau) = \sum_{j=1}^{i-1} t^f_{a_j}\big(\tau + T^f_{a_j}(\tau)\big). \tag{7.2.14b}$$

Each of (7.2.13) thus expresses the dynamic flow propagation along a path.

The *dynamic model of network flow propagation* is obtained by substituting, for each network link, either of the two (7.2.13) into (7.2.1b), that is, summing all the flows that, following different paths, enter a certain link at the same time. The model may be solved once we know the time profile of the path flow vector entering the network ($\boldsymbol{h}(\tau)$, $\forall\tau > 0$), and the time profile of forward or backward link travel times ($t^f_a(\tau) \circ t^b_a(\tau), \forall\tau > 0$). The profile of path flows $\boldsymbol{h}(\tau)$ is known to constitute model input data. Link travel times may instead be calculated using the link performance functions presented in the section below. We show that link travel time is generally a function of user load on the same link and hence of user flows entering and exiting from the link. Thus, as anticipated in Sect. 7.2, a circular dependence is configured between the network flow propagation model and the link performance model. The solution of the two models comes via the solution to the fixed-point problem known in the literature as *dynamic network loading*. This allows us to obtain the time trajectory of each of the system state variables (u, w, n, k) for each network link.

7.2.1.3 Link Performance and Travel Time Functions

The functions that express link travel time according to link flows are fundamental both for dynamic supply models and their static counterparts. Most of the dynamic models proposed in the literature adopt functions that explicitly (*travel time functions*), or implicitly (*exit time functions*), measure link travel time against the number of users on the link. Exit time functions directly express the flow exiting from a certain link according to its load $w_a(\tau) = w_a(n_a(\tau))$. However, such functions lead to a series of theoretical incongruences and therefore are not covered below.

Travel time functions express the travel time $t_a(\tau)$ of a vehicle arriving at the start of link a at time τ, according to traffic conditions at that instant. Most of the models proposed in the literature adopt separable travel time functions, in other words, functions that express the travel time on link a, $t_a^f(\tau)$ (or, $t_a^b(\tau)$), as a function only of the link load a, $n_a(\tau)$:

$$t_a^f(\tau) = t_a\big(n_a(\tau)\big). \tag{7.2.15}$$

Determination of the backward travel time as a function of the exit time function requires solution to the fixed-point problem that is unique only if the FIFO rule holds. Indeed, from (7.2.2b) we have:

$$t_a^b(\tau^*) = t_a\big(n_a\big(\tau - t_a^b(\tau^*)\big)\big).$$

Although various functional forms have been proposed for (7.2.15), not all lead to results consistent with the FIFO rule.

In applications two distinct link types are generally considered: *running links* represent the real movement of the vehicle, such as that of a vehicle traveling on an urban or motorway road section and *queuing or waiting links* represent waiting at intersections, toll barriers, and the like (see Fig. 7.11).

It can be shown that for running links, a linear travel time function as follows,

$$t_a\big(n_a(\tau)\big) = t_a^0 + \frac{1}{Q_a} \cdot n_a(\tau) \tag{7.2.16a}$$

where t_a^0 is the free flow travel time, means that the flow exiting link a, w_a, never exceeds link capacity Q_a, guarantees respect of the FIFO condition, and ensures the model's congruence.

Figure 7.12 illustrates the flow exiting a link according to the number of vehicles on the link itself, for a function of type (7.2.16a).

A similar function derived from deterministic queuing models may be applied for queuing links. Indeed, all the concepts introduced up to this point are applied to queuing models (see Sect. 2.2.2). The only difference is that, in the case of queuing, travel time is spent waiting rather than in movement along the link. The flow entering and that exiting from the link correspond to the rates of arrival and departure from the queue; the load is equivalent to the number of users in the queue, and so

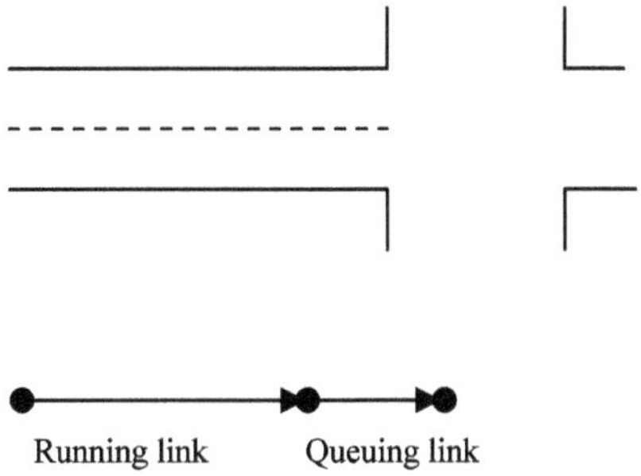

Fig. 7.11 Diagram of a road intersection with running and queuing links

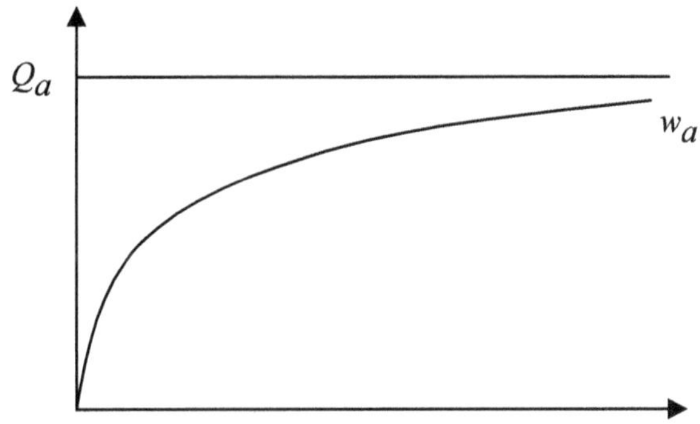

Fig. 7.12 Exit flow corresponding to a linear function of travel time (7.2.16a)

forth. In this case, (7.2.16a) may be rewritten as

$$tw_a\big(n_a(\tau)\big) = \frac{1}{Q_a} + \frac{1}{Q_a} n_a(\tau) \tag{7.2.16b}$$

where the "zero load" time is equal to the average service time; that is, $tw_a = 1/Q_a$.

7.2.1.4 Dynamic Network Loading

Dynamic network loading may be expressed by combining the network flow propagation model with that of link performance, and imposing time and space congruence of network flows. In the case of forward travel times, for example, a possible formulation is obtained by combining (7.2.13b), (7.2.14b), (7.2.15), (7.2.8), and (7.2.1d), (7.2.16b), as reported below for the reader's convenience.

$$u^k_{a_i}\big(\tau + T^f_{a_i}(\tau)\big) = \frac{h^k(\tau)}{\prod_{j=1}^{i-1}\big(1 + \frac{dt^f_{a_j}(\tau + T^f_{a_j}(\tau))}{d\tau}\big)} \quad \forall i, k, \tau \tag{7.2.17a}$$

$$T^f_{a_i}(\tau) = \sum_{j=1}^{i-1} t^f_{a_j}\big(\tau + T^f_{a_j}(\tau)\big) \tag{7.2.17b}$$

$$t_{a_i}^f(\tau) = t_{a_i}\left(n_{a_i}(\tau)\right) \quad (7.2.17c)$$

$$n_{a_i}(\tau) = \sum_k \delta_{a_i k}\left(\int_0^\tau u_{a_i}^k(t)\,dt - \int_0^\tau w_{a_i}^k(t)\,dt + n_{a_i}^k(0)\right) \quad (7.2.17d)$$

$$w_{a_i}^k(\tau) = u_{a_{i+1}}^k(\tau) \quad (7.2.17e)$$

to which initial conditions (e.g., $u_{a_i}(0) = w_{a_i}(0) = n_{a_i}(0) = h^k(0) = 0,\ \forall i, k)$ should be added.

Note that if link travel times do not vary in time ($dt_a/d\tau = 0,\ \forall a, \tau$), as, for example, occurs in the case of a noncongested network in which the supply is constant, the time profile of the flow entering any link along path k is equal to the time profile of path flow h^k, shifted of the time required to reach the link itself (as was seen for flow propagation on a link). In other words, (7.2.17a) (i.e., (7.2.13b)) becomes:

$$u_{a_i}^k\left(\tau + T_{a_i}^f(\tau)\right) = h^k(\tau) \quad \forall i, k, \tau.$$

Regardless of whether the network is congested, if the system functions under steady-state conditions (i.e., $h^k(\tau) = h^k = \text{const.}\ \forall k, \tau$ and $dt_a/d\tau = 0,\ \forall a, \tau$), we may write:

$$u_{a_i}^k\left(\tau + T_{a_i}^f\right) = h^k = u_{a_i}^k \quad \forall i, k, \tau$$

that is, the partial flow that enters a link following a certain path is constant and equal to the corresponding path flow.[5] Hence, as in any instant τ it holds that $w_{a_i}^k(\tau) = u_{a_{i+1}}^k(\tau)$, then we also obtain:

$$w_{a_i}^k = u_{a_i}^k = h^k = f_{a_i}^k.$$

Substituting h^k to $f_{a_i}^k$ into (7.2.1a), it is possible to calculate the link flow as the sum of (stationary) path flows which cross it:

$$f_a = \sum_k \delta_{ak} h^k.$$

The latter, as noted in Chap. 2, describes the network flow propagation model in the static case, and is indeed equivalent to (2.3.1).

7.2.1.5 Path Performance and Travel Time Functions

Path performance and travel time functions may be calculated directly by link performance and travel time functions.

[5] Flow constancy in time gives rise to the statement (although formally not correct) that flow propagation in the static case is "instantaneous," given that at each instant each partial link flow is equal to the corresponding path flow entering the network at the same time.

Equation (7.2.3a), applied recursively from the first to the ith link of path k, gives rise to the summation (7.2.14b). The latter, expressed as a function of only link travel times, is resolved in a nested sum of the same times:

$$T^f_{a^k_i}(\tau) = t^f_{a^k_1}(\tau) + t^f_{a^k_2}\big(\tau + t^f_{a^k_1}(\tau)\big) + t^f_{a^k_3}\big(\tau + t^f_{a^k_1}(\tau) + t^f_{a^k_2}\big(\tau + t^f_{a^k_1}(\tau)\big)\big)$$
$$+ \cdots + t^f_{a^k_{i-1}}\big(\tau + t^f_{a^k_1}(\tau) + \cdots + t^f_{a^k_{i-2}}\big(\tau + t^f_{a^k_1}(\tau) + \cdots\big)\big). \quad (7.2.18)$$

Similarly, for backward travel times, (7.2.14a) may be expressed as

$$T^b_{a^k_i}(\tau) = t^b_{a^k_{i-1}}(\tau) + t^b_{a^k_{i-2}}\big(\tau - t^b_{a^k_{i-1}}(\tau)\big) + t^b_{a^k_{i-3}}\big(\tau - t^b_{a^k_{i-1}}(\tau) - t^b_{a^k_{i-2}}\big(\tau - t^b_{a^k_{i-1}}(\tau)\big)\big)$$
$$+ \cdots + t^b_{a^k_1}\big(\tau - t^b_{a^k_{i-1}}(\tau) - \cdots - t^b_{a^k_2}\big(\tau - t^b_{a^k_{i-1}}(\tau)$$
$$- \cdots - t^b_{a^k_3}\big(\tau - t^b_{a^k_{i-1}}(\tau) - \cdots\big)\big)\big). \quad (7.2.19)$$

The previous equations may be easily used to express, as a function of link travel times, the total travel time on the whole path, the extra-costs of the path (assuming additive link attributes) and generalized costs. Thus let:

$TT^f_k(\tau)$ be the *total forward path travel time* function, in other words the time required to travel the whole path k, from origin to destination, starting at time τ

$TT^b_k(\tau)$ be the *total backward path travel time* function, in other words the time required to travel the whole path k, from origin to destination, completing the path at instant τ

$EC_k(\tau)$ be the *generalized extra-cost of path* k starting at time τ

$g_k(\tau)$ br the *total generalized cost* along path k starting at time τ

In the case of forward performances, for example, then:

$$TT^f_k(\tau) = t^f_{a^k_1}(\tau) + t^f_{a^k_2}\big(\tau + t^f_{a^k_1}(\tau)\big) + \cdots + t^f_{a^k_{n_k}}(\tau + \cdots) \quad (7.2.20a)$$
$$= T^f_{a^k_{n_k}}(\tau) + t^f_{a^k_{n_k}}(\tau + \cdots)$$

$$EC_k(\tau) = ec_{a^k_1}(\tau) + ec_{a^k_2}\big(\tau + t^f_{a^k_1}(\tau)\big) + \cdots + ec^f_{a^k_{n_k}}(\tau + \cdots) \quad (7.2.20b)$$

$$g_k(\tau) = \beta_t TT^f_k(\tau) + EC_k(\tau) \quad (7.2.20c)$$

where $a^k_{n_k}$ is the last link of path k. The relations between the vectors of forward path travel time functions $\boldsymbol{TT}^f(\tau)$ (with a component for each path of the network) and the forward link travel time functions $\boldsymbol{t}(\tau)$ (with a component for each link of the network) may be expressed as

$$\boldsymbol{TT}^f(\tau) = \Gamma\big(\boldsymbol{t}(\tau'), \tau' \geq \tau\big). \quad (7.2.21)$$

Equations (7.2.20) constitute the dynamic equivalent of the cost composition expressed by supply model (2.2.5) and (5.2.1) for static networks. In the static case, the order in which different link performance costs or attributes are summed to obtain path costs is unimportant. This no longer holds for within-day supply models for which link times and costs must be summed in their topological order along path k to respect the time sequence in which the links are crossed.

7.2.1.6 Formalization of the Whole Supply Model

The equations introduced in the previous sections express the dependence of entry and exit flows, of loads, and link and path times and costs in one instant, upon path flows starting from origins at previous times. The equations that define the complete supply model for congested networks, in respect of the FIFO rule, may be expressed symbolically as

$$\boldsymbol{f} = \Phi\left[\boldsymbol{t}(\tau), \boldsymbol{h}(\tau)\right] \tag{7.2.22a}$$

$$\boldsymbol{t}(\tau) = \boldsymbol{t}\left(\boldsymbol{f}(\tau'), \tau' \le \tau\right) \tag{7.2.22b}$$

$$\boldsymbol{TT}^f(\tau) = \Gamma\left(\boldsymbol{t}(\tau'), \tau' \ge \tau\right) \tag{7.2.22c}$$

where:

$\boldsymbol{t}(\tau)$ is the vector of link travel times at time τ
$\boldsymbol{TT}^f(\tau)$ is the vector of forward path travel times at time τ
$\boldsymbol{f}(\tau)$ is the vector of flow or load variables which are relevant to travel time functions at time τ
$\boldsymbol{h}(\tau)$ is the vector of path flows at time τ
Γ symbolically expresses the relation between link and path travel times (see (7.2.21))
Φ symbolically expresses the dynamic loading model of the network (see (7.2.17))

In the context of within-day dynamics of continuous flow, these equations are equivalent to the static equations:

$$\boldsymbol{f} = \Delta \boldsymbol{h}$$

$$\boldsymbol{c} = c(\boldsymbol{f})$$

$$\boldsymbol{g} = \Delta^T \boldsymbol{c}.$$

Note that (7.2.22) reflect the condition by which, in congested networks, the time for covering a link in a certain instant τ, depends on the load on all the network links that precede it along all the paths that lead to it, in the instants prior to τ $(\tau' \le \tau)$. Link flows and load in any one time depend on the link travel time profiles of all the links that precede it, up to that moment.

The solution to the dynamic supply model described above is based on the discretization of the integrodifferential equations that describe it. Given the large number of equations involved, what is important is also the sequence in which they are processed.

Note that this formulation of the supply model assumes that the only variables that affect link travel times are link loads, only at the moment of arrival at the link. This assumption, albeit convenient in solution terms as well as being close to the static case, is only appropriate for deterministic queuing links and very short running links.

As hinted in the introductory Sect. 7.2, other supply models for continuous flows (namely *space-continuous models*) are based on the application of systems of differential equations derived from space-continuous traffic flow models for each link, together with equations that guarantee flow conservation at each node.

The solution to such models, at least in theory, ensures the definition of variables such as flow, speed, and density at each point s and each instant τ. However, their solution requires discretization of space Δs, and hence in computational terms they may be considered similar to space-discrete models with an appropriate definition of link length (i.e., $\Delta s = L_a$).

7.2.2 Mesoscopic Models

Discrete flow models assume that users are discrete units; these units can be either individual vehicles or groups of vehicles moving together over the network and experiencing the same trip conditions. Discrete flow units are referred to below as *packets*, which includes the special case of single-vehicle packets.

As mentioned in Sect. 7.2, *mesoscopic models* simulate network performance at an aggregated level; as in discrete-space continuous-flow models, aggregated variables of capacity, flows, and occupancy are used. Traffic, however, is represented discretely by tracing the trips of individual packets; each packet is characterized by a departure time and by a path to its destination. It is often assumed that packets are concentrated at a point (concentrated or vertical packets); the smaller the size of the packets, the more realistic is this assumption. Mesoscopic models can be applied to general networks and extended to simulate queue-formation and spill-backs with reasonable computing times. On the other hand, they do not allow detailed simulation of the behavior of individual vehicles (overtaking, lane-changing, etc.).

Most discrete flow models are based on some form of time discretization, that is, a division of the reference period into intervals $[j]$ (which, for the sake of simplicity, are assumed below to be of equal duration DT). These models often assume that relevant flow variables are averaged over time intervals. They also assume that users begin their trips at a representative time instant τ_j in interval $[j]$, for example, its beginning or midpoint (see Fig. 7.13). In principle, the duration of departure intervals can differ from the duration of averaging intervals; for example, some models use very short departure intervals while averaging the variables over longer intervals. To

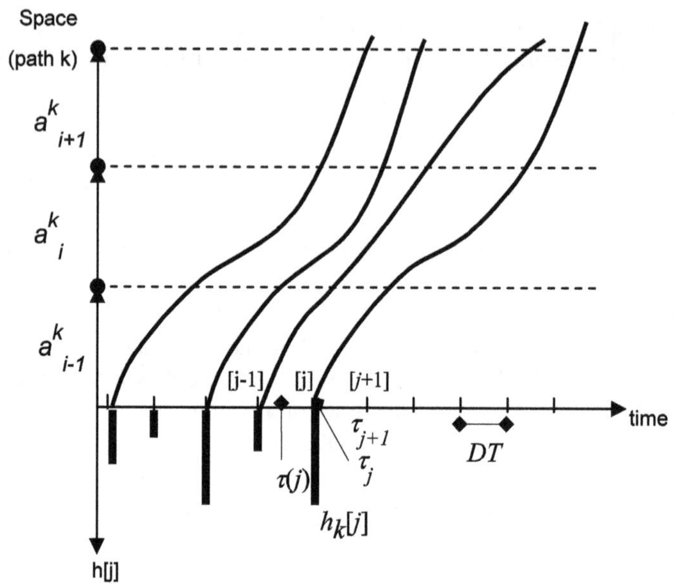

Fig. 7.13 Path flows and trajectories in discrete time: discrete flow models

simplify notation in the following discussion, only the single-interval case is considered; the generalization to multiple intervals is straightforward. Furthermore, it is assumed that the representative time instant of each interval is its final one; that is, $\tau_j = [j] \cdot DT$.

A general framework is more difficult to formalize for discrete time-discrete flow models than for continuous models, because there are several possible ways to discretize the relevant variables. The framework proposed in the following is general enough to include a number of models that have been presented in the literature.

7.2.2.1 Variables and Consistency Conditions

Discrete model variables and their structural relationships must first be defined.

Time Variables The discretization of time requires the introduction of time variables in addition to the absolute time τ. Let:

$\tau(j)$ be an arbitrary instant in time interval $[j]$, $\tau(j) \in ([j-1] \cdot DT, [j] \cdot DT)$

τ_j be the representative instant in time interval $[j]$, here assumed to be its end-point, $\tau_j = [j] \cdot DT$

Topological Variables Topological variables are the same as in the continuous-flow continuous-time case and are not restated.

Flow and Occupancy Variables The flow variables have the same definitions as in the continuous case, but in discrete flow models they represent "counts," that is, the number of users in an interval $[j]$, rather than flows, that is, temporal densities, as shown in Fig. 7.13. In the following, however, they are referred to equivalently as units (in a time interval) or flows, to simplify the notation and facilitate the extension of continuous flow results. Let:

k_j be an arbitrary packet, identified by its path k (which implies its O-D pair) and its departing interval $[j]$ from the origin; it is assumed that only one packet can leave on a given path in any time interval

$d_{od}[j]$ be the number of users departing from origin o in the representative instant of interval $[j]$ and traveling to destination d

$h_k[j]$ be the number of users starting their trip along path $k, k \in K_{od}$, in the representative instant of interval $[j]$; $h_k[j]$ can be thought of as the size of the packet k_j

$f^k_{a,s}[j], u^k_a[j], w^k_a[j]$ be, respectively, the number of users traveling on path k who cross section s of link a, and the number of users on path k who enter and leave link a during interval $[j]$

$f_{a,s}[j], u_a[j], w_a[j]$ be, respectively, the total (over all paths) number of users who cross section s of link a, and the total number of users who enter and leave link a during interval $[j]$. Note that they correspond to the variables $m(s|\tau_{j-1}, \tau_j)$, introduced in Sect. 2.2.2, with symbols modified to parallel those used for continuous models

Flow variables can also be defined with respect to any subinterval of interval j, for example, the interval $[\tau_{j-1}, \tau(j)]$ from the interval's beginning up to time $\tau(j)$. In this case the variables have as argument the specific instant in which they are calculated $\tau(j)$, for example, $f^k_{as}[\tau(j)]$ and so on. Let:

$n_a(\tau_j), n_a(\tau(j))$ be the link occupancy in time instants τ_j and $\tau(j)$, respectively

$\hat{n}_a[j]$ be the average load on link a during interval $[j]$. Clearly:

$$\hat{n}_a[j] = \frac{1}{DT} \int_{[j-1]\cdot DT}^{[j]\cdot DT} n_a(\tau(j))\, d\tau(j)$$

$U_a[\tau_j] U_a[\tau(j)], W_a[\tau_j], W_a[\tau(j)]$ be the cumulative in-flows and out-flows of link a up to the representative instant of interval $[j]$, τ_j, and up to an arbitrary time instant within that interval, $\tau(j)$, respectively. Cumulative in-flows and out-flows are related to interval-specific values by:

$$U_a[\tau_j] = \sum_{j' \leq j} u_a[j'] \tag{7.2.23a}$$

$$W_a[\tau_j] = \sum_{j' \leq j} w_a[j'] \tag{7.2.23b}$$

Equations (7.2.1), expressing the total flows as sums of path flows, and (7.2.6b), expressing flow conservation at nodes, also hold in the discrete case. In-flows and out-flows are also related to link occupancy through link conservation equations analogous to (7.2.7) and (7.2.8):

$$n_a(\tau_j) - n_a(\tau_{j-1}) = u_a[j] - w_a[j] \tag{7.2.24a}$$

$$n_a(\tau_j) = U_a[\tau_j] - W_a[\tau_j] \tag{7.2.24b}$$

Travel Time and Cost Variables In general, link and path travel times are continuous variables that vary with time τ as in the continuous case. In the discrete case, however, not all instants τ are meaningful because not all correspond to the arrival (or departure) of a packet (see Fig. 7.13). Below, time and cost variables are defined in terms of an arbitrary instant τ. Let:

$t_a^f(\tau), t_a^b(\tau)$ be the forward and backward travel time on link a for a packet that, respectively, enters or leaves the link at time τ. Forward and backward link travel times are related through mutual consistency equations identical to (7.2.5), which are restated here for convenience:

$$t_a^f(\tau) = t_a^b\left(\tau + t_a^f(\tau)\right)$$

$$t_a^b(\tau) = t_a^f\left(\tau - t_a^b(\tau)\right)$$

Because, in discrete flow models, users are individually identifiable units (packets k_j), it is possible to define temporal variables associated with a specific packet. Let:

$\tau_a^u[k_j], \tau_a^w[k_j]$ be, respectively, the entrance and exit times on link a of packet k_j. Consistency of travel times requires that (see Fig. 7.14):

$$\tau_a^w[k_j] = \tau_a^u[k_j] + t_a^f\left(\tau_a^u[k_j]\right) \tag{7.2.25a}$$

$$\tau_a^u[k_j] = \tau_a^w[k_j] - t_a^b\left(\tau_a^w[k_j]\right) \tag{7.2.25b}$$

The FIFO discipline also applies to discrete models if it is assumed that packets cannot overtake each other, or if no explicit overtaking mechanism is introduced. The formal representation of the FIFO rule is identical to that for continuous flow models in terms of forward and backward travel time, respectively:

$$\tau' + t_a^f(\tau') < \tau'' + t_a^f(\tau'') \quad \forall \tau' < \tau''$$

$$\tau' - t_a^b(\tau') < \tau'' - t_a^b(\tau'') \quad \forall \tau' < \tau''$$

Alternative conditions for the FIFO rule, analogous to those introduced in Sect. 7.2.1.1 for continuous models, can be stated. It should be observed, however, that, for discrete models, this condition is not so important because a packet is identified by the very nature of the model rather than implicitly through the trajectory crossing a given point at a given time.

As for the continuous case, the general discrete dynamic supply model can be formalized through link and path performance functions and the network flow propagation model.

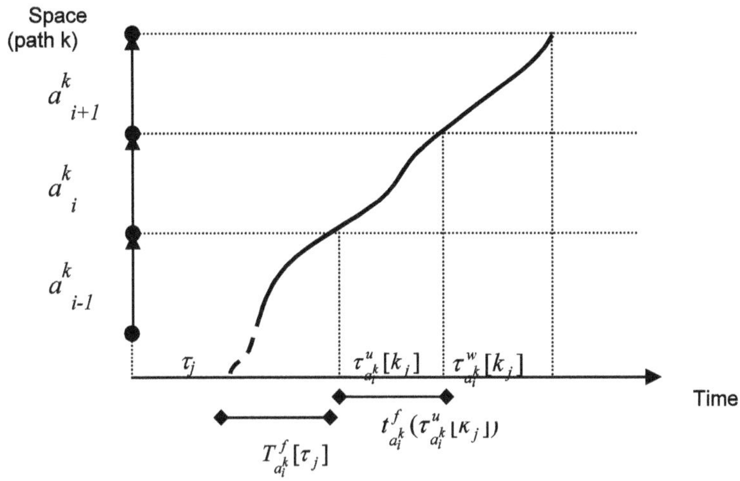

Fig. 7.14 Relationship among link entrance, exit, and travel times of a packet on a link

7.2.2.2 Link Performance and Travel Time Functions

The dependence of link travel time on link "flow" variables for congested networks can be expressed through a variety of models. It is possible to specify separable and nonseparable cost functions, the latter possibly allowing for spill-back effects from downstream links. The simpler separable travel time functions are similar to the functions described in Sect. 7.2.1.2 for running and queuing links.

Forward travel time on running link a can be expressed as a linear function of arrival time. It can thus vary for different time instants $\tau(j)$ within interval j:

$$t_a^f\left(\tau(j)\right) = t_a^0 + \frac{1}{Q_a} \cdot n_a\left(\tau(j)\right) \tag{7.2.26}$$

Other models express the travel time via the average speed computed as a function of link density, as in the fundamental diagram of traffic flow described in Chap. 2:

$$t_a^f\left(\tau(j)\right) = \frac{L_a}{V_a(n_a(\tau(j))/L_a)} \tag{7.2.27}$$

Given the discrete nature of the models, various assumptions can be made regarding the computation of travel times for packets entering the link in a given interval.

Some models proposed in the literature assume that the travel times are equal for all packets that enter the link in a given interval. In this case occupancy variables in (7.2.26) and (7.2.27) correspond to a representative time of interval j, typically its start-point τ_{j-1}, and are constant for all users entering the link during the interval. Alternatively, travel times can be computed as functions of the average link occupancy during the previous interval $\hat{n}_a[j-1]$, or the same interval $\hat{n}_a[j]$. In the latter case, however, link travel time for users entering the link during the interval depends on the number of users who enter the link later in the same interval; this may cause inconsistencies and counterintuitive results, and should be avoided. Other more accurate models compute travel times for each packet, for example, as a function of the instantaneous link occupancy at the entrance time.

7.2.2.3 Path Performance and Travel Time Functions

The concepts of the forward and backward travel time needed to reach link a_i^k along path k when leaving or arriving in a given instant can be immediately extended to discrete supply models. These variables are denoted by $T^f_{a_i^k}(\tau_j)$ and $T^b_{a_i^k}(\tau(j))$, respectively, to stress the fact that departures can occur only at the representative time of each interval τ_j, whereas arrivals can be at any time during the interval $\tau(j)$; see Fig. 7.13.

Therefore, the relationships between forward and backward travel times in the discrete case become:

$$T^f_{a_i^k}(\tau_j) = T^b_{a_i^k}\left(\tau + T^f_{a_i^k}(\tau_j)\right)$$

$$T^b_{a_i^k}\left(\tau(j)\right) = T^f_{a_i^k}\left(\tau(j) - T^b_{a_i^k}\left(\tau(j)\right)\right)$$

As with the continuous case, the forward (backward) total travel time on path k for a given departure (arrival) time can be defined for the discrete case, denoting the variables by $TT_k^f(\tau_j)$ and $TT_k^b(\tau(j))$, respectively. The FIFO rule for partial and total path travel times can also be extended to discrete flow models, as shown by (7.2.5).

Similarly the relationship between link and path travel times is analogous to (7.2.3) and, when applied recursively, leads to a "nested" structure:

$$\begin{aligned} T^f_{a_i^k}(\tau_j) &= t^f_{a_1^k}(\tau_j) + t^f_{a_2^k}\left(\tau_j + t^f_{a_1^k}(\tau_j)\right) + t^f_{a_3^k}\left(\tau_j + t^f_{a_1^k}(\tau_j) + t^f_{a_2^k}\left(\tau_j + t^f_{a_1^k}(\tau_j)\right)\right) \\ &\quad + \cdots + t^f_{a_{i-1}^k}\left(\tau_j + t^f_{a_1^k}(\tau_j) + \cdots + t^f_{a_{i-2}^k}\left(\tau_j + t^f_{a_1^k}(\tau_j) + \cdots\right)\right) \end{aligned} \tag{7.2.28}$$

In the discrete flow case, however, (7.2.28) can be expressed more straightforwardly using $\tau^u_{a_i^k}[k_j]$, the packet arrival time at link a_i^k

$$T^f_{a_i^k}(\tau_j) = t^f_{a_1^k}(\tau_j) + t^f_{a_2^k}\left(\tau^u_{a_2^k}[k_j]\right) + t^f_{a_3^k}\left(\tau^u_{a_3^k}[k_j]\right) + \cdots + t^f_{a_{i-1}^k}\left(\tau^u_{a_{i-1}^k}[k_j]\right) \tag{7.2.29}$$

The same construct applies to total path travel time $TT_k^f(\tau_j)$, to other path-additive attributes $EC_k(\tau_j)$ and finally to the total path cost $g_k(\tau_j)$:

$$\begin{aligned} TT_k^f(\tau_j) &= t^f_{a_1^k}(\tau_j) + t^f_{a_2^k}\left(\tau^u_{a_2^k}(\tau_j[k_j]\right) + t^f_{a_3^k}\left(\tau^u_{a_3^k}[k_j]\right) + \cdots + t^f_{a_{n_k}^k}\left(\tau^u_{a_{n_k}^k}[k_j]\right) \\ &= T^f_{a_{n_k}^k}(\tau_j) + t^f_{a_{n_k}^k}\left(\tau^u_{a_{n_k}^k}[k_j]\right) \end{aligned} \tag{7.2.30a}$$

$$EC_k(\tau_j) = ec^f_{a_1^k}(\tau_j) + ec^f_{a_2^k}\left(\tau^u_{a_2^k}[k_j]\right) + ec^f_{a_3^k}\left(\tau^u_{a_3^k}[k_j]\right)$$

$$+ \cdots + ec^f_{a_{n_k}^k}\left(\tau^u_{a_{n_k}^k}[k_j]\right) \tag{7.2.30b}$$

$$g_k(\tau_j) = \beta_t TT^f_k(\tau_j) + EC_k(\tau_j) \tag{7.2.30c}$$

Formally, the relationship between the vector of total path travel time $TT^f(\tau_j)$ for a given departure time τ_j, and travel times on the links making up each path, can be expressed symbolically as

$$\boldsymbol{TT}^f(\tau_j) = \Gamma\left(\boldsymbol{t}(\tau'), \tau' \geq \tau_j\right) \tag{7.2.31}$$

Equation (7.2.31) is the equivalent of (7.2.21) in the continuous-flow case.

7.2.2.4 Dynamic Network Loading

Unlike the continuous-flow case, the DNL model for discrete flows can easily be formulated explicitly because packets can be identified as they move across the network. In this case, the in-flow on link a in interval $[j]$ can be expressed as

$$u_a[j] = \sum_k \sum_{l \leq j} \delta_{ak}[l,j] \cdot h_k[l] \tag{7.2.32}$$

where the $\delta_{ak}(l, j)$ are zero/one variables analogous to the elements of the static link-path incidence matrix; they are equal to one if the packet k_l (of intensity $h_k[l]$) enters link a during interval j, and zero otherwise:

$$\delta_{ak}[l,j] = \begin{cases} 1 & \text{if } \tau^u_a[k_l] \in ([j-1]DT, [j]DT) \\ 0 & \text{otherwise} \end{cases}$$

Obviously the $\delta_{ak}[l, j]$ are all equal to zero if link a does not belong to path k (compare (7.2.32) with (7.2.17a)).

Equation (7.2.32) can also be formulated using matrix notation as

$$\boldsymbol{u}[j] = \sum_{l \leq j} \Delta[l,j] \cdot \boldsymbol{h}[l] \tag{7.2.33}$$

which is close to its static counterpart $\boldsymbol{f} = \Delta\boldsymbol{h}$.

Similar equations can be stated for the out-flow $w_a[j]$ from link a in time interval j:

$$w_a[j] = \sum_k \sum_{l \leq j} \delta'_{ak}[l,j] \cdot h_k[l] \tag{7.2.34}$$

where $\delta'_{ak}[l, j]$ is equal to one if packet k_l (of intensity $h_k[l]$) leaves link a during interval j, 0 otherwise:

$$\delta'_{ak}[l, j] = \begin{cases} 1 & \text{if } \tau_a^w[k_l] \in ([j-1]DT, [j]DT) \\ 0 & \text{otherwise} \end{cases}$$

and in matrix terms:

$$\boldsymbol{w}[j] = \sum_{l \le j} \Delta'[l, j] \cdot \boldsymbol{h}[l] \tag{7.2.35}$$

Note that the elements of dynamic incidence matrices depend on link travel times and, for congested networks, on link flows and occupancies. In this respect they should be denoted as

$$\delta_{ak}[l, j] = \delta_{ak}[l, j]\big(\boldsymbol{t}(\tau'); \tau' \in (\tau_l, \tau_j)\big)$$

The overall DNL model that relates link flows and occupancies to path flows can be expressed by combining the previous equations:

$$n_a(\tau_j) - n_a(\tau_{j-1}) = u_a[j] - w_a[j] \tag{7.2.36a}$$

$$u_a[j] = \sum_{l \le j} \Delta[l, j] \cdot \boldsymbol{h}[l] \tag{7.2.36b}$$

$$w_a[j] = \sum_{l \le j} \Delta'[l, j] \cdot \boldsymbol{h}[l] \tag{7.2.36c}$$

$$\tau_{ak}^u[k_l] = \tau_l + T_{a_i^k}^f(\tau_l) \tag{7.2.36d}$$

$$\tau_{ak}^w[k_l] = \tau_{ak}^u[k_l] + t_{ak}^f\big(\tau_{ak}^u[k_l]\big) \tag{7.2.36e}$$

$$t_a^f(\tau(j)) = \frac{L_a}{V_a(n_a(\tau_{j-1})/L_a)} \tag{7.2.36f}$$

The above set of equations has been specified under the assumption that link travel time functions depend on link occupancy at the beginning of each interval; the model can be expressed in a similar form with reference to an arbitrary time instant $\tau(j)$.

7.2.2.5 Formalization of the Whole Supply Model

Equations (7.2.36) can be expressed symbolically as nonlinear vector functions that relate link flows (in-flows and out-flows) and occupancies for an interval j, to the vector of path flows that depart in intervals from l to j and to the link travel times in intervals between τ_l and the end of interval j, τ_j

$$\boldsymbol{f}[j] = \Phi\big(\boldsymbol{h}[l], \boldsymbol{t}(\tau'); l \le j, \tau' \in [\tau_l, \tau_j]\big) \tag{7.2.37a}$$

Expression (7.2.37a) can be further combined with the equation relating link travel times to link occupancies for congested dynamic network loading models:

$$f[j] = \Phi\big(h[l], t(n(\tau')); l \leq j, \tau' \in [\tau_l, \tau_j]\big) \tag{7.2.37b}$$

The global supply model is completed by the symbolic relationships relating path travel times to link travel times:

$$TT^f(\tau_l) = \Gamma\big(t(n(\tau')); \tau' \geq \tau_l\big) \tag{7.2.38}$$

and path generalized transportation costs to travel times and other link costs:

$$g(\tau_l) = \beta_t TT^f(\tau_l) + EC(\tau_l) \tag{7.2.39}$$

7.3 Demand Models for Continuous Service Systems

Demand models used in dynamic assignment express the relationship between path flows and path costs. The "minimal" demand model in this context relates to path and departure time choice; it is included in some form in all dynamic assignment models, and is described in this section. Other models that simulate users' learning and choice adjustment mechanisms are needed for dynamic process assignment; they are briefly described in the next section on demand–supply interaction.

The flow $h_k(\tau)$ of users who depart at time τ on path k connecting O-D pair *od* can be represented with elastic demand profile models; these simulate not just path choice but also departure time choice given either the desired arrival time at destination τ_d, or the desired departure time from the origin τ_o.

The continuous time-continuous flow model is discussed first. Let:

$d_{od}(\tau_d)$ be the flow of trips between the O-D pair *od* with desired arrival time τ_d

$p_{od,k}(\tau/\tau_d)$ be the probability of choosing time τ and path k, given the O-D pair *od* and the desired arrival time τ_d

$V_k(\tau/\tau_d)$ be the systematic utility of path k and departure time τ, given the desired arrival time τ_d

$\boldsymbol{V}_{od}(\tau/\tau_d)$ be the vector of systematic utilities of all paths connecting O-D pair *od* for a given departure time τ and desired arrival time τ_d

The demand conservation condition over the whole reference interval [0, T] can be formally expressed as (compare with $h_k = p_{od,k} \cdot d_{od}$ in the static case):

$$h_k(\tau) = \int_o^T p_{od,k}(\tau/\tau_d) \cdot d_{od}(\tau_d) d\tau_d \tag{7.3.1}$$

Choice probabilities of departure time τ and path k are usually expressed with random utility models that depend on the systematic utilities of available path-departure time alternatives:

$$p_{od,k}(\tau/\tau_d) = p_{od,k}\big(V_{od}(\tau'/\tau_d),\ \forall\tau'\big) \tag{7.3.2a}$$

Such models are usually single-level random utility models with mixed continuous (departure time)/discrete (path) alternatives, as, for example, multinomial logit:

$$P_{od,k}(\tau/\tau_d) = \frac{\exp(V_k(\tau/\tau_d))}{\sum_{j\in K_{od}} \int_o^T \exp(V_j(\theta/\tau_d))\, d\theta} \tag{7.3.2b}$$

They can be partial share models as well. The combined choice probability is sometimes expressed as the product of path choice probability given the departure time, and the departure time choice probability:

$$p_{od,k}(\tau/\tau_d) = p_{od}(\tau/\tau_d) \cdot p_{od}[k/\tau, \tau_d]$$

Some empirical results on elasticities of demand with respect to changes in departure time and path seem to suggest a different sequence:

$$p_{od,k}(\tau/\tau_d) = p_{od}[k] \cdot p_{od}(\tau/k, \tau_d) \tag{7.3.2c}$$

Some dynamic assignment models proposed in the literature assume deterministic utility departure time and path models. In this case, as for static systems, choice probabilities cannot be expressed in closed form, because there may exist several departure time–path alternatives with equal systematic disutilities. Indirect expressions similar to the static models described in Chap. 4 can be adopted in this case:

$$p_{od,k}(\tau/\tau_d) > 0 \quad \Rightarrow \quad V_{od,k}(\tau/\tau_d) \geq V_{od,k'}(\tau'/\tau_d) \quad \forall \tau', k'$$

Deterministic choice models, however, are arguably less realistic when applied to continuous departure times than they are in the static case.

Systematic utility functions proposed for the simulation of combined path–departure time choice typically include, in addition to path attributes, the *schedule delay*, that is, the penalty for arriving early or late with respect to the desired arrival time (see Fig. 7.15). For desired arrival time τ_d, we have:

$$\begin{aligned} V_k(\tau/\tau_d) = {} & \beta_t TT_k^f(\tau) + EC_k(\tau) + \beta_e EAP_k\big(\tau, \tau_d, TT_k(\tau)\big) \\ & + \beta_l LAP_k\big(\tau, \tau_d, TT_k(\tau)\big) \end{aligned} \tag{7.3.3a}$$

where

$EAP_k(\tau, \tau_d, TT_k^f(\tau))$ is the penalty for arriving earlier than τ_d when departing at time τ and following path k. This penalty is usually considered only if the early arrival is above a minimum threshold Δ_e:

$$EAP_k\big(\tau, \tau_d, TT_k^f(\tau)\big) = \begin{cases} \tau_d - \Delta_e - (\tau + TT_k^f(\tau)) & \text{if } \tau_d - \Delta_e - (\tau + TT_k^f(\tau)) > 0 \\ 0 & \text{otherwise} \end{cases}$$

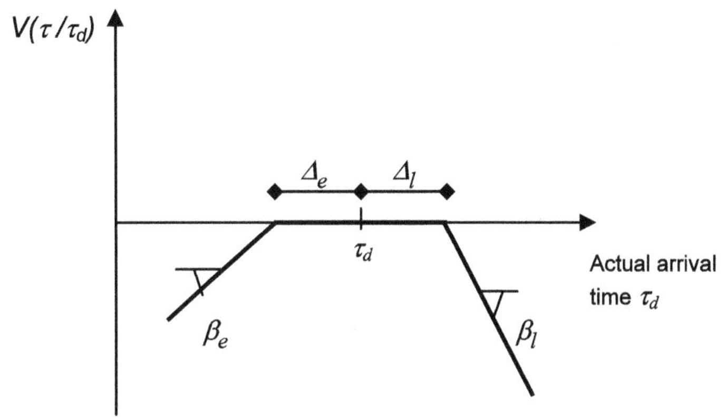

Fig. 7.15 Systematic utility function with respect to desired arrival time

$LAP_k(\tau, \tau_d, TT_k^f(\tau))$ is the penalty for arriving later than τ_d when departing at time τ and following path k. This penalty is usually considered only if the delay is above a minimum threshold Δ_l:

$$LAP_k\big(\tau, \tau_d, TT_k^f(\tau)\big) = \begin{cases} \tau + TT_k^f(\tau) - \tau_d - \Delta_l & \text{if } \tau + TT_k^f(\tau) - \tau_d - \Delta_l > 0 \\ 0 & \text{otherwise} \end{cases}$$

When users have a desired departure time from the origin (τ_o), rather than a desired arrival time at the destination (τ_d), the expression for the systematic utility is still a function of path travel time and schedule delay, but in this case the schedule delay does not depend on the path travel time $TT_k^f(\tau)$:

$$V_k(\tau/\tau_o) = \beta_t TT_k^f(\tau) + EC_k(\tau) + \beta_e EDP(\tau, \tau_o) + \beta_l LDP(\tau, \tau_o) \tag{7.3.3b}$$

where

$EDP(\tau, \tau_o)$ is the penalty for departing at a time τ that is earlier than τ_o; it is usually considered only if the early departure is above a minimum threshold Δ_e:

$$EDP(\tau, \tau_0) = \begin{cases} \tau_0 - \Delta_e - \tau & \text{if } \tau_0 - \Delta_e - \tau > 0 \\ 0 & \text{otherwise} \end{cases}$$

$LDP(\tau, \tau_o)$ is the penalty for departing at a time τ that is later than τ_o, usually considered only if the delay is above a minimum threshold Δ_l:

$$LDP(\tau, \tau_0) = \begin{cases} \tau - \tau_0 - \Delta_l & \text{if } \tau - \tau_0 - \Delta_l > 0 \\ 0 & \text{otherwise} \end{cases}$$

All the coefficients β in (7.3.3) are negative. Furthermore, the schedule early/delay penalties should have coefficients β_e and β_l with absolute values greater than the travel time coefficient ($|\beta_e| > |\beta_t|, |\beta_l| > |\beta_t|$) in order to avoid unrealistic user behavior, for example, large probabilities for alternatives with very high early/delay arrival penalties but with smaller travel times. Empirical results for work-related trips show that the disutility of late arrivals is larger than that for early arrivals ($|\beta_e| < |\beta_l|$), as shown in Fig. 7.15.

The global within-day dynamic demand model with elastic demand profile is expressed by (7.3.1) to (7.3.3) relating path flows to path travel times, extra costs, and schedule early/delay penalties for different departure times.

In *fixed demand profile* models, it is assumed that the distribution of demand flows over departure times is known and independent of variations in travel times; that is, the probabilities $p_{od}(\tau/\tau_d)$ or $p_{od}(\tau/\tau_o)$ are given. It follows that, for a given departure time, path is the only choice dimension considered:

$$h_k(\tau) = d_{od}(\tau) \cdot p_{od,k}\left(\mathbf{V}_{od}(\tau)\right) \tag{7.3.4}$$

where

$d_{od}(\tau)$ is the O-D demand flow leaving at time τ
$p_{od,k}(\tau)$ is the probability that trips starting at time τ will choose path k
$\mathbf{V}_{od}(\tau)$ is the vector of the systematic utilities $V_k[\tau]$ of the different paths, $k \in K_{od}$ connecting the O-D pair *od*

Path choice models in this case are analogous to those described in Sect. 4.3.3; the systematic utility of a path k can be expressed as a function of the path-related attributes introduced previously by:

$$V_k(\tau) = \beta_t TT_k^f(\tau) + EC_k(\tau) \tag{7.3.5}$$

The within-day dynamic demand model with a fixed demand profile is expressed by (7.3.4) and (7.3.5) connecting path flows to path travel times for a given departure time τ.

Considering now discrete time dynamic demand models, the necessary modifications to the previous discussion are straightforward. The only difference is that alternative departure times are the discrete intervals $[j-1]$, $[j]$, $[j+1]$, or their representative instants τ_{j-1}, τ_j, τ_{j+1}. Simultaneous departure time and path choice probabilities are thus expressed as $p_{od,k}[\tau_j/\tau_d]$. A multinomial logit specification can be:

$$p_{od,k}[\tau_j/\tau_d] = \frac{\exp(V_k[\tau_j/\tau_d])}{\sum_{\tau j'} \sum_{k' \in K_{od}} \exp(V_{k'}[\tau_{j'}/\tau_d])}$$

Alternatively the probability could be expressed using a partial share specification similar to (7.3.2c) introducing a correlation structure among adjacent departure intervals, for example, with a cross-nested logit model.

The previous results for choice models and systematic utility specifications apply also to the discrete departure time case. Discrete departure time models can be adopted for the continuous flows. In fact, some specifications of continuous departure time choice models assume that travelers do not choose among an infinite number of departure instants, but rather among a finite number of time intervals (e.g., five minutes long), and that actual departure times are uniformly distributed within the chosen interval. In this case, the multinomial logit probability of leaving at time $\tau(j)$ following path k would be:

$$p_{od,k}\big(\tau(j)/\tau_d\big) = \frac{1}{DT} \frac{\exp(V_k[j/\tau_d])}{\sum_{j'} \sum_{k' \in K_{od}} \exp(V_{k'}[\tau_{j'}/\tau_d])}$$

7.4 Demand–Supply Interaction Models for Continuous Service Systems

Demand–supply interaction models for within-day dynamic continuous service systems are conceptually analogous to those described for the equivalent static systems.

In the following sections, formal results are given for both uncongested and congested network assignment. These can be approached either through equilibrium or through dynamic process models. Both the continuous and discrete flow cases are discussed for uncongested and user equilibrium assignment models; on the other hand, dynamic process models, with and without information, are formulated only for the discrete flow case.

Dynamic Traffic Assignment (DTA) models are rather complex and few operational formulations have been developed (one of these is presented in Sect. 7.5). Furthermore, compared to the static case, few theoretical results on the existence and uniqueness of DTA solutions are currently available.

For simplicity, the following considers only (within-day dynamic) demand models with desired departure time τ_o. Extension to the case of desired arrival time is straightforward.

7.4.1 Uncongested Network Assignment Models

Dynamic assignment models for uncongested networks can be represented schematically as in Fig. 7.16. In this case link travel times do not depend on link occupancies.

In the continuous-flow case, the assignment model can be specified as

$$\boldsymbol{t}^f(\tau) = \boldsymbol{t}^0(\tau) \tag{7.4.1a}$$

$$\boldsymbol{TT}^f(\tau) = \Gamma\big(\boldsymbol{t}^0(\tau)\big) \tag{7.4.1b}$$

$$\begin{aligned} \boldsymbol{V}_{od}(\tau/\tau_o) =& \beta_t \boldsymbol{TT}(\tau) + \boldsymbol{EC}(\tau) + \beta_e \boldsymbol{EDP}(\tau, \tau_o) \\ & + \beta_l \boldsymbol{LDP}(\tau, \tau_o) \end{aligned} \tag{7.4.1c}$$

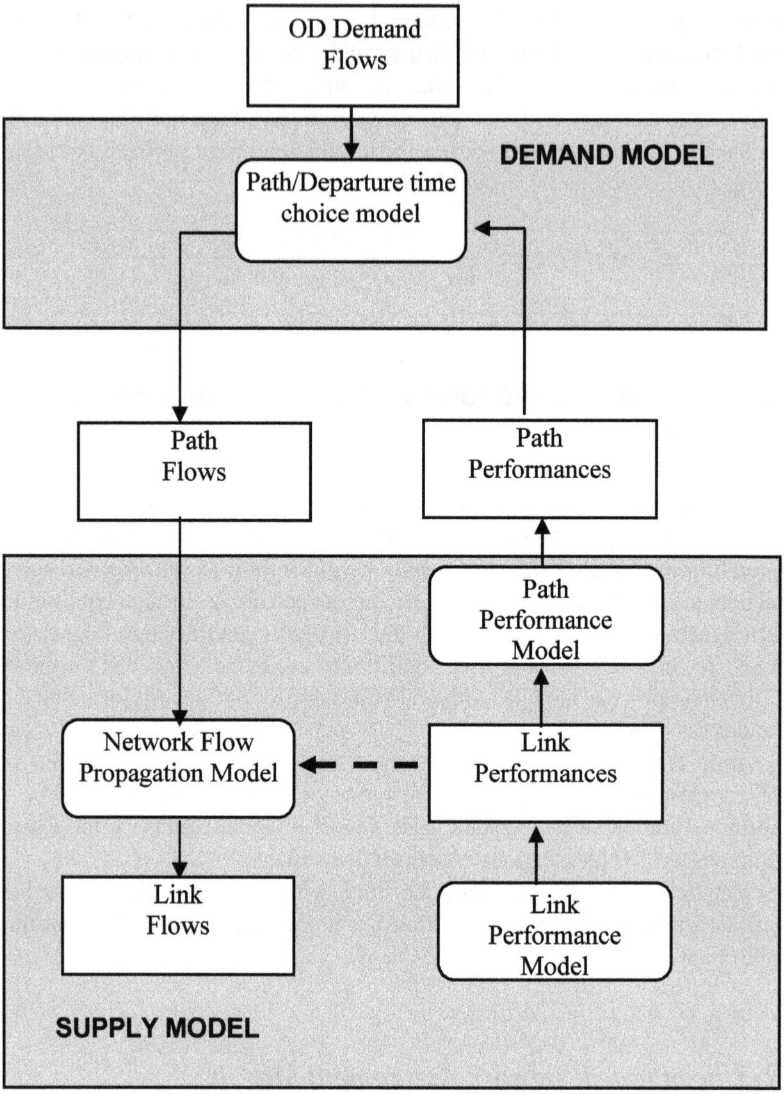

Fig. 7.16 Within-day dynamic traffic assignment for uncongested networks

$$\boldsymbol{h}(\tau) = \int_0^T \boldsymbol{P}\big(\boldsymbol{V}_{od}(\tau/\tau_o)\big) \cdot \boldsymbol{d}(\tau_o)\, d\tau_o \tag{7.4.1d}$$

$$\boldsymbol{f}(\tau) = \Phi\big(\boldsymbol{h}(\tau), \boldsymbol{t}^0(\tau)\big) \tag{7.4.1e}$$

Equations (7.4.1c) and (7.4.1d) represent the within-day dynamic demand models. On the other hand, (7.4.1a), (7.4.1b), and (7.4.1e) make up the supply model consisting of the link performance model, the path performance model, and the dynamic network flow propagation model, respectively. The uncongested dynamic assignment model (UND) can be deterministic (DUND) or stochastic (SUND) depending on the path choice model used in (7.4.1d).

The Dynamic Network Loading model (DNL) has been formulated symbolically in terms of an unspecified link flow vector $\boldsymbol{f}$, because, if FIFO holds, the different formulations in terms of in-flow, out-flow, or link occupancy are equivalent. For instance, (7.4.1e) can be stated in terms of in-flows on the link a as (see (7.2.1b) and (7.2.13a))

$$u_a(\tau) = \sum_k \delta_{ak} \cdot h^k\big(\tau - T^b_{a,k}(\tau)\big) \cdot \prod_{j=1}^{i-1}\left(1 - \frac{dt^b_{a^k_j}(\tau^L_{a^k_j})}{d\tau}\right)$$

where, for each path k that passes through link a, the second term product is extended to all the links that precede a along k (i.e., it is extended until the $(i-1)$th link of path k, where $a^k_i = a$, $\forall k \supseteq a$) and the backward travel time $T^b_{a,k}(\tau)$ are independent of link flows (the network being uncongested) but, in general, dependent on time τ:

$$T^b_{a,k}(\tau) = T^0_{a,k}\big(t^0(\tau)\big) \tag{7.4.1f}$$

From (7.4.1), in principle, both demand and link travel times vary with τ. However, in the absence of congestion, (7.4.1) can be solved sequentially to obtain path performances and link flows. In uncongested networks it is usually assumed that link travel times are constant over time; that is, $t^f_a(\tau) = t^0_a$. Thus the system of equations (7.4.1) becomes:

$$\boldsymbol{t}^f(\tau) = \boldsymbol{t}^0 \tag{7.4.2a}$$

$$\boldsymbol{TT}^f(\tau) = \Gamma(\boldsymbol{t}^0) \tag{7.4.2b}$$

$$\begin{aligned}\boldsymbol{V}_{od}(\tau/\tau_o) = {} & \beta_t \boldsymbol{TT}(\tau) + \boldsymbol{EC}(\tau) + \beta_e \boldsymbol{EDP}(\tau, \tau_o) \\ & + \beta_l \boldsymbol{LDP}(\tau, \tau_o)\end{aligned} \tag{7.4.2c}$$

$$\boldsymbol{h}(\tau) = \int_0^T \boldsymbol{P}\big(\boldsymbol{V}_{od}(\tau/\tau_o)\big) \cdot \boldsymbol{d}(\tau_o) d(\tau_o) \tag{7.4.2d}$$

$$\boldsymbol{f}(\tau) = \Phi\big(\boldsymbol{h}(\tau), \boldsymbol{t}^0\big) \tag{7.4.2e}$$

Here the only exogenous dynamic elements are the demand flows, which induce time-varying path and link flows. In particular, (7.4.2b) becomes:

$$TT^f_k(\tau) = \sum_k \delta_{ak} \cdot t^0_d \quad \forall \tau$$

or

$$\boldsymbol{TT}^f(\tau) = \Delta^T \cdot \boldsymbol{t}^0 \quad \forall \tau$$

In the discrete-flow case, the uncongested network assignment model can be formally specified as

$$\boldsymbol{t}^f(\tau_j) = \boldsymbol{t}_j^0 \tag{7.4.3a}$$

$$\boldsymbol{TT}^f(\tau_j) = \Gamma(\boldsymbol{t}_j^0) \tag{7.4.3b}$$

$$\begin{aligned} \boldsymbol{V}_{od}(\tau_j/\tau_o) =& \beta_t \boldsymbol{TT}^f(\tau_j) + \boldsymbol{EC}(\tau_j) + \beta_e \boldsymbol{EDP}(\tau_j, \tau_o) \\ & + \beta_l \boldsymbol{LDP}(\tau_j, \tau_o) \end{aligned} \tag{7.4.3c}$$

$$\boldsymbol{h}(\tau_j) = \sum_{\tau_o} \boldsymbol{P}\big(\boldsymbol{V}_{od}(\tau_j/\tau_o)\big) \cdot \boldsymbol{d}(\tau_o) \tag{7.4.3d}$$

$$\boldsymbol{f}[j] = \Phi\big(\boldsymbol{h}(\tau_j), \boldsymbol{t}_{j'}^0; j' < j\big) \tag{7.4.3e}$$

Note that time dependency in the above equations can be expressed equivalently in terms of the representative time instant of interval j, τ_j, or simply as $[j]$.

Equations (7.4.3c) and (7.4.3d) represent the within-day dynamic demand models and (7.4.3a), (7.4.3b), and (7.4.3e) represent, respectively, the link performance, path performance, and dynamic network flow propagation components of the overall supply model. The DNL can also be stated as

$$\boldsymbol{f}[j] = \sum_{l \le j} \Delta[l, j] \cdot \boldsymbol{h}[l]$$

Note that, if link travel times are constant for all time intervals of the simulation period (i.e., $t_a^f(\tau) = t_a^0$) the matrix Δ does not depend on the starting interval l, but only on the difference between j and l.

7.4.2 User Equilibrium Assignment Models

Dynamic equilibrium assignment on congested networks can be specified through fixed-point models by combining supply and demand models. For within-day dynamic systems, the dependency of travel times on link flows (loads) introduces two feedback loops (see Fig. 7.17): a path cost and flow loop that exists in static models, and a link flow and link travel time loop that is unique to dynamic models.

In the continuous-flow case, user-equilibrium models can be formally stated as a fixed-point problem in travel times, costs, and flows. The problem is derived from the following system of nonlinear equations.

$$\boldsymbol{t}^f(\tau) = \boldsymbol{t}^f\big(\boldsymbol{f}(\tau)\big) \tag{7.4.4a}$$

$$\boldsymbol{TT}^f(\tau) = \Gamma\big(\boldsymbol{t}^f(\tau'); \tau' \le \tau\big) \tag{7.4.4b}$$

$$\begin{aligned} \boldsymbol{V}_{od}(\tau/\tau_o) =& \beta_t \boldsymbol{TT}^f(\tau) + \boldsymbol{EC}(\tau) + \beta_e \boldsymbol{EDP}(\tau, \tau_o) \\ & + \beta_l \boldsymbol{LDP}(\tau, \tau_o) \end{aligned} \tag{7.4.4c}$$

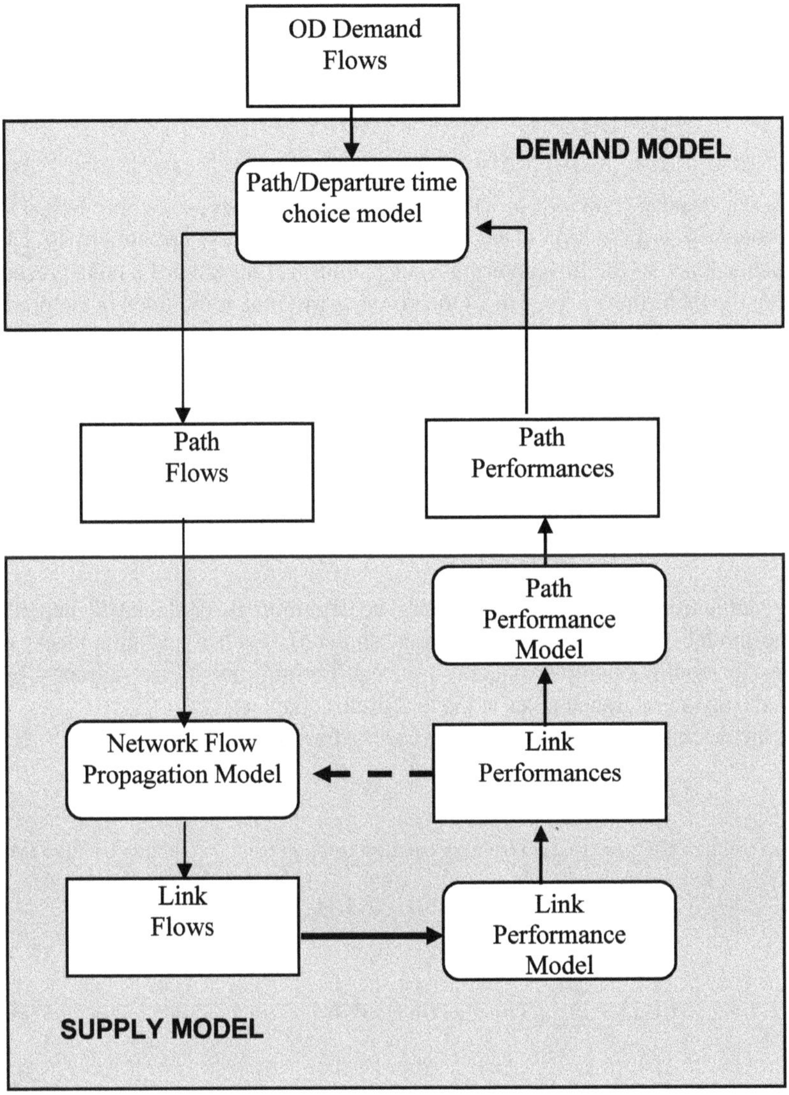

Fig. 7.17 Dynamic user equilibrium traffic assignment

$$\boldsymbol{h}(\tau) = \int_0^T \boldsymbol{P}\big(\boldsymbol{V}_{od}(\tau/\tau_o)\big) \cdot \boldsymbol{d}(\tau_o) d(\tau_o) \tag{7.4.4d}$$

$$\boldsymbol{f}(\tau) = \Phi\big(\boldsymbol{h}(\tau), \boldsymbol{t}^f(\tau); \tau' \leq \tau\big) \tag{7.4.4e}$$

Equation (7.4.4e) expresses the dependency of $\boldsymbol{f}(\tau)$, the link flow vector at time τ, on the path flow vectors $\boldsymbol{h}$ and on link travel time vectors $\boldsymbol{t}$ in all previous time instants $\tau' < \tau$. This can be more explicitly stated, for instance, in terms of in-flows on a link a, as (see Sect. 7.4.1)

$$u_a(\tau) = \sum_k \delta_{ak} \cdot h^k\left(\tau - T^b_{a,k}(\tau)\right) \cdot \prod_{j=1}^{i-1} \left(1 - \frac{dt^b_{a^k_j}(\tau^L_{a^k_j})}{d\tau}\right)$$

where the instants (previous to τ) in which path flows leave, reaching link a in τ, are expressed as a function of backward travel time $T^b_{a,k}(\tau)$ (defined by (7.2.19)), and dependence of the flow entering a, in τ, on travel times on the links preceding it along the paths that cross a, in all the previous instants, is included in the product.

Below is a formal fixed-point specification of dynamic user-equilibrium continuous-flow models in terms of link flows:

$$\boldsymbol{f}^*(\tau) = \Phi\Bigg(\sum_{\tau_o} \boldsymbol{P}\big(\beta_t \Gamma\big(\boldsymbol{t}\big(\boldsymbol{f}^*(\tau')\big) + \boldsymbol{EC}\big(\boldsymbol{t}\big(\boldsymbol{f}^*(\tau')\big) + \beta_e \boldsymbol{EDP}(\tau, \tau_o)$$

$$+ \beta_l \boldsymbol{LDP}(\tau, \tau_o)\big) \cdot \boldsymbol{d}_{\tau_o},\ \boldsymbol{t}^f\big(\boldsymbol{f}^*(\tau')\big);\ \tau' < \tau\Bigg)$$

Dynamic user equilibrium models may be deterministic or stochastic depending on the model of path and departure time choice. Existence and uniqueness conditions for continuous-flow dynamic user equilibrium models are currently being studied (see the reference notes at the end of this chapter).

In the discrete-flow case, the models can be formulated as follows.

$$\boldsymbol{t}^f(\tau_j) = \boldsymbol{t}^f\big(\boldsymbol{f}(\tau_j)\big) \tag{7.4.5a}$$

$$\boldsymbol{TT}^f(\tau_j) = \Gamma\big[\boldsymbol{t}^f(\tau_{j'});\ j' = 1, \ldots, j\big] \tag{7.4.5b}$$

$$\begin{aligned} \boldsymbol{V}_{od}(\tau_j/\tau_o) = {} & \beta_t \boldsymbol{TT}^f(\tau_j) + \boldsymbol{EC}(\tau_j) + \beta_e \boldsymbol{EDP}(\tau_j, \tau_o) \\ & + \beta_l \boldsymbol{LDP}(\tau_j, \tau_o) \end{aligned} \tag{7.4.5c}$$

$$\boldsymbol{h}(\tau_j) = \sum_{\tau_o} \boldsymbol{P}\big(\boldsymbol{V}_{od}(\tau_j/\tau_o)\big) \cdot \boldsymbol{d}(\tau_o) \tag{7.4.5d}$$

$$\boldsymbol{f}(\tau_{j'}) = \Phi\big(\boldsymbol{h}(\tau_{j'}), \boldsymbol{t}^f(\tau_{j'});\ j' = 1, \ldots, j\big) \tag{7.4.5e}$$

Equation (7.4.5e) is analogous to (7.4.1e) for the uncongested network case. It can be stated more explicitly as

$$\boldsymbol{f}[j] = \sum_{l \le j} \Delta[l, j] \cdot \boldsymbol{h}[l] \tag{7.4.5f}$$

The difference with respect to the uncongested network is that, in this case, Δ is a function of link travel times $\boldsymbol{t}$ in all the previous intervals up to interval j:

$$\Delta[l, j] = \Delta[l, j]\big(\boldsymbol{t}^f(\tau_i);\ i = l, \ldots, j\big) \tag{7.4.5g}$$

Below is a formal fixed-point specification of a dynamic user equilibrium model.

$$\boldsymbol{f}^*[\tau_j] = \sum_{\tau_o} \sum_{l=1,\dots,j} \Delta[l,j]\big(\boldsymbol{t}^f(\boldsymbol{f}^*[i];\ i=l,\dots,j)\big)$$

$$\boldsymbol{P}\big[\beta_t \Gamma\big(\boldsymbol{t}^f(\boldsymbol{f}^*[i];\ i=l,\dots,j)\big) + \boldsymbol{EC}\big(\boldsymbol{t}^f(\boldsymbol{f}^*[i];\ i=l,\dots,j)\big) + \beta_e \boldsymbol{EDP}(\tau_l,\tau_o) + \beta_l \boldsymbol{LDP}(\tau_l,\tau_o)\big] \cdot \boldsymbol{d}_{\tau_o}$$

Existence and uniqueness conditions for the fixed-point formulation have not been stated; however, in this case it is more difficult to arrive at general conditions, if indeed it is possible at all, given the discreteness of time and flows (i.e., packets).

7.4.3 Dynamic Process Assignment Models

Dynamic process models require models that simulate the mechanisms of learning (utility updating) and choice updating (see Fig. 7.18). These models can be seen as doubly dynamic assignment models.

As in the static case, to formalize a dynamic process model we need to distinguish between *expected* (or *anticipated*) and *actual* path performance attributes on day t. The former are the attributes (e.g., the travel time on a given path) that users expect to encounter on the network on a given day t; the latter are what they actually experience. Recall that, because of inertia and/or habit, users do not necessarily reconsider their choices every day t.

In the discrete-flow case, let us consider, for the sake of simplicity, that path travel time is the only attribute updated from one day to the next and let:

$\boldsymbol{TT}^{f,t}_{\text{exp}}(\tau_j)$ be the (forward) travel time that users expect to experience on day t if they depart at representative time instant τ_j

$\boldsymbol{TT}^{f,t}_{\text{act}}(\tau_j)$ be the (forward) travel time that users actually experience on day t when they depart at representative time instant τ_j

A deterministic dynamic process model, in which the travel time and choice updating models are simple exponential filters, can be formally stated as follows.

$$\boldsymbol{t}^{f,t-1}(\tau_j) = \boldsymbol{t}^{f,t-1}\big(\boldsymbol{f}(\tau_j)\big) \tag{7.4.6a}$$

$$\boldsymbol{TT}^{f,t-1}_{\text{act}}(\tau_j) = \Gamma\big(\boldsymbol{t}^{f,t-1}(\tau_{j'});\ j'=1,\dots,j\big) \tag{7.4.6b}$$

$$\boldsymbol{TT}^{f,t}_{\text{exp}}(\tau_j) = \beta \boldsymbol{TT}^{f,t-1}_{\text{act}}(\tau_j) + (1-\beta)\boldsymbol{TT}^{f,t-1}_{\text{exp}}(\tau_j) \tag{7.4.6c}$$

$$\boldsymbol{V}^t_{od}(\tau_j/\tau_o) = \beta_t \boldsymbol{TT}^{f,t}_{\text{exp}}(\tau_j) + \boldsymbol{EC}(\tau_j) + \beta_e \boldsymbol{EDP}(\tau_j,\tau_o) + \beta_l \boldsymbol{LDP}(\tau_j,\tau_o) \tag{7.4.6d}$$

$$\boldsymbol{h}^t(\tau_j) = \alpha \sum_{\tau_o} \boldsymbol{P}\big(\boldsymbol{V}^t_{od}(\tau_j/\tau_o)\big) \cdot \boldsymbol{d}(\tau_o) + (1-\alpha)\cdot \boldsymbol{h}^{t-1}(\tau_j) \tag{7.4.6e}$$

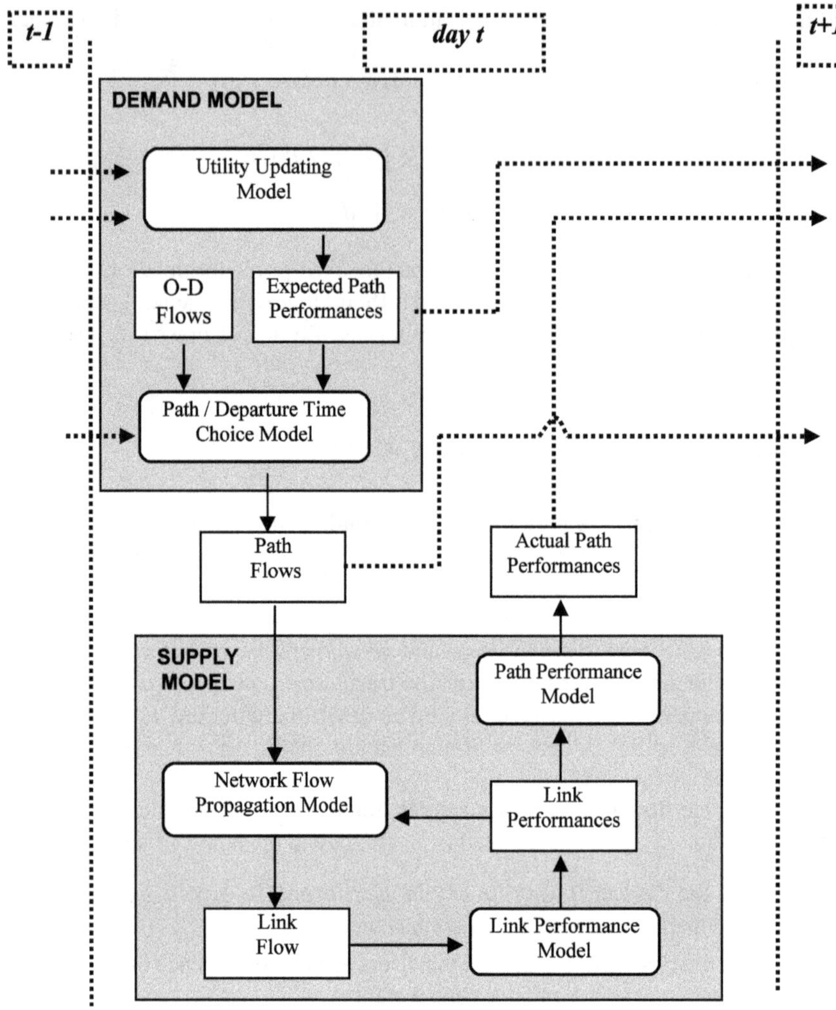

Fig. 7.18 Dynamic process assignment model (without information)

$$f^t[j] = \Phi\big(h^t(\tau_{j'}), t^{f,t}(\tau_{j'}); j' = 1, \dots, j\big) \tag{7.4.6f}$$

where β and α are, respectively, the weight given to the experience of the previous day $t-1$ and the fraction of users reconsidering their choice (assumed here to be constant for each day t).

Note that given the mesoscopic nature of the model, models to update individual packets can be easily implemented. In this case, for instance, it is possible to update expectations based only on the travel time experienced in the actual prior day trip.

Dynamic process models for within-day dynamic systems can be expanded to include real-time information that may be available to some users. This class of

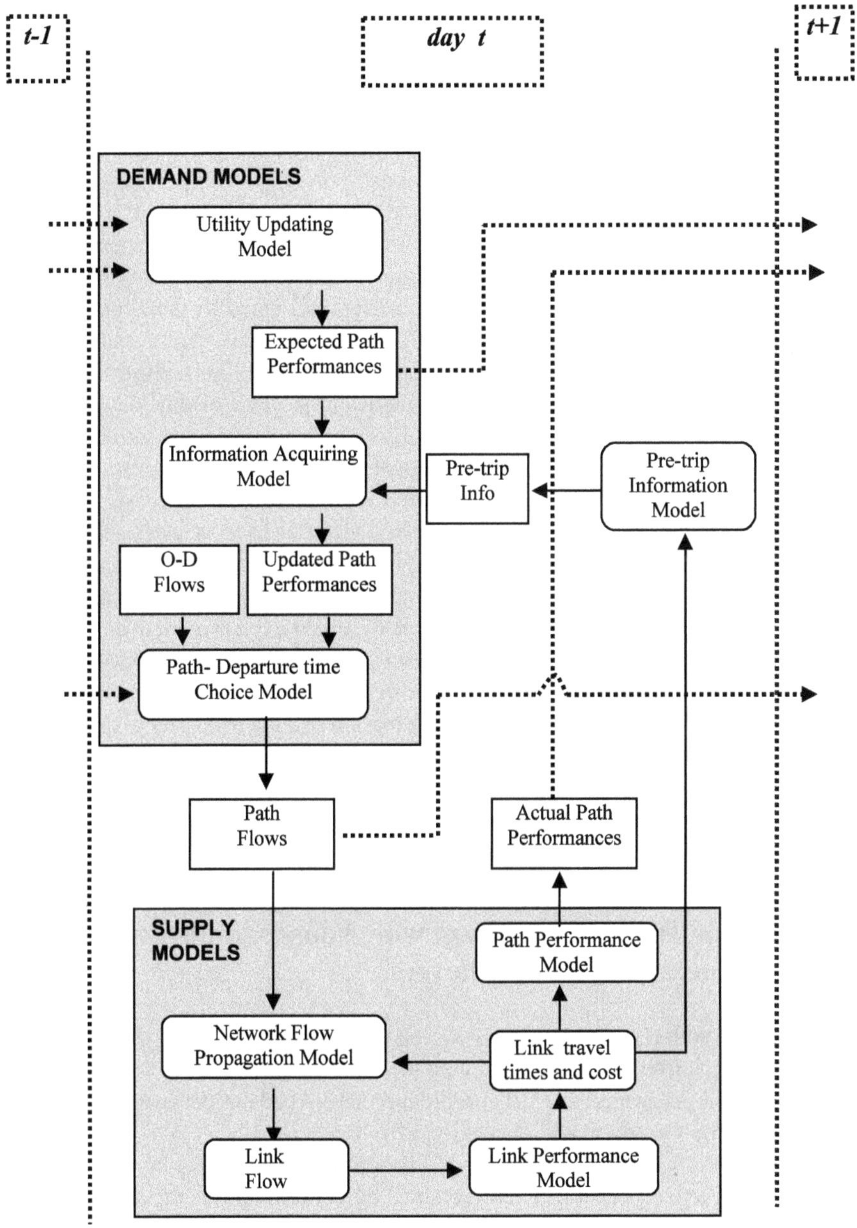

Fig. 7.19 Dynamic process assignment model (with pre-trip information)

assignment model is currently the subject of active development due to growing interest in Advanced Traveler Information Systems (ATIS). Two cases may be distinguished: information is available only before starting a trip (i.e., pre-trip information) and information is available during the journey (i.e., en-route information). The former case requires other demand models that represent the information acquisition process (see Fig. 7.19); the latter requires, in addition, models that simulate users' decisions at diversion nodes to comply with prescriptive information or to reconsider prior choices based on updated information (see Fig. 7.20)

Dynamic process models may be deterministic or stochastic, just as within-day static models, depending on assumptions made about the variables involved (average or deterministic variables or random variables). Full specification of these models requires assumptions on the type of information given and the information strategy, that is, how the information disseminated to users is related to the actual system state (see Fig. 7.21). In general, several information strategies are possible: ATIS can provide, for instance, *historical* information based on network performances in previous time periods with similar characteristics (e.g., time of day, day of week, weather conditions, etc.), real-time information on prevailing network conditions, or forecasts of what is going to happen on the network (i.e., *predictive* information). It is worth noting that predictive information is derived from forecasts of future conditions, but these conditions are themselves affected by how users react to the predictions that they receive. In other words, there is a circular dependency between predictive information and network performance; this can again be seen as a fixed-point problem. Furthermore, based on the type of information provided, any of these information systems can be described as *descriptive* (i.e., travel or congestion phenomena) or *prescriptive* (i.e., route guidance or turning movements).

Due to the multiple possible types of information and the necessity to distinguish between user categories (e.g., informed and noninformed, regular and nonregular, etc.) it is not possible to develop a general formulation for dynamic assignment models with ATIS; for this reason these models are not described here.

7.5 Dynamic Traffic Assignment with Nonseparable Link Cost Functions and Queue Spillovers[6]

In this section, with respect to the formulation described in Sect. 7.4.2, two main improvements are introduced, thus achieving the possibility of solving the within-day Dynamic Traffic Assignment (DTA) problem on large road networks while simulating explicitly the formation and dispersion of vehicle queues.

In Sect. 5.4 it was shown that the equilibrium flow pattern can be expressed as the solution of a fixed-point problem obtained by combining: (a) the supply model with the demand model; or (b) the uncongested network assignment map[7] and flow-dependent link cost functions, thereby making it possible to use an implicit path enumeration approach. In the static case the equivalence of the two formulations

[6] Guido Gentile and Natale Papola are the co-authors of this section.

[7] Under the assumption of probabilistic path choice behavior, the one-to-many map becomes a one-to-one function.

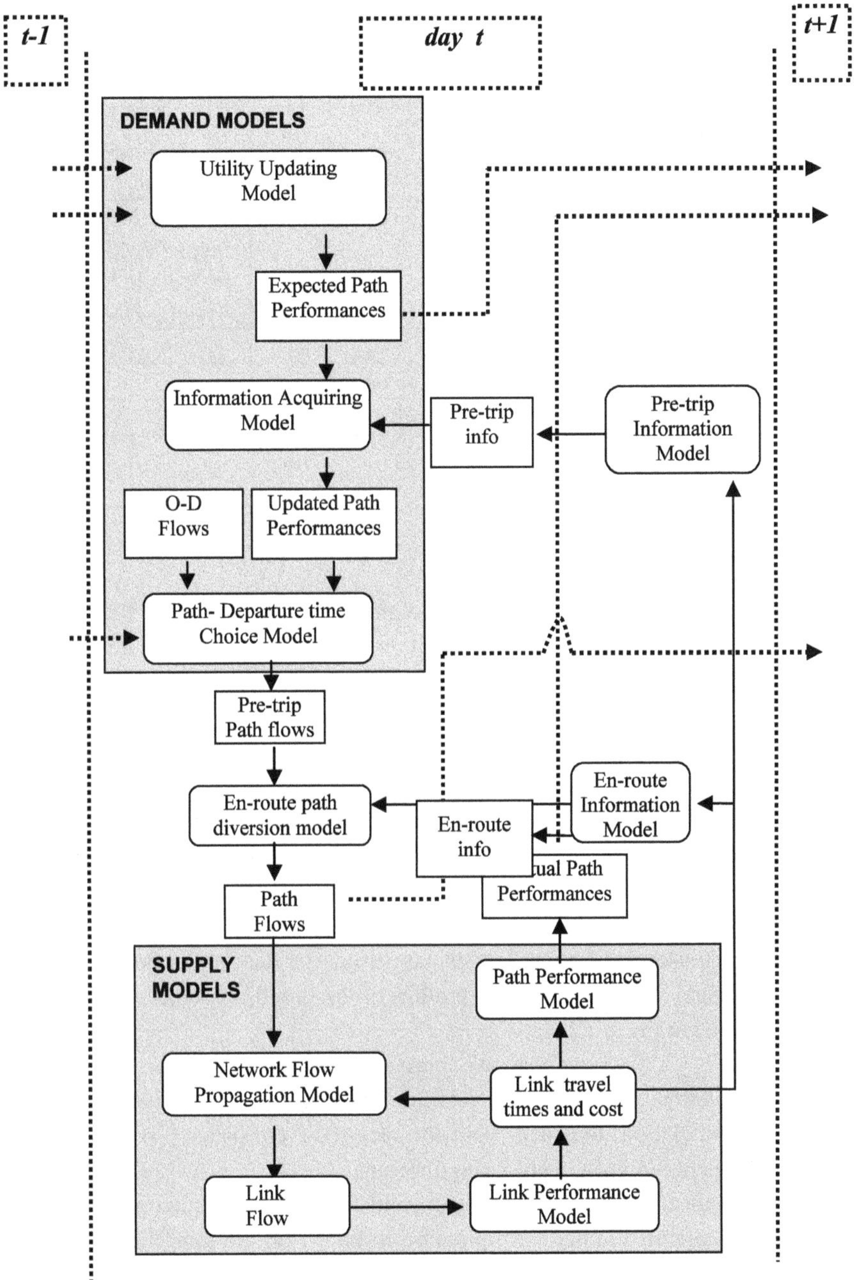

Fig. 7.20 Dynamic process assignment model (with pre-trip/en-route information)

<table>
<tr><th></th><th>Classification</th><th>Example</th></tr>
<tr><td rowspan="2">Information type</td><td>Descriptive</td><td>“Congestion ahead”
“Travel time to airport 5 min”</td></tr>
<tr><td>Prescriptive</td><td>“Turn left”</td></tr>
<tr><td rowspan="2">Information availability</td><td>Pre-trip</td><td>Information available via the Internet, or television</td></tr>
<tr><td>En-route</td><td>Variable Message Signs (VMS) or In-Vehicle Navigation Systems (IVNS)</td></tr>
<tr><td rowspan="3">Information time-dimension</td><td>Historical</td><td></td></tr>
<tr><td>Real-time (or prevailing)</td><td></td></tr>
<tr><td>Predictive (or self-consistent)</td><td></td></tr>
</table>

Fig. 7.21 Classification of information types

is proved and the uncongested network assignment map, also called the Network Loading Map (NLM), is available without requiring the explicit enumeration of path alternatives for each of the route choice models generally utilized in practice (deterministic, logit, probit). The first improvement consists then in extending approach (b) to the dynamic case, thus paving the way for the implementation of robust solving algorithms.

The second improvement consists in extending the continuous formulation of the DTA developed in the previous sections, so as to reproduce spill-back congestion within the Link Performance Model (LPM), which is a crucial step towards satisfactory simulation of highly congested networks.

The dynamic user equilibrium is then expressed as a fixed-point problem where the current variables are the temporal profiles of the link flows, consistent with the scheme depicted in Fig. 7.22.

Note that Fig. 7.22 shows that the approach followed in this section does not involve the solution of a Dynamic Network Loading (DNL) problem within the fixed-point formulation, thus achieving the reciprocal consistency between flows and travel times only jointly with the equilibrium.

Finally, none of the models presented in this section (unlike many others proposed in the literature) requires a limitation to be set on the time intervals introduced for solving the continuous formulation. In practice, this enables us to define a few long intervals of five to ten minutes to cover the simulation period, instead of many short intervals of a few seconds, thus making a decisive step towards the implementation of efficient DTA algorithms.

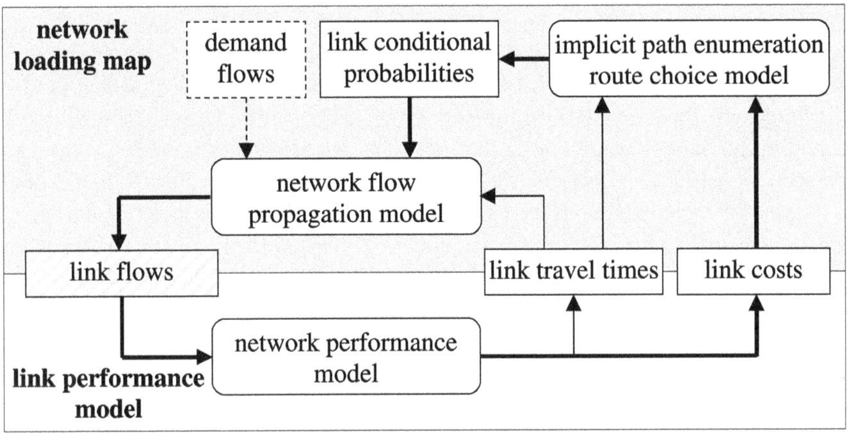

Fig. 7.22 Scheme of the fixed-point formulation for the DTA with spill-back congestion without explicit path enumeration

7.5.1 Network Performance Model

We now introduce a particular link performance model capable of reproducing queue spillovers, which is the main traffic phenomenon occurring on highly congested road networks. The prevalent nonseparability of this link cost function has suggested the term Network Performance Model (NPM). Because the NPM can be easily plugged into any dynamic model requiring an LPM, its relevance goes beyond the specific formulation of DTA presented in this section.

Proper simulation of spill-back congestion requires the formation and dispersion of vehicle queues to be explicitly represented under the condition that the queue length never exceeds the link length. To this end, any interaction among the flows on adjacent links will be translated in terms of time-varying link entry and exit capacities. The spill-back phenomenon is then modeled as a hypercritical flow state, either propagating backwards from the endpoint of a link until its initial point, or originating on the latter, which reduces the capacities of the links belonging to its backward star and eventually affects their flow states.

The key idea here is to introduce the spill-back representation directly in the LPM, without affecting the network flow propagation model internal to the NLM. On this basis the DTA can still be formulated as the system of a NLM based on implicit path enumeration and of a suitable LPM. The latter will be provided by the NPM, which is a system of spatially nonseparable macroscopic flow models specifically aimed at simulating the propagation of congestion due to queue spillovers among adjacent links.

To represent the spill-back phenomenon, we assume that each link is characterized by two time-varying bottlenecks, one located at the initial point and the other located at the end point, called "entry capacity" and "exit capacity," respectively.

The entry capacity, bounded from above by the physical capacity which is typically related to the number of road lanes, is meant to reproduce the effect of queues propagating backwards from the endpoint of the link itself, which can reach the initial point, thereby inducing spill-back conditions on the upstream links. In this case the entry capacity is set to limit the current inflow at a value that keeps the number of vehicles on the link equal to the storage capacity currently available, which is related to the queue density along the link. The latter changes dynamically in time and space as a function of the outflows at previous instants. Specifically, any change in the rate of the space freed by vehicles exiting the link at the head of the queue takes some time to become actually available at the tail of the queue, whereas the jam density multiplied by the length is just the upper bound of the storage capacity, which can be reached only if the queue is not moving.

The exit capacity, bounded from above by the saturation capacity which is typically related to the regulation of the road intersection, is meant to reproduce the effect of queue spillovers propagating backwards from the downstream links, which in turn may generate hypercritical flow states on the link itself. For given inflows, outflows, and intersection priorities,[8] the exit capacities are obtained as a function of the entry capacities based on flow conservation at the node.

The NPM is specified as a circular chain of three models, namely the "exit capacity model," the "exit flow and travel time model," and the "entry capacity model," whose system can be formulated and solved through a fixed-point problem to determine the temporal profiles of the bottleneck capacities and the link exit flows, for given inflows and outflows, and the link travel times and costs are determined accordingly. The three models, described separately in the following sections, are synthesized in Fig. 7.23, which shows how the entry capacities may be taken as current variables in the fixed-point formulation of the NPM.

It is worth pointing out that the exit flows, which are derived from the forward propagation of the inflows, are by definition different from the outflows, although the two coincide at the solution of the DNL[9] which in the proposed formulation is reached jointly with equilibrium.

To keep focusing on the extreme points of the link and avoiding its spatial discretization into many short segments, a wave model is assembled as the composition of three elements: the initial bottleneck, the running segment, and the final bottleneck. The general properties of bottlenecks and segments are analyzed in the context of the Simplified Theory of Kinematic Waves (STKW) based on cumulative flows in Appendix 7.A.

The initial bottleneck keeps the flow entering the running segment below its physical capacity, specified by the fundamental diagram, and reproduces the effects of queue spillovers coming from the initial point of the link itself.

The running segment aims at simulating the movement of vehicles along the link when no queue is present (i.e., in hypocritical conditions), and the effect of spillback on the entry capacity when the queue reaches the initial point of the link.

The final bottleneck keeps the flow exiting the running segment below its saturation capacity, which is usually lower than the physical capacity due to the pres-

[8] Intersection priorities are usually assumed proportional to the saturation capacities.

[9] The DNL guarantees that travel times and flows are reciprocally consistent.

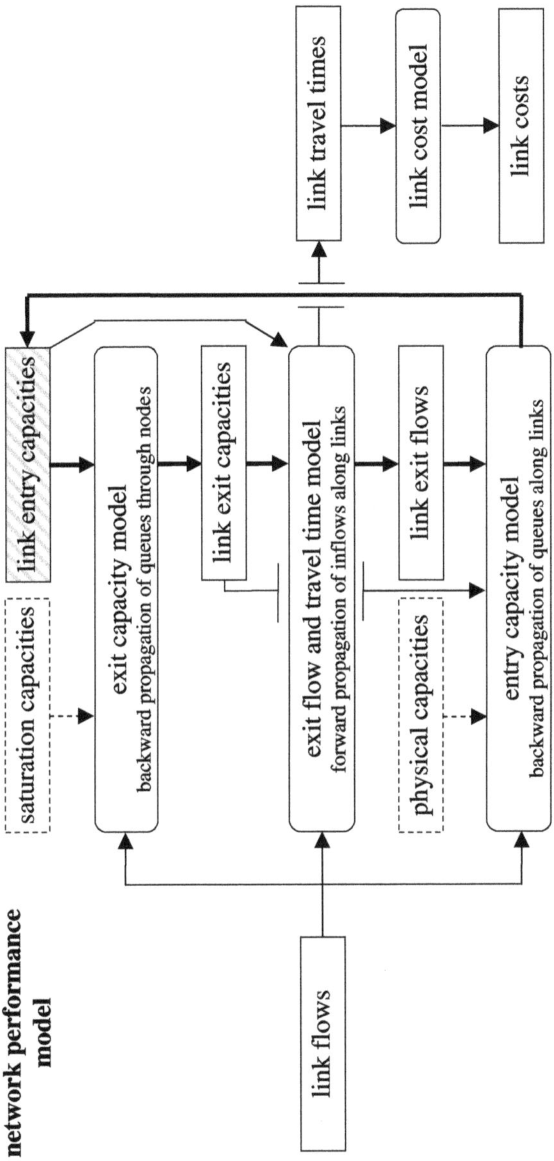

Fig. 7.23 Scheme of the fixed point formulation for the NPM

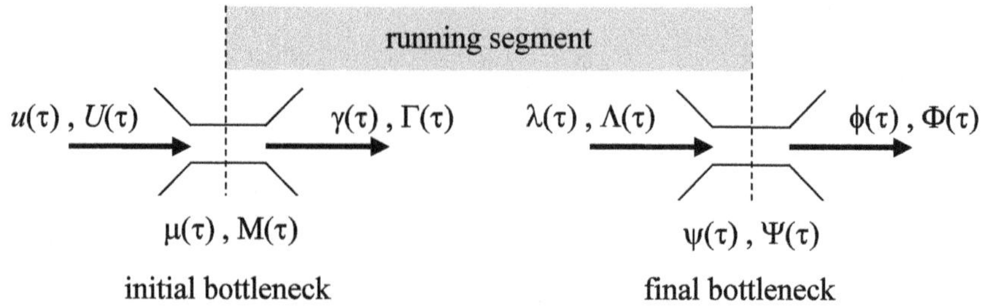

Fig. 7.24 Link model and flow notation

ence of an intersection at the end of the link, and reproduces the effects of queue spillovers coming from the links exiting from such intersection.[10]

Because the above three elements are in series, the leaving flow of one element corresponds to the arriving flow of the subsequent one. Thus, we deal with five distinct flow temporal profiles and two time-varying capacity constraints, as depicted in Figs. 7.24 and 7.25, where the link and node models are sketched, respectively[11]:

$u(\tau)$ the inflow, that is, the arriving flow to the initial bottleneck for each time τ, with cumulative $U(\tau)$

$\mu(\tau)$ the entry capacity of the initial bottleneck, with cumulative $M(\tau)$

$\gamma(\tau)$ the leaving flow from the initial bottleneck, which is equal to the arriving flow to the running segment, with cumulative $\Gamma(\tau)$

$\lambda(\tau)$ the leaving flow from the running segment in hypocritical condition, which is equal to the potential arriving flow to the final bottleneck, with cumulative $\Lambda(\tau)$

$\psi(\tau)$ the exit capacity of the final bottleneck, with cumulative $\Psi(\tau)$

$\phi(\tau)$ the exit flow, that is, the leaving flow from the final bottleneck, with cumulative $\Phi(\tau)$

$w(\tau)$ the outflow, with cumulative $W(\tau)$, which unlike the exit flow satisfies flow conservation at nodes when coupled with inflows[12]

Performances are denoted as follows.

[10]The saturation capacity can be assumed to be time-varying, so as to simulate the alternations in a traffic light between green and red; however, in many applications these are insignificant with respect to the within-day dynamic of traffic, so that the saturation capacity is often taken as constant in time, thus aiming at reproducing only the average effect of the junction regulation.

[11]The index referring to the link is omitted whenever unambiguous.

[12]Inflows and outflows are also referred to for short as link flows, because they are the current variable of the fixed-point formulating the DTA, whereas all other flow and capacity variables are internal to the NPM.

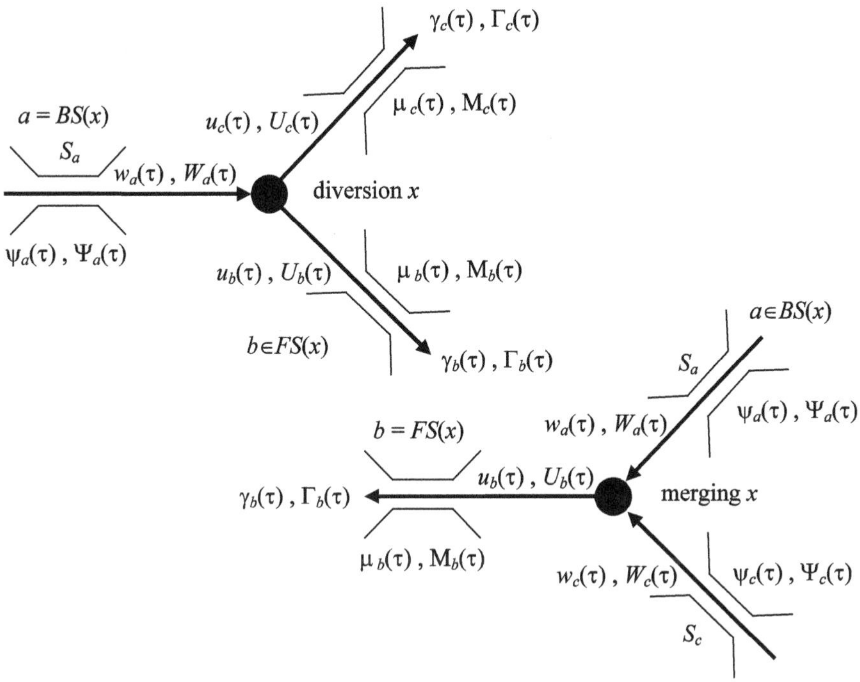

Fig. 7.25 Node model and flow notation

$t(\tau)$ the exit time from the final bottleneck, for vehicles arriving at the running segment at time τ[13]

$c(\tau)$ the link cost, for vehicles arriving at the running segment at time τ

Finally, we introduce below the notation for the main link characteristics:

S saturation capacity

C physical capacity

L length

$\omega^\circ(q), \omega^+(q)$ hypocritical and hypercritical wave speed as a function of flow q

$v^\circ(q), v^+(q)$ hypocritical and hypercritical vehicular speed as a function of flow q

When "mergings" and "diversions" are separated at the graph level, as often occurs to represent turn penalties and prohibitions, the maneuver flows, which play a role when the available entry capacity at a node is split among its upstream links,

[13]At the solution of the DNL, the vehicles entering the link arrive immediately at the running segment, because by definition no vehicle queues at the initial bottleneck, otherwise meaning that spill-back conditions are violated.

coincide with the link inflows and outflows, respectively.[14] Therefore, to simplify exposition, in the following we consider only these two types of nodes.[15]

7.5.1.1 Exit Capacity Model

In this section, the exit capacities of upstream links are determined on the basis of the entry capacities of the downstream links, and of the link flows at the node.

When considering a merging x (i.e., an intersection with a single exiting link) the problem is to split the entry capacity $\mu_b(\tau)$ of the link $b = FS(x)$ available at time τ among the links belonging to its backward star, whose outflows compete to get through the intersection. In principle, it is assumed that the available capacity is distributed proportionally to the saturation capacity S_a of each link $a \in BS(x)$.[16]

However, in this way it may happen that for some link a the outflow $w_a(\tau)$ is lower than the share of entry capacity assigned to it, so that only a lesser portion of the latter is actually exploited. Let $\Omega_b(\tau) \subseteq BS(x)$ be the set of such links. The rest of the entry capacity $\mu_b(\tau) - \sum_{a\in\Omega b(\tau)} w_a(\tau)$ shall then be distributed among the links making up the complementary set $BS(x)\backslash\Omega_b(\tau)$ with the same partition criterion. Moreover, when no spill-back phenomenon is active, that is, $\sum_{a\in BS(x)} w_a(\tau) < \mu_b(\tau)$, the exit capacity $\psi_a(\tau)$ of each link $a \in BS(x)$ shall be set equal to its saturation capacity S_a.

On these bases, we have:

$$\psi_a(\tau) = S_a \cdot \xi_b\big(\tau, \Omega_b(\tau)\big) \tag{7.5.1}$$

$$\Omega_b(\tau) = \big\{a \in BS(x) : w_a(\tau) < \psi_a(\tau)\big\} \tag{7.5.2}$$

where we denoted for any given set of links $\Omega \subseteq BS(x)$:

$$\xi_b(\tau, \Omega) = \begin{cases} \frac{\mu_b(\tau) - \sum_{a\in\Omega} w_a(\tau)}{\sum_{a\in BS(x)\backslash\Omega} S_a} & \text{if } \Omega \subset BS(x); \\ 1 & \text{otherwise.} \end{cases} \tag{7.5.3}$$

Note that a set $\Omega_b(\tau)$ satisfies jointly (7.5.1) and (7.5.2) if and only if every link $a \in BS(x)$ with a saturation ratio $w_a(\tau)/S_a < \xi_b(\tau, \Omega_b(\tau))$ belongs to $\Omega_b(\tau)$ itself and every link a with $w_a(\tau)/S_a \geq \xi_b(\tau, \Omega_b(\tau))$ does not. Because it is based on (7.5.3) $\xi_b(\tau, \Omega)$ decreases adding to Ω links for which $w_a(\tau)/S_a > \xi_b(\tau, \Omega)$,

---

[14] The extension of the exit capacity model to intersections with both mergings and diversions requires the DTA to be formulated in terms of maneuver flows at nodes.

[15] This leads to overlooking the phenomenon of performance deterioration due to a misuse of intersection capacity, which occurs when at a real node working as several separate mergings some users occupy the intersection although they cannot cross it due to the presence of a queue on their successive link. In this case, we should assume "polite behavior" where users wait until the necessary space becomes available.

[16] More general partition criteria require the introduction of priority coefficients that scale opportunely the saturation capacities.

whereas it increases removing from Ω links for which $w_a(\tau)/S_a < \xi_b(\tau, \Omega)$, and vice versa. The partition set $\Omega_b(\tau)$ can be easily proved to be unique, and it can be simply obtained by iteratively adding to an initially empty set Ω^* each link $a \in BS(x) \backslash \Omega^*$ such that $w_a(\tau)/S_a < \xi_b(\tau, \Omega^*)$. Finally, we prove that (7.5.1) yields $\psi_a(\tau) \leq S_a$ for each link $a \in BS(x)$. Assume by contradiction that $\xi_b(\tau, \Omega_b(\tau)) > 1$. Based on (7.5.1), we obtain $\psi_a(\tau) > S_a$; moreover, by definition $S_a \geq w_a(\tau)$. Based on (7.5.2) we then have $\Omega_b(\tau) = BS(x)$, which, considering (7.5.3), contradicts the hypothesis. The fact that (7.5.1) also holds for links belonging to $\Omega_b(\tau)$ enhances the continuity of the model.

When considering a diversion x, that is, an intersection with a single entering link, the problem is to determine at the generic time τ the most severe reduction to the outflow from the link $a = BS(x)$ among those produced by the entry capacities of the links belonging to its forward star. Again, when no link is spilling back, the exit capacity is set at the saturation capacity. When only one link $b \in FS(x)$ is spilling back, that is, $u_b(\tau) \geq \mu_b(\tau)$, the exit capacity $\psi_a(\tau)$ scaled by the share of vehicles turning on link b is set equal to the entry capacity in order to ensure capacity conservation at the node while satisfying the FIFO rule applied to the vehicles exiting from link a: $\psi_a(\tau) \cdot u_b(\tau)/w_a(\tau) = \mu_b(\tau)$. When more than one link $b \in FS(x)$ is spilling back, the exit capacity is the most penalizing among the above values. On this basis, we have:

$$\psi_a(\tau) = \min\{S_a; \mu_b(\tau) \cdot w_a(\tau)/u_b(\tau) : b \in FS(x), u_b(\tau) \geq \mu_b(\tau)\} \qquad (7.5.4)$$

Combining the solution of system (7.5.1) to (7.5.3) with (7.5.4), we can express the exit capacity model in the following compact form.

$$\boldsymbol{\psi} = \psi(\boldsymbol{u}, \boldsymbol{w}, \boldsymbol{\mu}; \boldsymbol{S}) \qquad (7.5.5)$$

Note that, in contrast with the models presented in the following two subsections, this model is spatially nonseparable, because the exit capacities of all the links belonging to the backward star of a given node are determined jointly, and temporally separable, because all relations refer to the same instant.

7.5.1.2 Exit Flow and Travel Time Model

The model input is the temporal profile of the inflow (i.e., the flow arriving at the initial bottleneck) and the temporal profile of the two bottleneck capacities, whereas the output of the model is the temporal profile of the exit flow (i.e., flow leaving the final bottleneck) and then the temporal profile of the exit time, for any given entry instant. However, as shown in Fig. 7.23, although the exit flow model is involved in the fixed-point formulation of the NPM, the travel time model is not, and link performances are therefore obtained only after mutually consistent entry and exit capacities have been found.

Applying (7.A.6) to the initial bottleneck, we determine the arriving flows to the running segment which are consistent with the time-varying entry capacity, corresponding to given inflows:

$$\Gamma(\tau) = \min\{U(\sigma) + M(\tau) - M(\sigma) : \sigma \le \tau\} \tag{7.5.6}$$

Applying (7.A.45) to the endpoint, we forward propagate the arriving flow to the running segment throughout the link as being hypocritical, thus obtaining the potential arriving flow at the final bottleneck:

$$\begin{aligned}\Lambda(\tau) = \min\big\{&\Gamma(\sigma) + \gamma(\sigma) \cdot L \cdot \big[1/\omega^\circ\big(\gamma(\sigma)\big) - 1/v^\circ\big(\gamma(\sigma)\big)\big] : \sigma \\ &+ L/\omega^\circ\big(\gamma(\sigma)\big) = \tau\big\}\end{aligned} \tag{7.5.7}$$

The above equation exploits the analytical solution of the STKW based on cumulative flows. It's worth noting that, indeed, any link performance model yielding exit flows for given entry flows can replace (7.5.7) to simulate the running segment.

Applying (7.A.6) to the final bottleneck, we determine the exit flows that are consistent with the time-varying exit capacity, corresponding to given arriving flows at the final bottleneck:

$$\Phi(\tau) = \min\big\{\Lambda(\sigma) + \Psi(\tau) - \Psi(\sigma) : \sigma \le \tau\big\} \tag{7.5.8}$$

A full understanding of the above equations requires thorough reading of Appendix 7.A, to which the reader is referred for any detail.

As shown in the scheme of Fig. 7.23, when the above exit flow model is applied, the exit capacities are consistent with the entry capacities to enable spill-back propagation through the nodes. This implies that the delay generated by the initial bottleneck is taken into account as the delay incurred at the final bottlenecks within the travel times of the upstream links. Indeed, this is exactly the main mechanism of the NPM, whose role is to transfer to backward links the excess travel time that any separable LPM would attribute to a link where spill-back conditions occur.

Therefore, the link exit time $t(\tau)$ at time τ is obtained, as depicted in Fig. 7.26, by applying (7.A.2) to the sequence of the sole running segment and final bottleneck, that is, without the initial bottleneck, through the following implicit expression:

$$\Phi\big(t(\tau)\big) = \Gamma(\tau) \tag{7.5.9}$$

This way we avoid computing the initial bottleneck delay twice; moreover, at the solution of the DNL (i.e., at equilibrium, in this case) the entry capacity constraint $u(\tau) \le \mu(\tau)$ is satisfied at any time τ, and thus such delay is null.

In presence of time intervals with null flow, (7.5.9) does not allow us to obtain a single value of exit time. To take these circumstances into account, once the cumulative exit flow temporal profile is known, the exit time temporal profile is calculated conventionally as

$$t(\tau) = \max\big\{\tau + L/v^\circ(0), \min\big\{\sigma : \Phi(\sigma) = \Gamma(\tau)\big\}\big\} \tag{7.5.10}$$

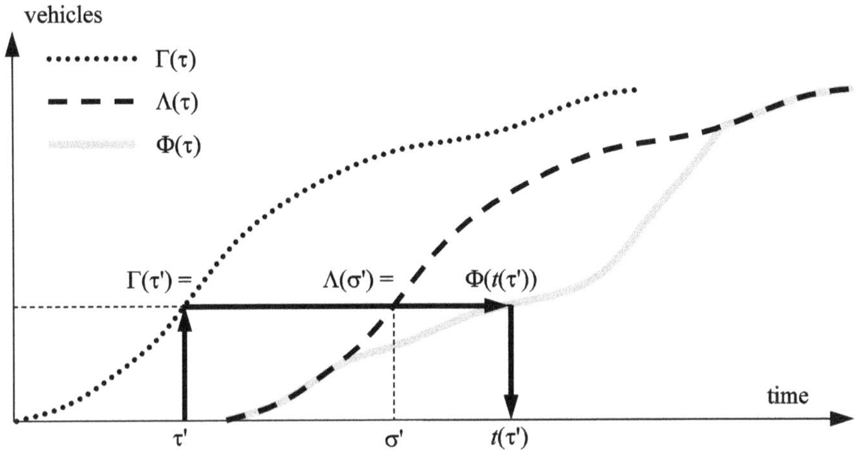

Fig. 7.26 Computation of the link exit time based on the cumulative leaving flow from the initial bottleneck and the cumulative exit flow from the link by applying the FIFO rule

where $L/v^{\circ}(0)$ is the free flow travel time of the running segment.

Combining (7.5.6) with (7.5.7) and the result with (7.5.8), we can express the exit flow model in the following compact form for all the links at once.

$$\boldsymbol{\Phi} = \Phi(\boldsymbol{u}, \boldsymbol{\mu}, \boldsymbol{\psi}) \tag{7.5.11}$$

where bold symbols denote temporal profiles of vector variables.

Combining (7.5.6) with (7.5.10), we can express the travel time model in the following compact form.

$$\boldsymbol{t} = t(\boldsymbol{u}, \boldsymbol{\mu}, \boldsymbol{\Phi}) \tag{7.5.12}$$

7.5.1.3 Entry Capacity Model

In this section, we represent the effect on the entry capacity of queues which, being generated at the endpoint of the link by the exit capacity, reach the initial link point, thus inducing spill-back conditions.

To better explain the proposed approach for modeling the phenomenon of queue spillovers, let us assume, for the moment, that the queue is incompressible; that is, only one hypercritical density exists. In this case, hypercritical kinematic waves have an infinite speed (see Fig. 7.A.3). Therefore, any hypercritical flow state occurring at the endpoint would propagate backward instantaneously, so that at any instant when the queue exceeds the link length, the entry capacity would be equal to the exit capacity. Note that also in this case the queue does not reach the initial point instantaneously, because there, consistent with the Newell Luke Minimum Principle (NLMP) presented in Appendix 7.A, the exiting hypercritical flow state does not prevail on the entering hypocritical flow state until the number of vehicles that have entered the link exceeds the number of vehicles that have exited the link plus the storage capacity, which in this case is constant in time and equal to the link length multiplied by the hypercritical density.

Actually, in the general case, hypercritical flow states may occur at different densities and their kinematic wave speeds are not only much lower, in absolute value, than the vehicle free flow speed, implying that the delay affecting the backward translation in space from the end to the initial point of the flow states produced by the exit capacity is not negligible, but also different from each other, which generates a distortion in their forward translation in time.

The spill-back effect on the entry capacity can be investigated by exploiting the analytical solution of the STKW based on cumulative flows, expressed by (7.A.46). Using this approach, we can avoid evaluating the queue length temporal profile, when the aim is only to determine the presence of spill-back. Indeed, this would be cumbersome, because the speed and density of the queuing vehicles vary over time and space as a function of the exit capacity. More simply, we just identify the time intervals when some leaving hypercritical flow state, propagating backward along the link, reaches the initial point and prevails on the arriving hypocritical flow state.

Applying (7.A.46) to the initial point, we backward propagate the exit flow from the running segment throughout the link, thus obtaining the potential leaving flow from the initial bottleneck:

$$\begin{aligned} G(\tau) = \min\{ & \Phi(\sigma) + \phi(\sigma) \cdot L \cdot \left[1/v^+(\phi(\sigma)) - 1/\omega^+(\phi(\sigma))\right] : \sigma + L/\omega^+(\phi(\sigma)) \\ & = \tau \phi(\sigma) = \psi(\sigma)\} \end{aligned} \tag{7.5.13}$$

where $G(\tau)$ is the maximum cumulative flow that can enter the running segment, consistent with the spill-back phenomenon.

According to the NLMP, the flow state consistent with the spill-back phenomenon occurring at the initial point is the one implying the lowest cumulative flow. Therefore, when at the generic time τ the cumulative inflow $U(\tau)$ equals or exceeds the maximum cumulative flow $G(\tau)$, such that spill-back actually occurs at that instant, the derivative $\mathrm{d}G(\tau)/\mathrm{d}\tau$ of the latter temporal profile may be interpreted as an upper bound to the inflow. This permits us to determine the proper value $\mu(\tau)$ of the entry capacity that maintains the queue length equal to the link length. When no spill-back is occurring, $\mu(\tau)$ is equal to the physical capacity C. Formally, we have:

$$\mu(\tau) = \begin{cases} \mathrm{d}G(\tau)/\mathrm{d}\tau, & \text{if } G(\tau) \le U(\tau) \\ C, & \text{otherwise} \end{cases} \tag{7.5.14}$$

Combining (7.5.13) with (7.5.14), we can express the entry capacity model in the following compact form.[17]

$$\boldsymbol{\mu} = \mu(\boldsymbol{u}, \boldsymbol{\psi}, \boldsymbol{\Phi}; \boldsymbol{C}) \tag{7.5.15}$$

---

[17]The dependency of $\boldsymbol{\mu}$ on $\boldsymbol{\psi}$ is solely due to the need for backward propagating only the hypercritical portions of the exit flow temporal profile.

7.5.1.4 Fixed-Point Formulation of the NPM

For given link flows the NPM allows us to determine (see Fig. 7.23 and the left-hand side of Fig. 7.27) link travel times and capacities consistent with the traffic flow theory that ensure the propagation of congestion through the network. It can be formulated by combining (7.5.11) and (7.5.5) with (7.5.15), yielding the following fixed-point problem in terms of entry capacity temporal profiles:

$$\boldsymbol{\mu} = \mu\big(\boldsymbol{u}, \psi(\boldsymbol{u}, \boldsymbol{w}, \boldsymbol{\mu}; \boldsymbol{S}), \Phi\big(\boldsymbol{u}, \boldsymbol{\mu}, \psi(\boldsymbol{u}, \boldsymbol{w}, \boldsymbol{\mu}; \boldsymbol{S})\big); \boldsymbol{C}\big) \qquad (7.5.16)$$

Although not formally proved, the above fixed-point problem behaves as a contraction and converges in a few iterations to a solution. However, when travel demand is very high, a solution may not exist due to the possible prevalence of gridlocks, which are queues spilling over intersections that generated them.[18] For given link flows, the solution to (7.5.16), if any, is denoted as follows.

$$\boldsymbol{\mu} = \mu^*(\boldsymbol{u}, \boldsymbol{w}) \qquad (7.5.17)$$

Combining (7.5.17) with (7.5.5), the result and (7.5.17) with (7.5.11), the result and (7.5.17) with (7.5.12), yields a performance function, expressing the link exit times in terms of the link flows:

$$\boldsymbol{t} = t\big(\boldsymbol{u}, \mu^*(\boldsymbol{u}, \boldsymbol{w}), \Phi\big(\boldsymbol{u}, \mu^*(\boldsymbol{u}, \boldsymbol{w}), \psi\big(\boldsymbol{u}, \boldsymbol{w}, \mu^*(\boldsymbol{u}, \boldsymbol{w}); \boldsymbol{S}\big)\big)\big) = t^*(\boldsymbol{u}, \boldsymbol{w}) \qquad (7.5.18)$$

The cost for users entering a link at any given time is assumed to depend on the travel time at that instant. Hence in compact form we have:

$$\boldsymbol{c} = \hat{\mathrm{c}}(\boldsymbol{t}) \qquad (7.5.19)$$

Finally, substituting (7.5.18) in (7.5.19), we obtain:

$$\boldsymbol{c} = \hat{\mathrm{c}}\big(t^*(\boldsymbol{u}, \boldsymbol{w})\big) = \mathrm{c}^*(\boldsymbol{u}, \boldsymbol{w}) \qquad (7.5.20)$$

which jointly with (7.5.18) expresses the LPM synthetically.

7.5.2 Network Loading Map and Fixed-Point Formulation of the Equilibrium Model

In the following, we briefly address both route choice and network flow propagation by adopting an implicit path enumeration approach. Referring to users traveling towards a single destination d, the formulation is based on the concepts of *link*

[18]This problem can be alleviated by a proper setting (raising) of priority coefficients to favor circulation in roundabouts and other close cycles of the graph.

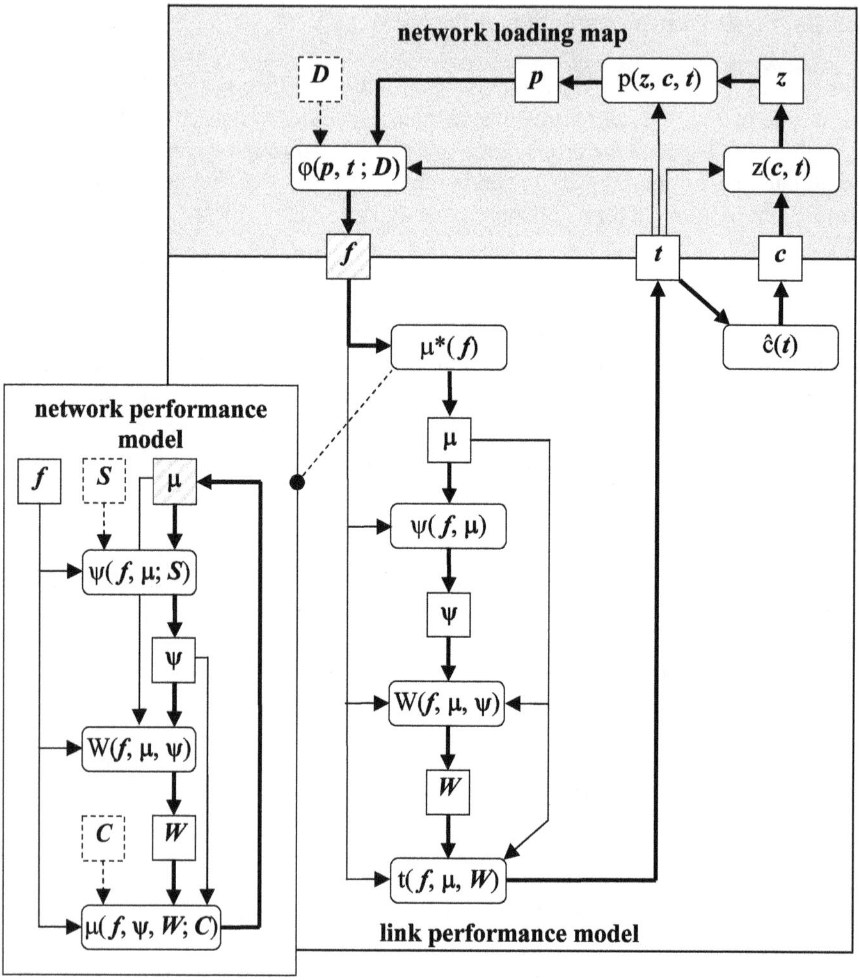

Fig. 7.27 Variables and models of the fixed-point formulations for the NPM (*left-hand side*) and for the DTA with spill-back (*right-hand side*) in terms of link flows $f = (u, w)$

conditional probability and *node satisfaction*, whose notation and definitions are introduced below.

$p_a^d(\tau)$ = probability of using link a, conditional on crossing node $TL(a)$ at time τ

$z_x^d(\tau)$ = expected value of the maximum perceived utility at time τ, relative to the paths K_{xd} connecting node x to d which are considered by the user

It can be proved that the following expressions of the node satisfaction and of the link conditional probability are consistent with a logit route choice model in which users consider all and only "efficient" paths (a path is efficient if each of its links is efficient):

$$z_x^d(\tau) = \begin{cases} \theta \cdot \ln\left(\sum_{a \in FS(x) \cap EA(d)} \exp\left(\frac{-c_a(\tau) + z^d_{HD(a)}(t_a(\tau))}{\theta}\right)\right), & \text{if } x \neq d; \\ 0, & \text{otherwise} \end{cases} \tag{7.5.21}$$

$$p_a^d(\tau) = \begin{cases} \exp\left(\frac{-c_a(\tau) + z^d_{HD(a)}(t_a(\tau)) - z^d_{TL(a)}(\tau)}{\theta}\right), & \text{if } a \in EA(d); \\ 0, & \text{otherwise} \end{cases} \tag{7.5.22}$$

where $EA(d)$ is the set of the efficient links that get closer to the destination with reference to a "distance" pattern on the network which is constant in time.

The solution of the triangular system formed by (7.5.21) in topological order, combined with (7.5.22), yields the route choice model, which can be expressed in compact form as

$$\boldsymbol{z} = \mathrm{z}(\boldsymbol{c}, \boldsymbol{t}) \tag{7.5.23}$$

$$\boldsymbol{p} = \mathrm{p}(\boldsymbol{z}, \boldsymbol{c}, \boldsymbol{t}) \tag{7.5.24}$$

Similar expressions can be derived for the deterministic case:

$$z_x^d(\tau) = \begin{cases} \max\{-c_a(\tau) + z^d_{HD(a)}(t_a(\tau)) : a \in FS(x) \cap EA(d)\}, & \text{if } x \neq d; \\ 0, & \text{otherwise} \end{cases} \tag{7.5.25}$$

$$\begin{cases} p_a^d(\tau) \cdot [z_x^d(\tau) + c_a(\tau) - z^d_{HD(a)}(t_a(\tau))] = 0, & \text{if } a \in EA(d); \\ 0, & \text{otherwise} \end{cases} \tag{7.5.26}$$

where by definition it is $p_a^d(\tau) \geq 0$ and

$$\sum_{a \in FS(x)} p_a^d(\tau) = 1. \tag{7.5.27}$$

Because the solution to the system (7.5.26) and (7.5.27) is nonunique, when more than one link exiting from node x yields the maximum utility $z_x^d(\tau)$, the symbol "=" in (7.5.24) should be replaced by the symbol "$\in$". The generalization of the deterministic route choice model to the case where the set of alternatives coincides with all acyclic paths is available but lies outside the scope of this outline. Moreover, the deterministic model can be exploited within a Monte Carlo simulation to address the case of probit route choice.

We assume that the origins and destinations are connected to the rest of the network by dummy links or infinitesimal length with infinite physical and saturation capacities, so that for all other nodes the flow conservation equation holds.

Therefore, the inflow $u_a^d(\tau)$ on the generic link a is given by the link conditional probability $p_a^d(\tau)$ multiplied by the flow exiting from node $TL(a)$. The latter is given, in turn, by the sum of the outflow $w_b^d(\tau)$ from each link $b \in BS(TL(a)) \cap EA(d)$ of its efficient backward star, whereas the inflow $u_{FS(o)}^d(\tau)$ on the dummy link $FS(o)$ exiting from origin o is instead equal to the demand flow $D_o^d(\tau)$ from o to d. Then we have:

$$u_a^d(\tau) = \begin{cases} D_{HD(a)}^d(\tau), & \text{if } TL(a) \text{ is an origin} \\ p_a^d(\tau) \cdot \sum_{b \in BS(TL(a)) \cap EA(d)} w_b^d(\tau), & \text{otherwise} \end{cases} \tag{7.5.28}$$

Based on (7.A.3), given the exit time temporal profile of link a, the outflow is related to the inflow temporal profile as follows.

$$w_a^d\left(t_a(\tau)\right) = u_a^d(\tau)/\left[\mathrm{d}t_a(\tau)/\mathrm{d}\tau\right] \tag{7.5.29}$$

where the weight $\mathrm{d}t_a(\tau)/\mathrm{d}\tau$ stems from the fact that users enter the link at a certain rate and exit it at a different rate, which is higher than the previous one, if the travel time is decreasing, and lower, otherwise.

Obviously, $u_a(\tau)$ and $w_a(\tau)$ are given by the sum for all destinations of $u_a^d(\tau)$ and $w_a^d(\tau)$, respectively.

The solution of the triangular system formed by (7.5.28) and (7.5.29) in reverse topological order, yields the network flow propagation model, which can be expressed in compact form as

$$(\boldsymbol{u}, \boldsymbol{w}) = \varphi(\boldsymbol{p}, \boldsymbol{t}; \boldsymbol{D}) \tag{7.5.30}$$

Combining (7.5.23) with (7.5.24) and the result with (7.5.30) yields a formulation based on implicit path enumeration of the NLM:

$$(\boldsymbol{u}, \boldsymbol{w}) = \varphi\left(\mathrm{p}(\mathrm{z}(\boldsymbol{c}, \boldsymbol{t}), \boldsymbol{c}, \boldsymbol{t}), \boldsymbol{t}; \boldsymbol{D}\right) = \varphi^*(\boldsymbol{c}, \boldsymbol{t}; \boldsymbol{D}) \tag{7.5.31}$$

On this basis the DTA can be formalized (see Fig. 7.27) as a fixed-point problem in terms of link flow temporal profiles by substituting into the NLM (7.5.31) the LPM (7.5.18) to (7.5.20):

$$(\boldsymbol{u}, \boldsymbol{w}) = \varphi^*\left(\mathrm{c}^*(\boldsymbol{u}, \boldsymbol{w}), t^*(\boldsymbol{u}, \boldsymbol{w}); \boldsymbol{D}\right) \tag{7.5.32}$$

The above fixed-point problem can be, as usual, solved by means of an MSA.

7.6 Models for Transport Systems with Scheduled Services[19]

Scheduled transportation services, such as those provided by airplanes, trains, and buses, can be considered discrete in both time and space: they can be accessed only

[19] Agostino Nuzzolo is the co-author of this section.

Run	Line	Service type	Initial station	Departure time	Intermediate stops	Terminal station
1	AA	Intercity	A	9.30	–	D
2	BB	Regional	A	9.50	B/C	D
3	AA	Intercity	A	10.30	–	D
4	CC	Intercity	A	11.30	D	E

Fig. 7.28 Time schedule, runs, and lines

at certain times and only at specific locations such as airports, rail stations, and bus stops. In a within-day dynamic context, supply, demand, and demand–supply interactions for scheduled service systems can be explicitly modeled by starting from the timetable, which defines runs and lines (see Fig. 7.28). A *run r* represents a connection with a given time schedule (e.g., a given train connection), whereas a *line ln*, as defined in Chap. 2, may be regarded as a set of runs of similar characteristics (e.g., stops, travel times, quality of service, etc.). Within-day dynamic models explicitly simulate supply and demand for runs rather than for lines, unlike static models for scheduled service systems described in previous chapters.

Dynamic models to simulate within-day dynamic scheduled service systems differ according to a number of factors related to system service characteristics. The main classification factors that apply to dynamic models are *frequency*, *regularity*, and *information* available to users.

Service frequency can be related directly to the frequency of the line in the reference period: the number of runs made on the line during such a period or, for overlapping lines, the sum of the frequencies of all attractive lines connecting the O-D pair. Service regularity is a measure of how closely the schedule is followed. Regularity, or rather its opposite, can be measured in different ways depending on the analysis purpose. If regularity is assumed to influence user behavior in line-based systems such as buses and trains, deviations from the schedule might, for example, be related to the average headway of runs belonging to the same line.

Regular services are usually associated with low frequencies, typical of systems that operate outside of urban areas, such as (intercity) rail or air. By contrast, irregular services generally correspond to high frequencies, such as bus or underground lines in urban or metropolitan areas. In any case, frequency and regularity are continuous variables and their segmentation in terms of "high" and "low" is conventional and somewhat arbitrary. In models, they correspond to different hypotheses on users' behavior and to different model systems. As such, they are at the analyst's discretion.

Information on services may be available to the user before a trip (i.e., at home) and/or en route (i.e., at stops). In both cases, the information might include data on waiting times, travel times, and on-board occupancy. Static information on run schedules is traditionally available from timetables. Intelligent Transportation Systems (ITS) have both significantly expanded the range of information available to the traveler, through Advanced Traveler Information Systems (ATIS), and also improved the performance of transit services, through Advanced Public Transportation Control Systems (APTCS).

Different supply and demand models are used to simulate scheduled service systems depending on their different characteristics. In the case of low frequencies and regular services, supply is modeled through deterministic dynamic networks. Users are assumed to have full information before starting their trip, and to choose a specific run based on expected performance attributes. Models analogous to those used to represent path choice in continuous service networks (see Sect. 4.3.3.1) can be applied to represent run choice.

On the other hand, supply models for high frequencies and irregular services are based on stochastic dynamic networks. Because users may not have full information before starting their trip, they are assumed to follow a mixed pre-trip/en-route choice behavior, as described in Sect. 4.3.3.2. It is commonly assumed that en-route choices occur at stops and involve the decision to board a particular run or to wait for a later and more suitable run. The choice of boarding stops is considered made before starting the trip, inasmuch as it is not influenced by unknown events.

As with other assignment models, dynamic assignment models for scheduled services can be decomposed into supply, demand, and supply–demand interaction models. A general framework for within-period dynamic assignment models for scheduled service systems is shown in Fig. 7.29.

In the following, the two cases of low-frequency regular services and high-frequency irregular services are addressed separately. It should be noted that dynamic traffic assignment for scheduled services is a newer and significantly less researched subject than DTA for continuous service systems. The models described here are thus somewhat less established than those that apply to the continuous case.

7.6.1 Models for Regular Low-Frequency Services

For regular low-frequency services, it is assumed that each run follows its scheduled departure and arrival times, that users have all relevant information before starting their trips, and that they choose access/egress terminals as well as runs according to their desired arrival or departure times.

In the following subsections, the within-day dynamic supply, demand, and demand–supply interaction models for this situation are discussed.

7.6.1.1 Supply Models

In general, within-period dynamic supply models of scheduled services consist of a network model (graph plus link performance and cost functions) and the network loading or flow propagation relationships that connect path costs to link costs and link flows to path flows. The main differences between dynamic supply models for scheduled and continuous service systems are in the graph model; the convenient linear loading relationships introduced in Chap. 2 for static systems remain applicable for scheduled service systems.

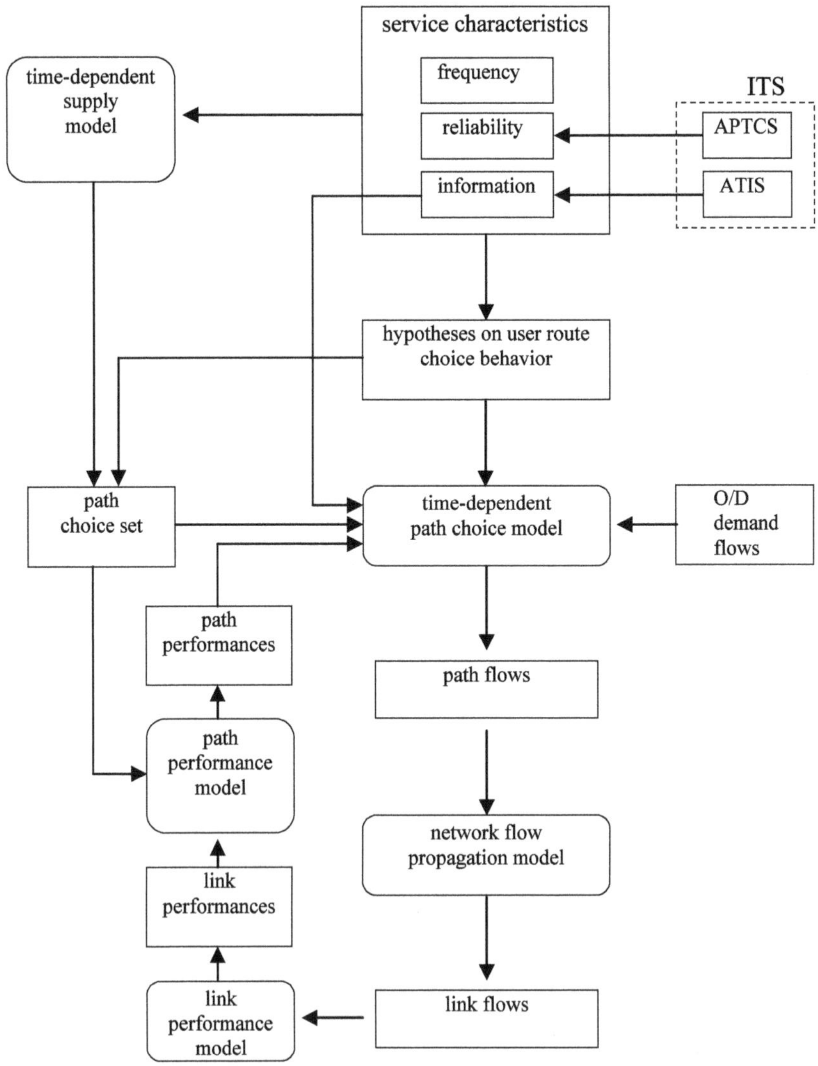

Fig. 7.29 Schematic representation of within-day dynamic transit assignment models

The graph model used for scheduled services is known as a space–time or *diachronic* graph. In this graph, some nodes represent events that take place at a given instant and therefore have an explicit time coordinate. Each run is described by means of a subgraph (Fig. 7.30) whose nodes represent the arrival and departure times of the vehicles (trains, planes, buses) at stations and whose links represent either travel from one station to another or dwelling at a given station. Other nodes represent the arrival or departure of users at the station to board or alight from each particular run. These nodes are connected, through boarding and alighting links, to the nodes representing the departure and arrival of that run. The arrival and depar-

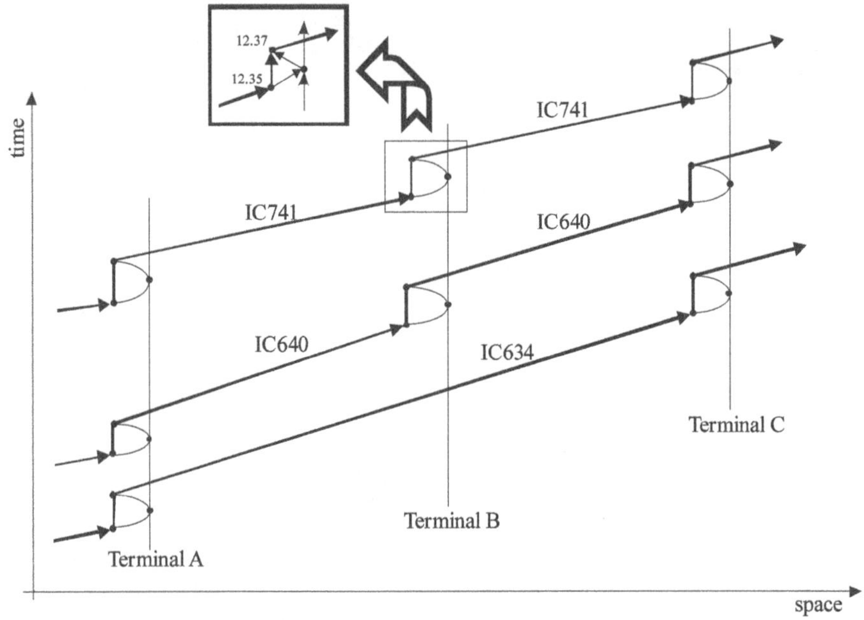

TIMETABLE

Run	Terminal A		Terminal B		Terminal C	
	Arr.	Dep.	Arr.	Dep.	Arr.	Dep.
IC634	08.25	08.30	–	–	12.00	12.05
IC640	08.55	09.00	10.10	10.15	13.15	13.18
IC741	10.58	11.00	12.35	12.37	14.00	14.02

Fig. 7.30 Diachronic graph representation of scheduled services

ture nodes of different runs at a station may also be connected by links that represent user transfers between the runs. This set of nodes and links is usually defined as a run subgraph.

Temporal centroid graphs are another kind of subgraph of a diachronic graph; they represent the times and locations of trip departures and arrivals. To simulate users' choices among different runs or sequences of runs, it is necessary to introduce the desired departure times from the origin τ_o, or the desired arrival times at the destination τ_d. Even if in principle these desired times are continuous variables, they are typically represented by discrete time intervals (e.g., five minutes long) in applications. Possible desired departure or arrival times are represented as temporal centroid nodes having the same spatial coordinates as the zone centroids introduced in Chaps. 1 and 2, and with time coordinates given by representative instants of the corresponding discrete time intervals (e.g., one node every five minutes). Nodes of the temporal centroid graph also represent the actual time of departure from the origin to the boarding terminal or the actual time of arrival at the destination from the alighting terminal. The difference between the desired and actual times of departure from the origin is modeled by a link that connects the temporal centroid (representing the user's desired departure time) to a temporal node representing the actual time the user leaves the origin to catch a particular run (Fig. 7.31). A similar subgraph represents the desired and actual times of arrival at the destination, and the difference between them.

The graph model for the overall system is usually completed with links that represent access (egress) from (to) the centroids, and that have the corresponding travel times and costs. Figure 7.31 shows a diachronic graph for a desired departure time situation; similar graphs can be built for a desired arrival time.

Diachronic graphs are very convenient because they exploit the intrinsically discrete service structure (the services being available only at certain time instants); this allows the use of very efficient network algorithms similar to those described for static continuous networks. Other models that represent regular services are based on timetable manipulations. These models are conceptually analogous to the graph representation, which we prefer because it is more consistent with the general approach to supply modeling followed throughout this book.

A trip is represented in a diachronic graph by a path k starting from the desired departure time on the temporal centroid subgraph and ending at the arrival time at the destination (see Fig. 7.31). Note that, unlike continuous service graphs, the desired departure time is uniquely associated with each path. The same sequence of runs for a different desired departure time corresponds to a different path k'. In the same way, a path k uniquely identifies the actual departure time (interval) τ_j.

In diachronic network models, performance variables and their relationships to flows are generally similar to those described above for static models. As in Chap. 2, link performance or level of service attributes r_{nl} are variables expressing average values of individual attributes perceived by users and associated with a given link. Examples of link attributes are monetary cost, access time, early or late schedule delay, on-board travel time, number of transfers, egress time, and so on. In the same way, the *average generalized transportation cost*, or simply the *link cost*, is the total disutility associated with each link. The link cost c_l is a (dis)utility function, typically linear, of link performance attributes that underlie travel-related choices and, in particular, path choices:

$$c_l = \sum_i \beta_i \cdot r_{il}$$

Depending on system characteristics (low frequency and regularity, booking of seats, etc.), it might be appropriate to assume that link performances and costs are independent of flows, and to model supply as in a noncongested network. In some cases, however, it can be appropriate to take into account congestion effects. Because of congestion, a link's performance attributes, and thus its average link cost, may depend on the number of users on the link and, possibly, on other links of the graph. In congested regional bus or rail systems, for example, passengers may not all have a seat and may even have difficulty boarding some runs. Referring to on-board links (see Fig. 7.32), separable cost functions similar to those introduced in Sect. 2.4.2.2 may be used to represent discomfort (2.4.32) and on-board travel time

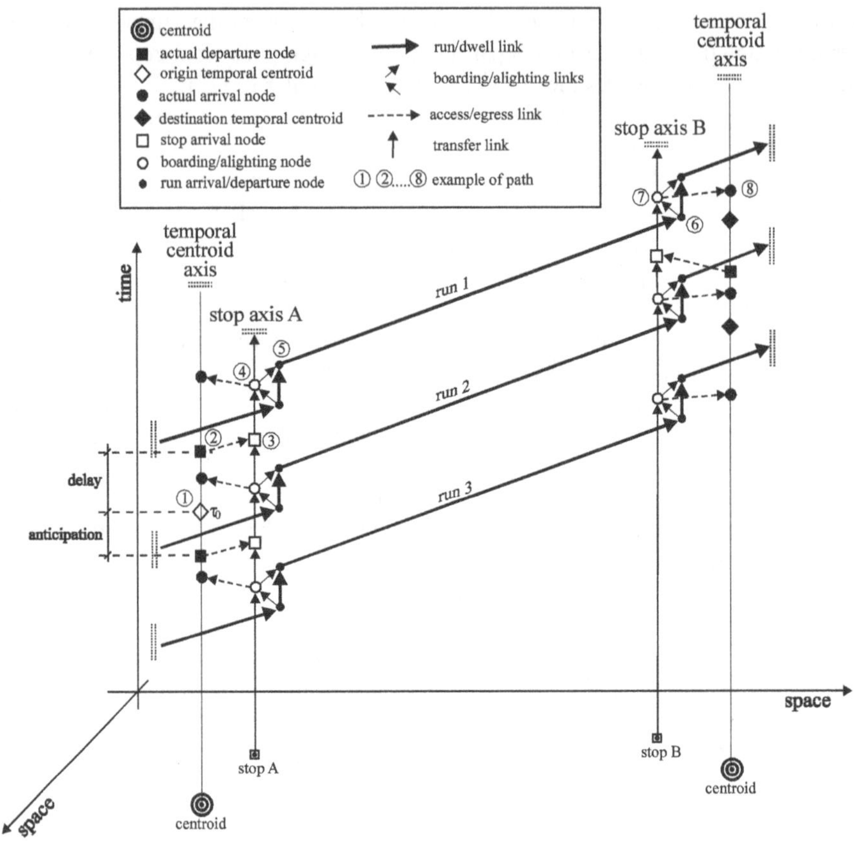

Fig. 7.31 Example of diachronic graph for low-frequency services

(2.4.27). Penalty functions can be adopted to represent the possibility of not being able to board a given run due to overcrowding.

Note that early or late schedule delay penalties, $EAP_k(\tau, \tau_d)$ or $EDP_k(\tau, \tau_o)$ and $LAP_k(\tau, \tau_d)$ or $LDP_k(\tau, \tau_o)$, introduced in Sect. 7.3 for continuous service dynamic demand models, can be represented as additive costs on the links in the temporal centroid graph that connect the two nodes corresponding to the desired and actual departure or arrival times; see Fig. 7.31.

Performance attributes and generalized transportation cost (disutility) can be extended from links to paths. The average generalized transportation cost g_k of a path k is defined as a scalar quantity that combines the different performance attributes perceived by users for the whole trip. As in Chap. 2, path cost in the most general case is made up of two parts: linkwise additive cost, g_k^{ADD}, and nonadditive cost, g_k^{NA}, assuming that they are commensurate:

$$g_k = g_k^{\text{ADD}} + g_k^{\text{NA}} = \sum_l \delta_{lk} c_l + g_k^{\text{NA}} \tag{7.6.1}$$

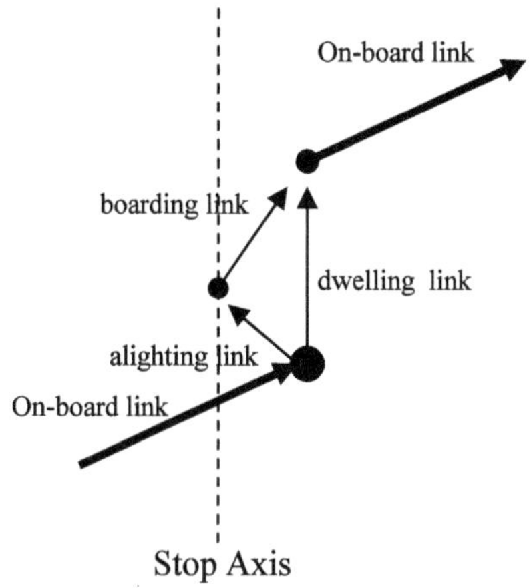

Fig. 7.32 Link classification at stops

or in matrix terms:

$$\boldsymbol{g} = \Delta^T \boldsymbol{c} + \boldsymbol{g}^{\mathrm{NA}} \tag{7.6.2}$$

where Δ is the link-path incidence matrix. Nonadditive costs must be introduced when the cost is nonlinear with respect to distance (e.g., fares based only on origin and destination, independently of the run or sequence of runs followed).

The average number of users (in a time unit) following path k is called the *path flow* h_k. The *link flow* f_l represents the average number of users on link l. Thus, the flow on a link that represents a connection between two successive stops of a particular run is the average number of travelers using that service segment. Following the terminology and notation of within-day static models, the number of users on a link or following a path in the diachronic network may be referred to as a *flow*, even though it is conceptually and dimensionally a number rather than a rate (users per time unit).

In within-day dynamic supply models for scheduled services, the flow on a link can be obtained by summing the flows on all the paths that include the link. This leads to a linear network loading model identical to the within-day static case:

$$f_l = \sum_k \delta_{lk} h_k \tag{7.6.3}$$

$$\boldsymbol{f} = \Delta \boldsymbol{h} \tag{7.6.4}$$

7.6.1.2 Demand Models

Demand models used in dynamic assignment for low-frequency regular scheduled service networks are analogous to those described above for discrete-time models of continuous services; they express the relationship between path flows and path costs.

The user flow on a path k connecting O-D pair od and departing in interval $[j]$ can be obtained with *elastic demand profile* models, which simulate departure interval choice as a function of the desired arrival time τ_d or the desired departure time τ_o. In this case, there is no need to model departure time choice separately from path choice because the former is implicitly included in each path alternative.

Path choice models for scheduled service systems determine the probability $p_{od,k}(\tau_j/\tau_o)$ of choosing path k and the related actual departure time τ_j, given O-D pair od and desired departure time τ_o (or alternatively desired arrival time τ_d). Pre-trip path choice models assume that users choose the path that minimizes the perceived disutility, taking into account attributes such as access and egress times and costs, travel time, number of transfers, monetary cost, comfort, and early or late schedule delay. These attributes are typically combined in a path cost variable as described in the previous section. Other attributes (e.g., socioeconomic variables) can be included in a V_{ok} term.

Most models proposed in the literature to simulate path choice also simulate choice set formation (see Sect. 4.3.3). It is typically assumed that only some of the topologically feasible paths belong to the choice set. Paths are selected by applying dominance rules such as:

- Runs that leave before and arrive after other runs in the choice set are not included in the set.
- Paths must satisfy criteria relative to maximum number of transfers, maximum time spent in transfers, maximum travel time, and so on.

The total systematic utility of a given path k can thus be expressed as

$$V_{od,k}(\tau_j/\tau_o) = g_k(\tau_j/\tau_o) + V_{ok} \tag{7.6.5}$$

Note that in (7.6.5) the departure time τ_j of the first run and the desired departure time τ_o, both associated with path k, have been made explicit in analogy with continuous service models.

A logit specification of the path choice model for desired departure time τ_o at the origin is:

$$p_{od,k}(\tau_j/\tau_o) = \frac{\exp(V_{od,k}(\tau_j/\tau_o))}{\sum_{k'} \exp(V_{od,k'}(\tau_j/\tau_o))} \tag{7.6.6}$$

If there are several service types (e.g., intercity and regional) and classes (e.g., first and second class), the interdependence of choice dimensions can be accounted for by assuming a positive correlation among the random residuals of the perceived utilities of paths that share the same service type, class, and so on. In this case, a multilevel hierarchical logit path choice model could be adopted.

The average flow h_k on path k can be expressed as

$$h_k = d_{od}(\tau_o) \cdot p_{od,k}(\tau_j/\tau_o) \tag{7.6.7}$$

Note that (7.6.7) is the equivalent of (7.3.1) for continuous-service continuous-flow models.

7.6.1.3 Demand–Supply Interaction Models

Given the supply and demand models described in the previous subsections, within-day dynamic assignment models for regular low-frequency scheduled service networks reduce to within-day static assignment on a diachronic network. It is also possible in this case to distinguish among uncongested network, user equilibrium, and dynamic process assignment models.

Because paths correspond to composite choice alternatives that include departure time, access–egress terminals, and runs, random utility choice modes are the only form that has been adopted and calibrated for this type of problem in practice. These give rise to analogues for scheduled service systems of static stochastic assignment models (SUN, SUE, etc.).

The general theoretical results on existence and uniqueness of solutions described in Chap. 5 can be applied to this case and are not repeated here.

7.6.2 Models for Irregular High-Frequency Services

For irregular high-frequency services, the complexity of the real system increases considerably with respect to both user behavior and performance variables. Different within-day dynamic models can be specified for these systems under different assumptions. In this section, one such model is described. We stress, once more, that this area is very little researched, and that further theoretical developments and applications are to be expected in the future.

In this model, users are assumed to make their choices at different times during their trips. The choice of the first boarding stop and the attractive line set is made before the trip begins (pre-trip choice). During the trip, users choose the runs to board at transfer points by adapting to the actual succession of run arrivals and to information given (if any) about waiting times. It is further assumed that, because of the high frequency and the irregularity of services, the actual departure time from the origin is equal to the desired departure time, so users arrive at stops independently of run departure times. Thus if τ_o is the (desired) departure time from the origin and $t_{a,os}$ the access time to stop s, the user arrives at the stop at the absolute time $\tau_{so} = \tau_o + t_{a,os}$.

In the following subsections, supply, demand, and demand–supply interaction models consistent with the above assumptions are described.

7.6.2.1 Supply Models

The diachronic network model described in Sect. 7.6.1 can also be adopted, with some differences, in the case of irregular services. Due to irregularity, the actual arrival and departure times of a run on day t can differ from the scheduled times and from the times on other days. This may be represented by a vector of random variables $\boldsymbol{b}$ whose elements are the arrival time $b_{a,rs}$ and the departure time $b_{p,rs}$ of each run r at each stop s. In the following, $\boldsymbol{b}^t$ indicates a realization of vector $\boldsymbol{b}$ representing day t and G^t is the corresponding diachronic graph (see Fig. 7.31). Equations (7.6.1) through (7.6.4), which express the relationships of path costs and flows with link costs and flows, can still be used once a link–path incidence matrix Δ^t for graph G^t is defined. It is usually assumed that the means of random variables $b_{a,rs}$ and $b_{pr,s}$ coincide with the scheduled arrival and departure times.

The vector $\boldsymbol{b}$ is related to another vector $\boldsymbol{y}$ with components y_{rl} and y_{rs} representing, respectively, the running time of run r on running link l and the dwelling time of run r at stop s (dwelling link). Due to service irregularity, $\boldsymbol{y}$ can also be modeled as a vector of random variables. The components of the two vectors $\boldsymbol{b}$ and $\boldsymbol{y}$ are related through the following recursive equations.

$$b_{a,rs} = b_{p,r(s-1)} + y_{r,l}, \qquad l \equiv \big((s-1), s\big), \qquad b_{p,rs} = b_{a,rs} + y_{r,s}$$

Thus, given the initial departure time of run r, for a given vector $\boldsymbol{y}^t$ it is possible to generate a vector $\boldsymbol{b}^t$ and vice versa. In applications, the random vector $\boldsymbol{y}$ is often modeled from empirical observations. One of the models proposed is a MultiVariate Normal (MVN) with mean $\bar{\boldsymbol{y}}$ (the scheduled running and dwelling times) and a variance–covariance matrix Σ_y whose elements can implicitly represent a variety of phenomena such as

- The propagation of delays between successive sections of the same line,

$$\mathrm{cov}(y_{r,l-1}, y_{rl}) > 0$$

- The persistence of perturbation factors on a given line section,

$$\mathrm{cov}(y_{r,l}, y_{r+1,l}) > 0$$

- The reduction in a run's dwelling time due to a longer dwelling time of the previous run at the same stop,

$$\mathrm{cov}(y_{r-1,s}, y_{rs}) < 0$$

From the algorithmic point of view, a configuration G^t of the diachronic network can be generated by sampling a vector $\boldsymbol{b}^t$ or $\boldsymbol{y}^t$ from the multivariate distribution assumed for $\boldsymbol{b}$ or $\boldsymbol{y}$. If $\boldsymbol{y}$ is assumed to be distributed $MNV(\bar{y}, \Sigma_y)$, the Monte Carlo method with a Cholesky factoring of the matrix Σ_y can be used for this purpose.

In any case, the resulting vector $\boldsymbol{y}^t$ must be modified to satisfy feasibility requirements. This might include ensuring correspondence between generated times and the allowed speeds for transit vehicles, preventing overtaking between successive runs, and so on.

7.6.2.2 Demand Models

In general, several different boarding stops s can be reached and many runs are available from a given origin temporal centroid (see Fig. 7.33). Path choice on a

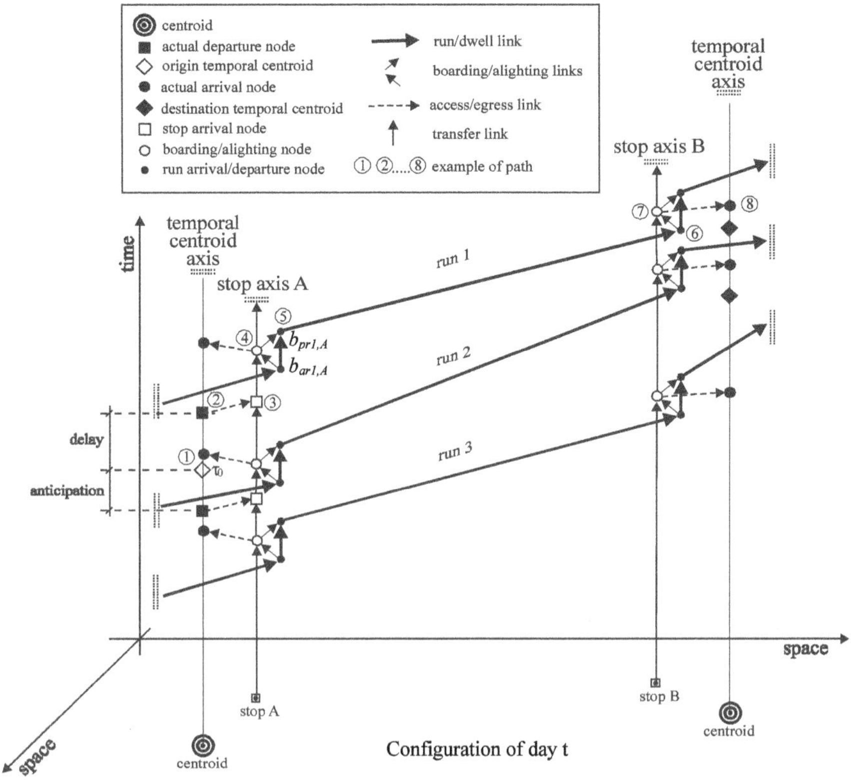

Fig. 7.33 Example of diachronic graph for high-frequency irregular services

realization G^t of the diachronic network thus implies choice of access stop and choice of the run(s) leading the user to the destination.

Path choice models give the probability $p_{od}[r, s|\tau_o]$ of choosing a path including run r at boarding stop s, given the O-D pair *od* and the desired departure time from the origin τ_o (or the arrival time at stop s, $\tau_{s,o}$). Because of the different choice behaviors assumed for pre-trip choices (stop s) and en-route choices (run r), this probability can be expressed as

$$p_{od}[r, s/\tau_o] = p_{od}[r/s, \tau_{s,o}]p_{od}[s/\tau_o] \tag{7.6.8}$$

This is the product of two probabilities: the probability of choosing run r at stop s, given the arrival time $\tau_{s,o}$; and the probability of choosing stop s, given the desired origin departure time τ_o.

Given the irregularity of services, some further assumptions have to be made on available information and the related choice set in order to model choice probabilities in (7.6.8).

If real-time information about waiting times is available at stops, the user can consider as choice alternatives the runs of different lines according to their actual

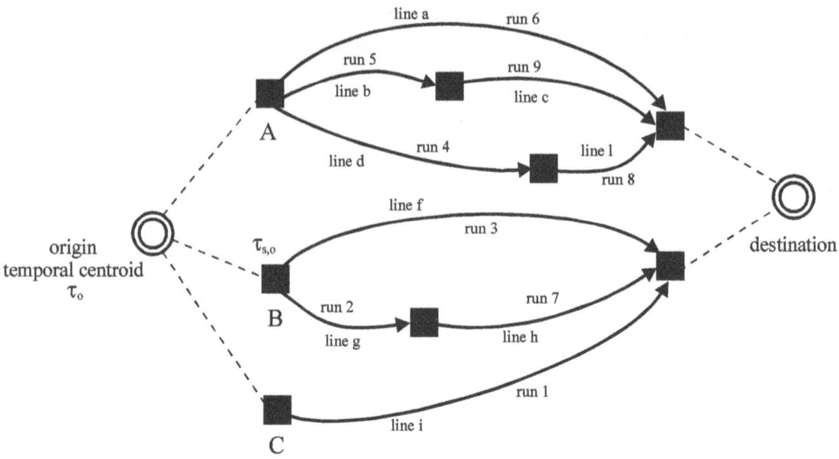

Fig. 7.34 Example of path choice set

arrival times on any particular day t. Thus an initial choice set of runs $K^s[\tau_{s,o}, \boldsymbol{b}^t]$ may be defined for users departing from origin o for destination d at time τ_o, arriving at stop s (where there is an ATIS providing information on run waiting or arrival times) at a time $\tau_{s,o}$ and finding a supply configuration $\boldsymbol{b}^t$. (Here and in the following, the index *od*, when not stated, is understood.) This set (see Fig. 7.34) is specified by line runs connecting stop s directly or indirectly to destination d and satisfying some feasibility rules, such as

- The set includes the first run of each line that leaves after the user's arrival at the stop at time $\tau_{s,o}$.
- The runs are not dominated (i.e., there are no runs leaving before and arriving after other runs of the choice set).
- The runs satisfy criteria such as the maximum number of transfers, maximum transfer time, maximum travel time, and so on.

The set $K^s[\tau_{s,o}, \boldsymbol{b}^t]$ depends on the user's arrival time $\tau_{s,o}$ at the stop, because different runs will be accessible to users at different times; it depends also on the system configuration $\boldsymbol{b}^t$ because, for the same arrival time on different days, different choice sets may be available due to random variations in system performance.

Furthermore, should an arriving run be too crowded to board, the set can be modified while the user waits at the stop. When a run of a specific line included in $K^s[\tau_{s,o}, \boldsymbol{b}^t]$ arrives and has no available places, the user can decide to extend the choice set, introducing the next run of the same line.

For example, with reference to Fig. 7.35, for a configuration $\boldsymbol{b}^t$ and a user arriving at τ_1, the run choice set consists of run 1 of line b, run 2 of line a, and run 1 of line c. This set will differ if the user arrives in τ_2 or if he arrives in τ_1 of day $t+1$ and finds a different supply configuration $\boldsymbol{b}^{t+1}$. In the latter case, if there is congestion (e.g., on run $b1$) the choice set may be extended to run 2 of line b.

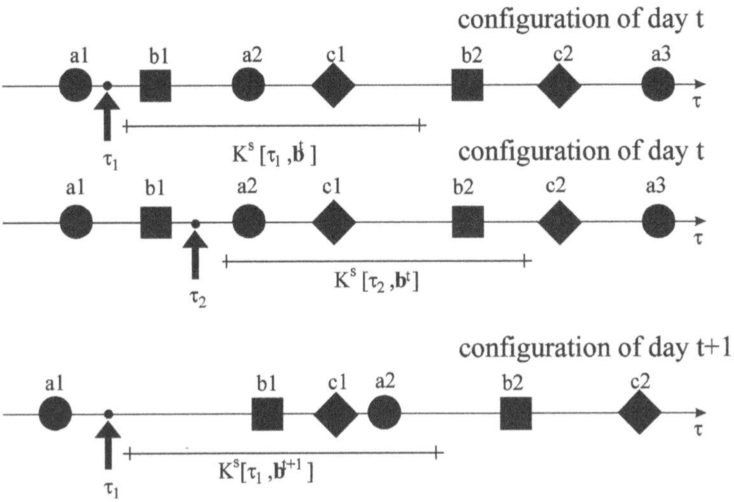

Fig. 7.35 Dependence of run choice set on configuration $\boldsymbol{b}^t$ and arrival time $\tau_{s,o}$

The choice set may change while the user waits at the stop not only because of congestion, but also because if an arrival is not boarded, the corresponding run is eliminated from the set. This point is clarified below.

A set of arrival times for the runs belonging to $K^s[\tau_{s,o}, \boldsymbol{b}^t]$ can be associated with each choice set $K^s[\tau_{s,o}, \boldsymbol{b}^t]$ for any arrival time τ^+ of run r^+. In the following, $K^s[\tau^+, \boldsymbol{b}^t]$ denotes the set available at time $\tau^+ > \tau_{s,o}$ of arrivals of run r^+ at the stop, with respect to which the user makes her choice.

A sequential mechanism can be assumed to simulate run choice. When a run r^+ of the path choice set $K^s[\tau^+, \boldsymbol{b}^t]$ arrives at time $\tau^+ > \tau_{s,o}$, the user chooses, in an intelligent adaptive way, to get on r^+ if the perceived utility U_{r^+} is greater than the utility U_{r^*} of all other runs $r^* \in K^s[\tau^+, \boldsymbol{b}^t]$ yet to arrive. In formal terms we have:

$$p_{od}[r^+/s, \tau^+] = \text{Prob}[U_{r^+} > U_{r^*}]$$
$$\forall r^* \neq r^+ \text{ with } \tau^* > \tau^+,\ r^+ \text{ and } r^* \in K^s[\tau^+, \underline{b}^t] \tag{7.6.9a}$$

As usual, perceived utilities can be specified as the sum of a systematic utility, expressed as a linear combination of attributes, and a random residual. A possible specification is:

$$U_{r^+} = V_{r^+}\varepsilon_{r^+} = \beta_{CFW}CFW_{r^+} + \beta_b Tb_{r^+} + \beta_c Tc_{r^+} + \beta_{CFB}CFB_{r^+}$$
$$+ \cdots + \beta_n Nn_{r^+} + \beta_p Tp_{r^+} + \varepsilon_{r^+} \tag{7.6.9b}$$

$$V_{r^*} = V_{r^*}\varepsilon_{r^*} = \beta_{CFW}CFW_{r^*} + \beta_b Tb_{r^*} + \beta_c Tc_{r^*} + \beta_{CFB}CFB_{r^*}$$
$$+ \cdots + \beta_n Nn_{r^*} + \varepsilon_{r^*} \tag{7.6.9c}$$

where

CFW_{r^+}, CFW_{r*} are the on-board comfort attributes (function of on-board crowding experienced at the stop)

Tw_{r^*} is the waiting time (equal to the difference between the arrival time of run r^+ and the arrival time of run r^*, provided by an information system)

Tb_{r^+} and Tb_{r^*} are on-board times

Tc_{r^+} and Tc_{r^*} are transfer times

Nn_{r^+} and Nn_{r^*} are the number of transfers

CFB_{r^+}, CFB_{r^*} are the "route" on-board comfort (a function of the amount of on-board crowding experienced in the following links)

Tp_{r^+} is the time already spent at the stop (equal to the difference between arrival time of run r^+ and the user arrival time τ_s at the stop) simulating a possible "impatience effect" ($\beta_p > 0$)

Note that in this model users cannot make their definitive choice upon arrival at stop at time τ_{so}, even if full information about waiting times is available, because the boarding comforts CFW of subsequent arrivals are not known. Of course, if the user does not choose run r^+, the choice is reconsidered when the subsequent run arrives and so on (sequential run choice behavior). Other more or less complex choice mechanisms can also be assumed.

If it is assumed that the random residuals ε in (7.6.9a) are i.i.d. Gumbel distributed, the choice probability $p_{od}[r^+/s, \tau^+]$ at time τ^+ of the arriving run r^+, conditional on not choosing previous runs and relative to the choice set $K^s[\tau^+, \boldsymbol{b}^t]$, can be expressed by a logit model:

$$p_{od}[r^+/s, \tau^+] - \frac{\exp(V_{r^+})}{\sum_{r \in K^s[\tau^+, b^t]} \exp(V_r)} \tag{7.6.10}$$

The total probability of choosing a given run r can be expressed as the product of the conditional probability (7.6.10) and the probability of not having chosen any previous run r belonging to the choice set $K_s[\tau^-, \boldsymbol{b}^t]$:

$$p_{od}[r/s, \tau_{s,o}] = \prod_{r^-=1,\ldots,r-1} \left(1 - p_{od}[r^-/s, \tau^-]\right) \cdot p[r/s, \tau] \tag{7.6.11}$$

where each conditional probability depends on the arrival time $\tau_{s,o}$ and may be computed through (7.6.9) and (7.6.10).

The probability $p_{od}[s/\tau_o]$ of choosing boarding stop s can be specified with a different model that refers to a choice set S_{od} of boarding stops. The choice set can be specified following different rules (e.g., by considering all stops within a certain distance from the origin). A perceived utility $U_s(\tau_o)$ can be associated with each stop in the choice set:

$$U_s(\tau_o) = V_s(\tau_o) + \varepsilon_s = \boldsymbol{\beta}^T \cdot \boldsymbol{X}_s + \beta_H \cdot H_s + \varepsilon_s \tag{7.6.12}$$

where $\boldsymbol{\beta}^T$ is the vector of the model parameters, $\boldsymbol{X}_s$ is a vector of stop-specific attributes (e.g., access time, presence of shops, etc.), and H_s is an "inclusive utility" expressing the average utility associated to all runs available at stop s. To model the inclusive utility, further assumptions have to be made on how travelers acquire and process information on system performance. This model is closely connected to the approach followed to simulate demand–supply interactions. One possible specification of H_s is based on the frequencies of the lines that are available at each stop and that belong to a feasible path on the line graph. This model is justified by the hypotheses of the lack of regularity (and information) and the high service frequencies of the system. Assuming a logit path choice model among the lines ln belonging to a set $Ln_s(o,d)$ of lines available at s to serve O-D pair od, the inclusive utility is proportional to the logsum variable H_s:

$$H_s = ln \sum_{ln \in Ln_s(o,d)} \exp(V_{ln,od})$$

with $V_{ln,od}$ depending on average (scheduled) level of service attributes of the line ln and given by:

$$V_{ln} = \beta_w T w_{ln} + \beta_b T b_{ln} + \beta_c T c_{ln} + \beta_n Nn$$

where the symbols have the same interpretation as in (7.6.9) but the coefficients are in principle different because they represent a different choice mechanism. Alternatively, the average cost of the minimum hyperpath connecting s to the destination d can be associated with each stop s. This model has the advantage of exploiting all the theoretical results and the computational algorithms described in Chaps. 5 and 7. In this case, it follows that $H_s \equiv x_{sd}^{\min}$.

Using a logit model, the stop choice probability can be expressed as

$$p_{od}[s/\tau_o] = \frac{\exp(V_s(\tau_o))}{\sum_{s' \in S_{od}} \exp(V_{s'}(\tau_o))} \tag{7.6.13}$$

Thus the total choice probability of a path k represented by departure time τ_o, boarding stop s, and run r (7.6.8) can be obtained through expressions (7.6.10), (7.6.11), and (7.6.13).

Finally, the average path flow h_k can be expressed as

$$h_k = d_{od}(\tau_o) \cdot p_{od}[r,s/\tau_o] = d_{od}(\tau_o) \cdot p_{od}[r,s/\tau_o] \cdot p_{od}[s/\tau_o]$$
$$k \equiv (\tau_o, s, r)$$

7.6.2.3 Demand–Supply Interaction Models

Given the irregularity of the system and the assumptions made about user behavior, especially at stops, demand–supply interactions should be modeled using a Stochastic Dynamic Process (SDP) approach. In this approach, service irregularities,

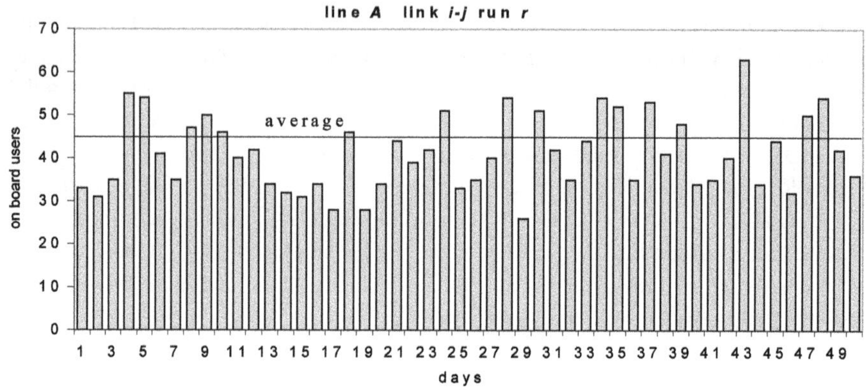

Fig. 7.36 Example of loads on the same section of the same run in different days

represented by random vectors $\boldsymbol{b}$ and $\boldsymbol{y}$, are simulated through a stochastic supply model. User choices at day t can be assumed to be independent multinomially distributed random variables with path choice probabilities given by (7.6.8). Figure 7.36 shows the number of users on the same section of the same run simulated on successive days for an urban transit network of realistic size under severe irregularity conditions.

The type of SDP model depends on a number of assumptions. First are the assumptions made about users' learning (cost-updating) mechanisms. If it is assumed that their pre-trip choices are based on average line attributes (see Sect. 7.6.2.2), stop choice probabilities $p_{od}[s/\tau_o]$ do not change over successive days, whereas run choice probabilities are affected by random events occurring at each day t but do not depend on previous days. Under these assumptions, the stochastic process is a renewal process; that is, the joint probability distribution of the variables describing the system state is independent of the states occupied in previous days. This assumption is reasonable for uncongested systems, where explicit utility updating mechanisms can be ignored and users base their choices on line frequencies because of the unreliability of the timetable.

Matters are further complicated by congestion effects. Given the randomness of the system, congestion levels vary over successive days. If users are assumed to choose the boarding stop based on uncongested attributes (as might be typical of infrequent users), congestion plays a role only in run choices at stops and the stochastic process is still a renewal one. Other (regular) users base their pre-trip choices on the congestion levels that they expect as a result of their previous experience. In this case, a utility updating filter similar to the ones described in Sect. 6.5 has to be introduced, and the process becomes Markovian.

7.A. The Simplified Theory of Kinematic Waves Based on Cumulative Flows: Application to Macroscopic Link Performance Models[20]

Macroscopic link performance models, aimed at reproducing travel times as a function of link inflows under the assumption of the fluid paradigm,[21] can be classified into two groups: space-continuous and space-discrete, as mentioned in Sect. 7.2.

The former are typically formulated as a system of differential equations in time and space that is solved through finite-difference methods. Such models yield accurate results, but require considerable computing resources, because their algorithmic implementation relies on a dense space discretization; for this reason they are also referred to as point-based. Altogether, they are very effective but somewhat inefficient.

The latter do not require any space discretization, and for this reason are also termed link-based. They can in turn be divided into whole link models and wave models.

The former yield link performances as a function of the space-average density (i.e., the number of vehicles on the link) without considering the propagation of flow states along the link. Such models are very simple and, for this reason, widely applied in DTA, but the representation of travel times becomes increasingly ineffective as the length of the link increases (see Sect. 7.2.1.3).

The latter, based on the Simplified Theory of Kinematic Waves (STKW), take (implicitly) into account the propagation of flow states, yielding link performances as a function of the traffic conditions that the vehicle encounters by traveling along the link. These models require minimal computing resources, and yield realistic results both in urban and extraurban contexts.

In this section we analyze the general properties of bottlenecks and segments under the STKW based on cumulative flows, because these are the main building blocks of any macroscopic link performance model.

In general, the bottleneck is defined as a gate with a null length and a constant or time-varying capacity, and a segment is assumed to be a homogeneous channel with a positive length and a time-constant capacity. Therefore, each element has a length L (for bottlenecks L is infinitesimal) and a capacity $\theta(\tau)$ (for segments $\theta(\tau) = C$ is constant in time).

With reference to any element two general properties hold true: the FIFO rule and the Newell–Luke Minimum Principle (NLMP). The latter states that:

- Among all possible states that may affect a given point of an element, bottleneck, or segment, the one yielding the minimum cumulative flow dominates the others.

Let $q(x, \tau)$ be the flow of vehicles crossing point $x \in [0, L]$ at time τ, and let $t(\tau)$ be the leaving time of a vehicle arriving to the element at time τ. The cumulative

[20] Guido Gentile and Natale Papola are the co-authors of this section.

[21] Vehicles are represented as particles of a mono-dimensional partially compressible fluid.

flow $Q(x,\tau)$ (see (7.2.8) and Fig. 7.9) is given by

$$Q(x,\tau) = \int_0^\tau q(x,\sigma) \cdot \mathrm{d}\sigma \tag{7.A.1}$$

Based on the fluid paradigm, the FIFO rule holds, and can be expressed formally as

$$Q(0,\tau) = Q\big(L, t(\tau)\big) \tag{7.A.2}$$

or equivalently as

$$q(0,\tau) = q\big(L, t(\tau)\big) \cdot \partial t(\tau)/\partial\tau \tag{7.A.3}$$

which is obtained by differentiating (7.A.2) with respect to time using the chain rule.

On this basis, once the cumulative flow temporal profiles at the initial and end-points of any element, or series of elements, are known, the exit time temporal profile can be easily determined, as depicted in Figs. 7.10 and 7.26.

The solution of (7.A.2) is based on the discretization of time in adjacent intervals $(\tau_{i-1}, \tau_i]$, with $i = 1, \ldots, n$. Under the classical numerical approximation that the flows are constant during each interval, we can apply the following algorithm, where we assume that $Q(0,\tau_n) = Q(L,\tau_n)$, $Q(0,\tau_0) = Q(L,\tau_0) = 0$, and T_0 is the free flow travel time of the element, or series of elements.

$t(\tau_0) = \tau_0 + T_0$

$j = 1$

for $i = 1$ to n do

  until $Q(L,\tau_j) \geq Q(0,\tau_i)$ do $j = j + 1$ (7.A.4)

if $Q(0,\tau_i) = Q(0,\tau_{i-1})$ then

  $t(\tau_i) = \tau_i + T_0$

  if $t(\tau_i) < t(\tau_{i-1})$ then $t(\tau_i) = t(\tau_{i-1})$

else

$$\begin{aligned} t(\tau_i) = {} & \tau_{j-1} + \big[Q(0,\tau_i) - Q(L,\tau_{j-1})\big] \cdot (\tau_j - \tau_{j-1})/\big[Q(L,\tau_j) \\ & - Q(L,\tau_{j-1})\big] \end{aligned} \tag{7.A.5}$$

The input of the algorithm is $Q(0,\tau_i)$ and $Q(L,\tau_i)$, and the output is $t(\tau_i)$, for $i = 0, \ldots, n$.

The cycle (7.A.4) aims to find, for each instant τ_i in chronological order, the earliest instant τ_j such that $Q(L,\tau_{j-1}) < Q(0,\tau_i) \leq Q(L,\tau_j)$. Because the leaving flow is by definition constant during the interval $(\tau_{j-1}, \tau_j]$, the cumulative leaving flow increases linearly with slope $[Q(L,\tau_j) - Q(L,\tau_{j-1})]/(\tau_j - \tau_{j-1})$. Therefore, in the general case where $Q(0,\tau_i) > Q(0,\tau_{i-1})$, the exit time $t(\tau_i)$ results from the simple proportion in (7.A.4). In the particular case where no flow arrives at the element in the interval $(\tau_{i-1}, \tau_i]$, the exit time $t(\tau_i)$ may be undetermined; it is thus set by definition as the maximum between the free flow exit time $\tau_i + T_0$ and the exit time $t(\tau_{i-1})$.

The following two sections are devoted to addressing the problem of determining the cumulative leaving flows for given arriving flows in the case of bottlenecks and segments, respectively.

7.A.1 Bottlenecks

Bottlenecks play a crucial role in modeling link performances in the context of DTA, because they allow explicit simulation of the formation and dispersion of vehicle queues, and hence evaluation of the delay due to the presence of intersections, which is an important part of the total travel time in highly congested urban networks.

A bottleneck can be conveniently formulated in terms of cumulative flows so as to yield the leaving flow by constraining the arriving flow below the bottleneck capacity, under the consideration that the former is stocked in a queue if it is not served at the moment, whereas the latter cannot be stocked if it is not utilized at the moment.

Based on the NLMP the cumulative flow leaving the bottleneck at time τ is the minimum among each cumulative outflow that would occur if the queue began at a previous instant $\sigma \le \tau$; that is;

$$Q(L,\tau) = \min\{Q(0,\sigma) + \Theta(\tau) - \Theta(\sigma) : \sigma \le \tau\} \tag{7.A.6}$$

where $\Theta(\tau)$ is the cumulative bottleneck capacity at time τ; that is;

$$\Theta(\tau) = \int_0^\tau \theta(\sigma) \cdot \mathrm{d}\sigma \tag{7.A.7}$$

The above expression (7.A.6) can be explained as follows. If there is no queue at a given time τ, the cumulative leaving flow $Q(L,\tau)$ is equal to the cumulative arriving flow $Q(0,\tau)$. If a queue arises at time $\sigma < \tau$, from that instant until the queue eventually vanishes, the outflow equals the bottleneck capacity, and then the cumulative leaving flow $Q(L,\tau)$ at time τ results from adding to the cumulative arriving flow $Q(0,\sigma)$ at time σ the integral of the bottleneck capacity between σ and τ; that is, $\Theta(\tau) - \Theta(\sigma)$. Note that, if there is no queue at time τ, the cumulative leaving flow is the same as the case when the queue arises exactly at $\sigma = \tau$.

Although not essential, below we illustrate in Fig. 7.A.1 the common case of bottlenecks with a time-constant capacity as an example of the above time-varying capacity model to facilitate the understanding of the NLMP. In this particular case, (7.A.6) becomes

$$Q(L,\tau) = \min\{Q(0,\sigma) + \theta \cdot (\tau - \sigma) : \sigma \le \tau\} \tag{7.A.8}$$

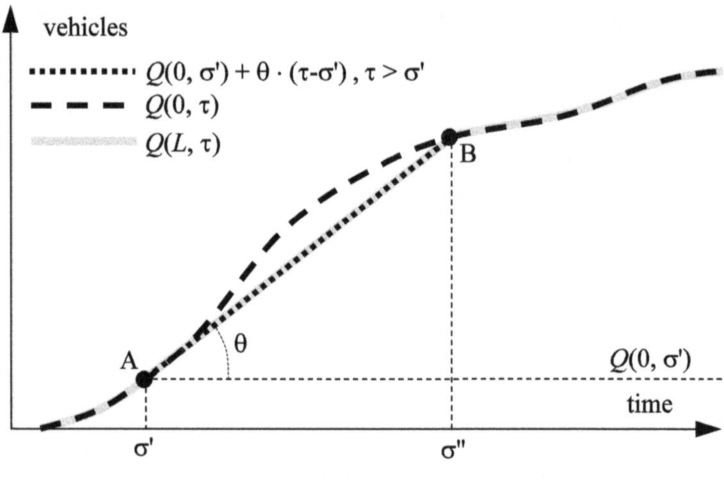

Fig. 7.A.1 Bottleneck with time-constant capacity

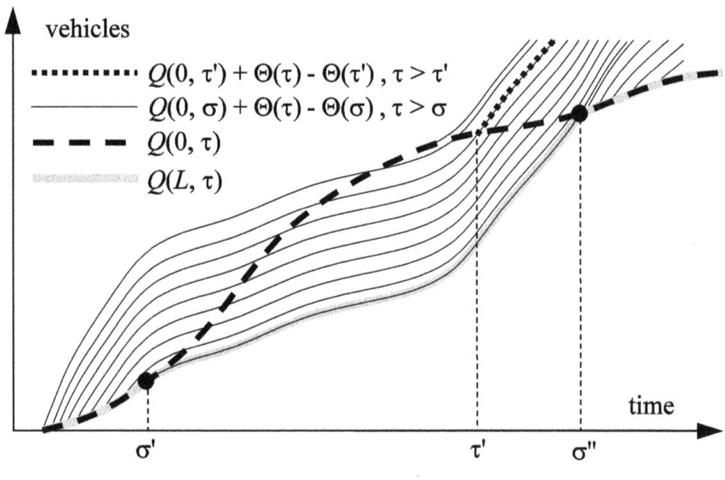

Fig. 7.A.2 Bottleneck with time-varying capacity

To explain (7.A.8), refer to Fig. 7.A.1, where the arriving and leaving cumulative flows are depicted. Let us consider a straight line with slope θ and let it translate vertically from the bottom upwards until it becomes tangent to a point where the temporal profile of the arriving flow is locally convex, like A in the figure. Just on the right of that point with time σ', as τ increases, the arriving flow becomes higher than the bottleneck capacity θ, meaning that the state $q(0, \sigma')$ prevails over the state θ, in terms of cumulative flows, until the point where the straight line intersects the cumulative arriving flow temporal profile, like B in the figure, where the queue disappears and $q(0, \sigma'') > \theta$. We see that the segment of straight line A–B belongs to the lower envelope of the possible flow states in the sense of the NLMP.

Figure 7.A.2 depicts instead a graphical interpretation of the general equation (7.A.6) for a bottleneck with a time-varying capacity, where the temporal profile $Q(L,\tau)$ of the cumulative leaving flow is the lower envelope of the following curves: (a) the cumulative arriving flow $Q(0,\tau)$; (b) the family of functions $Q(0,\sigma)+\Theta(\tau)-\Theta(\sigma)$ with $\tau>\sigma$, for every time σ, each obtained as the vertical translation of the temporal profile relative to the cumulative bottleneck capacity that goes through the point $(\sigma, Q(0,\sigma))$. No queue is present when curve (a) prevails; therefore, the queue arises at time σ' and vanishes at time σ''.

Let $N(\tau)$ be the number of vehicles queuing to exit the bottleneck at time τ; that is,

$$N(\tau)=Q(0,\tau)-Q(L,\tau) \tag{7.A.9}$$

Equation (7.A.6) can be numerically solved by means of the following algorithm.

$N(\tau_0)=0$
for $i=1$ to n do
  $N(\tau_i)=N(\tau_{i-1})+Q(0,\tau_i)-Q(0,\tau_{i-1})$
  if $N(\tau_i)\le\Theta(\tau_i)-\Theta(\tau_{i-1})$ then
    $Q(L,\tau_i)=Q(L,\tau_{i-1})+N(\tau_i)$
    $N(\tau_i)=0$
  else
    $Q(L,\tau_i)=Q(L,\tau_{i-1})+\Theta(\tau_i)-\Theta(\tau_{i-1})$
    $N(\tau_i)=N(\tau_{i-1})-\Theta(\tau_i)+\Theta(\tau_{i-1})$

The input of the algorithm is $Q(0,\tau_i)$ and $\Theta(\tau_i)$, and the output is $Q(L,\tau_i)$ and $N(\tau_i)$, for $i=0,\ldots,n$.

The number of vehicles desiring to leave the bottleneck during the interval $(\tau_{i-1},\tau_i]$, for short called here the demand, is given by the number of vehicles $N(\tau_{i-1})$ queuing to exit the bottleneck at the beginning of the interval, plus the number of vehicles $Q(0,\tau_i)-Q(0,\tau_{i-1})$ that arrive at the bottleneck during the interval. But, due to the capacity constraint, only the number of vehicles $\Theta(\tau_i)-\Theta(\tau_{i-1})$, for short called here the supply, can at most exit the bottleneck. If the supply is higher than the demand, then all such vehicles will actually leave the bottleneck during the interval, and no vehicle will be queuing to exit the bottleneck at the end of the interval; otherwise, only a number of vehicles equal to the supply will leave the bottleneck during the interval, and the rest of the demand will be queuing to exit the bottleneck at the end of the interval.

7.A.2 Segments

Segments aim at simulating the movement of vehicles along the links. Thus they are the main elements of moderately congested extraurban networks, where queues are not a prevalent phenomenon, and most links are so long that the presence of intersections plays a negligible role in the representation of travel times.

The modeling of a segment can be conveniently addressed in the framework of the STKW based on cumulative flows, of which a brief review is given below.

The STWK is founded on the following assumptions.

(a) The segment is a homogeneous channel.
(b) The vehicles change their speed, whenever needed, with infinite decelerations and accelerations.
(c) The fundamental diagram of traffic flow described in Sect. 2.2.2.2 for stationary conditions still holds for nonstationary traffic.

Specifically, based on (c) we have:

$$q(x,\tau) = k(x,\tau)\cdot v(x,\tau) \tag{7.A.10}$$

$$v(x,\tau) = \mathrm{v}\big(k(x,\tau)\big) \tag{7.A.11}$$

where $k(x,\tau)$ and $v(x,\tau)$ are respectively the density and speed at point x and time τ.

Based on (7.A.10), (7.A.11) also defines a relation between flow and density, called the fundamental diagram:

$$q(x,\tau) = \mathrm{q}\big(k(x,\tau)\big). \tag{7.A.12}$$

In the following we assume that the fundamental diagram is strictly concave; that is, it has only one maximum. Thus the (critical) density KC at which it takes the maximum flow (capacity) C divides the flow states in hypocritical (denoted by apex $^\circ$) and hypercritical (denoted by apex $+$), so that it is possible to derive the following inverse one-valued functions:

$$k(x,\tau) = \mathrm{k}^\circ\big(q(x,\tau)\big) \tag{7.A.13}$$

$$k(x,\tau) = \mathrm{k}^+\big(q(x,\tau)\big) \tag{7.A.14}$$

$$v(x,\tau) = \mathrm{v}^\circ\big(q(x,\tau)\big) \tag{7.A.15}$$

$$v(x,\tau) = \mathrm{v}^+\big(q(x,\tau)\big) \tag{7.A.16}$$

Dealing with nonstationary traffic, we state a conservation condition ensuring that vehicles are not created or lost along the segment. Let us consider an infinitesimal time interval $[\tau, \tau + \mathrm{d}\tau]$ and an infinitesimal portion of the segment $[x, x + \mathrm{d}x]$. Under the assumption that the flow remains constant during the infinitesimal time interval, but varies along the segment, the number of vehicles crossing point x is $q(x,\tau)\cdot \mathrm{d}\tau$, whereas those crossing point $x + \mathrm{d}x$ is $[q(x,\tau) + \partial q(x,\tau)/\partial x \cdot \mathrm{d}x]\cdot \mathrm{d}\tau$. Under the assumption that the density remains constant along the infinitesimal portion of the segment, but varies during the time, the number of vehicles present at time τ is $k(x,\tau)\cdot \mathrm{d}x$, whereas those present at time $\tau + \mathrm{d}\tau$ is $[k(x,\tau) + \partial k(x,\tau)/\partial\tau \cdot \mathrm{d}\tau]\cdot \mathrm{d}x$. The increase in vehicles present on the infinitesimal portion of the segment that occurs during the infinitesimal time interval must be equal to the number of vehicles that entered the infinitesimal portion

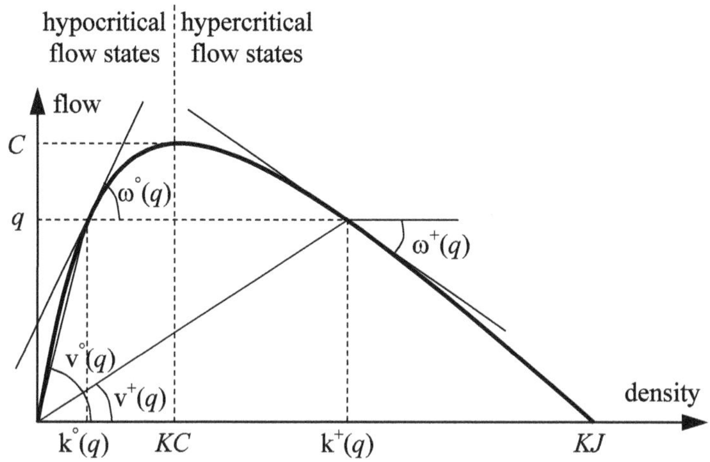

Fig. 7.A.3 The fundamental diagram

of the segment during the infinitesimal time interval minus those that exited it:

$$\begin{aligned}\left[k(x,\tau)+\partial k(x,\tau)/\partial\tau\cdot \mathrm{d}\tau\right]\cdot \mathrm{d}x-k(x,\tau)\cdot \mathrm{d}x&=q(x,\tau)\cdot \mathrm{d}\tau\\ -\left[q(x,\tau)+\partial q(x,\tau)/\partial x\cdot \mathrm{d}x\right]\cdot \mathrm{d}\tau&\end{aligned}\tag{7.A.17}$$

Therefore we have (see (2.2.3)):

$$\partial k(x,\tau)/\partial\tau+\partial q(x,\tau)/\partial x=0\tag{7.A.18}$$

Moreover, let us analyze the function $q(x,\tau)$ yielding the flow state at a given point (x,τ) in the time–space plane, looking for the points in the neighborhood of (x,τ) that are affected by its same flow state. If we consider the flow as the elevation of the point, this is like aiming to determine the contour line passing through (x,τ). Therefore, we are formally seeking a direction $\mathrm{d}x/\mathrm{d}\tau$ in the time–space plane such that:

$$\mathrm{d}q=\partial q(x,\tau)/\partial x\cdot \mathrm{d}\tau+\partial q(x,\tau)/\partial\tau\cdot \mathrm{d}x=0\tag{7.A.19}$$

Based on (7.A.18), it is:

$$\omega(x,\tau)=\mathrm{d}x/\mathrm{d}\tau=1/\left[\partial k(x,\tau)/\partial q(x,\tau)\right]\tag{7.A.20}$$

where we have introduced the notation $\omega(x,\tau)$ for such a direction, which in the time–space plane is a speed.

Because each point in the neighborhood of (x,τ) belonging to the straight line in the time–space plane with slope $\omega(x,\tau)$ passing through that point is affected by a same flow state, the latter will propagate as a wave keeping the same direction, which is therefore referred to as the wave speed.

This way we have shown that the solution in terms of flows to the system defined by (7.A.18), (7.A.13), and (7.A.14) is such that the generic hypocritical flow state propagates forward along the segment at a constant speed:

$$\omega(x,\tau) = \omega^{\circ}\big(q(x,\tau)\big) = 1/\big[\mathrm{dk}^{\circ}(q)\big)/\mathrm{d}q\big] \tag{7.A.21}$$

and the generic hypercritical flow state propagates backward along the segment at a constant speed:

$$\omega(x,\tau) = \omega^{+}\big(q(x,\tau)\big) = 1/\big[\mathrm{dk}^{+}(q)\big)/\mathrm{d}q\big] \tag{7.A.22}$$

One of the most simple specifications for (7.A.11) is the Greenshields linear model, already introduced in Sect. 2.2.2.2:

$$\mathrm{v}(k) = V \cdot (1 - k/KJ) \tag{7.A.23}$$

where V is the free flow speed and KJ is the jam density. The resulting capacity C is $0.25 \cdot V \cdot KJ$ and the critical density KC is $0.5 \cdot KJ$. In this case (7.A.21)–(7.A.22) and (7.A.15)–(7.A.16) become, respectively:

$$\omega^{\circ}(q) = V \cdot (1 - q/C)^{0.5} \tag{7.A.24}$$

$$\omega^{+}(q) = -\omega^{\circ}(q) \tag{7.A.25}$$

$$\mathrm{v}^{\circ}(q) = 0.5 \cdot \big[V + \omega^{\circ}(q)\big] \tag{7.A.26}$$

$$\mathrm{v}^{+}(q) = 0.5 \cdot \big[V + \omega^{+}(q)\big] \tag{7.A.27}$$

Another interesting model is the triangular fundamental diagram, which can be obtained from a simple car following approach.

The speed v is a function of the density k due to the need to keep a sufficient distance from the vehicle ahead taking into account: (a) the reaction time RT and (b) the length of the vehicle LV, including a safety margin. Indeed, if the vehicle ahead starts a break while traveling at a stationary state with speed v, the space run by the vehicle behind during the reaction time is: $RS = v \cdot RT$. Under the assumption that the braking power of the two vehicles is alike, the distance required to avoid collision is: $1/k = RS + LV$. Moreover we need to consider the speed limit or the free flow speed V of the road, so that for a given density k it is:

$$v = \min\big\{V, 1/(k \cdot RT) - LV/RT\big\} \tag{7.A.28}$$

The jam density, i.e. the maximum density, is obtained for the second case of (7.A.28) at $v = 0 : KJ = 1/LV$. Using the stationary flow equation $q = k \cdot v$, we then obtain the flow as a function of the density:

$$q = \min\{k \cdot V, 1/RT - k \cdot LV/RT\} \tag{7.A.29}$$

The capacity, i.e. the maximum flow, is obtained when $k \cdot V = 1/RT - k \cdot LV/RT$, that is at the critical density $KC = 1/(V \cdot RT + LV)$. Hence: $C = KC \cdot V = 1/(RT + LV/V)$.

The hypercritical wave speed is given by the ratio of the capacity and the difference between the jam density and the critical density: $\omega = Q/(KJ - KC) = LV/RT$. Therefore we have:

$$\omega^{\circ}(q) = V \tag{7.A.30}$$

$$\omega^{+}(q) = LV/RT \tag{7.A.31}$$

$$\mathrm{v}^{\circ}(q) = V \tag{7.A.32}$$

$$\mathrm{v}^{+}(q) = LV/(1/q - RT) \tag{7.A.33}$$

In the case of *NL* lanes, densities and flows are scaled by *NL*, so that (7.A.33) becomes:

$$\mathrm{v}^{+}(q) = LV/(NL/q - RT) \tag{7.A.34}$$

Flow states can disappear along the segment. In fact, where two kinematic waves with speed $\omega_1 = \omega(q_1)$ and $\omega_2 = \omega(q_2)$ collide, an interface (or shockwave) emerges separating the two flow states q_1 and q_2, whose speed ω_{12} is given by the change in flow across the interface over the change in density, that is,

$$\omega_{12} = (q_1 - q_2)/(k_1 - k_2) \tag{7.A.35}$$

where $k_1 = \mathrm{k}(q_1), k_2 = \mathrm{k}(q_2)$ and, for the sake of brevity, we denoted functions $\omega(q)$ and $\mathrm{k}(q)$, regardless of the flow state q being hypocritical or hypercritical, because (7.A.35) holds in any of the possible cases. Moreover, where two shockwaves with speed ω_{12} and ω_{23} collide, a new shockwave emerges with speed ω_{13} separating the two flow states q_1 and q_3, whereas the flow state q_2 disappears.

On the other hand, because the segment is a homogeneous channel, flow states can arise only at the initial and endpoints.

Equation (7.A.35), which is a consequence of the conservation equation, can be easily derived on the basis of purely geometric considerations. We address in the following the case where two different flow states, say q_1 and q_2, hold at two adjacent subspaces of the time–space plane as depicted in Fig. 7.A.4, where, without loss of generality, it is $q_1 < q_2$.

Considering the triangle BDE, the two similar triangles BDF and GEF, and the two similar triangles ABC and CDE, we get the following system in the three unknown ω_{12}, t and s, after denoting CD and DE as t and s, respectively:

$$\omega_{12} = s/(1/q_1 + t) \tag{7.A.36}$$

$$(1/k_2 + s)/(1/q_1 + t) = q_2/k_2 \tag{7.A.37}$$

$$s/t = q_1/k_1 \tag{7.A.38}$$

By substituting $t = s \cdot k_1/q_1$, obtained from (7.A.38), into (7.A.37), we have:

$$s = (q_2 - q_1)/(k_2 \cdot q_1 - k_1 \cdot q_2) \tag{7.A.39}$$

$$t = k_1/q_1 \cdot (q_2 - q_1)/(k_2 \cdot q_1 - k_1 \cdot q_2) \tag{7.A.40}$$

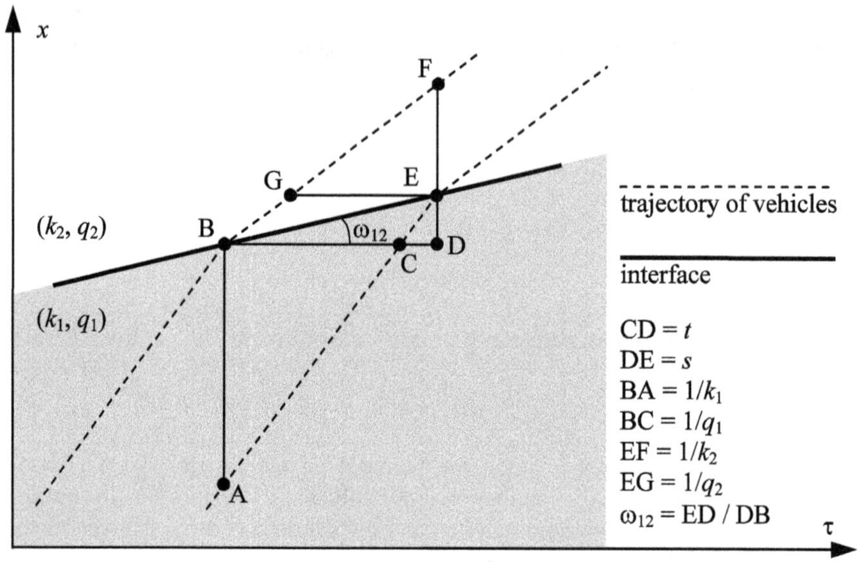

Fig. 7.A.4 Speed of an interface on the time–space plane

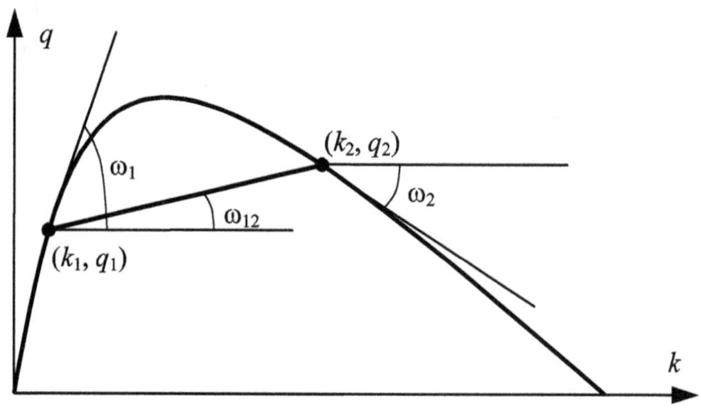

Fig. 7.A.5 Speed of an interface on the fundamental diagram

Finally, using (7.A.39) and (7.A.40) into (7.A.36) yields (7.A.35).

As this equation shows, the slope of the interface ω_{12} in Fig. 7.A.5 is equal to the slope of the segment joining the two points, (k_1, q_1) and (k_2, q_2), on the fundamental diagram in Fig. 7.A.5. When point 2 tends to point 1, the slope of the interface tends to the derivative $\omega_1 = 1/[\mathrm{dk}(q_1))/\mathrm{d}q]$, as stated by (7.A.21) and (7.A.22).

The theory of the NLMP allows us to solve the wave model in terms of cumulative flows with reference to any specific point of the segment as a function of the boundary conditions, which are the flow states arising at the initial and end points, specifically:

(a) Hypocritical flows arriving at the segment when the queue does not reach the initial point. That is, no spilling back is occurring from the segment, formally $q(0,\tau) < \mu(\tau)$, where $\mu(\tau)$ is the entry capacity at time τ.
(b) Hypercritical flows leaving the segment when a queue is present at the endpoint of the segment, formally $q(L,\tau) = \psi(\tau)$, where $\psi(\tau)$ is the exit capacity at time τ.

We now address the forward propagation of hypocritical inflows and the backward propagation of hypercritical outflows separately.

The instant $e^\circ(x,\tau) \geq \tau$ when the forward kinematic wave generated at time τ on the initial point of the segment by the hypocritical inflow $q(0,\tau) < \mu(\tau)$ reaches the generic point x is given by

$$e^\circ(x,\tau) = \tau + x/\omega^\circ\big(q(0,\tau)\big) \tag{7.A.41}$$

In general, $e^\circ(x,\tau)$ is not invertible, because more than one kinematic wave generated on the initial point may reach point x at the same time.

By definition, all the points in the time–space plane constituting the straight line trajectory produced by a kinematic wave are characterized by the same flow state. Figure 7.A.5 shows that the number of vehicles traveling at speed $\mathrm{v}^\circ(q)$ that pass an observer traveling at speed $\omega^\circ(q)$ along the hypocritical wave relative to flow q for any infinitesimal space ds moved in the same direction is equal to the time interval $\mathrm{d}s \cdot [1/\omega^\circ(q) - 1/\mathrm{v}^\circ(q)]$ multiplied by that flow. Therefore, integrating along the segment[22] from the initial point to point x, we obtain the cumulative flow $H^\circ(x,\tau)$ that may be observed at time $e^\circ(x,\tau)$ in that point:

$$H^\circ(x,\tau) = Q(0,\tau) + q(0,\tau)\cdot x \cdot \big[1/\omega^\circ\big(q(0,\tau)\big) - 1/\mathrm{v}^\circ\big(q(0,\tau)\big)\big] \tag{7.A.42}$$

The instant $e^+(x,\tau) \geq \tau$ when the backward kinematic wave generated at time τ on the endpoint of the segment by the hypercritical outflow $q(L,\tau) = \psi(\tau)$ reaches the generic point x is given by[23]

$$e^+(x,\tau) = \tau - (L - x)/\omega^+\big(q(L,\tau)\big) \tag{7.A.43}$$

As above, $e^+(x,\tau)$ is not invertible, because more than one kinematic wave generated on the endpoint may reach point x at the same time.

Figure 7.A.6 shows that the number of vehicles traveling at speed $\mathrm{v}^+(q)$ encountered by an observer traveling at speed $-\omega^+(q)$ along the hypercritical wave relative to the flow q for any infinitesimal space $\mathrm{d}s$ moved in the opposite direction is equal to the time interval $\mathrm{d}s \cdot [1/\mathrm{v}^+(q) - 1/\omega^+(q)]$ multiplied by that flow. Therefore, integrating along the segment from the endpoint to point x, we obtain the cumulative flow $H^+(x,\tau)$ that may be observed at time $e^+(x,\tau)$ in that point:

$$H^+(x,\tau) = Q(L,\tau) + q(L,\tau)\cdot(L - x)\cdot\big[1/\mathrm{v}^+\big(q(L,\tau)\big) - 1/\omega^+\big(q(L,\tau)\big)\big] \tag{7.A.44}$$

---

[22] The flow state along the wave is constant.

[23] Recall that $\omega^+(q)$ is negative.

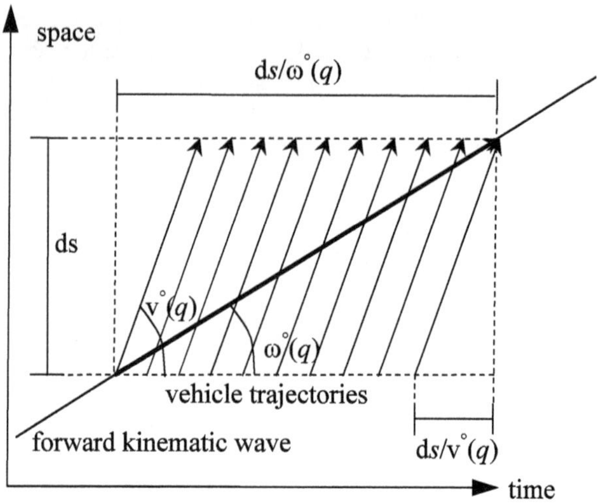

Fig. 7.A.6 Flow traversing a hypocritical kinematic wave

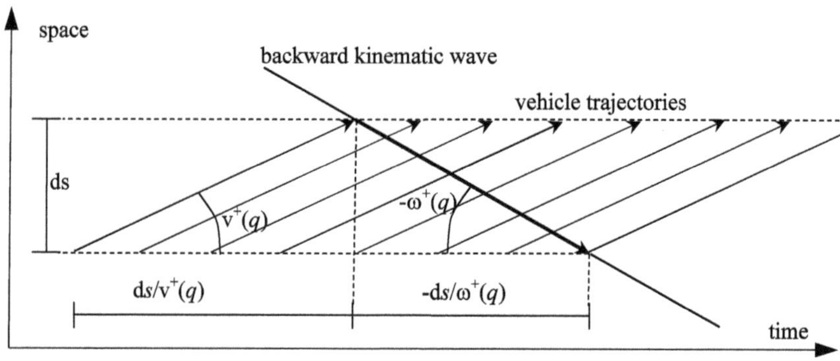

Fig. 7.A.7 Flow traversing a hypercritical kinematic wave

Based on the NLMP, of all kinematic waves that pass through a given point in the time–space plane the one yielding the minimum cumulative flow dominates the others.

Because the minimum operator is associative, we can separate the hypocritical flow states coming from upstream and the hypercritical flow states coming from downstream. This way the cumulative flow on point x at time τ is given by

$$Q^{\circ}(x,\tau) = \min\{H^{\circ}(x,\sigma) : e^{\circ}(x,\sigma) = \tau, q(0,\sigma) < \mu(\sigma)\} \quad (7.A.45)$$

$$Q^{+}(x,\tau) = \min\{H^{+}(x,\sigma) : e^{+}(x,\sigma) = \tau, q(L,\sigma) = \psi(\sigma)\} \quad (7.A.46)$$

$$Q(x,\tau) = \min\{Q^{\circ}(x,\tau), Q^{+}(x,\tau)\} \quad (7.A.47)$$

where $Q^{\circ}(x,\tau)$ or $Q^{+}(x,\tau)$ is not defined; that is, no hypocritical or hypercritical wave reaches point x at time τ, respectively. Their value resulting from the minimum operator is conventionally set equal to infinity.

Because hypocritical speeds are always higher than hypercritical speeds, forward propagating as hypocritical a hypercritical flow does not affect the overall model (7.A.47), therefore the condition $q(0,\sigma) < \mu(\sigma)$ in (7.A.45) can be omitted.

In conclusion, the flow state occurring on the generic point of the segment is the result of the interaction among hypocritical flow states coming from upstream and hypercritical flow states coming from downstream. With reference to the endpoint, the flow states coming from downstream are the hypercritical leaving flows generated by the exit capacity when a queue is present, whereas the flow states coming from upstream can be determined by forward-propagating the temporal profile of the cumulative arriving flows as hypocritical. With reference to the initial point, the flow states coming from upstream are the arriving flows, and the flow states coming from downstream can be determined by back-propagating the hypercritical portion of the temporal profile of the cumulative leaving flows.

The numerical solution of (7.A.45) and (7.A.46) can be easily addressed under the assumption that the arriving flows and leaving flows, respectively, are constant in each time interval. In this case, to the constant flow q_i, hypocritical or hypercritical, at the extreme point during the interval $(\tau_{i-1},\tau_i]$ a linear cumulative flow at point x corresponds, that is, a segment in the time-vehicles plane between the points $(\mathrm{e}(x,\tau_{i-1},q_i), \mathrm{H}(x,\tau_{i-1},q_i)) - (\mathrm{e}(x,\tau_i,q_i), \mathrm{H}(x,\tau_i,q_i))$, where functions $\mathrm{e}(x,\tau,q)$ and $\mathrm{H}(x,\tau,q)$ express (7.A.41) to (7.A.43) and (7.A.42) to (7.A.44), respectively, for the two cases. If we connect these segments, for $i=1,\ldots,n$, through additional segments between the points $(\mathrm{e}(x,\tau_i,q_i), \mathrm{H}(x,\tau_i,q_i)) - (\mathrm{e}(x,\tau_i,q_{i+1}), \mathrm{H}(x,\tau_i,q_{i+1}))$, for $i=1,\ldots,n-1$, then $Q(x,\tau_i)$ can be obtained as the minimum number of vehicles among the values taken at time τ_i by the segments that are defined at such instant. It is worth pointing out that connecting the segments through straight lines implies an approximation, because the points $(\mathrm{e}(x,\tau_i,q), \mathrm{H}(x,\tau_i,q))$ for $q \in [q_i,q_{i+1}]$ actually form a curve in the time–vehicles plane. The following algorithm can be applied to determine efficiently the cumulative flow at point x, where q_0 is assumed to be null, and $\tau_{n+1}=\infty$.

for $i=0$ to n do $Q(x,\tau_i)=\infty$
$j=0$
until $\tau_j > \mathrm{e}(x,\tau_0,q_0)$ do
  $Q(x,\tau_j)=0$
  $j=j+1$
for $i=1$ to n do
  if $\mathrm{e}(x,\tau_{i-1},q_i) > \mathrm{e}(x,\tau_{i-1},q_{i-1})$ then
    until $\tau_j \ge \mathrm{e}(x,\tau_{i-1},q_{i-1})$ do $j=j+1$
    until $\tau_j > \mathrm{e}(x,\tau_{i-1},q_i)$
     $\mathrm{H} = \mathrm{H}(x,\tau_{i-1},q_{i-1}) + [\tau_j - \mathrm{e}(x,\tau_{i-1},q_{i-1})]\cdot[\mathrm{H}(x,\tau_{i-1},q_i) - \mathrm{H}(x,\tau_{i-1},q_{i-1})]/[\mathrm{e}(x,\tau_{i-1},q_i) - \mathrm{e}(x,\tau_{i-1},q_{i-1})]$
    if $Q(x,\tau_j) > H$ then $Q(x,\tau_j)=H$

$j = j + 1$
else
until $\tau_j \leq \mathrm{e}(x, \tau_{i-1}, q_{i-1})$ do $j = j - 1$
until $\tau_j < \mathrm{e}(x, \tau_{i-1}, q_i)$
$H = \mathrm{H}(x, \tau_{i-1}, q_{i-1}) + [\tau_j - \mathrm{e}(x, \tau_{i-1}, q_{i-1})] \cdot [\mathrm{H}(x, \tau_{i-1}, q_i) - \mathrm{H}(x, \tau_{i-1}, q_{i-1})]/[\mathrm{e}(x, \tau_{i-1}, q_i) - \mathrm{e}(x, \tau_{i-1}, q_{i-1})]$
if $Q(x, \tau_j) > H$ then $Q(x, \tau_j) = H$
$j = j - 1$
until $\tau_j \geq \mathrm{e}(x, \tau_{i-1}, q_i)$ do $j = j + 1$
until $\tau_j > \mathrm{e}(x, \tau_i, q_i)$
$H = \mathrm{H}(x, \tau_{i-1}, q_i) + [\tau_j - \mathrm{e}(x, \tau_{i-1}, q_i)] \cdot [\mathrm{H}(x, \tau_i, q_i) - \mathrm{H}(x, \tau_{i-1}, q_i)] /[\mathrm{e}(x, \tau_i, q_i) - \mathrm{e}(x, \tau_{i-1}, q_i)]$
if $Q(x, \tau_j) > H$ then $Q(x, \tau_j) = H$
$j = j + 1$

The input of the algorithm is q_i and the output is $Q(x, \tau_i)$, for $i = 0, \ldots, n$.

The algorithms scans in chronological order, for $i = 1, \ldots, n$, first the additional segment between the points $(\mathrm{e}(x, \tau_{i-1}, q_i), \mathrm{H}(x, \tau_{i-1}, q_{i-1})) - (\mathrm{e}(x, \tau_{i-1}, q_i), \mathrm{H}(x, \tau_{i-1}, q_i))$ searching for any τ_j such that $\mathrm{e}(x, \tau_{i-1}, q_{i-1}) \leq \tau_j \leq \mathrm{e}(x, \tau_i, q_i)$, and then the segment between the points $(\mathrm{e}(x, \tau_{i-1}, q_i), \mathrm{H}(x, \tau_{i-1}, q_i)) - (\mathrm{e}(x, \tau_i, q_i), \mathrm{H}(x, \tau_i, q_i))$ searching for any τ_j such that $\mathrm{e}(x, \tau_{i-1}, q_i) \leq \tau_j \leq \mathrm{e}(x, \tau_i, q_i)$. While scanning, if the value taken at τ_j by the segment under analysis is lower than the current estimate of the cumulative flow $Q(x, \tau_j)$, then the latter is updated to the former; this way, at the end of the procedure, $Q(x, \tau)$ is at each τ_i the lower envelope of all above segments, and elsewhere a piecewise linear approximation is considered.

Reference Notes

Although dynamic traffic assignment is a relatively new research subject, a wide body of literature has been produced over the last 15 years (and only some of this output is cited here).

The first to propose DTA as a research subject of its own in a form similar to the present formulations were Ben-Akiva et al. (1984). The framework adopted in this chapter to present supply, demand, and supply–demand interaction models is original. The formulation reported in Sect. 7.5 is taken from Bellei et al. (2005) and Gentile et al. (2004, 2007), where solution algorithms are also devised.

Continuous-flow models were first investigated by the scientific community.

Continuous-flow link performance models can be classified into two major groups: space-continuous and space-discrete. Among the models belonging to the first group we recall METANET, proposed by Messmer and Papageorgiou (1990), which derives from a second-order approximation to the Theory of Kinematic Waves (TKW), and the Cell Transmission Model proposed by Daganzo (1994,

1995). The latter is consistent with the Simplified Theory of Kinematic Waves (STKW), that is, the first-order approximation of the TKW, independently proposed by Lighthill and Witham (1955) and Richards (1956). The models belonging to the second group do not require any spatial discretization. Specifically, whole link models (e.g., Astarita 1996; Ran et al. 1997) do not take into account the propagation of flow states along the link, because performances are assumed to depend on a space-average state variable, such as density (Heydecker and Addison 1998). This yields a poor representation of travel times, which gets worse as the link length increases (Daganzo 1995). Despite this major deficiency, these models are widely used in DTA because of their simplicity (e.g., Friesz et al. 1993; Tong and Wong 2000). Wave models, based on the simplified theory of kinematic waves, implicitly take into account the propagation of flow states along the link, yielding link performances as a function of the traffic conditions encountered while traveling throughout the link. For a theoretical analysis of the "Newell–Luke minimum principle" reference can be made to Newell (1993) and to Daganzo (1997), where the relevant aspect of the STKW can also be found. These models were first developed for bottlenecks with constant capacity, that is, when only two speeds may occur on the link: the free-flow speed and the queue speed (Newell 1993; Arnot et al. 1990; Ghali and Smith 1993; Bellei et al. 2005). Recently, these models have been extended to the case of long links and time-varying capacity (Gentile et al. 2005, 2007).

Continuous-flow models were adopted in the seminal work of Merchant and Nemhauser (1978), who addressed system-optimal DTA with a single origin. The first to identify the dynamic network loading model as a component of any DTA model were Cascetta and Cantarella (1991). The continuous-flow supply model described is based on the work of Friesz et al. (1989), who introduced the travel time link flow propagation model and equivalent conditions for respect of the FIFO rule. Recently more general equivalent conditions for FIFO were stated by Chabini and Kachani (1999), who also investigated the properties of uniqueness and existence of the continuous-flow single-link network DNL problem. Some heuristic algorithms have been proposed in the literature for solving the supply model for general networks (Astarita 1996; Wu et al. 1998; Xu et al. 1999).

In the literature there are several papers proposing discrete flow supply models, both at the mesoscopic level (Cascetta and Cantarella 1991; Jayakrishnan et al. 1994; Ben-Akiva et al. 1984; Cantarella et al. 1999), and at the microscopic level (Yang and Koutsopoulos 1996). In general it can be said that little or no effort has been made either to propose a general formulation of discrete flow models or to investigate their theoretical properties as for continuous models. In this respect, the general framework proposed here for discrete flow models is original.

Demand models of departure-time choice were first proposed by Abkowitz (1981) and Small (1982); a joint departure time-path choice model for urban networks was proposed by Cascetta et al. (1992). More complex departure-time and path switching models were proposed by Mahamassani and Liu (1999).

Most models proposed in the literature for demand–supply interactions are either deterministic or stochastic user equilibrium models.

Papers on continuous flow models have frequently proposed ways to extend to time-varying demand and link flows the deterministic (Wardrop's) user equilibrium equivalent formulations (i.e., optimization or variational inequalities). Among these, the papers of Boyce et al. (1991), Janson (1989), Vythoulkas (1990), Friesz et al. (1993), Wie et al. (1990), and the book by Ran and Boyce (1994) can be referred to. Static stochastic user equilibrium was formulated as a fixed-point problem by Daganzo (1983) and Cantarella (1997). However, the general formulation of continuous and discrete flow within-day dynamic, fixed-point models is original. Examples of dynamic process models are those proposed by Cascetta and Cantarella (1991), Cantarella et al. (1999), Jayakrishnan et al. (1994), and Jha et al. (1998).

Dynamic assignment for transit or other scheduled services has received considerably less attention in the literature. The idea of representing the schedule by a diachronic network can be credited to Nuzzolo and Russo (1996). Some examples of dynamic assignment models for low-frequency regular services can be found in Cascetta et al. (1996) and Nuzzolo et al. (2000) for multiple-service rail networks, and in Cascetta and Papola (2003) for mode-service choice simulation on multimodal bus and rail networks. Dynamic assignment for irregular scheduled services is a still newer area. The papers by Hickman and Wilson (1995), Hickman and Bernstein (1997), and Nuzzolo et al. (1999) are among the first presented in the literature. The books by Wilson and Nuzzolo (2004, 2008) provide an up-to-date and complete review of assignment models to scheduled service networks, as well as an overview of their applications.

Chapter 8
Estimation of Travel Demand Flows

8.1 Introduction

The analysis and design of transportation systems require the estimation of present demand and the forecasting of (hypothetical) future demand. These estimates and forecasts can be obtained using a variety of information sources and statistical procedures.

To estimate present demand, surveys can be conducted, typically by interviewing a sample of system users; direct estimates of the demand can then be derived using results from sampling theory.

Alternatively, demand (present or future) can be estimated using models similar to those described in Chap. 4. Model estimation requires that the models be specified (i.e., the functional form and the variables are defined), calibrated (i.e., the unknown coefficients are estimated), and validated (i.e., the ability to reproduce the available data is tested). These operations can be performed using disaggregate information about a sample of individuals. The type of survey and the size of the sample involved in model estimation are often different from those used for direct demand estimation. Once the models have been specified and calibrated, they can be applied to the present configuration of the activity and transportation systems in order to derive estimates of present demand; they can also be applied to hypothetical configurations representing possible future states of these systems (scenarios) in order to derive forecasts of future demand.

Aggregate data can also be used for direct demand estimation and for the specification and calibration of demand models. Flows measured on network links are the most sophisticated form of aggregate data and can complement other disaggregate data and the corresponding estimation methods.

The different types of survey and estimation methodologies are covered in this chapter as follows. Section 8.2 analyzes surveys and methods for direct demand estimation. Section 8.3 describes disaggregate estimation methods for the specification, calibration, and validation of demand models based on traditional revealed preference surveys. Section 8.4 describes some theoretical and operational aspects of stated preference survey and calibration techniques, based on the information elicited from a sample of individuals in hypothetical scenarios. Section 8.5 describes methods using traffic counts to estimate present demand, and Sect. 8.6 explores methods using traffic counts for aggregate calibration of demand models. Sections 8.7 and 8.8 extend some of the methods discussed in previous sections to deal with within-period dynamic estimation. Finally, Sect. 8.9 summarizes methodologies to estimate the different components of travel demand and discusses their fields of application. The topics listed are discussed for passenger travel demand; extensions to freight demand are relatively straightforward.

E. Cascetta, *Transportation Systems Analysis*,
Springer Optimization and Its Applications 29,
DOI 10.1007/978-0-387-75857-2_8,

8.2 Direct Estimation of Present Demand

Travel demand is the aggregation of individual trips made during a reference period (e.g., part or all of a typical day) by the users of a given system under study. Full knowledge of demand would therefore require information on the characteristics of the trips undertaken by all the users in the reference period. Furthermore, as noted in Chap. 1, such information should extend over several reference periods in order to compute average values. This complete census-like knowledge of travel demand is neither practicable nor necessary. The economic and organizational costs of the associated data collection effort would make the operation practically infeasible in most cases. For these reasons, present travel demand is typically estimated through sampling estimators, that is, estimators based on information about a sample of system users.

In Sect. 8.2.1, sampling surveys often used for direct demand estimation are described; estimators derived from sampling theory are covered in Sect. 8.2.2.

8.2.1 Sampling Surveys

The basic idea of sampling techniques is to estimate the population-level values of variables of interest from values observed in a relatively small group of individuals (a sample) belonging to the population.

Several types of sample surveys can be used for direct estimation of travel demand; these surveys, sometimes referred to generically as origin–destination surveys, may differ in their statistical characteristics and in the quality of information obtained. A comprehensive description of the various surveys is beyond the scope of this book; some typologies are briefly described below as examples.

With "*on-board*" surveys, a sample of users of one or more transportation modes is interviewed. The interviews can be conducted either at the roadside for car drivers and their passengers, or on board vehicles or at terminals (stations, airports, etc.) for scheduled transportation services. The sample of users is obtained by randomly interviewing a pre-determined fraction of the users of the chosen mode. In the case of surveys conducted at specific locations (road sections, stations, etc.), this requires counting the total number of travelers passing the point (count of the universe) and interviewing a given number of travelers selected through a random mechanism. When on-board surveys are conducted to estimate internal–external and external–external demand, they are also referred to as *cordon surveys*. In general, the information that can be gathered in these surveys is relatively simple because the interview has to be done in a short period of time during the trip, and it usually refers only to the trip or journey under way.

With household surveys, a sample of families or persons living within the study area is interviewed. For family-based surveys, the sample is extracted randomly from the set of all resident families (simple random sample) or from the set of families living in each traffic zone (stratified random sample). The family members in the sample are interviewed about the trips taken in a given reference period. The same approach can be used for individual- rather than for family-based surveys. The method of interviewing travelers in their homes usually is rather expensive but precise information is generally obtained because of the direct interaction between the interviewee and the interviewer. Household telephone surveys are becoming more and more popular; they have lower costs, although they usually yield less precise interview results.

There are several other types of sample surveys such as destination surveys, in which travelers are interviewed at trip destinations (workplaces, schools, shops, etc.); and e-mail surveys, in which travelers are interviewed by e-mail. These surveys, although less costly than household surveys, may produce biased results because of the systematic lack of information from some market segments.

The number of persons to be interviewed depends on the aims of the survey and the precision required for the estimates. Surveys aiming at direct estimation of present demand usually require larger samples than those needed to calibrate demand models.

In applications, different types of survey are employed to estimate different components of travel demand: for example, cordon surveys for internal–external and external–external trips, and household surveys for internal trips.

Whatever the type of survey, the statistical design of a sampling survey for demand estimation consists of several standard phases:

- Definition of the sampling unit (person, family, vehicle, etc.) and of the method for enumerating the population universe (e.g., lists of residents or counts of passing vehicles);
- Definition of the sampling strategy, that is, the method for extracting the sample of individuals to be interviewed;
- Definition of the estimator, that is, the function used to estimate the unknown quantities from the information obtained by the survey;
- Definition of the number of units in the sample (sample size).

The definition of the sampling unit is largely influenced by practical matters such as the type of survey (household, on-board, etc.) and the availability of information about the universe. For example, if a list of families living in a given area is available, but that of individuals is not, the logical sampling unit will be the family rather than the individual. In the case of on-board surveys, the sampling unit will be the vehicle if the survey is carried out at the roadside, or the passenger if the interviews are at the terminals.

For the choice of sampling strategy, almost all surveys utilize *probabilistic sampling*, that is, methods of sample extraction that define a priori the possible outcomes, assign a probability to each outcome, and randomly extract the elements of the sample according to this probability. In applications, the most commonly used probabilistic sampling strategies are as follows.

- *Simple random sampling*: all the elements of the population have an equal probability of belonging to the sample.

- *Stratified random sampling*: the population is divided into nonoverlapping, exhaustive groups (strata). A simple random sample of elements is drawn from each stratum, but the selection probabilities in different strata may be different.
- *Cluster sampling*: sampling units (e.g., people) are grouped in clusters (e.g., families or the passengers of a vehicle). Clusters are then extracted randomly with a pre-determined probability (simple random cluster sampling) or subdivided into strata and sampled with different probabilities (stratified random cluster sampling). A further possibility is two-stage cluster sampling, in which a sample of clusters (e.g., a sample of families) is first selected, and then a sample of individuals within each cluster is extracted. In this case, the probability that an individual belongs to the sample is the product of the probability of selecting the cluster to which he or she belongs and the probability that the individual is then extracted within the cluster.

The *choice of the estimator*, that is, the function of sample results, obviously depends on the variables to be estimated and on the sampling strategy adopted. In fact it can be demonstrated that an estimator that is statistically efficient for one strategy might not be for another.

The choice of the estimator and definition of the sample size involve a more substantial methodological content, and are discussed in the next subsection.

8.2.2 Sampling Estimators

Present travel demand can be estimated starting from the results of the sampling surveys described in the previous section. The problem of estimating origin–destination demand flows with certain characteristics (e.g., trip purpose and transport mode) and the main statistical properties of some sampling strategies is addressed below.

Simple random sampling. In this case, a sample of n elements is drawn at random from a universe of N users. For example, in a household survey the sample of n families is obtained from the universe of the N families living in the study area. Let d_{od}[1] be the demand flow between origin o and destination d with the characteristics concerned, and n^i_{od} be the number of these trips undertaken by the ith element (e.g., family) of the sample. Estimates of demand flows with given characteristics but not distinguished by origin–destination zone can be obtained in the same way. Let n_{od} be the total of trips obtained from the sample. Clearly,

$$n_{od} = \sum_{i=1,\ldots,n} n^i_{od} \tag{8.2.1}$$

[1]For simplicity of notation, demand flow characteristics such as user class, trip purpose (s), time of day (h), and mode (m) are generally taken as understood and so omitted.

The sample estimate $\hat{d}_{od}$ of the demand flow for the whole universe can be obtained as follows,

$$\hat{d}_{od} = (N/n)n_{od} = (1/\alpha)n_{od} = N\bar{n}_{od} \tag{8.2.2}$$

where $\alpha = n/N$ is the *sampling rate* and $\bar{n}_{od} = n_{od}/n$ is the average number of trips per element with the desired characteristics.

Sampling theory shows that the variance[2] of $\hat{d}_{od}$ can be estimated as:

$$\mathrm{Var}[\hat{d}_{od}] = N^2\hat{s}^2(1-\alpha)/n \tag{8.2.3}$$

where $\hat{s}^2$ is the sample estimate of the variance of the random variable n^i_{od}:

$$\hat{s}^2 = 1/(n-1) \sum_{i=1,\ldots,n} \left(n^i_{od} - \bar{n}_{od}\right)^2 \tag{8.2.4}$$

In some surveys, a sample element (e.g., a car driver for cordon surveys) undertakes at most one trip with the required characteristics (e.g., for a given purpose and/or in a given time period). In other surveys, the required information is whether the sample element has a given characteristic (e.g., made a trip with a given purpose or in a given time period). In such cases n^i_{od} is either zero or one, and $\bar{n}_{od}$ is the sample estimate of the percentage of travelers who have undertaken a trip with the given characteristics. This percentage is indicated below by $\hat{P}_{od}$.

$$\hat{P}_{od} = \sum_{i=1,\ldots,n} n^i_{od}/n \tag{8.2.5}$$

In this case the sampling estimate of the variance of n^i_{od} given by (8.2.4) can be expressed as the variance of a Bernoulli random variable:

$$\hat{s}^2 \cong \hat{P}_{od}(1-\hat{P}_{od}) \tag{8.2.6}$$

In fact, from (8.2.4), bearing in mind that in this case $n^{i2}_{od} \equiv n^i_{od}$, we have:

$$\hat{s}^2 = [1/(n-1)] \sum_{i=1,\ldots,n} \left(n^{i2}_{od} + \bar{n}^2_{od} - 2n^i_{od}\bar{n}_{od}\right) \cong \hat{P}_{od} + \hat{P}^2_{od} - 2\hat{P}^2_{od} = \hat{P}_{od}(1-\hat{P}_{od})$$

where the "almost equal" ($\cong$) results from assuming n equal to $(n-1)$. In the case under study, the estimate of the variance of $\hat{P}_{od}$ is given by

$$\mathrm{Var}[\hat{P}_{od}] = \hat{P}_{od}(1-\hat{P}_{od})(1-\alpha)/n$$

[2] The coefficient $(1-\alpha)$, known as the finite population correction coefficient, accounts for the fact that the population has a finite number of members; therefore, if a census (complete enumeration) were conducted, $\alpha = 1$ and the estimate would be the "true" value with zero variance. The weight of the correction coefficient, however, is negligible for the sampling rates used in practice for direct demand estimation.

Stratified random sampling. In this case, the total population is divided into K groups of users, or strata; stratum k has a population of N_k members and n_k elements are drawn at random from it. This type of sampling is the most widely used in practical demand surveys. In cordon surveys, the strata include users traveling through the different survey sections, whereas in household surveys the strata are often comprised of the families living in each zone (geographical stratification). In the first case, the sample is "structurally" stratified because the users can only be reached in this way; in the second, the stratification is a choice made to obtain a predetermined coverage of each zone.

If n_{od}^{ik} denotes the number of trips with the required characteristics undertaken by the ith element in the sample of stratum k, an estimate of the total number of trips can be obtained as follows.

$$\hat{d}_{od} = N \sum_{k=1,\dots,K} w_k \sum_{i=1,\dots,n_k} n_{od}^{ik}/n_k = N \sum_{k=1,\dots,K} w_k \bar{n}_{od}^{k} \qquad (8.2.7)$$

where n_{od}^{k} is the average number of trips observed in the kth stratum, and $w_k = N_k/N$ is the weight of the stratum k with respect to the universe.

The variance of $\hat{d}_{od}$, the stratified sampling estimate, can be estimated as follows,

$$\mathrm{Var}[\hat{d}_{od}] \approx N^2 \sum_{k=1,\dots,K} w_k^2 \hat{s}_k^2 (1-\alpha_k)/n_k \qquad (8.2.8)$$

where $\hat{s}_k^2$ is the sample estimate of the variance of n_{od}^{ik}:

$$\hat{s}_k^2 = 1/(n-1) \sum_{i=1,\dots,n_k} \left(n_{od}^{ik} - \bar{n}_{od}^{k}\right)^2$$

and α_k is the sampling rate in the kth stratum.

It can be shown that the sampling estimators (8.2.2), (8.2.5), and (8.2.7) are unbiased and consistent estimators of the unknown demand if the interviews do not contain systematic distortions of the information provided (e.g., underreporting of trips). The same can be said of the variance estimators (8.2.3) and (8.2.8). Variance estimates can be used to calculate the confidence limits of $\hat{d}_{od}$. If the sample is large enough to apply the central limit theorem, it can be assumed that the estimator $\hat{d}_{od}$ follows a normal distribution. The upper and lower confidence limits of the estimate, $L_{1-\gamma}^{S}(d_{od})$ and $L_{1-\gamma}^{I}(d_{od})$, define an interval which, with probability $(1-\gamma)$, includes the true value of d_{od}. On the assumption of a sufficiently large sample, these limits can be obtained as

$$L_{1-\gamma}^{S}(d_{od}) = \hat{d}_{od} + z_{1-\gamma/2}\,\mathrm{Var}[\hat{d}_{od}]^{1/2}$$

and

$$L_{1-\gamma}^{I}(d_{od}) = \hat{d}_{od} + z_{\gamma/2}\,\mathrm{Var}[\hat{d}_{od}]^{1/2}$$

where $z_{1-\gamma/2}$ and $z_{\gamma/2}$ are the $1-\gamma/2$ and $\gamma/2$ percentiles of the normal standard variable. For $\gamma = 0.05$, these percentiles are 1.96 and -1.96 and the confidence limits are the extremes of the interval which with a probability of 0.95 contains the true value.

The ratio $IR(1-\gamma)$ between the width of the confidence interval and the value to be estimated is called the *relative confidence interval* at $(1-\gamma)$ percent of the estimate $\hat{d}_{od}$:

$$IR(1-\gamma) = \left[L^S_{1-y}(d_{od}) - L^I_{1-y}(d_{od})\right]/d_{od} \tag{8.2.9}$$

Expressions for the estimators and their variances for sampling strategies other than simple and stratified random sampling are more complex. However, the latter can still be used as first approximations. For exact expressions of the estimators and their variances in more complex sampling schemes, specialized texts in sampling theory should be consulted.

In principle, the sample size could be calculated according to the level of precision required by using expression (8.2.9) and substituting tentative values obtained from other studies for the variances $\hat{s}^2$ and $\hat{s}^2_k$ and for the variable d_{od}. For example, in the case of simple random sampling, if an $IR(1-\gamma)$ relative confidence interval of the estimate $\bar{n}_{od}$ is required at a given confidence level and the coefficient of variation ($CV = s/\bar{n}_{od}$) of the variable n^i_{od} is known, the sample size n can be obtained by combining expressions (8.2.2), (8.2.3), and (8.2.9) as follows.

$$n \approx 4\frac{CV^2 z^2_{1-\gamma/2}(1-\alpha)}{IR(1-\gamma)^2} \tag{8.2.10}$$

A similar expression can be obtained for a given relative confidence interval of the O-D demand flows $\hat{d}_{od}$.

In practice, the computation of the sample size theoretically required by a survey is rarely possible because several parameters for the computation are obtained from the survey itself. Furthermore, the sample size required for sufficiently precise estimates of some parameters, and especially of the single elements of an O-D matrix, would be too large to be feasible. The usual practice is to choose a sample size used with other "successful" surveys, verifying that some aggregate estimates (e.g., the level of demand or the number of trips in each zone for each purpose) have a satisfactory minimum precision.

As an example, Fig. 8.1 shows the sampling rate for urban household origin–destination surveys recommended by the U.S. Bureau of Public Roads as a function of the resident population.

Finally, it should be noted that the use of models as estimators of present demand is becoming increasingly widespread (see Sect. 8.9). This is due not only to the generally low level of precision that results from direct estimation but also to the effectiveness of specification and calibration techniques of demand models.

Fig. 8.1 Sampling rates for household surveys in relation to resident population (BPR-USA)

Resident population		Sampling rate	
		Recommended	Minimum
		0.200	0.100
50,000	150,000	0.125	0.050
150,000	300,000	0.100	0.030
300,000	500,000	0.067	0.020
500,000	1,000,000	0.050	0.015
More than 1,000,000		0.040	0.010

8.3 Disaggregate Estimation of Demand Models

Estimation of travel demand by means of mathematical models, whether they are applied to the present situation or to hypothetical scenarios, requires the specification, calibration, and validation of such models. In other words, it is necessary to define the functional form and the variables included in the model, to estimate the model's coefficients or parameters, and to verify its statistical quality. A good demand model is usually the outcome of a trial and error process in which the specification–calibration–validation cycle is repeated several times until a satisfactory result is obtained. In this process the modeler's judgment and experience play a central role.

These operations, which together are called *model estimation*, can be performed starting from information on the travel behavior of a sample of users. This approach is called disaggregate estimation[3] of demand models. In general, surveys used to gather basic information are of two different classes: surveys of actual travel behavior in a real context (*Revealed Preference* or *RP* surveys) or surveys of hypothetical travel behavior in fictitious scenarios (*Stated Preference* or *SP* surveys). The traditional method of revealed preference is based on surveys analogous to those described in Sect. 8.2.1. They provide information on users' actual choices in situations relevant for the model to be calibrated (e.g., on the transportation mode chosen for the calibration of a mode choice model). Survey design therefore consists of the definition of the sample size, the questionnaire, and the sampling strategy. Stated preference (SP) surveys, on the other hand, are conceptually equivalent to a laboratory experiment designed with a larger number of degrees of freedom. Given the complexity of the subject, SP survey designs and their use for the calibration of demand models are covered separately in Sect. 8.4. The remainder of the present section considers the specification, calibration, and validation of demand models from RP survey results.

Irrespective of their interpretation (behavioral or descriptive) and functional form, demand models can be viewed as mathematical relationships giving the probability that a user i chooses a particular travel option from among those available.

[3] A different method, which can be called aggregate calibration, uses aggregate and indirect information on users' travel behavior, usually traffic counts, to specify and calibrate demand models. There are also mixed methods, which use both disaggregate and aggregate information. Aggregate and mixed estimators of demand model parameters are covered in Sect. 8.6.

Thus a mode choice model $p[m/ods]$ expresses the probability that a user, randomly selected from those who undertake a trip for purpose s between zones o and d, uses mode m. This section addresses the problem of building demand models or model systems, referring generically to a generic choice model that expresses the probability $p^i[j]$ that a user i chooses travel option j from among those available.

Section 8.3.1 discusses some general considerations relevant to model specification. Section 8.3.2 covers calibration methods, and finally Sect. 8.3.3 describes some validation methods.

8.3.1 Model Specification

The specification of a demand model can be defined as the complete identification of its mathematical structure, that is, the definition of its functional form and of the dependent and independent (explanatory) variables that it incorporates.

The choice of a model's functional form (e.g., multinomial logit or hierarchical logit) depends on many factors such as its computational tractability, the results obtained in similar cases, or the a priori expectations regarding the correlation of random residuals. In general, the assumptions can be tested a posteriori using the statistical tests described in Sect. 8.3.3.

The choice of the explanatory variables clearly depends on the specific type of model. However, there are some rules that should be observed to avoid problems in the calibration phase. In general, variables that are *collinear* (i.e., linearly dependent on each other) should be avoided. Indeed, if the systematic utility function is linear with respect to collinear attributes, infinitely many combinations of their coefficient values give equal values of systematic utilities and choice probabilities. This makes it impossible to estimate (identify) separately those coefficients during model calibration. A typical example of collinearity might be introduced when one attribute is derived from another. This would happen, for example, if travel time were derived from distance by assuming a constant speed; travel time and distance should not then be included in the model specification as two distinct variables.

It must be also taken into account that, for invariant random utility models, as described in Chap. 3, the choice probabilities depend on the differences between the perceived utilities of the alternatives and not on the absolute level of same perceived utilities. It is thus possible to add a constant to the perceived utility of each alternative, without changing the corresponding choice probabilities. This property gives rise to a correct specification rule according to which in an invariant random utility model with m alternatives at most $m-1$ independent ASA may be introduced. Indeed, if β_j^{ASA} is the coefficient of ASA relative to alternative j and V_j' the remaining part of systematic utility, from (3.2.4) we may write:

$$p[j] = Pr\left[V_j' + \beta_j^{\mathrm{ASA}} + \varepsilon_j \geq V_i' + \beta_i^{\mathrm{ASA}} + \varepsilon_i \ \forall i \in 1 \ldots m\right]$$

Consequently, by adding to the systematic utilities of all alternatives a constant equal to β_k^{ASA} of alternative k chosen arbitrarily, the previous formula may take the following form,

$$p[j] = Pr\left[V'_j + \left(\beta_j^{\text{ASA}} - \beta_k^{\text{ASA}}\right) + \varepsilon_j \geq V'_i + \left(\beta_i^{\text{ASA}} - \beta_k^{\text{ASA}}\right) + \varepsilon_i \ \forall i \in 1 \ldots m\right]$$

and this shows that the calculated choice probabilities, introducing the ASA into all the alternatives, are equal to those that would be obtained by introducing them into all the alternatives except one, chosen arbitrarily.

With similar steps it may be demonstrated that the user's socioeconomic attributes, such as income or car ownership, may be introduced at most in the utility function of all the alternatives except for one and not through specific variables of the alternative. For example, two "high income" variables cannot be introduced into systematic utilities of car and taxi alternatives with different coefficients.

As mentioned in Sect. 3.2, in the model specification phase it is necessary to define the expression of systematic utilities of alternatives as a function of explicative variables (attributes) identified in accordance with the rules described up to here. Usually one resorts to a functional dependence that is linear in the coefficients and may be linear (3.2.5a) or not (3.2.5b) in the attributes. At times it is worth introducing a nonlinearity also in respect of the coefficients; an interesting functional parametric transformation for nonnegative variables is that of Box–Cox:

$$x_k \rightarrow \left(x_k^{\lambda_k} - 1\right)/\lambda_k \quad \text{if } \lambda_k \neq 0$$

$$x_k \rightarrow \log(x_k) \quad \text{if } \lambda_k = 0$$

where λ_k is the parameter of transformation. It defines a family of functions that includes, as special cases, linear ($\lambda_k = 1$), exponential ($\lambda_k \neq 0$) and logarithmic transformation ($\lambda_k = 0$). It must also be stressed that the Box–Cox transformation introduces some difficulties into the estimation process due to the nonlinearity in λ_k of the utility functions. These difficulties may be avoided by estimating the model iteratively for different values of λ_k and choosing the model and λ_k values that supply the best results.

8.3.2 Model Calibration

Random utility models can be viewed as mathematical relationships expressing the probability $p^i[j](\boldsymbol{X}, \boldsymbol{\beta}, \boldsymbol{\theta})$ that individual i chooses alternative j as a function of the vector $\boldsymbol{X}$ of attributes for all the available alternatives; of the systematic utility parameter vector $\boldsymbol{\beta}$; and of the joint probability function of the random residuals $\boldsymbol{\theta}$. Choice probabilities depend on $\boldsymbol{X}$ and $\boldsymbol{\beta}$ through systematic utility functions, which are usually specified as linear combinations of the attributes $\boldsymbol{X}$ (or their transformations) with coefficients given by the parameters $\boldsymbol{\beta}$:

$$V_j\left(\boldsymbol{X}^i_j\right) = \sum_z \beta_z X^i_{zj} = \beta^T \boldsymbol{X}^i_j \tag{8.3.1}$$

Structural parameters θ include all parameters related to the random residual probability distribution function. Thus, in the case of the multinomial logit models, $\boldsymbol{\theta}$ is the scale parameter of the Gumbel random variables. In the hierarchical logit, $\boldsymbol{\theta}$ consists of scale parameters $\boldsymbol{\theta}_o$ and $\boldsymbol{\theta}_r$ associated with structural nodes. In the probit model, $\boldsymbol{\theta}$ consists of all the elements of the variance–covariance matrix, and so on.

Calibrating the model requires estimating the vectors $\boldsymbol{\beta}$ and $\boldsymbol{\theta}$ from data on the choices made by a sample of users. It should be observed that in general not all the coefficients can be identified, that is, estimated separately. We return to this point in greater detail later with reference to specific examples.

The Maximum Likelihood method. Maximum Likelihood (*ML*) is the method most widely used for estimating model parameters. In maximum likelihood estimation, the probability of observing the choices made by a sample of users (the sample likelihood) is expressed as a function of the unknown model parameters, and the parameter estimates are those which maximize that probability. The sample likelihood depends not only on the model and its parameters, but also on the sampling strategy adopted. The cases of simple and stratified random sampling are considered in the following.

In the case of *simple random sampling* of n users, the observations are statistically independent and the probability of obtaining the observed choices is the product of the probabilities that each user i chooses $j(i)$, that is, the alternative actually chosen by him or her. The probabilities $p^i[j(i)](\boldsymbol{X}^i;\boldsymbol{\beta},\boldsymbol{\theta})$ are computed by the random utility model and therefore depend on the coefficient vectors. Thus, the probability L of observing the whole sample is a function (the likelihood function) of the unknown parameters:

$$L(\boldsymbol{\beta},\boldsymbol{\theta}) = \prod_{i=1,\ldots,n} p^i\left[j(i)\right](\boldsymbol{X}^i;\boldsymbol{\beta},\boldsymbol{\theta}) \tag{8.3.2}$$

The maximum likelihood estimate $[\boldsymbol{\beta},\boldsymbol{\theta}]_{\mathrm{ML}}$ of the vectors of parameters $\boldsymbol{\beta}$ and $\boldsymbol{\theta}$ is obtained by maximizing (8.3.2) or, more conveniently, its natural logarithm (the log-likelihood function):

$$[\beta,\theta]_{\mathrm{ML}} = \arg\max \ln L(\boldsymbol{\beta},\boldsymbol{\theta}) = \arg\max \sum_{i=1,\ldots,n} \ln p^i\left[j(i)\right](\boldsymbol{X}^i,\boldsymbol{\beta},\boldsymbol{\theta}) \tag{8.3.3}$$

Figure 8.2 shows an elementary example of the maximum likelihood estimation of a single parameter.

In the calibration of some models, the n users may naturally be grouped in sets of n_i users, with all users in a set choosing the same alternative and having the same attributes. A typical example is an aggregate distribution model in which the users traveling between the same O-D pair possess the same attributes, namely the trip costs between zone pairs and the attraction variables of each destination. In this case, the likelihood function and its logarithm can be expressed as

$$L(\boldsymbol{\beta},\boldsymbol{\theta}) = \prod_i p^i[j(i)]^{n_i}(\boldsymbol{X}^i,\boldsymbol{\beta};\boldsymbol{\theta})$$

$$\ln L(\boldsymbol{\beta},\boldsymbol{\theta}) = \sum_i n_i \ln p^i[j(i)](\boldsymbol{X}^i;\boldsymbol{\beta},\boldsymbol{\theta})$$

In *stratified random sampling*, n_h users are sampled randomly from the N_h members of each stratum ($h = 1,\dots,H$) with a sampling rate $\alpha_h = n_h/N_h$. The probability of observing the sample choices and therefore the likelihood function, depends on the method used to identify the strata.

If the population is stratified using, either directly or indirectly, the attributes $\boldsymbol{X}$ but not the choices to be modeled, the strategy is known as *exogenous stratified sampling*. Typical examples are geographical stratification (the level-of-service attributes depend on the zone or zone pair on which the stratification is carried out) and/or income stratification.

For samples obtained through exogenous stratified sampling, it can be demonstrated that the log-likelihood function is:

$$\ln L(\boldsymbol{\beta},\boldsymbol{\theta}) = \sum_{h=1,\dots,H}\ \sum_{i=1,\dots,n_h} \ln p^i[j(i)](\boldsymbol{X}^i;\boldsymbol{\beta},\boldsymbol{\theta}) + \text{const.} \tag{8.3.4}$$

which, apart from a constant term, coincides with the function (8.3.3) obtained for a simple random sample with size $n = \sum_{h=1,\dots,h} n_h$.

If the stratification is based on the choices made by the users, the sampling strategy is known as *choice-based stratified sampling*. This is the case, for example, if the sample used to calibrate a mode choice model is obtained by randomly selecting a sample of users of each transport mode; the population of each stratum is comprised of all users choosing each mode. The exact closed form log-likelihood function is rather complex for this sampling strategy. As an approximation, the maximum likelihood estimator with *exogenous weights* can be adopted; in this case the function $\ln L(\boldsymbol{\beta},\boldsymbol{\theta})$ is expressed as

$$\ln L(\boldsymbol{\beta},\boldsymbol{\theta}) = \sum_{h=1,\dots,H} \left(\frac{w_h}{\alpha_h}\right) \sum_{i=1,\dots,n_h} \ln p^i[j(i)](\boldsymbol{X}^i_j;\boldsymbol{\beta},\boldsymbol{\theta}) \tag{8.3.5}$$

which, apart from the weights w_h and α_h, coincides with (8.3.4) and therefore with (8.3.3). To apply the maximum likelihood estimator with exogenous weights to a choice-based stratified sample, it is therefore necessary to have an estimate of the weight of each stratum, that is, of the fraction of the total population choosing each alternative. This information can be obtained from official statistics, or estimated from another simple random sample with smaller or less detailed questionnaires.

Under rather general assumptions, maximum likelihood estimators have many desirable asymptotic statistical properties such as consistency, efficiency, and normality, regardless of the model used to express the probabilities $p^i[j]$. Furthermore,

$$n = 3 \quad j = A, B \quad p[A] = \frac{\exp(-\beta C_A)}{\exp(-\beta C_A) + \exp(-\beta C_B)}$$

User	$j(i)$	C_A^i	C_B^i
1	A	3	5
2	A	2	1
3	B	4	3

$$L(\beta) = \frac{\exp(-3 \cdot \beta)}{\exp(-3 \cdot \beta) + \exp(-5 \cdot \beta)} \cdot \frac{\exp(-2 \cdot \beta)}{\exp(-2 \cdot \beta) + \exp(-1 \cdot \beta)} \cdot \frac{\exp(-3 \cdot \beta)}{\exp(-3 \cdot \beta) + \exp(-4 \cdot \beta)}$$

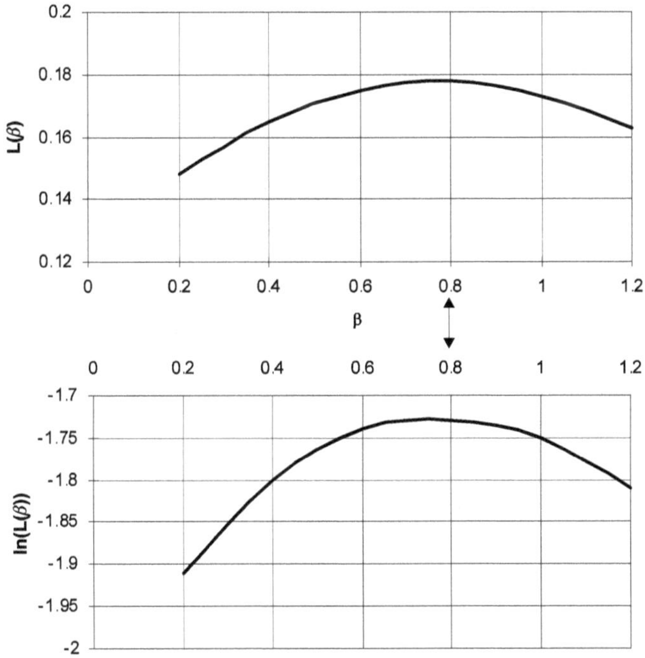

β	$p^1[A]$	$p^2[A]$	$p^3[B]$	$L(\beta)$	$\ln(L(\beta))$
0.20	0.60	0.45	0.55	0.148	−1.91
0.40	0.69	0.40	0.60	0.165	−1.80
0.60	0.77	0.35	0.65	0.175	−1.74
0.80	**0.83**	**0.31**	**0.69**	**0.178**	**−1.73**
1.00	0.88	0.27	0.73	0.173	−1.75
1.20	0.92	0.23	0.77	0.163	−1.81

Fig. 8.2 Maximum likelihood estimation of a single parameter

it is possible to obtain approximate estimates of the variances and covariances of the components of $\boldsymbol{\beta}_{\text{ML}}$, because its variance–covariance matrix $\boldsymbol{\Sigma}$ is asymptotically equal to the negative inverse of the log-likelihood function's Hessian, evaluated at the point $(\beta, \theta)_{\text{ML}}$:

$$\sum_{\beta,\theta} = -\left[\frac{\partial^2 \ln L(\beta,\theta)}{\partial(\beta,\theta)\partial(\beta,\theta)^T}\right]^{-1}_{(\beta,\theta)_{\text{ML}}} \tag{8.3.6}$$

If the sample is sufficiently large, expression (8.3.6) can be used to estimate variances and confidence limits for the coefficients.

From the algorithmic point of view, maximum likelihood estimation requires the solution of an unconstrained maximization problem, like (8.3.3). This problem can be solved by applying a gradient algorithm of the type described in Appendix A. The gradient of the objective function can be calculated either analytically or numerically, depending on the functional form of the model $p^i[j(i)]$ to be calibrated.

Maximum Likelihood Estimators for some random utility models. The explicit formulation of the functions $\ln L(\boldsymbol{\beta}, \boldsymbol{\theta})$ in expressions (8.3.3) through (8.3.5), the possibility of estimating the coefficients, as well as the properties of the unconstrained optimization problem all depend on the type of model used. Estimators for multinomial and hierarchical logit models can be derived analytically and are described below.

If the probabilities $p^i[j](\boldsymbol{X}^i; \boldsymbol{\beta}, \boldsymbol{\theta})$ are obtained with a multinomial logit model having a systematic utility linear in the coefficients β_k, the objective function (8.3.3) can be expressed analytically:

$$\ln L(\boldsymbol{\beta}, \boldsymbol{\theta}) = \sum_{i=1,\ldots,n}\left[\sum_{k=1,\ldots,K} \beta_k X^i_{kj(i)}/\theta - \ln\sum_{j\in I_i}\exp\left(\sum_{k=1,\ldots,K}\beta_k X^i_{kj}/\theta\right)\right]$$

or in vector form:

$$\ln L(\boldsymbol{\beta}, \boldsymbol{\theta}) = \sum_{i=1,\ldots,n}\left[\boldsymbol{\beta}^T \boldsymbol{X}^i_{j(i)}/\theta - \ln\sum_{j\in I_i}\exp(\boldsymbol{\beta}^T \boldsymbol{X}^i_j/\theta)\right] \tag{8.3.7}$$

In this case, the parameters to be estimated are the N_β coefficients β_k, plus a single scale parameter θ. As previously noted, not all parameters can be estimated separately because the values of the log-likelihood function (8.3.7) do not depend on the $N_\beta + 1$ single parameters but rather on the N_β ratios β_k/θ. It can be immediately verified, in fact, that the two vectors $[\beta_1, \beta_2, \ldots, \theta]$ and $[\alpha\beta_1, \alpha\beta_2, \ldots, \alpha\theta]$ give the same value of the function (8.3.7). Thus, it would be impossible to estimate $\boldsymbol{\beta}_k$ and θ separately, because there are infinitely many combinations of them giving the same choice probabilities and therefore the same log-likelihood function value. If the ratio

β_k/θ is denoted by β'_k, the vector $\boldsymbol{\beta}'$ is:

$$\boldsymbol{\beta}' = \boldsymbol{\beta}/\theta = [\beta_1/\theta, \beta_2/\theta, \ldots]$$

and expression (8.3.7) becomes[4]:

$$\ln L(\boldsymbol{\beta}') = \sum_{i=1\ldots n} \left[\boldsymbol{\beta}'^T \boldsymbol{X}^i_{j(i)} - \ln \sum_{j \in I_i} \exp(\boldsymbol{\beta}'^T \boldsymbol{X}^i_j) \right] \tag{8.3.8}$$

The first-order partial derivatives of (8.3.8) with respect to the generic parameter β'_k define the gradient of the objective function; they can be expressed in closed form:

$$\frac{\partial \ln L(\boldsymbol{\beta}')}{\partial \beta'_k} = \sum_{i=1\ldots n} \left[X^i_{kj(i)} - \sum_{j \in I_i} X^i_{kj} \frac{\exp(\boldsymbol{\beta}'^T \boldsymbol{X}^i_j)}{\sum_{h \in I_i} \exp(\boldsymbol{\beta}'^T \boldsymbol{X}^i_h)} \right]$$

or in more compact notation:

$$\frac{\partial \ln L(\boldsymbol{\beta}')}{\partial \beta'_k} = \sum_{i=1\ldots n} \left[X^i_{kj(i)} - \sum_{j \in I_i} X^i_{kj} p^i[j](\boldsymbol{X}^i, \boldsymbol{\beta}') \right] \tag{8.3.9}$$

The second-order partial derivatives of $\ln L(\boldsymbol{\beta}')$ can also be expressed in closed form:

$$\frac{\partial^2 \ln L(\boldsymbol{\beta}')}{\partial \beta'_k \partial \beta'_l} = - \sum_{i=1\ldots n} \sum_{j \in I_i} p^i[j](\boldsymbol{\beta}') \cdot \left(X^i_{jk} - \sum_{h \in I_i} X^i_{hk} p^i[h] \right) \cdot \left(X^i_{jl} - \sum_{h \in I_i} X^i_{hl} p^i[h] \right) \tag{8.3.10}$$

These derivatives are used in some algorithms to solve the optimization problem (8.3.3), and can also be used to obtain a sample estimate of the variance–covariance matrix $\boldsymbol{\Sigma}_{\mathrm{ML}}$ of the estimator $\boldsymbol{\beta}'_{\mathrm{ML}}$ given by (8.3.6).

Under rather general assumptions, it can be shown that the Hessian matrix of the objective function (8.3.8), whose components are given by the second derivatives (8.3.10), is definite negative; this implies that the function $\ln L(\boldsymbol{\beta}')$ is strictly concave. There is therefore a unique vector $\boldsymbol{\beta}'_{\mathrm{ML}}$ that maximizes the function $\ln L(\boldsymbol{\beta}')$, and the algorithms described in Appendix A converge to this value.

---

[4]Note that the terms of the summations in (8.3.7) and (8.3.8) are the difference between the systematic utility for the chosen alternative, $V_{j(i)}(\boldsymbol{X}, \boldsymbol{\beta})$, and the satisfaction associated with all the available alternatives:

$$\ln L(\boldsymbol{\beta}') = \sum_{i=1\ldots n} \left[V_{j(i)}(\boldsymbol{X}^i, \boldsymbol{\beta}) - s(\boldsymbol{X}, \boldsymbol{\beta}) \right]$$

and this difference, as can be seen from Sect. 3.4, is always less than zero.

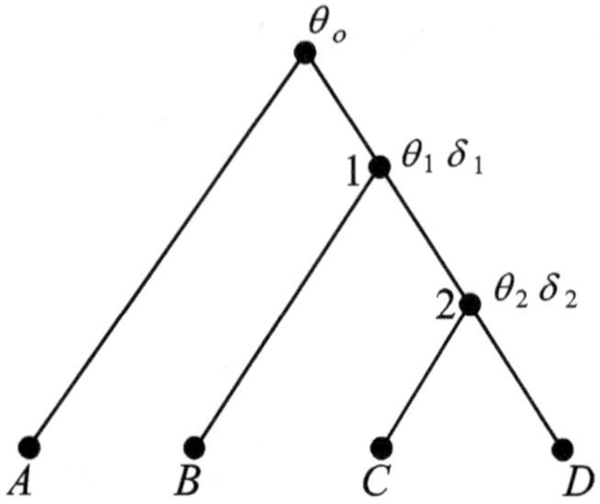

Fig. 8.3 Choice tree structure for a nested hierarchical model

These results can be extended to the case of functions $\ln L(\boldsymbol{\beta}')$ given by (8.3.4) and (8.3.5) for stratified samples.

In the case of hierarchical logit models, the choice probabilities depend on the structure of the choice tree. For the sake of simplicity, the discussion refers to the example in Fig. 8.3, in which the structural nodes have the indicated parameters θ and δ. The results can be extended to any choice tree structure.

In this case we have:

$$p[A] = \frac{e^{V_A/\theta_o}}{e^{V_A/\theta_o} + e^{\theta_1 Y_1/\theta_o}} = \frac{e^{V_A/\theta_o}}{e^{V_A/\theta_o} + [e^{V_B/\theta_1} + (e^{V_C/\theta_2} + e^{V_D/\theta_2})^{\theta_2/\theta_1}]^{\theta_1/\theta_o}}$$

Substituting the expressions for the systematic utilities, we have:

$$p[A] = \frac{e^{\sum_k \beta_k X_{kA}/\theta_o}}{e^{\sum_k \beta_k X_{kA}/\theta_o} + [e^{\sum \beta_k X_{kB}/\theta_1} + (e^{\sum_k \beta_k X_{kC}/\theta_2} + e^{\sum_k \beta_k X_{kD}/\theta_2})^{\theta_2/\theta_1}]^{\theta_1/\theta_o}} \tag{8.3.11}$$

The choice probabilities and the log-likelihood function depend not only on the N_β coefficients β_k, but also on the N_θ parameters θ_r, one for each intermediate node plus one (θ_o) for the root. It can also be observed that the structural coefficients always appear in (8.3.11) as ratios. Each coefficient β_{kj} in the systematic utility of an alternative j is divided by the parameter $\theta_{a(j)}$ of its parent node, whereas each parameter θ_r of an intermediate node r is divided by the parameter $\theta_{a(r)}$ of its parent node, which may be an intermediate node or the root.

In hierarchical logit models, the $N_\beta + N_\theta - 1$ ratios rather than the individual $N_\beta + N_\theta$ parameters are estimated. In fact, it can be verified immediately that a vector $[\beta_1, \beta_2, \ldots, \beta_{N_\beta}, \theta_1, \theta_2, \ldots, \theta_{N_\theta}]$, and a vector $[\alpha\beta_1, \alpha\beta_2, \ldots, \alpha\beta_{N_\beta}, \alpha\theta_1, \alpha\theta_2, \ldots, \alpha\theta_{N_\theta}]$ substituted in expression (8.3.11) give the same value of $p[A]$. All the parameters can therefore be identified but one. The parameters usually identified are the ratios $\beta^i_{kj} = \beta_{kj}/\theta_{a(j)}$ and $\delta_r = \theta_r/\theta_{a(r)}$.[5] From the previous expressions it

[5]Note that the estimates of a given coefficient β_k obtained with different specifications of the random utility model (multinomial logit, hierarchical logit, probit) are usually different because they contain different scale coefficients.

can also be deduced that the coefficients β_k of a generic attribute appearing in the utilities of alternatives belonging to different nests, for example β_{kA} and β_{kC}, must satisfy a consistency relationship:

$$\beta_{kA} = \beta_{kC} \quad \Rightarrow \quad \beta'_{kA} = \beta'_{kC}\delta_1\delta_2$$

From these considerations, if the vector of the ratios $\beta_{kj}/\theta_{a(j)}$ is denoted by $\boldsymbol{\beta}'$ and the vector of the ratios $\theta_r/\theta_{a(r)}$ by $\boldsymbol{\delta}$, the log-likelihood function becomes $\ln L(\boldsymbol{\beta}', \boldsymbol{\delta})$. It can be shown that, for a given δ, this function is concave with respect to the vector $\boldsymbol{\beta}'$, but that it is not concave with respect to the vector $\boldsymbol{\delta}$. Figure 8.4 shows the graph of the objective function $\ln L(\beta', \delta)$ for a simple hierarchical logit model as a function of a single parameter δ, where the vector $\boldsymbol{\beta}'$ is equal to the (unique) value that maximizes the log-likelihood function for the value of δ in the abscissa. The figure shows the nonconcavity of the function and two local maxima. For this reason, the problem (8.3.3) is sometimes solved using heuristic algorithms that maximize the log-likelihood function with respect to the vector $\boldsymbol{\beta}'$ for a set of fixed values of $\boldsymbol{\delta}$, and subsequently search for the overall maximum within the limited set of trial vectors δ (e.g., grid search). Other algorithms solve the problem (8.3.3) directly, with an appropriate definition of the ascent direction.

Another possibility for the calibration of hierarchical logit models is the sequential estimation of the parameters of multinomial logit models corresponding to each node of the choice tree associated with the decision process. The calibration process is started from the intermediate nodes that include only elemental alternatives. Parameters calibrated at one stage are kept fixed in the following stages. This type of calibration is known as *limited information estimation*, because the only information incorporated in each estimation concerns users who have chosen elemental alternatives (leaves) of the tree and/or compound alternatives (structural nodes) connected to the intermediate node under study. There are, however, both theoretical and practical problems connected with limited information maximum likelihood estimation. The method is theoretically suboptimal because it can produce an objective function value that is lower than the global maximum. Furthermore, the objective function values are sometimes even lower than those obtained from calibrating a multinomial logit model with equal systematic utilities, which is clearly a contradiction because the latter is a special case of the hierarchical logit model with all δs equal to one. From the practical point of view, it is very difficult in this method to estimate the coefficient of a generic attribute if the attribute is included (with the same coefficient value) in the systematic utilities of alternatives belonging to different groups. Each group is calibrated separately and there is no convenient way to impose equality constraints between coefficient values common to multiple groups. For these reasons the sequential estimation method is not to be recommended.

The same considerations reported for the hierarchical logit model can be easily extended to the cross-nested logit model, which requires estimation of the degrees of membership in addition to the variance parameters (see Sect. 3.3.4). It should be noted that the mathematical properties of the maximum likelihood for the cross-nested logit model, as well as the corresponding identification issues, are still under analysis.

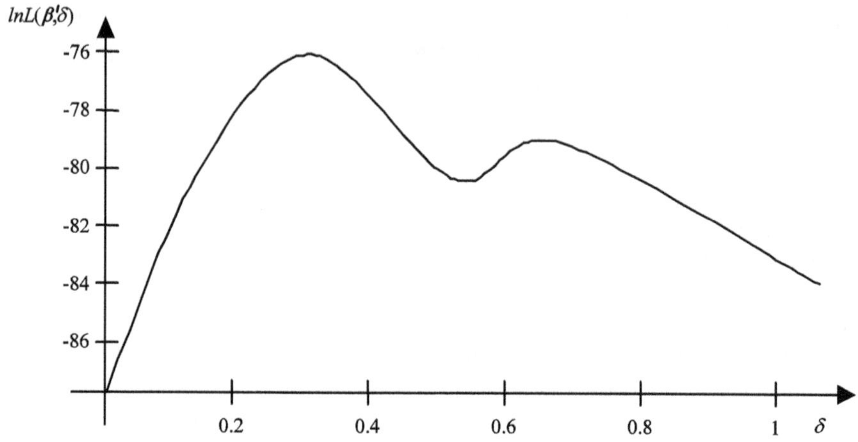

Fig. 8.4 Log-likelihood for a hierarchical logit model as a function of the parameter δ

Estimation of nonclosed-form random utility models (e.g., probit, mixed logit) can be theoretically carried out by introducing a *Maximum Simulated Likelihood* (MSL) estimation, that is, introducing in expressions (8.3.3) through (8.3.5) an estimate $\hat{p}^i[j(i)]$ of the choice probability for the alternative $j(i)$ provided through one of the methods described in Chap. 3. As a consequence, statistical properties of MSL estimators depend also on the procedure adopted for simulating choice probabilities. From an operational standpoint, the maximum simulated likelihood estimation faces some problems: the reader can refer to the specific literature for a deeper analysis.

Concerning the identification of the probit model parameters, as mentioned in Sect. 3.3.6, the covariance matrix related to an m-alternatives choice set is characterized at most by $m(m+1)/2$ distinct elements. In fact, under the hypothesis of the invariant utility model, choice behavior depends on the joint distribution function of the differences of perceived utilities with respect to an alternative chosen as reference. That is, the corresponding covariance matrix has order $m-1$ and therefore exhibits $m(m-1)/2$ distinct values. Taking into account the further degree of freedom provided by the choice of the scale factor, the covariance matrix of an invariant m-alternatives probit model is uniquely defined by $m(m-1)/2-1$ distinct values.

8.3.3 Model Validation

Once a demand model has been specified and calibrated, it must be validated. In this phase the reasonableness and the significance of estimated coefficients are verified, as well as the model's ability to reproduce the choices made by a sample of users. In addition, the assumptions underlying the functional form assumed by the model are tested. All of these activities can be completed with appropriate tests of hypotheses for a sample of users.

Informal tests on coefficients. These tests are based on expectations regarding the signs of the calibrated coefficients and the relationships between their values.

Wrong signs of the coefficients are likely indicators of attribute errors in the survey results database, or of model misspecification. For example, in a road path choice model, it may happen that paths including toll motorway sections are chosen, even though they have approximately the same average travel time and are more expensive than untolled paths. If the model specification does not somehow account for the greater driving comfort on a motorway, the calibration procedure may result in a positive cost coefficient in an attempt to increase the systematic utility, and therefore the choice probability, of motorway alternatives. A different model specification, for example, introducing an attribute equal to the length of the motorway section on each path, should adjust the cost coefficient to the expected negative value.

Other checks can be performed on the ratios of the coefficients of different attributes. As stated in Chap. 4 (see (4.3.15)), the ratio between time and monetary cost coefficients can be interpreted as a Value of Time (VOT), and can be compared with the results of other calibrations and with expectations about users' willingness to pay. The parameters of attributes corresponding to different components of travel time (e.g., waiting and on-board time) should have increasing absolute values for more onerous components. In general, the results reported in the scientific and technical literature are very helpful in these analyses.

As an example, consider the mode choice model described in Fig. 8.5; the time and cost coefficients are negative, and the availability coefficients (car, motorcycle, and bicycle) are positive. Furthermore, the perceived value of time is about 5 €/h. It can also be seen that the disutility associated with time on foot is about five times that of time on board and so on.

Formal tests on coefficients. For sufficiently large samples, the asymptotic properties of maximum likelihood estimates can be exploited to test different assumptions on $\boldsymbol{\beta}^{\mathrm{ML}}$.[6]

Student t-Tests on Particular Coefficients

These tests check the null hypothesis that the true value of a coefficient β_k is equal to zero and its estimate $\boldsymbol{\beta}_k^{\mathrm{ML}}$ differs from zero due to sampling errors (H_0 : $\beta_k = 0$). They are based on the Student t statistic:

$$t = \frac{\beta_k^{\mathrm{ML}}}{\mathrm{Var}[\beta_k^{\mathrm{ML}}]^{1/2}} \tag{8.3.12}$$

[6]For simplicity of notation, no distinction is made in what follows between the vector $\boldsymbol{\beta}$ of coefficients in the utility function and the vector $\boldsymbol{\theta}$ of structural coefficients, or more precisely between the vectors $\boldsymbol{\beta}'$ and $\boldsymbol{\delta}'$ of identifiable parameters. The vector $\boldsymbol{\beta}$ is to be understood as the set of all the coefficients to be estimated.

$$
\begin{aligned}
V_{\text{Car}} &= \beta_T \cdot T_a \quad + \beta_{\text{CA}} \cdot CA \quad + \beta_{\text{HF}} \cdot HF \quad + \beta_{\text{CAR}} \cdot CAR \\
V_{\text{motorbike}} &= \beta_T \cdot T_m \quad + \beta_{\text{MAN}} \cdot MAN \quad + \beta_{21-35} \cdot 21-35 \\
V_{\text{bus}} &= \beta_T \cdot T_b \quad + \beta_{T\,accb<10} \cdot T_{accb<10} \\
V_{\text{walking}} &= \beta_{T\,wlk} \cdot T_{wlk} + \beta_{\text{WLK}} \cdot WLK
\end{aligned}
$$

T_c, T_m, T_b = Travel times of the modes car, motorcycle, bus;
T_{wlk} = Walking travel time;
$T_{accb<10}$ = Dummy variable = 1 if access time to bus is less than ten minutes; 0 otherwise;

CA = Car availability (no. cars/no. licenses in the household);
MAN = Dummy variable = 1 if the user is male, 0 otherwise;
HF = Dummy variable = 1 if the user is head of family, 0 otherwise;
21–35 = Dummy variable = 1 if the user is aged between 21 and 35, 0 otherwise
CAR, WLK = Alternative Specific Attributes (ASA);

Coefficients	β_T	$\beta_{T\,wlk}$	$\beta_{T\,accb<10}$	β_{CA}	β_{HF}	β_{MAN}	β_{21-35}	β_{CAR}	β_{WLK}
Estimate	−0.748	−4.560	1.247	1.758	0.452	0.990	1.684	1.411	3.929
Std. dev.	0.338	0.431	0.472	0.384	0.225	0.532	0.466	0.560	0.548
t	−2.213	−10.59	2.642	4.573	2.012	1.962	3.616	2.519	7.168

Test	H	Test statistic	95th percentile
t student	$\beta_t = \beta_{T\,wlk}$	7.53	1.96
$LR(\mathbf{0})$	$\boldsymbol{\beta} = \mathbf{0}$	588.01	16.92
$LR(\boldsymbol{\beta}_{\text{ASA}})$	$\boldsymbol{\beta} = \boldsymbol{\beta}_{\text{ASA}}$	285.83	14.06

Goodness of fit test	
ρ^2	0.424
$\bar{\rho}^2$	0.411

Fig. 8.5 Parameters and statistics for a mode choice logit model

Alternatively, the Student t statistic can be used to test the hypothesis that two coefficients β_k and β_j are equal ($H_0 : \beta_k = \beta_j$):

$$t = \frac{\beta_k^{\text{ML}} - \beta_j^{\text{ML}}}{(\text{Var}[\beta_k^{\text{ML}}] + \text{Var}[\beta_j^{\text{ML}}] - 2\text{cov}[\beta_j^{\text{ML}} \beta_k^{\text{ML}}])^{1/2}}$$

In both cases, under the null hypothesis the statistic t is distributed as a Student t variable with degrees of freedom equal to the sample size minus the number of estimated coefficients. Given typical sample sizes, it is usually assumed that the t statistic is distributed as an $N(0, 1)$ standard normal variable, which is the limit distribution of the Student t-variable as the sample size increases. Sample estimates of variances and covariances can be computed through expression (8.3.6). As is well known, the null hypothesis is rejected with a probability α of making a Type I error (e.g., rejecting a true assumption) if the t statistic value is outside the interval $(z_{\alpha/2}, z_{1-\alpha/2})$, which for $\alpha = 0.95$ is equal to ± 1.96. The values of the Student t statistics (8.3.12) for the coefficients of the model reported in Fig. 8.5 show that all the coefficient estimates are significantly different from zero with $\alpha = 0.95$. The reader can check the significance of the coefficients of the different models described in Chap. 4.

Chi-Square Tests on Vectors of Coefficients

To test the null hypothesis that the true coefficient vector $\boldsymbol{\beta}$ or one of its subvectors is equal to a given vector $\boldsymbol{\beta}^*(H_0 : \boldsymbol{\beta} = \boldsymbol{\beta}^*)$, the following statistic can be used.

$$\text{chi}^2(\boldsymbol{\beta}^*) = (\boldsymbol{\beta}^{\text{ML}} - \boldsymbol{\beta}^*)^T \sum_{\beta}^{-1} (\boldsymbol{\beta}^{\text{ML}} - \boldsymbol{\beta}^*) \tag{8.3.13}$$

If the null hypothesis is true, the χ^2 statistic is asymptotically distributed as a chi-square variable with degrees of freedom equal to the number of components of $\boldsymbol{\beta}$.

Note that expressions (8.3.12) and (8.3.13) can be used to obtain the confidence interval for a single coefficient as well as the confidence region for a vector of coefficients.

Likelihood ratio tests on vectors of coefficients. The likelihood ratio test is similar to the previous one in that it tests the null hypothesis that the vector $\boldsymbol{\beta}$, or one of its subvectors, is equal to a vector $\boldsymbol{\beta}^*$. The vector $\boldsymbol{\beta}^*$ may be defined implicitly by imposing some constraints on $\boldsymbol{\beta}$, for example, by specifying a feasibility set B (with $\boldsymbol{\beta} \in B$). In both the implicit and the explicit case, $\boldsymbol{\beta}^*$ can be seen as the vector that maximizes the log-likelihood function under the constraints:

$$\boldsymbol{\beta}^* = \arg\max_{\beta \in B} \ln L(\boldsymbol{\beta})$$

For instance, one can test the hypothesis that $\boldsymbol{\beta}$ is null or that only some of its components are null; in the latter case the other components of $\boldsymbol{\beta}^*$ will be estimated by solving the constrained maximization problem.

The null hypothesis $H_o : \boldsymbol{\beta} = \boldsymbol{\beta}^*$ can be tested using the Likelihood Ratio statistic *LR*:

$$LR(\boldsymbol{\beta}^*) = -2\big[\ln L(\boldsymbol{\beta}^*) - \ln L(\boldsymbol{\beta}^{\text{ML}})\big] \tag{8.3.14}$$

which, under the null hypothesis, is asymptotically distributed as a chi-square variable with degrees of freedom equal to the number of constraints imposed in estimating β^*.

The *LR* statistic is always greater than zero because the unconstrained maximum $\ln L(\boldsymbol{\beta}^{\text{ML}})$ of the function $\ln L(\boldsymbol{\beta})$ is not smaller than the constrained maximum of the same function, $\ln L(\boldsymbol{\beta}^*)$. Note that the *LR* test is equivalent, but not equal from the numerical point of view, to the chi-square test described above when the constraints completely identify the vector $\boldsymbol{\beta}^*$. For example, in the case $\boldsymbol{\beta}^* = \mathbf{0}$ it yields:

$$LR(\mathbf{0}) = -2\big[\ln L(\mathbf{0}) - \ln L(\boldsymbol{\beta}^{\mathrm{ML}})\big] \tag{8.3.15}$$

The null hypothesis $\boldsymbol{\beta}^* = \mathbf{0}$ corresponds to assuming a "true" model with all coefficients equal to zero, which therefore predicts equal probabilities for all alternatives ($V_j = 0\ \forall j \Rightarrow p[j] = 1/J$). The larger the difference between the likelihood of observing the users' choices with the calibrated model ($\ln L(\beta^{\mathrm{ML}})$) and the corresponding probability with a zero coefficients model ($\ln L(\mathbf{0})$), the less likely is this hypothesis. Under the null hypothesis, the statistic $LR(\mathbf{0})$ will be distributed as a chi-square variable with degrees of freedom equal to N_β.

A more challenging specification of the test is obtained by comparing the calibrated model with a model whose only parameters are the alternative specific attributes β_{ASA}. The vector $\boldsymbol{\beta}^* = \boldsymbol{\beta}^{\mathrm{ML}}_{\mathrm{ASA}}$ is obtained by maximizing the log-likelihood function $\ln L(\boldsymbol{\beta})$ with all the other coefficients constrained to be equal to zero: the number of ASA and their coefficients, N_{ASA}, can at most be equal to one less than the number of alternatives; that is, $N_{\mathrm{ASA}} \leq (J - 1)$. In this case the LR statistic becomes:

$$LR(\boldsymbol{\beta}_{\mathrm{ASA}}) = -2\big[\ln L\big(\boldsymbol{\beta}^{\mathrm{ML}}_{\mathrm{ASA}}\big) - \ln L(\boldsymbol{\beta}^{\mathrm{ML}})\big] \tag{8.3.16}$$

Figure 8.5 shows the statistics $LR(\mathbf{0})$ and $LR(\boldsymbol{\beta}_{\mathrm{ASA}})$ with their respective degrees of freedom. These statistics far exceed the 95th percentile of the corresponding chi-square variables with N_β and $N_\beta - N_{\mathrm{ASA}}$ degrees of freedom, and therefore the assumptions that the "true" model has either null coefficients or mode-specific constants only can be rejected with a very low probability of error.

Statistics and tests on goodness of fit. The model's ability to reproduce the choices made by a sample of users[7] can be measured by using the *rho-square statistic*:

$$\rho^2 = 1 - \frac{\ln L(\boldsymbol{\beta}^{\mathrm{ML}})}{\ln L(\mathbf{0})} \tag{8.3.17}$$

This statistic is a normalized measure in the interval [0, 1]. It is equal to zero if $L(\boldsymbol{\beta}^{\mathrm{ML}})$ is equal to $L(\mathbf{0})$ (i.e., the model has no explanatory capability); it is equal to one if the model yields the probability one of observing the choices actually made by each user in the sample (i.e., the model has perfect ability to reproduce observed choices).

[7] In theory, the model's goodness of fit should be tested on a sample of observations different from the sample used for the calibration (a hold-out sample). In practice, this procedure is not always followed to make the best use of all the available information, given the limited size of many available samples.

Alternatively, it is possible to use an adjusted value of the rho-square statistic, sometimes named the rho-square bar, which replaces the log-likelihood function $\ln L(\boldsymbol{\beta}^{\mathrm{ML}})$ by its unbiased estimate $\ln L(\boldsymbol{\beta}^{\mathrm{ML}}) - N_\beta$, where N_β is the number of parameters estimated in the model:

$$\bar{\rho}^2 = 1 - \frac{\ln L(\boldsymbol{\beta}^{\mathrm{ML}}) - N_\beta}{\ln L(\mathbf{0})} \tag{8.3.18}$$

Expression (8.3.18) attempts to eliminate the effect of the number of parameters in the model's specification, in order to allow comparison of models with different numbers of parameters.

The adjusted rho-square statistic can be used to compare two models (model 1 and model 2) whose vectors $\boldsymbol{\beta}_1$ and $\boldsymbol{\beta}_2$ cannot be obtained as a special case of each other.[8] In this case, under the null hypothesis that model 1 is "true," the probability that the statistic $\bar{\rho}_2^2$ of model 2 is for sampling reasons larger by some z than that of model 1, is less than the value of the probability distribution function of an $N(0, 1)$ standard normal variable computed for the value

$$\bar{z} = -\left[-2z \ln L(\mathbf{0}) + (N_1 - N_2)\right]^{1/2} \tag{8.3.19}$$

or

$$Pr\left(\bar{\rho}_2^2 - \bar{\rho}_1^2 > z\right) \leq \phi(\bar{z}); \quad z > 0 \tag{8.3.20}$$

where $\phi(\bar{z})$ is the value of the p.d.f. of $N(0, 1)$ and N_1 and N_2 are the number of parameters in models 1 and 2 respectively.

In addition to the statistics ρ^2 and $\bar{\rho}^2$, other informal statistics are used to assess the goodness of fit of a model. One of these statistics (% right) relates to the percentage of observations in the sample for which the alternative actually chosen is that with maximum probability as predicted by the model. Other synthetic statistics are the choice percentage observed and predicted by the model for each alternative. The former is given by the ratio between the number of users choosing each alternative and the total number of users to whom it is available. The latter is obtained as the average of choice probabilities given by the model for the users to whom the alternative is available.

Tests on the functional form. The statistical tests described above examine different hypotheses on the coefficients $\boldsymbol{\beta}^{\mathrm{ML}}$ obtained from the calibration of a model, where the model specification is assumed given. This section describes some statistical tests that compare different hypotheses on the functional form of the model itself.

Two generic alternative specifications can be compared using the $\bar{\rho}^2$ test in (8.3.20). Alternatively, specific tests related to particular functional forms can be

[8]This type of assumption is known as "nonnested".

used. For example, in Chap. 3 it was shown that the multinomial logit model is a special case of a single-level hierarchical logit if $\delta = 1$ (expression (3.3.24)), and of the multilevel hierarchical logit if $\delta_r = 1$ for each intermediate node r of the choice tree (Sect. 3.3.3). The hypothesis that the "true" model is a multinomial logit can be tested by calibrating hierarchical logit models and testing the null hypothesis that the estimates $\boldsymbol{\delta}^{\text{ML}}$ are equal to one. These tests can be conducted using the statistics described previously for testing hypotheses on single or multiple parameters.

For the multinomial logit model, the Independence of Irrelevant Alternatives (IIA) property discussed in Sect. 3.3.1 can be tested directly. Under the IIA hypothesis, the choice model for any subset I' of alternatives (a partial choice set) contained in I (the universal choice set), $I' \subseteq I$, is still a multinomial logit model:

$$p^i[j/I'] = \exp(\bar{\beta}^T X^i_j) \Big/ \sum_{h \in I'} \exp \bar{\beta}^T X^i_h \tag{8.3.21}$$

where $\bar{\beta}$ indicates the subvector of coefficients included in the systematic utilities of the alternatives contained in I' (e.g., $\bar{\beta}$ will not contain the ASA coefficients of alternatives not belonging to I'). The number of these coefficients will be $N_{\bar{\beta}} \leq N_\beta$. The maximum likelihood estimator $\bar{\beta}^{\text{ML}}_{1'}$ for the model (8.3.21) can be obtained for the subsample of observations choosing the alternatives in I'. If the IIA hypothesis is true, the vector $\bar{\beta}^{\text{ML}}_I$ of the $N_{\bar{\beta}}$ coefficients obtained by calibrating the model for all the alternatives over the whole sample and the vector $\bar{\beta}^{\text{ML}}_{1'}$ described previously must be statistically equivalent. This hypothesis can be tested using the statistic:

$$(\bar{\boldsymbol{\beta}}^{\text{ML}}_I - \bar{\boldsymbol{\beta}}^{\text{ML}}_{1'})^{\text{T}} (\boldsymbol{\Sigma}_{\bar{\beta}_I} - \boldsymbol{\Sigma}_{\bar{\beta}_{1'}})^{-1} (\bar{\boldsymbol{\beta}}^{\text{ML}}_I - \bar{\boldsymbol{\beta}}^{\text{ML}}_{1'}) \tag{8.3.22}$$

which under the null hypothesis is distributed as a chi-square variable with $N_{\bar{\beta}}$ degrees of freedom. The matrices $\boldsymbol{\Sigma}_{\beta_I}$ and $\sum_{\bar{\beta}_{1'}}$ are the variance–covariance matrices of the estimates $\bar{\beta}^{\text{ML}}_I$ and $\bar{\beta}^{\text{ML}}_{1'}$ of the $N_{\bar{\beta}}$ common components. To test the IIA hypothesis, the test should be carried out on different subsets I' of the universal choice set I.

8.4 Disaggregate Estimation of Demand Models with Stated Preference Surveys*

The information on travel behavior needed to specify and calibrate demand models can also be obtained using *Stated Preference* (*SP*) surveys. This term refers to a set of techniques that use statements made by interviewees about their preferences in hypothetical scenarios. SP techniques are based on the possibility of "controlling the experiment" by designing the choice context rather than recording choices in a given (generally uncontrolled) choice context, which was the case with *Revealed Preference* (RP) surveys described in the previous section. SP surveys have several advantages over RP surveys, which can be summarized as follows.

- They allow the investigation of choice alternatives not available at the time of the survey (e.g., new modes or services in a mode choice context).
- They can control the variation of relevant attributes outside the presently observed range to obtain better estimates of the corresponding coefficients. For example, the monetary cost of travel in urban areas usually falls within a limited range of values.
- They can introduce new attributes not present in the real choice context (e.g., passenger information, vehicle air-conditioning, and other on-board services).
- They can collect more information, that is, larger samples, per unit cost because each interviewee is usually asked about several choice contexts.

These advantages are obtained at the price of introducing some distortion in the results and in the models calibrated. Distortions stem from the possible differences between stated and actual choice behavior: if the user experienced a real situation, her behavior might be different from that stated during the SP survey. These differences in behavior may be due to a variety of factors. For example, the context suggested might be or appear to be unrealistic, some attributes of the suggested alternative relevant to the decision-maker might be missing, or there may be fatigue and justification bias effects. Analysis of the possible causes of distortion and of their remedies is outside the scope of this book. However, it should be noted that some of these problems are inherent to the SP survey technique, whereas others can be solved by careful design and execution of the surveys, bringing them as close as possible to real choice contexts.

From the above, it is clear that SP surveys, in spite of their considerable application potential, should be seen as complementary to, rather than competing with, RP techniques. The advantages and disadvantages of the two techniques compensate for each other and, as shown in the following, the techniques can be used jointly to build demand models.

Different SP techniques and approaches are appropriate for different aims. In the following, reference is made to the SP techniques most widely used for the specification and calibration of travel demand models. In particular, Sect. 8.4.1 introduces some definitions and the main types of surveys, Sect. 8.4.2 describes some aspects of SP survey design, and Sect. 8.4.3 deals with model calibration methods using the combined results of RP and SP surveys.

8.4.1 Definitions and Types of Survey

A stated preference experiment is fully characterized by a number of elements: the composition of the choice contexts proposed to the decision-maker interviewee, the selection of the choice contexts proposed, the type of preference response elicited from the decision-maker and the way in which the interview is conducted.

During the interview, the decision-maker is usually presented with different *scenarios* or *choice contexts*. A scenario is defined by the set of *alternative options*[9]; each option is associated with some *attributes* or *factors* defining its characteristics. Figure 8.6 shows two choice contexts (scenario A and scenario B), each consisting of two alternative modes and their attributes.

In the choice contexts proposed, the attributes vary between a predetermined number of values, or *levels*. These levels can be defined in absolute terms, for example, specific travel times and costs, or obtained as percentage variations with respect to the values of the attributes for a real context known to the decision-maker (e.g., times and costs relative to current values for certain origin–destination pairs).

The decision-maker can be asked to express his *preference* in different ways:

- *Choice*, that is, an indication of which option he would choose in that context;
- *Ranking*, that is, ordering of the available options according to his preference;
- *Rating*, that is, the assignment of a vote of preference on a predefined scale for each alternative option.

Note that the three types of preference provide an increasing quantity of information but require increasing involvement of the decision-maker. Furthermore, "choice" and "ranking" coincide when the choice context consists of only two alternative options.

The number of possible scenarios depends on the number of combinations of the design elements introduced, namely the number of options, the number of attributes, and the number of levels for each attribute. Because the total number of scenarios might be very large and not all the combinations are equally "useful," one of the elements in the design of an SP experiment will be the selection of the scenarios to be proposed to the decision-maker(s).

Finally, the interviews can be conducted using different procedures. In traditional methods, the decision-maker is asked to fill in pre-printed paper forms. In more sophisticated computer-aided techniques, the scenarios are generated in real-time, taking previous answers into account.

8.4.2 Survey Design

Designing an SP survey requires the definition of all the elements described above. It must be recalled that, in spite of the operational guidelines and the theoretical

[9]Choice alternatives in any scenario depend on the functional form of the model to be calibrated. With multinomial logit models, due to the IIA property, estimates of the systematic utility coefficients do not depend on the number of alternatives proposed, so that the scenario might include any subset of the alternatives included in the model. In the case of models for which the IIA is not valid, for example, hierarchical logit and probit, scenarios must be designed to account explicitly for the structure of the model. For example, alternatives belonging to different groups, as well as multiple alternatives for the same group, must be included in some scenarios for hierarchical logit models.

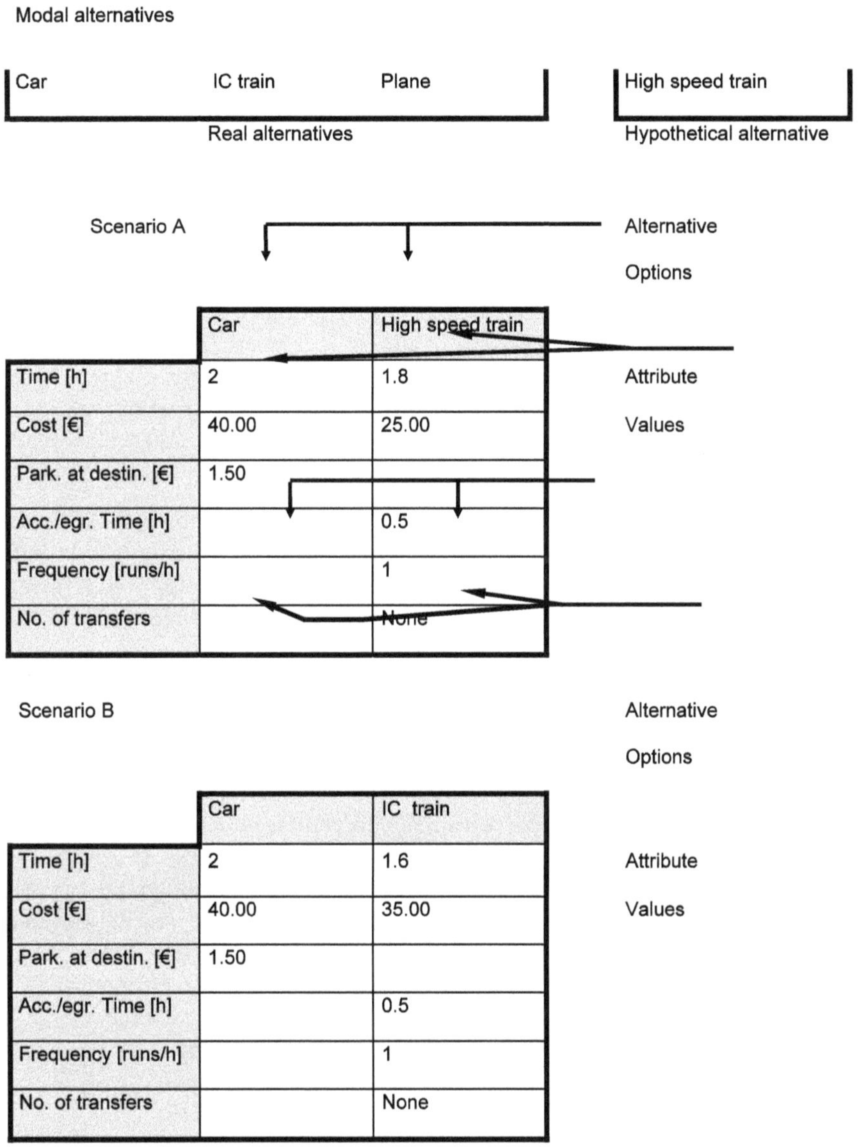

Scenario A

	Car	High speed train
Time [h]	2	1.8
Cost [€]	40.00	25.00
Park. at destin. [€]	1.50	
Acc./egr. Time [h]		0.5
Frequency [runs/h]		1
No. of transfers		None

Scenario B

	Car	IC train
Time [h]	2	1.6
Cost [€]	40.00	35.00
Park. at destin. [€]	1.50	
Acc./egr. Time [h]		0.5
Frequency [runs/h]		1
No. of transfers		None

Fig. 8.6 Hypothetical scenarios for an SP survey

analyses, SP survey design, even more than with traditional surveys, is a synthesis based on the analyst's experience and sensitivity. The main operational suggestions resulting from many years of research and experimentation are summarized below.

– *Scenario realism*: results of SP surveys are significantly better if choice scenarios are in the direct experience of the decision-maker. For example, in a survey for the calibration of a mode choice model, an RP interview can be carried out on an actual journey of a certain type, and then SP scenarios can be obtained from that journey by varying the attributes or by introducing a new mode. In this way the distortion effects described above can be reduced considerably. It is obvious that this type of survey requires more preparation. Portable computers can generate in real-time the level-of-service attributes for the different modes.

- *"Choice" rather than "ranking" and "rating"*. It seems that greater simplicity and less ambiguity of preference statements compensate for the smaller amount of information produced by this type of experiment. In addition, it is possible to use results and estimation techniques analogous to those applied to RP surveys.
- *Scenario simplification*. It seems that proposing a limited number of alternative options defined by a reduced number of attributes produces better results.
- *Limitation of the number of scenarios* proposed to each decision-maker in order to avoid fatigue effects that deteriorate the quality of results. Experience suggests that each decision-maker should be confronted with no more than nine or ten scenarios.

The latter aspect is strictly connected to the most theoretical phase of survey design, namely scenario selection. In most cases the number of scenarios theoretically possible is very large; in fact, subdividing the n factors into k groups of n_i elements $(i = 1, 2, \ldots, k)$ taking on m_i levels, the total number N of possible scenarios will be:

$$N = \prod_{i=1}^{k} m_i^{n_i}$$

The number of factors must be computed taking into account the fact that an attribute present in p alternatives counts for p different factors.

A full factorial design considers all possible scenarios. There are many techniques[10] for reducing the number of scenarios in a full factorial design, generating a subset of scenarios with desirable properties. Some results for the case of two levels per factor are given below; the case of several levels can be reduced by decomposing a multilevel factor into many two-level factors and introducing some compatibility constraints on the combinations of levels that the new factors can assume.

Figure 8.7 lists all the possible scenarios, indicating with $+$ and $-$ the two levels of each factor for an experiment with three factors and two levels $(N = 8)$; factors are time and cost for the car (*TC* and *CC*) and time for the bus (*TB*).

It is also assumed that the experiment associated with the ith scenario (ith row of the matrix in Fig. 8.7) yields an observation of the variable U_i not known a priori. In the example, this variable could be the difference in the perceived utility between the two alternatives (rating), or a binary indicator of the alternative preferred by the decision maker (ranking and choice). Let l_{ij} indicate the level of the jth factor in the

[10] These techniques are derived from multivariate statistical analysis and, in particular, from experimental design techniques. They are designed to allow the analysis of direct and indirect effects of relevant variables by means of linear models, and therefore do not correspond exactly to the case of demand models, which are typically nonlinear with respect to explanatory variables (attributes).

Scenario No.	Average	Factors			Interactions				Result of choice
		TC	*CC*	*TB*	*TC, CC*	*TC, TB*	*CC, TB*	*TC, CC, TB*	
1	+	−	−	−	+	+	+	−	U_1
2	+	+	−	−	−	−	+	+	U_2
3	+	−	+	−	−	+	−	+	U_3
4	+	+	+	−	+	−	−	−	U_4
5	+	−	−	+	+	−	−	+	U_5
6	+	+	−	+	−	+	−	−	U_6
7	+	−	+	+	−	−	+	−	U_7
8	+	+	+	+	+	+	+	+	U_8
Divisor	8	4	4	4	4	4	4	4	

Fig. 8.7 Example of full factorial design with levels and main interaction effects

ith scenario of the matrix in Fig. 8.7. Under the assumptions made here, l_{ij} assumes the values $+1$ and -1 corresponding to the "high" and "low" level of the factor. The complete experiment is called a contrast for factor j if:

$$\sum_{i=1,\dots,N} l_{ij} = 0 \tag{8.4.1}$$

that is, if the number of high levels (+) is equal to the number of low levels (−) in the N scenarios making up the experiment. Two contrasts involving factors j and h are said to be orthogonal if:

$$\sum_{i=1,\dots,N} l_{ij} l_{ih} = 0 \tag{8.4.2}$$

that is, if the numbers of scenarios in which the levels of the two factors are concordant (++, −−) is equal to the number in which they are discordant (+−, −+).

The variation (total variance) of the variables U_i can be explained in terms of the "main effects" and "interaction effects" of the factors considered in the experiment.

The main effect of factor j, $P_{(j)}$, is defined as the difference between the two averages $\bar{U}_+$ and $\bar{U}_-$ of the variable U calculated, respectively, from the (+) and (−) values of the factor. If the vector $\boldsymbol{l}_j$ is a contrast, it therefore follows that:

$$P_{(j)} = \frac{2}{N} \sum_{i=1,\dots,N} l_{ij} U_i \tag{8.4.3}$$

For the example in Fig. 8.7, the main effect of factor TC is therefore:

$$P_{(TC)} = \frac{1}{4}(U_2 + U_4 + U_6 + U_8) - \frac{1}{4}(U_1 + U_3 + U_5 + U_7)$$

The interaction effect between the factors j and h, $I_{(j,h)}$, is defined as the difference between the averages of the variable U obtained for the concordant values, (+) (+) or (−) (−), and the discordant values, (+) (−) or (+) (−), of the two factors.

If the two vectors $\boldsymbol{l}_j$ and $\boldsymbol{l}_h$ are contrasts, we have:

$$I_{(j,h)} = \frac{2}{N} \sum_{i=1,\dots,N} l_{ij} l_{ih} U_i \tag{8.4.4}$$

For the example of Fig. 8.7, the interaction effect of factors *TC* and *CC* is therefore:

$$I_{(T_C,C_C)} = \frac{1}{4}(U_1 + U_4 + U_5 + U_8) - \frac{1}{4}(U_2 + U_3 + U_6 + U_7)$$

Furthermore, from (8.4.4) it follows that, analogously to the levels of a factor, the level of interaction between two factors (j,h) for the ith scenario, $l_{i(j,h)}$, can be defined as $l_{ij} \cdot l_{ih}$; the interaction effect between the two factors can therefore be expressed as

$$I_{(j,h)} = \frac{2}{N} \sum l_{i(j,h)} U_i \tag{8.4.5}$$

Analogously, the interaction effect of three factors (j,h,k) can be defined as

$$I_{(j,h,k)} = \frac{2}{N} \sum_{i=1,\dots,N} l_{ij} l_{ih} l_{ik} U_i \tag{8.4.6}$$

and the interaction level of three factors can be expressed as $l_{i(j,h,k)}$.

Figure 8.7 shows the two-factor interaction levels and the unique three-factor interaction level as well as the average column, I, for which all variables m_i are equal to $(+1)$. These variables allow the average of U to be expressed as

$$\overline{U} = \frac{1}{N} \sum_{i=1,\dots,N} m_i U_i$$

Under the assumption of orthogonal contrasts, the N values taken by the variable U can be entirely explained as a linear combination, with coefficients a_i, of the average, the main effects and the interaction effects between the different factors. In the case of the example in Fig. 8.7, we have:

$$\begin{aligned} U_i &= \alpha_1 m_i + \alpha_2 l_{i(TC)} + \alpha_3 l_{2(CC)} + \alpha_4 l_{i(TB)} + \alpha_5 l_{i(TC\,CC)} \\ &\quad + \alpha_6 l_{i(TC\,TB)} + \alpha_7 l_{i(CC\,TB)} + \alpha_8 l_{i(TC\,CC\,TB)} \end{aligned}$$

Many experiments, however, lead to the conclusion that most of the overall variance of the variable U is explained by the main effects (approximately 80%), and the two-factor interaction effects explain a limited fraction of the global variance (3–6%). Furthermore, the variance explained by the interactions of more than two factors is usually negligible. In other words, if the variable U were expressed as a linear combination of the average and of the main effects, the variance explained by the model would be around 80% of the total variance observed for the variable U and so on. Based on these results, techniques have been developed for selecting less than the full factorial design. Examples are techniques to reduce the number of scenarios presented to each decision-maker, while retaining the orthogonality of the contrasts and the possibility of evaluating at least the main effects of the factors considered.

The first technique, known as *block decomposition* of the full factorial design, is based on the principle of subdividing the set of alternative scenarios into groups (blocks) that are presented to different decision-makers. In order to obtain blocks satisfying the properties (8.4.1) (contrasts) and (8.4.2) (orthogonality between contrasts) one or more "block variables" is selected and the scenarios corresponding to the same value of the block variable, or concordant (discordant) values of many block variables, are grouped together. The block variables normally used are high-level interactions, because the effects on the variance of the block variables and their interactions can be estimated only approximately on the basis of the observation of all the interviewees. Figure 8.8 shows two subdivisions into blocks of the full design in Fig. 8.7. In the first case, the eight scenarios are divided into two blocks of 4 $(8/2)$ scenarios, using the interaction level of the three factors (*TC*, *CC TB*) as the block variable. In the second case, 4 blocks $(8/2 \times 2)$ of two scenarios each are obtained by using as block variables the interaction level of the two factors (*TC*, *CC*) and of the two factors (*TC*, *TB*).

Another partialization technique of the full factorial design, known as *fractional factorial design*, eliminates some scenarios completely while retaining orthogonal contrasts that allow the estimation of the main effects. If the resulting number of scenarios is still too high to be presented to a single decision-maker, they can be further broken down into blocks by using the method described previously. A fractional factorial design can be obtained from the full design through a "defining relationship." The simplest case is that in which the level of a given factor is obtained from those of all the others resulting from a full design which excludes the factor to be obtained. The level of the "derived" factor is assumed equal to the level of a higher-level interaction effect. For example, in the case of Fig. 8.9 it is assumed that the level of the factor *TB* is equal to that of the interaction effect (*TC*, *CC*), where the levels of *TC* and *CC* are those defined in a two-factor full design $(N = 2^2)$. In this case the following "defining relationship" is adopted.

$$TB = (TC, CC); \quad \text{that is,} \ l_{iTB} = l_{iTC} \cdot l_{iCC} \tag{8.4.7}$$

The design in Fig. 8.9 is thus obtained as follows.

- Development of the full factorial design for the two factors *TC* and *CC*
- Calculation of the interaction effect level (*TC*, *CC*)
- Definition of the level of factor *TB* using (8.4.7)

With a fractional factorial design, the possibility of estimating some interaction effects is lost, as these effects get confounded with the retained ones. Confounded effects can be identified by manipulating the defining relationship of the fractional factorial design. Thus, recalling that the product of the levels of the same factor is equal to the average factor I, relationship (8.4.7) gives:

$$TB \times TB = TB \times TC \times CC = I$$

Sc. No.	Factors			Block var.	Alternatives organized by block					Sc. No.
	TC	*CC*	*TB*	*TC, CC, TB*	Block		*TC*	*CC*	*TB*	
1	−	−	−	−	I	Block I	−	−	−	1
2	+	−	−	+	II		+	+	−	4
3	−	+	−	+	II		+	−	+	6
4	+	+	−	−	I		−	+	+	7
5	−	−	+	+	II	Block II	+	−	−	2
6	+	−	+	−	I		−	+	−	3
7	−	+	+	−	I		−	−	+	5
8	+	+	+	+	II		+	+	+	8

Sc. No	Factors			Block var.		Alternatives organized by block					Sc. No
	TC	*CC*	*TB*	*TC, CC*	*TC, TB*	Block		*TC*	*CC*	*TB*	
1	−	−	−	+	+	IV	Block I	+	−	−	2
2	+	−	−	−	−	I		−	+	+	7
3	−	+	−	−	+	II	Block II	−	+	−	3
4	+	+	−	+	−	III		+	−	+	6
5	−	−	+	+	−	III	Block III	+	+	−	4
6	+	−	+	−	+	II		−	−	+	5
7	−	+	+	−	−	I	Block IV	−	−	−	1
8	+	+	+	+	+	IV		+	+	+	8

Note: Sc. = scenario

Fig. 8.8 Construction of two and four blocks from the full factorial design in Fig. 8.7

Scenario No.	Factors		Interaction	Factor
	TC	*CC*	*TC, CC*	*TB*
1	−	−	+	+
2	+	−	−	−
3	−	+	−	−
4	+	+	+	+

Fig. 8.9 Example of fractional factorial design for the full factorial design in Fig. 8.7

$$TC \times TB = TC \times TC \times CC = CC \qquad (8.4.8)$$
$$CC \times TB = CC \times CC \times TC = TC$$

that is, the three-factor interaction effect (*TC*, *CC*, *TB*) gets confounded with the average and the two-factor interaction effects (*TC*, *TB*) and (*CC*, *TB*) get confounded with the primary effects of the factors *CC* and *TC*, respectively. Obviously, the two-factor interaction effect (*TC*, *CC*) is confounded with the primary effect *TB* by construction.

The "length" of the defining relationship (i.e., the number of factors in it) is known as the resolution of a fractional factorial design. The resolution of (8.4.7) is equal to 3. The number of scenarios in a fractional factorial design depends on the number of defining relationships; for each defining equation, under the assumption of two levels for each factor, the number of scenarios halves. Obviously, the choice of defining relationships must be based on the analyst's expectations concerning the particular effects that should not be confounded to explain the observed behaviors.

To give a more detailed example, suppose there are seven factors generically indicated by A, B, C, D, E, F, G with two levels each. The full factorial design has $2^7 = 128$ scenarios, and a fractional factorial design with $2^{7-1} = 64$ scenarios can be obtained with a single defining relationship. For example:

$$G = (ABCDEF)$$

A design with $2^{7-4} = 8$ scenarios can be obtained with 4 defining relationships, for example:

$$D = (A, B); \qquad E = (A, C); \qquad F = (B, C); \qquad G = (A, B, C)$$

and so on.

Note the difference between the two methods described for partializing the full factorial design. With the block variables method, the whole full factorial design is used, even if it is presented to several decision-makers; with the fractional design; on the other hand, some scenarios are completely eliminated. In the former case many scenarios are generated but, given the number of decision-makers in the sample, less information (preference statements) is obtained for each scenario; with the fractional factorial design, the opposite occurs.

It should be pointed out that SP surveys are often aimed at the calibration of random utility models whose systematic utility function includes the values of individual attributes or their functional transformation. This specification assumes that the interactions between the attributes (or factors) can be disregarded in explaining the choice behaviors of decision-makers. Thus, the SP survey design should allow at least the evaluation of all the main effects of the factors considered.

8.4.3 Model Calibration

The results of an SP survey can be used to calibrate demand models involving the choice dimensions proposed to the decision-makers. The estimation methods used in practice are analogous to those described for revealed preferences in Sect. 8.3.2. In fact, each scenario i presented to a decision-maker can be seen as an element of a sample of observations of choice behaviors. The final size of the SP sample is thus equal to:

$$n_{\text{SP}} = \sum_{z=1,\dots,N_{\text{SP}}} n_z$$

where n_z is the number of scenarios presented to the zth decision-maker and N_{SP} is the number of decision-makers included in the SP survey.

The attributes proposed for the different alternatives can be associated with each scenario i. The chosen alternative is the one explicitly chosen by the decision-maker in choice surveys, or the one with greatest attractiveness in ranking or rating surveys. Under the approximate assumption that the n_{SP} observations are statistically independent,[11] it is possible to formulate likelihood and log-likelihood functions for the SP sample that formally coincide with expressions (8.3.2) and (8.3.3), and all the results described previously can be extended to the estimation of SP-based models.

As stated in the introduction to this section, SP surveys should be considered complementary to traditional RP surveys and the combined use of the two can balance their respective merits and shortcomings. From the point of view of demand modeling, it is therefore useful to carry out joint calibrations using RP and SP surveys on the same sample or on different samples of users. Random utility models explaining RP and SP choices should be specified separately because their attributes, random residuals, variances, and, in principle, even functional forms might be different. Possible specifications of the perceived utilities in both models are formalized below.

RP MODEL:

$$U_{ji}^{\text{RP}} = \boldsymbol{\beta}^T \boldsymbol{X}_{ji}^{\text{RP}} + \eta^T \mathbf{W}_{ji}^{\text{RP}} + \varepsilon_{ji}^{\text{RP}} = V_{ji}^{\text{RP}} + \varepsilon_{ji}^{\text{RP}} \quad i = 1, \ldots, n_{\text{RP}} \tag{8.4.9}$$

where

- U_{ji}^{RP} is the perceived utility associated with alternative j by decision-maker i in the RP context
- $\boldsymbol{X}_{ji}^{\text{RP}}$ is the vector of common RP/SP attributes of alternative j for decision-maker i; these attributes appear in the specification of the SP model with the same coefficients
- $\mathbf{W}_{ji}^{\text{RP}}$ is the vector of the RP-specific attributes of alternative j for decision-maker i
- $\varepsilon_{ji}^{\text{RP}}$ is the random residual of alternative j for decision-maker i
- V_{ji}^{RP} is the systematic utility of the RP model associated with alternative j for decision-maker i
- $\boldsymbol{\beta}$ and $\boldsymbol{\eta}$ are the vectors of unknown parameters to be estimated

SP MODEL:

$$U_{ji}^{\text{SP}} = \boldsymbol{\beta}^T \boldsymbol{X}_{ji}^{\text{SP}} + \gamma^T \boldsymbol{Z}_{ji}^{\text{SP}} + \varepsilon_{ji}^{\text{SP}} = V_{ji}^{\text{SP}} + \varepsilon_{ji}^{\text{SP}} \quad i = 1, \ldots, n_{\text{SP}} \tag{8.4.10}$$

where

- U_{ji}^{SP} is the perceived utility associated with alternative j in the hypothetical scenario i
- $\boldsymbol{X}_{ji}^{\text{SP}}$ is the vector of common RP/SP attributes of alternative j for scenario i

[11] In practice, it would be more correct to assume the existence of a correlation between observations of the choices of each individual. This can be obtained, for example, by introducing a common random residual for all the observations of a same respondent within a mixed logit formulation.

$\boldsymbol{Z}_{ji}^{SP}$ is the vector of the SP-specific attributes of alternative j for scenario i
ε_{ji}^{SP} is the random residual of alternative j for scenario I
V_{ji}^{SP} is the systematic utility of the SP model associated with alternative j for scenario i
$\boldsymbol{\beta}$ and $\boldsymbol{\gamma}$ are the vectors of unknown parameters to be estimated

Specific attributes of the RP model may involve variables not included among the SP factors. Specific attributes of the SP model may include quality attributes, such as on-board comfort, services contemplated in the SP survey, or ASA for alternatives not available in the RP context (e.g., new transport modes or services).

State dependence or state inertia is an SP-specific attribute often included in the specification (8.4.10). This attribute represents the conditioning of the SP decision-maker by the alternative actually chosen in the RP context. Inertia is often modeled as a dummy variable equal to one if user i chooses alternative j in the RP context, zero otherwise. Its coefficient is usually found to be statistically significant and positive indicating, given the values of other attributes, a larger perceived utility and choice probability for the alternative that is actually chosen in the real context. Obviously the state dependence attribute can be used only if the RP and SP surveys relate to the same sample of decision-makers.

The definition of the choice probabilities $p_{RP}^i[j]$ and $p_{SP}^i[j]$ obviously depends on the assumptions about the distribution of the random vectors $\boldsymbol{\varepsilon}^{RP}$ and $\boldsymbol{\varepsilon}^{SP}$. Assuming that ε_{ji}^{SP} and ε_{ji}^{RP} are i.i.d. Gumbel variables with parameters θ_{SP} and θ_{RP}, respectively, the probability of choosing alternative j in observation (decision-maker or scenario) i assumes the form of a multinomial logit model for both the RP and the SP models:

$$p_{RP}^i[j] = \frac{\exp(V_{ji}^{RP}/\theta_{RP})}{\sum_h \exp(V_{hi}^{RP}/\theta_{RP})}; \qquad p_{SP}^i[j] = \frac{\exp(V_{ji}^{SP}/\theta_{SP})}{\sum_h \exp(V_{hi}^{SP}/\theta_{SP})} \tag{8.4.11}$$

A scale factor taking into account the possibility that the variances of the vectors $\boldsymbol{\varepsilon}^{RP}$ and $\boldsymbol{\varepsilon}^{SP}$ might be different is usually introduced for joint RP/SP calibration. In fact, as stated in Sect. 8.3, for logit family models it is not possible to estimate the parameter θ separately from the coefficients β_k, so that the estimates $\hat{\beta}_k^{ML}$ are in reality ratios $\beta_k' = \beta_k/\theta$. To take into account the possible difference in the variances of the residuals $\boldsymbol{\varepsilon}^{RP}$ and $\boldsymbol{\varepsilon}^{SP}$, a scale factor μ, equal to the ratio between the parameters θ of the two random vectors, is introduced:

$$\mu^2 = \frac{\mathrm{Var}[\varepsilon_{RP}]}{\mathrm{Var}[\varepsilon_{SP}]} = \frac{\theta_{RP}^2}{\theta_{SP}^2}; \quad \text{that is,} \quad \mu = \frac{\theta_{RP}}{\theta_{SP}} \tag{8.4.12}$$

The log-likelihood function for the RP and SP samples can therefore be expressed including the parameter θ_{RP} in all the other coefficients:

$$\ln L^{RP}(\boldsymbol{\beta}', \boldsymbol{\eta}') = \sum_{i=1,\ldots,n_{RP}} \ln p_{RP}\big[j(i)\big]\big(\boldsymbol{X}_i^{RP}, \boldsymbol{W}_i^{RP}, \boldsymbol{\beta}', \boldsymbol{\eta}'\big) \tag{8.4.13}$$

$$\ln L^{\mathrm{SP}}(\boldsymbol{\beta}', \boldsymbol{\gamma}', \boldsymbol{\mu}') = \sum_{i=1,\ldots,n_{\mathrm{SP}}} \ln p_{\mathrm{SP}}[j(i)](\boldsymbol{X}_i^{\mathrm{SP}}, \boldsymbol{Z}_i^{\mathrm{SP}}, \boldsymbol{\beta}', \boldsymbol{\gamma}', \boldsymbol{\mu}') \quad (8.4.14)$$

where the probabilities $p[j(i)]$ are obtained by using the following systematic utilities:

$$V_{ij}^{\mathrm{RP}} = \boldsymbol{\beta}'^T X_{ij}^{\mathrm{RP}} + \boldsymbol{\eta}'^T W_{ij}^{\mathrm{RP}} \qquad \boldsymbol{\beta}' = \boldsymbol{\beta}/\theta_{\mathrm{RP}} \qquad \boldsymbol{\eta}' = \boldsymbol{\eta}/\theta_{\mathrm{RP}}$$

$$V_{ij}^{\mathrm{SP}} = \mu\boldsymbol{\beta}'^T X_{ij}^{\mathrm{SP}} + \boldsymbol{\gamma}'^T Z_{ij}^{\mathrm{SP}} \qquad \boldsymbol{\gamma}' = \boldsymbol{\gamma}/\theta_{\mathrm{SP}} \qquad \mu\boldsymbol{\beta}' = \boldsymbol{\beta}/\theta_{\mathrm{SP}}$$

The combined estimate of the parameters $(\boldsymbol{\beta}', \boldsymbol{\eta}', \boldsymbol{\gamma}', \mu)$ can therefore be obtained by maximizing the log-likelihood function of the joint sample, which is the sum of expressions (8.4.13) and (8.4.14) under the assumption that the RP and SP samples are independent:

$$\begin{aligned}(\boldsymbol{\beta}', \boldsymbol{\eta}', \boldsymbol{\gamma}', \mu') &= \arg\max\left[\ln L^{\mathrm{RP+SP}}(\boldsymbol{\beta}', \boldsymbol{\eta}', \boldsymbol{\gamma}', \mu)\right] \\ &= \arg\max\left[\ln L^{\mathrm{RP}}(\boldsymbol{\beta}', \boldsymbol{\eta}') + \ln L^{\mathrm{SP}}(\boldsymbol{\beta}', \boldsymbol{\gamma}', \mu)\right] \quad (8.4.15)\end{aligned}$$

Note also that under the hypothesis that the two choice models $p_{\mathrm{RP}}[j(i)]$ and $p_{\mathrm{SP}}[j(i)]$ are multinomial logit (such as (8.4.11)), the global log-likelihood function (8.4.15) is concave in the parameters $\boldsymbol{\beta}'$, $\boldsymbol{\eta}'$, and $\boldsymbol{\gamma}'$, but not in the scale factor μ as is the case with the hierarchical logit model. This implies that the maximization of (8.4.15) cannot use the gradient algorithms described in Appendix A for the multinomial logit model. A possible solution uses, on the one hand, the gradient algorithms to maximize the function $\ln L(\boldsymbol{\beta}', \boldsymbol{\delta}', \boldsymbol{\gamma}', \mu^k)$ for a predefined value μ^k of the scale factor and a one-dimensional line search algorithm that explores different values of μ (see Appendix A).

Experimental evidence indicates that the combined use of RP and SP data for estimating the parameters usually results in an improvement in statistical precision and in more reasonable parameter values. Furthermore, it is not possible to define a priori whether the scale factor μ must be greater or less than one. In fact, there are reasons both for a larger variance of RP random residuals (less precise attributes used for calibration, omitted attributes, etc.) and for the opposite (less realism of the choice context, fatigue effect, etc.). As an example, Fig. 8.10 reports the results of the calibrations of a multinomial logit mode choice model using RP and SP data separately and jointly.

The joint calibration on RP and SP data of more complex random utility models is further complicated if the joint density function of the vectors $\boldsymbol{\varepsilon}^{\mathrm{RP}}$ and $\boldsymbol{\varepsilon}^{\mathrm{SP}}$ involves more than one parameter. If the two models $p_{\mathrm{RP}}[j]$ and $p_{\mathrm{SP}}[j]$ were hierarchical logit models with the same tree structure (the same vector of parameters $\boldsymbol{\delta}$), it would still be possible to introduce a scale factor μ relating the variances of the two random vectors. For different correlation structures or other functional forms it would be more complicated to synthesize the different structures of the variance–covariance matrices of $\boldsymbol{\varepsilon}^{\mathrm{RP}}$ and $\boldsymbol{\varepsilon}^{\mathrm{SP}}$ with few parameters.

Parameters (attributes)	RP model	SP model	RP/SP model
β_1 (travel time)	−3.277 (−2.2)	−2.585 (−3.9)	−2.82 (−3.9)
β_2 (cost)	−2.863 (−3.5)	−1.336 (−5.9)	−1.371 (−2.4)
V.O.T.	0.591 €/h	0.999 €/h	1.062 €/h
β_3 (access time)	−6.606 (−1.2)	−3.176 (−3.5)	−4.776 (−4.8)
β_4 (waiting time)	−10.40 (−2.3)	−19.62 (−4.1)	−20.86 (−4.0)
β_5 (no. of motorbikes)	5.391 (3.9)	2.831 (3.6)	2.848 (5.5)
β_6 (no. of cars)	3.175 (2.5)	1.933 (4.4)	1.528 (3.9)
β_7 (chain)	−1.399 (−1.2)	−1.730 (−2.3)	−0.4545 (−1.7)
η_1 (ASA car RP)	−1.370 (−1.2)		−4.271 (−4.9)
η_2 (ASA motorbike RP)	−4.492 (−3.3)		−6.076 (−6.8)
γ_1 (ASA car SP)		−9748 (−1.8)	−1.923 (−3.1)
γ_2 (ASA motorbikes SP)		−1.499 (−2.3)	−2.480 (−3.4)
γ_3 (Inertia)			2.603 (4.4)
Scale factor μ			0.786 (4.0)
	STATISTICS		
$\ln L(0)$	−105.4668	−408.6838	−514.1506
$\mathrm{Ln}\, L(\beta)$	−55.4268	−210.4182	−282.2376
LR	100.08	396.53	463.826
RHO2	0.4745	0.4851	0.6612

RP Model

$$V_{\text{car}}^{\text{RP}} = \beta_1 Tb + \beta_2 Mc + \beta_6 Nc \beta_7 CH + \eta_1 CAR^{\text{RP}}$$
$$V_{\text{Motorbike}}^{\text{RP}} = \beta_1 Tb + \beta_2 Mc + \beta_5 Nm + \eta_2 MOTORBIKE^{\text{RP}}$$
$$V_{\text{Bus}}^{\text{RP}} = \beta_1 Tb + \beta_2 Mc + \beta_3 Ta + \beta_4 Tw$$

SP Model

$$V_{\text{car}}^{\text{SP}} = \beta_1 Tb + \beta_2 Mc + \beta_6 Nc \beta_7 CH + \gamma_1 CAR^{\text{SP}} + \gamma_3 IN$$
$$V_{\text{Motorbike}}^{\text{SP}} = \beta_1 Tb + \beta_2 Mc + \beta_5 Nm + \gamma_2 MOTORBIKE^{\text{SP}} + \gamma_3 IN$$
$$V_{Bus}^{\text{SP}} = \beta_1 Tb + \beta_2 Mc + \beta_3 Ta + \beta_4 Tw + \gamma_3 IN$$

Attributes

Tb = Travel time on board $[h]$

Mc = Monetary cost $[€ \cdot 10^3]$

Nm, Nc = No. of motorbikes and cars in household

Ta = Access time $[h]$

Tw = Waiting time $[h]$

CH = Dummy variable $(0/1)$, 1 if the trip belongs to a chain (sequence of more than 2 trips)

Car, Motorbike = Alternative Specific Attributes (ASA)

IN = Inertia variable $(0/1)$, 1 if mode was chosen in the RP survey

Fig. 8.10 Separate and joint RP/SP calibrations of a multinomial logit mode choice model

8.5 Estimation of O-D Demand Flows Using Traffic Counts

This section covers methods aimed at improving estimates of present origin–destination demand flows by combining direct and/or indirect (model) estimators with other aggregate information related to the flows. This aggregate information will be considered here to be traffic counts, that is, counts of user flows, on some elements (links) of the transportation supply system (network).[12] The problem of estimating O-D flows by combining traffic counts with all the other available information is sometimes referred to in the literature as the origin–destination count based estimation (ODCBE) problem.

From a certain point of view, the problem of estimating O-D flows by using traffic counts can be considered as the inverse assignment problem. Chapter 5 posed the assignment problem as that of calculating link flows starting from O-D flows, and network and path choice models. Conversely, the problem under study here is that of calculating O-D flows starting from measured link flows, using network and path choice models (see Fig. 8.11).

Estimation of O-D matrices using traffic counts has received considerable attention in recent years both from theoretical and empirical points of view. This can be easily explained given the cost and complexity of sampling surveys, as well as the lack of precision of direct and model estimators of O-D flows. In contrast to this, user flows on network links (traffic counts) are cheaply and easily obtainable, often automatically. Furthermore, in many transportation engineering applications, O-D flow estimates are primarily used to predict traffic flows resulting from changes in the supply system (network). The focus is on estimating and predicting aggregate implications of the O-D matrix, that is, the total link flows, rather than individual O-D flows, and it is expected that a matrix capable of reproducing such aggregates with good precision will also give good predictions following network changes.

Before solving the O-D estimation problem, it is necessary to express formally the relationship between the vector of observed flows and the unknown O-D demand flows, by reformulating some of the relationships presented in the previous chapters. As stated in Chaps. 2 and 5, the link flow f_l in the reference period can be expressed as the sum of reference period flows on the paths that include link l:

$f_l = \sum_k \delta_{lk} h_k$ where δ_{lk} is the element of the link-path incidence matrix $\boldsymbol{\Delta}$.

Path flows can be expressed as the product of the total O-D demand flow by the percentage (fraction) of users following each O-D path:

$$f_l = \sum_k \delta_{lk} h_k = \sum_k \delta_{lk} \sum_i p_{ki} d_i \tag{8.5.1}$$

[12]This is both the most frequent and most complex case. Other aggregate information sources can be easily represented as particular cases of link counts by properly specifying the "assignment equation" (8.5.2). Total generated and/or attracted flows, average trip length, distribution of trip lengths, and total flows crossing internal cordons are examples of other aggregate information about O-D flows that can be seen as special cases. In the following it is also assumed that flow-counting locations are given; that is, the links are given as input to the problem of O-D demand estimation. Although this is sometimes the case, count locations should be determined based on the information they provide. The problem of determining optimal counting locations can be formulated as a network design problem similar to those described in Chap. 9.

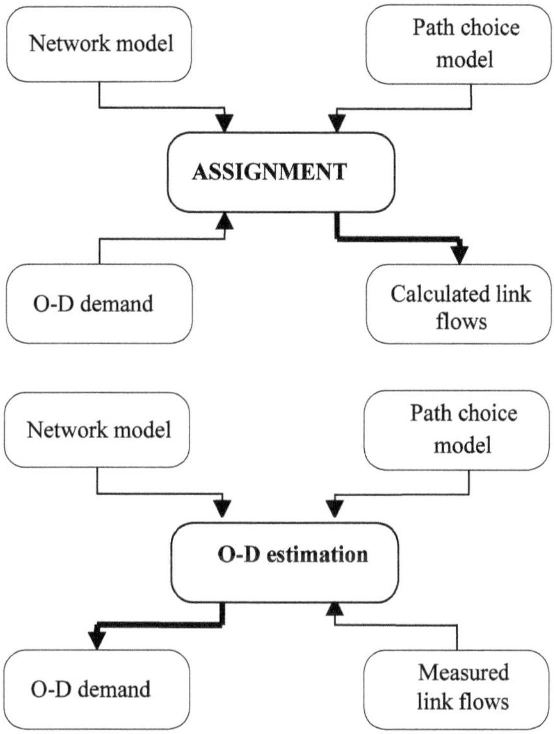

Fig. 8.11 Relationship between traffic assignment and estimation of O-D flows from traffic counts

where p_{ki} is the fraction of the flow d_i[13] between the ith O-D pair using path k. Note that, in this expression, the variables (link flows, O-D flows, path flows, and path fractions) indicate the "true" values for the system and the reference period under study.

Equation (8.5.1) can be expressed differently as

$$f_l = \sum_i d_i \sum_k \delta_{lk} p_{ki} = \sum_i m_{li} d_i \tag{8.5.2}$$

or

$$f_l = \boldsymbol{m}_l^T \boldsymbol{d}$$

where $m_{li} = \sum_k \delta_{lk} p_{ki}$ is the assignment fraction, that is, the fraction of the flow d_i using link l, and $\boldsymbol{m}_l$ is the column vector obtained by ordering these fractions.[14]

---

[13]For simplicity of notation in this section, the generic element of the demand vector is denoted as d_i, i.e., using a single index for the O-D pair as in Chap. 5, instead of the double index d_{od} used previously.

[14]Note that values of p_{ki}, and therefore m_{li}, are "true" values, that is, the actual fractions of users who use a given path or a given link in the reference period.

Using matrix notation, expression (8.5.2) becomes

$$\boldsymbol{f} = \boldsymbol{\Delta h} = \boldsymbol{\Delta P d} = \boldsymbol{M d} \tag{8.5.3}$$

All the variables introduced here refer to the links for which traffic counts are available (n_l being their number), to the paths using them and to the O-D flows using those paths (n_{OD} being their number). Thus the matrix $\boldsymbol{M}$, or *assignment matrix*, has dimensions ($n_l \times n_{\text{OD}}$). The relationship (8.5.3) between link flows and O-D demand flows is known as the assignment relationship or map; Fig. 8.12 shows an example of the assignment map for an elementary network.

When several paths are available between an O-D pair, the elements m_{li} of the assignment relationship are not uniquely defined, and therefore must be estimated. Path choice and network assignment models described in Chaps. 4 and 5 provide methods for obtaining estimates $\hat{p}_{ki}$ of the fractions p_{ki} and estimates $\hat{m}_{li}$ of the fractions m_{li}.

In the case of pre-trip, deterministic, or probabilistic path choice models, fractions $\hat{p}_{ki}$ can be expressed as probabilities of choosing each path k connecting the ith O-D pair as a function of the path cost vector $\boldsymbol{g}$ (see Sect. 4.3.3.1):

$$\hat{p}_{ki} = p[k/i](\boldsymbol{g}) \tag{8.5.4}$$

In the case of mixed pre-trip/en-route path choice models (often used for high-frequency public transport networks), the probability of choosing path k for O-D pair i can be obtained from the choice probability $q[j/i]$ of hyperpath j, which depends on the vector of hyperpath costs $\boldsymbol{x}$, and from the probabilities ω_{kj} of following path k within hyperpath j (see Sect. 4.3.3.1):

$$\hat{p}_{ki} = \sum_j \omega_{kj} q[j/i](\boldsymbol{x}) \tag{8.5.5}$$

To underline the dependence of assignment matrix estimates $\hat{m}_{li}$ on the path choice model and, through this, on the link costs $\boldsymbol{c}$, the matrix $\hat{\boldsymbol{M}}$ can be formally expressed as

$$\hat{\boldsymbol{M}} = \boldsymbol{\Delta} \hat{\boldsymbol{P}}(\boldsymbol{c}) \tag{8.5.6}$$

$$\hat{\boldsymbol{M}} = \boldsymbol{\Delta \Omega}\, \boldsymbol{Q}(\boldsymbol{c}) \tag{8.5.7}$$

leaving as understood the relationship between additive path costs and link costs.

If link and path costs are known,[15] an estimate $\hat{\boldsymbol{M}}$ of the "true" assignment matrix $\boldsymbol{M}$ can be calculated through path choice models (8.5.4) and (8.5.5). It is to be expected that $\hat{\boldsymbol{M}}$ may differ from the true assignment matrix $\boldsymbol{M}$ because of the approximations implicit in any assignment model (network extraction, cost functions,

[15]The calculation of the assignment matrix $\hat{\boldsymbol{M}}$ in the case of congested networks for which the link costs are not known is covered in Sect. 8.5.4 on solution methods.

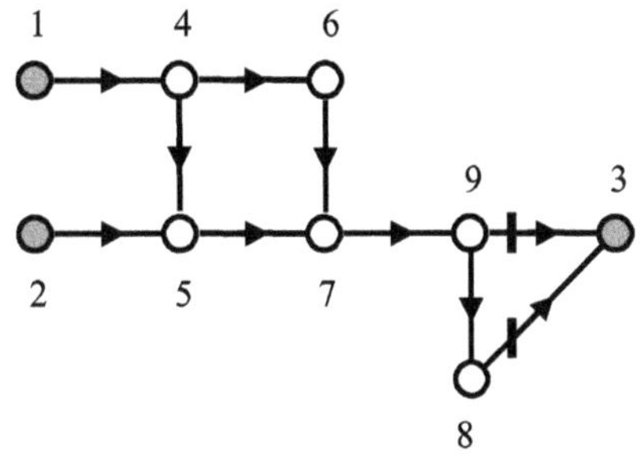

$$d = \begin{bmatrix} d_{13} \\ d_{23} \end{bmatrix}$$

O-D pair	Path k								p_{ki}
1-3	1)	1	4	6	7	9	3		0.30
	2)	1	4	5	7	9	3		0.30
	3)	1	4	5	7	9	8	3	0.20
	4)	1	4	6	7	9	8	3	0.20
2-3	5)	2	5	7	9	3			0.70
	6)	2	5	7	9	8	3		0.30

$N = 2$ (*link* 9-3 *and link* 8-3)

$\boldsymbol{\Delta} \bullet \boldsymbol{P} = \boldsymbol{M}$

Δ

	1	2	3	4	5	6
9-3	1	1			1	
8-3			1	1		1

•

P

	1-3	2-3
1	0.3	
2	0.3	
3	0.2	
4	0.2	
5		0.7
6		0.3

=

M

	1-3	2-3
9-3	0.6	0.7
8-3	0.4	0.3

Fig. 8.12 Assignment map for an elementary network

path choice model, etc.). Thus, a vector $\varepsilon^{\mathrm{SIM}}$[16] of assignment-related errors should be added when substituting $\hat{\boldsymbol{M}}$ for $\boldsymbol{M}$ in (8.5.3):

$$\boldsymbol{f} = \boldsymbol{M}\boldsymbol{d} = (\hat{\boldsymbol{M}} + \boldsymbol{E}^{\mathrm{SIM}})\boldsymbol{d} = \hat{\boldsymbol{M}}\boldsymbol{d} + \boldsymbol{\varepsilon}^{\mathrm{SIM}} \tag{8.5.8}$$

where $\boldsymbol{E}^{\mathrm{SIM}}$ is the matrix of differences between the true assignment matrix and that obtained with the assignment model; and $\varepsilon^{\mathrm{SIM}}$ is the vector of differences, or assignment errors, between the flows resulting from the assignment of "true" demand and "true" flows. In other words, if the "true" vector of demand flows $\boldsymbol{d}$[16]

[16]It should be remembered that the components of the "true" vectors $\boldsymbol{d}$ and $\boldsymbol{f}$ are the flows between each O-D pair and on each link, averaged over different observation periods.

were known, its assignment to the network would produce a flow vector $\boldsymbol{v}$:

$$\boldsymbol{v} = \hat{\boldsymbol{M}}\boldsymbol{d} = \boldsymbol{v}(\boldsymbol{d}) \tag{8.5.9}$$

which is different from the "true" link flow vector $\boldsymbol{f}$. These differences are the components of the vector ε^{SIM}:

$$\boldsymbol{f} = \boldsymbol{v} + \boldsymbol{\varepsilon}^{\text{SIM}}$$

Flow counts are a further source of error. Like all measures, traffic counts are affected by errors that depend, among other things, on the technique used (manual, automatic, etc.). Furthermore, the counts are usually conducted over several days, sometimes on different days for different links, whereas the "true" demand vector $\boldsymbol{d}$ represents the average O-D flows in periods with similar characteristics (e.g., peak hour of the average weekday). Thus, if $\hat{\boldsymbol{f}}$ is the vector of measured flows, it will differ from the "true" vector $\boldsymbol{f}$ by a vector ε^{OBS} of measurement errors:

$$\hat{\boldsymbol{f}} = \boldsymbol{f} + \boldsymbol{\varepsilon}^{\text{OBS}} \tag{8.5.10}$$

By combining (8.5.8) and (8.5.10), it is possible to express the relationship between the vector of counts $\hat{\boldsymbol{f}}$, the assignment matrix $\hat{\boldsymbol{M}}$ and the "true" O-D demand flow vector $\boldsymbol{d}$ as

$$\hat{\boldsymbol{f}} = \hat{\boldsymbol{M}}\boldsymbol{d} + \boldsymbol{\varepsilon}^{\text{SIM}} + \boldsymbol{\varepsilon}^{\text{OBS}} = \boldsymbol{v}(\boldsymbol{d}) + \boldsymbol{\varepsilon} \tag{8.5.11}$$

where the vector $\boldsymbol{\varepsilon}$ is the algebraic sum of the vectors ε^{SIM} and ε^{OBS}. It is usually assumed that the assignment model and the counts are unbiased estimators of the "true" flows, that is, that the vector $\boldsymbol{\varepsilon}$ is a zero mean random vector $E(\boldsymbol{\varepsilon}) = \boldsymbol{0}$. Empirical evidence seems to support this assumption.

Usually the information on O-D flows contained in traffic counts, represented by the system of stochastic equations (8.5.11), is not sufficient to estimate the vector $\boldsymbol{d}$. Indeed, even assuming that $\boldsymbol{\varepsilon}$ is null, the number of independent equations in the linear system (8.5.3) is usually much less than the number of unknown O-D flows to be estimated. The example in Fig. 8.13 shows that even for an elementary network with a single path for each O-D pair, there are many O-D matrices that, when assigned to the network, exactly reproduce the flows observed on the links.

Furthermore, because in general the vector ε differs from zero, the system of linear equations $\hat{\boldsymbol{f}} = \hat{\boldsymbol{M}}\boldsymbol{d}$ may not have a solution. In summary, the information contained in the counts must be combined with that from other sources to estimate the unknown O-D demand flows.

The additional information can be of two types: sampling or experimental information derived from demand surveys, and nonexperimental information representing the a priori knowledge of the analyst. In the former case, the classic theory of statistical interference can be applied; in the latter, Bayesian estimators should be used. The two methods, whose statistical foundations are described in the following sections, give rise to several estimators, some of them having similar formal representations.[17] In fact, if $\hat{d}$ is the vector representing the initial information (i.e. the information on O-D-demand not given by the counts), the ODCBE problem can be expressed in general form as

$$\boldsymbol{d}^* = \arg\min_{x \geq 0} \left[z_1(\boldsymbol{x}, \hat{\boldsymbol{d}}) + z_2(\boldsymbol{v}(\boldsymbol{x}), \hat{\boldsymbol{f}}) \right] \tag{8.5.12}$$

where $\boldsymbol{x}$ is the unknown demand vector. The two functions $z_1(\boldsymbol{x}, \hat{\boldsymbol{d}})$ and $z_2(\boldsymbol{v}(\boldsymbol{x}), \hat{\boldsymbol{f}})$ can be considered as different "distance" measures: z_1 measures the "distance" of the unknown demand $\boldsymbol{x}$ from the a priori estimate $\hat{\boldsymbol{d}}$ and z_2 measures the "distance" of the flows $\boldsymbol{v}(\boldsymbol{x})$ obtained by assigning $\boldsymbol{x}$ to the network from the traffic counts $\hat{\boldsymbol{f}}$. An intuitive interpretation of the problem (8.5.12) is that it searches the vector $\boldsymbol{d}^*$ that is closest to the a priori estimate $\hat{\boldsymbol{d}}$, and, once it is assigned to the network, produces the flows $\boldsymbol{v}(\boldsymbol{d}^*)$ closest to the counts $\hat{\boldsymbol{f}}$.

In general, the functional form of the two terms $z_1(\cdot)$ and $z_2(\cdot)$, depends on the type of information available (experimental or nonexperimental) and on the probability laws associated with such information. The statistical bases of the various estimators and their resulting functional forms are described in the following sections.

8.5.1 Maximum Likelihood and GLS Estimators*

Classic estimators of $\boldsymbol{d}$ can be derived from Maximum Likelihood (ML) or Generalized Least Squares (GLS) theory, depending on whether explicit assumptions on the probability distribution of the random residuals ε^{SIM} and ε^{OBS} are made.

Maximum likelihood estimators $\boldsymbol{d}^{\text{ML}}$ are obtained by maximizing the probability of observing both the additional sampling survey results and the counted flows. Under the usually acceptable assumption that these two probabilities are independent, the maximum likelihood estimator can be expressed as

$$\boldsymbol{d}^{\text{ML}} = \arg\max_{x \in S} \left[\ln L(\boldsymbol{n}/\boldsymbol{x}) + \ln L(\hat{\boldsymbol{f}}/\boldsymbol{x}) \right] \tag{8.5.13}$$

where

- $\boldsymbol{x}$ is the "unknown" demand vector, of dimension $(n_{\text{OD}} \times 1)$, whose components x_{od} are the trip flows between the O-D pair *od* (which from now on are denoted with a double index)
- $\boldsymbol{n}$ is the vector of O-D demand counts, with dimension $(n_{\text{OD}} \times 1)$. The generic component of $\boldsymbol{n}$, n_{od}, is the number of trips in the sample that were observed between the O-D pair *od*

[17] This can be seen as confirmation of the essentially interpretative nature of the difference between the objective and subjective approaches in probability theory and statistical inference.

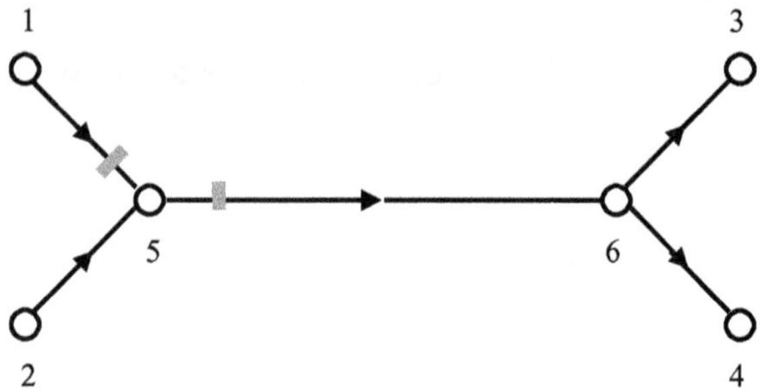

$\hat{f}_{1,5} = 15 \quad \hat{f}_{5,6} = 20$

$\hat{d}$

	3	4
1	12	7
2	3	2

O-D matrix	Distance Measures: $\sum_i (d_i - \hat{d}_i)^2$	Distance Measures: $\sum_i \frac{(d_i - \hat{d}_i)^2}{\hat{d}_i}$
d_1: 1→(3: 5, 4: 10); 2→(3: 5, 4: 0)	66	8.702
d_2: 1→(3: 7, 4: 8); 2→(3: 3, 4: 2)	26	(2.226)
d_3: 1→(3: 12, 4: 3); 2→(3: 3, 4: 2)	(16)	2.286

Fig. 8.13 O-D matrices corresponding to the same link flows vector

$\hat{f}$ is the vector of observed flows, or traffic counts, with dimension $(n_l \times 1)$

$\ln L(\boldsymbol{n}/\boldsymbol{x})$ is the log-likelihood function of demand counts, that is, the logarithm of the probability of observing the sampling vector $\boldsymbol{n}$ if $\boldsymbol{x}$ is the (true) demand vector

$\ln L(\hat{\boldsymbol{f}}/\boldsymbol{x})$ is the log-likelihood function of the traffic counts, that is, the logarithm of the probability of observing the vector of the counts $\hat{\boldsymbol{f}}$ if $\boldsymbol{x}$ is the (true) demand vector

S is the feasibility set of the (true) demand vector, usually taken to be the nonnegative orthant; that is, $S = \{\boldsymbol{x} : \boldsymbol{x} \geq 0\}$

Maximum likelihood estimators can therefore be obtained by solving the constrained maximization problem expressed by (8.5.13) once the log-likelihood functions $\ln L(\boldsymbol{n}/\boldsymbol{x})$ and $\ln L(\hat{\boldsymbol{f}}/\boldsymbol{x})$ have been specified. This requires the formulation of hypotheses on the probability laws of O-D demand counts $\boldsymbol{n}$ and of traffic counts $\hat{\boldsymbol{f}}$, conditional on the demand vector $\boldsymbol{x}$.

It is usually assumed that traffic counts are random variables with means given by the flows $\boldsymbol{v}(\boldsymbol{x})$ obtained by assigning the demand $\boldsymbol{x}$. By (8.5.11), this implies that the vector $\boldsymbol{\varepsilon}^{\text{SIM}}$ has a zero mean. The probability laws most widely used for count data are the Poisson and the multivariate normal. If it is assumed that the traffic counts on each link l are independently distributed as Poisson random variables with mean equal to $v_l(\boldsymbol{x})$, that is,

$$E[\hat{f}_l] = v_l(\boldsymbol{x}) = \hat{\boldsymbol{m}}_l^T \boldsymbol{x} \tag{8.5.14}$$

then the probability of observing $\hat{\boldsymbol{f}}$ is given by the product of the probabilities of observing its individual components:

$$L(\hat{\boldsymbol{f}}/\boldsymbol{x}) = \prod_{l=1,\dots,n_L} \frac{\exp(-v_l(\boldsymbol{x}))v_l(\boldsymbol{x})^{\hat{f}_l}}{\hat{f}_l!} \tag{8.5.15}$$

and the log-likelihood function becomes[18]

$$\ln L(\hat{\boldsymbol{f}}/\boldsymbol{x}) \cong \sum_{l=1\dots n_L} \left(\hat{f}_l \ln v_l(\boldsymbol{x}) - v_l(\boldsymbol{x})\right) + \text{const.} \tag{8.5.16}$$

where the constant denotes other terms that do not depend on the unknown demand vector $\boldsymbol{x}$, and which are therefore irrelevant for the maximization problem (8.5.13).

If the traffic counts are jointly distributed according to a multivariate normal random variable with mean vector $\boldsymbol{v}(\boldsymbol{x})$ and variance–covariance matrix $\boldsymbol{W}$, the likelihood of observing the vector $\hat{\boldsymbol{f}}$ is proportional to:

$$L(\hat{\boldsymbol{f}}/\boldsymbol{x}) \propto \exp\left[-\frac{1}{2}\left(\hat{\boldsymbol{f}} - \boldsymbol{v}(\boldsymbol{x})\right)^T \boldsymbol{W}^{-1}\left(\hat{\boldsymbol{f}} - \boldsymbol{v}(\boldsymbol{x})\right)\right] \tag{8.5.17}$$

and the log-likelihood function becomes:

$$\ln L(\hat{\boldsymbol{f}}/\boldsymbol{x}) = -\frac{1}{2}\left(\hat{\boldsymbol{f}} - \boldsymbol{v}(\boldsymbol{x})\right)^T \boldsymbol{W}^{-1}\left(\hat{\boldsymbol{f}} - \boldsymbol{v}(\boldsymbol{x})\right) + \text{const.} \tag{8.5.18}$$

The log-likelihood function of *O-D demand counts* depends on the type of sampling adopted (see Sect. 8.2). In the simplest case of stratified random sampling by zone of origin, a simple random sample of n_o trips is extracted from the $d_{o.}$ trips originating from each zone o (e.g., sampling at the entrances of a motorway network or at the cordon sections of the study area). Here it can be assumed that the number of trips sampled from each origin to all destinations is distributed as a multinomial random variable. It can further be assumed that the probability of observing the

[18]Equation (8.5.16) is obtained from (8.5.15) by using Stirling's approximation: $\ln(x!) \cong x \ln x - x$.

whole vector $\boldsymbol{n}$ is the product of the probability functions of these variables over all origins:

$$L(\boldsymbol{n}/\boldsymbol{x}) = \prod_o \left[\left(n_o!/\prod_d n_{od}! \right) \prod_d (x_{od}/x_o)^{n_{od}} \right] \tag{8.5.19}$$

where x_{od}/x_o is the unknown probability of observing a trip with destination d.

From (8.5.19) the log-likelihood function can be obtained:

$$\operatorname{Ln} L(\boldsymbol{n}/\boldsymbol{x}) = \sum_{od} n_{od} \ln x_{od} + \text{const.} \tag{8.5.20}$$

with the further constraint that the number of trips generated in each zone o is equal to the number of those counted $d_{o.}$, or:

$$S = \left\{ \boldsymbol{x} : \sum_d x_{od} = d_{o.}; \boldsymbol{x} \geq 0 \right\}$$

If the number of trips sampled at each origin is sufficiently large (a few dozen or more), the multinomial variable can be closely approximated by the product of independent Poisson variables (one for each O-D pair), with parameters equal to the means $\alpha_o x_{od}$, where α_o is the sampling rate for origin o:

$$\alpha_o = \frac{n_o}{d_{o.}}$$

In this case the functions $L(\boldsymbol{n}/\boldsymbol{x})$ and $\ln L(\boldsymbol{n}/\boldsymbol{x})$ given by (8.5.19) and (8.5.20) can be approximated by:

$$L(\boldsymbol{n}/\boldsymbol{x}) = \prod_{od} \frac{\exp(-\alpha_o x_{od})(\alpha_o x_{od})^{n_{od}}}{n_{od}!} \tag{8.5.21}$$

and

$$\ln L(\boldsymbol{n}/\boldsymbol{x}) = \sum_{od} \left(n_{od} \ln(\alpha_o x_{od}) - \alpha_o x_{od} \right) + \text{const.} \tag{8.5.22}$$

Analogous expressions can be obtained for more complex sampling methods; in applications, however, expressions (8.5.21) and (8.5.22) are often used as reasonable approximations.

In conclusion, the general maximum likelihood estimator $\boldsymbol{d}^{\mathrm{ML}}$ given by expression (8.5.13) is made more detailed by substituting expression (8.5.16) or (8.5.18) for the log-likelihood function of traffic counts, and expression (8.5.20) or (8.5.22) for the log-likelihood function of the O-D demand counts (see Fig. 8.14).

Generalized Least Squares (GLS) is the other estimator derived within the classic theory of statistical inference. The GLS estimator provides the estimate of an unknown vector, in this case the O-D demand flow vector, starting from a system of linear stochastic equations. The latter can be obtained by combining the demand-related information contained in the traffic counts, expressed by (8.5.11), and in the direct estimate $\hat{d}$, obtained from O-D demand counts:

$$\begin{aligned} \hat{\boldsymbol{f}} &= \hat{\boldsymbol{M}}\boldsymbol{x} + \boldsymbol{\varepsilon} \qquad E(\boldsymbol{\varepsilon}) = \boldsymbol{0} \qquad \mathrm{Var}[\boldsymbol{\varepsilon}] = \boldsymbol{W} \\ \hat{\boldsymbol{d}} &= \boldsymbol{x} + \boldsymbol{\eta} \qquad E(\boldsymbol{\eta}) = \boldsymbol{0} \qquad \mathrm{Var}[\boldsymbol{\eta}] = \boldsymbol{Z} \end{aligned} \tag{8.5.23}$$

where $\hat{\boldsymbol{d}}$ is the O-D demand vector whose components $\hat{d}_{od}$ are the sample estimates, obtained using the methods described in Sect. 8.2. For example, in the case of simple random sampling with rate α, these estimates will be:

$$\hat{d}_{od} = \frac{n_{od}}{\alpha}$$

The vector $\boldsymbol{\eta}$ in expression (8.5.23) is the vector of sampling errors whose components are the differences between the true unknown demand $\boldsymbol{x}$ and the sample estimates $\hat{\boldsymbol{d}}$. If the estimator adopted is unbiased the vector $\boldsymbol{\eta}$ has zero mean. The elements of the variance–covariance matrix $\boldsymbol{Z}$ can be estimated using the appropriate expressions for variances and covariances of sample estimates.

The *GLS* estimator of the demand vector can therefore be expressed as

$$\boldsymbol{d}^{\mathrm{GLS}} = \underset{x \in S}{\arg\min}\left[(\hat{\boldsymbol{d}} - \boldsymbol{x})^T \boldsymbol{Z}^{-1} (\hat{\boldsymbol{d}} - \boldsymbol{x}) + (\hat{\boldsymbol{f}} - \hat{\boldsymbol{M}}\boldsymbol{x})^T \boldsymbol{W}^{-1} (\hat{\boldsymbol{f}} - \hat{\boldsymbol{M}}\boldsymbol{x})\right] \tag{8.5.24}$$

Expression (8.5.24) is often applied assuming that the matrices $\boldsymbol{Z}$ and $\boldsymbol{W}$ are diagonal, that is, ignoring the covariances between the components of vectors $\varepsilon^{\mathrm{SIM}}$ and η. This is done both because these covariances are difficult to express and also to reduce memory requirements and computing times. Under this simplified assumption, expression (8.5.24) becomes:

$$\boldsymbol{d}^{\mathrm{GLS}} = \underset{x \geq 0}{\arg\min}\left[\sum_{od} \frac{(\hat{d}_{od} - x_{od})^2}{\mathrm{Var}[\eta_{od}]} + \sum_{l} \frac{(\hat{f}_l - \sum_{od} \hat{m}_{l,od} x_{od})^2}{\mathrm{Var}[\varepsilon_l]}\right] \tag{8.5.25}$$

The intuitive interpretation given for (8.5.12) can be extended to (8.5.25): the demand vector $\boldsymbol{d}^{\mathrm{GLS}}$ minimizes the sum of squared differences between it and the O-D flows estimated from the sample, and between the flows resulting from its assignment and the counted link flows. Furthermore, the squared differences have weights inversely proportional to the variances of their respective errors. In other words, poorer estimates of differences from a component $\hat{d}_{od}$ (those with higher $\mathrm{Var}[\eta_{od}]$) will be weighted less, and the same is true for the flows.

Note also the role of information on the vector $\boldsymbol{d}$ contained in traffic counts. If this information did not exist, the second term of (8.5.24) and (8.5.25) would disappear, and the estimate of $\boldsymbol{d}^{\mathrm{GLS}}$ would coincide with $\hat{\boldsymbol{d}}$ because the latter minimizes the quadratic objective function (setting it to zero). A similar observation can be made for the maximum likelihood estimators.

From the formal point of view, the *GLS* estimator coincides with the maximum likelihood estimator if both the demand estimates and the traffic counts are assumed to be distributed as multivariate normal random variables with means $\boldsymbol{x}$ and $\hat{\boldsymbol{M}}\boldsymbol{x}$, and with variance–covariance matrices $\boldsymbol{Z}$ and $\boldsymbol{W}$, respectively.

8.5.2 Bayesian Estimators*

Bayesian methods estimate unknown parameters by combining experimental (or sampling) information with nonexperimental (a priori or "subjective") information.[19] In the particular case of O-D demand estimation, the experimental information comes from traffic counts, whereas nonexperimental information may come from old O-D estimates to be updated, from estimates obtained with demand models, or simply from analyst "expectations." In each case, $\hat{\boldsymbol{d}}$ will indicate the demand vector derived from nonexperimental information. Bayesian estimators are obtained from the a posteriori probability function $h(\boldsymbol{x}/\hat{\boldsymbol{f}}, \hat{\boldsymbol{d}})$ of the unknown demand vector $\boldsymbol{x}$ conditional on a priori information $\hat{\boldsymbol{d}}$ and on experimental information $\hat{\boldsymbol{f}}$. According to Bayes' theorem, the a posteriori probability is proportional to the product of two factors: the a priori probability function $g(\boldsymbol{x}/\hat{\boldsymbol{d}})$, which expresses the distribution of subjective probability attributed to the unknown vector given the a priori estimate $\hat{\boldsymbol{d}}$; and the probability, or likelihood, function $L(\hat{\boldsymbol{f}}/\boldsymbol{x})$, which expresses the probability of observing the traffic counts $\hat{\boldsymbol{f}}$ conditional on the unknown demand vector $\boldsymbol{x}$:

$$h(\boldsymbol{x}/\hat{\boldsymbol{f}}, \hat{\boldsymbol{d}}) \propto L(\hat{\boldsymbol{f}}/\boldsymbol{x}) g(\boldsymbol{x}/\hat{\boldsymbol{d}}) \tag{8.5.26}$$

A family of Bayesian estimators for demand flows $\boldsymbol{d}^B$ can be obtained by maximizing[20] the a posteriori probability (8.5.26) or its natural logarithm:

$$\boldsymbol{d}^B = \underset{\boldsymbol{x} \in S}{\arg\max} \left[\ln g(\boldsymbol{x}/\hat{\boldsymbol{d}}) + \ln L(\hat{\boldsymbol{f}}/\boldsymbol{x}) \right] \tag{8.5.27}$$

The detailed specification of a Bayesian estimator depends on the assumptions made about the probability functions $L(\hat{\boldsymbol{f}}/\boldsymbol{x})$ and $g(\boldsymbol{x}/\hat{\boldsymbol{d}})$. With respect to the function $L(\hat{\boldsymbol{f}}/\boldsymbol{x})$, (8.5.16) and (8.5.18), corresponding to the assumptions of independent Poisson and multivariate normal random variables respectively, can be used.

---

[19] Bayesian estimators coincide with "classic" estimators under the assumption that the subjective estimates are obtained from sampling surveys. This shows that "classic" estimators can be obtained as special cases in the context of Bayesian statistics.

[20] In theory, different Bayesian estimators can be derived from different properties of the a posteriori probability function. The maximizing estimator given by (8.5.27) corresponds to the mode of the a posteriori probability function (8.5.26). Another estimator could be obtained by finding the expected value of the a posteriori function:

$$\boldsymbol{d}^B = E[x/\hat{\boldsymbol{f}}, \hat{\boldsymbol{d}}] = \int_{\boldsymbol{x} \in S} \boldsymbol{x} h(\boldsymbol{x}/\hat{\boldsymbol{f}}, \hat{\boldsymbol{d}})\, d\boldsymbol{x}$$

In practice, however, the calculation of this expected value estimator would be very complex because it is not usually possible to solve analytically the multiple integral that defines it.

The a priori probability function, $g(\boldsymbol{x}/\hat{\boldsymbol{d}})$, can be specified in different ways; the formulations proposed in the literature are described below.

If it is assumed that the unknown demand vector is a multinomial random variable resulting from the distribution of total demand d, among all possible O-D pairs, with probabilities π_{od} derived from the matrix $\hat{\boldsymbol{d}}$:

$$\pi_{od} = \frac{\hat{d}_{od}}{\hat{d}..} \quad \hat{d}.. = \sum_{od} \hat{d}_{od}$$

the function $g(\boldsymbol{x}/\hat{\boldsymbol{d}})$ can be written as

$$g(\boldsymbol{x}/\hat{\boldsymbol{d}}) = \frac{(\sum_{od} x_{od})!}{\prod_{od} x_{od}!} \prod_{od} (\hat{d}_{od}/\hat{d}..)^{x_{od}} \tag{8.5.28}$$

Using Stirling's approximation ($\ln x! \cong x \ln x - x$), the logarithm of (8.5.28) can be expressed as

$$\ln g(\boldsymbol{x}/\hat{\boldsymbol{d}}) = \sum_{od} x_{od} \ln\left(\sum_{od} x_{od}\right) + \sum_{od} x_{od} \ln(\hat{d}_{od}/\hat{d}..x_{od}) \tag{8.5.29}$$

Furthermore, if the total number of trips ($\sum_{od} x_{od} = \hat{d}..$) is assumed to be known, expression (8.5.29) further simplifies to:

$$\ln g(x/\hat{d}) = -\sum_{od} x_{od} \ln(x_{od}/\hat{d}_{od}) + \text{const.} \tag{8.5.30}$$

The negative of function (8.5.30) is known as the entropy function of the unknown vector $\boldsymbol{x}$.

Alternatively, it may be assumed that the components x_{od} are independently distributed Poisson random variables, with mean (parameter) equal to $\hat{d}_{od}$. In this case the function $g(\boldsymbol{x}/\hat{\boldsymbol{d}})$ becomes:

$$g(\boldsymbol{x}/\hat{\boldsymbol{d}}) = \prod_{od} \frac{\exp(-\hat{d}_{od})^{x_{od}}}{x_{od}!} \hat{d}_{od}^{x_{od}} \tag{8.5.31}$$

The latter, using Stirling's approximation, can be expressed as

$$\ln g(\boldsymbol{x}/\hat{\boldsymbol{d}}) = -\sum_{od} x_{od}\left[\ln(x_{od}/\hat{d}_{od}) - 1\right] + \text{const.} \tag{8.5.32}$$

The negative of function (8.5.32) is known as the information function of the unknown vector $\boldsymbol{x}$.

Finally, it may be assumed that the vector $\boldsymbol{x}$ is a multivariate normal random variable with mean $\hat{\boldsymbol{d}}$ and variance–covariance matrix $\boldsymbol{Z}_B$; in this case the probability function is proportional to:

$$g(\boldsymbol{x}/\hat{\boldsymbol{d}}) \propto \exp\left[-\frac{1}{2}(\boldsymbol{x}-\hat{\boldsymbol{d}})^T \boldsymbol{Z}_B^{-1}(\boldsymbol{x}-\hat{\boldsymbol{d}})\right]$$

and its logarithm becomes:

$$\ln g(\boldsymbol{x}/\hat{\boldsymbol{d}}) = -\frac{1}{2}(\boldsymbol{x}-\hat{\boldsymbol{d}})^T \boldsymbol{Z}_B^{-1}(\boldsymbol{x}-\hat{\boldsymbol{d}}) + \text{const.} \tag{8.5.33}$$

If the a priori probability function $g(\boldsymbol{x}/\hat{\boldsymbol{d}})$ and the traffic counts probability function $L(\hat{\boldsymbol{f}}/\boldsymbol{x})$ are both assumed to be multivariate normal variables, expressions (8.5.18) and (8.5.33) are substituted in the general expression (8.5.27) and the resulting Bayesian estimator is formally analogous to the generalized least squares estimator $\hat{\boldsymbol{d}}^{\text{GLS}}$. However, the similarity between the two estimators is only formal, because the vector $\hat{\boldsymbol{d}}$ and the variance–covariance matrices $\boldsymbol{Z}$ and $\boldsymbol{Z}_B$ have different interpretations. In the *GLS* estimator, the vector $\hat{\boldsymbol{d}}$ is a direct demand estimate from sampling surveys, and the matrix $\boldsymbol{Z}$ includes its sampling variances and covariances. In Bayesian estimators, $\hat{\boldsymbol{d}}$ is an a priori estimate of the O-D demand vector, and $\boldsymbol{Z}_B$ is made up of variances and covariances that summarize the analyst's confidence in the estimate.

The formal analogy of the two estimators should, however, be considered an advantage because it allows the use of the same model and algorithm in very different estimation situations. This generality of the *GLS* estimator has contributed to its widespread use in applications.

8.5.3 Application Issues

We stated that different estimators combining traffic counts $\hat{\boldsymbol{f}}$ and other information $\hat{\boldsymbol{d}}$ can be expressed in a general form as the vector $\boldsymbol{d}^*$ solving the constrained minimization problem[21];

$$\boldsymbol{d}^* = \arg\min_{\boldsymbol{x}\in S}\left[z_1(\boldsymbol{x},\hat{\boldsymbol{d}}) + z_2\big(\boldsymbol{v}(\boldsymbol{x}),\hat{\boldsymbol{f}}\big)\right] \tag{8.5.34}$$

Figure 8.14 summarizes the functional forms of $z_1(\cdot)$ and $z_2(\cdot)$ described above, together with the corresponding assumptions.

The application of these methods in practice poses a number of problems that are briefly addressed below.

The choice of functional form from among the various possibilities obviously depends on the type of available information about the O-D flows and therefore on

[21] The problem (8.5.34) can be easily applied to maximization problems by changing the sign of the objective function.

GENERAL ESTIMATION MODEL
$\boldsymbol{d}^* = \arg\min_{\boldsymbol{x}\in S}\left[z_1(\boldsymbol{x},\hat{\boldsymbol{d}}) + z_2(\boldsymbol{v}(\boldsymbol{x}),\hat{\boldsymbol{f}})\right]$

Distance from the initial estimate	Distance from flow counts
$z_1(\boldsymbol{x},\hat{\boldsymbol{d}})$	$z_2(\boldsymbol{v}(\boldsymbol{x}),\hat{\boldsymbol{f}})$
Generalized Least Squares (GLS)	Generalized Least Squares (GLS)
$(\hat{\boldsymbol{d}}-\boldsymbol{x})^T\boldsymbol{Z}^{-1}(\hat{\boldsymbol{d}}-\boldsymbol{x})$ or $\sum_{od}(x_{od}-\hat{d}_{od})^2/\,\mathrm{Var}[\eta_{od}]$	$(\hat{\boldsymbol{f}}-\boldsymbol{v}(\boldsymbol{x}))^T\boldsymbol{W}^{-1}(\hat{\boldsymbol{f}}-\boldsymbol{v}(\boldsymbol{x}))$ or $\sum_{l\in M}(\hat{f}_l-v_l(\boldsymbol{x}))^2/\,\mathrm{Var}[\varepsilon_l]$
Maximum Likelihood (ML)	Maximum Likelihood (ML)
Poisson $-\sum_{od}(n_{od}\ln(\alpha_{ods}x_{od})-\alpha_{od}x_{od})$	Poisson $-\sum_{l\in M}(\hat{f}_l\ln v_l(\boldsymbol{x})-v_l(\boldsymbol{x}))$
Multinomial $-\sum_{od}n_{od}\ln x_{od}$	MVN $(\hat{\boldsymbol{f}}-\boldsymbol{v}(\boldsymbol{x}))^T\boldsymbol{W}^{-1}(\hat{\boldsymbol{f}}-\boldsymbol{v}(\boldsymbol{x}))$ or $\sum_{l\in M}(\hat{f}_l-v_l(\boldsymbol{x}))^2/\,\mathrm{Var}[\varepsilon_l]$
Bayes	Bayes
Poisson $\sum_{od}x_{od}\ln[(x_{od}/\hat{d}_{od})-1]$	Poisson $-\sum_{l\in M}(\hat{f}_l\ln v_l(\boldsymbol{x})-v_l(\boldsymbol{x}))$
MVN $(\hat{\boldsymbol{d}}-\boldsymbol{x})^T\boldsymbol{Z}^{-1}(\hat{\boldsymbol{d}}-\boldsymbol{x})$ or $\sum_{od}(x_{od}-\hat{d}_{od})^2/\,\mathrm{Var}[\eta_{od}]$	MVN $(\hat{\boldsymbol{f}}-\boldsymbol{v}(\boldsymbol{x}))^T\boldsymbol{W}^{-1}(\hat{\boldsymbol{f}}-\boldsymbol{v}(\boldsymbol{x}))$ or $\sum_{l\in M}(\hat{f}_l-v_l(\boldsymbol{x}))^2/\,\mathrm{Var}[\varepsilon_l]$
Multinomial $\sum_{od}x_{od}\ln(x_{od}/\hat{d}_{od})$	

Fig. 8.14 Functional forms of the terms $z_1(\cdot)$ and $z_2(\cdot)$

the estimation context (classic or Bayesian). The generalized least squares estimator is "robust" because it can be adopted in both cases and, as a classic estimator, does not require explicit assumptions on the probability, or likelihood, function of traffic and demand counts. Obviously this robustness is paid for in terms of statistical properties that are less satisfactory than those of other estimators if probability distributions are known for traffic and demand counts.

The literature presents a number of studies comparing the statistical performance of various estimators. Statistical performance can be measured by the "divergence" between the estimates $\boldsymbol{d}^*$ obtained for different specifications of the model (8.5.34) and the true demand vector $\boldsymbol{d}$. The mean square error between the two demand vectors, $MSE(\boldsymbol{d}^*,\boldsymbol{d})$, is one of the most popular divergence measures:

$$MSE(\boldsymbol{d}^*,\boldsymbol{d}) = \frac{1}{n_{\mathrm{OD}}}\sum_{od}\left(d^*_{od}-d_{od}\right)^2$$

where n_{OD} is the number of O-D pairs.

An alternative measure is the ratio between the square root of the mean square error and the average demand, which is analogous to the coefficient of variation of a random variable:

$$RMSE\% = \frac{MSE(\boldsymbol{d}^*, \boldsymbol{d})^{1/2}}{d_{..}/n_{\mathrm{OD}}}$$

Obviously, the lower the *MSE* and *RMSE%* are, the better is the estimator $\boldsymbol{d}^*$. Numerical results seem to confirm the theoretical indications and suggest that, under a wide range of hypotheses on the information contained in $\hat{\boldsymbol{d}}$ and $\hat{\boldsymbol{f}}$, the *GLS* estimator gives more stable results compared with other estimators.

The use of *GLS* estimators requires the definition of variance–covariance matrices $\boldsymbol{Z}$ and $\boldsymbol{W}$. This issue arises only in the case of *GLS* estimators and should be seen as a further degree of freedom because variances and covariances are implicitly defined by the distributions underlying the other functional forms of $z_1(\cdot)$ and $z_2(\cdot)$. For example, expression (8.5.16) for $z_2(\boldsymbol{v}(\boldsymbol{x}), \hat{\boldsymbol{f}})$ implies the assumptions that traffic counts are independent Poisson variables, their deviations from the mean $\boldsymbol{v}(\boldsymbol{x})$ are independent ($\mathrm{Cov}(\varepsilon_l \varepsilon_m) = 0$) and their variance is equal to the mean ($\mathrm{Var}[\varepsilon_l] = v_l(x)$).

In applications, covariances among the components of $\boldsymbol{\varepsilon}$ and $\boldsymbol{\eta}$ are usually ignored; that is, matrices $\boldsymbol{Z}$ and $\boldsymbol{W}$ are assumed to be diagonal. If $\hat{\boldsymbol{d}}$ is a sample estimate, the variance of the sampling error η_{od} depends on the sampling strategy and can be computed, for example, by using formulas (8.2.3) and (8.2.8). In Bayesian estimation, variances are a measure of analyst "confidence" in the a priori estimates and therefore cannot be objectively defined. The variances of the residuals $\varepsilon_l^{\mathrm{OBS}}$ can be obtained through empirical relationships expressing the coefficient of variation CV of assignment errors for different assignment models as a function of measured flows. An example of this type of result was shown in Fig. 5.29 of Chap. 5.

8.5.4 Solution Methods

The main computational problem in solving the ODCBE models is the calculation of the assignment map $\boldsymbol{v}(\boldsymbol{x})$, that is, the assignment matrix $\hat{\boldsymbol{M}}$ expressed by (8.5.6) and (8.5.7). The elements $\hat{m}_{li}$, depend on path choice probabilities, (8.5.4) and (8.5.5), which in turn are functions of path (or hyperpath) costs and hence of link costs, as formally expressed by (8.5.6) and (8.5.7).

In general, given the path or hyperpath choice model, computation of the assignment matrix for given costs $\boldsymbol{c}$ can be carried out with relatively straightforward modifications to the network loading algorithms described in Sect. 7.3. Furthermore, in the case of congested networks, link costs depend on the link flow vector $\boldsymbol{f}$. Solution of the ODCBE problem has two levels of complexity according to whether the link cost vector is known.

Link costs known. Let us assume that an estimate $\hat{\boldsymbol{c}}$ of link costs is available. This is the case if the network is uncongested or moderately congested, and link costs can be estimated independently of flows. Alternatively, link costs can be estimated for congested networks either directly through network (travel time) surveys or indirectly through cost functions and flow counts on congested links $\hat{c}_l = c_l(\hat{f}_l)$. Direct network surveys can be carried out automatically with surveillance systems based on vehicle location and remote transmission technologies.

If link costs are known, the assignment matrix $\hat{\boldsymbol{M}}(\hat{\boldsymbol{c}})$ can be estimated independently of the demand vector; thus $\boldsymbol{d}^*$ can be estimated by applying model (8.5.34):

$$\boldsymbol{d}^* = \arg\min_{\boldsymbol{x}\in S}\left[z_1(\boldsymbol{x}, \hat{\boldsymbol{d}}) + z_2\left(\hat{\boldsymbol{M}}(\hat{\boldsymbol{c}})\boldsymbol{x}, \hat{\boldsymbol{f}}\right)\right] \tag{8.5.35}$$

Model (8.5.35) is a constrained minimization problem that can be solved with different algorithms, depending on the constraints defining the set S. Often the feasibility set S is defined by nonnegativity constraints on the demand flows ($x_i \geq 0\ \forall i$). The projected gradient algorithm, described in Appendix A, can be used in this case. It is usually possible to formulate explicitly the gradient of the objective function. For the *GLS* estimator, under the assumption that the matrices $\boldsymbol{Z}$ and $\boldsymbol{W}$ are diagonal, the ith component of the gradient can be expressed as

$$\begin{aligned} Gr_i &= \frac{\partial}{\partial x_i}\left[\sum_i \frac{(x_i - \hat{d}_i)^2}{\mathrm{Var}[\eta_i]} + \sum_l \frac{(\hat{f}_l - \sum_i \hat{m}_{li}x_i)^2}{\mathrm{Var}[\varepsilon_l]}\right] \\ &= 2\left[\frac{(x_i - \hat{d}_i)}{\mathrm{Var}[\eta_i]} + \sum_l \frac{\hat{m}_{li}(\sum_j \hat{m}_{lj}x_j - \hat{f}_l)}{\mathrm{Var}[\varepsilon_l]}\right] \end{aligned}$$

Figure 8.15 reports the main variables of an application of the projected gradient algorithm for the calculation of $\boldsymbol{d}^{\mathrm{GLS}}$ on a test network.

Link costs unknown. Estimates of link costs might not be available for all links. This is typically the case with congested networks, for which the information described above is often not available. In this case a problem of circular dependence arises, because it is possible to estimate link flows $\boldsymbol{v}(\boldsymbol{d}^*)$ and costs $\boldsymbol{c}(\boldsymbol{v}(\boldsymbol{d}^*))$ by assigning the demand $\boldsymbol{d}^*$ that solves problem (8.5.35), which is expressed in terms of link flows and costs. The estimation problem can be formalized as a fixed-point problem as described below.

Let $\boldsymbol{d} = \delta(\hat{\boldsymbol{M}})$ be the solution of the estimation problem (8.5.35) for a given assignment matrix $\hat{\boldsymbol{M}}$:

$$\boldsymbol{d} = \delta(\hat{\boldsymbol{M}}) = \arg\min_{\boldsymbol{x}\in S}\left[z_1(\boldsymbol{x}, \hat{\boldsymbol{d}}) + z_2(\hat{\boldsymbol{M}}\boldsymbol{x}, \hat{\boldsymbol{f}})\right]$$

If the above problem has only one solution, the relationship $\boldsymbol{d} = \delta(\hat{\boldsymbol{M}})$ can be considered to be a function that associates with each assignment matrix $\hat{\boldsymbol{M}}$ an estimate of the demand vector $\boldsymbol{d}$. The assignment matrix $\hat{\boldsymbol{M}}$ can be expressed as a

TRUE O/D MATRIX

-	25	25	25
25	-	25	25
25	25	-	25
25	25	25	-

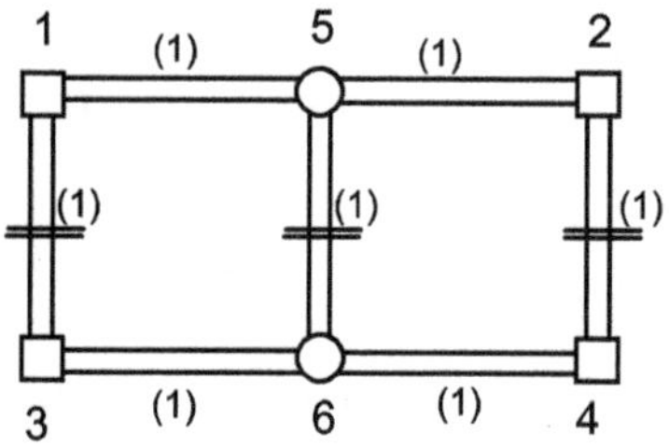

(.) link costs $\hat{c}_l$

TRUE DEMAND VECTOR

$$d^T = \begin{bmatrix}25 & 25 & 25 & 25 & 25 & 25 & 25 & 25 & 25 & 25 & 25 & 25\end{bmatrix}$$

INITIAL DEMAND VECTOR

$$\hat{d}^T = \begin{bmatrix}0 & 50 & 0 & 0 & 50 & 50 & 50 & 0 & 50 & 0 & 100 & 0\end{bmatrix}$$

COUNTED FLOWS

$$\hat{f}^T = \begin{bmatrix}49 & 49 & 33 & 33 & 49 & 49\end{bmatrix}$$

PATH CHOICE MODEL

$$\hat{p}_{ki} = \frac{exp(-C_i)}{\sum_k exp(-C_k)}$$

ASSIGNMENT MATRIX *M*

O/D pairs		1	2	3	4	5	6	7	8	9	10	11	12
links		1-2	1-3	1-4	2-1	2-3	2-4	3-1	3-2	3-4	4-1	4-2	4-3
1	1-3	20	86	33		33	2						20
2	3-1				20			86	33	20	33	2	
3	5-6	10	12	33	10	33	12			10			10
4	6-5	10			10			12	33	10	33	12	10
5	2-4		2	33	20	33	86			20			
6	4-2	20						2	33		33	86	20

Fig. 8.15a Application of the projected gradient algorithm for the computation of $\boldsymbol{d}^{\mathrm{GLS}}$ (input data)

function of demand flows: if we combine the relationship connecting the assignment matrix to link costs, $\hat{\boldsymbol{M}} = \hat{\boldsymbol{M}}(\boldsymbol{c})$, with the cost functions $\boldsymbol{c} = \boldsymbol{c}(\boldsymbol{f})$, and introduce the relationship between link and demand flows through the assignment model $\boldsymbol{f} = \boldsymbol{v}(\boldsymbol{d})$, we get:

$$\hat{\boldsymbol{M}} = \hat{\boldsymbol{M}}\big(\boldsymbol{c}\big(\boldsymbol{v}(\boldsymbol{d})\big)\big)$$

	$z^k(\cdot)$	$\|h\|$		1	2	3	4	5	6	7	8	9	10	11	12
1	425	3.658	**d**	0	50	0	0	50	50	50	0	50	0	100	0
			$-\nabla$	-0.905	-1.195	-1.089	-0.254	-1.089	-1.352	-0.081	-0.823	-0.254	-0.823	-2.978	-0.905
			h	0	-1.195	0	0	-1.089	-1.352	-0.084	0	-0.254	0	-2.978	0
2	92	1.347	**d**	0	35	0	0	36	33	49	0	47	0	63	0
			$-\nabla$	0.194	0.340	0.502	0.379	0.517	0.347	0.248	0.444	0.383	0.444	-0.487	0.194
			h	0.194	0.340	0.502	0.379	0.517	0.347	0.248	0.444	0.383	0.444	-0.487	0.194
3	62	0.817	**d**	2	38	5	4	41	36	51	4	50	4	58	2
			$-\nabla$	-0.078	-0.282	-0.215	-0.018	-0.201	-0.329	-0.281	0.036	-0.015	0.056	-0.544	-0.078
			h	-0.078	-0.282	-0.215	-0.018	-0.201	-0.329	-0.281	0.056	-0.015	0.056	-0.544	-0.078
4	47	0.647	**d**	1	35	2	3	39	33	48	5	50	5	53	1
			$-\nabla$	0.145	0.108	0.182	0.162	0.206	0.072	-0.077	0.314	0.166	0.314	-0.151	0.145
			h	0.145	0.108	0.182	0.162	0.206	0.072	-0.077	0.314	0.166	0.314	-0.151	0.145
9	9.92	0.159	**d**	5	33	4	9	43	29	33	19	56	19	37	5
			$-\nabla$	-0.025	-0.049	-0.054	-0.020	-0.042	-0.053	-0.082	-0.020	-0.017	-0.020	-0.078	-0.025
			h	-0.025	-0.049	-0.054	-0.020	-0.042	-0.053	-0.082	-0.020	-0.017	-0.020	-0.078	-0.025
10	9.45	0.099	**d**	5	32	4	8	42	29	33	18	56	18	36	5
			$-\nabla$	0.019	0.021	0.021	0.025	0.033	0.018	0.008	0.052	0.028	0.052	-0.005	0.019
			h	0.019	0.021	0.021	0.025	0.033	0.018	0.008	0.052	0.028	0.052	-0.005	0.019

	1	2	3	4	5	6	7	8	9	10	11	12
True O/D vector	25	25	25	25	25	25	25	25	25	25	25	25
Initial O/D vector	0	50	0	0	50	50	50	0	50	0	100	0
Estimated O/D vector	5	32	4	8	42	29	33	18	56	18	36	5

STATISTICS	
MSE (true – estimate)	3120.894
MSE (true – initial)	12500.0
Percentage reduction of MSE	0.7503

Fig. 8.15b Application of the projected gradient algorithm for the computation of $\boldsymbol{d}^{\mathrm{GLS}}$ (main variables and comparison statistics)

Thus the ODCBE problem can be expressed as a fixed-point model that is obtained by combining the two functions $\boldsymbol{d} = \delta(\hat{\boldsymbol{M}})$ and $\hat{\boldsymbol{M}} = \hat{\boldsymbol{M}}(\boldsymbol{c}(\boldsymbol{v}(\boldsymbol{d})))$:

$$\boldsymbol{d}^* = \delta\big(\hat{\boldsymbol{M}}(\boldsymbol{d}^*)\big)$$

or

$$\boldsymbol{d}^* = \arg\min_{\boldsymbol{x}\in S}\big[z_1(\boldsymbol{x}, \hat{\boldsymbol{d}}) + z_2\big(\hat{\boldsymbol{M}}\big(\boldsymbol{c}\big(\boldsymbol{v}(\boldsymbol{d}^*)\big)\big)\boldsymbol{x}, \hat{\boldsymbol{f}}\big)\big] \qquad (8.5.36)$$

Alternatively, the ODCBE problem for congested networks can be stated as a bilevel optimization problem. This is the case when the equilibrium assignment map is expressed through an optimization model, as described in Sects. 5.4 and 5.A for DUE and SUE, respectively. In this case, the problem can be stated formally as

$$\boldsymbol{d}^* = \arg\min_{\boldsymbol{x}\in S}\big[z_1(\boldsymbol{x}, \hat{\boldsymbol{d}}) + z_2\big(\boldsymbol{v}(\boldsymbol{x}), \hat{\boldsymbol{f}}\big)\big] \qquad (8.5.37)$$

$$\boldsymbol{v}(\boldsymbol{x}) = \arg\min_{\boldsymbol{f}\in S_f(\boldsymbol{x})} z(\hat{\boldsymbol{f}})$$

where $z(\cdot)$ is the objective function corresponding to the DUE or SUE equivalent optimization problem and the dependence of the link flow feasibility set on the demand vector has been stated explicitly. Obviously, the bilevel optimization approach requires that the assignment problem can be expressed by an optimization model; that is, it must satisfy the mathematical properties stated in Chap. 5: continuous cost functions with symmetric Jacobian. If this is the case, the two formulations (8.5.36) and (8.5.37) are equivalent.

Problems (8.5.36) and (8.5.37) are computationally more complex than problem (8.5.35) because it is necessary simultaneously to solve the constrained optimization problem (8.5.35) given the demand estimate, and the equilibrium assignment problem that gives the link flows and costs.[22] The fixed-point problem (8.5.36) can be solved by using fixed-point iterative algorithms, which alternately solve the demand estimation and the assignment problems by averaging the results until convergence. For example, the MSA algorithm described in Appendix A and applied in Chap. 5 to calculate SUE equilibrium flows can be adopted here. Given an estimate $\boldsymbol{d}^{k-1}$ from iteration $k-1$, the main steps of the algorithm in step k are as follows.

- Calculation of assignment matrix $\hat{\boldsymbol{M}}^k$ corresponding to demand $\boldsymbol{d}^{k-1}$:
  - Assignment of demand $\boldsymbol{d}^{k-1}$ to the network and computation of the corresponding flows

$$\boldsymbol{v}^k = \boldsymbol{v}(\boldsymbol{d}^{k-1});$$

  - Estimation of the link costs and assignment matrix from the obtained flows

$$\boldsymbol{c}^k = \boldsymbol{c}(\boldsymbol{v}^k)$$
$$\hat{\boldsymbol{M}}^k = \hat{\boldsymbol{M}}(\boldsymbol{c}^k)$$

  - Estimation of the auxiliary demand vector $\boldsymbol{y}^k$

$$\boldsymbol{y}^k = \arg\min_{x \in S} \left[z_1(\boldsymbol{x}, \hat{\boldsymbol{d}}) + z_2(\hat{\boldsymbol{M}}^k \boldsymbol{x}, \hat{\boldsymbol{f}}) \right]$$

  - Updating the demand estimate with a weighted average of $\boldsymbol{d}^{k-1}$ and $\boldsymbol{y}^k$:

$$\boldsymbol{d}^k = \frac{k-1}{k} \boldsymbol{d}^{k-1} + \frac{1}{k} \boldsymbol{y}^k$$

This procedure is repeated until a suitable termination test ($\boldsymbol{y}^k \cong \boldsymbol{d}^{k-1}$) is satisfied. The MSA algorithm could be applied to other variables such as link costs or assignment fractions.

[22] In the literature, the fixed-point formulation has been proposed mainly for SUE assignment, where the equilibrium assignment map is defined uniquely, and the bilevel formulation has been proposed for DUE assignment.

8.6 Aggregate Calibration of Demand Models Using Traffic Counts

The aggregate information on travel demand contained in traffic counts[23] can also be used to estimate the parameters of (calibrate) demand models. As stated in Chap. 4, demand models can be viewed as functions relating the demand flows to variables that characterize the activity system $\boldsymbol{SE}$ and the transport system $\boldsymbol{T}$ through a vector of unknown parameters $\boldsymbol{\beta}$.[24]

$$\boldsymbol{d} = \boldsymbol{d}(\boldsymbol{SE}, \boldsymbol{T}; \boldsymbol{\beta}) \tag{8.6.1}$$

For a given specification of the model and given values of $\boldsymbol{SE}$ and $\boldsymbol{T}$, expression (8.6.1) can be considered a relationship between demand flows and the unknown vector $\boldsymbol{\beta}$. This section discusses the problem of combining traffic counts with other information (experimental or not) to estimate the unknown parameters $\boldsymbol{\beta}$. As in the case of O-D flow estimation, the problem can be formulated following either classical or Bayesian approaches.

The classical approach is appropriate when other experimental information is available, typically from RP or SP sampling surveys, for the calibration of demand models. The estimates $\hat{\boldsymbol{\beta}}$ resulting from the methods described in Sects. 8.3 and 8.4 can be viewed as realizations of random variables; thus the estimate of a component $\hat{\beta}_i$ diverges from the "true" value by an unknown quantity σ_i:

$$\hat{\beta}_i = \beta_i + \sigma_i \tag{8.6.2}$$

If $\hat{\beta}_i$ is a maximum likelihood estimate, the variance of σ_i can be calculated as the inverse of the Hessian matrix of the log-likelihood function; see (8.3.6). Furthermore, the estimator is (asymptotically) unbiased: $E(\sigma_i) \to 0$.

In a Bayesian approach, $\hat{\boldsymbol{\beta}}$ can include a priori expectations on the parameters, such as values obtained in a similar study area. In this case, expression (8.6.2) can be viewed as a relationship between the "true" parameter and an initial value and the variance of σ_i is a measure of the analyst's "confidence" in the initial estimate. The two approaches coincide if the a priori estimates are obtained from sampling surveys.

To use traffic counts to estimate $\boldsymbol{\beta}$, it is necessary to express the relationship that links these counts to the unknown parameters of a demand model. In general,

---

[23]The methods described in this section, although presented in terms of traffic counts, can easily be extended to mixed (aggregate/disaggregate) or purely aggregate calibration, using other kinds of aggregate data. For example, the model parameters can be calibrated from estimates $\hat{d}_{od}$ of demand flows derived from different sources (data from transport companies or sampling estimates). In this case the assignment matrix $\hat{\boldsymbol{M}}$ relating the aggregate counts to the demand vector is the identity matrix. Other aggregate data can complement or substitute for traffic counts.

[24]The vector $\boldsymbol{\beta}$ denotes all the identifiable parameters of the specific demand model system, including those that characterize the random residual probability density function. It will include, for example, the coefficients β_i' of a multinomial logit model and the coefficients β_i' and δ_j of a hierarchical logit model.

to calibrate a direct demand model, that is, a single model that encompasses the sequence generation, distribution, and mode choice, it is necessary to have counts on the different modal networks. Let $\hat{\boldsymbol{f}}_m$ be the vector of flows (measured) on the mode m network and $\hat{\boldsymbol{f}}$ be the vector obtained by sequentially ordering all the vectors $\hat{\boldsymbol{f}}_m$. The relationship between the traffic counts $\hat{\boldsymbol{f}}_m$ and the "true" mode m demand vector $\boldsymbol{d}_m$ is basically analogous to (8.5.11), which now becomes:

$$\hat{\boldsymbol{f}}_m = \hat{\boldsymbol{M}}_m \boldsymbol{d}_m + \boldsymbol{\varepsilon}_m^{\mathrm{SIM}} + \boldsymbol{\varepsilon}_m^{\mathrm{OBS}} \tag{8.6.3}$$

The vector of O-D mode m flows, as obtained from the direct demand model, can be expressed as $\boldsymbol{d}_m(\boldsymbol{\beta})$ where, for the sake of simplicity, the vectors $\boldsymbol{SE}$ and $\boldsymbol{T}$ are understood.[25] Even if the "true" parameter vector $\boldsymbol{\beta}$ were known, the demand obtained from the model would diverge from the "true" demand by a vector of errors $\boldsymbol{\varepsilon}_m^{\mathrm{MOD}}$:

$$\boldsymbol{d}_m = \boldsymbol{d}_m(\boldsymbol{\beta}) + \boldsymbol{\varepsilon}_m^{\mathrm{MOD}} \tag{8.6.4}$$

and by substituting (8.6.4) in (8.6.3) we have:

$$\hat{\boldsymbol{f}}_m = \hat{\boldsymbol{M}}_m \boldsymbol{d}_m(\boldsymbol{\beta}) + \boldsymbol{\varepsilon}_m \tag{8.6.5}$$

where the vector ε_m is the sum of all the error components:

$$\boldsymbol{\varepsilon}_m = \boldsymbol{\varepsilon}_m^{\mathrm{SIM}} + \boldsymbol{\varepsilon}_m^{\mathrm{OBS}} + \hat{\boldsymbol{M}}_m \boldsymbol{\varepsilon}_m^{\mathrm{MOD}}$$

and has zero mean if the vectors $\boldsymbol{\varepsilon}_m^{\mathrm{SIM}}, \boldsymbol{\varepsilon}_m^{\mathrm{OBS}}, \boldsymbol{\varepsilon}_m^{\mathrm{MOD}}$ have zero mean.

The relationship (8.6.5) can be extended to the set of counting links belonging to different modal networks:

$$\hat{\boldsymbol{f}} = \hat{\boldsymbol{M}} \boldsymbol{d}(\boldsymbol{\beta}) + \boldsymbol{\varepsilon} \tag{8.6.6}$$

where the vectors $\hat{\boldsymbol{f}}, \boldsymbol{d}(\boldsymbol{\beta})$, and $\boldsymbol{\varepsilon}$ are obtained by sequentially ordering the vectors of the different modes for which traffic counts are available. Similarly the assignment matrix $\hat{\boldsymbol{M}}$ is obtained by sequentially ordering the modal assignment matrices.

The two sources of information about $\boldsymbol{\beta}$, expressed, respectively, by (8.6.2) and (8.6.6), can be combined in different ways, leading to different estimators of $\boldsymbol{\beta}$ that result from the classic or Bayesian interpretation of the initial estimate $\hat{\boldsymbol{\beta}}$ and from the assumptions about the probability distribution of vectors $\boldsymbol{\sigma}$ and $\boldsymbol{\varepsilon}$. It is possible to specify maximum likelihood, generalized least squares, and Bayesian estimators of $\boldsymbol{\beta}$ analogous to those described in Sects. 8.5.1 and 8.5.2. Most estimators can be expressed in the general form:

$$\beta^* = \underset{b \in S_B}{\arg\min} \left[z_1(\boldsymbol{b}, \hat{\boldsymbol{\beta}}) + z_2\left(\hat{\boldsymbol{M}} \boldsymbol{d}(\boldsymbol{b}), \hat{\boldsymbol{f}}\right) \right] \tag{8.6.7}$$

---

[25] Note that the O-D demand on a given mode m usually depends on the level of service attributes of all the competing modes. For this reason, the vector $\boldsymbol{d}_m$ has been expressed as a function of the vector $\boldsymbol{T}$, which includes the attributes of all transport modes. Furthermore, because it is assumed that the assignment matrix $\hat{\boldsymbol{M}}$ is known, the vector of the unknown parameters $\boldsymbol{\beta}$ does not include those involved with path choice. This assumption is relaxed in what follows.

Note that the unknown parameter vector $\boldsymbol{b}$ has significantly fewer components than the O-D demand vector $\boldsymbol{x}$ (dozens of components instead of hundreds or thousands). Thus, problem (8.6.7) has a smaller dimensionality than the ODCBE problem (8.5.12). Conversely, the optimization problem (8.6.7) is "more nonlinear" than in the direct demand estimation case, because the nonlinearity of demand models as a function of unknown parameters is added to the nonlinearity of functions $z_1(\cdot)$ and $z_2(\cdot)$. The feasibility set S_B may coincide with the entire Euclidean space, as in the case of the maximum likelihood estimation dealt with in Sect. 8.3, or alternatively constraints may be imposed on the "expected" signs of the coefficients (e.g., negative cost coefficients).

Model (8.6.7) can also be specified when only aggregate traffic counts or other information sources are available. In this case, the aggregate estimator results from the minimization of the "distance" $z_2(\cdot)$ between the observed traffic counts and the link flows obtained by assigning the O-D flows generated by the demand model. Unlike the ODCBE problem, it is possible to use only traffic counts because the number of independent counts is in general much larger than the number of unknown model parameters.

$$\beta^* = \arg\min_{b \in S_B} z_2\big(\hat{\boldsymbol{M}}\boldsymbol{d}(\boldsymbol{b}), \hat{\boldsymbol{f}}\big)$$

NonLinear Generalized Least Squares (NLGLS) is one of the most widely used specifications of problem (8.6.7). This, in its simplified form, becomes:

$$\beta^* = \arg\min_{\boldsymbol{b} \in S_B} \left[\sum_i \frac{(b_i - \hat{\boldsymbol{\beta}}_i)^2}{\mathrm{Var}[\sigma_i]} + \sum_l \frac{(\hat{f}_l - \sum_i \hat{m}_{li} d_i(\boldsymbol{b}))^2}{\mathrm{Var}[\varepsilon_l]} \right] \tag{8.6.8}$$

Problem (8.6.8), can be solved by a gradient or a projected gradient algorithm similar to those described in Appendix A, according to whether constraints on the components of $\boldsymbol{b}$ have been imposed. The kth component of the gradient for objective function (8.6.8) can be expressed as

$$\begin{aligned} GR_k &= \frac{\partial}{\partial b_k} \left[\sum_i \frac{(b_i - \hat{\beta}_i)^2}{\mathrm{Var}[\sigma_i]} + \sum_l \frac{(\hat{f}_l - \sum_i \hat{m}_{li} d_i(\boldsymbol{b}))^2}{\mathrm{Var}[\varepsilon_l]} \right] \\ &= \frac{2(b_k - \hat{\beta}_k)}{\mathrm{Var}[\sigma_k]} + 2\sum_l \frac{(\sum_i \hat{m}_{li} d_i(\boldsymbol{b}) - \hat{f})}{\mathrm{Var}[\varepsilon_l]} \cdot \sum_i \hat{m}_{li} \frac{\partial d_i(\boldsymbol{b})}{\partial b_k} \end{aligned} \tag{8.6.9}$$

The calculation of the partial derivative of the demand function for the ith O-D pair with respect to the generic parameter β obviously depends on the specification adopted for the demand models being calibrated. Analytical calculation of these derivatives can be very cumbersome, or even impossible (e.g., for probit models); in these cases recourse is had to numerical differentiation methods.

The methods described have been applied to rather simple aggregate demand models, such as traditional four-stage models.[26] The results obtained are generally satisfactory. Figure 8.16 shows an application of estimator (8.6.8) to the coefficients of a four-stage demand model for the city of Reggio Calabria, starting from two different initial parameter vectors. There are no systematic comparisons in the literature of alternative specifications for $z_1(\cdot)$ and $z_2(\cdot)$.

From the statistical point of view, model (8.6.7) can be considered a two-stage mixed (disaggregate/aggregate) estimator of parameters β if it uses disaggregate information (choices made by a sample of users) to estimate $\hat{\boldsymbol{\beta}}$, as well as aggregate information (traffic counts $\hat{\boldsymbol{f}}$) to correct this initial estimate. It is also possible to formulate a "simultaneous" mixed estimator, such as a maximum likelihood estimator that maximizes the probability of observing both the choices $j(i)$ of a sample of users and also the traffic counts $\hat{f}_l$. In this case, assuming that the observations are independent and that users' choices $j(i)$ are obtained with a simple random sample, the estimate $\boldsymbol{\beta}^{\text{ML}}$ can be obtained by combining the log-likelihood function (8.3.3) with one of the functions $z_2(\cdot)$ described in Fig. 8.14, expressing the log-likelihood of observing the counts as a function of the assignment matrix and of the parameter vector $\boldsymbol{\beta}$:

$$\boldsymbol{\beta}^{\text{ML}} = \arg\max_{\boldsymbol{b}\in S_b}\left[\sum_i \ln p^i\left[j(i)\right](\boldsymbol{b}) + z_2\left(\hat{\boldsymbol{M}}\boldsymbol{d}(\boldsymbol{b}), \hat{\boldsymbol{f}}\right)\right]$$

Little is currently known about the simultaneous mixed estimator or about its properties compared with those of the sequential estimator.

A final consideration relates to path choice parameters. In all previous analyses, it has been assumed that the path choice model providing the elements $\hat{p}_{ki}$ of matrix $\hat{\boldsymbol{P}}$, and therefore the matrix $\hat{\boldsymbol{M}}$, was given. In other words, it was assumed that the parameters $\boldsymbol{\beta}^{\text{PATH}}$ in the systematic utility and in the random residual distribution were known. Consequently, these parameters were not included in the vector $\boldsymbol{\beta}$ to be estimated. However, the estimation problem can be specified to improve an initial estimate $\hat{\boldsymbol{\beta}}^{\text{PATH}}$ of these parameters by using traffic counts. In this case, the general expression of the model (8.6.7) becomes:

$$\boldsymbol{\beta}^* = \arg\min_{\boldsymbol{b}\in S_b}\left[z_1(\boldsymbol{b}, \hat{\boldsymbol{\beta}}) + z_2\left(\hat{\boldsymbol{M}}(\boldsymbol{b})\boldsymbol{d}(\boldsymbol{b}), \hat{\boldsymbol{f}}\right)\right] \tag{8.6.10}$$

where the vector $\boldsymbol{\beta}^{\text{PATH}}$ has been included in the general parameter vector $\boldsymbol{\beta}$ and in the variable vector $\boldsymbol{b}$. Comparing expressions (8.6.7) and (8.6.10), the latter is even more nonlinear because the elements of the assignment matrix now depend on unknown parameters.

A similar approach can be followed for the specification of joint estimators of O-D demand flows and path choice parameters. This case results in the following

---

[26] In the case of disaggregate demand models, and sample enumeration aggregation techniques, all the previous expressions still hold. However, each calculation of the demand flow vector requires the application of the entire aggregation procedure, which might be quite burdensome.

	Model	Purpose	Attributes	$\hat{\beta}_1$	$\boldsymbol{\beta}^*$	$\hat{\beta}_2$	β^*
β_1	Gener.	H-WPL	Workers	0.46	0.604	0.230	0.602
β_2	Gener.	H-SC	Students	0.86	0.902	1.015	0.900
β_3	Distrib.	H-WPL	Distance	1.02	0.346	1.103	0.347
β_4	Distrib.	H-WPL	Workplaces	0.70	0.570	1.008	0.550
β_5	Distrib.	H-SC	Distances	0.93	0.900	0.335	0.908
β_6	Distrib.	H-SC	School places	0.35	0.272	0.346	0.269
β_7	Mod. ch.	H-WPL	Walking time	1.19	1.424	1.848	1.649
β_8	Mod. ch.	H-WPL	On-board time	0.54	0.628	0.466	0.559
β_9	Mod. ch.	H-WPL	Cost car/bus	1.80	0.100	1.541	0.100
β_{10}	Mod. ch.	H-WPL	ASA car	2.54	2.543	3.536	3.352
β_{11}	Mod. ch.	H-WPL	ASA bus	2.29	2.330	2.116	3.179
β_{12}	Mod. ch.	H-SC	Walking time	2.18	2.207	3.436	2.737
β_{13}	Mod. ch.	H-SC	On-board time	0.39	0.506	0.349	0.642
β_{14}	Mod. ch.	H-SC	Cost bus	1.58	1.713	1.315	1.980
β_{15}	Mod. ch.	H-SC	ASA bus	1.53	1.544	0.796	2.632

Demand model

$$d_{odm}(H-WPL) = \beta_1\, \text{Work}_o \frac{\exp[\beta_3 \ln \text{dist}_{od} + \beta_4 \ln \text{WPL}_d]}{\sum_{d'} \exp[\beta_3 \ln \text{dist}_{od'} + \beta_4 \ln \text{WPL}_{d'}]} \cdot \frac{\exp[V_{m/od}]}{\sum_{m'} \exp[V_{m'/od}]}$$

$$d_{odm}(H-SC) = \beta_2\, \text{Stud}_o \frac{\exp[\beta_5 \ln \text{dist}_{od} + \beta_6 \ln \text{ScPL}_d]}{\sum_{d'} \exp[\beta_5 \ln \text{dist}_{od'} + \beta_6 \ln \text{ScPL}_{d'}]} \cdot \frac{\exp[V_{m/od}]}{\sum_{m'} \exp[V_{m'/od}]}$$

Mode choice models

H-WPL
$$V_{\text{walk}} = \beta_7 T_w$$
$$V_{\text{car}} = \beta_8 T_c + \beta_9 Mc + \beta_{10} Car$$
$$V_{\text{bus}} = \beta_8 T_b + \beta_9 Mc + \beta_{11} Bus$$

Number of counts	
Road	: 30
Public transport	: 6
Pedestrians	: 26

H-SC
$$V_{\text{walk}} = \beta_{12} T_w$$
$$V_{\text{bus}} = \beta_{13} T_b + \beta_{14} Mc + \beta_{15} Bus$$

Fig. 8.16 Example of demand model calibration with traffic counts

formulation.

$$\boldsymbol{\beta}^*, \boldsymbol{d}^* = \underset{\substack{b \in S_b \\ x \in S_d}}{\arg\min} \left[z_1(\boldsymbol{x}, \hat{\boldsymbol{d}}) + z_2(\boldsymbol{b}, \hat{\boldsymbol{\beta}}) + z_3\left(\hat{\boldsymbol{M}}(\boldsymbol{b})\boldsymbol{x}, \hat{\boldsymbol{f}}\right) \right] \qquad (8.6.11)$$

where the vector $\boldsymbol{\beta}$ coincides with $\boldsymbol{\beta}^{\text{PATH}}$. Problem (8.6.11) simultaneously estimates the path choice model parameters and demand flows that minimize the "distances" from their respective initial estimates and from the observed traffic counts.

Other combined estimators of model parameters and/or demand flows can be specified along the lines described thus far. It should be observed that investigations of the statistical properties and computational issues of these estimators are at a very early stage of research.

8.7 Estimation of Within-Period Dynamic Demand Flows Using Traffic Counts

The O-D flow estimators discussed in Sect. 8.5 were specified under the usual assumption of a within-day static system, that is, assuming that on average all relevant variables are constant within the reference period. In this section the statistical framework proposed for the static problem is generalized and extended to the dynamic O-D estimation case. This problem can be formally stated as that of combining time-varying traffic counts with other available information to estimate time-varying O-D demand flows. The problem is conceptually analogous to the one discussed in Sect. 8.5. The main difference lies in the further complexity introduced by the within-day dynamic framework, as discussed in Chap. 7. In this section some models developed for solving the Dynamic O-D Count Based Estimation (DODCBE) problem are presented, starting with formal relationships between traffic counts and O-D flows. The DODCBE problem has been recently formulated in conjunction with its inverse problem of dynamic traffic assignment (see Chap. 7), and is much less studied than its static counterpart.

Relationships Between Demand and Counts

Relationships between link flows and O-D flows are expressed in terms of discrete time intervals, because this is how flows are counted in practice.

Let the total study period J be divided into n_j intervals $j = 1, \dots, n_j$, of equal duration T, so that $J = n_j \cdot T$. Let $d_{od}[j]$ represent the number of users moving between O-D pair od and leaving the origin during the interval j, and let $\boldsymbol{d}[j]$ be the column vector obtained by arranging the O-D flows for this interval. Let $\hat{d}_{od}[j]$ denote a priori information (an initial estimate of the true demand $d_{od}[j]$), and let $\hat{\boldsymbol{d}}[j]$ be the corresponding vector.

For each interval j a link flow $f_l[j]$ can be associated with each link l of the network,[27] or more precisely to each section of a link: it is the number of users crossing the section in that interval. In general, link counts over an interval are affected by measurement errors $\varepsilon_l^{\mathrm{OBS}}[j]$; the measured flow $\hat{f}_l[j]$ is therefore only an estimate of the actual flow $f_l[j]$. In vector form:

$$\hat{\boldsymbol{f}}[j] = \boldsymbol{f}[j] + \boldsymbol{\varepsilon}^{\mathrm{OBS}}[j] \tag{8.7.1}$$

The link flow $f_l[j]$ is comprised of O-D flows that depart during the same or earlier intervals, and that reach link l in interval j. This can be formally expressed by defining the quantity $m_{lj}^{od,t} \in [0, 1]$ as the fraction of O-D flow $d_{od}[t]$ contributing

[27] As noted in Chap. 7, user flows in a dynamic network model may differ at different cross-sections of the same link. The generic flow at section s of link l was denoted by $f_{l,s}[0]$. In the following, it is assumed that only one counting section is associated to a counted link, and the link flow relates to that counting section. Furthermore, to be consistent with the notation used for the static case, a generic link is denoted l, unlike in Chap. 7.

to the flow on link l in interval j, which gives:

$$f_l[j] = \sum_{t=1}^{j} \sum_{od} m_{lj}^{od,t} d_{od}[t] \tag{8.7.2}$$

Equation (8.7.2) can be expressed in matrix form by introducing the $(n_l \times n_{od})$ assignment fraction matrices $\boldsymbol{M}[t, j]$, analogous to the within-day static counterpart defined in (8.5.11):

$$\boldsymbol{f}[j] = \sum_{t=1}^{j} \boldsymbol{M}[t, j]\boldsymbol{d}[t]$$

This equation assumes that demand flows and counts before the first interval are negligible, which introduces a positive bias in O-D estimates for the first interval. The assumption can be easily relaxed if an estimate is available of O-D demand leaving before the study period.

Let $h_k[j]$ be the path flow, that is, the average number of travelers per time unit following path k between O-D pair od who leave from their origin during period j. Path flows can also be expressed as the product of the O-D demand $d_{od}[t]$ and the probability (average fraction) $p[k/t]$ of choosing path k given the departure interval t:

$$h_k[t] = d_{od}[t] \cdot p[k/t] \tag{8.7.3}$$

In order to express assignment fractions $m_{lj}^{od,t}$ in terms of path choice probabilities, the formal dependence of link flows on path flows must be introduced:

$$f_l[j] = \sum_{od} \sum_{k \in K_{od}} \sum_{t=1}^{j} b_{lj}^{kt} h_k[t] \tag{8.7.4}$$

where the summation is extended to all paths belonging to the set K_{od} of paths connecting O-D pair *od*.

In the above expression, b_{lj}^{kt} is the crossing fraction, that is, the fraction of path flow $h_k[t]$ crossing a section of link l in interval j; the above fractions depend on how link flows are defined, when each path flow reaches link l, and how it moves on it.

By combining (8.7.3) and (8.7.4), and comparing with (8.7.2), we obtain:

$$m_{lj}^{od,t} = \sum_{k \in K_{od}} b_{lj}^{kt} p[k/t] \tag{8.7.5}$$

Equation (8.7.4) can be expressed in matrix form:

$$\boldsymbol{f}[j] = \sum_{t=1}^{j} \boldsymbol{B}[t, j]\boldsymbol{h}[t]$$

where $\boldsymbol{B}[t, j]$ is the crossing fraction matrix $\boldsymbol{B}[t, j] = \{b_{lj}^{kt}\}$.

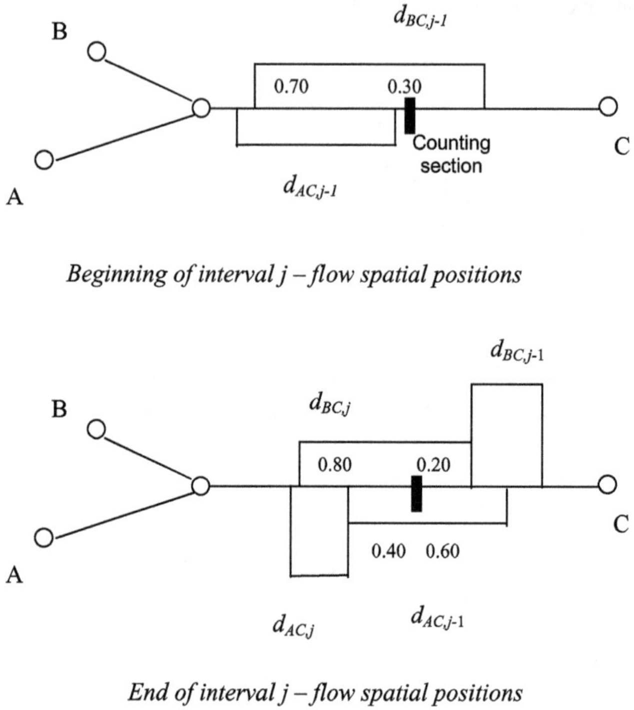

$m_{lj}^{\mathrm{AC},j-1} = 0.60 \qquad m_{lj}^{\mathrm{BC},j-1} = 0.70$

$m_{lj}^{\mathrm{AC},j} = 0 \qquad m_{lj}^{\mathrm{BC},j} = 0.20$

$f_l[j] = 0.60 d_{\mathrm{AC},j-1} + 0.70 d_{\mathrm{BC},j-1} + 0.20 d_{\mathrm{BC},j}$

Fig. 8.17a Relationship between within-day dynamic traffic counts and O-D flows: continuous path flow representation

In practice, path choice and Dynamic Network Loading (DNL) models only provide estimates $\hat{p}[k/t]$ and $\hat{b}_{lj}^{kt}$ of the true values $p[k/t]$ and b_{lj}^{kt}; see Chap. 7. Estimates of assignment fractions can thus be formally expressed as

$$\hat{m}_{lj}^{od,t} = \sum_{k \in I_{od}} \hat{b}_{lj}^{kt} \hat{p}[k/t] \tag{8.7.6}$$

If path flows $h_k[j]$ are modeled as space-continuous packets, crossing fractions may take any value in the interval [0, 1]. More commonly, path flows are modeled as space-discrete packets, and for these models the crossing fractions are either 0 or 1, depending on whether packet $[k, j]$ crosses the counting section on link l during interval t.

Figure 8.17 shows an elementary example of the relationship between within-day dynamic traffic counts and O-D demand flows, for both space-continuous and space-discrete path flows.

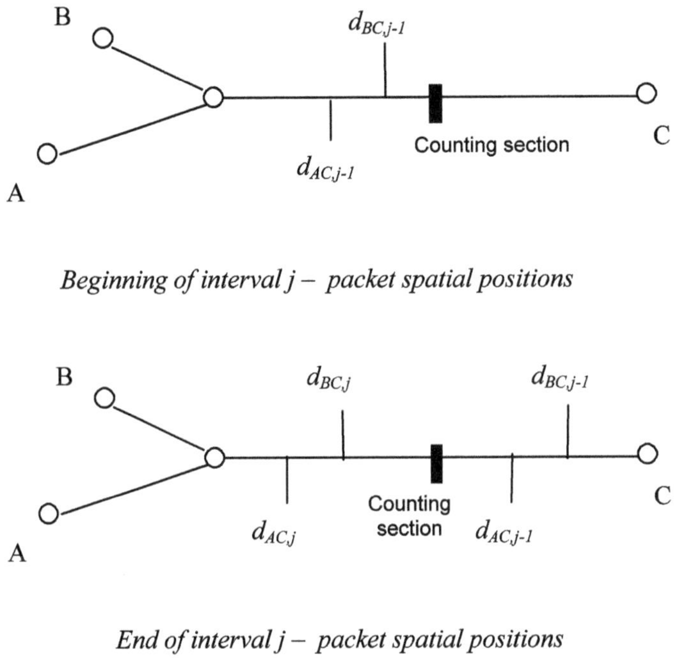

$m_{lj}^{\mathrm{AC},j-1} = 1 \qquad m_{lj}^{\mathrm{BC},j-1} = 1$

$m_{lj}^{\mathrm{AC},j} = 0 \qquad m_{lj}^{\mathrm{BC},j} = 0$

$f_l[j] = d_{\mathrm{AC},j-1} + d_{\mathrm{BC},j-1}$

Fig. 8.17b Relationship between within-day dynamic traffic counts and O-D flows: discrete path flow representation

The estimated values of crossing fractions $\hat{b}_{lj}^{kt}$, path choice probabilities $\hat{p}[k/t]$, and resulting assignment fractions $\hat{m}_{lj}^{od,t}$ are expected to be different from the true ones. As already seen in the static context, this implies that, even if the true demand vector $\boldsymbol{d}[t]$ were known and assigned to the network substituting $\hat{m}$ instead of m in (8.7.2), the resulting link flows would differ from the actual ones by a random error term $\varepsilon^{\mathrm{SIM}}$ (modeling error):

$$f_l[j] = \sum_{t=1}^{j} \sum_{od} \hat{m}_{lj}^{od,t} d_{od}[t] + \varepsilon^{\mathrm{SIM}}[j] \tag{8.7.7}$$

or in matrix form:

$$\boldsymbol{f}[j] = \sum_{t=1}^{j} \hat{\boldsymbol{M}}[t,j]\boldsymbol{d}[t] + \varepsilon^{\mathrm{SIM}} \tag{8.7.8}$$

Equations (8.7.1) and (8.7.8) can be combined into:

$$\hat{f}[j] = \sum_{t=1}^{j} \hat{M}[t, j]d[t] + \varepsilon \tag{8.7.9}$$

where the random vector ε is the sum of the two (independent) vectors $\varepsilon^{\mathrm{OBS}}$ and $\varepsilon^{\mathrm{SIM}}$.

Within-Period Dynamic Estimators of O-D Demand Flows Using Traffic Counts

In Sect. 8.5, it was shown that most O-D demand static estimators can be obtained by solving a constrained optimization problem of the form:

$$d^* = \arg\min_{x \in S} \left[z_1(x, \hat{d}) + z_2\big(v(x), \hat{f}\big) \right] \tag{8.7.10}$$

In this section the estimators previously proposed for the static context are extended to the dynamic estimation problem.

The problem here is to estimate O-D demand flows $d[t]$ for each interval from counts $\hat{f}[j]$. Two alternative approaches are possible. The *simultaneous* approach applies an estimator that gives, in a single step, the entire O-D demand pattern $(d[1], \ldots, d[n])$ for all intervals by processing all available count data. The *sequential* approach involves multiple steps; each step estimates the O-D demand vector for one period only, using counts for that period and the previous one and if possible, the O-D demand estimates developed for previous periods. The following subsections describe these two estimators.

8.7.1 Simultaneous Estimators

Static estimators can be extended in a straightforward manner to the simultaneous estimation framework. In this case, however, the single unknown demand vector has to be replaced by the n_j vectors $(x[1], \ldots, x[j], \ldots, x[n_j])$. Likewise, the counted flow vector is replaced by $(\hat{f}[1], \ldots, \hat{f}[j], \ldots, \hat{f}[n_j])$. The general form of the estimator then becomes:

$$\begin{aligned} d^*[1] \ldots d^*[n_j] = \arg\min_{x[1] \geq 0, \ldots, x[n_j] \geq 0} & \Big[z_1\big(x[1], \ldots, x[n_j]; \hat{d}[1], \ldots, \hat{d}[n_j]\big) \\ & + z_2\big(x[1], \ldots, x[n_j]; \hat{f}[1], \ldots, \hat{f}[n_j]\big) \Big] \end{aligned} \tag{8.7.11}$$

All specifications of objective functions $z_1(\cdot)$ and $z_2(\cdot)$ reported in Sects. 8.5.1 and 8.5.2 can be extended and substituted in (8.7.11), thus obtaining ML, GLS, or Bayesian estimators depending on the distribution assumptions made about the random residuals ε_j. For example, for a GLS estimator they become:

$$z_1 = \sum_{j=1}^{n} (\boldsymbol{x}[j] - \hat{\boldsymbol{d}}[j])^T Z^{-1}[j](\boldsymbol{x}[j] - \hat{\boldsymbol{d}}[j])$$

$$z_2 = \sum_{j=1}^{n} \left(\sum_{t=1}^{j} \hat{\boldsymbol{M}}[t,j]\boldsymbol{x}[t] - \hat{\boldsymbol{f}}[j] \right)^T W^{-1} \left(\sum_{t=1}^{j} \hat{\boldsymbol{M}}[t,j]\boldsymbol{x}[t] - \hat{\boldsymbol{f}}[j] \right)$$

8.7.2 Sequential Estimators

In this method, the O-D demand vector is estimated one interval j at a time. There are two advantages to this approach. The first is the reduction of computational complexity that results from decomposing a large optimization problem into a number of smaller and more manageable ones; the second is that the estimates obtained for one interval can be used as the initial estimates for a subsequent interval.

The main idea in this approach is to express the counts in a given period as a linear (stochastic) function of the unknown demand of the same period only. This is achieved by equating the demand in earlier periods to the corresponding estimates $\boldsymbol{d}^*[t]$, which will have already been computed:

$$\hat{\boldsymbol{f}}[j] = \sum_{t=1}^{j-1} \hat{\boldsymbol{M}}[t,j]\boldsymbol{d}^*[t] + \hat{\boldsymbol{M}}[j,j]\boldsymbol{x}[j] + \boldsymbol{\varepsilon}[j] \tag{8.7.12}$$

The general formulation of the static estimation problem can be adapted to this context, leading to

$$\boldsymbol{d}^*[j] = \arg\min_{x[j] \geq 0} \left[z_1(\boldsymbol{x}[j], \hat{\boldsymbol{d}}[j]) + z_2(\boldsymbol{x}[j]/\boldsymbol{d}^*[1], \ldots, \boldsymbol{d}^*[j-1]; \hat{\boldsymbol{f}}[j]) \right] \tag{8.7.13}$$

where $\hat{\boldsymbol{f}}[j]$ is given by (8.7.12).

In the case of a GLS estimator, the objective functions z_1 and z_2 become:

$$z_1 = (\boldsymbol{x}[j] - \hat{\boldsymbol{d}}[j])^T Z^{-1}[j](\boldsymbol{x}[j] - \hat{\boldsymbol{d}}[j])$$

$$z_2 = \left(\sum_{t=1}^{j-1} \hat{\boldsymbol{M}}[t,j]\boldsymbol{d}^*[t] + \hat{\boldsymbol{M}}[j,j]\boldsymbol{x}[j] - \hat{\boldsymbol{f}}[j] \right)^T \times \boldsymbol{W}^{-1} \left(\sum_{t=1}^{j-1} \hat{\boldsymbol{M}}[t,j]\boldsymbol{d}^*[t] + \hat{\boldsymbol{M}}[j,j]\boldsymbol{x}[j] - \hat{f}[j] \right)$$

8.8 Real-Time Estimation and Prediction of Within-Period Dynamic Demand Flows Using Traffic Counts

By virtue of their distinctive characteristics, simultaneous estimators covered in Sect. 8.7.1 and sequential estimators from Sect. 8.7.2 may be optimally used in different application contexts. Indeed, given an observation period T subdivided into n_t intervals of length t, a simultaneous estimator is run once, fed with the traffic counts for all n_t intervals, producing an estimate of O-D flows for all n_t intervals. Hence such estimators are computationally demanding and may only be applied downstream of period T (i.e., when data for the whole period are made available). These two circumstances make this type of estimator optimal for offline, typically planning, applications.

Instead, a sequential estimator is run n_t times (once for each interval of length t), fed each time with traffic counts for the interval being estimated and producing an estimate of O-D flows for that one interval. Hence such estimators are computationally much more efficient and may be applied downstream of each interval t (in other words, as soon as counts for that interval are made available) without waiting for the end of period T. These two circumstances make this type of estimator optimal for applications in which O-D flows have to be predicted/estimated in real-time (as in certain online traffic or infomobility management applications), where traffic count data are used to correct O-D flows as they are made available (i.e., in real-time).

Of the more efficient procedures proposed in the literature on this point, mention should be made of those based on using the Kalman filter. The Kalman filter for the state estimation problem is a recursive estimator able to combine information from a model with information derived from measurements made on the system. The filter is applied to a linear state–space model of a real system[28]:

$$\begin{aligned} \boldsymbol{x}[j+1] &= \boldsymbol{A}[j]\cdot\boldsymbol{x}[j] + \boldsymbol{B}[j]\cdot\boldsymbol{u}[j] + \boldsymbol{D}[j]\cdot\boldsymbol{w}[j] \\ \boldsymbol{y}[j] &= \boldsymbol{C}[j]\cdot\boldsymbol{x}[j] + \boldsymbol{v}[j] + \boldsymbol{\varepsilon}[j] \end{aligned} \tag{8.8.1}$$

where $\boldsymbol{x}[j]$ are the state variables, $\boldsymbol{u}[j]$ and $\boldsymbol{v}[j]$ the control and measurement inputs, $\boldsymbol{w}[j]$ and $\boldsymbol{\varepsilon}[j]$ the model and measurement errors, and $\boldsymbol{A}$, $\boldsymbol{B}$, $\boldsymbol{C}$, and $\boldsymbol{D}$, their respective coefficient matrices, which are time-dependent in the case of a nonstationary system. The errors $\boldsymbol{w}[j]$ and $\boldsymbol{\varepsilon}[j]$ due to the model and the measurements, respectively, are white stochastic processes with zero mean and known covariance matrices $\boldsymbol{Q}[j]$ and $\boldsymbol{R}[j]$:

$$\begin{aligned} E\left[\boldsymbol{w}(j)\boldsymbol{w}(i)^T\right] &= \begin{cases} \boldsymbol{Q}[j] & \text{if } i=j \\ \boldsymbol{0} & \text{otherwise} \end{cases} \\ E\left[\boldsymbol{\varepsilon}(j)\boldsymbol{\varepsilon}(i)^T\right] &= \begin{cases} \boldsymbol{R}[j] & \text{if } i=j \\ \boldsymbol{0} & \text{otherwise} \end{cases} \end{aligned} \tag{8.8.2}$$

[28] In the case in which the system is nonlinear we may resort to an extended Kalman filter.

The first equation of (8.8.1), called the state equation, links the system state at instant $j+1$, to the state, the inputs, and noise at instant j. The second equation, known as the output or measurement equation, defines the system output at instant j.

Estimation of the state variables made by the Kalman filter entails at each step j a prediction phase in which the unknown vector of the state variables is calculated from the model, and an updating or adjustment phase in which the prediction is adjusted on the basis of the measurements made. In the adjustment phase, the model information and the measurements are weighted with their respective variances. The algorithm supplies the value of the state variables that minimize the estimation error, that is, the difference between estimation of the state and the actual state.

The filter estimates are obtained by solving the following equation,

$$\hat{\boldsymbol{x}}_F[j] = \hat{\boldsymbol{x}}_P[j] + \boldsymbol{K}[j] \cdot \left[\boldsymbol{y}[j] - \boldsymbol{C}[j] \cdot \hat{\boldsymbol{x}}_P[j] - \boldsymbol{v}[j]\right] \tag{8.8.3}$$

where $\hat{\boldsymbol{x}}_P$ is the one step prediction at j, obtained by solving the following recursive equation,

$$\hat{\boldsymbol{x}}_P[j] = \boldsymbol{A}[j-1] \cdot \hat{\boldsymbol{x}}_F[j-1] + \boldsymbol{B}[j-1] \cdot \boldsymbol{u}[j-1] + \boldsymbol{D}[j-1] \cdot \boldsymbol{w}[j-1] \tag{8.8.4}$$

The matrix $\boldsymbol{K}[j]$ is called the gain matrix:

$$\boldsymbol{K}[j] = \boldsymbol{P}_P[j] \cdot \boldsymbol{C}[j]^T + \left[\boldsymbol{C}[j] \cdot \boldsymbol{P}_P[j] \cdot \boldsymbol{C}[j]^T + \boldsymbol{R}[j]\right]^{-1} \tag{8.8.5}$$

and represents a compromise between two distinct requirements: the need to use measurements available to adjust the model estimate of the future state and not to downgrade this estimate because of errors in measurements. The gain matrix is proportional to the covariance matrix of the estimate error $\boldsymbol{P}_P[j]$, which must be updated for every step through the following formula.

$$\boldsymbol{P}_P[j+1] = \boldsymbol{A}[j] \cdot \boldsymbol{P}_F[j] \cdot \boldsymbol{A}[j]^T + \boldsymbol{D}[j] \cdot \boldsymbol{Q}[j] \cdot \boldsymbol{D}[j]^T \tag{8.8.6}$$

where $\boldsymbol{P}_F[j]$ is given by:

$$\boldsymbol{P}_F[j] = \left[\boldsymbol{I} - \boldsymbol{K}[j] \cdot \boldsymbol{C}[j]\right] \cdot \boldsymbol{P}_P[j] \tag{8.8.7}$$

and $\boldsymbol{I}$ is the identity matrix.

In applying the Kalman filter to the estimate/forecast of O-D demand flows $\boldsymbol{x}[j]$, because there are no control inputs the state equations (8.8.1) are expressed as follows.

$$\boldsymbol{x}[j+1] = \boldsymbol{A}[j] \cdot \boldsymbol{x}[j] + \boldsymbol{D}[j] \cdot \boldsymbol{w}[j] \tag{8.8.8}$$

In general, the measurement equations correspond to assignment equations (8.7.12) that relate link flows measured at instant j to unknown demand flows to instant j and to demand flows already estimated for previous instants, and may be boiled down similarly to the general expression (8.8.1) as follows.

$$\hat{\boldsymbol{f}}[j] = \boldsymbol{C}[j] \cdot \boldsymbol{x}[j] + \boldsymbol{v}[j] + \boldsymbol{\varepsilon}[j] \tag{8.8.9}$$

where $\boldsymbol{C}[j]$ represents the assignment matrix $\hat{\boldsymbol{M}}[j, j]$ and $\boldsymbol{v}[j]$ the contribution to link flows from the O-D assignment flows for instants prior to j and already estimated at previous steps $(\sum_{t=1}^{j-1} \hat{\boldsymbol{M}}[t, j]\boldsymbol{d}^*[t])$.

Note that the model described above is independent of an a priori estimate of demand flows. Alternatively, if historical data are available on demand flows, this important additional information may be allowed for, considering as state variables no longer the demand flows but the deviations of such flows from the historical data. In this case the state equations (8.8.8) become:

$$\boldsymbol{x}[j+1] - \boldsymbol{d}^H[j+1] = \boldsymbol{A}[j] \cdot \left[\boldsymbol{x}[j] - \boldsymbol{d}^H[j]\right] + \boldsymbol{D}[j] \cdot \boldsymbol{w}[j] \tag{8.8.10}$$

where $\boldsymbol{d}^H[j]$ represents the historical flow demand vector for period j and the measurement equations become:

$$\hat{\boldsymbol{f}}[j] - \boldsymbol{f}^H[j] = \boldsymbol{C}[j] \cdot \left[\boldsymbol{x}[j] - \boldsymbol{d}^H[j]\right] + \boldsymbol{v}'[j] + \boldsymbol{\varepsilon}[j] \tag{8.8.11}$$

where

$$\boldsymbol{v}'[j] = \sum_{t=1}^{j-1} \hat{\boldsymbol{M}}[t, j]\left[\boldsymbol{d}^*[t] - \boldsymbol{d}^H[t]\right] \tag{8.8.12}$$

It is worth pointing out that if the variables are not flows but their deviations from historical data, the hypothesis that the errors of model $\boldsymbol{w}[j]$ may be normally distributed proves even more convincing.

8.9 Applications of Demand Estimation Methods

The methods described in this chapter can be used to estimate demand flows for an existing system or for hypothetical transportation and/or activity system scenarios. These estimates can in turn be used to determine link flows and performances with an assignment model and/or to analyze the structure of the travel demand in a given area. Obviously, different techniques, or combinations of techniques, can be used for different applications and for different demand components. Below, the main application areas and corresponding demand estimation methodologies are described, with the results of the previous sections being summarized (see Fig. 8.18).

8.9.1 Estimation of Present Demand

Estimation of average demand flows in the reference period can be performed either by using sampling surveys and direct estimation methods, or by applying a system of demand models to the present configuration of the system. In the former case, the sampling methods described in Sect. 8.2 are used. From a practical point of view,

Area of application	Estimation method	Input data	Complementary techniques
Estimation of present demand	Direct estimation	Sampling surveys	Estimation of O-D matrices with traffic counts
	Model estimation	• Model parameters • Attributes of the activity systems SE^P • Attributes of the transport system T^P	
Estimation of demand variations (forecast)	Model estimation	• Model parameters • Attributes of the activity systems (scenarios) SE^F • Attributes of the transport system (projects) T^F	Pivoting on the present demand

Fig. 8.18 Application of demand estimation methods

it should be noted that different types of sampling surveys are typically used for the estimation of different demand components. In particular, on-board or en-route surveys are often used to estimate internal–external and external–external flows, whereas household surveys are used to estimate internal demand flows.

Demand models can be used as estimators of present demand by applying them with present values of the attributes of the activity system $\boldsymbol{SE}^P$ and of the transportation supply system $\boldsymbol{T}^P$. Model estimation of present demand can be formally expressed as

$$\hat{\boldsymbol{d}}^P_{\text{MOD}} = \boldsymbol{d}(\boldsymbol{SE}^P, \boldsymbol{T}^P; \hat{\boldsymbol{\beta}}) \tag{8.9.1}$$

where $\hat{\boldsymbol{\beta}}$ indicates the estimate of the parameter vector. Expression (8.9.1) can be applied to estimate demand flows with different levels of aggregation, for example, by origin, destination, and mode.

The model-based estimation of present demand deserves a few comments.

- The rationale of the method is that, for a given sample size, estimates of the parameters $\hat{\boldsymbol{\beta}}$ are significantly more precise than direct sampling estimates of $\hat{\boldsymbol{d}}$. The underlying assumption of the method is that deviations between true demand flows and model-based estimates are less dispersed than deviations between direct estimates and the true demand flows. This assumption has received some limited empirical validation.
- Application of demand models requires the aggregation of the results. The different aggregation techniques described in Sect. 3.7 can be used to obtain estimates of the trip flows between the different origin–destination pairs. Aggregation by categories (aggregate models) and sample enumeration (disaggregate models) are the most common options.
- Models used for present demand estimation might be different and less sophisticated than those used to predict demand variations following changes in the ac-

tivity or transportation system. In the former case, the sole function of the model may be to describe the observed phenomenon. However, it is reasonable to assume that a model that captures underlying behavioral decision-making should be a better predictor. Again, models of various levels of complexity can be used to estimate different components of present demand based on their importance. In particular, internal–external demand can be estimated with simpler models that require less information than those used to estimate demand flows within the study area.
- Model specification, calibration and validation can be conducted using the disaggregate methodologies described in Sects. 8.3 and 8.4, if possible integrated with the mixed aggregate/disaggregate estimation method using traffic counts, as described in Sect. 8.6.

The two methods (direct estimation and model-based estimation) are generally used to estimate different components of present demand. For example, it is quite common to use direct estimation for internal–external and external–external demand (for which it is both easier to conduct direct cordon surveys and more complicated to formulate demand models) and model-based estimation for internal demand. Finally, present demand can be estimated by combining direct estimation and/or model-based estimation with aggregate information on traffic counts, using the methods described in Sect. 8.5.

8.9.2 Estimation of Demand Variations (Forecasting)

The classic use of demand models is to represent demand variations following changes in the activity system and/or the transportation supply system. There is obviously a close interdependence between the characteristics of demand models and the project under study, inasmuch as the model must be "elastic" with respect to variables describing the changes whose effects are to be evaluated. For example, when developing the circulation plan of an urban road network, it is sometimes assumed that all aspects of travel demand will remain unchanged by the plan except for user path choices. This implies that the present O-D demand matrix for the "car" mode can be used to measure the impacts of alternative plans, and that the only demand model necessary for this application is the path choice model used for fixed demand assignment. On the other hand, if the same plan is included in a wider project aimed at modifying the current modal split, for example, by introducing parking charges, it would be necessary to use both mode and path choice models, which could be applied to present O-D matrices of total demand.

For short-term projects it is generally assumed that socioeconomic variables of the activity system remain unaffected and the transportation performance variables are modified by the project. These variations may affect travel choices in several dimensions (path, mode, destination, frequency). In this case, the application of the demand models can be formally expressed as

$$\hat{\boldsymbol{d}}^F_{\text{MOD}} = \boldsymbol{d}(\boldsymbol{SE}^P, \boldsymbol{T}^F; \hat{\boldsymbol{\beta}}) \tag{8.9.2}$$

where $\hat{\boldsymbol{d}}^F_{\text{MOD}}$ indicates the vector of the model-based estimates of "future" demand flows and $\boldsymbol{T}^F$ indicates the vector of level-of-service attributes corresponding to the project.

Medium- to long-term projects usually require quantifying their effects over a correspondingly long period. In this case, it is necessary to forecast the evolution of these variables. In general, it is very difficult to develop reliable medium- to long-term forecasts of the evolution of significant activity system variables such as resident population and income levels, the organization of economic production, family lifestyles, and the location of manufacturing and service activities. Even if some activity system variables can be considered endogenous to the model system, particularly transport–land use interaction models, the evolution of other exogenous variables must still be forecast. In practice, for long-term applications, different scenarios[29] for the evolution of the variables $\boldsymbol{SE}^F$ are used. Demand models are applied to each scenario and the resulting ranges of variation of the key variables can be used to design and evaluate project alternatives, as shown in Chap. 10. Estimation of demand flows over long periods can therefore be formally expressed as

$$\hat{\boldsymbol{d}}^F_{\text{MOD}} = \boldsymbol{d}(\boldsymbol{SE}^F, \boldsymbol{T}^F; \hat{\boldsymbol{\beta}}) \tag{8.9.3}$$

The comments on model calibration and aggregation techniques can be extended to both of the applications (8.9.2) and (8.9.3).

Another method, known as pivoting, forecasts future demand by estimating the change that it represents relative to present demand. This approach assumes that it is possible to obtain estimates, $\hat{\boldsymbol{d}}^P$, of present demand that are better than those obtained using only demand models. This may be the case if other sources of information on present demand are available (e.g., traffic counts), so that direct or model-based estimates of present demand are improved with such information. In this case, modeling errors can be reduced by using the models as predictors of relative demand changes and, therefore, obtaining "future" demand estimates as

$$\hat{d}^F_{od} = \hat{d}^P_{od} \cdot \frac{d_{od}(\boldsymbol{SE}^F, \boldsymbol{T}^F; \hat{\boldsymbol{\beta}})}{d_{od}(\boldsymbol{SE}^F, \boldsymbol{T}^P; \hat{\boldsymbol{\beta}})} \tag{8.9.4}$$

The general form (8.9.4) must be specialized for the particular demand dimensions to which it is applied. For example, by applying the method to modal O-D matrices and leaving to network assignment the determination of path choice probabilities and flows (see Fig. 8.19), expression (8.9.4) becomes:

$$d^F_{od}[shm] = \hat{d}^P_{od}[shm] \cdot \frac{d_{od}[shm](\boldsymbol{SE}^F, \boldsymbol{T}^F; \hat{\boldsymbol{\beta}})}{d_{od}[shm](\boldsymbol{SE}^P, \boldsymbol{T}^P; \hat{\boldsymbol{\beta}})} \tag{8.9.5}$$

[29] A scenario can be defined as a set of internally consistent assumptions regarding the exogenous variables of a model system. In some applications, scenarios are obtained from separate macroeconomic models that require fewer input variables (e.g., population and economic growth rates). The outputs of these models are used to generate consistent sets of disaggregate input variables for travel demand models.

OD Pair	$\hat{d}^{P}_{od\,Car}$	$\hat{d}^{P}_{od\,Train}$	$d^{P(MOD)}_{od\,Car}$	$d^{P(MOD)}_{od\,Train}$	$d^{F(MOD)}_{od\,Car}$	$d^{F(MOD)}_{od\,Train}$	$\hat{d}^{F}_{od\,Car}$	$\hat{d}^{F}_{od\,Train}$
1,2	100	30	92	31	85	40	92.4	38.7
1,3	30	15	26	11	22	25	25.4	34.1
1,4	70	25	73	22	60	31	57.5	35.2
2,1	120	46	116	47	103	53	106.6	51.9
2,3	50	22	47	19	49	29	52.1	33.6
2,4	60	18	55	20	53	31	57.8	27.9
3,1	85	32	88	27	76	39	73.4	46.2
3,2	70	27	71	30	68	46	67.0	41.4
3,4	23	5	20	6	18	11	20.7	9.2
4,1	58	24	56	22	52	30	53.9	32.7
4,2	65	26	66	24	60	35	59.1	37.9
4,3	90	32	87	33	70	48	72.4	46.5

Fig. 8.19 Application of the pivoting method

The use of the pivoting method in the form of (8.9.4) requires a double application of the model to both present $(\boldsymbol{SE}^P, \boldsymbol{T}^P)$ and future $(\boldsymbol{SE}^F, \boldsymbol{T}^F)$ scenarios. Furthermore, the method must be adapted for practical applications; for example, (8.9.5) would not allow the estimation of demand associated with the introduction of a new mode of transport for which no present demand exists. These distortions can be corrected in various ways, for example, by applying the pivoting method partially to foresee changes in present demand in some dimensions and then directly applying the models to the other dimensions.

Reference Notes

Direct demand estimation is based on the application of sampling surveys and estimators. A description of "classical" travel demand surveys can be found in manuals such as the one from RRL (1965) and EPA (1996). For statistical sampling theory, refer to the texts by Cochran (1963) and Yates (1981). Applications to travel demand estimation are covered in several articles, such as those of Smith (1979) and Brog and Ampt (1982), as well as in the volume by Ortuzar and Willumsen (2001).

The literature on specification, calibration, and validation of demand models is quite substantial. The books by Domencich and McFadden (1975) and Ortuzar and Willumsen (2001), as well as the articles by Horowitz (1981, 1982) and Manski and McFadden (1981) address various statistical aspects of the calibration of disaggregate models. A review of the field as of the early 1980s is contained in Gunn and Bates (1982). The work of Manski and Lerman (1977) studies model calibration based on nonrandom samples. A detailed and systematic discussion of the subjects in Sect. 8.3 is contained in the volume by Ben-Akiva and Lerman (1985), and the reader is referred to its comprehensive bibliography.

Recent advances in random utility model estimation are reported in the book by Train (2003), who also provides an exhaustive insight into model estimation in the presence of repeated observations (panel data), and reports theoretical and operational details on simulated log-likelihood estimation for nonclosed random utility models.

Stated preference survey techniques have been the subject of growing interest over the last 10 to 15 years and are an area in continuous evolution from both the theoretical and application points of view. An exhaustive review on this subject is provided by Louviere et al. (2000). A discussion of the theoretical aspects of SP techniques can be found in the works of Hensher et al. (1988), Louviere (1988), and Ortuzar (1992), whereas practical aspects are covered in Pearmin et al. (1991). The statistical bases of factorial survey designs are described in greater detail in texts on experimental design such as Box et al. (1978). The calibration of demand models from combined SP–RP surveys is dealt with in Ben-Akiva and Morikawa (1990) and Bradley and Daly (1992). An application to mode choice modeling is described in Biggiero and Postorino (1995), from which the example in Fig. 8.10 is taken.

Estimation of demand flows using traffic counts is a subject that has been intensely researched over the last two decades. A state of the art and literature review can be found in Cascetta and Improta (1999). The general statistical bases are addressed in Cascetta and Nguyen (1986). For estimation of O-D demand flows using traffic counts, there are several papers on particular estimators or specific applications. The papers by Van Zuylen and Willumsen (1980) on the maximum entropy estimator, Maher (1983) on Bayesian estimators, Cascetta (1984) proposing the GLS estimator, Bell (1991) on applications of the GLS method, and Di Gangi (1988) on a numerical comparison of the statistical performances of different estimators are all of interest.

The problem of estimating O-D flows using traffic counts in congested networks is more recent; it has been studied by a number of authors, typically as a bilevel programming problem for DUE assignment: see Florian and Chen (1995) and Yang (1995). The fixed-point formulation and the MSA algorithm described in Sect. 8.5, with some variants, are described in Cascetta and Postorino (2001).

Estimation of model parameters using traffic counts and other sources is a well-established heuristic practice, but has received relatively limited theoretical attention. Among the first papers proposing methods for aggregate estimation of coefficients using traffic counts, those by Cascetta (1986) proposing GLS estimators and by Willumsen and Tamin (1989) describing an estimator for gravity type models are of note. The paper by Cascetta and Russo (1997) describes the general statistical framework discussed in Sect. 8.6. The combined estimation (both aggregate and disaggregate) of model parameters and O-D flows using traffic counts is original.

In the literature, various methods have been proposed to estimate time-varying O-D flows using traffic counts. Among others, Cremer and Keller (1987) propose sequential estimators for a simple network using traffic counts alone. Cascetta et al. (1993) propose dynamic estimators obtained by optimizing a two-term objective function as described in Sect. 8.7. Nguyen et al. (1989) proposed different simultaneous estimators on a general transit network. The Kalman filtering approach was first used for real-time prediction of time-varying O-D flows by Okutani and Stephanades (1984) and Ashok and Ben-Akiva (2002) propose to use deviation with respect to historical O-D flows instead of O-D flows as state variables. A recent contribution on the dynamic O-D matrix estimation was provided by the doctoral dissertation of Lindveld (2003). Overall validation of the effectiveness of the procedure for estimating O-D flows using traffic counts both in static and in dynamic contexts was recently provided by Marzano et al. (2008).

Chapter 9
Transportation Supply Design Models

9.1 Introduction

This chapter outlines a wide range of methods and mathematical models that may assist the transportation systems engineer in designing projects or other interventions. It should be stated at the outset that supply design models[1] are not meant to "automate" the complex task of design, especially when the proposed actions can significantly alter the performance of the transportation system. In this case, as we have seen, a project may have structural effects ranging from changes in land use to modifications in the level and structure of travel demand. On the other hand, the elements of the transportation supply to be designed may assume a very large number of possible configurations: circulation directions in an urban road network or the lines and frequencies of a transit system are two cases in point. In the presence of such a large number of possibilities, it is practically impossible to explore and compare all the feasible configurations in order to identify the optimum with respect to a given set of objectives and constraints.

From the modeling perspective, supply design models belong to a different class from the models described thus far and, in some respects, can be considered as extensions or generalizations of these models. The mathematical models described in the previous chapters aim at simulating the relevant aspects of a transportation system under the assumption that the supply (facilities, services, and prices) and activity systems are exogenously given. These models can be used as "design tools" by simulating the main effects of exogenously specified projects, verifying their technical compatibility and evaluating their "convenience" as shown in Chap. 10. This approach is known as "what if." By contrast, supply design models provide "what to" indications, that is, how to alter supply in order to optimize given objectives while satisfying given constraints (see Fig. 9.1). Clearly, in order to identify solutions for the design problem, it is necessary to evaluate the system responses (demand, flows, and performances) to the possible actions; therefore the simulation model is a component of the design model. The cost of this generalization is not only the simplification of the real design problem, but also the simplification of the simulation models, which now become submodels of a wider model.

An interesting interpretation of the differences between simulation and design models can be given in terms of game theory. The design problem can be seen as a Stackelberg game. One of the two players (or groups of players), called the leader,

---

[1] In the literature, supply design problems and their corresponding models are often called network design problems (NDP). This definition, as shown, applies to a wide subset of the entire range of supply design problems, specifically those that refer to the definition of network elements.

E. Cascetta, *Transportation Systems Analysis*, Springer Optimization and Its Applications 29, DOI 10.1007/978-0-387-75857-2_9, © Springer Science+Business Media, LLC 2009

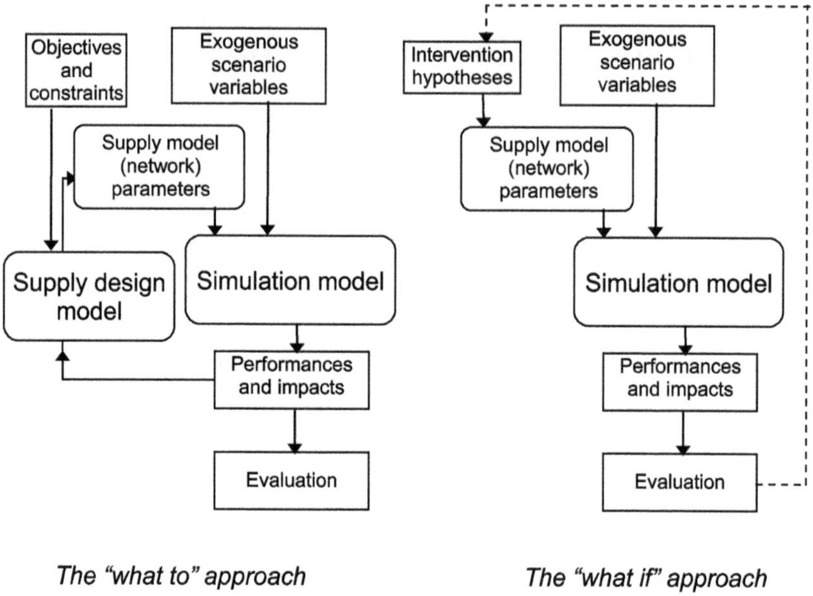

Fig. 9.1 Two approaches to transportation supply design

knows in advance the reactions of the other player (or group of players), called the follower, to her actions. In this case, the leader is the designer (or manager) of the supply system and the followers are the users of the transportation system. The designer is able to anticipate user reactions and exploits this information to achieve his objectives.[2] In this context, the simulation models represent the tools to predict user reactions, whereas supply design models provide the leader with the "winning" strategy. On the other hand, within the context of game theory, the simulation model can be interpreted as a description of a Nash game, in which the generic player (i.e., the user) does not know the possible reactions of the other players.

Supply design models typically simplify the actual design problem, accounting for only some control variables and simulating the relationships between these variables and the system through simplified models. In general, the design problem is expressed as one of optimizing an objective function under certain constraints; the solution, or solutions, of this problem are then used as starting points for successive extensions and comparative evaluation described in Chap. 10. Obviously, the more "elementary" is the intervention to be designed, the closer the formulation of the

[2]The model described corresponds to a monopoly market situation. In reality, the situation is often more complex. For example, in the transportation market, there might be multiple operators (e.g., air service, railway, and road managers), each with their own objectives and constraints and with the ability to forecast the demand reactions to their own actions as well as those of their competitors. Supply design models currently available are not yet capable of simulating this type of oligopolistic market situation.

Fig. 9.2 Classification of supply design models

Design (control) variables	Network topology Performances Prices and fares
Objectives	Society's Operator's
Constraints	External Technical Demand/flow/cost consistency
Simulation model	Assignment model

supply design model is likely to be to the real problem. Thus, the problem of designing traffic signal control parameters at an isolated intersection can be expressed by an optimization model that, among all possible values, searches for those minimizing the total delay, or maximizing the total capacity of the intersection. The resulting optimal control parameters can be directly used in the real world.

On the other hand, if the problem is to design the transportation system of an entire region, it is practically impossible to represent the complexity of the objectives and constraints. In this case, one or more simplified design models can be formulated, for example, to define the road network, the public transport network, and the pricing structure, which jointly or separately minimize the total generalized user costs under budget, technical, and environmental constraints. In any case, the solution or solutions of the partial problems will only be the starting point for the further phases of design, evaluation, and negotiation, which will lead eventually to choices by society. The design models proposed in the literature and most often used in applications can be classified on the basis of some elements described below and summarized in Fig. 9.2.

Design (control) variables. Design problems can be divided into three groups with respect to the control variables: *network topology* or layout (e.g., of the road network or public transport lines), *performances* of supply elements (e.g., transit line frequencies or traffic signal control parameters), and *pricing* (e.g., air, rail, parking, or motorway fares). The design variables may be discrete (topology and performances) or continuous (prices and performances) according to the specific problem. Obviously a model can, and often does, aim at defining the optimal combination of different types of variables.

Objectives. The design can be developed from different perspectives; in other words, the design model can be defined to optimize (maximize or minimize) different objective functions. Design models can account for society's objectives, such as minimization of total generalized user costs, and/or the operator's objectives, such as minimization of investment and/or management costs or maximization of net revenues from traffic. The social objectives underlying larger projects are significantly simplified. The objective function may be mixed, that is, a combination of society's and the operator's objectives as in benefit-cost analysis, described in

Sect. 10.3.1. Other objective functions correspond to multiattribute utility functions in multi-criteria analysis, as described in Sect. 10.3.3.

Constraints. Most supply design models can be formulated as constrained optimization problems and, as is often the case in modeling, some objectives can be introduced as constraints (and vice versa) for computational convenience. Some of these constraints may be external, for example, the maximum available budget or the maximum concentration levels of pollutants. In the former case, the implicit objective is to minimize the cost; in the latter, it is to reduce air pollution. Technical constraints relate to aspects of the system such as maximum flow-capacity ratios, and minimum and maximum frequencies of bus lines. Some specifications of the design model use a third category of constraints representing the consistency between demand, flows, design variables, and system performances. These constraints represent the system simulation model and are considered in the next section.

Simulation model. The simulation model that is usually most relevant to design problems is the assignment (or demand–supply interaction) model. As shown in previous chapters, such models can be based on within-day static or dynamic system representations, on deterministic or stochastic path choice models, and may or may not account for congestion effects. Furthermore, the assignment model may assume fixed or elastic demand according to whether demand flows are considered constant with respect to the values of the design variables.

Although transportation supply design models have received considerable scientific and professional attention, they have not reached a level of theoretical completeness and/or breadth of applications comparable to those described in previous chapters for simulation models. Furthermore, design problems have not been studied at the same level of detail as these other models. It is difficult to present general results for all supply models, as they are specific to the design problem and to a number of assumptions that can be made in connection with each of them. A systematic review of all the supply models presented in the literature and of their transportation engineering implications would require a book on its own. Rather, this chapter briefly analyzes this broad application area. General formulation of the supply design models is described first in Sect. 9.2; some specialization of the general model to the most common design problems is introduced in Sect. 9.3 without analyzing either the specific models proposed or the implications of related results. Finally Sect. 9.4 describes some algorithms that can be applied to solve various design problems.

9.2 General Formulations of the Supply Design Problem

The supply design problem (SDP) can be formulated as a constrained optimization model, maximizing or minimizing an objective function $w(\cdot)$ that depends on design variables $\boldsymbol{y}$ and link flows $\boldsymbol{f}$. Representation of the system and its variables can be within-day static or dynamic. Although some SDP models for dynamic systems are covered in the literature, most specifications refer to static systems and assignment models. This is not surprising, given both the recent development and computational complexity of dynamic assignment models which, as shown later, are used repeatedly in an SDP. For these reasons, the following deals with static models.

As stated in Chap. 5, link flows resulting from a static assignment model can be expressed as a function of the O-D demand flows (vector $\boldsymbol{d}$), network topology (link-path incidence matrix $\boldsymbol{\Delta}$), and path choice probabilities (matrix $\boldsymbol{P}$). In general, both the network topology and path choice probabilities depend on the supply configuration, either directly or through link costs and cost functions. Demand flows are constant if the assignment model assumes fixed demand, and depend on supply performances if demand is elastic. The general supply design model can be formulated as

$$\boldsymbol{y}^* = \arg\underset{\boldsymbol{y}}{\text{opt}}\, w(\boldsymbol{y}, \boldsymbol{f}^*) \tag{9.2.1a}$$

subject to the constraints:

$$\boldsymbol{f}^* = \boldsymbol{\Delta}(\boldsymbol{y})\boldsymbol{P}\big[\boldsymbol{y}, \boldsymbol{g}(\boldsymbol{f}^*, \boldsymbol{y})\big]\boldsymbol{d}\big[\boldsymbol{g}(\boldsymbol{f}^*, \boldsymbol{y})\big] \tag{9.2.1b}$$

$$\boldsymbol{y}, \boldsymbol{f}^* \in E \tag{9.2.1c}$$

$$\boldsymbol{y}, \boldsymbol{f}^* \in T \tag{9.2.1d}$$

where $\boldsymbol{y}^*$ is the optimal solution of the supply design problem and $\boldsymbol{f}^*$ is the equilibrium flow vector; (9.2.1b) expresses the consistency constraint among supply performances, demand, and flows (i.e., the equilibrium assignment); (9.2.1c) identifies the set of supply parameters satisfying the external constraints; and (9.2.1d) expresses the system of technical constraints. Furthermore, the notation $\boldsymbol{\Delta}(\boldsymbol{y})$ indicates that, in the case of design variables influencing the network topology, both the paths and the link–path incidence matrix depend on the values of the design variables; the same holds for the path choice probabilities, as expressed by $\boldsymbol{P}(\boldsymbol{y}, \boldsymbol{g})$, where $\boldsymbol{g}$ is the path cost vector.

The formulation (9.2.1) is based on explicit representation of the assignment model as a fixed-point model. As was seen in Chap. 5, this formulation presents some mathematical problems for deterministic user equilibrium (DUE) assignment. In this case, the consistency constraint (9.2.1b) is usually replaced by a variational inequality, which for fixed demand becomes (see Sect. 5.4.3):

$$\boldsymbol{c}(\boldsymbol{f}^*, \boldsymbol{y})^T(\boldsymbol{f} - \boldsymbol{f}^*) \geq 0 \quad \forall \boldsymbol{f} \in S(\boldsymbol{y}, \boldsymbol{d}) \tag{9.2.1e}$$

Expression (9.2.1e) makes explicit the dependence of the link flow feasibility set S on demand and design parameters. For elastic demand, the analogous expression is given in Sect. 6.3.1.2.

The design model can be formulated differently if the assignment model is formulated as an optimization problem. In this case, model (9.2.1) can be expressed as a bilevel optimization model where the value of the first-level objective function $w(\cdot)$ depends on the solution of a second-level optimization problem, usually with a different objective function $z(\cdot)$:

$$y^* = \arg\underset{y}{\operatorname{opt}}\, w\big(y, f(y)\big) \tag{9.2.2a}$$

$$y \in E$$

$$y \in T$$

$$f(y) = \arg\min_{f \in S_f} z(f, y, d) \tag{9.2.2b}$$

The specific form of the objective function $z(\cdot)$ in (9.2.2b) depends on the particular assignment model (see Chap. 5 for DUE and the SUE specifications). For an uncongested network assignment model, the link cost vector depends exclusively on the design variables, $c = c(y)$, simplifying the specification and the solution of the design model.

The actual specification of the supply design model, whether in the form (9.2.1) or in (9.2.2), comes from the particular design problem and the assumptions. Examples of specifications are given in the next section. As mentioned earlier, the design variables can be divided into three categories: topological or network layout variables, usually discrete, that are denoted in the following by the vector y^{TOP}; supply performance variables, continuous or discrete, denoted by y^{PER}; and price variables, usually continuous, indicated by y^{PRI}. Consequently, in the general case the vector y of design variables can be decomposed into three subvectors:

$$y^T = (y^{\mathrm{TOP}}, y^{\mathrm{PER}}, y^{\mathrm{PRI}})^T \tag{9.2.3}$$

The objective function can assume different forms depending on the goal of the project. Social objective functions $w_1(\cdot)$ usually correspond to the network indicators described in Sect. 5.2.4. The most common specification is in terms of the total actual cost, which in the absence of nonadditive path costs can be expressed as

$$w_1(y, f) = \sum_l c_l(y, f) f_l \tag{9.2.4}$$

The total Expected Maximum Perceived Utility (EMPU) with respect to path (and possibly mode) choice is seldom adopted as an objective function[3] because of its computational complexity, even if it is a more appropriate measure of user's surplus, as shown in Sect. 10.2.3:

$$w_1(y, f) = \sum_{od} d_{od} s_{od}\big(-\Delta^T(y) c(y, f)\big) \tag{9.2.5}$$

[3]The two objective functions (9.2.4) and (9.2.5) coincide only in the case of deterministic path choice models.

Operator objective functions $w_2(\cdot)$ express the total investment and operations and maintenance (O&M) costs in terms of the design parameters $\boldsymbol{y}$ or functional transformations of them:

$$w_2(\boldsymbol{y}) = \sum_j b_j(y_j) y_j \tag{9.2.6}$$

where b_j is the unit cost related to each design variable y_j. For example, if the design variables are 0/1 topological variables that indicate whether to include the connection j, b_j is the investment and/or O&M cost for that connection. Another type of operator objective function includes traffic revenues, which depend on the design price variables, and which may be associated either with individual links (vector $\boldsymbol{y}_L^{\mathrm{PRI}}$) or with O-D pairs (vector $\boldsymbol{y}_{\mathrm{OD}}^{\mathrm{PRI}}$).

$$w_2(\boldsymbol{y}_L^{\mathrm{PRI}}, \boldsymbol{f}) = \sum_l y_l^{\mathrm{PRI}} f_l \tag{9.2.7}$$

$$w_2(\boldsymbol{y}_{\mathrm{OD}}^{\mathrm{PRI}}, \boldsymbol{d}) = \sum_{od} y_{od}^{\mathrm{PRI}} d_{od} \tag{9.2.8}$$

In the case of multiobjective optimization, objective functions are usually expressed as linear combinations of two or more of the above functions. For example, the total user and operator cost is usually obtained by adding (9.2.4) and (9.2.6) with coefficients representing the relative weight of the two objectives. Furthermore, expression (9.2.6) can also be used to specify an overall (external) budget constraint:

$$\sum_j b_j(y_j) y_j \leq B \tag{9.2.9}$$

where B represents the maximum available budget.

Little can be said about the mathematical properties of supply design models in general, or about the existence and uniqueness of the solution $\boldsymbol{y}^*$ in particular, since the solution depends on the particular specification adopted. In most cases neither the objective function nor the constraints have convexity properties sufficient to guarantee solution uniqueness. In fact, many models have shown multiple solutions, or local optima, corresponding to similar values of the objective function. This may have significant practical implications because nearly equivalent solutions can be generated, among which the best solution can be chosen on the basis of a wider set of objectives and criteria. Similar comments can be made regarding the existence of solution $\boldsymbol{y}^*$, which obviously depends on the definition of the constraints; erroneous or incompatible specifications could lead to problems without any feasible solution.

9.3 Applications of Supply Design Models

Supply design models have been studied in greater detail for certain classes of "partial" problems, which are described below. For more complex projects, the actions to be jointly designed may involve many elements of the supply system and many modes. In the case of tactical urban transportation planning, for example, actions may include the directions and traffic signal control of the road network, the availability of parking areas on- and off-street, the structure and frequency of the transit lines, parking and transit pricing, and so on. Similarly, for a railway system program, design variables may include the structure of the lines, the timetables of individual runs, and the fare structure. Design problems of this complexity are usually solved by formulating separate design models for one or more individual components, following a sequence related to the (implicit) hierarchy of the objectives.

9.3.1 Models for Road Network Layout Design

Design problems in this class identify the road connections to be built or the optimal circulation scheme for a given network of facilities. The design variables for these models are discrete topological variables represented by the vector $\boldsymbol{y}^{\mathrm{TOP}}$, with a component for each possible road connection. These variables are a subset of the expanded road network links that include the existing connections as well as possible future connections to be designed.

Typically in the optimal infrastructure layout problem, roads are assumed to be bidirectional and the design variables are binary variables $y_j^{\mathrm{TOP}} = 0/1$, indicating that the link j is to be excluded (zero) or included (one) in the solution. Figure 9.3 shows an example of the initial configuration and some possible alternative configurations with the corresponding values of the design variables for a small test network. This SDP is often associated with extraurban road networks.

For this problem, the objective function is usually specified as a linear combination of the total transportation cost (9.2.4) and the total construction and O&M cost (9.2.6), where $b_j(y_j^{\mathrm{TOP}})$ is the cost to build, maintain, and operate the road connection represented by link j. To ensure comparability of the two terms, transportation and construction/operation costs should be expressed in monetary units and cover the same period, for example, a generic average year. This can be accomplished by "projecting" into a given year (typically the first year of operation) the values of O-D flows and the annual user transportation costs. Similarly, the operator cost will be the equivalent annual amount of the total investment cost and the yearly O&M cost.

External constraints usually include a budget constraint (9.2.9) and, in some cases, constraints on the total level of pollutants emitted. Some specifications may include network constraints that ensure the connection of all origin–destination pairs, the conservation of flow at nodes, and so on, as described in Sect. 5.2. It should be noted, however, that network constraints are necessary only if the assignment model is deterministic (DUN or DUE) and is expressed by a variational inequality or an optimization model formulated in terms of link variables. Recall that, in stochastic assignment models, these constraints are implicit in the relationships between demand and link flows as expressed by (9.2.1b). Many specifications

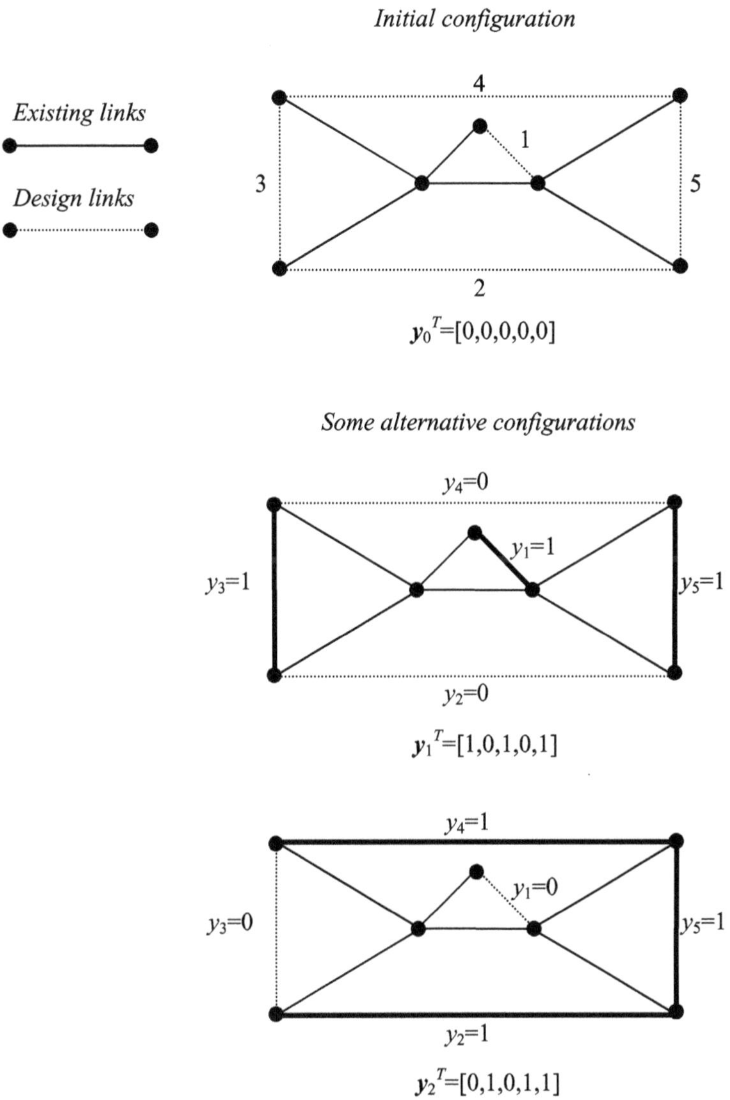

Fig. 9.3 Design variables for an optimal infrastructure layout problem

of this model consider fixed demand, using the modal O-D matrices established for the reference year.

A simplified specification of the design problem is:

$$\boldsymbol{y}^{*\,\mathrm{TOP}} = \arg\min_{\boldsymbol{y}^{\mathrm{TOP}}} \sum_{l} c_l\left(\boldsymbol{y}^{\mathrm{TOP}}, \boldsymbol{f}^*\right) \cdot f_l^* \tag{9.3.1}$$

subject to the constraints:

$$y_j^{\mathrm{TOP}} = 0/1$$

$$\boldsymbol{f}^* = \boldsymbol{\Delta}(\boldsymbol{y}^{\mathrm{TOP}})\boldsymbol{P}\left[\boldsymbol{y}^{\mathrm{TOP}}, \boldsymbol{g}(\boldsymbol{f}^*, \boldsymbol{y}^{\mathrm{TOP}})\right]\boldsymbol{d}$$

$$\sum_j y_j^{\mathrm{TOP}} \cdot b_j \leq B$$

The optimal functional layout problem considers the optimal circulation scheme, that is, the optimal configuration of traffic directions for a road network, typically an urban network. Optimal circulation schemes may be needed for two conflicting reasons. The single-direction use of a road increases the available width for this direction and, in turn, the saturation flow at its intersections. This reduces the waiting time for a given flow. On the other hand, two-way roads generally reduce the distance between an O-D pair and increase the number of conflict points at intersections. The design variables are still discrete variables $\boldsymbol{y}^{\mathrm{TOP}}$ associated with each link, and can assume different values (e.g., 0, 1 or 2), according to whether the link is used in both directions or in each of the two ways (see Fig. 9.4).

The cost functions of each link j depend on the variable y_j; furthermore, the objective function usually includes only the users' generalized cost (9.2.4), the construction cost of existing roads being null and the difference in O&M costs being negligible. The link constraints are analogous to those described for the infrastructure layout problem and the same considerations hold.

The model is sometimes specified by introducing external constraints that limit flow/capacity ratios for particular links. These constraints express the need for both technical functionality (flows near capacity induce instability phenomena and possible spill-backs at intersections) and pollution reduction (emissions are higher for low commercial or average speeds). Another type of external constraint requires that the distance between each O-D pair on the shortest path not exceed the shortest feasible distance, that is, the minimum distance on a fully bidirectional road configuration, by more than a specified amount. In this case, the implicit "equity" objective is to distribute penalties among users.

The optimal urban road network layout problem, discussed below, is usually associated with the control of intersections that determine road link capacity. As was seen in Sect. 2.4.1.2, the capacity of a signalized intersection movement is given by the product of the saturation flow and the ratio of the movement's effective green time to the cycle length.

9.3.2 Models for Road Network Capacity Design

Capacity design models optimize link capacity in a road network of a given topology. The design variables are generally continuous, expressed by a vector $\boldsymbol{y}^{\mathrm{PER}}$ whose components are link capacities. The problem may assume two different forms, typically corresponding to interurban (or rural) and urban road networks.

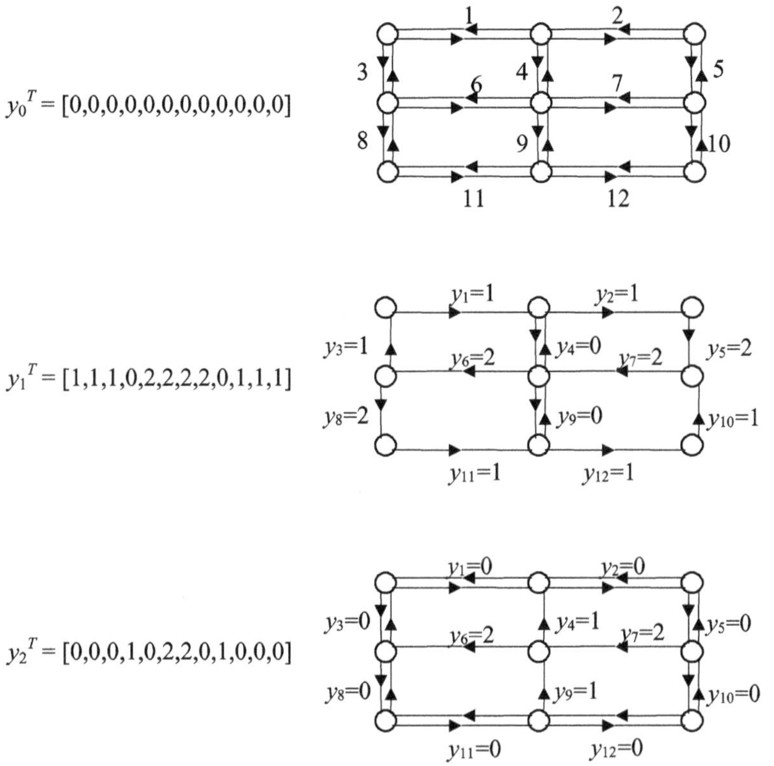

Fig. 9.4 Design variables for an optimal functional layout problem

For interurban or rural road network capacity design, the decision variables are the link capacities, usually constrained to be between specified minimum and maximum values. Formulation of the model is substantially similar to that described for the optimal network layout problem; the objective function can be expressed as the sum of user costs and construction costs. Budget and congestion-level constraints (maximum flow/capacity ratios) are also typically included.

The capacity of an interurban or rural road depends, to a first approximation, on its cross-section (lane number and width, lateral clearance, etc.). In practice, capacities do not range over all possible values, but rather take values from a finite set, corresponding to the different section types. From this point of view, the design variables should be discrete even though, in the literature, they are often approximated as continuous. In the discrete capacity case, the problem is analogous to that described in the preceding section, with the difference that the design variables can assume several discrete values corresponding to the different section types.

Urban road network capacity design often addresses the problem of finding optimal traffic signal control parameters for a subset of intersections (the traffic signal setting problem). In the most simplified formulations, it is assumed that intersections are "isolated"; that is, the traffic signal coordination between adjacent intersections

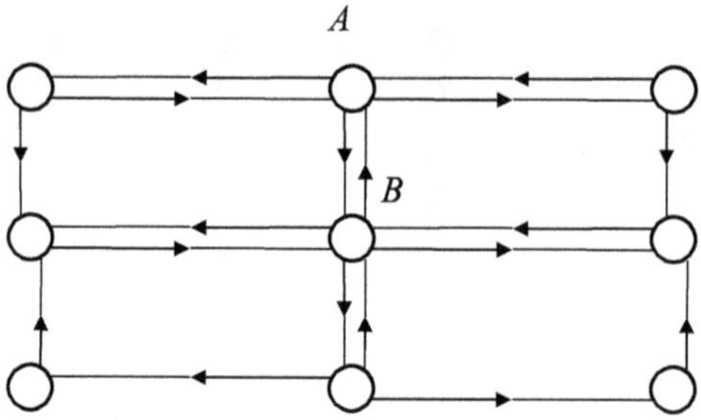

Intersection A

Cycle	T_{cA}
Total lost time	LT_A
Phases	$J_A = \{1_A, 2_A, 3_A\}$
Variables	$y_{1A}^{\mathrm{PER}} = G_{1A}/(T_{cA} - LT_A)$
	$y_{2A}^{\mathrm{PER}} = G_{2A}/(T_{cA} - LT_A)$
	$y_{3A}^{\mathrm{PER}} = G_{3A}/(T_{cA} - LT_A)$

Phase plan

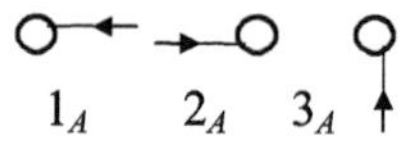

Constraint on total cycle length

$G_{1A} + G_{2A} + G_{3A} = T_{cA} - LT_A$

Intersection B

Cycle	T_{cB}
Total lost time	LT_B
Phases	$J_B = \{1_B, 2_B\}$
Variables	$y_{1B}^{\mathrm{PER}} = G_{1B}/(T_{cB} - LT_B)$
	$y_{2B}^{\mathrm{PER}} = G_{2B}/(T_{cB} - LT_B)$

Phase plan

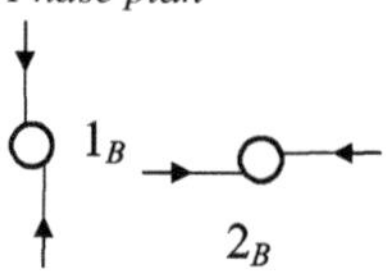

Constraint on total cycle length

$G_{1B} + G_{2B} = T_{cB} - LT_B$

Vector of variables

$\mathbf{y}^{\mathrm{PER}} = [y_{1A}^{\mathrm{PER}}, y_{2A}^{\mathrm{PER}}, y_{3A}^{\mathrm{PER}}, y_{1B}^{\mathrm{PER}}, y_{2B}^{\mathrm{PER}}]^T$

Constraints

$y_{1A}^{\mathrm{PER}} + y_{2A}^{\mathrm{PER}} + y_{3A}^{\mathrm{PER}} = 1$

$y_{1B}^{\mathrm{PER}} + y_{2B}^{\mathrm{PER}} = 1$

Fig. 9.5 Design variables and constraints for an optimal signal setting problem

has no effect. This assumption implies that offsets between the green times of different intersections are not relevant control variables. It can also be assumed that, for each intersection, the overall duration of the cycle and the structure of the traffic signal phases are known. This implies that for each node (or group of nodes) n, representative of a signalized intersection, the set J_n of phases j_n and the set of links $I(j_n)$ for which flows receive green in the same phase is known. In this case, the design variables $\mathbf{y}^{\mathrm{PER}}$ can be identified as the effective green to cycle length ratios, with the latter reduced by the lost times for each phase. The design variables are therefore continuous over the interval (0, 1) (see Fig. 9.5).

Note the difference between capacity design of signalized intersections for the entire network and for a single intersection. In the former, as the green/cycle ratios vary, capacities also vary and, because of the effect of assignment constraints (9.2.1b), link flows vary as well. In the latter, it is assumed that flows are known and invariant with respect to capacity parameters.

The specification usually adopted for the capacity design model is analogous to that presented for the road network layout problem. The objective function to be minimized is the total generalized cost (usually time) spent on the network. Construction costs are not taken into consideration and the external constraints might include maximum levels of congestion and pollution. Technical constraints set the maximum and minimum duration of each phase and require that the sum of green/cycle ratios over all phases is equal to one for each intersection.

Two different approaches can be followed to optimize the traffic signal control parameters: local and global. In the *global* approach, the control parameters of all intersections are jointly optimized to minimize the total travel time on the network. In the *local* approach, each signalized intersection is optimized to minimize the total user delay at the intersection. In this case, a circular dependence among flows, costs, and control parameters arises and the resulting problem can be seen as a fixed-point problem. This problem can be modeled as an asymmetric user assignment problem.

A possible simplified formulation of the global optimal signal setting problem is:

$$\boldsymbol{y}^{*\mathrm{PER}} = \arg\min_{y^{\mathrm{PER}}} \sum_{l} c_l\left(\boldsymbol{y}^{\mathrm{PER}}, \boldsymbol{f}^*\right) \cdot f_l^* \tag{9.3.2}$$

subject to:

$$0 \le y_{jn}^{\mathrm{PER}} \le 1$$

$$\sum_{jn \in J_n} y_{jn}^{\mathrm{PER}} = 1 \quad \forall n$$

$$y_{jn}^{\mathrm{PER}} T_{cn} \ge T_{\min} \quad \forall n$$

$$\boldsymbol{f}^* = \boldsymbol{\Delta P}\left[\boldsymbol{y}^{\mathrm{PER}}, \boldsymbol{g}(\boldsymbol{f}^*, \boldsymbol{y}^{\mathrm{PER}})\right]\boldsymbol{d}$$

where T_{cn} is the duration of the cycle at intersection n and $T_{\min}$ is the minimum value for a green time interval.

More complex traffic signal control problems introduce other design variables, including offsets between green times in nearby intersections, cycle length for each intersection, and sequence and number of phases. In the first case, the link delay models described in Chap. 2 should account for the effects of platoon dispersion between coordinated intersections.

9.3.3 Models for Transit Network Design

It is usually assumed that the relevant supply variables for high-frequency urban transit systems are service frequencies rather than actual timetables (see Sect. 2.4.2). Under this assumption, the design problem identifies the optimal layout for the lines as well as their service frequencies in the reference period (e.g., rush hour). In this case, the design model simultaneously identifies the topological configuration and the optimal performances of the supply system.

The design variables are the discrete layout variables y_{ln}^{TOP}, equal to one if the physical link l (e.g., a road or railway section) belongs to line n, and zero otherwise, and the continuous performance variables y_n^{PER}, representing the service frequency of each line n; see Fig. 9.6. The layout variables are equivalent to the duplication of physical links in line links, that is, to the implicit construction of the line network model described in Chap. 2. For this reason, link variables in this model are written with a double index.

The objective function usually includes the user generalized cost and the operator cost, expressed in commensurate terms. For urban transit systems, the function $w_1(\cdot)$ expressing user costs is different from (9.2.4) under the usual assumptions on mixed preventive/adaptive path choice behavior. In this case, alternative travel strategies are represented by hyperpaths on the lines, and average path costs include a nonadditive component associated with waiting times at stops (see Sect. 4.3.3.2).

Formally, the objective function $w_1(\cdot)$ can be expressed as

$$w_1(\mathbf{y}^{\text{TOP}}, \mathbf{y}^{\text{PER}}) = \sum_n \sum_{l \notin Jw} c_{ln} f_{ln}(\mathbf{y}^{\text{TOP}}, \mathbf{y}^{\text{PER}}) + \sum_{l \in Jw} \sum_k tw_l^k(\mathbf{y}^{\text{TOP}}, \mathbf{y}^{\text{PER}}) f_l^k(\mathbf{y}^{\text{TOP}}, \mathbf{y}^{\text{PER}}) \tag{9.3.3}$$

where c_{ln} is the generic additive cost associated with link l and line n (e.g., on-board or access travel times); J_w is the set of waiting links; k is the generic hyperpath; tw_l^k is the waiting time (cost) associated with link l and hyperpath k; f_{ln} is the user flow on on-board link l belonging to line n; and f_{lk} is the flow on waiting link l belonging to hyperpath k; see Sect. 4.3.3.2.

The overall operator cost $w_2(\cdot)$ is usually expressed using the unit running cost CE_n for each journey (bus, train, etc.) of line n, expressed in monetary units per distance or time unit:

$$w_2(\mathbf{y}^{\text{TOP}}, \mathbf{y}^{\text{PER}}) = \sum_n \sum_l y_{ln}^{\text{TOP}} CE_n L_{ln} y_n^{\text{PER}} \tag{9.3.4}$$

where L_{ln} is the length (or round trip time) of link l on line n.

The assignment constraints can be expressed using the formulation introduced in Chap. 6 as

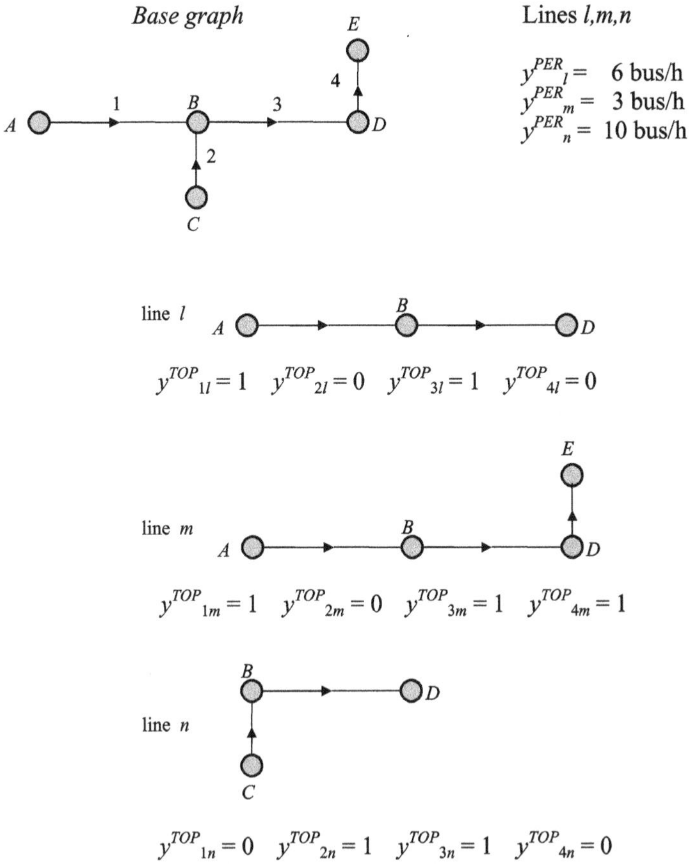

Fig. 9.6 Design variables for an optimal line layout and frequency problem

$$
\begin{aligned}
\boldsymbol{f}(\boldsymbol{y}^{\text{TOP}}, \boldsymbol{y}^{\text{PER}}) = \boldsymbol{\Lambda}(\boldsymbol{y}^{\text{TOP}}, \boldsymbol{y}^{\text{PER}})\boldsymbol{Q}\big(\boldsymbol{\Lambda}^{T}(\boldsymbol{y}^{\text{TOP}}, \boldsymbol{y}^{\text{PER}})\boldsymbol{c}(\boldsymbol{y}^{\text{TOP}}) \\
+\boldsymbol{x}^{\text{NA}}(\boldsymbol{y}^{\text{TOP}}, \boldsymbol{y}^{\text{PER}})\big)\boldsymbol{d}
\end{aligned}
\tag{9.3.5}
$$

where it is implicitly assumed that the network is not congested and that the link crossing probability matrix $\boldsymbol{\Lambda}$, and the nonadditive hyperpath costs $\boldsymbol{x}^{\text{NA}}$, both depend on the topological configuration of the lines ($\boldsymbol{y}^{\text{TOP}}$) and on the respective frequencies ($\boldsymbol{y}^{\text{PER}}$).

The technical constraints of the problem usually restrict the flow f_{ln} on each line link l to the capacity of line n, which can be expressed as the product of the capacity Cap_n of each vehicle and the frequency of line n:

$$f_{ln} \le Cap_n y_n^{\text{PER}} \quad \forall l$$

Furthermore, the frequencies must be nonnegative, and equal to zero if the line is not active and below a maximum technically feasible value $y_{\max}^{\mathrm{PER}}$:

$$0 \le y_n^{\mathrm{PER}} \le y_{\max}^{\mathrm{PER}}$$

Another possible technical constraint is a budget constraint on the vehicle stock. This constraint can be expressed as a function of the travel time t_l of each line link, because the number of vehicles necessary for a line of frequency y_n^{PER} is equal to the product of the frequency for the total travel time of the line:

$$\sum_n \sum_l t_l y_n^{\mathrm{PER}} \cdot y_{ln}^{\mathrm{TOP}} \le N_{\max} \tag{9.3.6}$$

with $N_{\max}$ equal to the maximum number of available vehicles.

Finally, a technical constraint sometimes introduced, although not easily expressed in formal terms, requires that lines must have their terminals in a given set of nodes.

A simplified version of the transit design problem assumes that the topological configuration of all the lines (the components of vector $\boldsymbol{y}^{\mathrm{TOP}}$) is given. In this case, the design problem is reduced to the calculation of optimal service frequencies, that is, the components of the vector $\boldsymbol{y}^{\mathrm{PER}}$, with a significant reduction in the number of variables and in computational complexity. A minimum required service frequency may be added to the technical constraints.

For interurban or rural services (with low frequency and high regularity), the supply design problem is quite different, as are the models used to simulate these services. As was seen in Sect. 7.6.1, the diachronic network models used to simulate regular low-frequency services are based on the explicit representation of the service schedule. Optimal schedule design models define the departure and arrival times of each run of a pre-defined set of lines. Furthermore, in the most general case, they jointly determine the lines and their departure and arrival times, under a set of technical constraints. The latter are the feasible range of travel times (feasible commercial speeds), the available vehicle stock, the range of acceptable connection times between different lines at intermediate stops, and so on. The problem of optimal service scheduling has not been extensively covered in the literature.

9.3.4 Models for Pricing Design

Pricing design models can be applied to different contexts. Prices, generally represented as continuous variables $\boldsymbol{y}^{\mathrm{PRI}}$, may be related to the different transportation supply elements: road tolls, parking, air and rail fares, and the like. Specification of the design variables $\boldsymbol{y}^{\mathrm{PRI}}$ will depend on the assumed pricing structure, that is, on how prices are computed and applied. If constant access prices are assumed, for example, constant road tolls at motorway entrance/exit points or constant parking fares, the components y_j^{PRI} of the vector can be associated with the network links j, representative of the toll points or parking facilities. If the price is proportional to the distance covered, for example, road tolls or rail fares proportional to the journey length, the price parameter y_l^{PRI} can be associated with each link l, corresponding to a section with a physical length.

The objectives of pricing design might also differ in different situations. If the pricing policy is meant to improve the efficiency of the transportation system, for example, by reducing the overall generalized cost of the system and/or the overall pollution level, the resulting pricing is known as *efficiency pricing*. A typical example of efficiency pricing design is *road pricing*, that is, charging a driver for the use of roads according to the social cost (typically the total travel time) that her use produces. In this case, the social objective function[4] is $w_1(y^{\mathrm{PRI}}) = \sum_l c_l(\boldsymbol{f}_l) f_l$. The efficiency road pricing design problem with fixed demand can therefore be formulated as

$$\boldsymbol{y}^{*\,\mathrm{PRI}} = \arg\min_{\boldsymbol{y}^{\mathrm{PRI}}} \sum_l c_l(\boldsymbol{f}^*) f_l^* \tag{9.3.7}$$

subject to the constraints:

$$\boldsymbol{y}^{\mathrm{PRI}} \geq \mathbf{0}$$

$$\boldsymbol{f}^* = \boldsymbol{\Delta P}\big(\boldsymbol{g}(\boldsymbol{f}^*, \boldsymbol{y}^{\mathrm{PRI}})\big)\boldsymbol{d}$$

In the special case of DUE assignment with separable cost functions, problem (9.3.7) is equivalent to the system optimal (SO) assignment problem described in Sect. 5.4.6. This problem is a single-level optimization problem and the optimum price $y_l^{*\,\mathrm{PRI}}$ can be calculated as

$$y_l^{*\,\mathrm{PRI}} = c_l'(f_l^*) f_l^* \quad \forall l \tag{9.3.8}$$

where $c_l'(f_l)$ is the first derivative of the cost function.

However, it should be noted that the price vector $\boldsymbol{y}^{*\,\mathrm{PRI}}$ given by (9.3.8) is not, in general, a unique solution to the problem (9.3.7). Under deterministic path choice, there may be other vector solutions to the general problem (9.3.7) with different operational impacts (e.g., less expense to users or the possibility of applying the price only to certain network links).

The formulation (9.3.7) of the road pricing problem assumes that O-D demand $\boldsymbol{d}$ is fixed. The resulting prices tend to reduce the total travel time by modifying path choices; this is achieved by increasing the generalized cost on a link as a function of its congestion level. However, many empirical results indicate that the most significant congestion reductions can be obtained by focusing on demand flows. To address this problem it is necessary to consider the O-D demand for the car mode

[4]The economic interpretation of this objective function, which differs from the individual user generalized cost, is that the monetary cost can be considered a transfer from users to system operators who, in principle, can return it to the users in another form (see Sect. 10.3.1 on benefit–cost analysis).

$\boldsymbol{d}^C$ as variable. For example, it may be assumed that demand is variable with respect to mode choice and that roads are the only congested mode, that is, that only road costs are dependent on link flow $\boldsymbol{f}^C$ and design prices $\boldsymbol{y}^{*\text{PRI}}$. It may be appropriate to impose further constraints on the problem (9.3.7), for example, requiring that road link flows are below a pre-determined fraction of the corresponding capacities.

Under these assumptions, the efficiency road pricing problem with variable demand can be formulated as

$$\boldsymbol{y}^{*\text{PRI}} = \arg\min_{\boldsymbol{y}^{\text{PRI}}} \sum_l c_l(\boldsymbol{f}^*) f_l^* \tag{9.3.9}$$

subject to:

$$\boldsymbol{y}^{\text{PRI}} \geq \boldsymbol{0}$$
$$\boldsymbol{f}^* = \boldsymbol{\Delta P}\big(\boldsymbol{g}^C(\boldsymbol{y}^{\text{PRI}}, \boldsymbol{f}^*)\big)\boldsymbol{d}^C\big(\boldsymbol{s}^C(\boldsymbol{y}^{\text{PRI}}, \boldsymbol{f}^*), \boldsymbol{s}^B\big)$$

where $\boldsymbol{s}^C(\cdot)$ and $\boldsymbol{s}^B$ are $(n_{od} \times 1)$ vectors of EMPU variables related to path choice for car and bus modes.

The pricing design model has a different form when maximizing traffic revenues or net profits (revenues minus costs). In the former case, for example, assuming a single operator in the market, the operator's objective function $w_2(\cdot)$ is the total revenue, and the problem can be formulated as

$$\boldsymbol{y}^{*\text{PRI}} = \arg\max_{\boldsymbol{y}^{\text{PRI}}} \sum_l y_l^{\text{PRI}} f_l^* \tag{9.3.10}$$

subject to:

$$\boldsymbol{f}^* = \boldsymbol{\Delta P}\big(\boldsymbol{g}^C(\boldsymbol{y}^{\text{PRI}}, \boldsymbol{f}^*)\big)\boldsymbol{d}^C\big(\boldsymbol{g}^C(\boldsymbol{y}^{\text{PRI}}, \boldsymbol{f}^*), \boldsymbol{g}^B\big)$$
$$\boldsymbol{y}^{\text{PRI}} \geq \boldsymbol{0}$$

where the modal O-D demand flows are considered price elastic.

Pricing design models for other types of transport infrastructure (e.g., rail lines or airport slots) and service (e.g., train or air connections) can be formulated in a similar way. Typically, optimal infrastructure use prices are computed with respect to social objectives (efficiency pricing) whereas service prices are computed with respect to operator objectives. The literature contains relatively few descriptions of applications of these methods.

9.3.5 Models for Mixed Design

Complex projects involving an areawide transportation system or multiple aspects of a transportation company's services would require design models that integrate two or more of the models described earlier. For example, a regional transportation plan usually involves the optimal design of road and rail infrastructure, rail and bus services, road and transit pricing, and so on. Similarly, definition of a road project financing scheme includes the optimal design of new infrastructure and pricing systems. Clearly, the computational complexity of these problems increases exponentially and the (few) examples published in the literature are based on a number of ad hoc simplifying hypotheses specific to the individual problem. Solution algorithms are generally based on the sequential solution of separate design problems corresponding to separate design variables.

9.4 Some Algorithms for Supply Design Models

Using mathematical programming terminology, supply design models can be characterized as discrete, continuous, or mixed optimization problems; such models are generally nonlinear with nonlinear constraints, or bilevel optimization models with ill-defined mathematical properties. For most of these problems, optimal algorithms – algorithms that can be proven to converge to global or local optimal solutions – do not exist.

For this reason, heuristic algorithms have generally been used in applications with, in many cases, satisfactory results. This is especially relevant given that the goal is to define interventions in the physical system with the help of design models rather than to solve a mathematical problem per se. In what follows, some examples of heuristic algorithms are briefly presented for discrete and continuous problems. These algorithms are applicable to a wide range of design models. A comprehensive review lies beyond the scope of this book.

9.4.1 Algorithms for the Discrete SDP

A number of algorithms have been proposed for solving discrete SDPs; most solve specific network design problems. These algorithms can be classified in two groups:

- *Exact algorithms* that yield globally optimal solutions, such as total enumeration and "branch and bound" algorithms
- *Heuristic algorithms* that yield suboptimal solutions (local optimum or near optimal solution), such as add-and-delete algorithms, neighborhood search algorithms, genetic algorithms, and simulated annealing algorithms

In general, exact algorithms can be applied only to small networks, whereas heuristic algorithms can be applied to relatively large networks. To facilitate comparisons here, the algorithms are applied to the small network of Fig. 9.7 for the uncongested road layout design problem with a deterministic route choice model. The algorithms can be extended to other discrete design problems (e.g., optimal layout of transit lines).

Add-and-delete algorithms. These perform a sequence of insertion and deletion operations starting from an initial solution. The insertion operation adds design links

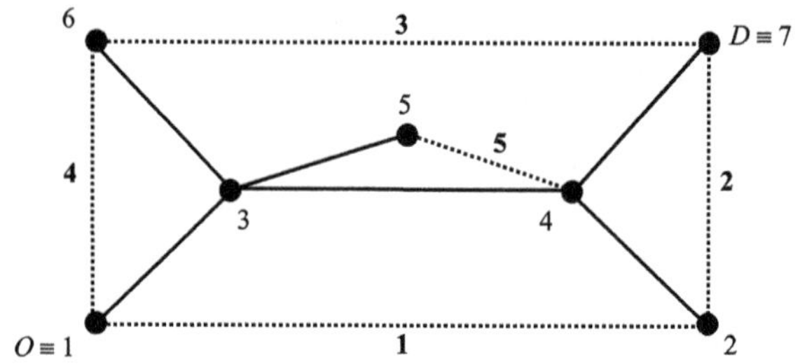

existing link

design link

Objective function:
$w = \mathbf{c}^{\mathrm{T}} \mathbf{f}(\mathbf{y}^{TOP}, \mathbf{d}) + \beta\, \mathbf{b}^{T} \mathbf{y}^{TOP}$
$\beta = 5$ time units/cost unit

Link number	Link	Travel time ($\boldsymbol{c}$)	Construction cost ($\boldsymbol{b}$)
		Design links	
1	1–2	8	800
2	2–7	12	1000
3	6–7	15	1500
4	1–6	10	600
5	4–5	20	500
		Existing links	
	1–3	10	–
	2–4	10	–
	3–4	15	–
	3–5	10	–
	3–6	10	–
	4–7	10	–

Demand OD (d_{1-7})	1000

Fig. 9.7 Test network (starting configuration)

sequentially to generate new solutions. For each possible addition, the resulting objective function value is calculated, and the link with the largest objective function improvement is added to the current configuration. The insertion routine continues to add links until no further link insertion improves the objective function. The deletion routine is then invoked. This operation deletes links from the current configuration, calculating the objective function for each possible deletion. The link with the largest objective function improvement is deleted from the current configuration. The deletion routine continues to delete links until no link deletion improves the objective function. If at least one link is deleted, the algorithm repeats the insertion tests; otherwise the algorithm stops. In the last generated network configuration, no link insertion or deletion could improve the objective function value.

Inclusion or deletion of link l in iteration it is indicated by a 1 in the corresponding component of the design vector $\mathbf{y}^{\mathrm{TOP}}_{it,l}$. The results of an application of the *add-and-delete* algorithm to the test network of Fig. 9.7 are summarized in Fig. 9.8.

Solution	User costs	Construction costs	Objective function value
Starting solution			
$y_0^{TOP} = [0, 0, 0, 0, 0]$	35000	0	35000
Insertion operation			
First insertion			
$y_{1,1}^{TOP} = [\mathbf{1}, 0, 0, 0, 0]$	28000	4000	32000
$y_{1,2}^{TOP} = [0, \mathbf{1}, 0, 0, 0]$	35000	5000	40000
$y_{1,3}^{TOP} = [0, 0, \mathbf{1}, 0, 0]$	35000	7500	42500
$y_{1,4}^{TOP} = [0, 0, 0, \mathbf{1}, 0]$	35000	3000	38000
$y_{1,5}^{TOP} = [0, 0, 0, 0, \mathbf{1}]$	35000	2500	37500
Best inserted link			
$y_1^{TOP} = [1, 0, 0, 0, 0]$	28000	4000	32000
Second insertion			
$y_{2,2}^{TOP} = [1, \mathbf{1}, 0, 0, 0]$	20000	9000	29000
$y_{2,3}^{TOP} = [1, 0, \mathbf{1}, 0, 0]$	28000	11500	39500
$y_{2,4}^{TOP} = [1, 0, 0, \mathbf{1}, 0]$	28000	7000	35000
$y_{2,5}^{TOP} = [1, 0, 0, 0, \mathbf{1}]$	28000	6500	34500
Best inserted link			
$y_2^{TOP} = [1, 1, 0, 0, 0]$	20000	9000	29000
Third insertion			
$y_{3,3}^{TOP} = [1, 1, \mathbf{1}, 0, 0]$	20000	16500	36500
$y_{3,4}^{TOP} = [1, 1, 0, \mathbf{1}, 0]$	20000	12000	32000
$y_{3,5}^{TOP} = [1, 1, 0, 0, \mathbf{1}]$	20000	11500	31500
No inserted link improves objective function			
Deletion operation			
First deletion			
$y_{3,1}^{TOP} = [\mathbf{0}, 1, 0, 0, 0]$	35000	5000	40000
$y_{3,2}^{TOP} = [1, \mathbf{0}, 0, 0, 0]$	28000	4000	32000
No deleted link improves objective function			
Optimal solution			
$y_{opt}^{TOP} = [1, 1, 0, 0, 0]$	20000	9000	29000

Fig. 9.8 Add and delete algorithm applied to the test network of Fig. 9.7

Neighborhood search algorithms. Starting from an initial solution, these algorithms generate the set of solutions that can be reached directly from the current solution by an elementary operation, called a *move*. Each such solution is termed a *neighbor* of the current solution, and the set of all neighbors is called the *neighborhood*. The next solution is chosen by selecting either the best solution (the descent/ascent method) or a random solution (the Monte Carlo method) from among all neighbors. The algorithm ends when no neighbor of the current solution improves the objective function in the descent method, or when the objective function does not significantly improve over the last m iterations in the Monte Carlo method. The results of an application of the neighborhood search algorithm to the test network of Fig. 9.7, using the descent method, are summarized in Fig. 9.9.

The neighborhood search algorithm is similar to the add-and-delete algorithm described previously; the main difference is the sequence in which insertions and deletions are performed. In add-and-delete algorithms, a link can be added to the current solution only in insertion routines and can be deleted only in deletion routines. In the neighborhood search, at each step all the links can be added or deleted. However, practical applications have shown that neighborhood search algorithms are better suited to find local optima close to their starting solution. Consequently, they are best used as second-stage algorithms, coupled with other algorithms that explore the entire feasible set.

Genetic algorithms. These algorithms, often used for combinatorial problems, mimic the mechanisms of genetics and natural selection. They are heuristic algorithms that start with an initial *population* (set of initial feasible solutions) and iteratively generate a new population with a higher probability of containing the optimal (or near-optimal) solution. Each feasible solution is an *element* (named a *chromosome*) of the population, and is composed of *genes*. A gene is a group of variables satisfying "local" constraints, such as the number of lanes to be allocated in each direction for an urban road network design problem. Future populations are generated with three operations: *reproduction*, *crossover*, and *mutation*.

The *reproduction* operation randomly generates a new population from the current population. When generating the new population, current solutions with a higher objective function values have a higher probability of surviving; thus only the fitter solutions will be submitted to the subsequent crossover and mutation routines. Survival probabilities are defined by the *fitness function*, a monotone increasing (decreasing) function of the objective function for maximization (minimization) problems. One possible specification of the fitness function is:

$$ff(i) = \exp(-\alpha w_i)$$

where i is an element of the current population (a feasible solution), α is a parameter and w_i is the corresponding value of the objective function. Reproduction probabilities can be computed as

$$p_r(i) = \frac{ff(i)}{\sum_j ff(j)} = \frac{\exp(-\alpha w_i)}{\sum_j \exp(-\alpha w_j)}$$

Solution	User costs	Construction costs	Objective function
Starting solution			
$y_0^{TOP} = [0, 0, 0, 0, 0]$	35000	0	35000
Neighborhood generation			
$y_{1,1}^{TOP} = [\mathbf{1}, 0, 0, 0, 0]$	28000	4000	32000
$y_{1,2}^{TOP} = [0, \mathbf{1}, 0, 0, 0]$	35000	5000	40000
$y_{1,3}^{TOP} = [0, 0, \mathbf{1}, 0, 0]$	35000	7500	42500
$y_{1,4}^{TOP} = [0, 0, 0, \mathbf{1}, 0]$	35000	3000	38000
$y_{1,5}^{TOP} = [0, 0, 0, 0, \mathbf{1}]$	35000	2500	37500
Next solution			
$y_1^{TOP} = [1, 0, 0, 0, 0]$	28000	4000	32000
Neighborhood generation			
$y_{2,1}^{TOP} = [\mathbf{0}, 0, 0, 0, 0]$	35000	0	35000
$y_{2,2}^{TOP} = [1, \mathbf{1}, 0, 0, 0]$	20000	9000	29000
$y_{2,3}^{TOP} = [1, 0, \mathbf{1}, 0, 0]$	28000	11500	39500
$y_{2,4}^{TOP} = [1, 0, 0, \mathbf{1}, 0]$	28000	7000	35000
$y_{2,5}^{TOP} = [1, 0, 0, 0, \mathbf{1}]$	28000	6500	34500
Next solution			
$y_2^{TOP} = [1, 1, 0, 0, 0]$	20000	9000	29000
Neighborhood generation			
$y_{3,1}^{TOP} = [\mathbf{0}, 1, 0, 0, 0]$	35000	5000	40000
$y_{3,2}^{TOP} = [1, \mathbf{0}, 0, 0, 0]$	28000	4000	32000
$y_{3,3}^{TOP} = [1, 1, \mathbf{1}, 0, 0]$	20000	16500	36500
$y_{3,4}^{TOP} = [1, 1, 0, \mathbf{1}, 0]$	20000	12000	32000
$y_{3,5}^{TOP} = [1, 1, 0, 0, \mathbf{1}]$	20000	11500	31500
No neighbor improves objective function			
Optimal solution			
$y_{opt}^{TOP} = [1, 1, 0, 0, 0]$	20000	9000	29000

Fig. 9.9 Neighborhood search algorithm applied to the test network of Fig. 9.7

where the summation is extended to all the elements of the current population. The *crossover* routine generates a new population by randomly exchanging parts (genes) between the feasible solutions (chromosomes). The *mutation* routine generates a

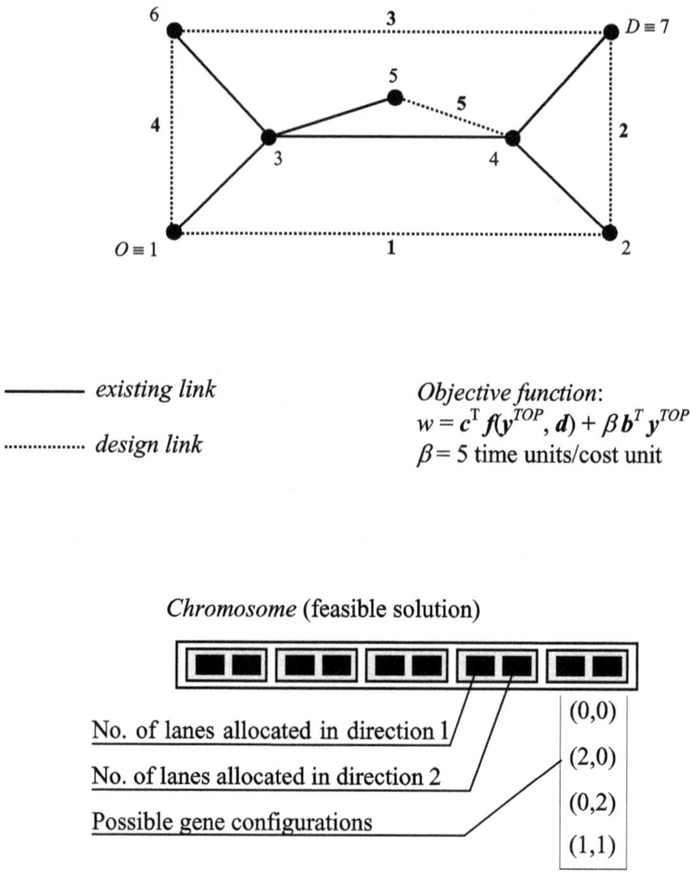

Fig. 9.10a Genetic algorithm for a discrete road network design problem: test network

new population by randomly "mutating" a gene (variable) of a "chromosome" (solution). The algorithm stops when the objective function no longer improves with new solutions, for example, in their average or min/max values over the most recent iterations.

One of the differences between genetic algorithms and those previously considered is that the outcome of the former is a population of feasible solutions with similar objective function values. Comparisons can be made among these values based on individual components of the objective function as well as on other variables. By contrast, optimization algorithms seek a single "best" solution.

The algorithm can be adjusted by setting the parameters of the fitness functions, as well as the number of crossover and mutation operations at each iteration. An example of a cycle of reproduction – crossover – mutation is provided in Figs. 9.10a and 9.10b for a road network design problem to determine the number of lanes in each direction. In this case, each gene represents the configuration of a given road and has two components, one for each direction.

Parameters of the algorithm	
Number of design links	5
Design variables	Lanes in each direction
Lanes in each design link	2
Population	3 elements
Fitness function	$\exp(-0.0001w_i)$
Number of crossovers	1
Number of mutations	1

Starting population		Objective function
Solution 1: Present configuration	$y_1 = [0, 0; 0, 0; 0, 0; 0, 0; 0, 0]$	35000
Solution 2: Random configuration 1	$y_2 = [1, 1; 1, 1; 1, 1; 0, 0; 0, 0]$	44000
Solution 3: Random configuration 2	$y_3 = [0, 0; 0, 0; 1, 1; 1, 1; 1, 1]$	46500
Reproduction		
Reproduction Probability (RP)	$\exp(-0.0001w_i)/\sum_j \exp(-0.0001w_j)$	
RP1 = 58.8%	range [0; 0.588)	
RP2 = 23.5%	range [0.588; 0.823)	
RP3 = 17.7%	range [0.823; 1]	
Random number extraction	New population	
0.456 → Solution 1	$y_1 = [0, 0; 0, 0; 0, 0; 0, 0; 0, 0]$	35000
0.672 → Solution 3	$y_2 = [0, 0; 0, 0; 1, 1; 1, 1; 1, 1]$	46500
0.089 → Solution 1	$y_3 = [0, 0; 0, 0; 0, 0; 0, 0; 0, 0]$	35000
Crossover		
Random solution selection		
Solution 1	$y_1 = [0, 0; 0, 0; 0, 0; 0, 0; 0, 0]$	
Solution 2	$y_2 = [0, 0; 0, 0; 1, 1; 1, 1; 1, 1]$	
Random cut points selection		
Point 1 → 2	$[x, x; x, x; \mid x, x; x, x; x, x]$	
Point 2 → 4	$[x, x; x, x; x, x; x, x; \mid x, x]$	
New population		
Solution 1 (crossed)	$y_1 = [0, 0; 0, 0; \mid 1, 1; 1, 1; \mid 0, 0]$	44000
Solution 2 (crossed)	$y_2 = [0, 0; 0, 0; \mid 0, 0; 0, 0; \mid 1, 1]$	37500
Solution 3	$y_3 = [0, 0; 0, 0; 0, 0; 0, 0; 0, 0]$	35000
Mutation		
Random solution selection		
Solution 3	$y_3 = [0, 0; 0, 0; 0, 0; 0, 0; 0, 0]$	
Random mutation link	Link 1 → $[\boldsymbol{x}, \boldsymbol{x}; x, x; x, x; x, x; x, x]$	
New random link configuration	$y_3 = [2, 0; 0, 0; 0, 0; 0, 0; 0, 0]$	32000
New population		
Solution 1	$y_1 = [0, 0; 0, 0; 1, 1; 1, 1; 0, 0]$	44000
Solution 2	$y_2 = [0, 0; 0, 0; 0, 0; 0, 0; 1, 1]$	37500
Solution 3 (Mutated)	$y_3 = [2, 0; 0, 0; 0, 0; 0, 0; 0, 0]$	32000

Fig. 9.10b Genetic algorithm for the discrete road network design problem of Fig. 9.10a

Fig. 9.11 General scheme of feasible direction algorithms for continuous SDP

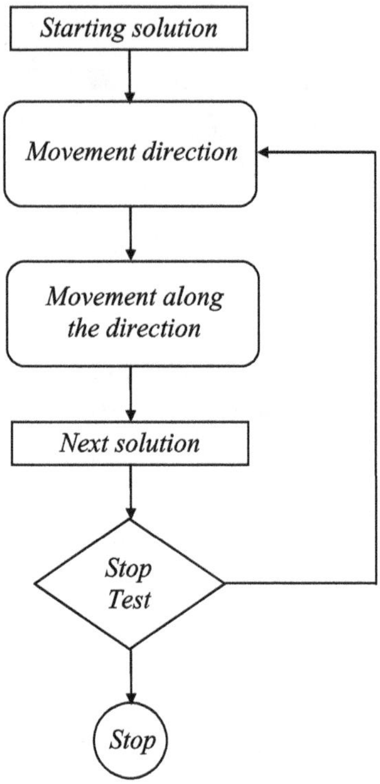

9.4.2 Algorithms for the Continuous SDP

Algorithms for continuous supply design problems are based on the principles of nonlinear optimization (see Appendix A). The optimal solution can be expressed in a closed form only for few simple problems (e.g., transit frequency optimization for a single line or cycle length, and green/cycle ratios for an isolated intersection with fixed flows). In general it is necessary to implement algorithms to perform a local search along a feasible direction, that is, a direction that moves towards a local optimum. The solution reached will be the global optimum only if the objective function is convex. If the objective function is convex, the solution reached will be the global optimum. However, it is impossible to demonstrate convexity of the objective function for most network design problems. The general scheme of a feasible direction nonlinear optimization algorithm is presented in Fig. 9.11.

Different algorithms can be specified according to their movement directions. If the movement direction can be shown to be an ascent or descent direction (e.g., the gradient or its opposite) the algorithm is exact, otherwise it is heuristic. The amount of movement along the direction can be determined by a linear search or by a fixed or variable step length according to the computational difficulty.

Below are two example applications of the gradient algorithm to continuous network design problems: traffic signal setting and transit line frequency optimization.

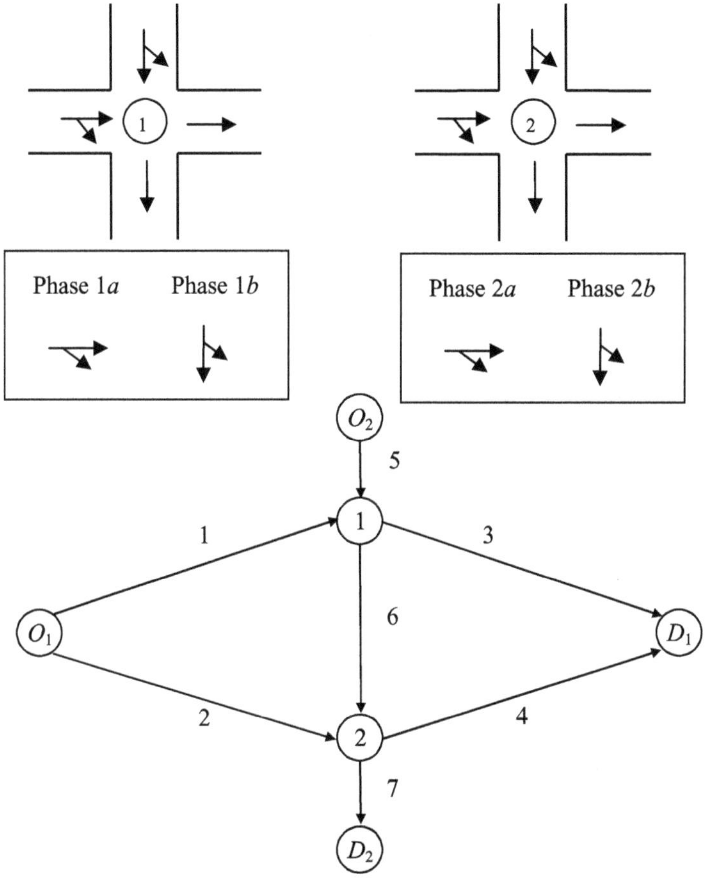

Link	Running time	Variable	Saturation flow
1	10	$y_1^{\text{PER}} = y_{1a}$	1000 vph
2	8		
3	13	$y_2^{\text{PER}} = y_{2a}$	1000 vph
4	10		
5	10	$y_1^{\text{PER}} = 1 - y_{1b}$	1000 vph
6	10		
7	4	$y_2^{\text{PER}} = 1 - y_{2b}$	1000 vph

Demand
$O_1 - D_1 = 700$ vph
$O_2 - D_2 = 900$ vph

Delay function	$tw(f_l, y_n^{\text{PER}}) = 3 + 5[f_i/(y_n^{\text{PER}} s_i)]^4$

Objective function	$w(\boldsymbol{y}) = [\boldsymbol{c}(\boldsymbol{y}, \boldsymbol{f}^*)]^T \boldsymbol{f}^*$

Fig. 9.12a Projected gradient algorithm for the optimal signal setting problem: test network

Iteration	1	2	3	4	5	6	7	8	9
y_{1it}	0.500	0.400	0.300	0.300	0.300	0.275	0.275	0.275	0.269
y_{2it}	0.500	0.401	0.306	0.306	0.306	0.285	0.285	0.285	0.287
f_1^*, f_3^*	292	309	327	327	327	330	330	330	326
f_2^*, f_4^*	408	391	373	373	373	370	370	370	374
$w(\mathbf{y}_{it})$	138,719	91,315	74,771	74,771	74,771	74,011	74,011	74,011	74,002
Step size	0.100	0.100	0.100	0.050	0.025	0.025	0.013	0.006	0.006
$\partial w/\partial y_1$	378,527	94,661	35,000	35,000	35,000	4,000	4,000	4,000	−1,000
$\partial w/\partial y_2$	373,340	88,501	29,000	29,000	29,000	−1,000	−1,000	−1,000	−2,000
$\max \lvert\partial w/\partial y_n\rvert$	378,527	94,661	35,000	35,000	35,000	4,000	4,000	4,000	2,000
Δy_1	−0.100	−0.100	−0.100	−0.050	−0.025	−0.025	−0.013	−0.006	−0.006
Δy_2	−0.099	−0.096	−0.083	−0.041	−0.021	+0.006	+0.003	+0.002	−0.003
y_{1it+1}	0.400	0.300	0.200	0.250	0.275	0.250	0.262	0.269	0.263
y_{2it+1}	0.401	0.306	0.223	0.265	0.285	0.291	0.288	0.287	0.284
$w(\mathbf{y}_{it+1})$	91,315	74,771	84,864	74,864	74,011	74,150	74,030	74,002	74,053
Step size red.	NO	NO	YES	YES	NO	YES	YES	NO	YES
Stop test	NO	NO	NO	NO	NO	NO	NO	NO	YES

VALID Steps | OPTIMAL Solution

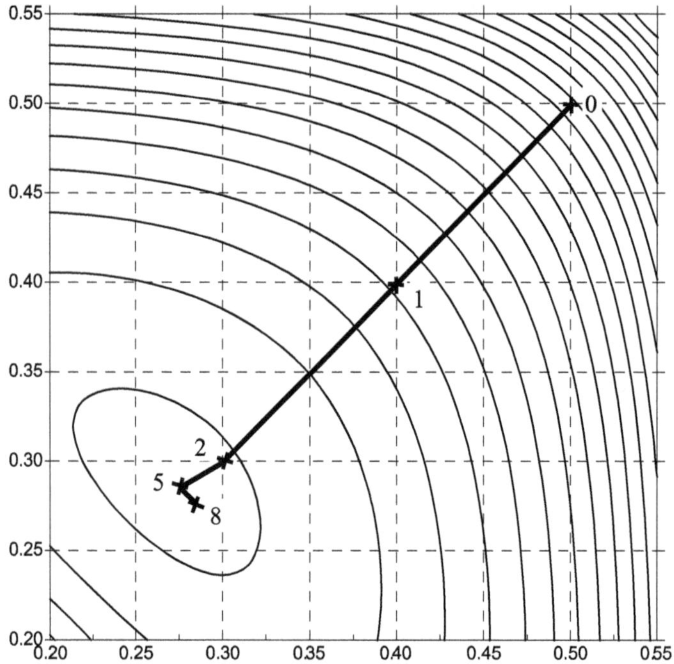

Fig. 9.12b Projected gradient algorithm for the optimal signal setting problem of Fig. 9.12a

An algorithm for optimal signal setting. An example of continuous SDP is the global optimization of traffic signal settings for urban networks. Under the assumption that each intersection (node n) has only two phases (a and b), the control vari-

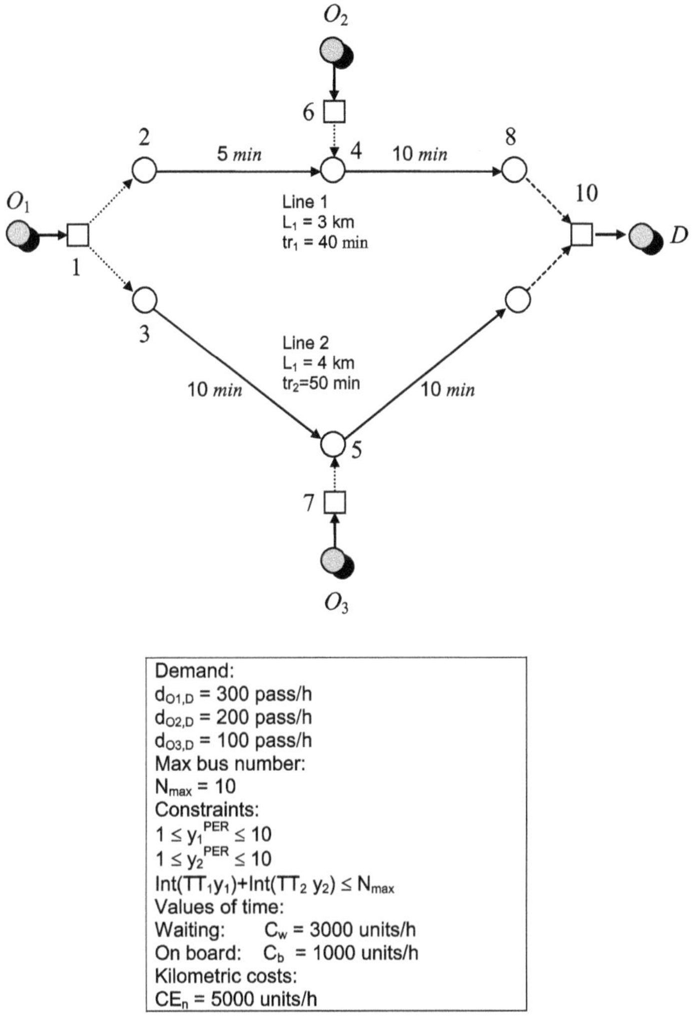

Fig. 9.13a Projected gradient algorithm for the optimal transit frequency problem: test network

ables are the ratios of effective green time to cycle length, one for each intersection:

$$y_n^{\text{PER}} = G_n^a / T_{cn} = 1 - \left(G_n^b / T_{cn}\right) \quad \forall \text{ intersection } n$$

where

G_n^a is the effective green time for phase a at intersection n
G_n^b is the effective green time for phase b at intersection n
T_{cn} is the cycle length for intersection n

For such a problem, the number of variables is equal to the number of signalized intersections, because there is only one decision variable for each of them. The

Numerical results					
Iterations →	1	2	3	4	5
y_{1it}	3	5	7	7	8
y_{2it}	3	3,92	4,78	4,78	4,41
f_{2-4}	150	168	178	178	193
f_{4-8}	350	368	378	378	393
f_{3-5}	150	132	122	122	107
f_{5-9}	250	232	222	222	207
Operator cost	105,000	153,387	200,506	200,506	208,268
Users cost	587,700	433,621	360,288	360,288	349,536
Total costs (w^k)	692,700	587,008	560,794	560,794	557,805
Bus number	6	8	9	9	10
Step size	2	2	2	1	
$\partial w/\partial y_1$	−78,500	−21,492	−4,580	−4,580	
$\partial w/\partial y_2$	−36,084	−9,198	1,657	1,657	
$\text{Max}\lvert\partial w/\partial y_i\rvert$	78,500	21,492	4,580	4,580	
Δy_{1it}	2	2	2	1	
Δy_{2it}	0.92	0.86	−0.72	−0.36	
y_{1it+1}	5	7	9	8	
y_{2it+1}	3.92	4.78	4.05	4.41	
Total costs (w_{it+1})	587,008	560,794	558,624	557,805	
Bus number	8	9	11	10	
Step size red.	NO	NO	YES	NO	
Stop test	NO	NO	NO	YES	

Fig. 9.13b Projected gradient algorithm for the optimal transit frequencies problem of Fig. 9.13a

social objective function to be minimized can be the total user cost on the network, as given by (9.3.2).

To solve this problem, a projected gradient algorithm with numerical calculation of derivatives and variable step length can be used; the algorithm follows the general framework reported in Fig. 9.11 and computes the descent direction as the opposite of the numerical gradient. The descent direction is projected back to the feasible region (i.e., some components being set to zero) if the next trial solution violates a constraint.

In order to find the step length, the descent direction can be normalized by dividing its components by the maximum absolute value. The algorithm proceeds with a fixed step length each time the objective function value is improved and the constraints on the variables are satisfied. The step length is reduced each time the objective function value worsens in an iteration, and the algorithm stops when the step length is less than a fixed value. A numerical example of the optimal signal setting problem was performed for the small network with two controlled intersections described in Fig. 9.12a. The assignment model used in the example is stochastic user equilibrium with multinomial logit path choice and an MSA algorithm to compute the equilibrium flows (see Sect. 5.4.2). The main variables generated by the projected gradient algorithm are presented in Fig. 9.12b.

An algorithm for optimal transit frequencies. Another example of a continuous network design problem is the optimization of transit frequencies.

This problem looks for the optimal frequencies y_j^{PER} for a transit network with given transit lines. The objective function is the sum of user and operator costs expressed by (9.3.3) and (9.3.4), respectively, taking into account only frequency control variables $\mathbf{y}^{\text{PER}}$. The constraints included in this model are the assignment constraints (9.3.5), minimum and maximum frequency constraints, and the vehicle budget constraint (9.3.6).

In Fig. 9.13 numerical results of an application of the projected gradient algorithm on a test network are shown. The step length can be reduced if the objective function increases, and/or if the budget constraint is violated.

Reference Notes

There is a large body of literature on network (supply) design models, many contributions considering only network features without considering price parameters. The general formulation proposed in this chapter is original. Almost all proposed models assume the demand as rigid: supply modifications do not influence user mobility choices except for route choice.

The numerous papers proposed in the literature can be classified in several ways: by transportation system (road, transit, multimodal), by kind of variable (continuous, discrete, mixed), by assumptions on demand (rigid, elastic), by solution approaches (simulation, optimization), and by solution algorithms (exact, heuristic, metaheuristic).

Design of topological variables of road networks has been tackled by discrete variable models in papers by Billheimer and Gray (1973), Boyce and Janson (1980), Poorzahedy and Turnquist (1982), and Solanki et al. (1998), which propose heuristic solution algorithms, and in papers by Le Blanc (1975), Los and Lardinois (1982), and Chen and Alfa (1991a, 1991b), advocating branch-and-bound solution algorithms.

Continuous variable models have been proposed in papers by Dantzing et al. (1979), Marcotte (1983), Le Blanc and Boyce (1986), Suwansirikul et al. (1987), and Meng et al. (2001), that propose heuristic solution algorithms; Abdulaal and Le Blanc (1979) and Davis (1994) propose, instead, descent algorithms; and Friesz et al. (1992) propose simulated annealing techniques. Models to solve network design and location problems jointly have been proposed in papers by Melkote and Daskin (2001) and Drezner and Wesolowsky (2003).

General formulation of the signal setting design problem, with the distinction between local and global approaches, is reported in papers by Marcotte (1983), Cantarella et al. (1991), Cantarella and Sforza (1991, 1995), and Cascetta et al. (1999, 2006). Abdelfatah and Mahmassani (1999) propose a dynamic approach to the problem.

The local approach has been widely studied in the literature; papers by Allsop (1977), Smith (1979), Dafermos (1980, 1982a, 1982b), Fisk and Nguyen (1982), Florian and Spiess (1982), Meneguzzer (1995), Smith and Van Vuren (1993), and

Cascetta et al. (1999) faced the problem from a static point of view. Dynamic models have been proposed in papers by Han (1996), Hu and Mahmassani (1997), and Lo et al. (2001). Cantarella and Improta (1991) consider also offsets as decisional variables. Real-time actuated traffic signals have been studied in a paper by Mirchandani and Head (2001). Wong and Yang (1999) proposed group-based methods.

The solution of the signal setting design problem from the global point of view has been covered in papers by Sheffi and Powell (1983), Yang and Yagar (1995), Heydecker (1996), Wong and Yang (1997), Ziyou and Yifan (2002), and Cascetta et al. (1998, 2006). Wong et al. (2002) propose group-based methods.

Joint design of topological variables and signal settings in an urban context has been tackled in papers by Cantarella and Vitetta (1994, 2006), Gallo (2002), Cantarella et al. (2006), and Gallo et al. (2009), which propose heuristic or meta-heuristic algorithms.

The transit network design problem is widely studied in the literature; a recent review is reported in Guihaire and Hao (2008). Because the literature is very extensive, below we refer to only more recent papers. Models and methods to design only line frequencies have been proposed in papers by Shih and Mahmassani (1991), Constantin and Florian (1995), Russo (1995) and Crisalli (1996). The design of the whole network is treated in papers by Baaj and Mahmassani (1991, 1995), Pattnaik et al. (1998), Iman (1998), Soehodo and Koshi (1999), and Montella and Gallo (2002). Design of feeder lines to rail systems is studied in papers by Rama Moorthy (1997), Martins and Pato (1998), Chien and Schonfeld (1998), and Chieng and Yang (2000).

Multimodal network design problems, under the assumption of variable demand, are introduced in papers by Montella et al. (2000, 2007); a multimodal model to optimize parking fares is proposed in D'Acierno et al. (2006); and in D'Acierno et al. (2003) the optimization of transit fares is studied.

Chapter 10
Methods for the Evaluation and Comparison of Transportation System Projects

10.1 Introduction

As stated in Chap. 1, transportation systems engineering can be defined as a discipline aimed at the functional design of physical and/or organizational interventions on transportation systems. A set of coordinated, internally consistent actions on a transportation system are referred to here as a project or plan. Transportation system modeling allows the prediction of the impacts of projects in order to assess their technical suitability and to support intermediate and final decision-makers through the process of project evaluation.

The range of transportation system projects is very diverse, and so also are the points of view from which their consequences can be evaluated. Projects might involve transportation facilities, control systems, services, and/or fares. Similarly, projects can be designed and evaluated from the perspective of the community that they will serve, or from the perspective of the service and/or facility operators.

Design and decision-making are two interdependent activities. Decision-making for transportation systems is often more complex than for systems in other sectors of engineering. This is especially true when the decision-maker must consider, either directly or indirectly, the effects of proposed actions on the larger community. Projects that involve decisions and/or points of view which are only relevant to an operator, such as the organization of freight distribution or the design of a traffic signal control system, may only require a relatively simple and straightforward decision process. On the other hand, it often happens that projects undertaken by a transportation agency, such as a reorganization of transit lines, lead to impacts that are external to the agency but that may influence its final decisions. For this reason, this chapter mainly considers complex projects with a wide range of impacts.

The decision-making process can address the evaluation of an individual project or the comparison of multiple alternative projects. The first case examines the economic, financial, and other impacts of carrying out a particular project, such as constructing a new road. In the second case, the decision-making process is intended to help choose the best among different possible solutions whose economic or financial impacts have been previously determined. An example is the choice between different possible alignments for a particular road.

The following sections focus on the activities relating to project evaluation and comparison. Section 10.2 identifies relevant project impacts and discusses their quantification using the models and methodologies described in the previous chapters. Section 10.3 then presents economic and financial analysis techniques used in project evaluation, and multicriteria analysis methods for the comparison of alternative projects.

E. Cascetta, *Transportation Systems Analysis*,
Springer Optimization and Its Applications 29,
DOI 10.1007/978-0-387-75857-2_10,

10.2 Evaluation of Transportation System Projects

Because it is an activity intended to support decision-making, evaluation necessarily depends on the decision-maker's perspective. A classic example of this dependency is the difference between the financial and the economic analysis of a project.

Financial analysis is traditionally associated with private operators that attempt to maximize profit under constraints such as regulations, service obligations, concessions, and the like. In this case benefits and costs have a natural expression in monetary terms: the former come from the revenues from service sales and subsidies, if any, and the latter from the financial costs of service production such as construction, maintenance, and operating costs, tolls, taxes, and so on.

Economic analysis is traditionally associated with a public decision-maker.[1] Alternative projects are evaluated taking into account their benefits and costs (general positive and negative impacts) with respect to the objectives of the community, or rather of different groups in the community that are homogeneous in terms of their socioeconomic characteristics and of project impacts on them. Indeed, some transportation system users may benefit from a particular project (reduced travel times and costs, increased accessibility, etc.) whereas others may receive lesser advantages or even disadvantages from it (increased travel times and costs, etc.). This might occur in an urban area, for example, as a result of the shifting of congestion from one zone to another due to traffic signal control strategies, reserved lanes for public transportation, limited access traffic zones, and the like. The contrast is even more evident if the benefits to system users are compared with the costs borne by some nonusers, for example the increase in noise and air pollution for residents in zones close to a new motorway or a new airport.

Quantitative techniques for the evaluation for public infrastructure projects have been the object of many theoretical studies and practical applications over the decades. Begun in the United States in the 1930s during the New Deal as a set of methods and criteria for the evaluation of water resource projects, both the theory and the practice of this discipline have rapidly evolved. In recent years the aim and scope of project evaluation have expanded as a result of major changes in the transportation arena. These developments include changes in the values and the level of participation of different stakeholder groups, deregulation of some sectors of the transportation market, and involvement of private capital in financing infrastructure construction and/or service operations. The systematic analysis of the results achieved in this field is well beyond the scope of this book. The following sections consider the role of quantitative methods in the overall activity of project evaluation.

[1] In reality, companies operating transportation services also have several objectives and/or must take into account the impacts of their decisions on different subjects. Economic analysis, in the broad sense, should be extended to all the main decision-makers who operate in a transportation system, although with different objectives and constraints.

10.2.1 *Identification of Relevant Impacts*

The impacts of a transportation project can be defined as the consequences of the project as they relate to specific groups that are affected by it (i.e., groups that are homogeneous with respect to an issue under consideration). It follows that the breadth of the evaluation task and the evaluation approach followed are determined by the selection of impacts to be considered in the evaluation. The range of impacts typically considered in transportation project evaluation has widened over time, hand in hand with improvements in models and in computing power. Similarly, there has been expanded recognition of the different and often contrasting objectives and goals of stakeholders and decision-makers. Indeed, the role of the decision-maker has also become increasingly complex as transportation system financing and management options have grown more diverse. In the past, the decision-maker was typically viewed as an anonymous and uninvolved public official who was expected somehow to reconcile the objectives of the general public and the transportation service providers. This has evolved to a more explicit recognition of the distinct nature of the different roles, in particular those of the public institutions, the major interest groups, and the transportation service providers independently of their public or private nature.

First-generation quantitative evaluation exercises were undertaken to support investment in motorways and were later extended to consider transportation system projects in the broader sense. These studies typically took into account only the monetary and monetarily quantifiable effects (benefits and costs) to the users of the planned facilities resulting from building and operating these facilities and services. The former included the changes[2] in level-of-service attributes such as travel time, in the monetary cost of tolls and vehicle operation, and sometimes in the expected number of accidents. The monetary costs for service and/or infrastructure operators (agency costs) included construction costs, investment costs in vehicles and technologies, changes in maintenance and operating costs, as well as changes in revenues from service sales. The effects for the operator sometimes included changes in transfer payments with other public authorities (e.g., reimbursements for service obligations, duties and taxes on gasoline and real estate, etc.).

With better understanding and modeling of the mechanisms underlying transportation systems, the range of effects considered for the users of the transportation system has gradually increased. More consideration is now given to impacts on all users, both current and project-induced, calculating the changes in generalized costs, both perceived and not perceived, for the different transportation modes. Analysts often distinguish the impacts on different user classes (market segments), that is, groups of users that are homogeneous in terms of their trip purpose, socioeconomic characteristics, and level-of-service attributes. As for the effects on operators, the construction, maintenance, and running costs calculated on the basis of market

[2] As shown more clearly in the following, effects on users are measured in terms of changes induced in their choices.

prices are increasingly disaggregated and expressed in terms of the resources consumed (manpower, materials, capital), because market prices do not always reflect the actual social value of these resources.

The perspective and range of effects considered in transportation project evaluation has been further widened by consideration of project external effects. These are impacts on members of society who are not directly involved in the use of the transportation system. Examples of nonuser impacts are given below, subdividing the external effects into economic, land use, social, and environmental. It should be noted that the classification of some impacts can be somewhat arbitrary, and that analysts do not always agree on the characterization of some of these.

Economic impacts can be defined as changes in the state of the economic system brought about by a project. This includes changes in residential and commercial property values and in economic production resulting from changes in accessibility; and changes in the economic consequences of accidents directly and indirectly related to the project. Economic externalities are directly measurable in monetary units, or at least can easily be translated into such units.

Land use impacts are related to land use and its quality. Examples of land use impacts are changes in land use type (e.g., from residential to commercial) and intensity, or more generally the relocation of housing and economic activities brought about by accessibility differentials. This category includes changes in the geographical structure of a region or in the urban quality of specific neighborhoods.

Social impacts can be defined as impacts on social values and changes in the relationships among people and social institutions such as the family, local communities, education, government bodies, and so on, brought about by the project. In this case too there are effects of different types: social effects of accidents, changes in accessibility to social activities (schools, public offices, parks, etc.), changes in cohesion and stability of local communities, as well as impacts on historic and cultural sites. Changes in social equity, that is, changes in the distribution of travel-related opportunities with respect to space (zone) and socioeconomic status (income class or age) can also be considered as social impacts.

Finally, *environmental impacts* can be defined as the effects of a project on the physical environment. These can be further classified as effects on the ecosystem, on noise and air pollution, and on visual perception. Transportation system projects, especially new infrastructure in rural areas, can alter the ecological balance of plant and animal populations. Furthermore, any transportation system generates noise and air pollution, and a project may significantly change their intensity and distribution in the affected areas. Lastly, transportation infrastructure and vehicles affect the viewscape over a potentially large area. The nature and severity of these impacts depend on the visibility of the transportation facility and its contrast with the background.

Figure 10.1 summarizes some of the potential impacts of a transportation system project on different groups. It is obvious that not all the impacts listed are relevant to the evaluation of all projects. In certain cases, particular effects might be absent or their magnitude might be considered negligible; in other cases some impacts may be present but deemed irrelevant to the analysis.

**Users (by class)*

- Differences of net utility (surplus) perceived by users
- Differences of costs not perceived by users

**Agencies and operators (for each subject involved)*

- Differences in resources (manpower, materials, capital) and costs needed for building transportation infrastructure, vehicles and control systems (investments)
- Differences in resources and costs for maintenance of the infrastructure and technologies
- Differences in resources and costs for the operation of transportation services
- Expropriation and relocation costs
- Differences in traffic revenues
- Changes in taxes paid by users (fuel, etc.) and nonusers (property, etc.)
- Differences in transfer payments between government agencies

**Nonusers of the transportation system (for each homogenous group)*

Economic impacts

- Differences in the production of different economic sectors
- Differences in the economic impacts of accidents
- Differences in property values

Land use impacts

- Differences in the location of households and economic activities
- Differences in urban structure and "quality"

Social impacts

- Impacts on the preservation of historic and cultural sites
- Difference in accessibility to social activities (school, social and religious centers, recreational activities, etc.)
- Modifications in the structure and cohesion of local communities
- Changes in the social effects of accidents
- Changes in the distribution of users' surplus by zone and socioeconomic group (impact on equity)
- Changes in visual and aesthetic impacts

Environmental impacts

- Changes in the ecosystem
- Changes in noise and air pollution

Fig. 10.1 Classification of impacts for the evaluation of transportation system projects

10.2.2 Identification and Estimation of Impact Indicators

The effects of a transportation project can be represented by a set of variables known as *impact indicators* or *measures of effectiveness* (MOE). Because, at a detailed level, a project can have a large number of different impacts, and it would be impractical to handle all the related variables, it is common practice in analysis to use a smaller number of performance indicators that are obtained by aggregating the detailed impacts and their measures.

Some impact indicators are quantitative variables such as travel time or tons of CO emissions; others are intrinsically qualitative and can only be expressed by descriptive variables (such as little, much, etc.) or on an arbitrary scale (such as from A to F).

The effects of a project are usually evaluated in differential terms, that is, as changes in the values of impact variables between the project (P) and nonproject (NP) states. The latter, sometimes known as the reference or baseline solution, is defined as the option to maintain the present state of the system, or to implement projects that have already been accepted and that are not subject to the evaluation. The definition of the baseline alternative may require considerable care.

The *time dimension* is an important factor in estimating impacts. Different project impacts generally follow different profiles over time. For example, construction and investment costs are spent in a relatively short period of time at the beginning of the project, whereas maintenance and operating costs continue throughout the entire life of the project. Furthermore, with the passing of time some effects change in intensity or even in direction: user travel costs may increase during the construction phase due to capacity reductions and other disturbances, and then decrease when the infrastructure is in service.

For analysis purposes, the total economic life[3] of the project is divided into a set of reference periods, with conditions assumed to be stationary during each period. Such periods might be, for example, different times during a year or different periods during a typical day, both during construction and during representative operation years. As was seen in Chap. 1, impact indicators are typically computed for these reference periods and then extrapolated to longer time periods. As shown in Fig. 10.2, many impacts can be predicted using the models described in previous chapters. The estimation of these impacts requires modeling the system in the Project (P) and NonProject (NP) states followed by calculation of differences (changes) in the variables measuring quantifiable impacts.

The resources needed for construction, maintenance, and operation and their corresponding costs can be estimated either (i) analytically, from the actual design of the proposed facilities and services, or (ii) synthetically, using statistical relationships known as *production functions*. These functions estimate the resources needed,

---

[3] The "economic life" of a project can be defined conventionally as its period of validity. For infrastructure, this may be considered to be the period for which no major extraordinary maintenance works are necessary. The arbitrariness of this definition is partly compensated for by the possibility of a residual value of the project at the end of the period under consideration.

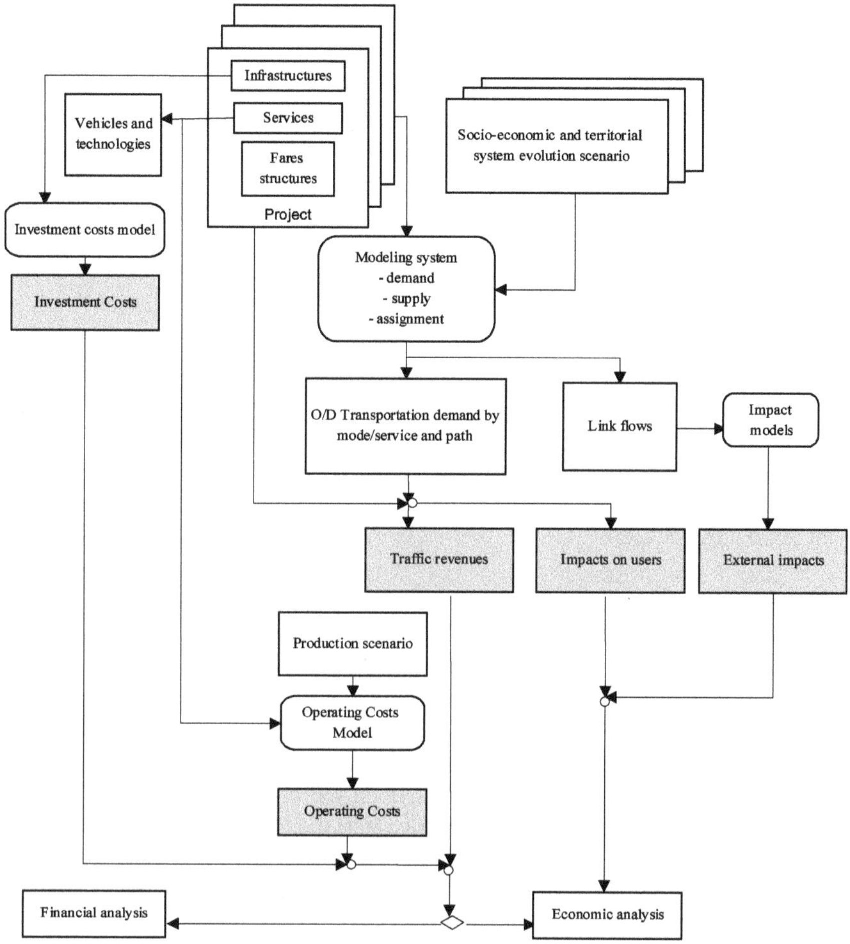

Fig. 10.2 Main components of an impact assessment process

for example, to build and equip a unit length of typical infrastructure, to produce vehicles and technologies with given characteristics, or to maintain infrastructure and operate a transportation service of a given type. Alternatively, construction, maintenance, and operation cost functions directly estimate the corresponding costs in monetary terms.

Traffic revenues can be calculated by multiplying the number of predicted users of tolled infrastructure and/or for-pay services by the corresponding charges.

A variety of other impacts can be calculated from the models described earlier in this book. For example, the probability of accidents and their consequences, fuel consumption, and noise and air pollution can all be evaluated through the relevant link impact functions described in Sect. 2.3.3. The ease of access to different services can be measured through accessibility variables derived from destination choice models (as discussed in Sect. 4.3.1.2) or developed in other ways described in the literature. In any case, generalized costs or level-of-service attributes play a key role in the measurement of accessibility.

The calculation of transportation system user benefits requires a further elaboration of the concepts and demand models described in previous chapters, and is covered in the following section.

10.2.3 Computation of Users' Surplus Changes

The impacts perceived by users can be calculated as a change in net perceived utility (or surplus) associated with the travel choices made in the project and nonproject situations. Either of two different calculation approaches can be applied, depending on whether the underlying demand model is a behavioral random utility model or a descriptive nonbehavioral model. The two approaches are analyzed and compared in the following sections.

(a) Random Utility Demand Models

Random utility demand models are based on explicit assumptions about the choice behavior of decision-makers or users. These assumptions can be used to estimate changes in average perceived utilities for the choice dimensions affected by a project. As an example, consider the classic choice sequence defined by the decision of user i to make x trips for a particular purpose from zone o to destination d by mode m and following path k.

In this case, the utility U_P^i perceived by user i for the sequence that would be chosen in the project situation P can be expressed as:

$$U_P^i = \sum \beta_k X_{kj(i)}^{iP} + \varepsilon_{j(i)}^i = V_{xodmk}^i\left(\boldsymbol{X}_{j(i)}^P\right) + \varepsilon_{xodmk}^i \tag{10.2.1}$$

where $j(i)$ indicates the specific sequence $(x\ o\ d\ m\ k)$ chosen and the vector of attributes $\boldsymbol{X}_{j(i)}^P$ includes level-of-service (times, costs, etc.) and other variables corresponding to $j(i)$ in the situation P. Because in general some attributes $X_{kj(i)}^i$ have positive coefficients (i.e., they represent utilities) whereas others have negative coefficients (costs), expression (10.2.1) represents the perceived net utility (utility minus cost), or *surplus*. A simple specification of the systematic utility V_{xodmk} for shopping trips might be:

$$V_{xodmk}(\boldsymbol{X}) = \beta_1 NOTRIP + \beta_2 SHP_d - \beta_3 t_{odmk} - \beta_4 mc_{odmk} \tag{10.2.2}$$

where x takes the value zero or one, *NOTRIP* is an alternative-specific variable for the choice not to make a trip, $(x = 0)$, SHP_d is the number of shops in zone d, t_{odmk} and mc_{odmk} are, respectively, the travel time and monetary cost to go to d

by mode m departing from origin o and following path k. The linear combination of travel time and monetary cost is often called the generalized path cost, $g_{odmk} = \beta_3 t_{odmk} + \beta_4 mc_{odmk}$.

In random utility models, the user is assumed to choose the alternative that maximizes the perceived utility (10.2.1), but this is unknown to the analyst and so is represented as a random variable. The impacts of a project on a homogeneous group of transportation system users can be expressed by the change of the expected value of the surplus perceived by group members. This corresponds to the mean value of the perceived utility (surplus) of the chosen alternative, that is, the alternative with maximum utility. The mean value of the perceived surplus thus coincides with the mean value of the maximum perceived utility among all the available alternatives, that is, with the Expected Maximum Perceived Utility (EMPU) variable discussed in Chap. 3.

Inasmuch as the model specification, the attributes considered, and the systematic utility coefficients β_k in (10.2.1) usually all depend on the trip purpose and the socioeconomic characteristics of the decision-maker, the EMPU variable s is calculated separately for a representative user of each class i[4] in each zone o:

$$s_P(o,i) = E\Big[\max_{xdmk} U_P^i(xdmk)\Big] \tag{10.2.3}$$

As shown in Chap. 3, if the residuals ε_{xodmk} are i.i.d. Gumbel variables with parameter $\theta = 1$, the EMPU (10.2.3) can be expressed in closed form as a logsum variable:

$$s_P(o,i) = \ln \sum_{xdmk} \exp\big[V_{xdmk}^i\big(\boldsymbol{X}_i^P\big)\big] \tag{10.2.4}$$

Similar closed-form expressions can be obtained for other models belonging to the logit family. The total utility of all the (homogeneous) users of class i in zone o in the project situation $S_P(o,i)$ can be estimated as

$$S_P(o,i) = N_o^P(i) s_P(o,i) \tag{10.2.5}$$

where $N_o^P(i)$ is the number of such users. Note that $N_o^P(i)$ includes both actual and potential users (potential users are those who could travel but choose not to).

The perceived surplus change brought about by the project for users of class i in zone o can then be expressed as

$$DS_P(o,i) = S_P(o,i) - S_{NP}(o,i) \tag{10.2.6}$$

where $S_{NP}(o,i)$ is the total perceived surplus in the nonproject situation, and is calculated in the same way as described above.

[4]As stated in Chap. 4, the class is a group of users sharing the same values of behavioral parameters that are relevant to the specific application. A user class is usually defined by the pair (socioeconomic category, trip purpose). In the limiting case of completely disaggregate models, the class i may coincide with a single individual.

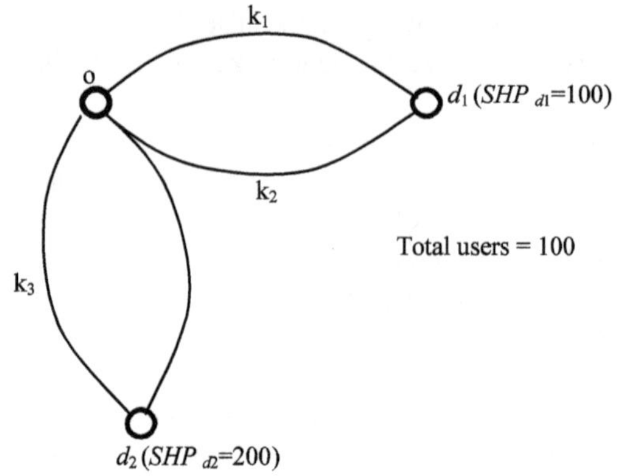

$V_{xdmk} = \beta_1 NOTRIP + \beta_2 SHP_d + \beta_3 T_{odk}$

$$S^S(o) = 100\ln\big[\exp(\beta_1 NOTRIP) + \exp(\beta_2 SHP_{d_1} + \beta_3 T_{od_1k_1}) + \exp(\beta_2 SHP_{d_1} + \beta_3 T_{od_1k_2}) + \exp(\beta_2 SHP_{d_2} + \beta_3 T_{od_2k_3}) + \exp(\beta_2 SHP_{d_2} + \beta_3 T_{od_2k_4})\big]$$

	Non-project		Project	
$\beta_1 = 1$	$T^{NP}_{od1k1} = 6$	$g^{NP}_{od,k1} = 1.2$	$T^{P}_{od1k1} = 5$	$g^{P}_{od,k1} = 1.0$
$\beta_2 = 0.015$	$T^{NP}_{od1k2} = 7$	$g^{NP}_{od,k1} = 1.4$	$T^{P}_{od1k2} = 5$	$g^{P}_{od,k1} = 1.0$
$\beta_3 = -0.2$	$T^{NP}_{od2k3} = 10$	$g^{NP}_{od,k1} = 2.0$	$T^{P}_{od2k3} = 6$	$g^{P}_{od,k1} = 1.2$
	$T^{NP}_{od2k4} = 10$	$g^{NP}_{od,k1} = 2.0$	$T^{P}_{od2k4} = 7$	$g^{P}_{od,k1} = 1.4$
	$S_{NP}(o) = 236.2$	$S_P(o) = 283.4$	$\Delta S(o) = 47.2$	

Fig. 10.3 Example of calculation of surplus change with a behavioral model

The monotonicity of the EMPU variable, discussed in Sect. 3.5, ensures that the net utility change is positive or negative according to whether the systematic utility of each alternative increases or decreases when passing from the nonproject to the project state. Reductions in cost attributes and/or increases in utility attributes will lead to increases in the total surplus, and vice versa. By the same property, the total surplus will increase if the number of available alternatives increases; this may happen if the project makes available new transportation modes or services. Figure 10.3 provides an example of the calculation of perceived surplus for a logit choice model over the sequence *xdmk* for shopping trips.

Assuming that it is valid to sum the utilities of different people, aggregate utilities of users in different classes and/or in different traffic zones can be calculated. The perceived impact of project *P* over all users is therefore given by:

$$DS_P = \sum_i \sum_o DS_P(o,i) \tag{10.2.7}$$

It should be noted, however, that in many applications the perceived surplus changes should be analyzed at the level of disaggregate user groups in order to highlight the distribution of project (dis)benefits among the different groups or zones in the study area.

Average perceived surplus $s_P(o,i)$ and total utility changes calculated by (10.2.6) and (10.2.7) are expressed in dimensionless measurement units, sometimes called *utils*. In order to compare them with other effects of the project P, these values can be expressed in monetary units by dividing them by the coefficient of the monetary cost coefficient β_c, which has units of (monetary units)$^{-1}$.

If perceived utility changes from the project do not influence trip frequencies, and therefore the total demand level for each class of users remains constant, the surplus of nontravelers does not change and the total perceived surplus of users of a given class and origin can be expressed as

$$S_P(o,i) = d_{o.}(i) E\left[\max_{dmk} U^i(dmk/os)\right] = d_{o.}(i) \cdot s_P(o,i) \tag{10.2.8}$$

where $d_o(i)$ is the number of trips with origin o undertaken by users of class i in the reference period. Similar simplified expressions can be derived for fixed origin–destination demand flows.

(b) Descriptive Demand Models

A different methodology is applied to evaluate user impacts when descriptive demand models are used. In this case the model can be interpreted as a "demand function" relating the number of users undertaking trips with given characteristics to the average generalized trip cost and other explanatory variables. This cost is defined, in analogy with Chap. 2, as a (linear) combination of the amount of resources spent by the user on a trip (time, money, stress, etc.), with weights reflecting the user's travel behavior. The cost parameters (weights) may vary according to trip purpose and socioeconomic category (i.e., user class), and are generally estimated together with other coefficients of the demand model; this is commonly done for path and mode choice models.

In the following, the generalized cost of a trip undertaken by a user of class i between o and d by mode m and following path k is indicated by $g_{odmk}(i)$. This is equivalent to the cost g_k on path k in the network for mode m; for uniformity of notation, the zone pair *od* that the path and mode (or mode combination) connect is explicitly noted.

A simplified specification of the generalized cost analogous to that implicit in expression (10.2.2) is:

$$g_{odmk}(i) = \beta_1(i) t^i_{odmk} + \beta_2(i) mc^i_{odmk} \tag{10.2.9}$$

where t and *mc* are, respectively, the travel time and the monetary cost. (The coefficients here have been explicitly associated with user class i.) As above, the gener-

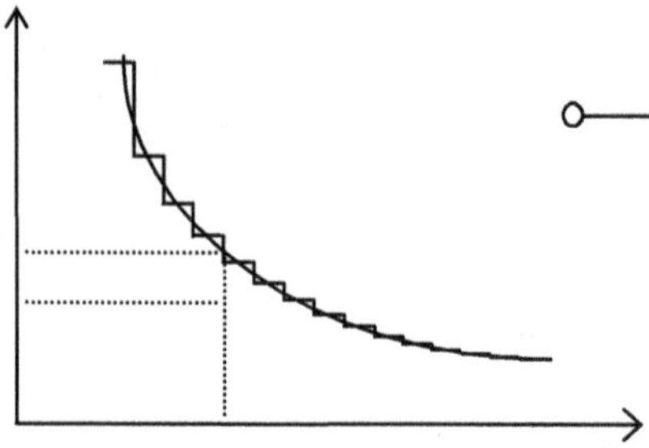

Fig. 10.4 Demand curve of a single O/D pair, mode, and path system

alized cost can be expressed in monetary units by dividing it by the cost coefficient $\beta_2(i)$.

To introduce the method for calculating perceived surplus changes with descriptive demand models, consider first a simple system consisting of a single O-D pair connected by a single mode and a single path, as shown in Fig. 10.4. Assume that all users belong to one class; that is, that they have the same behavioral parameters.

In this case, the demand model can be formally written as $d_{od} = d_{od}(g_{od})$, which gives the average number of users undertaking a trip for each value of the generalized average cost in the reference period.

The relationship $d_{od}(g_{od})$ can be represented in a two-dimensional graph and usually has a form similar to that shown in Fig. 10.4. (Strictly speaking, the figure represents the *inverse* demand curve, which is itself an important analytical construct. Alternatively, the figure can be considered to represent the demand curve itself, but using transposed axes.) The demand curve, in its traditional neoclassical interpretation, represents the ordering of individual trips by users on the basis of the generalized cost that they are willing to pay to undertake the trip; this is a measure of the utility of the trip to the user. In other words, the marginal trip corresponding to each point on the horizontal axis has a total utility (or willingness to pay) equal to the corresponding value of the generalized cost on the vertical axis. An increase in the cost would discourage this marginal user from making the trip and therefore reduce the value of the demand d_{od}.

All users of a given class incur the same generalized cost. Let g_{od}^{NP} be this cost and $d_{od}(g_{od}^{NP})$ be the number of users traveling in the nonproject situation. For all trips undertaken, except the marginal one, there is a net utility, or surplus, given by the difference between the amount that the user would be willing to pay to make the trip, and the cost that is actually paid (see Fig. 10.4). If as a result of project P the generalized cost is reduced to g_{od}^{P}, the number of users traveling increases to $d_{od}(g_{od}^{P})$, as shown in Fig. 10.5.

To calculate the total surplus change resulting from project P, a distinction should be made between trips undertaken in the situation *NP* and the new trips that are only undertaken because of the cost reduction (trips generated or induced by the

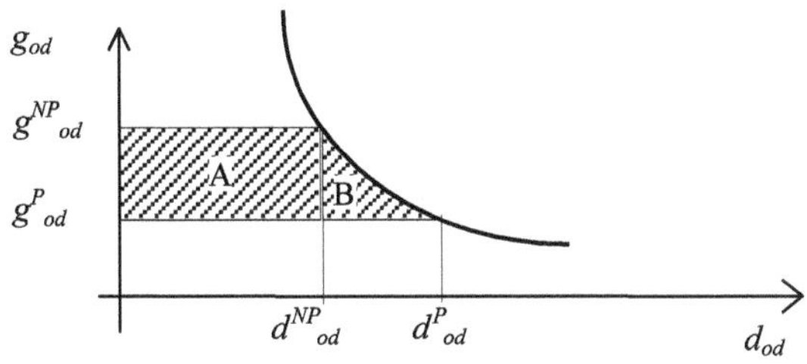

Fig. 10.5 Surplus change between project (P) and nonproject (NP) states: case of cost reduction

project).[5] For a trip by user i in the first group, the surplus change will be given by:

$$DS = \left(U^i - g^P_{od}\right) - \left(U^i - g^{NP}_{od}\right) = g^{NP}_{od} - g^P_{od} \tag{10.2.10}$$

that is, by the difference between the generalized cost in the nonproject and project situations. The total surplus change DS'_p for all the trips/users of this group is therefore:

$$DS'_p = d_{od}\left(g^{NP}_{od}\right) \cdot \left(g^{NP}_{od} - g^P_{od}\right) \tag{10.2.11}$$

and is represented by the area A in Fig. 10.5.

A trip i that is generated by the cost reduction brought about by project P will have a surplus $U^i - g^P_{od}$ in the project situation, and zero in the nonproject situation. The total surplus change for the $d^*_{od} = d_{od}(g^P_{od}) - d_{od}(g^{NP}_{od})$ project generated trips is therefore given by the area B in Fig. 10.5. Typically it is assumed that all generated trips d^*_{od} experience identical utility, given by the average value of the interval $[g^{NP}_{od}, g^P_{od}]$ (i.e., $U^i = (g^{\mathrm{NP}}_{od} + g^P_{od})/2$), and therefore the total surplus for the generated demand can be calculated as

$$DS^*_p = d^*_{od}\left[\frac{g^{NP}_{od} + g^P_{od}}{2} - g^P_{od}\right] = \frac{1}{2} d^*_{od}\left(g^{NP}_{od} - g^P_{od}\right) \tag{10.2.12}$$

Equivalently, if the demand curve is approximated by a line between $d_{od}(g^{NP}_{od})$ and $d_{od}(g^P_{od})$, the change in surplus for generated trips results from the formula for the area of the (approximate) triangle B.

The total surplus change is given by the sum of the terms (10.2.11) and (10.2.12):

$$\begin{aligned} DS_p = DS'_p + DS^*_p &= d_{od}\left(g^{NP}_{od}\right)\left(g^{NP}_{od} - g^P_{od}\right) \\ &\quad + \frac{1}{2}\left[d_{od}\left(g^P_{od}\right) - d_{od}\left(g^{NP}_{od}\right)\right]\left(g^{NP}_{od} - g^P_{od}\right) \\ &= \frac{1}{2}\left[d_{od}\left(g^P_{od}\right) + d_{od}\left(g^{\mathrm{NP}}_{od}\right)\right] \cdot \left(g^{NP}_{od} - g^P_{od}\right) \end{aligned} \tag{10.2.13}$$

Expression (10.2.13) can be interpreted as the product of the average demand between situations P and NP by the change in the corresponding generalized cost.

[5]Extra trips undertaken because of the effect of the generalized cost reduction are sometimes called the demand generated or induced by the project P.

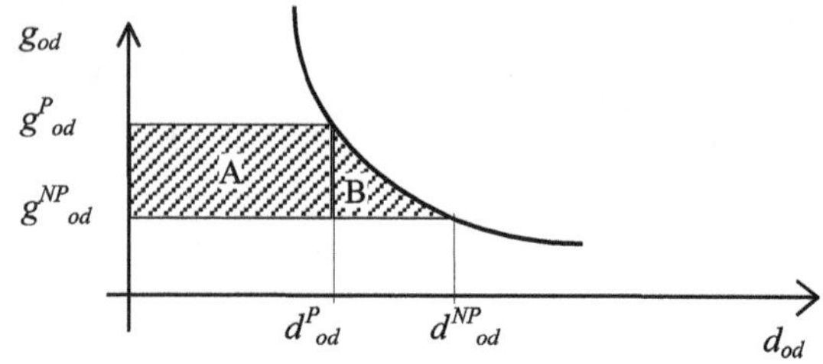

Fig. 10.6 Surplus change between project (P) and nonproject (NP) states for a cost increase

Equivalently, with the linear approximation mentioned above, this expression can be interpreted as the area of the (approximate) trapezoid consisting of the two parts A and B.

The exact expression for the surplus change can be obtained by calculating the hatched area in Fig. 10.5 as the integral of the demand function $d(g)$:

$$DS_p = -\int_{g^{NP}}^{g^P} d(g)\,dg \tag{10.2.14}$$

The results described still hold if the project increases the generalized cost ($g^P_{od} > g^{NP}_{od}$), as shown in Fig. 10.6. In this case, there will clearly be a reduction of surplus and a decrease in the number of trips. The surplus change can also be obtained through the algebraic sum of (10.2.11) and (10.2.12), which in this case are both negative.

The concept of surplus change and expressions (10.2.13) and (10.2.14) can be generalized to the case of multiple cost "dimensions" (e.g., multiple destinations and/or modes and/or paths). However, this generalization is neither straightforward nor universal. Consider, in fact, a slightly more complex case with two possible alternatives, for example, two paths with costs g_1 and g_2 (see Fig. 10.7); the two demand curves can be defined as $d_1(g_1 g_2)$ and $d_2(g_1 g_2)$. The demand, that is, the number of trips, on each path depends on the cost of both paths with a graph similar to that shown in Fig. 10.7. (The demand function can be obtained by combining trip generation and path choice models, e.g.) In this case the integral (10.2.14) can be replaced by:

$$DS = -\int_{(g_1^{NP}, g_2^{NP})}^{(g_1^P, g_2^P)} \sum_{i=1,2} d_i(g_1, g_2)\,dg_1\,dg_2 \tag{10.2.15}$$

However, the value of this integral usually depends on the path of integration between the two limits.[6]

To determine the surplus change, two heuristic approaches can be followed, corresponding to two approximate methods for the evaluation of integral (10.2.15).

[6]The integral (10.2.15) depends only on the extremes of integration if the Jacobian of demand functions is symmetrical with respect to generalized path costs: $\frac{\partial d_1}{\partial g_2} = \frac{\partial d_2}{\partial g_1}$. This condition is seldom, if ever, met by usual demand models.

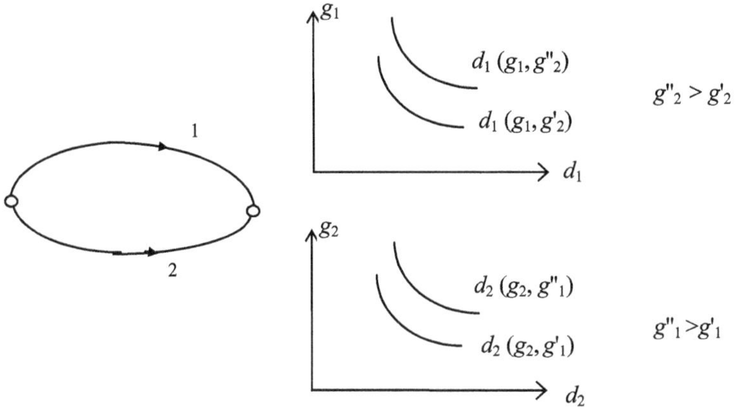

Fig. 10.7 Demand curves for a system with two paths

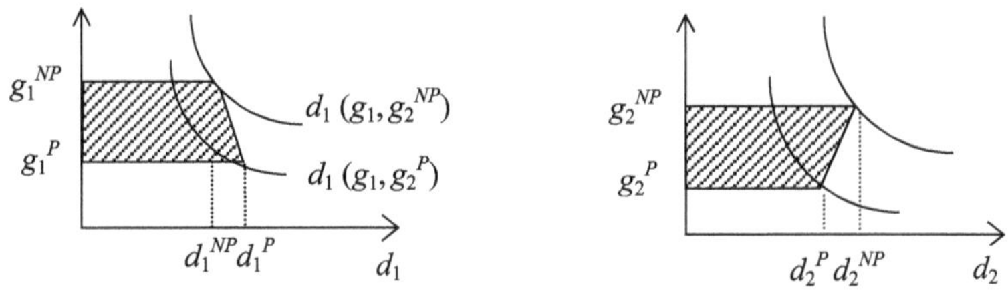

Fig. 10.8 Calculation of the surplus change with the average demand method

The first approach, which can be called the *average demand method*, calculates the surplus change as

$$DS_P = \frac{1}{2} \sum_{i=1,2} (d_i^{NP} + d_i^P)(g_i^{NP} - g_i^P) \tag{10.2.16}$$

where d_i^{NP} and d_i^P are, respectively, equal to $d_i(g_1^P, g_2^P)$ and $d_i(g_1^{NP}, g_2^{NP})$. The expression (10.2.16) can be interpreted as the summation extended to all relevant dimensions (in this case, the two paths) of the product of the average demand between states *P* and *NP*, and the cost change in that dimension.

Expression (10.2.16) corresponds to the sum of the two hatched areas in Fig. 10.8.

The alternative approach, which can be called the *average cost method*, reduces the problem to a single choice dimension by considering an average trip cost $\bar{g}$ given by the weighted average of the costs in each dimension:

$$\begin{aligned} \bar{g}^P &= p_1(g_1^P, g_2^P)g_1^P + p_2(g_1^P, g_2^P)g_2^P \\ \bar{g}^{NP} &= p_1(g_1^{NP}, g_2^{NP})g_1^{NP} + p_2(g_1^{NP}, g_2^{NP})g_2^{NP} \end{aligned} \tag{10.2.17}$$

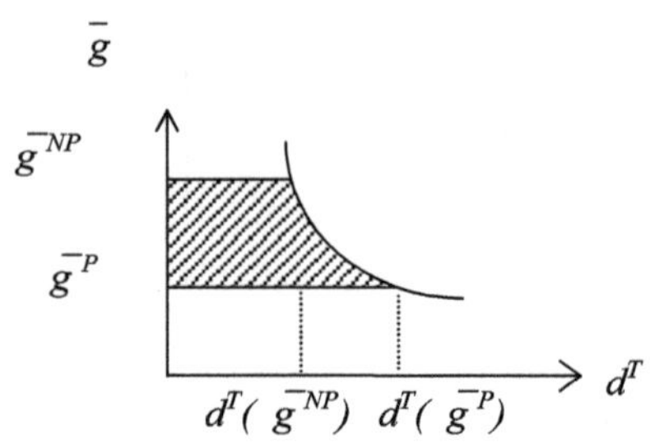

Fig. 10.9 Total demand curve as a function of the average trip cost

where p_1 and p_2 are the demand shares of each dimension, $p_i = d_i/(d_1+d_2)$. In this approach, the demand curve expresses the total demand $d^T = d_1 + d_2$ as a function of the average cost $\bar{g}$ (see Fig. 10.9). The surplus change can therefore be calculated from expression (10.2.13):

$$DS_P = \frac{1}{2}\left[d^T(\bar{g}^P) + d^T(\bar{g}^{NP})\right](\bar{g}^{NP} - \bar{g}^P) \tag{10.2.18}$$

The surplus change expressed by (10.2.18) can be interpreted intuitively as the product of the average of the total demand between the states *P* and *NP* and the change in average cost between the two states. Comparing expressions (10.2.16) and (10.2.18), it is clear that the two approaches give different results, as can be seen in the example presented in Fig. 10.10.

The partial share demand model can be conveniently expressed as the product of the demand level and the fraction of trips with given characteristics:

$$d^i_{odmk} = d^i_{o.}(\boldsymbol{SE}^i\boldsymbol{g}^i)\,p^i_{dmk/o}(\boldsymbol{SE}^i\boldsymbol{g}^i) \tag{10.2.19}$$

where $d^i_{o.}$ is the number of trips from zone o undertaken by users of class i, and $p^i_{dmk/o}$ is the fraction of these trips with the characteristics dmk.

As noted in Chaps. 4 and 5, both d^i_o and $p^i_{dmk/o}$ depend on a vector of socioeconomic and activity system attributes $\boldsymbol{SE}$, as well as on a vector of level of service attributes, expressed by the perceived generalized costs for all destinations, by all modes and on all paths $\boldsymbol{g}^i$. (In the following, the dependence of demand on the *SE* variables is implicit.) The surplus change for user class i resulting from the passage from state *NP* with costs $\boldsymbol{g}^{NPi}$ to state *P* with costs $\boldsymbol{g}^{Pi}$, can be calculated by extending the two previous approximate expressions to the general case. The average demand method, expressed by (10.2.16), therefore yields:

$$DS_p(o,i) = \frac{1}{2}\sum_{dmk}\left[d^i_{odmk}(\boldsymbol{g}^{NPi}) + d^i_{odmk}(\boldsymbol{g}^{Pi})\right]\cdot\left(g^{NPi}_{odmk} - g^{Pi}_{odmk}\right) \tag{10.2.20}$$

On the other hand, the average cost method, expressed by (10.2.18) yields:

$$DS_p(o,i) = \frac{1}{2}\left[d^i_{o.}(\boldsymbol{g}^{Pi}) + d^i_{o.}(\boldsymbol{g}^{NPi})\right]\cdot(\bar{g}^{NPi} - \bar{g}^{Pi}) \tag{10.2.21}$$

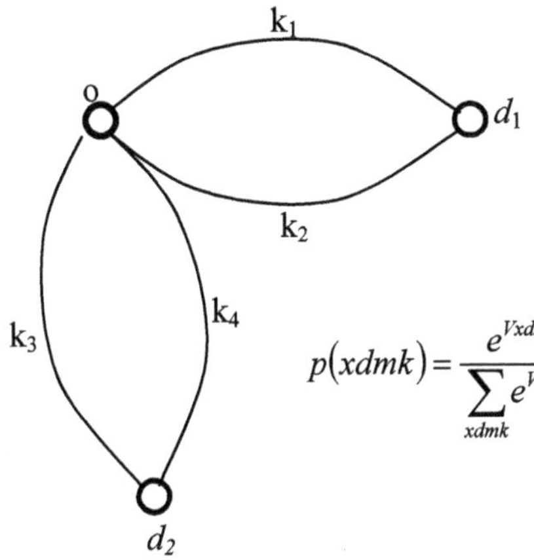

$$p(xdmk) = \frac{e^{Vxdmk}}{\sum_{xdmk} e^{Vxdmk}}$$

Nonproject (NP)

$$g^{NP}_{od_1k_1} = 1.2 \; d_{od_1k_1}(\mathbf{g}^{NP}) = 100 \cdot p^{NP}_{od_1k_1} = 36$$
$$d_o(\mathbf{g}^{NP}) = 100$$
$$g^{NP}_{od_1k_2} = 1.4 \; d_{od_1k_2}(\mathbf{g}^{NP}) = 100 \cdot p^{NP}_{od_1k_2} = 30$$
$$g^{NP}_{od_2k_3} = 2.0 \; d_{od_2k_3}(\mathbf{g}^{NP}) = 100 \cdot p^{NP}_{od_2k_3} = 17$$
$$g^{NP}_{od_2k_4} = 2.0 \; d_{od_2k_4}(\mathbf{g}^{NP}) = 100 \cdot p^{NP}_{od_2k_4} = 17$$

Project (P)

$$g^{P}_{od_1k_1} = 1.0 \; d_{od_1k_1}(\mathbf{g}^{P}) = 100 \cdot p^{P}_{od_1k_1} = 29$$
$$d_o(\mathbf{g}^{P}) = 100$$
$$g^{P}_{od_1k_2} = 1.0 \; d_{od_1k_2}(\mathbf{g}^{P}) = 100 \cdot p^{P}_{od_1k_2} = 29$$
$$g^{P}_{od_2k_3} = 1.2 \; d_{od_2k_3}(\mathbf{g}^{P}) = 100 \cdot p^{P}_{od_2k_3} = 23$$
$$g^{P}_{od_2k_4} = 1.4 \; d_{od_2k_4}(\mathbf{g}^{P}) = 100 \cdot p^{P}_{od_2k_4} = 19$$

Average demand method

$$DS_P(o) = \frac{1}{2}\sum_{dmk}\left[d_{odmk}(\mathbf{g}^{NP}) + d_{odmk}(\mathbf{g}^{P})\right] \cdot \left(g^{NP}_{odmk} - g^{P}_{odmk}\right)$$
$$= 0.5 \cdot \left[(36+29) \cdot 0.20 + (30+29) \cdot 0.4 + (17+23) \cdot 0.8 + (17+19) \cdot 0.6\right] = 45.1$$

Average cost method

$$DS_P(o) = \frac{1}{2}\left[d_{o.}(\mathbf{g}^{P}) + d_{o.}(\mathbf{g}^{NP})\right] \cdot (\bar{g}^{NP} - \bar{g}^{P})$$
$$= 0.5 \cdot (100+100) \cdot \left[(1.2 \cdot 0.36 + 1.4 \cdot 0.30 + 2.0 \cdot 0.17 + 2.0 \cdot 0.17)\right.$$
$$\left. - (1.0 \cdot 0.29 + 1.0 \cdot 0.29 + 1.2 \cdot 0.23 + 1.4 \cdot 0.19)\right] = 41.0$$

Fig. 10.10 Calculation of surplus change applying the average demand and average cost methods to the system of Fig. 10.3

with

$$\bar{g}^{Pi} = \sum_{dmk} p^i_{dmk/o}(\mathbf{g}^{Pi}) g^{Pi}_{odmk}$$

and

$$\bar{g}^{NPi} = \sum_{dmk} p^i_{dmk/o}(\mathbf{g}^{NPi}) g^{NPi}_{odmk}$$

Expressions (10.2.20) and (10.2.21) are the equivalent of expression (10.2.6) for descriptive demand models. The surplus change for all system users can be calculated by adding the results of expressions (10.2.20) or (10.2.21) for all user classes, all zones, and all trip purposes. However, because the surplus changes resulting from a project may be positive for some user classes, zones, or phases of the project and negative for others, it is helpful to keep these values separate, just as for behavioral demand models.

Figure 10.10 provides an example of the calculation of DS_p for users in a single market segment with two alternative destinations and two modes (or paths) to each destination.

(c) Comparison Between Calculation Methods

The change of perceived net utility (surplus) for transportation system users can be calculated either by following the behavioral interpretation of random utility models or by treating the model as a descriptive demand function. In the second case, the exact calculation poses some definition problems and two different simplified approaches have been proposed. The behavioral approach is certainly more consistent and elegant because it is based on an explicit theory of behavior. It also has two further advantages[7] in practice.

The first advantage stems from the possibility of taking into account surplus changes resulting from changes in attributes that are not traditionally considered as components of the generalized cost. This allows the evaluation of surplus changes produced for example by improvements in the availability of transportation services (e.g., new connections) or in travel comfort, or by the provision of travel information to users. To do this obviously requires that the corresponding variables be included as explicit or implicit attributes (e.g., alternative specific constants) in the systematic utility functions.

Such effects could not be assessed with the descriptive method because they do not correspond to a reduction in the generalized cost (which usually only incorporates attributes such as time, monetary cost, etc.). Paradoxical results could be obtained when increased demand for a mode of superior quality, but with greater

[7]In spite of the advantages of the behavioral approach to the calculation of surplus change, in applications descriptive models are often adopted, even when demand models have logit or other random utility specifications. This can be explained, at least partly, by the persistence of tradition.

generalized cost, yields a negative surplus change, that is, a seeming disbenefit for its users.

The other advantage arises from the possibility of computing changes in users' surplus corresponding to the introduction of alternatives that are not available in the nonproject situation, again avoiding obvious paradoxes.

This point can be clarified with the example in Fig. 10.11. In the *NP* situation, the system offers a single alternative (e.g., a single path) for the pair (o, d). In the *P* situation, a second path with a higher generalized cost is added. The total demand is assumed to be constant and the split between the two paths is obtained with the binomial logit model shown in the figure. With the behavioral method, the surplus change can be calculated by computing the surplus change (10.2.6) from the logsum variable (10.2.4). Because the expected maximum perceived utility function is monotone increasing with respect to the number of available alternatives (see Sect. 3.5), the users' surplus in situation *P* is greater than in situation *NP*.

Calculation of surplus change with descriptive methods, on the other hand, poses some problems in this case. First, the average demand method corresponding to expression (10.2.20) cannot be used because it is not possible to define a cost g_2^{NP} for the new path. The average cost method corresponding to expression (10.2.21) can be used because it only requires computation of the total demand d^{NP} and d^P and the weighted average of path costs for situations *P* and *NP*. However, because of the increase in the average cost, the method gives a negative surplus change, that is, a reduction in the net utility for the system's users. This outcome is clearly paradoxical inasmuch as an increase in capacity should correspond to an increase in users' surplus if some users are using the new path. The explanation is to be found in the difference between the assumptions underlying the demand model and the calculation of the surplus.

The logit model, beyond its behavioral interpretation, assigns a positive probability to alternatives with greater generalized cost, implying that the cost perceived by the users is different from the average cost. Conversely, the average cost method assumes that users perceive the average cost of the alternative chosen.

The two methods give the same outcome only for deterministic utility choice models. In this case, all demand would choose path 1 in both the *NP* and *P* situations, the average cost would be equal to g_1, and the surplus change would be equal to zero.

This result can be generalized because, as discussed in Sect. 3.4, the EMPU variable for deterministic choice coincides with the maximum utility (minimum cost) value. However, if the demand model were a deterministic utility model, it would be a behavioral model and the previous discussion of such models would apply.

The surplus change can also be calculated using a mixed approach in which the average cost descriptive method (10.2.21) is applied but replacing the average costs $\bar{g}^P$ and $\bar{g}^{NP}$ by the corresponding EMPU values $\bar{s}^P$ and $\bar{s}^{NP}$, calculated for the choice dimensions for which a behavioral model is used. With a multinomial logit model in three dimensions $d\ m\ k$ with parameter $\theta = 1$, for example, the surplus

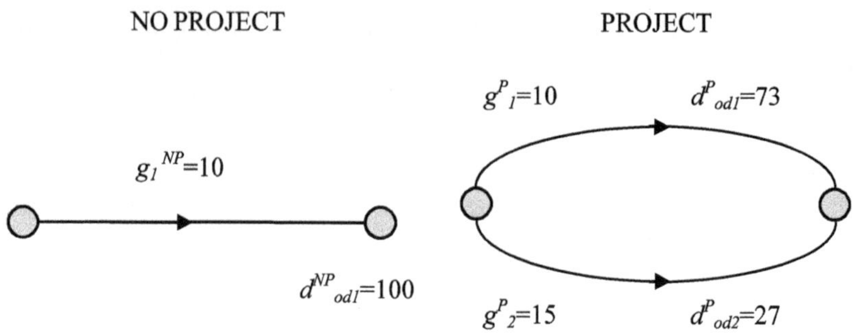

DESCRIPTIVE APPROACH

$$DS = 100 \cdot \left[10 - (0.73 \cdot 10 + 0.27 \cdot 15)\right] = -135$$

BEHAVIORAL APPROACH

$$V_k = -0.2 \cdot g_k$$

$$S^{NP} = 100 \ln\left[\exp(-0.2 \cdot 10)\right] = -200$$

$$S^{P} = 100 \ln\left[\exp(-0.2 \cdot 10) + \exp(-0.2 \cdot 15)\right] = -169$$

$$DS = +31$$

Fig. 10.11 Calculation of the surplus change applying descriptive and behavioral approaches

would be given by:

$$s_P(o) = \ln \sum_{d'm'k'} \exp[V_{d'm'k'}] = Y_{o.}$$

which is the *accessibility* from zone o to all destinations via all available modes and paths.

In this case, the curve expressing the demand level $d_o(s)$ as a function of the EMPU variable on choice dimensions d, m, and k can be interpreted as the ordering of the trips with respect to the corresponding average perceived net utility. The number of users making a trip increases with s. Figure 10.12 shows a diagram of the demand function with respect to the inclusive utility, with the shaded area corresponding to the surplus change for an increase in the EMPUs.[8] A linear approximation can again be used for the calculation of surplus change:

$$DS = 1/2(d^P + d^{NP})(s^P - s^{NP})$$

bearing in mind that EMPU and cost have opposite signs.

From the previous discussion it follows that, as far as possible, changes in surplus should be computed using expected maximum perceived utility variables, especially

[8]A surplus increase can result both from a reduction of generalized costs and from an increase of attractiveness of some zones.

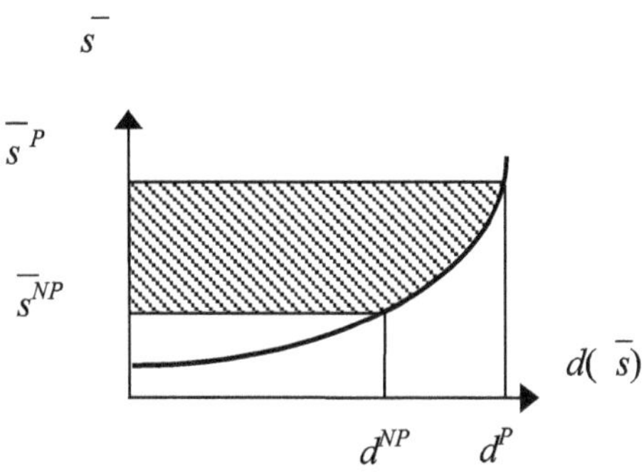

Fig. 10.12 Demand function with respect to EMPU values

when the project increases the number of available alternatives.[9] If the EMPU approach is not used, the surplus change should be computed using a mixed approach based on the EMPU variables for the choice dimensions that are most closely related to the changes in the transportation system, such as mode and path choice.

10.3 Methods for the Comparison of Alternative Projects*

A number of methods are available for comparing alternative transportation system projects. This section briefly presents the quantitative methods that are most frequently used in practice. In Sect. 10.3.1 the focus is on benefit-cost analysis as an example of an economic evaluation method; Sect. 10.3.2 describes revenue-cost analysis methods used for financial evaluation; and Sect. 10.3.3 deals with Multi-Criteria analysis methods.

Basic elements of these methods are described here to provide an overview of the various approaches used to compare projects, summarizing their assumptions and possibilities. A systematic presentation is well beyond the scope of this book, and the interested reader should refer to the extensive literature in this area for further information.

10.3.1 Benefit-Cost Analysis

Benefit-Cost (B/C) analysis expresses the impacts of each alternative project in monetary units. For each alternative, a single aggregate measure of economic worth is formed by algebraically adding the different impacts, considering benefits as positive and costs (disbenefits) as negative, and taking account of the time when these

[9]The descriptive approach was introduced to deal with the case in which the project results in the reduction of generalized transportation costs, particularly for road systems. Furthermore, the implicit demand models were often deterministic and this, as it has been shown, implies that there will be no "pathological" results. These conditions, however, are not necessarily met by all the applications of the method.

occur. (The definition and quantification of benefits and costs depend, of course, on the stakeholders for whom the analysis is performed.) The alternative with the highest worth is preferred.

Applications of the B/C method from the viewpoint of a single public decision-maker (typically a government agency) may consider for each year t of the economic life of project P_i all or some of the following effects.

- *CC* Difference between the construction costs of the project and the costs of construction and other major works (reconstruction, rehabilitation), if any, required for the nonproject alternative. (Recall that already-committed investments should be included in the *NP* situation.) In some applications, a negative construction cost *CC* (i.e., a benefit) is assumed for the final year: this corresponds to the residual value of the project at the end of the analysis period. In this way it is possible to mitigate the unavoidable arbitrariness in the definition of the technical life of the project
- *CVT* Difference between investment costs in vehicles and technologies for the project and nonproject situations. Here again, the *np* situation might require such investments
- *CMO* Difference between maintenance and operating costs for the project and nonproject situations
- *REV* Difference between direct (sale of transportation services) and indirect (commercial activities) revenues in the project and nonproject situations
- *TR* Difference between government revenues from taxes and duties in the project and non-project situations
- *DS* Change in transportation system user perceived surplus in the project and nonproject situations, expressed in monetary units. This is typically obtained by adding up the changes in perceived surplus for different user classes
- *UNPB* Change in benefits not perceived by the users between the project and nonproject situations. These benefits might include costs changes due to accidents or vehicle operations (lubricants, tires, etc.). And other non out-of-pocket costs not perceived by the users in their travel-related choices. All these benefits are expressed in monetary units; the variable has a positive sign if there is a reduction in these costs
- *NUI* Change in the nonuser impacts between the project and nonproject situations. Impacts on the environment (e.g., reduction of pollutant emissions) and on the economy and land use system can be included in this variable after conversion to monetary units. These impacts are sometimes referred to as indirect benefits and are positive if the benefits increase

The above variables are usually evaluated using market prices, when available. Transfer payments (VAT, sales, income and fuel taxes, etc.) are sometimes excluded from the market price because these do not correspond to actual consumption of economic resources; for example, construction, maintenance, and operation costs may be computed by evaluating the corresponding resource costs at market prices minus VAT and other taxes. In some applications, market prices are replaced by shadow prices or opportunity costs, which are considered more accurately to reflect the marginal value of a particular resource to the community. Shadow prices may be assigned when there is no market price for a resource, when the price is distorted by market imperfections, or when it is felt desirable to take into account social objectives or constraints that are not adequately reflected by market mechanisms. The opportunity cost of labor, for example, might be lower than the market price of manpower when there is a high level of unemployment, and unemployment reduction is one of the objectives of the project. In this case, the opportunity cost could be obtained as the difference between the net market price and the unemployment subsidy for each category of workers.

The apparent monetary value of costs or benefits that occur in the future may be influenced by the effects of inflation. It is common in economic B/C analysis to ignore these effects by measuring all monetary values in constant (real) monetary units.

It is important to stress that the variables considered and the way they are computed both depend on the viewpoint from which B/C analysis is performed.

Whatever point of view is adopted, the evaluation must avoid double counting of an individual project effect by quantifying its impacts with different variables having the same sign. An example of such double counting is given by increased accessibility (reduction of the generalized transportation cost) of one zone compared with others. This effect often leads to an increase in real estate values in the zone as a consequence of the willingness of residents and/or firms to pay for the greater accessibility. If the change in user surplus and the change in real estate values were both counted as benefits, the accessibility effect of the project would be accounted for twice and, in this particular case, the overall benefits would be overestimated. In this example the change in real estate values should not be considered, or it should be accounted for with opposite signs for those who benefit from them (i.e., the property owners) and those who incur costs (i.e., renters or buyers). It is clear that these impacts will be very different for the different stakeholders, and their distribution among different groups in society should be taken into account.

Some effects may be present with different signs in two or more variables: for example, user fares may be counted as revenue with a positive sign (benefit) and in the perceived surplus change with a negative sign (cost). The same effect occurs with gasoline taxes, and other variables. Effects of this kind could even be excluded from the B/C analysis, as proposed by some analysts for traffic revenues. Their exclusion, however, is acceptable only in the special case in which the effects count linearly in all terms. In the previous example, this would be the case if the monetary cost appears linearly in generalized user costs and if surplus change is computed through descriptive methods (10.2.20) and (10.2.21) with fixed demand. However, if this variable appears nonlinearly in different terms, for example, if the monetary cost is used in the EMPU variable (10.2.4) and (10.2.6) for the evaluation of user surplus changes, and/or if there are changes in the level of demand, the variable must necessarily be accounted for twice. Figure 10.13 shows graphically the difference between user surplus and revenue changes in the case where generalized cost coincides with monetary cost; the two changes would coincide only in the special case of fixed demand (areas B and C equal zero).

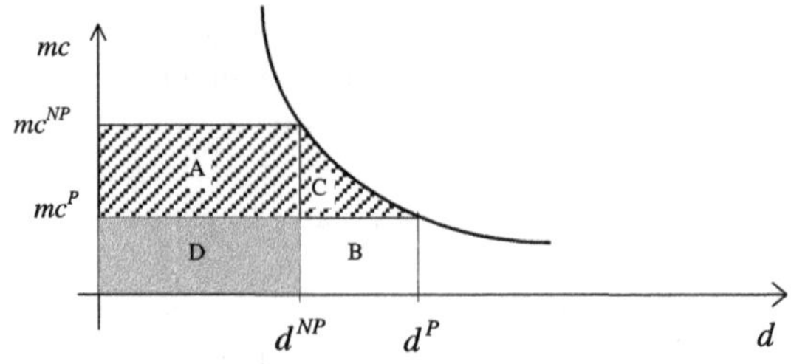

$$\Delta_{\text{rev.}} = d_P \cdot mc^P - d_{\text{NP}} \cdot mc^{NP} = (A + D) - (B + D) = A - B$$

$$\Delta_{\text{surplus}} = A + C$$

$$|\Delta_{\text{rev.}}| \neq |\Delta_{\text{surplus}}|$$

Fig. 10.13 Difference between surplus and revenue changes

Once the relevant effects have been defined and measured in monetary units, different alternative projects are then compared using aggregate indicators of project worth. Benefits and costs that occur in different years are compared by applying the interest or discount rate r. This is defined as the relative increase in the value of a monetary amount M after one year:

$$r = \frac{M^{t+1} - M^t}{M^t}$$

$$M^{t+1} = (1 + r)M^t$$

The discount rate is generally assumed to remain constant over time, so the value M^t, after t years of an amount M_o available today, can therefore be calculated as

$$M^t = M_o(1 + r)^t$$

from which it follows that the value today (the present value) M_o of an amount M^t occurring t years in the future is:

$$M_o = \frac{M^t}{(1 + r)^t} \tag{10.3.1}$$

With this formula, the various annual benefits and costs occurring over a project's life can be converted into equivalent amounts in year 0 as needed. This allows projects with different benefit and cost time streams to be directly compared.

Several synthetic indicators have been proposed for comparing the time streams of benefits and costs of different projects P_i. The *Net Present Value* (*NPV*) is the equivalent value in year 0 of the time stream of annual project costs and benefits:

$$NPV_i(r) = \sum_{t=1}^{T} \frac{(DS_i^t + UNPB_i^t + NUI_i^t + TR_i^t + REV_i^t - CC_i^t - CVT_i^t - CMO_i^t)}{(1 + r)^t} \tag{10.3.2}$$

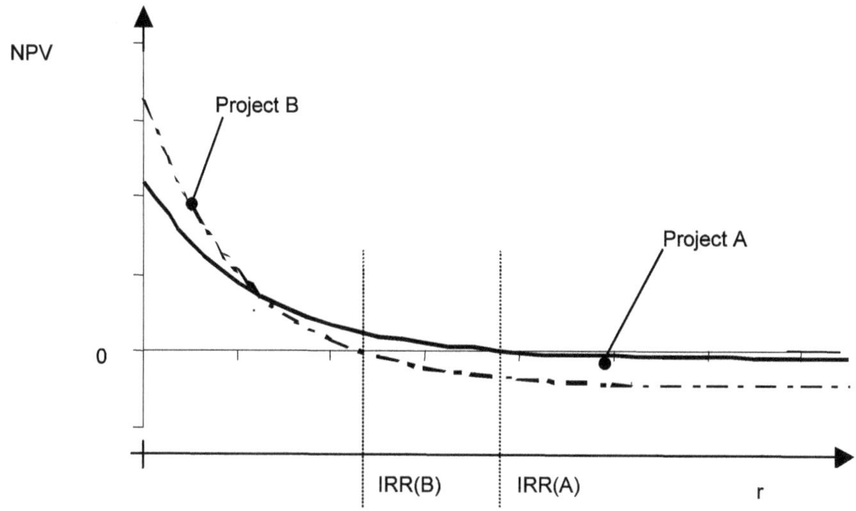

Fig. 10.14 Net present value as a function of the discount rate

where T is the number of years included in the time stream and r is the applicable discount rate per year. The *Internal Return Rate* (*IRR*) is defined as the value of the discount rate r_o such that the *NPV* calculated over a period of T years is equal to zero:

$$IRR_i = r_o; \qquad NPV_i(r_o) = 0 \tag{10.3.3}$$

Using the net present value criterion, a project P_i is preferable to the nonproject alternative *NP* if its *NPV* is positive, and project P_i is preferable to project P_j if $NPV_i > NPV_j$. The relative ranking of project P_i compared to P_j may depend significantly on the discount rate r used for the *NPV* calculation, as shown in Fig. 10.14. Projects with lower investment costs and fewer benefits are usually favored by higher values of r (project P_2 in Fig. 10.14), whereas lower discount rates tend to favor more costly projects with greater future benefits (project P_1 in Fig. 10.14). This is because higher discount rates tend to reduce the present value of project benefits, which usually occur some years after the investment is made; conversely, the present value of project investment costs, which are typically incurred in the early years, is less sensitive to the discount rate.

In terms of the internal rate of return criterion, a project P_i is preferable to the nonproject alternative *NP* if its *IRR* is above the social discount rate, and is preferable to project P_j if $IRR_i > IRR_j$.

The discount rate used for project evaluation can be selected in several different ways. One possibility is to use an appropriate prevailing market or government interest rate. If costs and benefits are measured in real (constant monetary units) terms, the inflation rate is subtracted from prevailing interest rates. Other more complex methods determine the social opportunity cost of capital from the returns potentially achievable from alternative uses, from the marginal social utility of consumption, or based on the risk connected to the project. This subject has been discussed at length in the economic literature, to which the interested reader is referred. Here it is worth mentioning that the choice of discount rate implies important value judgments regarding present versus future impacts; these should be explicitly stated and will generally depend on the point of view of the analysis.

Proper application of benefit/cost analysis for transportation project evaluation can require considerable care. The most frequent criticism raised in practical applications of B/C analysis is that the evaluation of a project's user and nonuser impacts is incomplete or inexact. In some applications, impacts are computed only for the users of the planned facilities, ignoring the effects on other parts of the transportation system and on other stakeholders. Although in some cases this approximation may be acceptable, in others it may significantly distort the results of the analysis. Because of the interdependencies inherent in a transportation system, it frequently happens that a project's impacts on those who do not use it are comparable in magnitude to its impacts on its users. This and other similar criticisms can be overcome by analyzing the transportation system and stakeholder impacts with the methods described in previous sections.

More fundamentally, however, B/C analysis as an evaluation method has been subjected to a number of criticisms of its theoretical foundations.

(i) The use of market prices as indicators of the social value of resources is theoretically correct only under key assumptions of socially optimal income distribution and perfectly competitive markets. In reality, both assumptions are almost always far from the truth. Use of market prices implies value judgments about the prevailing income distribution and is inconsistent with the theory of welfare economics that underlies B/C analysis. As an alternative, shadow prices can be used if the project is expected to contribute to the pursuit of "social" objectives. For example, objectives such as reduction of unemployment, air pollution, or energy consumption can be reflected in shadow prices of labor and fuel. However, the rigorous calculation of shadow prices is extremely complex, and in practice rough estimates are often used.

(ii) The aggregation of project impacts across different groups implies value judgments regarding the optimality of the prevailing income distribution, and indifference to the income redistribution that may be caused by the project. For example, the method assumes that a generalized cost increase for some users can be exactly offset by a cost reduction of the same magnitude for other users. Furthermore, the quantification of perceived user impacts may depend on the user's income; for example, if the value of time is related to income, travel time savings of the same amount would produce larger perceived surplus changes for higher-income groups.

(iii) The implicit decision criterion (Kaldor criterion) underlying B/C analysis is that a project is worthwhile if the individuals receiving benefits from the project could adequately compensate those incurring disbenefits and still be better off than without the project. In reality, such compensations rarely take place. Nonetheless, in B/C analysis one project is considered preferable to another if the equivalent monetary value remaining after such compensations is greater, regardless of the actual occurrence of these transactions.

(iv) Benefit/cost analysis only considers impacts that are, or can be, expressed in monetary units, thus ignoring those that cannot reasonably be measured in this way. This implicitly favors the objective of economic efficiency over other social and environmental objectives.

On the basis of these criticisms, many economists argue that, from the point of view of a public decision-maker, B/C analysis should have a role that is essentially normative and/or conventional. As a normative tool, the main elements of the analysis, that is, types of impacts to be considered, prices, discount rates, and the like, should be fixed by the agencies that fund public projects. This allows a variety of alternative project proposals to be evaluated and compared on a common and equitable basis. As a conventional tool, the parameters used in B/C analysis should be derived and consolidated from practical applications of the method in specific sectors.

Put differently, B/C analysis should not be considered as a comprehensive evaluation method, but rather as a tool for evaluating economic impacts for some stakeholders of the decision process, taking into account only the monetary or monetarily quantifiable costs and benefits of projects. In this interpretation, B/C analysis can be considered as a way to evaluate projects' economic efficiency impacts for the involved stakeholders. Thus several B/C indicators could be computed, representing different stakeholders such as users, service operators, and public agencies. These indicators could then be used together with others in the context of a wider multicriteria analysis discussed below.

10.3.2 Revenue-Cost Analysis

The financial evaluation of a project involves an assessment of the investments and subsequent revenues of all operators, bodies, and enterprises (public and/or private) involved in planning and/or implementing the project. It can be seen as a special case of cost-benefit analysis, where the only costs and benefits considered are those that give rise to financial inputs and outputs. It is conducted using market prices (actual and/or expected). Thus, the result of an evaluation is expressed in monetary terms; it measures the overall financial impacts of a proposed project on its stakeholders. The goal of revenue-cost analysis is to identify the financially preferred alternative: generally the one that maximizes net revenues.

As in benefit/cost analysis, the reference situation (which is not necessarily the status quo, but may include already committed projects and/or other baseline actions) is considered a possible alternative. The criterion commonly used in the evaluation of the financial worth of a project is the present value of its net revenues, defined in incremental terms in comparison with the nonproject situation. The project's Net Present Value (*NPV*) is calculated as

$$NPV_i(r) = \sum_{t=1}^{T} \frac{(REV_i^t - CC_i^t - CVT_i^t - CMO_i^t)}{(1+r)^t} \tag{10.3.4}$$

where the variables *CC*, *CVT*, and *CMO* were defined in the previous section. Here *REV* is the difference between direct revenues in the project and nonproject situations. In addition to the sales of transportation services, public contributions or other transfers should be considered as revenues as well. It is common in revenue-cost analysis to express values in current monetary terms, that is, incorporating the effects of inflation on the magnitude of future revenues and costs. In this case the discount rate (e.g., the market interest rate) need not be reduced by the inflation rate.

In order to accept a project, its NPV must be positive; when there are multiple alternatives, the one with the highest NPV is preferable.

Two other criteria are also often used:

– The ratio of discounted revenues to discounted costs
– The *internal rate of return (IRR)*, defined as the discount rate for which the project's NPV is 0

Although the approaches and methods of financial analysis derive from investment analyses carried out in the private sector, financial analysis can also be of use in the analysis of public sector investments. There are, however, significant differences between the two approaches. For example:

– The private operator is interested in maximizing profit, that is, the difference between revenues and costs; the public operator may only aim to cover part of its costs, because it pursues goals related to equity criteria among different social groups and minimization of net costs
– The private operator generally considers a short-term analysis period (e.g., 5 years); the public operator often considers a long-term period (20 to 40 years)
– The private operator usually considers only the direct users (the producers and providers of production factors); the public operator generally considers all users (direct and indirect), that is, the citizens
– The private operator compares the profits deriving from an investment with more favorable opportunities; the public operator considers social welfare in addition to economic criteria when deciding if a project should be implemented

10.3.3 Multi-criteria Analysis

Transportation system projects may produce a variety of impact types, and decision-makers generally have multiple goals. Each impact affects one or multiple stakeholders and, as such, can be transformed into an objective. Thus, increasing users' surplus, reducing costs, increasing revenues, increasing social equity and accessibility, increasing the efficiency of the transportation system, reducing environmental impacts, and inducing economic growth in a given area are all objectives that an individual decision-maker or stakeholder may want simultaneously to pursue. However, these objectives often conflict with each other; the maximization of users' surplus might, for example, conflict with the reduction of noise and air pollution and with the minimization of capital investment costs.

MultiCriteria (MCA) (or MultiObjective (MOA)) analysis establishes preferences between options by reference to an explicit set of objectives that have been identified, and for which performance indicators (that measure the degree to which an objective is attained) have been defined. Good decisions need clear objectives. These should be specific, measurable, agreed, realistic, and time-dependent.

It is sometimes useful to classify objectives according to their timeframe, for example, ultimate, intermediate, and immediate objectives; it is particularly useful to distinguish between ultimate and immediate ones. Ultimate objectives are usually framed in terms of strategic or higher-level variables, such as the target level of economic growth, or sustainable development. Immediate objectives are those that can be directly linked with the outputs of a policy, program, or project.

To compare the contribution of different options towards given objectives, it is necessary to have criteria that reflect the options' performance in meeting those objectives. In simple situations, the process of identifying and assessing objectives and criteria may alone provide enough information for decision-makers. However, where a level of detail broadly akin to cost-benefit analysis is required, multicriteria analysis offers a number of ways of aggregating the data on individual criteria to provide indicators of the overall performance of each available option.

A key feature of multicriteria analysis is its emphasis on the judgment of the decision-making team in establishing objectives and criteria, in estimating relative importance weights, and, to some extent, in judging the contribution of each option towards each evaluation criterion. Its foundation, in principle, is the decision-maker's own choice of objectives, criteria, weights, and her assessments of the options' performance towards achieving the objectives, although "engineering" data such as times and costs can of course also be incorporated in this process. Multicriteria analysis can bring a degree of structure, analysis, and openness to classes of decisions that lie beyond the practical reach of cost-benefit analysis.

In general, a multicriteria method consists of the following main steps (see Fig. 10.15)[10]: (i) establish the decision context; (ii) identify the options (alternative projects or more generally alternative courses of action) to be appraised; (iii) identify objectives and criteria; (iv) assign scores; (v) assign weights; (vi) combine the weights and the scores for each option to derive a limited set of aggregate values; (vii) examine the results; and (viii) conduct a sensitivity analysis. Each of these steps is discussed in the paragraphs below.

(i) A first step is always to establish a shared understanding of the decision context. This consists of the entire complex of administrative, political, and social structures that surround the decision being made. Central to this context are the objectives of the decision-making body, the administrative and historical environment, the set of people who may be affected by the decision, and an identification of those responsible for the decision.

It is crucial to have a clear understanding of objectives. Towards what overall ambition is this decision seeking to contribute? Multicriteria analysis is about multiple conflicting objectives: there are ultimately trade-offs to be made. Nonetheless, in applying multicriteria analysis it is important to identify a number of high-level objectives, for which there will usually be subobjectives. To establish objectives it

[10] Figure 10.15 has been taken from the DTLR multicriteria manual.

Steps in a Multi-Criteria Analysis

1. Establish the decision context. What are the aims of the MCA, and who are the decision makers and other key players?

2. Identify the options.

3. Identify the objectives and criteria that reflect the value associated with the consequences of each option.

4. Describe the expected performance of each option against the criteria; 'score' the options, i.e. assess the value associated with the consequences of each option.

5. 'Weighting'. Assign weights for each of the criteria to reflect their relative importance to the decision.

6. Combine the weights and scores for each of the options to derive aggregate evaluation value.

7. Examine the results.

8. Conduct a sensitivity analysis of the results to changes in scores or weights.

Fig. 10.15 Applying multicriteria analysis: detailed steps

is essential to establish who the decision-makers are and who may be affected by the decision.

The aims of transportation systems decision-making typically involve general public goals such as providing accessibility for different parts of a geographic area, increasing equity, or reducing pressure on the environment. They may also include specific objectives more related to the particular interests of decision-makers, for example, the overall objective of increasing political support for the local government may include subobjectives such as improving the relationships with its constituents, with relevant interest groups, with the media and so on.

(ii) It is unlikely that the decision-making group will arrive at the stage of formally structuring the multicriteria analysis without having some intuition about the decision options. As discussed in Sect. 1.5, options may include infrastructure, control systems, services, and/or fares. Sometimes the initial problem statement will include an unmanageably large number of possibilities, and it will be an early function of multicriteria analysis to provide a structured screening of alternatives to identify a short list, using basic data and quick response procedures.

(iii) Objectives are the general goals that are pursued, and towards which the proposed options are expected to contribute. Objectives can be very broad, for example, improve the environment, and may be further disaggregated into a variety of subobjectives, for example, reduce air pollution and noise pollution. Objectives should be defined taking into account the perspectives of different interest groups.

Criteria and subcriteria are the performance measures by which the options will be judged. Whether in a decision-making team or as an individual, an effective way to start the process of identifying criteria is first to recapitulate briefly the general aims and more specific objectives and then to brainstorm responses to the question, "What would distinguish between a good choice and a bad one in this decision problem?"

The viewpoint of stakeholder groups is generally important. One way to include them is directly to involve the affected parties in some or all steps of the multicriteria analysis. This might be appropriate, for example, in local planning situations. A second approach is to examine policy statements that reflect their concerns. A third possibility is to encourage one or more of the members of the decision-making team to role-play the position of key stakeholder groups, to ensure that this perspective is not overlooked when criteria are being derived. The number of criteria should be kept as low as is consistent with making a well-founded decision. There is no rule to guide this judgment and it will certainly vary from application to application.

The different objectives of the decision-makers are transformed into evaluation criteria or performance indicators; examples are given below. The performance indicator corresponding to the objective of increasing users' utility might be the difference between the total users' surplus in the project and nonproject situations. Values of *NPV* and *IRR* may correspond to the objective of increasing economic efficiency; the indicator corresponding to the objective of reducing air pollution might be the change in total pollutant emissions; and so on. Some objectives may involve criteria expressed qualitatively (e.g., with terms such as little, much, etc.). This could be appropriate for objectives such as preserving the historical identity of an area or minimizing the visual impact of new infrastructure. In these cases qualitative criteria can be transformed into quantitative variables by indirect quantitative determination techniques.

(iv) In many multicriteria techniques, the various performance indicators are first transformed to allow their comparison. Suppose that M evaluation criteria corresponding to the objectives of the project have been identified, and that the value of the mth performance indicator for the jth project is represented by the variable x_{mj}. Variables x_{mj} are usually all expressed on a scale having values that increase with the level of satisfaction, so that better levels of performance lead to higher value scores. This may mean a reversal of the natural units. When x_{mj} measures a disbenefit, for example, the quantity of emitted pollutants or distance to the nearest transit service, the reduction with respect to the maximum value taken by the indicator can be substituted for it:

$$x'_{mj} = \left(\max_k x_{mk}\right) - x_{mj}$$

Let s_{mj} be the score of option j relative to criterion m. It is conventional to assign a value score to each criterion using an interval scale between 0 and 100. The advantage of an interval scale is that differences in scores are consistent within each criterion; note, however, that such scales do not permit conclusions that, for example, a score of 80 represents a performance that on some absolute standard is five times as good as a score of 16 (statements of this type would require the use of a ratio scale).

The first step in establishing an interval scale for a criterion is to define the levels of performance corresponding to any two reference points on the scale; minimum and maximum scores of 0 and 100 are used in the examples below. One possibility (*global scaling*) is to assign a score of 0 to represent the worst level of performance that is likely to be encountered in a decision problem of the type being addressed, and 100 to represent the best level. Another option (*local scaling*) associates 0 with the performance level of the poorest performing project in the currently considered set, and 100 with the best performing one. Formally:

$$s_{mj}^{\text{GLOB}} = \frac{x_{mj} - x_m^{\text{WORSE}}}{x_m^{\text{BEST}} - x_m^{\text{WORSE}}} \times 100 \qquad \text{(a)}$$

$$s_{mj}^{\text{LOC}} = \frac{x_{mj} - \min_k x_{mk}}{\max_k x_{mk} - \min_k x_{mk}} \times 100 \qquad \text{(b)} \qquad (10.3.5)$$

The choice between local and global scaling should make no difference to the ranking of options. An advantage of global scaling is that it more easily accommodates new projects if these exhibit performance values that lie outside those of the original set.

Once the endpoints are established for each criterion, there is a way in which scores may be established for the individual projects under consideration. Specifically, it uses a value function to translate a project's measure of performance with respect to a criterion into a value score on the chosen interval scale. The value functions used in many multicriteria analysis applications can for practical purposes be

Fig. 10.16 Regional jobs

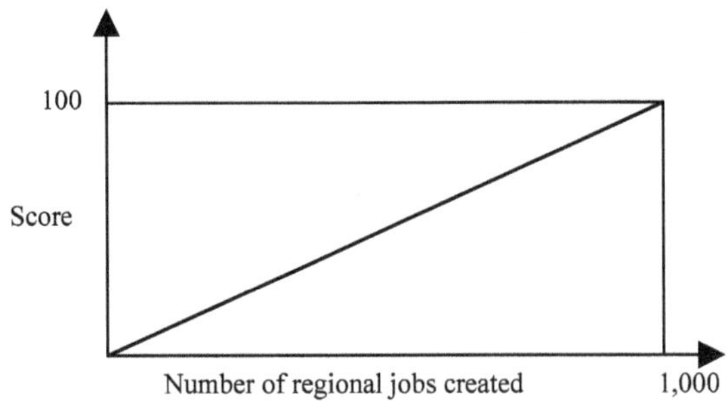

Fig. 10.17 Distance to public transport

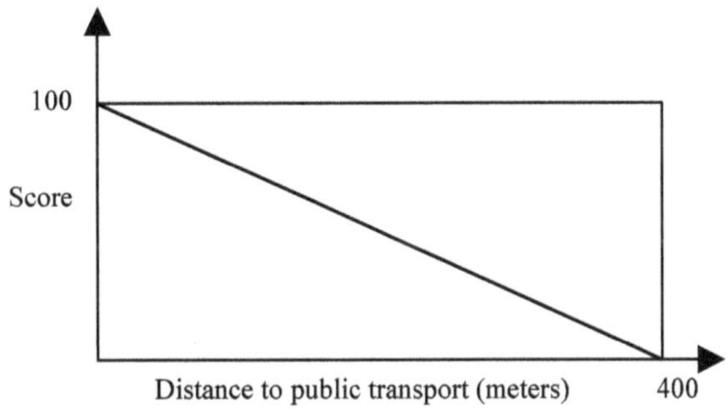

assumed to be linear.[11] For example, if one criterion corresponds to number of regional full-time jobs created, with a minimum likely level of 200 and a maximum of 1000, then a simple graph allows conversion from the natural scale of measurement to a 0–100 range. This is shown in Fig. 10.16.

Any project's score with respect to the regional job creation criterion is assessed by simply locating on the horizontal axis the number of jobs created, and reading on the vertical axis the corresponding score. Thus a project that creates 600 jobs, say, would receive a score of 50.

Where higher measurements on the scale of natural units correspond to worse rather than better performance, the slope of the function that maps performance onto the interval scale score is simply reversed, as in Fig. 10.17.

(v) Weighting is another fundamental stage in multicriteria analysis. A weight $w_m \geq 0$ is attributed to each criterion m; it measures the importance of the corresponding objective compared to the others. Obviously, in defining weights decision-makers must express value judgments. In principle, different sets of weights can be

[11] However, on same occasions it may be desirable to use a nonlinear function. For example, it is well known that human reaction to changes in noise levels measured on a decibel scale is nonlinear. Alternatively, there are sometimes thresholds of achievement above which further increments are not greatly appreciated. For example, in valuing office area, initial increments above the absolute minimum space lead to high estimates of the marginal value of increased room size, but after an acceptable amount of space is available, further marginal increments are valued much less highly.

associated to the same set of objectives in order to express the point of view of different stakeholders in the decision process.

Many methods have been proposed to estimate the unknown weights of an individual decision-maker as well as to reach a compromise weighting among multiple decision-makers. In the most direct approach for the latter, known as the DELPHI method, each decision-maker is independently asked to express the weight that he or she would apply to each objective or criterion. The interviews are subsequently repeated, but in this second round each interviewee is told the weights previously stated by the other decision-makers before being asked again to assign weights to the objectives or criteria. The process is repeated until a compromise agreement is reached.

When weights cannot be obtained directly from decision-makers, other procedures can be used. For example, it is possible to infer the implicit weights that would justify a posteriori the choices previously made in similar contexts for projects of the same type and size. In a different approach, the decision-maker is asked to express preferences between pairs of alternative hypothetical projects; the implicit set of weights can be estimated so as to reproduce as closely as possible the stated choices. Yet other methods estimate the weights by asking the decision-maker to perform explicit trade-offs, for example, to choose between solutions that improve the attainment of some objectives while worsening others.[12]

The *evaluation matrix* or *impact tableau* presents the evaluation indicators (x_{mj}, x'_{mj}, or s_{mj}). The number of rows of this matrix is equal to the number of evaluation criteria, and the number of columns is equal to the number of alternative projects under consideration. Figure 10.18 presents an example of such a matrix.

A project j is *dominated* if there exists at least one project h satisfying all the objectives better than, or at least as well as, project j:

$$x_{mj} \leq x_{mh} \quad \forall m = 1, \ldots, M \tag{10.3.6}$$

with at least one of the inequalities (10.3.6) holding strictly.

A nondominated project is also called *efficient*. The set of nondominated projects that satisfy any additional imposed constraints (e.g., budget constraints) is called the *project efficiency frontier* or *boundary*. It can be shown that each point on this boundary is an optimal solution to the decision problem for some specific set of objectives/criteria weights.

The *Dominance* method can be applied to eliminate all the inefficient alternatives. The procedure is as follows.

- The first two alternatives are compared; if one of the two is dominated by the other, then it is eliminated
- The noneliminated alternatives are compared with a third alternative; any dominated alternatives are eliminated

[12]Notice that this approach is equivalent to the calibration of implicit decision-maker utility functions based on revealed and/or stated preferences. It is conceptually analogous to the calibration of demand models with utility functions for transportation-related choices and can be addressed using the parameter estimation techniques described in Chap. 8.

Objectives	Evaluation criteria	Project						Weights
		1	2	...	J	...	J	
1	1	x_{11}	x_{12}	...	x_{1j}	...	x_{1J}	w_1
2	2	x_{21}	x_{22}	...	x_{2j}	...	x_{2J}	w_2
...	...	...	...	...	...	...	...	...
i	m	x_{m1}	x_{m1}	...	x_{mj}	...	x_{mJ}	w_m
...	...	...	...	...	...	...	...	...
...	...	...	...	...	...	...	...	...
N	M	x_{M1}	x_{M1}	...	x_{Mj}	...	x_{MJ}	w_M

Fig. 10.18 Evaluation matrix of J alternative projects with respect to M criteria

Evaluation criteria	Alternatives		
	A	B	C
Reduction of km of congested network	89	72	68
Reduction of total travel time on network	1606	1500	1100
Veh-km	130	140	98

Fig. 10.19 Evaluation matrix

- At each step a new alternative is introduced and compared with the others, after which the dominated alternatives are eliminated
- At step $J - 1$ the comparison among the J alternatives is finished and the set of all nondominated alternatives has been identified

Note that, after application of this method, multiple nondominated alternatives may still remain.

Following is a simple example of this method. The objective is to identify the best among three road alignments, taking into account criteria such as the impact that the road construction has on the entire network, measured in terms of:

- Reduction of the extent of congested network (expressed in km)
- Reduction of total travel time on the network (expressed in equivalent veh-hr)
- Traffic usage of each road (expressed in thousands of veh-km/day)

From Fig. 10.19, it follows that alternative C is dominated by alternatives A and B because the impacts of C are all lower. However, neither A nor B dominates the other. Therefore the nondominated alternatives are A and B.

(vi) Multicriteria analysis techniques proposed in the literature generate a set of nondominated solutions (projects) and assist the decision-maker in selecting a reasonable compromise between contrasting objectives.

In some application areas, there exist techniques that generate a continuous set of nondominated projects, defined by continuous decision variables with explicit relationships (preferably linear) between these variables and their impacts.[13] Trans-

[13]The continuous variable supply design problem discussed in Chap. 9 is an example of this. The main difference with the problems described in Chap. 9, though, is that there are multiple objective functions (indicators) rather than a single objective function.

portation system projects rarely meet these requirements, however, because of the discrete ("lumpy") nature of many projects (new infrastructure, e.g.), the intrinsic nonlinearity of the system (cost functions and demand models) and the complexity of the relationships between control variables and effects (e.g., changes in equilibrium flows and costs resulting from a transportation network project). Alternative projects are assumed to be nondominated because dominated ones, by the assumption of monotonicity of preferences, could never be optimal choices under any set of weights.

The roles of the analyst and the decision-maker vary greatly among the different techniques proposed. According to some authors, the analyst's task should end after informing the decision-maker about the list of nondominated projects and their characteristics and impacts, presented in a way that facilitates understanding by nonspecialists. Other methods assume that interaction between the analyst and the decision-maker continues throughout the decision-making process.

The following paragraphs describe multicriteria methods that are representative of the different approaches. In general, MCA methods can be divided into two classes: *compensatory* and *noncompensatory*.

Noncompensatory methods (see Sect. 10.3.3.1) do not permit trade-offs between attributes. An unfavorable value in one attribute (performance indicator) cannot be offset by a favorable value in other attributes. Each attribute must stand on its own. Hence comparisons are made on an attribute-by-attribute basis. The multicriteria methods in this category are noted for their simplicity, which is why they are also called *elementary* methods.

Compensatory methods permit trade-offs between attributes. A decline in one attribute may be acceptable if it is offset by an enhancement in one or more other attributes.

When compensation is acceptable, most multicriteria methods involve implicit or explicit aggregation of each option's performance across all the criteria to form an overall assessment of the option, on the basis of which the alternative options are compared. The principal difference between the main families of multicriteria methods is the way in which this aggregation is done.

The multicriteria analysis methods described in this section concentrate on prescriptive[14] approaches to decision making, that is, approaches that give support to the decision-maker. There is no universally accepted normative model of how individuals should make multicriteria choices.

The compensatory methods covered below are briefly summarized here. Multiattribute utility theory methods (Sect. 10.3.3.2), which are the ones that come closest to universal acceptance, are based on three building blocks: the evaluation matrix, methods to define criteria independence, and specification and calibration of a utility function U. The latter expresses the decision-maker's overall assessment of an

[14] Note that the term *prescriptive* may be interpreted in either a strong or a weak sense. The discussion here is not meant to imply the strong sense of giving an order or instruction to the decision-maker. Rather, the procedures reported here should be interpreted in the weaker sense as providing support or advice for the decision-maker.

option in terms of the value of its performance with respect to each of the separate criteria.

If it can be either proved or reasonably assumed that the criteria are independent of each other, and if uncertainty is not formally accounted for by the multicriteria analysis method, then the simpler linear additive evaluation method (Sect. 10.3.3.3) is applicable. The linear model shows how an option's values with respect to multiple criteria can be combined into one overall value.

The Analytic Hierarchy Process (AHP), discussed in Sect. 10.3.3.4, also develops a linear additive method but, in its usual form, derives the criterion's weights and the alternative scores based, respectively, on pairwise comparisons between criteria and between alternatives. Thus, for example, in assessing weights, the decision-maker is asked a series of questions, each of which ascertains how important one particular criterion is relative to another for the decision being addressed.

A rather different approach from any of those mentioned so far depends upon the concept of *outranking* (Sect. 10.3.3.5). Outranking methods seek to eliminate alternatives that are, in a particular sense, dominated. However, weights are used to give more influence to some criteria than to others when identifying the dominated options.

One option is said to outrank another if it outperforms the other on a sufficient number of important criteria and, conversely, if it is not outperformed by the other in the sense of having a significantly inferior performance on any individual criterion. Each option is then assessed in terms of the extent to which it sufficiently outranks the full set of options being considered, as measured against a pair of threshold parameters.

An interesting feature of outranking methods is that it is possible, under certain conditions, to determine that two options cannot be compared, in which case they are classified as incomparable. Incomparability of two options is not the same as indifference between them. It might, for example, be associated with missing information at the time the assessment is made; this is not an unlikely occurrence in many decision-making exercises. Building this possibility into the mathematical structure of outranking allows formal analysis of a problem to continue without having either to declare (without justification) indifference between the options or to drop the options entirely, simply because full information is not currently at hand.

Finally Sect. 10.3.3.6 introduces the constrained optimization method. This method is based on the observation that any multiobjective and multidimensional decision-making problem can be changed into a single-objective constrained optimization problem and solved as a linear programming problem in binary variables.

(vii) Whatever comparison technique is used in multicriteria analysis, *sensitivity analysis* is of considerable importance. Sensitivity analysis explores how dependent the analysis outcome is on the parameter assumptions. In other words, it attempts to establish whether the solution obtained (the ordering of the alternatives) is robust with respect to changes in the parameters, which are intrinsically arbitrary. Sensitivity analyses can be carried out by different methods having different levels of sophistication. The description of these methods is beyond the scope of this book and can be found in the specialized literature.

10.3.3.1 Noncompensatory Methods*

This section outlines some of the procedures applied to establish preferences between options when:

- Each option is evaluated against a common set of criteria that are set out in the evaluation matrix
- The decision-maker is not willing to allow compensation, that is, strong performance on one criterion to compensate for weak performance on some other criterion

Use of *noncompensatory* evaluation methods severely restricts the extent to which, in practice, overall preferences between options can be established.

Noncompensatory methods generally involve the introduction of performance thresholds for one of more criteria. This reflects a judgment that the selected criteria should be prioritized in this way compared to others for which no thresholds are given. Thresholds can be of three different types:

- "Acceptable minimum": the worst value, in terms of achievement of the objective, that the decision-maker considers acceptable
- "Desirable minimum": the worst value, in terms of achievement of the objective, that the decision-maker considers satisfactory
- "Target": the optimum value of the objective

Once the thresholds have been defined, it is possible to use a series of noncompensatory techniques to evaluate the alternatives. Common techniques include the following methods.

Disjunctive
Conjunctive
Maxmin
Maxmax
Lexicographic
Elimination by aspects

The *Disjunctive* constraint method evaluates an alternative on the basis of its best attribute, regardless of all other attributes. The procedure is:

- The decision-maker establishes a desirable threshold value for each attribute
- For a given alternative, each attribute is compared with its threshold
- If at least one attribute value is better than or equal to the threshold, the alternative is selected, otherwise it is eliminated

Let x^*_m be the desirable threshold value of the mth attribute; then alternative j is selected if:

$$\exists m: \quad x_{mj} \geq x^*_m \quad \forall m = 1, 2, \ldots, M \tag{10.3.7}$$

assuming that higher values of the attribute are better; the inequality is reversed if lower values are better.

Evaluation criteria	Thresholds		
	Acceptable threshold	Desirable threshold	Target
Reduction of km of congested network	70	75	80
Reduction of total travel time on network	1200	1300	1600
Veh-km	100	130	150

Fig. 10.20 Attribute thresholds

Evaluation criteria	Alternatives		
	A	B	C
Reduction of km of congested network	1	0	0
Reduction of total travel time on network	1	1	0
Veh-km	1	1	0

(1 = threshold attained; 0 = threshold not attained)

Fig. 10.21 Disjunctive method

Evaluation criteria	Alternatives		
	A	B	C
Reduction of km of congested network	1	0	0
Reduction of total travel time on network	1	1	0
Veh-km	1	1	0

(1 = threshold attained; 0 = threshold not attained)

Fig. 10.22 Conjunctive method

From the comparison between the desirable threshold values in Fig. 10.20 and the performance indicators, it follows that alternatives A and B are both selected because each alternative attains the threshold for at least one attribute (see Fig. 10.21).

In the *Conjunctive* constraint method, a minimum acceptable threshold is established for each attribute, and each attribute of an alternative is required to meet or exceed this standard. If the threshold is an accurate reflection of the decision-maker's expectations, the solutions obtained by this method will be acceptable.

The procedure is:

- The decision-maker establishes a minimum acceptable threshold value $\min_m$ for each attribute m
- For a given alternative, each attribute is compared with the corresponding threshold
- Alternative j is selected if:

$$x_{mj} \geq \min_m \quad \forall m = 1, 2, \ldots, M \tag{10.3.8}$$

again assuming that higher values of the attribute are better.

According to this method alternative A is selected (see Fig. 10.22).

The Maxmin method finds the lowest attribute value (min) of each alternative and then chooses the alternative having the highest (max) of these values. The logic is that a chain is as strong as its weakest link. This method is applicable only when attribute values are comparable with each other: either measured with the same units or transformed to a common scale.

In contrast, the Maxmax method selects an alternative by its best attribute value. It too is applicable only when attributes are comparable.

These techniques may have application domains in which they are reasonable, but they may not be very useful for general decision-making.

Another approach to noncompensatory choice requires the decision-maker to provide supplementary information about the ranking of individual criteria in terms of their perceived importance. This approach then considers each criterion in turn and works as a sequential elimination method.

Specifically, in *lexicographic* elimination, all alternatives are first compared in terms of the criterion deemed most important. If there is a unique best performing alternative in terms of this criterion, then it is selected as the most preferred. If there is a tie, then the selection process moves on to the second most important criterion and, considering only the alternatives that previously tied for first, seeks the ones that score best on the second criterion. Again, if this leads to a unique selection, then this alternative is designated as the preferred one. If not, the process is repeated, applying the third most important criterion to the alternatives that tied for both the first and second criteria, and so on until a unique option is identified or all criteria have been considered.

The *elimination by aspects* (*EBA*) method combines elements of both lexicographic ordering and the conjunctive/disjunctive methods. Alternative attributes are compared against corresponding thresholds. They are examined criterion by criterion and, for each criterion, alternatives that do not attain the threshold are eliminated. The criteria are examined, not in order of their importance to the decision-maker, but rather in an order that attempts to maximize the number of alternatives that fail to pass. This process is continued until only one alternative remains.

Neither the lexicographic elimination nor the elimination by aspects method has contributed much to the practice of public sector decision-making.

10.3.3.2 Multiattribute Utility Theory Method (MAUT)*

Multiattribute utility theory methods explicitly compute the utility of the different options under consideration. The options are then compared with each other on the basis of their utility values, and the preferred alternative is the one with the highest utility. In the context of choice among different projects, the utility of a project represents the decision-maker's level of satisfaction with it, which derives from the project's performance with respect to the different points of view considered in the evaluation. Therefore if m criteria for each alternative j are considered, a utility function $u(j) = f(x_{1j}, x_{2j}, \ldots, x_{mj})$ needs to be determined, where element x_{ij} is project j's attribute value for the ith criterion.

The core of MAUT is the procedure for estimating the decision-maker's utility function. This is done by first determining the value of u for a set of attribute values $x_{i,}$. The utility function is then obtained either by interpolation or by using a curve-fitting method to obtain a continuous function. The curve-fitting method uses known functional forms that satisfy the qualitative characteristics of the decision-maker's utility. The coefficients of the utility function are then estimated using quantitative information pertaining to the case under study. This procedure can be recommended when the number of attributes and options is high. However, whatever estimation technique is adopted, the value of the utility function u must first be determined for a number of points.

Determination of the value of u at selected points is carried out using a lottery mechanism. A lottery is defined as a list of possible outcomes $(x_1, x_2, \ldots, x_m)$ with a probability $(p_1, p_2, \ldots, p_m)$ associated with each. A binary lottery has two possible outcomes x_i and x_j with probability p_i and $p_j = 1 - p_i$, respectively; it is usually denoted as $(x_i, p_i; x_j)$.

For the sake of simplicity, consider the case of a single attribute x, so that $u = f(x)$. Suppose that the utility of an alternative i is a monotone increasing function of the attribute x so that $x_i > x_j \Rightarrow u_i > u_j$ in the interval $[x_w, x_b]$, whose endpoints are, respectively, the worse and the best values that the decision-maker assigns to attribute x. The method described here determines the utility of alternative a by reference to a lottery $\tilde{x}^a$ having n possible outcomes x_i^a with probabilities p_i. The utility of this lottery is equal to:

$$u^a = \sum_{i=1}^{n} p_i \cdot u\left(x_i^a\right) = E\left[u(\tilde{x}^a)\right] \tag{10.3.9}$$

Two lotteries that have the same utility are considered equivalent, so alternative a is preferred to alternative b if

$$E\left[u(\tilde{x}^a)\right] \geq E\left[u(\tilde{x}^b)\right] \tag{10.3.10}$$

The following definitions are very important for the direct determination of the utility functions. Consider a lottery $\tilde{x}$; its *certainty equivalent* is the value $\hat{x}$ for which:

$$u(\hat{x}) = E\left[u(\tilde{x})\right] \tag{10.3.11}$$

Two utility functions u_1 and u_2 are *strategically equivalent* if they produce the same preference order between two given lotteries. Therefore two utility functions that are strategically equivalent have the same certainty equivalent.

In order to estimate the value of u at points in the interval $[x_w, x_b]$, either of two distinct methods can be used, one based on the concept of the certainty equivalent and the other on equivalent lotteries. In both cases the decision-maker should be interviewed in order to understand her preferences. Initially the decision-maker is asked to identify the endpoints of the interval within which she thinks that the values of x can vary. After the worse and best values x_w and x_b have been identified, the utility function $u(x)$ is constrained so that $u(x_w) = 0$ and $u(x_b) = 1$.

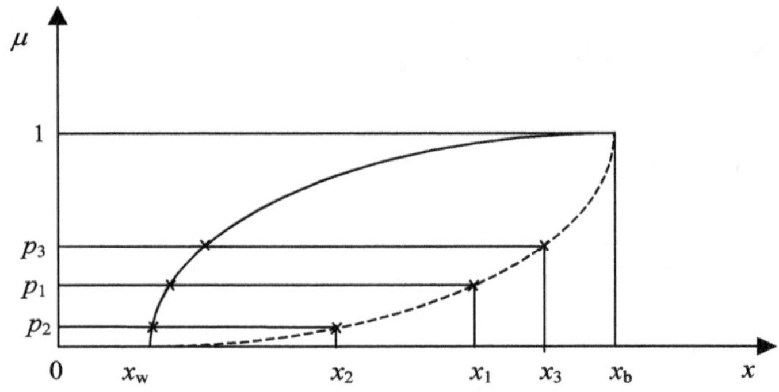

Fig. 10.23a Example of the certainty equivalent method

With the certainty equivalent method, the decision-maker is presented with the binary lottery $(x_b, p; x_w)$; the p value is usually taken equal to 0.5. The decision-maker is asked to indicate the value $\hat{x}$ of the lottery's certainty equivalent; that is, she is asked to evaluate the quantity $\hat{x}$ such that, if it were available as a certain outcome, it would have the same utility as the given lottery. This defines one point of the utility function:

$$u(\hat{x}_1) = p_1 \cdot u(x^b) + (1 - p_1) \cdot u(x^w) = p_1 \quad \text{with } p_1 = 0.5 \qquad (10.3.12)$$

In an analogous way, it is possible to determine other points on the function by considering two new lotteries derived from the preceding one: $(\hat{x}, p_2; x_w)$ and $(x_b, p_3; \hat{x})$; see Fig. 10.23a.

However, this method presents some limits: (i) if different p values are used, different $u(x)$ functions may be obtained; (ii) errors can be propagated in the estimation procedure, as the utility in one step is derived from the utility computed in the previous step; (iii) the decision-maker compares a certain outcome with a lottery, that is, with uncertain outcomes, therefore the method is not precise.

The equivalent lottery method eliminates problems (ii) and (iii). The decision-maker only compares uncertain outcomes and the utility is calculated step by step independently of the values computed before. The decision-maker is asked to compare two lotteries, one $(x_b, p; x_w)$ defined by the interval endpoints, the other defined by x_w and by an intermediate value x_i, which has probability p_i (usually 0.5) assigned by the interviewer. Therefore she is asked to determine the probability p that satisfies the following equation.

$$u(x_b, p; x_w) = u\left(x^i, p_i; x_w\right) \qquad (10.3.13)$$

With probability $p = 0.5$, the utility function value at x_i is obtained as follows.

$$u(x_i) = \frac{p_i}{p} \cdot u(x_b) = \frac{p_i}{p} = 2p_i \qquad (10.3.14)$$

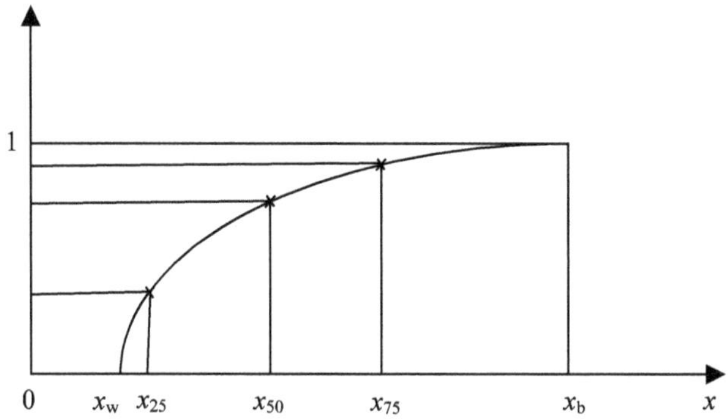

Fig. 10.23b Example of the equivalent lottery method

By asking the decision-maker to make the comparison described for m intermediate points, the values $u(x_i)$ of the utility function at m different points $x_i, I = 1, \ldots, m$ can be obtained (see Fig. 10.23b).

Once several points $(x_i, u_i;)$ on the utility function have been obtained, a continuous function $u(x)$ that passes through them can be estimated using, for example, least squares methods. The functional forms used for this purpose should obey some theoretical properties. For instance, the utility function of a risk averse (neutral, prone) decision-maker should be concave (linear, convex). Classes of risk averse and risk neutral functions are:

$$u(x) = a + b(-e^{-cx}) \tag{10.3.14a}$$

$$u(x) = a + b(cx) \tag{10.3.14b}$$

respectively, where a and $b > 0$ are constants to ensure that u is scaled appropriately (e.g., from zero to one) and c is positive for increasing utility functions and negative for decreasing ones. The parameter c indicates the degree of the decision-maker's risk aversion. For the linear case (10.3.14b), parameter c can be set at $+1$ or -1 for the increasing and decreasing cases, respectively.

So far only one-dimensional utility functions of the type $\boldsymbol{u} = f(x)$ have been discussed. However, the choice among alternative projects requires the decision-maker to evaluate those projects with respect to a variety of criteria such as noise, cost, and travel time. In this case the decision-maker's utility function is a function that aggregates all the criteria considered in the decision-making process. More specifically, it is a function of the type $\boldsymbol{u} = f(x)$, where $\boldsymbol{X} = (x_1, x_2, \ldots, x_n)$ represents the vector of project attributes.

The estimation of this function can be onerous. However, in practice the problem is simplified by assuming that the utility components related to different attributes are additively independent; that is,

$$u(X) = \sum_{i=1}^{n} k_i \cdot u_i(x_i) \tag{10.3.15}$$

The total utility function is the weighted average of the individual functions $u_i(x_i)$. The weights (i.e., the k_i coefficients) are scaling constants and are determined by asking the decision-maker appropriate questions concerning his preferences regarding the different attributes. Successive k_i values can be obtained via a lottery mechanism, applying a procedure similar to the one described above for the determination of points on the utility function $u(x)$. This procedure is outlined next.

In order to compute the k_i values, attribute vectors $\boldsymbol{X}_w = (x_{1w}, x_{2w}, \ldots, x_{mw})$ and $\boldsymbol{X}_b = (x_{1b}, x_{2b}, \ldots, x_{mb})$ are first constructed. $\boldsymbol{X}_w$ represents the situation in which each attribute has the worst value in the previously determined interval and $\boldsymbol{X}_b$ is the situation where each attribute has the best value. These vectors have utility equal to

$$U_w(X) = \sum_{i=1}^{n} k_i u_i(x_i) = 0 \quad \text{and} \quad U_b(X) = \sum_{i=1}^{n} k_i u_i(x_i) = 1.$$

Moreover, the vector $\boldsymbol{X}_i = (x_{1w}, x_{2w}, \ldots, x_{i-1w}, x_{ib}, x_{i+1b}, \ldots, x_{mb})$ having utility equal to

$$U(X_i) = \sum_{i=1}^{n} k_i u_i(x_i) = k_i$$

is defined as well. The decision-maker is asked to provide the probability p for which he is indifferent between the lottery $(X_b, p; X_w)$ and the certain outcome X_i. This means that the decision-maker is asked to estimate the impact of a variation in each u_i on the global utility U. Because the decision-maker is indifferent between $(X_b, p; X_w)$ and X_i, they have the same utility. It follows that the k_i values can be obtained from the expression $k_i = U(X_i) = pU(X_b) + (1-p)U(X_w) = p$.

Following is an example of this procedure. The objective is to choose the best location for intersections or ramps on limited access divided highways. Eight different scenarios (alternative projects), identified as a_1 to a_8, are under consideration. The criteria corresponding to the different objectives are as follows: public financial costs (c_1), users' nonmonetary costs (c_2), energy consumption (c_3), pollution (c_4), residential impact (c_5), and safety (c_6). With respect to these criteria, the following impact variables have been defined: $x_1(a_i)$, the monetary cost of project construction in millions of Euro; $x_2(a_i)$, the travel time spent on the network in hours per day; $x_3(a_i)$, the fuel consumption in kilograms per day; $x_4(a_i)$, the pollutant emissions in kilograms per day; $x_5(a_i)$, the weighted average of the volume/capacity ratio on urban roads; and $x_6(a_i)$, the weighted average of the percentage of time spent waiting to pass on two-lane rural roads.

Figure 10.24 presents the evaluation matrix.

Suppose that the utility function $u = f(x_1, x_2, \ldots, x_6)$ is expressed by equation (10.3.15). The k_i weights shown in Fig. 10.25 satisfy the condition $\sum_{i=1}^{6} k_i = 1$. These were determined by interviewing the decision-maker with questions such as "Would you prefer bringing x_1 and x_6 from x_{1w} and x_{6w} to x_{1b} and x_{6b}, rather than increasing all the other four attributes from the worse value to the

Alternative projects	Evaluation criteria					
	M€	h/day	kg/day	kg/day	V/C	%
1	10.25	80500	251000	23845	0.80	0.68
2	9.50	81800	255000	24225	0.85	0.73
3	10.75	82000	250000	23750	0.86	0.70
4	10.00	80600	253000	24035	0.87	0.64
5	10.50	79100	249000	23655	0.79	0.65
6	10.25	80900	252000	23940	0.82	0.73
7	11.50	78500	249000	23655	0.78	0.62
8	10.75	79400	251500	23892	0.83	0.67

Fig. 10.24 The evaluation matrix

k_1	k_2	k_3	k_4	k_5	k_6
0.2540	0.0745	0.0745	0.1350	0.1750	0.2870

Fig. 10.25 Weights

best?" As the answer is positive it follows that $k_1 + k_6 > 0.5$. By asking analogous questions involving pairwise comparisons between criteria, relations of the type $k_2 = k_3, k_1 > k_6$, and so on are obtained. To obtain numerical values, the decision-maker is asked to evaluate the probability p for which he is indifferent between: the certain outcome $(x_{1b}, x_{2w}, x_{3w}, x_{4w}, x_{5w}, x_{6w})$, whose utility is $u(x_{1b}, x_{2w}, x_{3w}, x_{4w}, x_{5w}, x_{6w}) = k_1u_1(x_{1b}) + k_2u_2(x_{2w}) + k_3u_3(x_{3w}) + k_4u_4(x_{4w}) + k_5u_5(x_{5w}) + k_6u_6(x_{6w}) = k_1$ and, on the other hand, the lottery $((x_{1b}, x_{2b}, x_{3b}, x_{4b}, x_{5b}, x_{6b}), p; (x_{1w}, x_{2w}, x_{3w}, x_{4w}, x_{5w}, x_{6w}))$, with utility equal to p. It follows that the value of k_1 is equal to p. For consistency, the computed numerical values should satisfy the relations introduced above; if they do not, the contradictions should be pointed out to the decision-maker, who should then be asked to revise his preferences on the basis of the previous results. After having determined for each attribute the endpoints of the interval within which the decision-maker thinks the values vary, three points are determined with the certainty equivalent method for each interval $x_{.25}$, $x_{.50}$, and $x_{.75}$. The utility values $u(x)$ equal to 0.25, 0.50, and 0.75 correspond to these intervals. These points are shown in Fig. 10.26. Utility functions u_i of the form $u(x) = Ax^2 + Bx + C$ are then fit through the points. The utility values for each alternative are reported in Fig. 10.27; they were computed with (10.3.14). From Fig. 10.27 it follows that the best alternative is x_5.

10.3.3.3 Linear Additive Methods*

The linear additive multicriteria method has a transparency and straightforward intuitive appeal that ensures it a central role in any discussion of multicriteria analysis.

The *weighted sum method* is the most commonly used linear additive method and it can be seen as a special case of MAUT with $w_i = k_i$ and $u(x_i) = x_i$. It assumes that the utility produced by an alternative can be expressed as a linear function

Utilities	Points for each alternative				
	$x_{0.0}$	$x_{.25}$	$x_{.50}$	$x_{.75}$	$x_{1.0}$
u_1	12.25	11.529	10.634	9.613	8.000
u_2	83500	82250	80121	78685	76500
u_3	256500	254360	251880	249670	246500
u_4	25000	24496	23770	23044	22000
u_5	0.88	0.862	0.825	0.771	0.700
u_6	0.75	0.734	0.701	0.659	0.600

Fig. 10.26 Points for each alternative

$u(a_1)$	$u(a_2)$	$U(a_3)$	$u(a_4)$	$u(a_5)$	$u(a_6)$	$U(a_7)$	$u(a_8)$
0.597	0.398	0.445	0.561	0.679	0.443	0.655	0.555

Fig. 10.27 Utilities of the alternatives

Evaluation criteria	Weights	Alternatives		
		A	B	C
Investment costs	0.200	1	1.25	2
Travel time	0.067	30	15	10
V/C ratio	0.333	20	30	40
CO emissions	0.333	1000	750	100
HC emissions	0.067	10	12	2

Fig. 10.28 Evaluation matrix with quantitative elements

of the normalized measures of the performance of the alternatives relative to each criterion:

$$u_i = \sum_m w_{mi} s_{mi} \tag{10.3.16}$$

The performance measure weights can be established by pairwise comparisons; this is described in detail in Sect. 10.3.3.4. The ranking of the alternatives is based on the utility values: the preferred alternative is the one that maximizes u_i.

Following is an example of this procedure. Consider the construction of a new road having three possible alignments. The criteria are investment costs (millions of Euros), travel time (minutes), volume/capacity ratio (V/C), CO emissions (tons), and HC emissions (tons). The evaluation matrix is shown in Fig. 10.28.

Figure 10.29 reports the normalized matrix obtained using (10.3.5).

The application of (10.3.16) to the elements of Fig. 10.29 gives the following values of the utilities for the three alternatives.

$$u_A = 0.345 \qquad u_B = 0.441 \qquad u_C = 0.711$$

It follows that the preferred alternative is C.

Evaluation criteria	Weights	Alternatives		
		A	B	C
Investment costs	0.200	0.500	0.375	0.000
Travel time	0.067	1.000	0.500	0.333
V/C ratio	0.333	0.500	0.750	1.000
CO emissions	0.333	0.000	0.250	0.900
HC emissions	0.067	0.167	0.000	0.833

Fig. 10.29 Normalized evaluation matrix

10.3.3.4 The Analytical Hierarchy Process (AHP)*

Procedures based on the approach named the *Analytic Hierarchy Process* (AHP) begin by organizing into a hierarchy the various elements that are involved in the decision-making process. Knowledge of the relationships between a level and the next one up in the hierarchy is essential. For example, a choice problem involving alternative transportation projects may involve three hierarchical levels: at the lowest level are the projects, at the next higher level are the different points of view on the basis of which the projects will be evaluated, and at the highest level is the ultimate objective to be achieved. Choice requires knowledge of the impacts of the projects with respect to the different points of view, and of the importance of the different points of view with respect to the ultimate objective. Sometimes other intermediate levels are present: for example, between the levels representing the ultimate objective and the points of view, there may be a further level defining different socioeconomic scenarios that affect the importance of the points of view. These individual scenarios would be related to the final objective by their probabilities of occurrence.

Consider a decision-making process and divide the set of elements of such a process into z subsets in an ordered sequence of levels, so as to impose a hierarchical structure on the process. The level 1 has only one element; n_r is the number of elements on level r; and y_i^r is the element that occupies position i in level r, as follows.

$$\begin{array}{l} y_1^1 \\ y_1^2\ y_2^2\ \dots\ y_{n_2-1}^2\ y_{n_2}^2 \\ \vdots \\ y_1^r\ y_2^r\ \dots\ y_{n_r-1}^r\ y_{n_r}^r \\ y_1^{z-1}\ y_2^{z-1}\ \dots\ y_{n_{z-1}-1}^{z-1}\ y_{n_{z-1}}^{z-1} \\ y_1^z\ y_2^z\ \dots\ y_{n_z-1}^z\ y_{n_z}^z \end{array} \tag{10.3.17}$$

Let $a_{j,k}^{i,r}$ be a number that measures the relative importance of element y_j^{r+1} compared to y_k^{r+1} on level $r+1$, where the comparison is made with respect to element

y_i^r on level r. The *AHP* method assumes that

$$a_{k,j}^{i,r} = 1/a_{j,k}^{i,r} \qquad a_{j,k}^{i,r} \cdot a_{k,h}^{i,r} = a_{j,h}^{i,r} \tag{10.3.18}$$

The value $a_{j,k}^{i,r}$ is the j,k element of a matrix $B_i^r(n_{r+1} \cdot n_{r+1})$, where the elements of the main diagonal are equal to 1 and the symmetric elements j,k, and k,j are inverses of each other. A matrix that satisfies this property is called *reciprocal*; if the second equation of (10.3.18) is also verified, it is also called *consistent*.

Matrix B_i^r has rank 1, because any row j can be obtained by multiplying row 1 by $a_{j,1}^{i,r}$:

$$a_{j,1}^{i,r} \cdot a_{1,k}^{i,r} = a_{j,k}^{i,r} \quad \forall k \in (1 \ldots n_{r+1}) \tag{10.3.19}$$

B_i^r has therefore a single nonzero eigenvalue λ_i^r, which is equal to the matrix trace n_{r+1}. Let the corresponding eigenvector be w_i^r.

It follows, as a consequence of (10.3.18), that the elements $a_{j,k}^{i,r}$ of B_i^r can be expressed as ratios between numbers drawn from a set of n_{r+1} numbers $\gamma_1^{i,r}, \gamma_2^{i,r} \ldots \gamma_{n_{r+1}}^{i,r}$:

$$a_{j,k}^{i,r} = \frac{\gamma_j^{i,r}}{\gamma_k^{i,r}} \quad \forall j,k \in (1 \ldots n_{r+1}) \tag{10.3.20}$$

Let γ_i^r be the vector with elements $\gamma_1^{i,r}, \gamma_2^{i,r} \ldots \gamma_{n_{r+1}}^{i,r}$; it follows that

$$B_i^r \gamma_i^r = n_{r+1} \gamma_i^r \tag{10.3.21}$$

and, because $\lambda_i^r = n_{r+1}$, it follows that γ_i^r is the same as the eigenvector w_i^r of B_i^r.

As the components of w_i^r are only defined up to a multiplicative factor, let us assume that the sum of these components is equal to 1, and let us interpret such components as normalized measures of the importance (weights) of the elements of level $r+1$ of the hierarchy with respect to the element y_i^r of level r.

Let $W_{r+1}^r(n_{r+1} \cdot n_r)$ be the matrix whose columns are the vectors w_i^r $\forall i \in (1 \ldots n_r)$, and let w_i^{r-1} be the normalized eigenvector of matrix B_i^{r-1}, whose elements are the weights of level r of the hierarchy with respect to y_i^{r-1} of level $r-1$.

Let $w_{i,r+1}^{r-1}$ be the product of matrix W_{r+1}^r by w_i^{r-1}:

$$W_{r+1}^r \cdot w_i^{r-1} = w_{i,r+1}^{r-1} \tag{10.3.22}$$

Let us assume that the components of $w_{i,r+1}^{r-1}$, whose sum is equal to 1, are the normalized weights of the elements of level $r+1$ of the hierarchy with respect to element y_i^{r-1} of level $r-1$.

Let W_{r+1}^{r-1} be the matrix whose components are the vectors $w_{i,r+1}^{r-1}, i \in (1 \ldots n_{r-1})$, and w_j^{r-2} be the normalized eigenvector of matrix B_j^{r-2}; then the product $W_{r+1}^{r-1} \cdot w_j^{r-2}$ provides the weights of level $r+1$ with respect to y_j^{r-2} of level $r-2$.

γ_i/γ_j	Outcome of the comparison
1	Equal importance
3	Moderately greater importance
5	Greater importance
7	Much greater importance
9	Extremely greater importance
2, 4, 6, 8	Intermediate values

Fig. 10.30 Pairwise comparison scale

In this way, by assigning a quantitative measure to the relative importance of each pair of elements in a given level $r+1$ with respect to each element of the next higher level r, the normalized list of weights of the elements of a given level with respect to any element of an upper level can be obtained. In particular, the normalized list of weights of the elements of level z (alternative projects) with respect to the single element of level 1 (the final objective) can be determined in this way.

The attributes of the elements can be *tangible*, meaning that they can be obtained through computation, or *intangible*, meaning that they can be deduced only in a qualitative way and are therefore subjective.

When the attributes are tangible, the best way of obtaining the relative importance of two elements y_j^{r+1}, y_k^{r+1}, expressed by $a_{j,k}^{i,r}$ in matrix B_i^r, is by considering the ratio of their attributes. Let $e_j^{i,r+1}$, $e_k^{i,r+1}$ be the attributes of y_j^{r+1} and y_k^{r+1}, respectively, with respect to y_i^r: if $e_j^{i,r+1} > e_k^{i,r+1}$ then y_j^{r+1} is more important than y_k^{r+1}, so the attributes are favorable, and we set $a_{j,k}^{i,r} = e_j^{i,r+1}/e_k^{i,r+1}$. On the other hand, if $e_j^{i,r+1} > e_k^{i,r+1}$ then y_j^{r+1} is less important than y_k^{r+1}, so the attributes are unfavorable, and we set $a_{j,k}^{i,r} = e_k^{i,r+1}/e_j^{i,r+1}$.

When the attributes are unfavorable, the relative importance of two elements is qualitatively assessed using a numerical scale to provide values corresponding to the different qualitative judgments. Figure 10.30 shows the pairwise comparison scale of Saaty.

When the attributes are tangible, matrix B_i^r satisfies both properties of (10.3.18), and it is easy to verify that

$$\gamma_j^{i,r} = \frac{e_j^{i,r+1}}{\sum_{k=1}^{n_{r+1}} e_k^{i,r+1}} \tag{10.3.23}$$

if the attributes $e_k^{i,r+1}$ $\forall k = 1 \dots n_{r+1}$ are favorable. On the other hand, if the attributes are unfavorable then:

$$\gamma_j^{i,r} = \frac{1/e_j^{i,r+1}}{\sum_{k=1}^{n_{r+1}} 1/e_k^{i,r+1}} \tag{10.3.24}$$

When the attributes are intangible, it is not a problem to satisfy the first condition of (10.3.18) (i.e., the reciprocal condition), but it is difficult to satisfy the second condition (i.e., consistency). In this case, if $\lambda_{\max}$ is the principal eigenvalue of matrix B_i^r and n_{r+1} is its rank, the consistency index CI of B_i^r is defined as

$$CI = \frac{\lambda_{\max} - n_{r+1}}{n_{r+1} - 1} \tag{10.3.25}$$

which represents the average value of B_i^r, which would be equal to zero if the matrix were consistent. The consistency ratio is $CR = CI/ACI$, where ACI is the mean value of CI for a random matrix of rank n_{r+1}. ACI values for typical values of n_{r+1} are shown below:

n_{r+1}	2	3	4	5	6	7	8	9	10
ACI	0.0000	0.4887	0.8045	1.0591	1.1797	1.2519	1.3171	1.3733	1.4055

The consistency of B_i^r can be accepted if $CR < 0.1$, and in this case the principal eigenvector B_i^r provides a good estimate of the weights $\gamma_j^{i,r}$, $j = 1, \ldots, n_{r+1}$.

Following is a detailed example application of the AHP method described above.[15] The purpose is to recommend the best transportation system management (TSM) strategy for reducing traffic congestion and emissions in a town.

Step 1: Setting Up the Hierarchy

The first step is to set up a decision process hierarchy (see Fig. 10.31). The first level denotes the overall goal of the decision-maker. In this example, this is to find the best TSM strategy for reducing congestion and pollution. The second level defines factors that contribute to this goal. The number of factors considered in the AHP depends on the problem; in the present case the factors are environment, technology, cost, and average vehicle speed (i.e., effectiveness for travelers). Environment and speed are obviously factors that relate to the goal of reducing congestion and pollution, but technology (e.g., construction and management methods) and cost are always highly important considerations in transportation planning. The last level of the hierarchy then describes the alternative TSM strategies, which are to be evaluated in terms of the criteria in the level above. The number of alternatives is not limited in the AHP, but this example focuses on two alternatives to make the process more clear. Let us consider only High Occupancy Vehicle lanes and an increase in parking charges as possible strategies for reducing congestion and pollution. The analytical hierarchy process will aid the process of choosing which alternative is the best for achieving this goal.

Step 2: Comparison of Characteristics

In the next step, the factors from the second level of the hierarchy are compared with each other to determine the relative importance of each factor in accomplishing the overall goal. The easiest way is to prepare a matrix with factors (i.e., environment, technology, cost, and speed) listed at the top and on the left. Based on the

[15] The example reported here is taken from Boulter (1999).

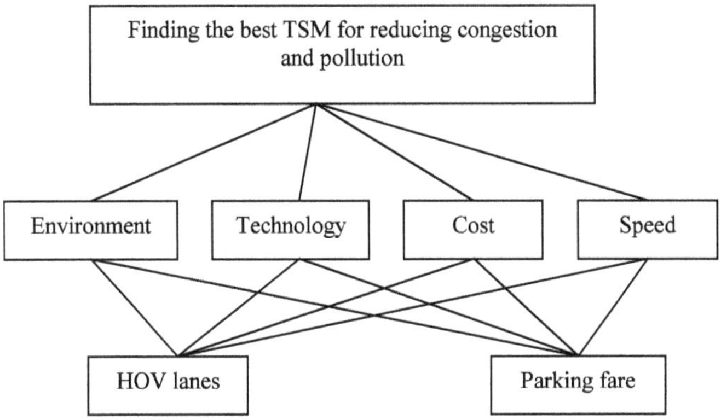

Fig. 10.31 A hierarchy of priorities

	Environment	Technology	Cost	Speed
Environment	1	4	3	2
Technology	1/4	1	1/2	1/3
Cost	1/3	2	1	1/2
Speed	1/2	3	2	1

Fig. 10.32 Comparison of characteristics

resulting judgment of the decision-maker, the matrix is then filled in with numerical values denoting the importance of the factor on the left relative to the importance of the factor on the top. A high value means that the factor on the left is relatively more important than the factor at the top. In Fig. 10.32 for example, environment is considered to be four times as important as technology, whereas average speed of vehicles is only half as important as the environment. When a factor is compared with itself the ratio of importance is obviously one. The eigenvalue of the comparison matrix is 4.026, whereas the *CI* is 0.009 and *CR* is 0.011.

In this example the priorities are clear. Because the overall goal is to reduce congestion and emissions, environment and speed are deemed the most important factors and thus are assigned the highest value in the matrix. Environment is slightly more important than speed. Cost is a less important factor compared to the positive effects achievable, that is, reduction of congestion and pollution. The technology involved in implementing the different alternatives is less important than the cost. Thus in the matrix, cost is assigned the value 2 relative to technology. Once the matrix has been filled in, the decision-maker can move to step 3, in which the priority vector is established.

Step 3: Establish the Priority Vector

In this step the decision-maker uses the numbers from the matrix in Fig. 10.33 to determine an overall priority weight for each factor. To do this, the decision-maker computes the sum of the values in each row of the matrix and divides each result by the sum of the results for all the rows.

The calculations are reported in Fig. 10.33.

Environment	$1+4+3+2$	$=10$
Technology	$1/4+1+1/2+1/3$	$=2.08$
Cost	$1/3+2+1+1/2$	$=3.83$
Speed	$1/2+3+2+1$	$=6.5$
		22.41 Total

$$\Rightarrow \begin{array}{l} 10\ :22.41=0.45 \\ 2.08:22.41=0.09 \\ 3.83:22.41=0.17 \\ 6.5\ :22.41=0.29 \end{array}$$

Fig. 10.33 The priority vector

Environment	HOV lanes	Parking
HOV lanes	1	2
Parking charge	1/2	1

Technology	HOV lanes	Parking
HOV lanes	1	1
Parking charge	1	1

Cost	HOV lanes	Parking
HOV lanes	1	1/3
Parking charge	3	1

Speed	HOV lanes	Parking
HOV lanes	1	4
Parking charge	1/4	1

Fig. 10.34 Comparison of alternative TSM strategies

Step 4: Comparison of Alternatives

The decision-maker moves from level 2 to level 3 of the hierarchy and makes a pairwise comparison of the two alternatives at the bottom of the hierarchy in Fig. 10.34. In this example the elements to be compared are the HOV lanes and the increase in parking charges, with the comparison to be made on the basis of how much one is better than the other in satisfying the factors from level 2.

In order to show how the numerical values have been assigned, it is useful to look at one or two examples in detail. The first case in Fig. 10.34 shows that, in terms of the impact on the environment, HOV lanes are between equally and moderately more important than parking charges (following Saaty's scale). This is because the carpooling that is necessary for the use of HOV lanes immediately reduces the number of vehicles on the road and thus reduces pollution. Numerical values, obtained by analyzing the system using mode choice and assignment models, could be used to assess the importance of the two options. In terms of cost, the increase in parking charges seems more desirable, as this is a much cheaper option than building additional lanes for HOVs, even when the cost of employing officers to enforce the parking charges is considered. The remaining two cases show similar evaluations of the two TSM strategies in terms of their technology and speed impacts.

Step 5: Establish the Priority Vectors for the Alternatives

This follows the same procedure as Step 3.

The results from Steps 3 and 5 can be summarized in the two priority matrices shown in Fig. 10.36.

Environment:	$1+2=3$ $0.5+1=1.5$, the sum is 4.5	Technology:	$1+1=2$ $1+1=2$, the sum is 4
Then	$3:4.5=0.67$ $1.5:4.5=0.33$	Then	$2:4=0.5$ $2:4=0.5$
Cost:	$0+0.33=1.33$ $3+1=4$, the sum is 5.33	Speed:	$1+4=5$ $0.25+1=1.25$, the sum is 6.25
Then	$1.33:5.33=0.25$ $4:5.33=0.75$	Then	$5:6.25=0.8$ $1.25:6.25=0.2$

Fig. 10.35 Priority vector for the alternatives

Env.	0.45
Tech.	0.09
Cost	0.17
Speed	0.29

Env.	Tech.	Cost	Speed
0.67	0.5	0.25	0.8
0.33	0.5	0.75	0.2

Fig. 10.36 Priority matrices

Step 6: Obtaining the Overall Ranking

The final step in determining which alternative best fulfills the goal of reducing traffic congestion and emissions is to calculate the overall ranking for both alternatives. This is accomplished by multiplying the criteria priority vector by the alternative priority vector. For example, the priority vector for the environment criterion is 0.45, and the priority vector for the HOV alternative is 0.67; the HOV alternative's rating with respect to the environment criterion is thus $0.45 * 0.67 = 0.30$. Results of the application of this method to each criterion and alternative are shown in Fig. 10.37. It can be seen that HOV lanes have received an overall rating of 0.62 whereas the increase in parking charges has received a rating of only 0.39. Therefore, HOV lanes are the best TSM strategy for reducing traffic congestion and emissions in this hypothetical case study.

10.3.3.5 Outranking Methods*

Partial aggregation methods assume that, for each pair of actions (or projects), the decision-maker can either express his preference or indifference between them or declare them to be not comparable. Preferences expressed in this situation will not necessarily be consistent and may not allow a complete ranking of alternatives. The *ELECTRE* (ELimination Et Choix Traduisant la REalité) decision methods are based on this approach.

Four different versions of the ELECTRE software are available. The first (ELECTRE I) was designed to select "the best" alternatives from among a given set. The second, third, and fourth versions (ELECTRE II, III, IV) were developed to rank a set of alternatives.

At the basis of the ELECTRE methods lies the concept of "outranking." Option A outranks option B if there are enough arguments to decide that A is at least as good as B, although there is no overwhelming reason to reject that statement. The heart

Alternatives	Criteria priority vector	Alternative priority vector	Product of vector
		Environment	
HOV	0.45	0.67	0.30
Parking charge	0.45	0.33	0.15
		Technology	
HOV	0.09	0.5	0.05
Parking charge	0.09	0.5	0.05
		Cost	
HOV	0.17	0.25	0.04
Parking charge	0.17	0.75	0.13
		Speed	
HOV	0.29	0.8	0.23
Parking charge	0.29	0.2	0.06
HOV total			**0.62**
Parking charge total			**0.39**

Fig. 10.37 Overall ranking of alternatives

of the evaluation procedure depends on the verification, for each pair of alternatives, of the outranking assumption. The latter is based on two conditions that should both be satisfied:

- A *concordance* condition (the majority of the criteria should be in favor of alternative a_i)
- A *discordance* condition (none of the criteria should be too much against a_i)

The satisfaction of both conditions is verified through concordance and discordance indices that are computed by comparing pairs of alternatives criterion by criterion. It is clear that if an alternative performs better with respect to all criteria, then it is definitely preferred to the others (dominating alternative). The opposite occurs if its performance is worse with respect to all the criteria (dominated alternative). In intermediate cases, a rule is defined that fixes the acceptable limits of the outranking assumption.

Whatever the version, ELECTRE methods require weights w_m and performance indicators s_{mi}. for each alternative i and criterion m.

ELECTRE I defines the *index of concordance* c_{ij} of the project i compared with project j as a standardized measure of the preference for project i compared to project j; the index is equal to one if i dominates j:

$$c_{ij} = \frac{\sum_{m \in S_{ij}} w_m}{\sum_{n=1 \ldots M} w_n} \tag{10.3.26}$$

where $S_{ij} \equiv \{m : s_{mi} \geq s_{mj}\}$ is the set of criteria for which project i is as good as or better than j and s_{mj} is the normalized performance indicator defined by (10.3.5). Obviously, the closer index c_{ij} is to one (i.e., the more project i is preferred to project j), the greater will be the weights w_m of the criteria for which i is superior to j.

The *discordance index* d_{ij} of project i compared to project j is a standardized measure of the "inferiority" of i compared to j. It is equal to one if the maximum weighted deviation in favor of j among all the criteria for which j is superior, coincides with the maximum absolute weighted deviation between i and j for all the criteria. In formal terms, it results:

$$d_{ij} = \frac{\max_{m \in I_{ij}}[w_m(l_{mj} - l_{mi})]}{\max_n[w_n|l_{ni} - l_{nj}|]} \tag{10.3.27}$$

where $I_{ij} \equiv \{m : l_{mi} < l_{mj}\}$ is the set of criteria indices for which project i is inferior to j.

Concordance and discordance indices can be used in different ways to compare available alternatives. The *mobile threshold method* calculates the concordance and discordance indices for all ordered pairs of alternative projects. This ordering of alternatives can be obtained by fixing two thresholds, $\bar{c}$ and $\bar{d}$ (with $\bar{c} \leq \bar{d}$), and rejecting all project pairs (i, j) such that c_{ij} is less than $\bar{c}$ (i.e., pairs for which i is not significantly superior to j) and/or d_{ij} is greater than $\bar{d}$ (i.e., i is clearly inferior to j). Pairs of alternative projects that meet both requirements satisfy:

$$c_{ij} \geq \bar{c} \quad \text{and} \quad d_{ij} \leq \bar{d}$$

These pairs are considered to give a significant indication of the superiority of alternative i over alternative j. If the resulting pairs still do not lead to a unique ordering (e.g., i is preferable to j, j is preferable to k but k is preferable to i), the values of the thresholds $\bar{c}$ and $\bar{d}$ are modified by increasing the former and reducing the latter until a set of project pairs expressing a unique preference ordering is obtained (see Fig. 10.38).

Some multicriteria analyses are intended for application to choice contexts where no precise numerical data are available. This may be the case when many criteria do not naturally lend themselves to numerical measurement or when there is insufficient budget to carry out a detailed modeling of project impacts. It should be clear that for these cases the somewhat arbitrary qualitative assessment of project impacts means that the criteria weights are themselves somewhat arbitrary. Within this range of methods, the *REGIME* method is outlined in detail.

The starting point of this method is to calculate, for each pair of alternatives i and j, a concordance index defined as in the *ELECTRE* methods.

Given two alternatives i and j, the *regime index* r_{ij} is then computed; it can be interpreted as an indicator of the preference for alternative i compared to j:

$$r_{ij} = c_{ij} - c_{ji} \tag{10.3.28}$$

Because the weights w_M are ordinal, it is impossible to compute a numerical value for r_{ij}. Therefore the *REGIME* analysis focuses on the sign of the regime index.

Alternatives A, B, C, D

$\bar{c} = 0.30 \quad \bar{d} = 0.70$

Pairs of alternatives	Concordance indices	Discordance indices
A-B	0.40	0.20
A-C	0.70	0.50
B-A	0.60	0.50
B-C	0.65	0.30
C-B	0.35	0.60
C-D	0.35	0.40

$\bar{c} = 0.45 \quad \bar{d} = 0.50$

Pairs of alternatives	Concordance indices	Discordance indices
A-C	0.70	0.50
B-A	0.60	0.50
B-C	0.65	0.30

Ordering B > A > C

Fig. 10.38 Example application of the *ELECTRE IV* mobile threshold method

Evaluation criteria	Weights	Alternatives		
		A	B	C
1	++	−	−−	−−−
2	+	+++	++	+
2	+++	+	++	+++
4	+++	−−−	−−	−
5	+	−−	−−−	−

Fig. 10.39 Evaluation matrix with ordinal elements

Ordinal information about weights and/or effects is enough to determine the sign of r_{ij}. In this sense the qualitative evaluations that define the importance of the different criteria are sufficiently clear to conclude with certainty that the criteria for which alternative i dominates j are more important than those for which j dominates i. A matrix can be filled in with all the possible pairs of alternatives i and j. Element *ij* is $+1$ if i is preferred to j and -1 otherwise.

An example clarifies this procedure. Let us interpret Fig. 10.39 expressed in ordinal terms.

The series $---+++$ should be interpreted as an increasing ordinal scale from negative impacts (disadvantages or costs with a maximum of negativity indicated as $---$) to positive impacts (advantages or benefits, with a maximum of positivity indicated as $+++$). The series of ordinal values of weights should be associated with a series of cardinal values. Each weight expressed in ordinal terms can be replaced by a score between 1 and 5 and then the values are normalized by dividing them by their sum. In this example the following correspondences have been accepted (see Fig. 10.40):

The concordance indices can be calculated according to the *ELECTRE* method. In Fig. 10.41 the concordance indices and the regime indices are reported for each pair of alternatives.

Qualitative weight	+	+	++	+++	+++	Total
Quantitative weight	1	1	3	5	5	15
Normalized weight	0.067	0.067	0.200	0.333	0.333	1.000

Fig. 10.40 Qualitative, quantitative, and normalized scores

Pairs of alternatives	Concordance index	Regime index
A-B	0.334	−0.332
A-C	0.267	−0.466
B-A	0.666	+0.332
B-C	0.267	−0.466
C-A	0.733	+0.466
C-B	0.733	+0.466

Fig. 10.41 Concordance and regime indices for each pair of alternatives

Alternatives	A	B	C	Total
A	−−	−1	−1	−2
B	+1	−−	−1	0
C	+1	+1	−−	+2

Fig. 10.42 Regime indices matrix

The pairwise comparison is reported in Fig. 10.42, where element *ij* is equal to $+1$ or -1 if the regime index is positive or negative and, then, if the alternative of the ith row is better or worse than the one of the jth column (or vice versa).

It follows that the classification is C-B-A without ambiguity.

10.3.3.6 Constrained Optimization Method*

The constrained optimization method is based on the principle that any multiobjective and multidimensional decision-making problem can be converted into a constrained single-objective optimization problem and solved using integer programming techniques with binary variables. This transformation can be done in the following way.

- By choosing an objective to maximize or minimize
- By considering the other objectives as constraints, whose minimum or maximum level of achievement should be respected
- By assigning to each project alternative j a binary variable x_j, which takes the value 1 if alternative j is to be implemented and 0 if not. (It is not possible to implement part of a project)

Following is an example of this method.

Suppose that one or more projects are to be chosen from among six candidate projects, identified by a code from x_1 to x_6. For all projects, it is assumed that the

Year	x_1	x_2	x_3	x_4	x_5	x_6
0	−100	−200	−200	−100	−200	−300
1	18	30	0	25	50	40
2	18	30	0	25	50	40
3	18	30	5	20	40	40
4	18	30	10	20	40	40
5	18	30	10	20	40	40
6	18	30	20	15	30	40
7	18	30	20	15	30	40
8	18	30	30	15	30	40
9	18	30	30	15	30	40
10	−82	30	40	10	20	40
11	18	30	40	10	20	40
12	18	30	40	10	20	40
13	18	30	50	10	20	40
14	18	30	50	10	20	40
15	18	30	50	5	10	40
16	18	30	50	5	10	40
17	18	30	50	5	10	40
18	18	30	50	5	10	40
19	18	30	50	5	10	40
20	18	30	100	5	10	40
NPV (6%)	50.62	144.10	116.75	68.06	136.12	158.80

Fig. 10.43 Annual benefits and NPV for each project

investment is concentrated at the end of year 0 and that the net benefits are concentrated at the end of each year of the economic project life. Figure 10.43 indicates the annual benefits and the NPV associated to each of the six investments.

Furthermore, assume that projects x_1 and x_2 are mutually exclusive (e.g., two links that connect the same O/D pair). Moreover, project x_3 cannot be implemented if project x_4 is not, whereas x_4 can be carried out by itself (x_3 is a project that depends on the implementation of x_4, whereas x_4 is an independent project). The projects do not otherwise interact, so their individual annual benefits and *NPV* are additive.

The choice of the investments to carry out is made in a context where the available financial resources K are exogenously defined and insufficient to finance all the projects.

To formalize the decision-making problem among the different objectives (e.g., NPV maximization, investment cost minimization, etc.) the function to maximize or minimize must be chosen. In this case, for example, the total NPV of the selected alternatives will be maximized, using a discount rate of 6%.

Three different exogenous constraints should be considered:

- The total cost of the selected projects cannot exceed K
- Some alternatives are complementary
- Some alternatives are mutually exclusive

The binary programming problem can be formulated in the following way:

$$\max \sum_j NPV_j(6\%)x_j \quad j = 1, \ldots, n \tag{10.3.29}$$

subject to:

$x_j(1 - x_j) = 0$ Variable x_j can only assume the values 0 or 1 if this constraint is to be respected

$x_3 - x_4 \leq 0$ If project 4 is implemented ($x_4 = 1$) then project 3 can be either implemented ($x_3 = 1; x_1 - x_3 = 0$) or not ($x_3 = 0$); if project 4 is not implemented ($x_4 = 0$), then project 3 cannot be ($x_3 = 0$ necessarily). This constraint expresses the dependency of project 3 on project 4

$x_1 + x_2 \leq 1$ Implementation of either project 1 or 2 precludes implementation of the other (e.g., $x_2 = 1$ necessarily implies $x_1 = 0$). The projects are mutually exclusive. It is possible that neither is implemented

$\sum_j q_{0j}x_j \leq K$ The total investment necessary to finance the selected projects should be less than the available budget K. (Here q_{0j} represents the investment cost of project j)

The solution of the above binary programming problem identifies the set of projects to be implemented, that is, those that maximize the objective function subject to all the constraints.

By solving the same problem with a different budget constraint, the list of projects to be implemented may change. This may also happen if constraints of a different nature are included: for example, constraints on the expected maximum number of annual accidents on the regional network (e.g., $\sum_j INC_jX_j \leq INC^*$, where INC^* is the maximum acceptable number of annual accidents).

Each alternative formulation of the decision-making problem produces an "*evaluation scenario*," characterized by the choice of a given objective function, by the set of technical, physical, and behavioral constraints and by the targets defined by the decision-maker/analyst for the other objectives and constraints.

The information derived from the different scenarios is used to generate new scenarios to test, so increasing the level of knowledge with respect to the impacts of the various objectives on the decision-making problem. Examples of the important conclusions that can be drawn from this procedure include the following.

- The contributions of the selected alternatives, in each scenario, to the criteria that are considered relevant by the decision-makers
- The robust alternatives, that is, those that tend to be selected a larger number of times in the different scenarios

The procedure ends when a satisfactory solution is obtained as a good compromise between the different relevant objectives.

Reference Notes

The method proposed for the calculation of surplus changes for transportation system users is original; it extends the "classical" results for aggregate models and those in Williams (1977) for behavioral models. The paper by Jara-Diaz and Friesz (1982) deals with the evaluation of user surplus changes using descriptive demand functions involving multiple service attributes.

The traditional approach of benefit-cost analysis is covered in Wohl and Martin (1967), Hutchinson (1974) and Stopher and Meybourg (1976). Alternative approaches such as cost-effectiveness analysis are described in Stopher and Meybourg (1976) and Meyer and Miller (2001).

The literature on welfare economics applied to investment analysis is quite substantial and a systematic analysis is well beyond the scope of this book. Among the many texts on the subject, reference can be made to the classic book of Mishan (1974), Trezza et al. (1978), and some chapters of the volume by Adorisio (1986).

Applications to the evaluation of transportation system investments can be found in almost all books on transportation planning. A critical review of financial analysis techniques is presented by Fleming and Giugale (2001).

Although there are some clear links back to earlier work, the works of Von Neumann and Morgenstern (1947) and later by Savage (1954) are generally considered to be the starting point for multicriteria analysis in terms of a normative theory of how individuals should rationally choose between competing options. More recent fundamental contributions to these techniques can be found in Chankong and Haimes (1983), Voogd (1983), Haimes and Chankong (1985), Nijkamp et al. (1990), and Rostirolla (1998). The DLTR multicriteria analysis manual by Dodgson et al. (2000) is an important reference. It provides guidance for government officials and other practitioners on how to undertake and make the best use of multicriteria analysis for the appraisal of options for policy and other decisions, including but not limited to those having implications for the environment. It covers a range of techniques that can be of practical value to public decision-makers, and is increasingly being used in the United Kingdom and in other countries. The paper by Shiftan et al. (2002) is also fundamental for the application of these methods to transportation projects.

Examples of elimination by aspects models can be found in Tversky (1972).

Keeney and Raiffa's (1993) book is the key guide to multiattribute utility applications. More recent references that cite examples and indicate the range of potential applications include Menichini (2003) (the example presented in this book is his) and Dyer (2005). Von Neumann and Morgenstern (1947) discussed some paradoxes that are at the base of the MAUT method. One of these is the Allais paradox.

Examples of linear additive multicriteria methods are presented by Edwards (1971) and Edwards and Barron (1994).

The analytic hierarchy process method was originally devised by Saaty (1990). It has proved to be one of the most widely applied multicriteria methods; see, for example, Zahedi (1986), Shim (1989), Boulter (1999), Chavarria (2002), and Ferrari (2003) for summaries of applications.

Outranking as a basis for multicriteria analysis originated in France in the work of Bernard Roy and colleagues in the mid-1960s and has continued to be applied and extended since that time. For a summary of the European school of multicriteria analysis thinking, see Roy and Vanderpooten (1996). Vinckie's book (1992) provides a clear introduction to the best-known outranking methods, of which there are several. Nijkamp and Van Delft (1977) and Voogd (1983) both suggest procedures for a qualitative outranking analysis. Examples of the regime method are reported in Hinloopen and Nijkamp (1990), de Luca (2000), and De Montis et al. (2000). Di Maio and Rostirolla (2002) present applications of the constrained optimization method.

A different response to the imprecision that surrounds much of the data on which public decision making is based has been to look to the developing field of fuzzy sets to provide a basis for decision-making models. The fundamentals of decision-making in a fuzzy environment are presented by Zadeh (1976) and Zimmerman (1995). See Munda (1995) for a detailed description of the NAIADE method. However, a detailed description of these methods is beyond the scope of this book.

Appendix A
Review of Numerical Analysis

This appendix contains an overview of numerical analysis for the formulation, analysis, and solution of the mathematical models described in the text.

A.1 Sets and Functions

A.1.1 Elements of Set Topology

In this section some properties of numerical sets are outlined, with reference to the n-dimensional Euclidean space E^n. Numerical sets are made up of points in E^n; that is, vectors (assumed to be column vectors) with n real components $\boldsymbol{x}^T = (x_1, \ldots, x_n)$, among which the Euclidean norm (or module) $\|\boldsymbol{x}\| = (\sum_i x_i^2)^{1/2} = (\boldsymbol{x}^T\boldsymbol{x})^{1/2}$ and the corresponding Euclidean distance are defined.

The sphere of radius δ and center $\boldsymbol{x}$ is defined as a neighborhood $N_\delta(\boldsymbol{x})$ of radius δ of the point $\boldsymbol{x} \in E^n$:

$$N_\delta(\boldsymbol{x}) = \left\{\boldsymbol{y} : \|\boldsymbol{y} - \boldsymbol{x}\| < \delta\right\}$$

A point $\boldsymbol{x} \in E^n$ is said to be interior to the set $S \subseteq E^n$, if there is at least a neighborhood of finite radius δ entirely contained in S. A point $\boldsymbol{x} \in E^n$ is at the boundary of the set S if all the neighborhoods of $\boldsymbol{x}$, however small the radius δ, contain points belonging and points not belonging to S. A nonempty set S is said to be *open* if all the points belonging to S are interior points (i.e., if no boundary point belongs to the set); S is closed if all the boundary points belong to the set. A set S is said to be *limited* if (for all the points belonging to it) a neighborhood of finite radius including all the points of the set can be found:

$$\forall \boldsymbol{x} \in S\ \exists \delta > 0,\ \delta \text{ finite: } S \subseteq N_\delta(\boldsymbol{x})$$

A closed and limited subset of E^n is *compact* (and vice versa).

For example, the set $S = \{(x_1, x_2) : x_1^2 + x_2^2 \leq 1\}$ of the points belonging to the circle with unitary radius and center in the origin is a closed and limited set, and the set $S_1 = \{(x_1, x_2) : x_1^2 + x_2^2 < 1\}$ is an open and limited set. The boundary of S and S_1 consists of the set $S_2 = \{(x_1, x_2) : x_1^2 + x_2^2 = 1\}$.

Given two points $\boldsymbol{x}_1$ and $\boldsymbol{x}_2$, the set of points $\boldsymbol{x}$ defined by:

$$\left\{\boldsymbol{x} : \boldsymbol{x} = \mu\boldsymbol{x}_1 + (1 - \mu)\boldsymbol{x}_2,\ \mu \in [0, 1]\right\}$$

Giulio Erberto Cantarella is co-author of this appendix.

E. Cascetta, *Transportation Systems Analysis*,
Springer Optimization and Its Applications 29,
DOI 10.1007/978-0-387-75857-2,

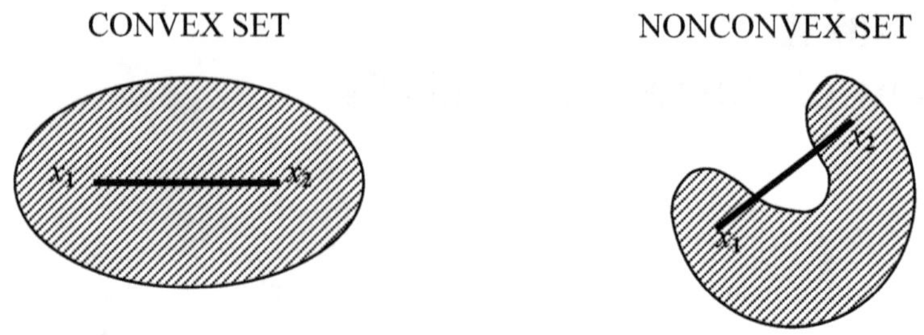

Fig. A.1 Illustration of convex and nonconvex sets

is called a segment of extremes $\boldsymbol{x}_1$ and $\boldsymbol{x}_2$.

A nonempty set S is said to be *convex* if all the points of the segment joining any two points belonging to the set, belong to the set itself (Fig. A.1).

$$\boldsymbol{x} = \mu \boldsymbol{x}_1 + (1-\mu)\boldsymbol{x}_2 \in S \quad \forall \mu \in [0,1],\ \forall \boldsymbol{x}_1, \boldsymbol{x}_2 \in S \tag{A.1.1}$$

The intersection of convex sets is a convex set. Sets defined by a system of linear equalities and/or inequalities, also known as *polyhedral sets*, are convex sets.

In fact, given $S = \{\boldsymbol{x} : \boldsymbol{A}\boldsymbol{x} \leq \boldsymbol{b}\}$ if $\boldsymbol{x}_1$ and $\boldsymbol{x}_2$ belong to S, let $\boldsymbol{x}$ be any point belonging to the segment with extremes $\boldsymbol{x}_1$ and $\boldsymbol{x}_2$, it yields:

$$\begin{aligned} \boldsymbol{x} &= \mu \boldsymbol{x}_1 + (1-\mu)\boldsymbol{x}_2 \quad \mu \in [0,1], \quad \text{hence} \\ \boldsymbol{A}\boldsymbol{x} &= \boldsymbol{A}\big(\mu \boldsymbol{x}_1 + (1-\mu)\boldsymbol{x}_2\big) = \mu \boldsymbol{A}\boldsymbol{x}_1 + (1-\mu)\boldsymbol{A}\boldsymbol{x}_2 \leq \mu \boldsymbol{b} + (1-\mu)\boldsymbol{b} \\ &= \boldsymbol{b} \mu \in [0,1] \end{aligned}$$

Thus, point $\boldsymbol{x}$ belongs to S. An analogous demonstration can be repeated for the set $S \equiv \{\boldsymbol{x} : \boldsymbol{A}\boldsymbol{x} = \boldsymbol{b}\}$.

Given a point $\boldsymbol{x}^*$, for each nonnull vector, call direction, $\boldsymbol{h} \neq \boldsymbol{0}$, the set of points lying on the half-line of origin $\boldsymbol{x}^*$ and direction defined by the vector $\boldsymbol{h}$ is *a ray* emanating from $\boldsymbol{x}^*$ along direction $\boldsymbol{h}$. This set is formally defined by:

$$\{\boldsymbol{x} : \boldsymbol{x} = \boldsymbol{x}^* + \mu \boldsymbol{h},\ \mu \geq 0\}$$

A vector $\boldsymbol{h}$ is a feasible direction at the point $\boldsymbol{x}^*$ for the set S, if it is possible to move along the direction of a finite quantity from the point $\boldsymbol{x}^*$ and remain within the set S:

$$\exists \mu^* > 0 : \boldsymbol{x} = \boldsymbol{x}^* + \mu \boldsymbol{h} \in S \quad \forall \mu < \mu^*, \mu \geq 0$$

Given a set S, the set $D(\boldsymbol{x}^*)$ of the feasible directions at a point $\boldsymbol{x}^*$ belonging to S (Fig. A.2) is formally defined as

$$D(\boldsymbol{x}^*) = \{\boldsymbol{h} \neq \boldsymbol{0} : \exists \mu^* > 0 : \boldsymbol{x} = \boldsymbol{x}^* + \mu \boldsymbol{h} \in S\ \forall \mu < \mu^*,\ \mu \geq 0\}$$

For a convex set S, the set of feasible directions at point $\boldsymbol{x}^*$ can also be defined as

$$D(\boldsymbol{x}^*) = \big\{\boldsymbol{h} = (\boldsymbol{x} - \boldsymbol{x}^*) : \boldsymbol{x} \in S\big\}$$

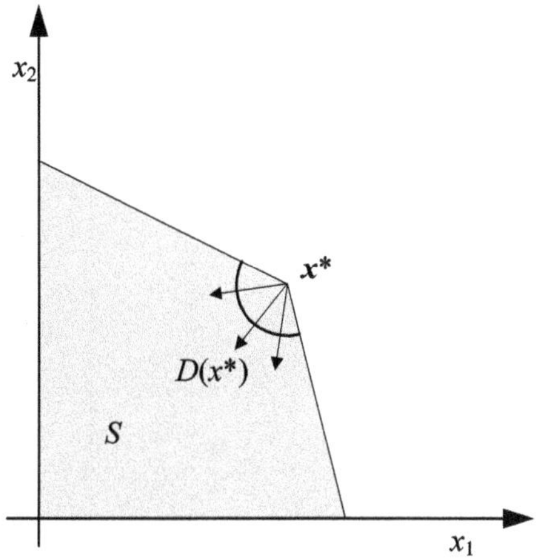

Fig. A.2 Illustration of feasible directions

A.1.2 Continuous and Differentiable Functions

A scalar-valued function of a vector $y = f(\boldsymbol{x})$, with values in E^1 and defined on an open set $S \subseteq E^n$ is said to be continuous at point $\boldsymbol{x}^* \in S$ if small variations of the variables $\boldsymbol{x}$ induce small variations in the variable y. Formally, the function $y = f(\boldsymbol{x})$ is said to be continuous at point $\boldsymbol{x}^*$ if for any neighborhood $N_\delta(y^*)$ of the point $y^* = f(\boldsymbol{x}^*)$, however small, there is a neighborhood $N_\varepsilon(\boldsymbol{x}^*)$ of point $\boldsymbol{x}^*$ such that the points $\boldsymbol{x}$ belonging to them have values $y = f(\boldsymbol{x})$ in the neighborhood of y^*:

$$\forall\delta > 0,\ \exists\varepsilon > 0 : y = f(\boldsymbol{x}) \in N_\delta\big(y^* = f(\boldsymbol{x}^*)\big) \quad \forall\boldsymbol{x} \in N_\varepsilon(\boldsymbol{x}^*)$$

A scalar function of vector $f(\boldsymbol{x})$ with values in E^1 and defined on a closed set $S \subseteq E^n$ is said to be *differentiable* at the point $\boldsymbol{x}^* \in S$ if there is a vector, known as the *gradient* of the function in the point and denoted by $\nabla f(\boldsymbol{x}^*)$, such that the difference between the value of the function at any point $\boldsymbol{x} \in S$ and its linear approximation in $\boldsymbol{x}^*$ along $\nabla f(\boldsymbol{x}^*)$, given by $f(\boldsymbol{x}^*) + \nabla f(\boldsymbol{x}^*)^T(\boldsymbol{x} - \boldsymbol{x}^*)$, is an infinitesimal of superior order with respect to the norm of the vector $(\boldsymbol{x} - \boldsymbol{x}^*)$:

$$\lim_{\boldsymbol{x}\to\boldsymbol{x}^*} \frac{f(\boldsymbol{x}) - f(\boldsymbol{x}^*) - \nabla f(\boldsymbol{x}^*)^T(\boldsymbol{x} - \boldsymbol{x}^*)}{\|\boldsymbol{x} - \boldsymbol{x}^*\|} = 0 \quad \forall\boldsymbol{x} \in S \tag{A.1.2}$$

The components of the vector $\nabla f(\boldsymbol{x}^*)$ are the partial derivatives of the function:

$$\nabla f(\boldsymbol{x}^*)^T = \left[\frac{\partial f(\boldsymbol{x}^*)}{\partial x_1}, \frac{\partial f(\boldsymbol{x}^*)}{\partial x_2}, \ldots, \frac{\partial f(\boldsymbol{x}^*)}{\partial x_n}\right] \tag{A.1.3}$$

A function with continuous first partial derivatives can be proved differentiable, and also continuous.

The gradient of a function can be represented in the space E^n with the same dimensionality of the definition set S. In the same space the level curves of the

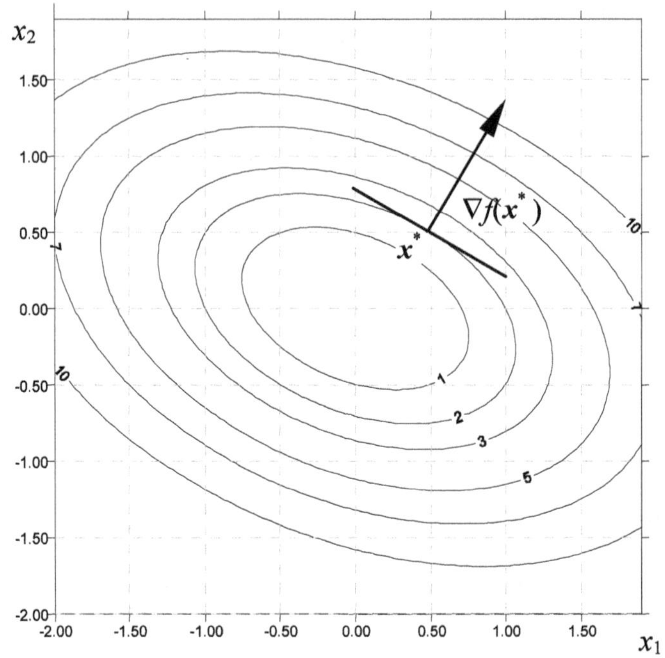

Fig. A.3 Level curves and gradient

function (loci of the points $\boldsymbol{x}$ to which the same value of $f(\boldsymbol{x})$ corresponds) can be defined. The gradient at each point $\boldsymbol{x}^*$ is a vector perpendicular to the tangent at the level curve $f(\boldsymbol{x}^*)$ and, as shown, points towards increasing values of the function (see Fig. A.3).

Given a scalar function $f(\boldsymbol{x})$, defined in S, a point $\boldsymbol{x}^* \in S$ and a direction vector $\boldsymbol{h}$ such that $\boldsymbol{x}^* + \mu\boldsymbol{h} \in S$ for values of μ less than μ^*, the *directional derivative* of the function in $\boldsymbol{x}^*$ along direction $\boldsymbol{h}$ can be defined as the limit:

$$f'(\boldsymbol{x}^*, \boldsymbol{h}) = \lim_{\mu \to 0} \frac{f(\boldsymbol{x}^* + \mu\boldsymbol{h}) - f(\boldsymbol{x}^*)}{\mu} \tag{A.1.4}$$

If $f(\boldsymbol{x})$ is differentiable in $\boldsymbol{x}^*$, it is rather easy to demonstrate that the directional derivative can be expressed in terms of the gradient:

$$f'(\boldsymbol{x}^*, \boldsymbol{h}) = \nabla f(\boldsymbol{x}^*)^T \boldsymbol{h} \tag{A.1.5}$$

A direction $\boldsymbol{h}$ along which it is possible to move by a finite quantity starting from $\boldsymbol{x}^*$, increasing the value of the function, at least in a neighborhood of $\boldsymbol{x}^*$, is known as an ascent direction. In other words, a direction $\boldsymbol{h}$ is an ascent direction if a positive scalar θ^* can be found such that for each $0 < \theta < \theta^*$ it results that:

$$f(\boldsymbol{x}^* + \theta\boldsymbol{h}) > f(\boldsymbol{x}^*) \tag{A.1.6}$$

It can be demonstrated (by using the property of the directional derivative (A.1.5) and the theorem of sign permanence) that a direction $\boldsymbol{h}$ is an ascent direction if and only if the directional derivative of $f(\boldsymbol{x})$ at point $\boldsymbol{x}^*$ along direction $\boldsymbol{h}$ is positive:

$$f'(\boldsymbol{x}^*, \boldsymbol{h}) = \nabla f(\boldsymbol{x}^*)^T \boldsymbol{h} > 0 \tag{A.1.7}$$

Similarly, the directions along which it is possible to move starting from $\boldsymbol{x}^*$, causing a decrease in the function value, are known as descent directions, and have negative directional derivative at $\nabla f(\boldsymbol{x}^*)^T \boldsymbol{h} < 0$.

The gradient of a differentiable function, at whatever point it differs from zero, is an ascent direction. In fact, under the assumptions made, it results that:

$$f'\left(\boldsymbol{x}^*, \nabla f(\boldsymbol{x}^*)\right) = \nabla f(\boldsymbol{x}^*)^T \nabla f(\boldsymbol{x}^*) = \left\| \nabla f(\boldsymbol{x}^*) \right\|^2 > 0 \tag{A.1.8}$$

Vice versa, the direction opposite to the gradient $-\nabla f(\boldsymbol{x}^*)$, if different from zero, is a descent direction of $f(\boldsymbol{x})$ in $\boldsymbol{x}^*$.

A scalar function $f(\boldsymbol{x})$ is said to be *doubly* or *twice differentiable* in $\boldsymbol{x}$ if there is a vector $\nabla f(\boldsymbol{x}^*)$ and a symmetric matrix $\boldsymbol{H}_f(\boldsymbol{x}^*)$ of dimensions $(n \times n)$ such that:

$$\lim_{x \to x^*} \frac{f(\boldsymbol{x}) - f(\boldsymbol{x}^*) - \nabla f(\boldsymbol{x}^*)^T (\boldsymbol{x} - \boldsymbol{x}^*) - 1/2 (\boldsymbol{x} - \boldsymbol{x}^*)^T \boldsymbol{H}_f(\boldsymbol{x}^*)(\boldsymbol{x} - \boldsymbol{x}^*)}{\|\boldsymbol{x} - \boldsymbol{x}^*\|^2} = 0$$
$$\forall \boldsymbol{x} \in S \tag{A.1.9}$$

Equation (A.1.9) expresses the condition that the difference between the value of the function and its quadratic approximation is an infinitesimal of superior order with respect to the square norm of the vector $(\boldsymbol{x} - \boldsymbol{x}^*)$. The matrix $\boldsymbol{H}_f(\boldsymbol{x}^*)$ is called the *Hessian matrix* of $f(\boldsymbol{x})$ at $\boldsymbol{x}^*$ and its components are the second-order partial derivatives of $f(\boldsymbol{x})$ at $\boldsymbol{x}^*$:

$$\boldsymbol{H}_f(\boldsymbol{x}^*) = \begin{vmatrix} \frac{\partial^2 f(\boldsymbol{x}^*)}{\partial \boldsymbol{x}_1^2} & \cdots & \frac{\partial^2 f(\boldsymbol{x}^*)}{\partial \boldsymbol{x}_1 \partial \boldsymbol{x}_n} \\ \cdots & & \cdots \\ \cdots & & \cdots \\ \cdots & & \cdots \\ \frac{\partial^2 f(\boldsymbol{x}^*)}{\partial \boldsymbol{x}_1 \partial \boldsymbol{x}_n} & \cdots & \frac{\partial^2 f(\boldsymbol{x}^*)}{\partial \boldsymbol{x}_n^2} \end{vmatrix} \tag{A.1.10}$$

A function can be proved doubly differentiable if it has continuous second partial derivatives. In this case the first partial derivatives are differentiable (because they have continuous partial derivatives), thus the function is differentiable and therefore continuous. Furthermore, the second partial derivatives do not depend on the order of derivation and the Hessian matrix is symmetric.

Taylor's formulae of the first- and second-order relative to the scalar function $f(\boldsymbol{x})$ around the point $\boldsymbol{x}^*$ are, respectively,

$$\exists \boldsymbol{x}^\circ \in (\boldsymbol{x}^*, \boldsymbol{x}) : f(\boldsymbol{x}) = f(\boldsymbol{x}^*) + \nabla f(\boldsymbol{x}^\circ)^T (\boldsymbol{x} - \boldsymbol{x}^*) \quad \forall \boldsymbol{x} \in S \tag{A.1.11}$$

$$\exists \boldsymbol{x}^\circ \in (\boldsymbol{x}^*, \boldsymbol{x}) : f(\boldsymbol{x}) = f(\boldsymbol{x}^*) + \nabla f(\boldsymbol{x}^*)^T(\boldsymbol{x} - \boldsymbol{x}^*) + 1/2(\boldsymbol{x} - \boldsymbol{x}^*)^T \times \boldsymbol{H}_f(\boldsymbol{x}^\circ)(\boldsymbol{x} - \boldsymbol{x}^*) \quad \forall \boldsymbol{x} \in S \tag{A.1.12}$$

where $\boldsymbol{x}^\circ$ is a point within the segment $(\boldsymbol{x}^*, \boldsymbol{x})$. Equations (A.1.11) and (A.1.12) obviously require $f(\boldsymbol{x})$ to be differentiable of the first- and second-order, respectively.

An m-vectorial function $\boldsymbol{g}(\boldsymbol{x})$ associates a vector of m components, that is, a point of E^m, to an n-dimensional vector, that is, a point of E^n; it is a vector in m functions:

$$\boldsymbol{g}(\boldsymbol{x}) = \left[g_1(\boldsymbol{x}), g_2(\boldsymbol{x}), \ldots, g_m(\boldsymbol{x})\right]^T$$

that associates with each n-dimensional vector $\boldsymbol{x} \in S$ an m-dimensional vector $\boldsymbol{g}(\boldsymbol{x})$.

The function $\boldsymbol{g}(\boldsymbol{x})$ is said to be differentiable at point $\boldsymbol{x}^*$ if all its component functions are differentiable. The *Jacobian matrix* of $\boldsymbol{g}(\boldsymbol{x})$ is a matrix of dimensions $(m \times n)$ that has the gradients of the component functions $g_i(\boldsymbol{x})$ as its rows:

$$\boldsymbol{Jac}\left[\boldsymbol{g}(\boldsymbol{x}^*)\right] = \begin{vmatrix} \frac{\partial g_1(\boldsymbol{x}^*)}{\partial x_1} & \cdots & \frac{\partial g_1(\boldsymbol{x}^*)}{\partial x_n} \\ \cdots & & \cdots \\ \cdots & & \cdots \\ \cdots & & \cdots \\ \frac{\partial g_m(\boldsymbol{x}^*)}{\partial x_1} & \cdots & \frac{\partial g_m(\boldsymbol{x}^*)}{\partial x_n} \end{vmatrix} \tag{A.1.13}$$

An n-vectorial function of vector $\boldsymbol{g}(\boldsymbol{x})$ (in this case $m = n$) defined in a set $S \subseteq E^n$ is strictly increasing monotone if for each pair of different points $\boldsymbol{x}_1 \neq \boldsymbol{x}_2 \in S$ it results that:

$$\left(\boldsymbol{g}(\boldsymbol{x}_1) - \boldsymbol{g}(\boldsymbol{x}_2)\right)^T (\boldsymbol{x}_1 - \boldsymbol{x}_2) > 0 \quad \forall \boldsymbol{x}_1 \neq \boldsymbol{x}_2 \in S \tag{A.1.14}$$

The function is said to be nondecreasing monotone if weak inequality holds (≥ 0). Similarly, functions can be denoted as strictly decreasing or nonincreasing monotone if the reversed inequalities hold. If the two points, $\boldsymbol{x}_1 \neq \boldsymbol{x}_2 \in S$, differ only in the ith component, that is, $x_{1,i} \neq x_{2,i}$, with $x_{1,j} = x_{2,j}$ $\forall j \neq i$, from inequality (A.1.14) it follows that:

$$\left(g_i(x_{1,i}) - g_i(x_{2,i})\right)^T (x_{1,i} - x_{2,i}) > 0$$

Hence all the component functions are increasing monotone functions of every component of the vector $\boldsymbol{x}$ for given values of all other components (scalar functions of scalar).

If the Jacobian matrix $\boldsymbol{Jac}[\boldsymbol{g}(\boldsymbol{x})]$ of the function $\boldsymbol{g}(\boldsymbol{x})$, assumed to be differentiable, is positive (negative) semidefinite over the whole set of definition S, the function $\boldsymbol{g}(\boldsymbol{x})$ is nondecreasing (nonincreasing monotone). If the Jacobian is positive (negative) definite, the function is monotone strictly increasing (decreasing).

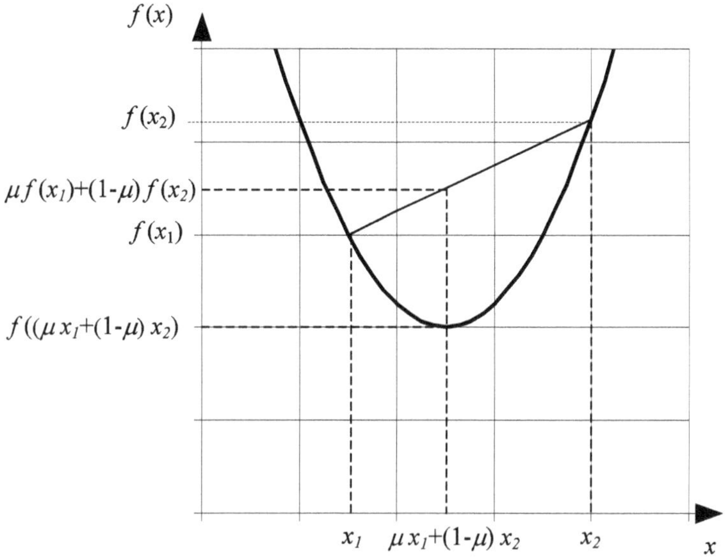

Fig. A.4 Geometrical interpretation of the definition of convex function

A.1.3 Convex Functions

A scalar function of vector $f(\boldsymbol{x})$ defined in the convex set $S \subseteq E^n$ is termed convex if for any pair of points $\boldsymbol{x}_1$ and $\boldsymbol{x}_2$ belonging to S the following relationship holds.

$$f\left(\mu \boldsymbol{x}_1 + (1-\mu)\boldsymbol{x}_2\right) \leq \mu f(\boldsymbol{x}_1) + (1-\mu) f(\boldsymbol{x}_2) \quad \forall \mu \in [0, 1] \qquad \text{(A.1.15)}$$

The geometrical interpretation of (A.1.15) is that the value of the function calculated at whatever point of the segment joining $\boldsymbol{x}_1$ and $\boldsymbol{x}_2$ is not greater than the linear combination of the values calculated at the endpoints (see Fig. A.4).

It can be demonstrated that a differentiable function is convex if and only if it satisfies the following condition,

$$f(\boldsymbol{x}_2) \geq f(\boldsymbol{x}_1) + \nabla f(\boldsymbol{x}_1)^T (\boldsymbol{x}_2 - \boldsymbol{x}_1) \quad \forall \boldsymbol{x}_2, \boldsymbol{x}_1 \in S \subseteq E^n \qquad \text{(A.1.16)}$$

that is, if the value of the function in $\boldsymbol{x}_2$ is not lower than the value of its linear extrapolation starting from $\boldsymbol{x}$ (see Fig. A.5). By inverting points $\boldsymbol{x}_1$ and $\boldsymbol{x}_2$ in (A.1.16) and summing the two expressions we also get:

$$\left(\nabla f(\boldsymbol{x}_1) - \nabla f(\boldsymbol{x}_2)\right)^T (\boldsymbol{x}_1 - \boldsymbol{x}_2) \geq 0 \qquad \text{(A.1.17)}$$

that is, the gradient of a convex differentiable function is a monotone nondecreasing vectorial function of the vector $\boldsymbol{x}$.

It can also be shown that the necessary and sufficient condition for a doubly differentiable function to be convex is that its Hessian matrix is positive semidefinite

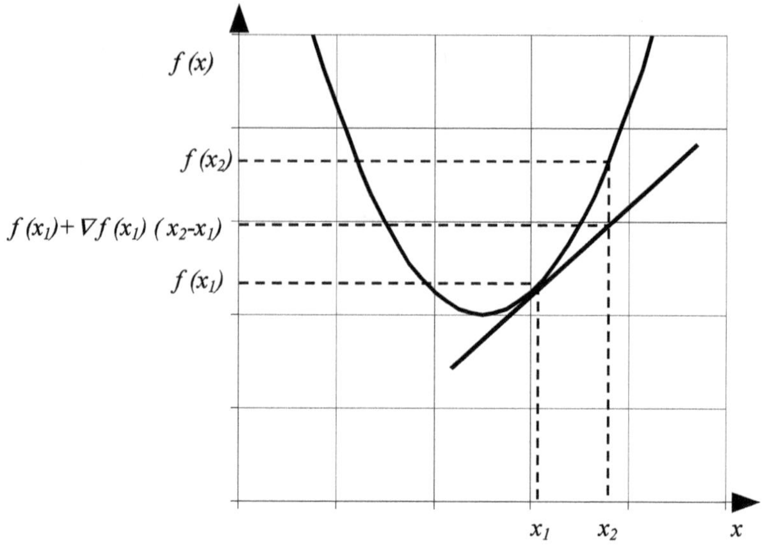

Fig. A.5 Geometrical interpretation of the convexity of a differentiable function

over the whole set of definition S:

$$\boldsymbol{x}^T \boldsymbol{H}(\boldsymbol{x}^*)\boldsymbol{x} \leq 0 \quad \forall \boldsymbol{x}, \boldsymbol{x}^* \in S \tag{A.1.18}$$

If the inequalities (A.1.15), (A.1.16), and (A.1.17) and (A.1.18) hold with the sign of strict inequality, the function is said to be strictly convex.

A function $f(\boldsymbol{x})$ given by a linear combination with positive coefficients of convex functions $f^i(\boldsymbol{x})$ is convex:

$$f(\boldsymbol{x}) = \sum_i \mu_i f^i(\boldsymbol{x}) \quad \mu_i > 0$$

It is also strictly convex if at least one of the component functions is strictly convex.

The function $f(\boldsymbol{x})$ is said to be (strictly) concave if $-f(\boldsymbol{x})$ is (strictly) convex. In this case $f(\boldsymbol{x})$ verifies (A.1.15) through (A.1.18) with the inequalities inverted. A linear function is both convex and concave because (A.1.15) holds with the sign of equality.

A.2 Solution Algorithms

A mathematical problem with a solution given by a vector $\boldsymbol{x}^* \in S$ is termed solvable in closed form if there is a relationship allowing the calculation of the solution (or solutions) of the problem as a function of the parameters of the problem itself.

Consider, for example, the problem of searching for the null points of a function $f(\boldsymbol{x})$ in the set S, that is, the problem of solving the equation $f(\boldsymbol{x}) = 0$, with the condition that the solutions belong to the set S. In the case of a second-order polynomial function, $ax^2 + bx + c$, the null points are a solution to the equation: $ax^2 + bx + c = 0$. The equation is known to have two solutions x_1 and x_2 (real or conjugate complex) within E, which can be calculated in closed form by means of the formula: $x_{1,2} = (-b \pm (b^2 - 4ac)^{1/2})/(2a)$; then it can be verified whether any of them belongs to set S.

More in general, when a closed-form solution cannot be found, recursive equations generating a succession of points $\{\boldsymbol{x}^1, \ldots, \boldsymbol{x}^k, \boldsymbol{x}^{k+1}, \ldots\}$ are adopted; that is,

$$\boldsymbol{x}^{k+1} = \varphi(\boldsymbol{x}^k) \tag{A.2.1}$$

Equation (A.2.1) defines an algorithm solving the problem, if the recursive equation stops in the solution being sought $\boldsymbol{x}^*$:

$$\boldsymbol{x}^* = \varphi(\boldsymbol{x}^*)$$

and vice versa if it is found that the point at which the equation stops $\boldsymbol{x}^* = \varphi(\boldsymbol{x}^*)$ is the solution sought.

An algorithm is said to be feasible if all the elements of the succession belong to the set of feasible solutions, $\boldsymbol{x}^k \in S$. An algorithm is *finitely convergent* if it can be demonstrated that there is a finite number n such that $\boldsymbol{x}^n = \boldsymbol{x}^*$ (a closed form solution therefore is an algorithm convergent in one step). A resolutive algorithm is said to be *asymptotically convergent* if it can be demonstrated that the succession of points converges to the solution sought: that is, $\lim_{k\to\infty} \boldsymbol{x}^k = \boldsymbol{x}^*$. If no form of convergence can be demonstrated, the issue remains open and the algorithm is to be considered heuristic, as the algorithms for which nonconvergence can be proven (at least for some instances of the problem).

A.3 Fixed-Point Problems

Let $\psi(\boldsymbol{x})$ be an n-vectorial function of a vector $\boldsymbol{x}$ defined in a set $S \subseteq E^n$, with values in the set $T = \psi(S) = \{\psi(\boldsymbol{x}) : \boldsymbol{x} \in S\} \subseteq E^n$; the point $\boldsymbol{x}^* \in S$ is called a fixed-point if the function has a value equal to the argument (see Fig. A.6):

$$\boldsymbol{x}^* = \psi(\boldsymbol{x}^*) \quad \boldsymbol{x}^* \in S \tag{A.3.1}$$

(Note that specifying a solution algorithm for any mathematical problem based on the recursive equation (A.2.1) is equivalent to defining a function $\varphi(x)$ having as its fixed-point the solution of the mathematical problem under study.) Fixed-point problems, found in various branches of engineering and economics, can easily be related to nonlinear systems of equations (and vice versa):

$$\boldsymbol{x}^* - \psi(\boldsymbol{x}^*) = 0 \quad \boldsymbol{x}^* \in S$$

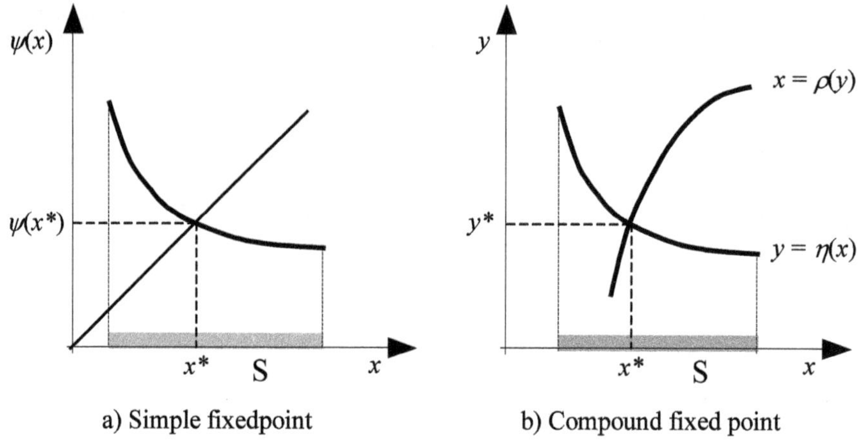

Fig. A.6 Simple and compound fixed-points

A particularly interesting case of the fixed-point problem, called the *compound fixed-point* problem, is identified in the search for equilibrium configurations between two vectors, $\boldsymbol{x} \in S_x \subseteq E^n$ and $\boldsymbol{y} \in S_y \subseteq E^m$ (also with $n \neq m$) which reciprocally influence each other (see Fig. A.6b); that is:

$$\begin{cases} \boldsymbol{y}^* = \eta(\boldsymbol{x}^*) & \boldsymbol{x}^* \in S_x \ \boldsymbol{y}^* \in S_y \\ \boldsymbol{x}^* = \rho(\boldsymbol{y}^*) & \boldsymbol{y}^* \in S_y \ \boldsymbol{x}^* \in S_x \end{cases} \tag{A.3.2}$$

In fact, by combining the previous relationships, a compound fixed-point problem in the variable $\boldsymbol{x}$ is obtained:

$$\boldsymbol{x}^* = \rho\big(\eta(\boldsymbol{x}^*)\big) \quad \boldsymbol{x}^* \in S_x \tag{A.3.3a}$$

with $\eta(S_x) \subseteq S_y$ and $\rho(\eta(S_x)) \subseteq S_x$. Similarly, an equivalent[1] fixed-point problem in the variable $\boldsymbol{y}$ can be defined:

$$\boldsymbol{y}^* = \eta\big(\rho(\boldsymbol{y}^*)\big) \quad \boldsymbol{y}^* \in S_y \tag{A.3.3b}$$

with $\rho(S_y) \subseteq S_x$ and $\eta(\rho(S_y)) \subseteq S_y$.

The properties of the nonlinear equations system (A.3.2) or of each of the two compound fixed-point problems (A.3.3a) and (A.3.3b) depend on the characteristics of the two functions involved, $\boldsymbol{y} = \eta(\boldsymbol{x})$ and $\boldsymbol{x} = \rho(\boldsymbol{y})$, and on the sets of definition of the variables, S_x and S_y.

[1] Two mathematical problems are said to be equivalent if the solutions of one problem are also solutions of the other and vice versa. In this case, analysis of the theoretical properties of the solutions such as their existence and uniqueness, and the convergence analysis of resolutive algorithms can be carried out for only one of the two problems.

A.3.1 Properties of Fixed-Points

Sufficient conditions for the existence and uniqueness of the solution of a fixed-point problem are given by the well-known Banach's theorem[2] which also allows the specification of an asymptotically convergent resolutive algorithm; only a restricted class of functions, however, satisfies these conditions. What follows, therefore, describes weaker conditions (some of these conditions can be extended with some mathematical complications).

Sufficient conditions for the existence of at least one solution of the fixed-point problem (A.3.1), that is, for the existence of at least one fixed-point of a function, are given by Brouwer's theorem stated below.

Brouwer's theorem *The fixed-point problem* (A.3.1) *has at least one solution; that is, the function* $\psi(\boldsymbol{x})$ *defined in the set* $S \subseteq E^n$ *with values in the set* $T = \psi(S) \subseteq E^n$ *has at least one fixed-point if:*

T	*is a subset of* $S, T \subseteq S$, *i.e.* $\psi(\boldsymbol{x}) \in S \;\forall \boldsymbol{x} \in S$
S	*is a nonempty compact and convex set*
$\psi(\boldsymbol{x})$	*is a continuous function*

Application of Brouwer's theorem to compound fixed-point problems, such as that defined by (A.3.3a), requires both the functions $\eta(\boldsymbol{x})$ and $\rho(\boldsymbol{y})$ to be continuous, the definition set to be a nonempty, compact, and convex set, and $S_x \subseteq \rho(\eta(S_x))$; that is, $\rho(\eta(\boldsymbol{x})) \in S_x \;\forall \boldsymbol{x} \in S_x$.

A graphic illustration of the relevance of some of the assumptions of Brouwer's theorem is given in Fig. A.7, for a simple case.

Sufficient conditions for the uniqueness of the solution of the fixed-point problem (A.3.1), that is, for the existence of at most one fixed-point of a function, are given by the simple theorem described below.

Theorem *The fixed-point problem* (A.3.1) *has at most one solution; that is, the function defined in the set* $\psi(\boldsymbol{x})$ *with values in the set* $S \subseteq E^n$ *has at most one fixed-point, if* $T = \psi(\boldsymbol{x}) \subseteq E^n$, *and* $\psi(\boldsymbol{x})$ *is a monotone nonincreasing*[3] *function, over the whole set* S:

$$\left(\psi(\boldsymbol{x}') - \psi(\boldsymbol{x}'')\right)^T (\boldsymbol{x}' - \boldsymbol{x}'') \le 0 \quad \forall \boldsymbol{x}', \boldsymbol{x}'' \in S$$

In fact, if there existed two different fixed-point vectors, $\boldsymbol{x}_1^* \neq \boldsymbol{x}_2^*$, that is, $\boldsymbol{x}_1^* = \psi(\boldsymbol{x}_1^*) \in S, \boldsymbol{x}_2^* = \psi(\boldsymbol{x}_2^*) \in S$ for the monotonicity of the function $\psi(\boldsymbol{x})$, it would

[2]Banach's theorem requires the function $\psi(\boldsymbol{x})$, defined over S with values in $T \subseteq S$, to be a contraction (implying monotonicity) over a complete set (a weaker property than that of compactness), or that the function $\psi(\boldsymbol{x})$ is a quasi-contraction (implying monotonicity) over a compact set. Note that in both cases the function is continuous.

[3]In general, it is sufficient to have: $(\psi(\boldsymbol{x}') - \psi(\boldsymbol{x}''))^T (\boldsymbol{x}' - \boldsymbol{x}'') < (\boldsymbol{x}' - \boldsymbol{x}'')^T (x' - x'') \;\forall \boldsymbol{x}' \neq \boldsymbol{x}'' \in S$.

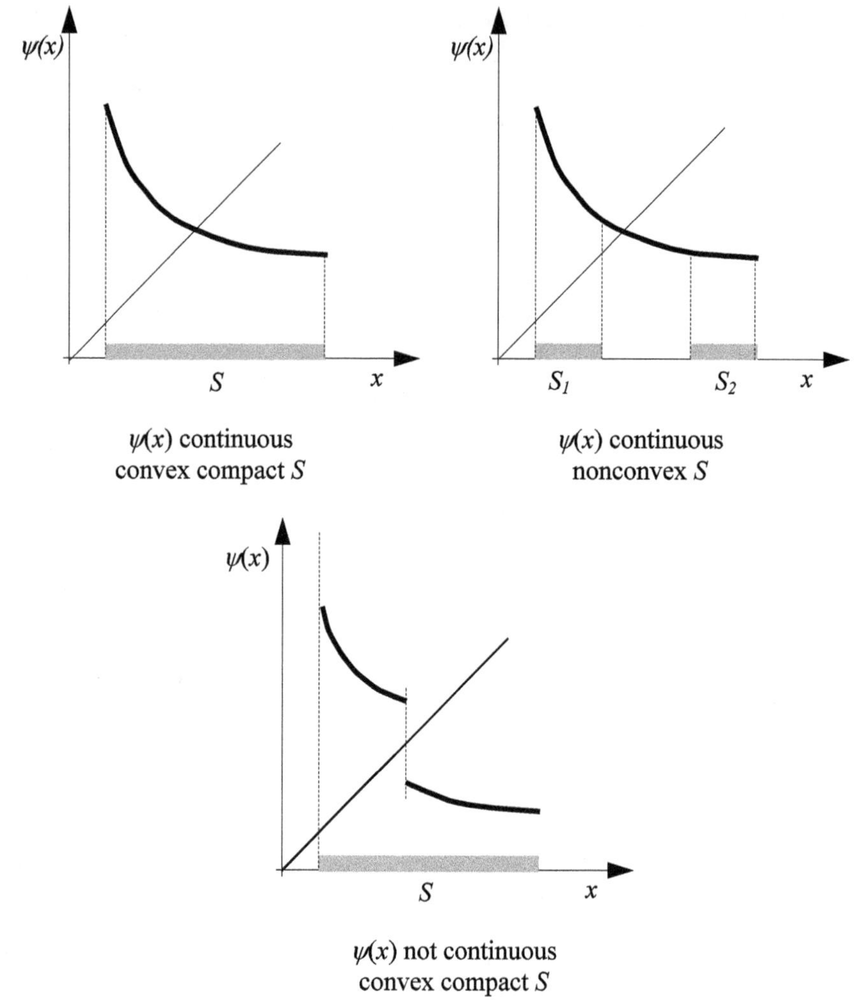

Fig. A.7 Illustration of the assumptions of Brouwer's theorem

follow that

$$\|(\boldsymbol{x}_1^* - \boldsymbol{x}_2^*)\|^2 = (\boldsymbol{x}_1^* - \boldsymbol{x}_2^*)^T(\boldsymbol{x}_1^* - \boldsymbol{x}_2^*) \leq 0$$

In contradiction to the condition $\|(\boldsymbol{x}_1^* - \boldsymbol{x}_2^*)\|^2 > 0$ for any $\boldsymbol{x}_1^* \neq \boldsymbol{x}_2^*$.

Uniqueness conditions can be extended to compound fixed-point problems, such as that defined by (A.3.3a), in the case of two monotone functions in the opposite direction (and at least one of the two strictly monotone), as shown in the theorem described below.

Theorem *The compound fixed-point problem* (A.3.3a) *has at most one solution, i.e. the compound function* $\psi(\mathbf{x}) = \rho(\eta(\mathbf{x}))$ *defined in the set* $S_x \subseteq E^n$ *with* $\eta(S_x) \subseteq S_y$ *and* $\rho(\eta(S_x)) \subseteq S_x$, *has at most one fixed-point if the two functions* $\rho(.)$ *and* $\eta(.)$ *are monotone in the opposite direction. For example*:

$\mathbf{y} = \eta(\mathbf{x})$ *is a strictly increasing function*; *that is,*

$$\left(\eta(\mathbf{x}') - \eta(\mathbf{x}'')\right)^T (\mathbf{x}' - \mathbf{x}'') > 0 \quad \forall \mathbf{x}' \neq \mathbf{x}'' \in S_x$$

$\mathbf{x} = \rho(\mathbf{y})$ *is a nonincreasing function*; *that is,*

$$\left(\rho(\mathbf{y}') - \rho(\mathbf{y}'')\right)^T (\mathbf{y}' - \mathbf{y}'') \leq 0 \quad \forall \mathbf{y}', \mathbf{y}'' \in \eta(S_x)$$

In fact, if there existed two different fixed-point vectors, $\mathbf{x}_1^* \neq \mathbf{x}_2^*$, that is, $\mathbf{x}_1^* = \rho(\eta(\mathbf{x}_1^*))$ and $\mathbf{x}_2^* = \rho(\eta(\mathbf{x}_2^*))$, denoted $\mathbf{y}_1^* = \eta(\mathbf{x}_1^*)$ and $\mathbf{y}_2^* = \rho(\mathbf{x}_2^*)$, from which $\mathbf{x}_1^* = \rho(\mathbf{y}_1^*)$ and $\mathbf{x}_2^* = \rho(\mathbf{y}_2^*)$, for the monotonicity of the function $\rho(\mathbf{y})$ it would follow that

$$\left(x_1^* - x_2^*\right)^T \left(y_1^* - y_2^*\right) = \left(\rho(y_1^*) - \rho(y_2^*)\right)^T \left(y_1^* - y_2^*\right) \leq 0$$

In contradiction to the monotonicity of function $\eta(\mathbf{x})$, for $\mathbf{x}_1^* \neq \mathbf{x}_2^*$:

$$\left(\eta(\mathbf{x}_1^*) - \eta(\mathbf{x}_2^*)\right)^T \left(\mathbf{x}_1^* - \mathbf{x}_2^*\right) = \left(\mathbf{y}_1^* - \mathbf{y}_2^*\right)^T \left(\mathbf{x}_1^* - \mathbf{x}_2^*\right) > 0$$

A.3.2 Solution Algorithms for Fixed-Point Problems

In general, solution algorithms for solving fixed-point problems are more recent and less developed than those for optimization problems, described in the next section.

Algorithms for fixed-point problems (A.3.1) are usually based on the explicit calculation of the Jacobian of the function $\psi(\mathbf{x})$, and eventually on the calculation of its eigenvalues, or an estimate of them. This approach is generally difficult to apply to large-scale problems; for this reason what follows describes some solution algorithms whose application requires only the calculation of the function $\psi(\mathbf{x})$. In particular, given a sequence $\{\mu_k\}_{k>0}$ satisfying the condition:

$$\sum_{k>0} \mu_k = \infty, \qquad \sum_{k>0} \mu_k^2 < \infty \tag{A.3.4}$$

an algorithm for the solution of a fixed-point problem can be specified by the following recursive equation,

$$\mathbf{x}^{k+1} = \mathbf{x}^k + \mu_k\left(\psi(\mathbf{x}^k) - \mathbf{x}^k\right) \quad \text{that is,} \quad \mathbf{x}^{k+1} = (1 - \mu_k)\mathbf{x}^k + \mu_k \psi(\mathbf{x}^k) \tag{A.3.5}$$

with $\mathbf{x}^1 \in S$.

By using Blum's theorem (not reported here because of its complexity), it can be demonstrated that if the function $\psi(\boldsymbol{x})$ has a unique fixed-point $\boldsymbol{x}^* = \eta(\boldsymbol{x}^*)$; the relationship (A.3.5) defines a sequence convergent[4] to the fixed-point $\boldsymbol{x}^*$. That is, $\lim_{k\to\infty} \boldsymbol{x}^k = \boldsymbol{x}^*$, if the function $\psi(\boldsymbol{x})$ is continuous and monotone nonincreasing and the set S is nonempty, compact, and convex (as required by the sufficient conditions of existence and uniqueness). From a practical point of view, the algorithm is stopped when $\boldsymbol{x}^k \cong \psi(\boldsymbol{x}^k)$, for example, when a norm value of the vector of components $(x_i^k - \psi_i(\boldsymbol{x}^k))/x_i^k$ is lower than a pre-assigned threshold. Stop tests based on the distance between values of the vector $\boldsymbol{x}$, between successive iterations (i.e., $\boldsymbol{x}^{k+1} \cong \boldsymbol{x}^k$) are to be avoided because this difference tends to zero because of the structure of the algorithm, regardless of the proximity to the solution of the fixed-point problem.

If the sequence $\{\mu_k\}_{k>0}$ also satisfies the condition

$$\mu_k \in (0, 1) \tag{A.3.6}$$

the elements of the sequence generated by the relationship (A.3.5) belong to the set $S, \boldsymbol{x}^k \in S$, S being convex. This property is especially useful in practical terms because it provides a feasible solution to the problem at whatever iteration the algorithm stops.

The sequence with the largest elements satisfying both conditions (A.3.4) and (A.3.6) is given by $\{\mu_k = 1/k\}_{k>0}$. In this case the relationship (A.3.5) leads to the so-called Method of Successive Averages or MSA:

$$\boldsymbol{x}^{k+1} = \boldsymbol{x}^k + (1/k)\big(\psi(\boldsymbol{x}^k) - \boldsymbol{x}^k\big) \in S \tag{A.3.7}$$

that is,

$$\boldsymbol{x}^{k+1} = \big((k-1)\boldsymbol{x}^k + \psi(\boldsymbol{x}^k)\big)/k \in S$$

with $\boldsymbol{x}^1 \in S$.

The above observations can also be applied to the compound fixed-point problem (A.3.3a). In this case the relationship (A.3.5) becomes:

$$\boldsymbol{x}^{k+1} = \boldsymbol{x}^k + \mu_k\big[\rho\big(\eta(\boldsymbol{x}^k)\big) - \boldsymbol{x}^k\big] \tag{A.3.8}$$

with $\boldsymbol{x}^1 \in S_x$.

If the function $\boldsymbol{x} = \rho(\boldsymbol{y})$ is continuous and monotone nonincreasing, the function $\boldsymbol{y} = \eta(\boldsymbol{x})$ is continuous and strictly monotone increasing, and the set S is nonempty, compact, and convex and the compound function $\rho(\eta(\boldsymbol{x}))$ has a unique fixed-point $\boldsymbol{x}^* = \rho(\eta(\boldsymbol{x}^*))$ (according to the sufficient conditions of existence and uniqueness). In this case by using Blum's theorem, it can be demonstrated that the relationship (A.3.8) defines a sequence convergent to the fixed-point $\boldsymbol{x}^*$ and if the function $\boldsymbol{y} = \eta(\boldsymbol{x})$ has a symmetrical and continuous Jacobian.

[4]If the function $\psi(\boldsymbol{x})$ is the realization of a random variable, and an unbiased estimate of its value is available, convergence is almost certain. (This is the most general of the cases originally analyzed in Blum's theorem.)

A.4 Optimization Problems

Optimal points $\boldsymbol{x}^*$ of a scalar function of a vector $f(\boldsymbol{x})$ are the points corresponding to minimum or maximum values of the function. For simplicity, what follows makes reference only to minimum points.[5] Formally, let $f(\boldsymbol{x})$ be a scalar function defined in a set $S \subseteq E^n$; the point $\boldsymbol{x}^*$ is called a local minimum point of the function if there is a neighborhood $N_\delta(\boldsymbol{x}^*)$ of radius δ such that the following condition holds.

$$f(\boldsymbol{x}) \geq f(\boldsymbol{x}^*) \quad \forall \boldsymbol{x} \neq \boldsymbol{x}^*,\ \boldsymbol{x} \in N_\delta(\boldsymbol{x}^*)$$

If this condition holds for all the points of S, point $\boldsymbol{x}^*$ is called a global minimum point of the function $f(\boldsymbol{x})$ over S. In general, a continuous function $f(\boldsymbol{x})$ over a compact set S always has at least one global minimum point. A function with a unique minimum point is termed *unimodal*; an example of this kind of function is given by the strictly convex functions defined previously.

The problem of the search for the minimum points $\boldsymbol{x}^*$ of a function is called a minimum or minimization problem, $f(\boldsymbol{x})$ the objective function, and S the feasibility set. The minimization problem is formally expressed as:

$$\begin{aligned} \boldsymbol{x}^* &= \operatorname{argmin} f(\boldsymbol{x}) \\ \boldsymbol{x} &\in S \end{aligned} \tag{A.4.1}$$

Minimization problems and fixed-point problems are related to each other. Indeed, the fixed-point problem (A.3.1) defined by the function $\psi(\boldsymbol{x})$ is equivalent to a minimum problem defined by the objective function with nonnegative values $f(\boldsymbol{x}) = (\psi(\boldsymbol{x}) - \boldsymbol{x})^T(\psi(\boldsymbol{x}) - \boldsymbol{x})$, $f(\boldsymbol{x})$ being 0 if and only if $\psi(\boldsymbol{x}) - \boldsymbol{x} = 0$.

The definition of local and global minimum points cannot be used in the search for such points inasmuch as it would require the calculation of $f(\boldsymbol{x})$ over all the points in S and comparison of their values. It is therefore essential to find necessary and/or sufficient conditions for the minimum points expressed in terms of "local" properties of the function. Such conditions are reported in the following by differentiating the case in which the minimum point is interior to an open set and that in which it may be on the boundary of a closed set.

A.4.1 Properties of Minimum Points

A.4.1.1 Properties of Minimum Points on Open Sets

A necessary condition for which point $\boldsymbol{x}^*$ is a local minimum for the differentiable function $f(\boldsymbol{x})$ defined in an open set S is that it is a point of stationarity of the objective function, that is, that in $\boldsymbol{x}^*$ we have $\nabla f(\boldsymbol{x}^*) = 0$. In fact, if $\boldsymbol{x}^*$ is a point interior

[5] The results reported and the algorithms described can easily be extended to maximum points, bearing in mind that the maximum points of a function correspond to the minimum points of the opposite function $-f(\boldsymbol{x})$.

to S, any direction is feasible. Furthermore, because $\boldsymbol{x}^*$ is a local minimum point, the directional derivative calculated in $\boldsymbol{x}^*$ must be nonnegative for any direction:

$$\nabla f(\boldsymbol{x}^*)^T \boldsymbol{h} \geq 0 \quad \forall \boldsymbol{h} \tag{A.4.2}$$

Because $-\nabla f(\boldsymbol{x}^*)$ is a feasible direction, the condition (A.4.2) holds only if the gradient is null.

Note that the nullity of the gradient in $\boldsymbol{x}^*$ is only a necessary condition for $\boldsymbol{x}^*$ to be a local minimum point. In particular, local maximum points also satisfy the same condition.

If $\boldsymbol{x}^*$ is a point of stationarity of the continuous and second-order differentiable function $f(\boldsymbol{x})$, with continuous first and second derivatives, the sufficient condition for $\boldsymbol{x}^*$ to be a local minimum is that the Hessian matrix in $\boldsymbol{x}^*$ is positive semidefinite. In fact, applying Taylor's second-order formula (A.1.12), it follows that

$$f(\boldsymbol{x}) = f(\boldsymbol{x}^*) + 1/2(\boldsymbol{x} - \boldsymbol{x}^*)^T \boldsymbol{H}_f(\boldsymbol{x}^\circ)(\boldsymbol{x} - \boldsymbol{x}^*) \tag{A.4.3}$$

where $\boldsymbol{x}^\circ$ is a point of the segment $(\boldsymbol{x}, \boldsymbol{x}^*)$. If $\boldsymbol{H}_f(\boldsymbol{x})$ is positive definite for the sign permanence theorem, a neighborhood of $\boldsymbol{x}^*$, $N_\delta(\boldsymbol{x}^*)$, can be found such that $\boldsymbol{H}_f(\boldsymbol{x})$ is positive definite at all points within this neighborhood. If $\boldsymbol{x}$ belongs to this neighborhood, all the points of the segment $(\boldsymbol{x}, \boldsymbol{x}^*)$ belong to it and so does $\boldsymbol{x}^\circ$. From this it follows that

$$1/2(\boldsymbol{x} - \boldsymbol{x}^*)^T \boldsymbol{H}_f(x^\circ)(\boldsymbol{x} - \boldsymbol{x}^*) \geq 0 \quad \Rightarrow \quad f(\boldsymbol{x}) \geq f(\boldsymbol{x}^*) \quad \forall \boldsymbol{x} \in N_\delta(\boldsymbol{x}) \tag{A.4.4}$$

If the function $f(\boldsymbol{x})$ is convex, the Hessian matrix is positive semidefinite in all S and from (A.4.4) it follows that the nullity of the gradient is a necessary and sufficient condition for $\boldsymbol{x}^*$ being a global minimum point. The minimum points of a convex function make up a convex set. Furthermore, if the function is strictly convex, a point of stationarity is also the unique global minimum point.

A.4.1.2 Properties of Minimum Points on Closed Sets

In general, the closed set S is defined by equality and/or inequality relationships, known as *constraints*. The case of m inequalities constraints is discussed below. Equality constraints can be reduced to two inequalities: $g_i(\boldsymbol{r}) = 0$ is equivalent to $g_i(\boldsymbol{x}) \leq 0$ and $-g_i(\boldsymbol{x}) \leq 0$ or used to reduce the number of variables:

$$S \equiv \left\{ \boldsymbol{x} : g_i(\boldsymbol{x}) \leq 0 \; i = 1, 2, \ldots, m \right\}$$

Using the m-vectorial function of vector $g(\boldsymbol{x})$, the constraints can be expressed as

$$\boldsymbol{g}(\boldsymbol{x}) \leq \boldsymbol{0}$$

With this notation the optimization problem can formally be expressed as

$$\min f(\boldsymbol{x})$$
$$\boldsymbol{g}(\boldsymbol{x}) \leq \boldsymbol{0}$$

Unlike the previous case, the minimum point might lie on the boundary of the set S. In this case, not all directions are feasible; in particular, the gradient may not be a feasible direction and the stationarity of the function in $\boldsymbol{x}^*$ does not have to be verified by a minimum point.

Denoting $D(\boldsymbol{x}^*)$, the set of feasible directions at the minimum point $\boldsymbol{x}^*$, because of the results (A.1.7) and (A.1.8) on directional derivatives, the function must have nonnegative directional derivatives in $\boldsymbol{x}^*$ for all the feasible directions:

$$\nabla f(\boldsymbol{x}^*)^T \boldsymbol{h} \geq 0 \quad \forall \boldsymbol{h} \in D(\boldsymbol{x}^*) \tag{A.4.5a}$$

If the set S is convex and $\boldsymbol{x}$ is a point belonging to S, the direction $(\boldsymbol{x} - \boldsymbol{x}^*)$ is feasible by definition and (A.4.5a) becomes:

$$\nabla f(\boldsymbol{x}^*)^T (\boldsymbol{x} - \boldsymbol{x}^*) \geq 0 \quad \forall \boldsymbol{x} \in S \tag{A.4.5b}$$

In general the two conditions (A.4.5a) or (A.4.5b) are only necessary for point $\boldsymbol{x}^*$ to be the minimum.

They are also sufficient if the objective function $f(\boldsymbol{x})$ is convex. Also in this case the minimum points of a convex function make up a convex set. In the case of a strictly convex function, there is a unique minimum point.

A.4.2 Solution Algorithms for Optimization Problems

This section describes some solution algorithms for particular optimization problems which have been mentioned in previous chapters.

A.4.2.1 Monodimensional Optimization Algorithms

These algorithms solve the problem of finding the minimum of a function $f(\theta)$ of a scalar variable θ. If the value θ of minimizing $f(\theta)$ in the interval $(\theta_{\min}, \theta_{\max})$ is indicated by θ^*, the monodimensional optimization problem can be expressed as follows.

$$\theta^* = \underset{\theta_{\min} \leq \theta \leq \theta_{\max}}{\operatorname{argmin}} \; f(\theta) \tag{A.4.6}$$

In practice, the problem (A.4.6) is rarely solved as such. However, it is an element common to many solution algorithms for more complex problems because, as shown, it allows us to obtain the minimum of a vector function along a direction $\boldsymbol{h}$ starting from a point $\boldsymbol{x}^*$. In this case, the points of the straight line passing from $\boldsymbol{x}^*$ oriented as the vector $\boldsymbol{h}$ are expressed as $\boldsymbol{x}^* + \theta \boldsymbol{h}$ and as θ varies, the points of the whole straight-line $(-\infty < \theta < +\infty)$ or of the half-line concordant with $\boldsymbol{h}$ $(\theta > 0)$ are described (see Sect. A.1.1).

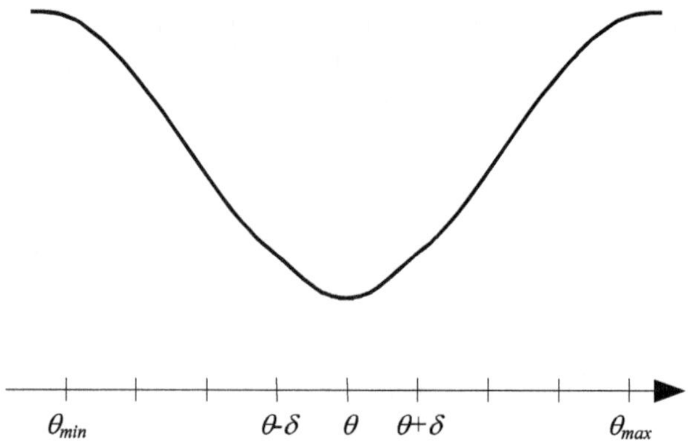

Fig. A.8 Uniform search algorithm

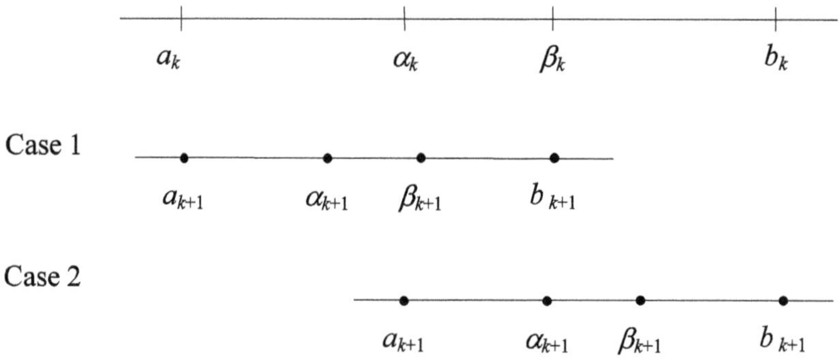

Fig. A.9 Illustration of the golden section algorithm

The most straightforward algorithm solving the problem (A.4.6) is the "uniform search." The interval $\theta_{\min}, \theta_{\max}$, is subdivided into subintervals of equal widths δ with extremes at the "grid points" $\theta_1 = \theta_{\min}, \theta_2, \ldots, \theta_n = \theta_{\max}$; the objective function is evaluated in each of the n points θ_k and θ^* is the point corresponding to the lowest value of the function (Fig. A.8). If the function is convex, the actual minimum point lies in the interval $\theta^* \pm \delta$.

More efficient algorithms for convex functions are based on the principle of "reduction of the uncertainty interval." At each iteration, an interval of extremes (a_k, b_k) is obtained which includes the minimum of the function, called the interval of uncertainty. The width of this interval is reduced at each iteration.

In the following, the main steps of one such algorithm, known as the "method of the golden section," is described. The name derives from its use of the property of the golden section of a segment to recompute the value of the $f(\theta)$ only once at each iteration (see Fig. A.9). The algorithm is asymptotically convergent if $f(.)$ is a convex function (even in a weaker sense than that described in Sect. A.1.3).

Golden Section Algorithm
(k is just the counter of iterations, irrelevant for computation)

The maximum width ε, allowed for the uncertainty interval is chosen.
The extremes of the initial interval are set:

$a_1 = \theta_{\min} b_1 = \theta_{\max}$

This gives the points:

$\alpha_1 = a_1 + 0.382(b_1 - a_1)$ and $\beta_1 = a_1 + 0.618(b_1 - a_1)$

The values of the function $f(\alpha_1)$ and $f(\beta_1)$ are calculated.

WHILE $(b_k - a_k) > \varepsilon$
  IF $f(\alpha_k)$ is greater than $f(\beta_k)$ (Case 2 in Fig. A.9) **THEN**
    Let $a_{k+1} = \alpha_k$ and $b_{k+1} = b_k$.
    By definition of the golden section, we have

$$\alpha_{k+1} = \beta_k,\ f(\alpha_{k+1}) = f(\beta_k), \text{ and } \beta_{k+1} = a_{k+1} + 0.618(b_{k+1} - a_{k+1}).$$

    Compute $f(\beta_{k+1})$.
OTHERWISE (Case 1 in Fig. A.9)
    Let $a_{k+1} = a_k$ and $b_{k+1} = \beta_k$.
    By definition of the golden section, we have

$$\beta_{k+1} = \alpha_k,\ f(\beta_{k+1}) = f(\alpha_k) \text{ and } \alpha_{k+1} = a_{k+1} + 0.382(b_{k+1} - a_{k+1}).$$

    Compute $f(\alpha_{k+1})$.
  END
END

The solution of the problem is $\theta^* = \frac{a_k + b_k}{2}$

As an example, Fig. A.10 reports the relevant variables of the golden section method for the following problem,

$$\min_{-3 \le \theta \le 5} (\theta^2 + 2\theta)$$

with stop threshold $\varepsilon = 0.2$.

The algorithms discussed thus far are based exclusively on evaluations of the objective function in different points without using derivatives. If the function is differentiable, and especially if its derivative $f'(\theta)$ is easily calculated, algorithms exploiting the "information" contained in the derivative can be used to solve the problem (A.4.6). The main steps of the *bisection algorithm* which halves the uncertainty interval at each iteration on the basis of the value assumed by the derivative at

k	a_k	b_k	α_k	β_k	$f(\alpha_k)$	$f(\beta_k)$	ε
1	−3.00	5.00	−0.104	2.104	−0.197184	8.634816	8.00
2	−3.00	2.10	−1.152	−0.104	−0.976789	−0.197184	5.10
3	−3.00	−0.10	−1.952	−1.152	−0.094366	−0.976789	2.90
4	−1.95	−0.10	−1.152	−0.773	−0.976789	−0.948402	1.85
5	−1.95	−0.77	−1.525	−1.152	−0.724456	−0.976789	1.18
6	−1.52	−0.77	−1.152	−1.045	−0.976789	−0.997966	0.75
7	−1.15	−0.77	−1.045	−0.910	−0.997966	−0.991941	0.38
8	−1.15	−0.91	−1.065	−1.045	−0.995813	−0.997966	0.24
9	−1.06	−0.91	−1.045	−0.966	−0.997966	−0.998854	0.15
10	−1.05	−0.91	−0.966	−0.959	−0.998854	−0.998323	0.13
11	−1.05	−0.96	−1.014	−0.966	−0.999805	−0.998854	0.09
12	−1.05	−0.97	−1.017	−1.014	−0.999727	−0.999805	0.08
13	−1.02	−0.97	−1.014	−0.984	−0.999805	−0.999756	0.05
14	−1.02	−0.98	−1.005	−1.014	−0.999976	−0.999805	0.03
15	−1.02	−1.01	−1.016	−1.005	−0.999757	−0.999976	0.00

Fig. A.10 Relevant variables of the golden section algorithm

the midpoint of the current uncertainty interval are described below. The algorithm is asymptotically convergent if the function $f(\theta)$ is convex (also in a weaker sense than that introduced in Sect. A.1.3).

Bisection Algorithm
(k is just the counter of iterations, irrelevant for computation)

The maximum width ε, allowed for the uncertainty interval is chosen.
If the function is convex, the condition $f'(\theta_{\min}) \geq 0$ implies that $\theta_{\min}$ is a minimum point; analogously $f'(\theta_{\max}) \leq 0$ implies that $\theta_{\max}$ is a minimum point. Otherwise, the extremes of the initial interval are set:

$$a_1 = \theta_{\min},\ b_1 = \theta_{\max}$$

REPEAT

The derivative $f'(\theta_k)$ is calculated at the midpoint of the uncertainty interval $\theta_k = 1/2(a_k + b_k)$.

IF $f'(\theta_k) = 0$ **THEN**

$$a_{k+1} = b_{k+1} = \theta_k$$

OTHERWISE

IF $f'(\theta_k) > 0$ **THEN**

Let $a_{k+1} = a_k,\ b_{k+1} = \theta_k$

OTHERWISE $f'(\theta_k) < 0$

Let $a_{k+1} = \theta_k,\ b_{k+1} = b_k$

END

END
UNTIL $b_{k+1} - a_{k+1} < \varepsilon$

The solution of the problem is $\theta^* = \frac{a_{k+1}+b_{k+1}}{2}$

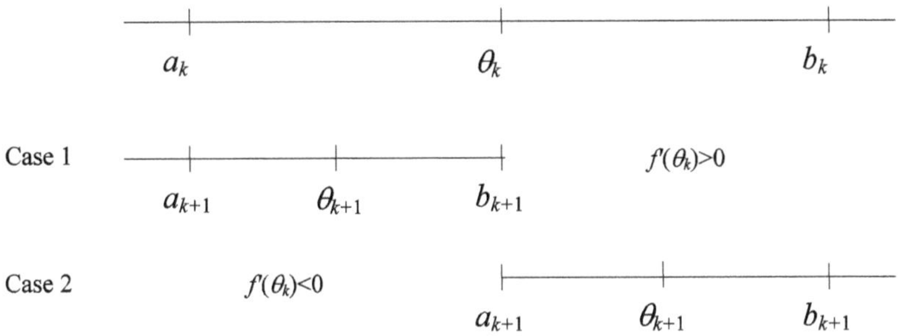

k	a_k	b_k	θ_k	$f(\theta_k)$	$f'(\theta_k)$	ε
1	−3.00	5.00	1.00	3.00	4.00	8.00
2	−3.00	1.00	−1.00	−1.00	0.00	4.00
3	−1.00	1.00	0.00	0.00	2.00	2.00
4	−1.00	0.00	−0.50	−0.75	1.00	1.00
5	−1.00	−0.50	−0.75	−0.94	0.50	0.50
6	−1.00	−0.75	−0.88	−0.98	0.25	0.25
7	−1.00	−0.88	−0.94	−1.00	0.13	0.13
8	−1.00	−0.94	−0.97	−1.00	0.06	0.06
9	−1.00	−0.97	−0.98	−1.00	0.03	0.03
10	−1.00	−0.98	−0.99	−1.00	0.02	0.02
11	−1.00	−0.99	−1.00	−1.00	0.01	0.01

Fig. A.11 Illustration of the bisection algorithm

Figure A.11 illustrates the bisection algorithm for the previous problem.

A.4.2.2 Unconstrained Multidimensional Optimization Algorithms

The unconstrained multidimensional optimization problem:

$$\begin{aligned} &\boldsymbol{x}^* = \operatorname{argmin} f(\boldsymbol{x}) \\ &\boldsymbol{x} \in E^n \end{aligned} \tag{A.4.7}$$

can be solved by using different algorithms, some of which are based exclusively on the calculation of the values of the objective function, others on the use of first and second-order derivatives.

A brief description of some *descent direction algorithms* follows. These algorithms make use of the results described in Sect. A.1 and, at each iteration k, search for the minimum of the function $f(\boldsymbol{x})$ along a direction of negative directional derivative $\boldsymbol{h}^k$ (linear minimization). The algorithms converge towards a null gradient point (stationarity point) of the function $f(\boldsymbol{x})$; they converge towards a global minimum point if the objective function is convex. The simplest of such algorithms, known as the gradient algorithm, assumes the opposite of the gradient as the descent direction. The main steps of the algorithm are given below.

Gradient Algorithm
(k is just the counter of iterations, irrelevant for computation)

The stop parameter ε is fixed. This can be either the maximum gradient module or the maximum deviation between the values of $f(\boldsymbol{x})$ in two successive iterations. An initial point $\boldsymbol{x}^1$ is chosen.

REPEAT

Calculation of the search direction

$$\boldsymbol{h}^k = -\nabla f(\boldsymbol{x}^k)$$

Monodimensional search. The value of the parameter θ minimizing the function of a single variable $f(\boldsymbol{x}_k + \theta \boldsymbol{h}_k)$ is sought

$$\theta^k = \text{argmin}_{0 \leq \theta \leq \theta^*} f(\boldsymbol{x}^k + \theta \boldsymbol{h}^k)$$

where θ^* is a prefixed, large enough value. The line search can be carried out by using one of the algorithms described in the previous section.
Calculation of the next point as

$$\boldsymbol{x}^{k+1} = \boldsymbol{x}^k + \theta^k \boldsymbol{h}^k$$

UNTIL

The module of the function gradient in $\boldsymbol{x}^{k+1}$ is less than the stop threshold:

$$\|\nabla f(\boldsymbol{x}^{k+1})\| < \varepsilon$$

or the relative difference of two successive values of the objective function is less than the stop threshold:

$$\frac{[f(\boldsymbol{x}^k) - f(\boldsymbol{x}^{k+1})]}{f(\boldsymbol{x}^k)} < \varepsilon$$

Figure A.12 describes an application of the gradient algorithm to the minimization of the function:

$$f = (x_1 - 2)^4 + (x_1 - 2x_2)^2$$

with stop parameter $\varepsilon = 0.10$.

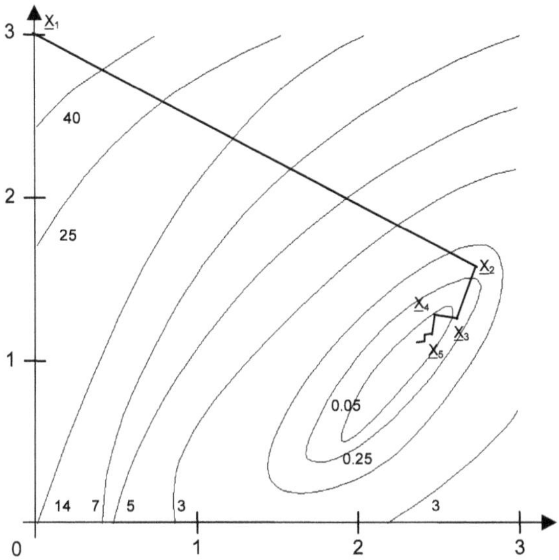

k	$\boldsymbol{x}_k\ f(\boldsymbol{x}_k)$	$\nabla f(\boldsymbol{x}_k)$	$\|\nabla f(\boldsymbol{x}_k)\|$	$h_k = -\nabla f(\boldsymbol{x}_k)$	θ^k	$\boldsymbol{x}_{k+1}$
1	(0.00, 3.00) 52.00	(−44.00, 24.00)	50.12	(44.00, −24.00)	0.062	(2.70, 1.51)
2	(2.70, 1.51) 0.34	(0.73, 1.28)	1.47	(−0.73, −1.28)	0.24	(2.52, 1.20)
3	(2.52, 1.20) 0.09	(0.80, −0.48)	0.93	(−0.80, 0.48)	0.11	(2.43, 1.25)
4	(2.43, 1.25) 0.04	(0.18, 0.28)	0.33	(−0.18, −0.28)	0.31	(2.37, 1.16)
5	(2.37, 1.16) 0.02	(0.30, −0.20)	0.36	(−0.30, 0.20)	0.12	(2.33, 1.18)
6	(2.33, 1.18) 0.01	(0.08, 0.12)	0.14	(−0.08, −0.12)	0.36	(2.30, 1.14)
7	(2.30, 1.14) 0.009	(0.15, −0.08)	0.17	(−0.15, 0.08)	0.13	(2.28, 1.15)
8	(2.28, 1.15) 0.007	(0.05, 0.08)	0.09			

Fig. A.12 Graphic representation and relevant variables for an application of the gradient algorithm

This figure shows a typical characteristic of the gradient algorithm: in the first iterations a rapid decrease in the objective function is observed, whereas successive iterations show smaller reductions and zigzagging towards the optimum value. The problem (A.4.7) can be solved with other algorithms whose structure is similar to that described above, apart from the calculation of the descent direction $\boldsymbol{h}_k$. These algorithms, in order to accelerate convergence, use directions obtained by "deflecting" the gradient and for this reason they are denoted "*deflected gradient*."

Fletcher and Reevers' conjugate gradient algorithm deflects the opposite gradient at each iteration, adding a positive multiple of the direction used in the previous iteration. In the case of quadratic objective function ($f(\boldsymbol{x}) = \boldsymbol{x}^T \boldsymbol{H}\boldsymbol{x}$), this algorithm generates a series of conjugate directions (from which it derives its name) with respect to the matrix $\boldsymbol{H}$, and converges at the optimum point in a finite number of iterations equal to the number of components of $\boldsymbol{x}$. In the general case, it usually converges more quickly than the gradient algorithm, and in particular solves the zigzagging problems in proximity of the minimum point typical of the gradient.

The description of the conjugate gradient algorithm is basically similar to that of the gradient algorithm. The only difference lies in the calculation of the descent direction which is substituted as follows.

$$\boldsymbol{h}^k = -\nabla f(\boldsymbol{x}^k) + \alpha_k \boldsymbol{h}^{k-1} \quad \text{with } \alpha_k = \frac{\|\nabla f(\boldsymbol{x}^k)\|^2}{\|\nabla f(\boldsymbol{x}^{k-1})\|^2}$$

A.4.2.3 Bounded Variables Multidimensional Optimization Algorithms

The problem of minimizing the objective function, imposing constraints on the lower and/or upper bounds of the components of the vector $\boldsymbol{x}$ is slightly more complex than that of unconstrained optimization (A.4.8). In this case the constraint $g_i(\boldsymbol{x})$ can be written as

$$x_i \geq c_i \quad \text{and/or} \quad x_i \leq c_i$$

The variables x_i can be easily modified so that the constraints are always expressed in the form $x_i \geq 0$. Therefore the problem of optimization with inequality constraints can be formally expressed as

$$\min_{x \geq 0} f(\boldsymbol{x}) \tag{A.4.8}$$

The problem (A.4.8) can be solved by using a feasible directions algorithm similar to those described previously. The main difference is that the descent direction used for the unconstrained problem (A.4.7) (i.e., the opposite gradient) is not necessarily a feasible direction with respect to the feasibility set defined by the constraints of the problem. To solve this inconvenience when it occurs, the descent direction can be "projected" over the feasibility set as in the projected gradient algorithm described below.

Projected Gradient Algorithm
(k is just the counter of iterations, irrelevant for computation)

The stop parameter ε is fixed, a feasible initial point $\boldsymbol{x}_1$ is chosen (e.g., $\boldsymbol{x}_1 = \boldsymbol{0}$), and the value of the objective function $f(\boldsymbol{x}_1)$ is calculated.

REPEAT

Calculation of the search direction. The components of the direction $\boldsymbol{h}^k$ are equal to the components of the gradient with changed sign if these components are feasible (i.e., if x_i^k is positive and/or the gradient component is negative). Vice versa if the jth component of the gradient changed of sign is not feasible, the corresponding component of h_i^k is set at zero; this corresponds to the projection of $-\nabla f(\boldsymbol{x}^k)$ over the hyperplane perpendicular to the jth axis.

$$h_i^k = -\frac{\partial f(x^k)}{\partial x_i} \text{ if } x_i^k > 0 \quad \text{and/or} \quad -\frac{\partial f(x^k)}{\partial x_i} \geq 0$$

$$h_i^k = 0 \quad \text{otherwise}$$

Monodimensional search. The minimum of the function $f(\boldsymbol{x}^k + \theta \boldsymbol{h}^k)$ is searched for in the interval $[0, \theta^*]$ where θ^* is the maximum value allowing nonexit from the feasibility set (i.e., ensuring the nonnegativity of all the components of x^{k+1}):

$$\theta^k = \text{argmin}_{0 \leq \theta \leq \theta^*} f(\boldsymbol{x}^k + \theta \boldsymbol{h}^k)$$

with $\theta^* = \max_i \frac{x_i^k}{-h_i^k}$ for $i : h_i^k < 0$, otherwise $\theta^* = \infty$
(Note that for $h_i^k < 0$, $x_i^k > 0$ must result because of previous step)

Calculation of the next point

$$\boldsymbol{x}^{k+1} = \boldsymbol{x}^k + \theta^k \boldsymbol{h}^k$$

UNTIL $\|\boldsymbol{h}^k\| < \varepsilon$
thus verifying the impossibility of any further move along the projected gradient or heuristically, on the percentage decrease of the objective function in the last two iterations:

$$\left|\frac{f(\boldsymbol{x}^{k+1}) - f(\boldsymbol{x}^k)}{f(\boldsymbol{x}^k)}\right| < \varepsilon$$

A.4.2.4 Linearly Constrained Multidimensional Optimization Algorithms

The problem of minimizing the objective function over a closed set defined by linear inequality and/or equality constraints can be stated formally as

$$\begin{aligned} &\min f(\boldsymbol{x}) \\ &\boldsymbol{A}\boldsymbol{x} \leq \boldsymbol{a} \\ &\boldsymbol{B}\boldsymbol{x} = \boldsymbol{b} \end{aligned} \tag{A.4.9}$$

This problem can be solved with different algorithms which differ in the way they generate the "feasible descent direction" $\boldsymbol{h}_k$, that is, a direction along which it is possible to move while reducing the objective function $f(\boldsymbol{x})$ and remaining within the set S defined by the constraints. A description of the Frank–Wolfe algorithm follows which at each iteration generates the direction $\boldsymbol{h}_k$, minimizing a linear approximation of $f(\boldsymbol{x})$.

Also in this case it is possible to demonstrate that if $f(\boldsymbol{x})$ is a convex function, the algorithm converges to the solution of the problem (A.4.9).

Figure A.13 illustrates the application of the Frank–Wolfe algorithm to the following optimum problem.

$$\min x_1^2 + 2x_2^2 - 2x_1x_2 - 10x_2$$
$$0 \le x_1 \le 4$$
$$0 \le x_2 \le 6$$

with stop parameter $\varepsilon = 0.10$.

Frank–Wolfe Algorithm
(k is just the counter of iterations, irrelevant for computation)

The stop parameter ε is fixed; a feasible initial point $\boldsymbol{x}^1$ is chosen.

REPEAT

Generation of the feasible direction. The linear programming problem is solved:

$$\boldsymbol{y}^k = \text{argmin}\, \nabla f(\boldsymbol{x}^k)^T \boldsymbol{y}$$
$$\boldsymbol{A}\boldsymbol{y} = \boldsymbol{a}$$
$$\boldsymbol{B}\boldsymbol{y} \le \boldsymbol{b}$$

The problem is equivalent to the minimization of the linear approximation of $f(\boldsymbol{x})$ at point $\boldsymbol{x}^k$ given by

$$f_L(\boldsymbol{y}) = f(\boldsymbol{x}^k) + \nabla f(\boldsymbol{x}^k)^T(\boldsymbol{y} - \boldsymbol{x}^k)$$

once the constant terms are eliminated. This problem can be solved with the simplex algorithm or one of its variants.[6]
The descent direction is $\boldsymbol{h}^k = \boldsymbol{y}^k - \boldsymbol{x}^k$

Monodimensional search. The linear minimum of the function $f(\boldsymbol{x}^k + \theta\boldsymbol{h}^k)$ is searched for:

$$\theta^k = \text{argmin}_{0\le\theta\le 1} f(\boldsymbol{x}^k + \theta\boldsymbol{h}^k)$$

for θ in the interval [0, 1]. Points $\boldsymbol{x}^k$ and $\boldsymbol{y}^k$ correspond to the extreme values of θ; because both are feasible by construction and the set S is convex, all the points of the segment that joining them are feasible.

---

[6]The solution of problem introduced is generally one of the vertices of the set defined by the linear equations and inequalities. Therefore the Frank–Wolfe algorithm can move only along directions pointing to the vertices and presents zigzagging problems in proximity of the minimum similar to those described for the gradient algorithm.

Calculation of the next point

$$\boldsymbol{x}^{k+1} = \boldsymbol{x}^k + \theta^k \boldsymbol{h}^k$$

UNTIL $\nabla f(\boldsymbol{x}^k)^T(\boldsymbol{x}^{k+1} - \boldsymbol{x}^k) > -\varepsilon$
or, more simply (but less effectively), $|f(\boldsymbol{x}^k) - f(\boldsymbol{x}^{k+1})|/f(\boldsymbol{x}^k) < \varepsilon$.

A.5 Variational Inequality Problems

Let $\boldsymbol{\varphi}(\boldsymbol{x})$ be a vectorial function of a vector defined in a convex set $S \subseteq E^n$, with values in the set $T = \boldsymbol{\varphi}(S) = \{\boldsymbol{\varphi}(\boldsymbol{x}) : \boldsymbol{x} \in S\} \subseteq E^n$; the mathematical problem, called variational inequality, with solution in the point $\boldsymbol{x}^* \in S$, is defined as

$$\boldsymbol{\varphi}(\boldsymbol{x}^*)^T(\boldsymbol{x} - \boldsymbol{x}^*) \geq 0 \quad \forall \boldsymbol{x} \in S \tag{A.5.1}$$

In other words, the problem of the variational inequality of a vectorial function of a vector consists in the search for point $\boldsymbol{x}^*$ at which the vector function $\boldsymbol{\varphi}(\boldsymbol{x}^*)$ has a nonnegative scalar product (i.e., angles $\leq \pi/2$) with all the vectors joining point $\boldsymbol{x}^*$ with every other point $\boldsymbol{x}$ of the set of definition S.

Variational inequality problems can be considered a generalization of minimization problems, in particular of the conditions of virtual minimum (A.4.5b), because the vectorial function of vector $\varphi(\boldsymbol{x})$ is not required to be the gradient of a scalar function of vector $f(\boldsymbol{x})$. To show this, let us consider the generic minimization problem:

$$\begin{aligned} \boldsymbol{x}^* &= \operatorname{argmin} f(\boldsymbol{x}) \\ \boldsymbol{x} &\in S \end{aligned} \tag{A.5.2}$$

If the function $f(\boldsymbol{x})$ is differentiable, its gradient $\nabla f(\boldsymbol{x})$ is a vectorial function of vector, and the virtual minimum conditions are given by

$$\nabla f(\boldsymbol{x}^*)^T(\boldsymbol{x} - \boldsymbol{x}^*) \geq 0 \quad \forall \boldsymbol{x} \in S \tag{A.5.3}$$

It then results that the variational inequality (A.5.1) in the function $\varphi(\boldsymbol{x}) = \nabla f(\boldsymbol{x})$ coincides with the expression of the necessary minimum conditions of the problem (A.5.2). Furthermore, if the gradient $\nabla f(\boldsymbol{x})$ exists, the minimization problem (A.5.2) can be reformulated as

$$\begin{aligned} \boldsymbol{x}^* &= \operatorname{argmin} z(\boldsymbol{x}) = \int_0^{\boldsymbol{x}} \nabla f(\boldsymbol{t})^T \, d\boldsymbol{t} \\ \boldsymbol{x} &\in S \end{aligned} \tag{A.5.4}$$

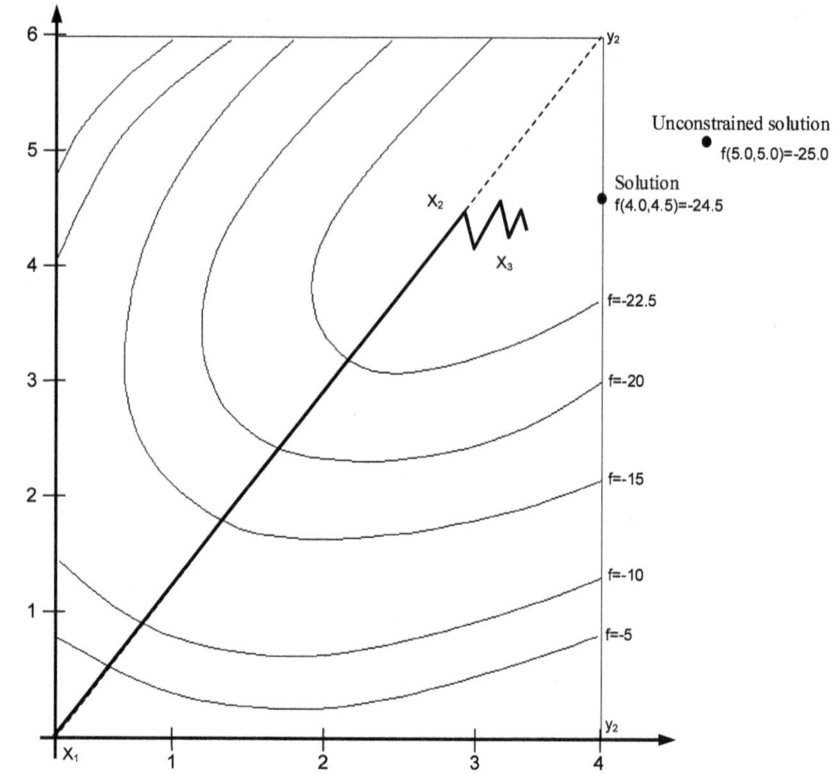

k	$\nabla f(\boldsymbol{x}^k)$	$\boldsymbol{y}^k$	θ^k	$\boldsymbol{x}^{k+1}$	$f(\boldsymbol{x}^{k+1})$	$f(\boldsymbol{x}^k) - f(\boldsymbol{x}^{k+1})$
0				0.000 0.000		
1	0.000 −10.000	(4, 6)	0.750	3.000 4.500	−22.500	22.500
2	−3.000 2.000	(4, 0)	0.119	3.119 3.969	−23.213	0.713
3	−1.700 −0.362	(4, 6)	0.206	3.301 4.385	−23.446	0.233
4	−2.168 0.938	(4, 0)	0.063	3.344 4.111	−23.622	0.176
5	−1.534 −0.244	(4, 6)	0.144	3.439 4.383	−23.728	0.106
6	−1.888 0.654	(4, 0)	0.045	3.464 4.186	−23.816	0.089

Fig. A.13 Graphic representation and significant variables for application of the Frank–Wolfe algorithm

On the other hand, given a vectorial function of vector $\varphi(\boldsymbol{x})$ with symmetric Jacobian $\boldsymbol{Jac}[\varphi(\boldsymbol{x})]$, a minimization problem can be defined:

$$\boldsymbol{x}^* = \operatorname{argmin} f(\boldsymbol{x}) = \int_0^{\boldsymbol{x}} \varphi(\boldsymbol{t})^T \, d\boldsymbol{t} \qquad \boldsymbol{x} \in S \tag{A.5.5}$$

In general, the value of the curvilinear integral appearing in (A.5.4) depends on the integration path. However, if the Jacobian $\boldsymbol{Jac}[\varphi(\boldsymbol{x})]$ of the integrating function $\varphi(\boldsymbol{x})$ is symmetric, the value of the integral is independent of the integration path, being the set of definition convex (Green's theorem). In other words, if and only if the integrating function $\varphi(\boldsymbol{x})$ has symmetrical Jacobian, the former can be the gradient of a function $f(\boldsymbol{x})$, that is, $\nabla f(\boldsymbol{x}) = \varphi(\boldsymbol{x})$, of which $\boldsymbol{Jac}[\varphi(\boldsymbol{x})]$ is the (symmetrical) Hessian matrix. If the equivalent minimization problem (A.5.5) is correctly defined, the necessary conditions of virtual minimum are given by

$$\nabla f(\boldsymbol{x}^*)^T (\boldsymbol{x} - \boldsymbol{x}^*) \geq 0 \quad \forall \boldsymbol{x} \in S \tag{A.5.6}$$

that is,

$$\varphi(\boldsymbol{x}^*)^T (\boldsymbol{x} - \boldsymbol{x}^*) \geq 0 \quad \forall \boldsymbol{x} \in S$$

It can immediately be seen that the condition (A.5.6) is formally coincident with the variational inequality (A.5.1).

If the function $\varphi(\boldsymbol{x})$ is continuous and differentiable with symmetrical and semidefinite positive Jacobian $\boldsymbol{Jac}[\varphi(\boldsymbol{x})]$, a vector $\boldsymbol{x}^*$ solving the constrained optimization model (A.5.5) solves the corresponding variational inequality (A.5.1) and vice versa.

In this case, the objective function of the problem (A.5.5) $f(\boldsymbol{x})$ is differentiable with continuous gradient and continuous positive semidefinite Hessian matrix, because $\nabla f(\boldsymbol{x}) = \varphi(\boldsymbol{x})$, and $\boldsymbol{Hess}[f(\boldsymbol{x})] = \boldsymbol{Jac}[\varphi(\boldsymbol{x})]$. Therefore $f(\boldsymbol{x})$ is convex, and so the conditions of virtual minimum (A.5.6) are necessary and sufficient.

A.5.1 Properties of Variational Inequalities

Sufficient conditions for the existence of at least one solution of the variational inequality (A.5.1) can be obtained by applying Brouwer's theorem, as follows.

Theorem *The variational inequality problem* (A.5.1) *has at least one solution if*:

S *is a nonempty, compact and convex set*
$\varphi(\boldsymbol{x})$ *is a continuous function*

Sufficient conditions for the uniqueness of the variational inequality solution are given by the following theorem.

Theorem *The variational inequality* (A.5.1) *has at most one solution if*:

$\varphi(\boldsymbol{x})$ *is a strictly monotone increasing function*; *that is*,

$$\left(\varphi(\boldsymbol{x}') - \varphi(\boldsymbol{x}'')\right)^T (\boldsymbol{x}' - \boldsymbol{x}'') > 0 \quad \forall \boldsymbol{x}', \boldsymbol{x}'' \in S$$

In fact, if there existed two different vectors solving the variational inequality, $\boldsymbol{x}_1^* \neq \boldsymbol{x}_2^* \in S$, we would have:

$$\varphi\left(\boldsymbol{x}_1^*\right)^T \left(\boldsymbol{x} - \boldsymbol{x}_1^*\right) \geq 0 \quad \forall \boldsymbol{x} \in S_x \tag{A.5.7a}$$

$$\varphi\left(\boldsymbol{x}_2^*\right)^T \left(\boldsymbol{x} - \boldsymbol{x}_2^*\right) \geq 0 \quad \forall \boldsymbol{x} \in S_x \tag{A.5.7b}$$

From (A.5.7a) for $\boldsymbol{x} = \boldsymbol{x}_2^*$:

$$\varphi\left(\boldsymbol{x}_1^*\right)^T \left(\boldsymbol{x}_2^* - \boldsymbol{x}_1^*\right) \geq 0 \tag{A.5.8a}$$

Furthermore, from (A.5.7a) for $\boldsymbol{x} = \boldsymbol{x}_1^*$, it results that:

$$\varphi\left(\boldsymbol{x}_2^*\right)^T \left(\boldsymbol{x}_1^* - \boldsymbol{x}_2^*\right) \geq 0 \tag{A.5.8b}$$

that is,

$$-\varphi\left(\boldsymbol{x}_2^*\right)^T \left(\boldsymbol{x}_2^* - \boldsymbol{x}_1^*\right) \geq 0$$

Adding (A.5.8a) and (A.5.8b), it would follow that:

$$\left(\varphi\left(\boldsymbol{x}_1^*\right) - \varphi\left(\boldsymbol{x}_2^*\right)\right)^T \left(\boldsymbol{x}_2^* - \boldsymbol{x}_1^*\right) \geq 0$$

that is,

$$\left(\varphi\left(\boldsymbol{x}_2^*\right) - \varphi\left(\boldsymbol{x}_1^*\right)\right)^T \left(\boldsymbol{x}_2^* - \boldsymbol{x}_1^*\right) \leq 0$$

which contradicts the monotonicity assumption.

A.5.2 Solution Algorithms for Variational Inequality Problems

Solution algorithms for the variational inequality problem (A.5.1), in the case of the function $\varphi(\boldsymbol{x})$ with symmetric Jacobian, are based on algorithms solving the equivalent minimization problem (A.5.5) described in Sect. A.4.2. Note that in this case the gradient $\nabla f(\boldsymbol{x})$ of the objective function of the minimization problem, used by the algorithm, is given by the function $\varphi(\boldsymbol{x})$ defining the variational inequality.

In the general case of a function $\varphi(\boldsymbol{x})$ with nonsymmetric Jacobian, various solution algorithms can be adopted; even though their convergence analysis usually requires conditions that are not easily verifiable. One of the simplest, called the diagonalization algorithm, generates a succession of vectors $\boldsymbol{x}^k$, starting from a feasible point $\boldsymbol{x}^0 \in S$, solving a succession of variational inequalities defined by functions with diagonal Jacobians approximating the problem (A.5.1). In particular, at a point $\boldsymbol{x}^* \in S$, the ith component function $\varphi_i(\boldsymbol{x})$ of the vectorial function $\varphi(\boldsymbol{x})$, can be approximated by a function $\varphi_i^*(x_i)$ obtained by fixing all the other components of $\boldsymbol{x}$ at their values x_j^* (i.e., by diagonalizing the Jacobian):

$$\varphi_i(x_1, \ldots, x_{i-1}, x_i, x_{i+1}, \ldots) \cong \varphi_i^*\left(x_1^*, \ldots, x_{i-1}^*, x_i, x_{i+1}^*, \ldots\right) = \varphi_i^*(x_i) \quad \forall i$$

Thus the variational inequality (A.5.1) can be approximated by a variational inequality defined by a function of $\varphi^*(\boldsymbol{x})$ with diagonal Jacobian:

$$\varphi(\boldsymbol{x}^*)^T(\boldsymbol{x} - \boldsymbol{x}^*) \cong \sum_i \varphi_i^*(x_i)\left(x_i - x_i^*\right) \geq 0 \quad \forall \boldsymbol{x} \in S_x \tag{A.5.9}$$

The solution of the approximate variational inequality (A.5.9) can be obtained by solving the equivalent minimization problem (A.5.5), with one of the algorithms described in Sect. A.4.2.

Index

E. Cascetta, *Transportation Systems Analysis*,
Springer Optimization and Its Applications 29,
DOI 10.1007/978-0-387-75857-2, © Springer Science+Business Media, LLC 2009

References

Abbe, E., Bierlaire, M., & Toledo, T. (2007). Normalization and correlation of cross-nested logit models. *Transportation Research B*, *41*, 795–808.

Abdelfatah, A., & Mahmassani, S. H. (1999). System optimal time-dependent path assignment and signal timing in traffic networks. *Transportation Research Record*, *1645*, 185–193.

Abdulaal, M., & Le Blanc, L. J. (1979). Continuous equilibrium network design models. *Transportation Research B*, *13*, 19–32.

Abkowitz, M. D. (1981). An analysis of the commuter departure time decision. *Transportation*, *10*, 283–297.

Adler, T., & Ben-Akiva, M. (1979). A theoretical and empirical model of trip chaining behaviour. *Transportation Research B*, *13*, 243–257.

Adorisio, I. (1986). *Ingegneria della produzione astratta*. Padova: CEDAM.

Aho, A. V., Hopcroft, J. E., & Ullman, J. D. (1983). *Data Structures and Algorithms*. Reading: Addison–Wesley.

Ahuja, R. K., Magnanti, T. L., & Orlin, J. B. (1993). *Networks flows: theory, algorithms, and application*. Englewood Cliffs: Prentice Hall.

Akcelik, R. (1988). The highway capacity manual delay formula for signalized intersections. *ITE Journal*, *58*(3), 23–27.

Alexander, E. R. (1997). *Introduzione alla pianificazione: teorie, concetti e problemi attuali*. Traduzione di Approaches to planning: introduction to current planning theories. Napoli: Ed. Clean.

Algers, S., Daly, A., & Widlert, S. (1993). *The Stockholm model system* (Technical report). The Hague: Hague Consulting Group.

Allsop, R. E. (1977). Some possibilities for using traffic control to influence trip distribution and route choice. In D. J. Buckley (Ed.), *Proceedings of the sixth international symposium on transportation and traffic theory*. New York: Elsevier.

Antonisse, R. W., Daly, A., & Gunn, H. (1986). The primary destination tour approach to modelling trip chains. In *Proceedings of seminar M on transportation planning methods at the 14th PTRC summer annual meeting*. England: University of Sussex.

Arnott, R., De Palma, A., & Lindsey, R. (1990). Departure time and route choice for the morning commute. *Transportation Research B*, *24*, 209–228.

Ashok, K., & Ben-Akiva, M. (2002). Estimation and prediction of time-dependent origin-destination flows with a stochastic mapping to path flows and link flows. *Transportation Science*, *36*(2), 184–198.

Astarita, V. (1996). A continuous time link model for dynamic network loading based on travel time functions. In *Proceeding of the 13th international symposium on the theory of traffic flow* (pp. 87–102). Lyon.

Baaj, M. H., & Mahmassani, H. S. (1991). An AI based approach for transit route system planning and design. *Journal of Advanced Transportation*, *25*, 187–210.

Baaj, M. H., & Mahmassani, H. S. (1995). Hybrid route generation heuristic algorithm for the design of transit networks. *Transportation Research C*, *3*, 31–50.

Barry, T. M., & Reagan, J. A. (1978). FHWA highway traffic noise prediction model (FHWA Report, RD 77-108). Washington: Transportation Department.

Bath, C. (1997). Recent methodological advances relevant to activity and travel behavior analysis. In *Proceedings of the VIII IATBR conference* (Resource papers), Austin.

Bayliss, B. (1988). *The Measurement of Supply and Demand in Freight Transport*. Hants: Avebury Grower.

Beckman, M., McGuire, C. B., & Winsten, C. B. (1956). *Studies in the economics of transportation*. New Haven: Yale University Press.

E. Cascetta, *Transportation Systems Analysis*,
Springer Optimization and Its Applications 29,
DOI 10.1007/978-0-387-75857-2, © Springer Science+Business Media, LLC 2009

Bell, M. (1991). The estimation of origin-destination matrices by constrained generalized least squares. *Transportation Research B*, *25*, 13–22.

Bell, M. (1995). Stochastic user equilibrium assignment in networks with queues. *Transportation Research B*, *29*, 125–137.

Bell, M., & Iida, Y. (1997). *Transportation networks analysis*. New York: Wiley.

Bell, M., Inaudi, D., Lam, W., & Ploss, G. (1993). Stochastic user equilibrium assignment and iterative balancing. In *Proceedings of the 12th international symposium on traffic and transportation theory* (pp. 427–440). Berkeley.

Bellei, G., Gentile, G., & Papola, N. (2005). A within-day dynamic traffic assignment model for urban road networks. *Transportation Research Part B*, *39*, 1–29.

Ben-Akiva, M., Bergman, M. J., Day, A. L., & Ramaswamy, R. (1984). Modelling interurban route choice behaviour. In *Proceedings of the 9th international symposium on transportation and traffic theory* (pp. 299–330). Utrecht: VNU Science Press.

Ben-Akiva, M., & Bierlaire, M. (1999). Discrete choice methods and their applications to short-term travel decisions. In R. Hall (Ed.), *Handbook of transportation science. International series in operation research and management science*. Hingham: Kluwer Academic.

Ben-Akiva, M., & Boccara, B. (1995). Discrete choice models with latent choice sets. *International Journal of Research in Marketing*, *12*, 9–24.

Ben-Akiva, M., & Bolduc, D. (1996). *Multinomial probit with a logit kernel and a general parametric specification of the covariance structure* (Working paper). Boston: Department of Economics, MIT.

Ben-Akiva, M., Bowman, J., & Gopinath, D. (1996). Travel demand model system for the information era. *Transportation*, *23*, 241–266.

Ben-Akiva, M., Cyna, M., & de Palma, A. (1984). Dynamic model of peak period congestion. *Transportation Research B*, *18*, 339–355.

Ben-Akiva, M., & de Palma, A. (1987). Dynamical models of transportation networks. In *Proceedings of the 15th PTRC summer annual meeting*, University of Bath, England.

Ben-Akiva, M., & Francois, B. (1983). *Homogeneous generalized extreme value model* (Working paper). Cambridge: Department of Civil Engineering, MIT.

Ben-Akiva, M., Koutsopoulos, H. N., & Mukundan, A. (1994). A dynamic traffic model system for ATMS/ATIS operations. *IVHS Journal*, *2*, 9–24.

Ben-Akiva, M., & Lerman, S. (1985). *Discrete choice analysis: theory and application to travel demand*. Cambridge: MIT Press.

Ben-Akiva, M., & Morikawa, T. (1990). Estimation of switching models from revealed preferences and stated intentions. *Transportation Research A*, *24*, 149–164.

Ben-Akiva, M. E., & Bowman, J. L. (1998). Activity based travel demand model systems. In P. Marcotte & S. Nguyen (Eds.), *Equilibrium and advanced transportation modelling*. Dordrecht: Kluwer Academic, Chapter 2.

Bergman, L. (1990). The development of computable general equilibrium modelling. In L. Bergman & W. Jorgenson (Eds.), *General equilibrium modelling and economic policy analysis*. Oxford: Blackwell.

Bernstein, D., & Smith, T. E. (1994). Equilibria for networks with lower semicontinuous costs: with an application to congestion pricing. *Transportation Science*, *28*, 221–235.

Bhat, C. R., Guo, J. Y., Srinivasan, S., & Sivakumar, A. (2004). A comprehensive econometric micro-simulator for daily activity-travel patterns. *Transportation Research Record*, *1894*, 57–66.

Bianco, L. (1986). The role of quantitative methods in urban transportation planning. In *Proceedings of the international seminar "The management and planning of urban transport systems: from theory to practice"*. Montreal: Canada.

Bifulco, G. N. (1993). A stochastic user equilibrium assignment model for the evaluation of parking policies. *European Journal of Operational Research*, *71*, 269–287.

Bifulco, G. N., Cartenì, A., & Papola, A. (2003). *Un modello comportamentale per la simulazione di fenomeni di mobilità complessi: un approccio activity-based*. Presented at the 2003 scientific seminar of the Italian Society of Transportation Engineering Professors (SIDT).

Biggiero, L. (1991). Un modello comportamentale per la generazione degli spostamenti non sistematici in area urbana. *Trasporti e Trazione*, *4*, 142–152.
Biggiero, L., & Postorino, M. N. (1995). La calibrazione di modelli di scelta modale mediante l'uso congiunto di dati RP e SP. In E. Cascetta & G. Salerno (Eds.). *Sviluppi della ricerca sui sistemi di trasporto*. Milano: Franco Angeli ed.
Billheimer, J. W., & Gray, P. (1973). Network design with fixed and variable cost elements. *Transportation Science*, *7*, 49–74.
Blum, J. R. (1954). *Ann. math. stat.: Vol. 25. Multidimensional stochastic approximation methods* (pp. 737–744). Berkeley: University of California.
Bolduc, D. (1999). A practical technique to estimate multinomial probit models in transportation. *Transportation Research B*, *33*, 63–99.
Bolduc, D., Fortin, B., & Fournier, M. A. (1996). The impact of incentive policies to influence practice location of general practitioners: a multinomial probit analysis. *Journal of Labor Economics*, *14*, 703–732.
Boulter, J. (1999). *The analytical hierarchy process* (Working paper). Champaign: University of Illinois of Urbana.
Bouzaiene-Ayari, B., Gendreau, M., & Nguyen, S. (1995). *On the modelling of bus stops in transit networks*. Centre de recherche sur les transports, Université de Montréal.
Bouzaiene-Ayari, B., Gendreau, M., & Nguyen, S. (1997). *Transit equilibrium assignment problem: a fixed-point simplicial-decomposition solution algorithm*. Internal report.
Bouzaiene-Ayari, B., Gendrau, M., & Nguyen, S. (1998). Passenger assignment in congested transit networks: a historical perspective. In P. Marcotte & S. Nguyen (Eds.). *Equilibrium and advanced transportation modelling*. Dordrecht: Kluwer Academic.
Bovy, P. H. L., & Jansen, G. R. M. (1983). Network aggregation effect upon equilibrium assignment outcomes: an empirical investigation. *Transportation Science*, *17*, 240–262.
Box, G., Hunter, W., & Hunter, G. (1978). *Statistics for experimenters*. London: Wiley.
Boyce, D. E., & Janson, B. N. (1980). A discrete transportation network design problem with combined trip distribution and assignment. *Transportation Research B*, *14*, 147–154.
Boyce, D. E., Ran, B., & Le Blanc, L. J. (1991). Dynamic user-optimal traffic assignment model: a new model and solution technique. In *Proceedings of the 1st TRISTAN*, Montreal, Canada.
Boyer, K. D. (1998). *Principles of transportation economics*. Longman: Addison–Wesley.
Bowman, J. L., & Ben-Akiva, M. E. (2001). Activity-based disaggregate travel demand model system with activity schedules. *Transportation Research A*, *35*, 1–28.
Bradley, M. A., & Daly, A. J. (1992). Estimation of logit choice models using mixed stated preferences and revealed preference information. In *Proceedings of the 20th PTRC summer annual meeting*, University of Manchester, England.
Branston, D. (1976). Link capacity functions: a review. *Transportation Research*, *10*, 223–236.
Brog, W., & Ampt, E. (1982). *State of the art in the collection of travel behaviour data*. In Travel behaviour for the 1980's (Special report 201). Washington: National Research Council.
Burrell, J. E. (1968). Multiple route assignment and its application to capacity restraint. In W. Leutzbach & P. Baron (Eds.). *Proceedings of the 4th international symposium on the theory of road traffic flow*, Karlsruhe: Germany.
Canale, S., Corriere, F., & Lo Bosco, D. Tesoriere, Jr G. (1990). *L'inquinamento acustico da traffico nelle aree urbane: l'indagine nella città di Palermo*, Accademia Nazionale di Scienze e Arti of Palermo.
Cantarella, G. E. (1997). A general fixed-point approach to multi-mode multi-user equilibrium assignment with elastic demand. *Transportation Science*, *31*, 107–128.
Cantarella, G. E. (2001). *Introduzione alla tecnica ed economia dei trasporti e del traffico con elementi di economia dei trasporti*. Torino: UTET.
Cantarella, G. E., & Binetti, M. (1998). Stochastic equilibrium traffic assignment with value-of-time distributed among user. *International Transactions of Operational Research*, *5*, 541–553.
Cantarella, G.E., Binetti, M. (2002). Stochastic assignment with gammit path choice models. In M. Patriksson & M. Labbé (Eds.), *Transportation planning*. Dordrecht: Kluwer Academic.
Cantarella, G. E., & Cascetta, E. (1995). Dynamic processes and equilibrium in transportation networks. *Transportation Science*, *29*, 305–329.

Cantarella, G. E., & Cascetta, E. (1998). Stochastic assignment to transportation networks: models and algorithms. In P. Marcotte & S. Nguyen (Eds.), *Equilibrium and advanced transportation modelling* (pp. 87–107). Boston: Kluwer Academic.

Cantarella, G. E., Cascetta, E., Adamo, V., & Astarita, V. (1999). A doubly dynamic traffic assignment model for planning applications. In *Proceedings of the 14th international symposium on transportation and traffic theory*, Jerusalem, Israel.

Cantarella, G. E., & Improta, G. (1991). Iterative procedure for equilibrium network traffic signal setting. *Transportation Research A*, *25*, 241–249.

Cantarella, G. E., Improta, G., & Sforza, A. (1991). Road network signal setting: equilibrium conditions. In M. Papageorgiou (Ed.), *Concise encyclopedia of traffic and transportation systems* (pp. 366–371). Oxford: Pergamon Press.

Cantarella, G. E., Pavone, G., & Vitetta, A. (2006). Heuristics for urban road network design: lane layout and signal settings. *European Journal of Operational Research*, *175*, 1682–1695.

Cantarella, G. E., & Sforza, A. (1991). Road network signal setting: equilibrium conditions. In M. Papageorgiou (Ed.), *Concise encyclopaedia of traffic and transportation systems* (pp. 366–371). Oxford: Pergamon Press.

Cantarella, G. E., & Sforza, A. (1995). Network design models and methods for urban traffic management. In N. H. Gartner & G. Improta (Eds.), *Urban traffic networks – dynamic flow modelling and control* (pp. 123–153). New York: Springer.

Cantarella, G. E., & Vitetta, A. (1994). Algoritmi genetici per il progetto di reti di trasporto. *Ricerca Operativa*, *24*, 33–55.

Cantarella, G. E., & Vitetta, A. (2000). Stochastic assignment to high frequency transit networks: models, algorithms and applications with different perceived cost distributions. In *Proceedings of the 7th meeting of the EURO working group on transportation*, Helsinki, Finland, August 1999.

Cantarella, G. E., & Vitetta, A. (2006). The multi-criteria road network design problem in an urban area. *Transportation*, *33*, 567–588.

Carey, M., & McCartney, M. (2002). Behaviour of a whole-link travel time model used in dynamic traffic assignment. *Transportation Research B*, *36*, 83–95.

Cartenì, A., & Punzo, V. (2007). Travel time cost functions for urban roads: a case study in Italy. *Urban transport XIII: urban transport and the environment in the 21st century*. Wessex Institute of Technology, Coimbra, Portugal, 3–5/09.

Cascetta, E. (1984). Estimation of trip matrices from traffic counts and survey data: a generalized least squares estimator. *Transportation Research B*, *18*, 289–299.

Cascetta, E. (1986). *CRT publication: Vol. 375. A class of travel demand estimators using traffic flows*. Montreal: Université de Montreal.

Cascetta, E. (1987). Static and dynamic models of stochastic assignment to transportation networks. In G. Szaego, L. Bianco, & A. Odoni (Eds.). *Flow control of congested networks*. Berlin: Springer.

Cascetta, E. (1989). A stochastic process approach to the analysis of temporal dynamics in transportation networks. *Transportation Research B*, *23*, 1–17.

Cascetta, E. (1993). L'ingegneria dei sistemi di trasporto: metodi quantitativi per la pianificazione e la gestione. Sistemi di trasporto, 1.

Cascetta, E. (1995). Modelli di scelta del percorso su reti di trasporto. In *Proceedings of seminar, assegnazioni alle reti di trasporto*, Dipartimento di Ingegneria Civile, Università di Roma "Tor Vergata".

Cascetta, E. (1998). *Teoria e metodi dell'ingegneria dei sistemi di trasporto*. Torino: UTET.

Cascetta, E. (2005). *La sfida dei trasporti in Campania: un sistema integrato per la mobilità sostenibile*. Napoli: Electa.

Cascetta, E., & Cantarella, G. E. (1991). A day-to-day and within-day dynamic stochastic assignment model. *Transportation Research A*, *25*, 277–291.

Cascetta, E., Di Gangi, M., & Conigliaro, G. (1996). A multi-regional input-output model with elastic trade coefficients for the simulation of freight transport demand in Italy, in transportation planning methods. In *Proceedings of the 24th PTRC summer annual meeting*, England.

Cascetta, E., Gallo, M., & Montella, B. (1998). Optimal signal setting on traffic networks with stochastic equilibrium assignment. In *Preprints of the 3rd TRISTAN meeting*, Portorico.

Cascetta, E., Gallo, M., & Montella, B. (1999). An asymmetric SUE model for the combined assignment-control problem. In *Selected proceedings of 8th WCTR* (Vol. 2, pp. 189–202). Oxford: Pergamon.

Cascetta, E., Gallo, M., & Montella, B. (2006). Models and algorithms for the optimization of signal settings on urban networks with stochastic assignment. *Annals of Operation Research, 144*, 301–328.

Cascetta, E., & Improta, A. A. (1999). Estimation of travel demand using traffic counts and other data sources. *Optimization days 1999*. Michael Florian's special session, Montreal, Canada.

Cascetta, E., Inaudi, D., & Marquis, G. (1993). Dynamic estimators of origin-destination matrices using traffic counts. *Transportation Science, 27*, 363–373.

Cascetta, E., & Montella, B. (1979). Modelli di arrivo dei passeggeri alle fermate di un sistema di trasporto collettivo urbano. *La Rivista della Strada B, 453*, 303–308.

Cascetta, E., & Nguyen, S. (1986). A unified framework for estimating or updating origin-destination matrices from traffic counts. *Transportation Research B, 22*, 437–455.

Cascetta, E., & Nuzzolo, A. (1982). Analisi statistica del processo delle velocità in autostrada. *Autostrade, 6*, 29–41.

Cascetta, E., & Nuzzolo, A. (1986). Uno schema comportamentale per la modellizzazione delle scelte di percorso nelle reti di trasporto pubblico urbano. In *Proceedings of the IV conference of PFT-CNR*, Torino.

Cascetta, E., & Nuzzolo, A. (1988). Un modello di equilibrio domanda/offerta per la simulazione dei sistemi di trasporto nelle aree urbane di medie dimensioni. In *Proceedings of the V conference of PFT-CNR* (Vol. I, pp. 141–176). Napoli.

Cascetta, E., Nuzzolo, A., & Biggiero, L. (1992). Analysis and modelling of commuters departure time and route choices in urban networks. In *Proceedings of the 2nd international seminar on urban traffic network*, Capri.

Cascetta, E., Nuzzolo, A., & Biggiero, L. (1995). A system of behavioural models for the simulation of intercity travel demand in Italy. In D. Hensher (Ed.), *Proceedings of the 7th WCTR*. Sidney: Elsevier.

Cascetta, E., Nuzzolo, A., Biggiero, L., & Russo, F. (1995). Passenger and freight demand models for the Italian transportation system. In D. Hensher (Ed.), *Proceedings of the 7th WCTR*. Sidney: Elsevier.

Cascetta, E., Nuzzolo, A., Biggiero, L., & Russo, F. (1996). A system of within-day dynamic demand and assignment models for scheduled intercity services. In *Proceedings of 24th European transportation forum seminar D-E part 2*, London.

Cascetta, E., Nuzzolo, A., & Rostirolla, P. (1989). Optimal railway pricing: results of a model for the Italian case. In *Proceedings of seminar D on "Public transport planning and operations" at the 17th PTRC summer annual meeting*, University of Sussex, England.

Cascetta, E., Nuzzolo, A., Russo, F., & Vitetta, A. (1996). A new route choice logit model overcoming IIA problems: specification and some calibration results for interurban networks. In J. B. Lesort (Ed.), *Proceedings of the 13th international symposium on transportation and traffic theory*. Oxford: Pergamon Press.

Cascetta, E., Nuzzolo, A., & Velardi, V. (1986). Un'analisi sperimentale dei modelli di assegnazione alle reti urbane di trasporto privato. In *Proceedings of the IV conference of PFT-CNR*, Torino.

Cascetta, E., Nuzzolo, A., & Velardi, V. (1994). A time of the day tour based trip chaining model system for urban transportation planning. In *Transportation planning methods: Vol. I. Proceedings of the 22nd PTRC summer annual meeting*, University of Warwick.

Cascetta, E., & Papola, A. (2001). Random utility models with implicit availability-perception of choice alternatives for the simulation of travel demand. *Transportation Research C, 9*, 249–263.

Cascetta, E., & Papola, A. (2003). A joint mode-run choice model to simulate the schedule influence at a regional level. *Transportation Research B, 37*, 595–614.

Cascetta, E., & Papola, A. (2009). Dominance among alternatives in random utility models. *Transportation Research A*, *43*, 170–179.

Cascetta, E., & Postorino, M. N. (2001). Fixed point models for the estimation of O-D matrices using traffic counts on congested networks. *Transportation Science*, *35*, 134–147.

Cascetta, E., & Rostirolla, P. (1989). Un modello matematico per il calcolo della tariffa tecnico-economica. *Ingegneria Ferroviaria*, *6*, 1–8.

Cascetta, E., & Russo, F. (1997). Calibrating aggregate travel demand models with traffic counts: estimators and statistical performances. *Transportation*, *24*, 271–293.

Catling, I. (1977). A time-dependent approach to junction delays. *Traffic Engineering Control*, *18*, 520–526.

Chabini, I. (2000). The analytical network loading problem: formulation, solution algorithms and computer implementation. In *Proceedings of the 79th annual meeting of the transportation research board*.

Chabini, I., & Kachani, S. (1999). Analytical dynamic network loading models: analysis of a single link network. *Transportation Research*, forthcoming.

Chankong, V., & Haimes, Y. Y. (1983). *Multi-objective decision making: theory and methodology*. New York: Elsevier/North-Holland.

Chavarria, S. (2002). *Analytic hierarchy process* (Internal report). Champaign: Department of Urban and regional planning, University of Illinois at Urbana.

Chen, M., & Alfa, A. S. (1991a). A network design algorithm using a stochastic incremental traffic assignment approach. *Transportation Science*, *25*, 215–224.

Chen, M., & Alfa, A. S. (1991b). Algorithms for solving Fisk's stochastic traffic assignment model. *Transportation Research B*, *25*, 405–412.

Chen, Y., & Florian, M. (1995). A coordinate descent method for the bi-level O-D matrix adjustment problem. *International Transactions in Operational Research*, *2*, 165–179.

Chenery, H. (1953). The structure and growth of the Italian economy. In H. Chenery & P. Clark (Eds.), *Regional analysis* (pp. 97–116). Rome: United States Mutual Security Agency.

Cherkassky, B. V., Goldberg, A. V., & Radzik, T. (1996). Shortest paths algorithms: theory and experimental evaluation. *Mathematical Programming*, *73*, 129–174.

Chien, S., & Schonfeld, P. (1998). Joint optimization of a rail transit line and its feeder bus system. *Journal of Advanced Transportation*, *32*, 253–284.

Chieng, S., & Yang, Z. (2000). Optimal feeder bus routes on irregular street networks. *Journal of Advanced Transportation*, *34*, 213–248.

Chu, C. (1989). A paired combinatorial logit model for travel demand analysis. In *Proceedings of the 5th WCTR conference* (pp. 295–309). Ventura.

Cochran, W. G. (1963). *Sampling techniques*. New York: Wiley.

Codina, E., & Barcelò, J. (1995). Dynamic traffic assignment: considerations on some deterministic modelling approaches. *Annals of Operations Research*, *60*, 1–58.

Constantin, I., & Florian, M. (1995). Optimizing frequencies in a transit networks: a nonlinear bi-level programming approach. *International Transportation Operational Research*, *2*, 149–164.

Costa, P., & Roson, R. (1988). Transport margins, transportation industry and the multiregional economy. Some experiments with a model for Italy. *Ricerche Economiche*, *2*, 237–287.

Cremer, M., & Keller, H. (1987). A new class of dynamic methods for the identification of origin-destination flows. *Transportation Research B*, *21*, 117–132.

Crisalli, U. (1996). Un algoritmo di tipo greedy per il progetto delle frequenze di una rete di trasporto collettivo. In *Proceedings of SIDT conference*, Napoli.

D'Acierno, L., Gallo, M., & Montella, B. (2003). Un modello per la determinazione delle tariffe ottimali per il Trasporto Pubblico Locale. In *Metodi e tecnologie dell'ingegneria dei trasporti – seminar 2001* (pp. 256–272). Milano: Franco Angeli ed.

D'Acierno, L., Gallo, M., & Montella, B. (2006). Optimisation models for the urban parking pricing problem. *Transport Policy*, *13*, 34–48.

Dafermos, S. (1971). An extended traffic assignment model with applications to two-way traffic. *Transportation Science*, *5*, 366–389.

Dafermos, S. C. (1972). The traffic assignment problem for multi-class user transportation networks. *Transportation Science*, *6*, 73–87.

Dafermos, S. (1980). Traffic equilibrium and variational inequalities. *Transportation Science*, *14*, 42–54.
Dafermos, S. (1982a). The general multimodal network equilibrium problem with elastic demand. *Networks*, *12*, 57–72.
Dafermos, S. (1982b). Relaxation algorithms for the general asymmetric traffic equilibrium problem. *Transportation Science*, *16*, 231–240.
Daganzo, C. F. (1977). *On achieving stochastic user equilibrium on a transportation network.* (Working paper UCB-ITS-PWP 7704). Berkeley: Institute of Transportation Studies, University of California.
Daganzo, C. F. (1979). *Multinomial probit: the theory and its application to demand forecasting.* New York: Academic Press.
Daganzo, C. F. (1982). Unconstrained extremal formulation of some transportation equilibrium problems. *Transportation Science*, *16*, 332–360.
Daganzo, C. F. (1983). Stochastic network equilibrium with multiple vehicle types and asymmetric, indefinite link cost Jacobians. *Transportation Science*, *17*, 282–300.
Daganzo, C. F. (1994). The cell transmission model 1: a dynamic representation of highway traffic consistent with the hydrodynamic theory. *Transportation Research B*, *28*, 269–287.
Daganzo, C. F. (1995). The cell transmission model 2: network traffic simulation. *Transportation Research B*, *29*, 79–93.
Daganzo, C. F. (1995a). Properties of link travel time functions under dynamic loads (Working paper WP, UCB ITS PWP-93-5). Berkeley: Institute of Transportation Studies, University of California.
Daganzo, C. F. (1997). *Fundamentals of transportation and traffic operations*. Oxford: Pergamon.
Daganzo, C. F., & Kusnic, M. (1993). Another look at the nested logit model: two properties of the nested logit model. *Transportation Science*, *27*, 395–400.
Daganzo, C. F., & Sheffi, Y. (1977). On stochastic models of traffic assignment. *Transporation Science*, *11*, 253–274.
Daganzo, C. F., & Sheffi, Y. (1982). Unconstrained extremal formulation of some transportation equilibrium problems. *Transportation Science*, *16*, 332–360.
Daly, A., & Zachary, S. (1978). Improved multiple choice models. In D. Hensher & M. Dalvi (Eds.), *Determinant of travel choice*. Sussex: Saxon House.
Damberg, O., Lundgren, J. T., & Patriksson, M. (1996). An algorithm for the stochastic user equilibrium problem. *Transportation Research B*, *30*, 115–131.
Dantzing, G. B., Harvey, R. P., Lansdowne, Z. F., Robinson, D. W., & Maier, S. F. (1979). Formulating and solving the network design problem by decomposition. *Transportation Research B*, *13*, 5–17.
Davis, G. A., & Nihan, N. L. (1993). Large population approximations of a general stochastic traffic assignment model. *Operations Research*, *41*, 169–178.
Davis, G. A. (1994). Exact local solution of the continuous network design problem via stochastic user equilibrium assignment. *Transportation Research B*, *28*, 61–75.
de la Barra, T. (1989). *Integrated land use and transport modelling*. Cambridge: Cambridge University Press.
de Luca, M. (2000). *Manuale di pianificazione dei trasporti*. Milano: Franco Angeli ed.
De Montis, A., De Toro, P., Droste-Franke, B., Omann, I., & Stagl, S. (2000). Criteria for quality qssessment of MCDA methods. In *Proceedings of third international conference of the European Society of Ecological Economics, Transition Towards a Sustainable Europe Ecology, Economy, Policy*, Vienna.
Del Bono, F., & Zamagni, S. (1996). *Lezioni di microeconomia*. Bologna: Il Mulino.
Di Gangi, M. (1988). Una valutazione delle prestazioni statistiche degli estimatori della matrice O/D che combinano i risultati di indagini e/o modelli con i conteggi di flussi di traffico. *Ricerca Operativa*, *51*, 23–59.
Di Maio, A., & Rostirolla, P. (2002). *Tecniche e supporti per la selezione dei progetti di investimento: un approccio per la selezione dei progetti*. Project NUVAL, La formazione per la rete dei nuclei di valutazione e verifica degli investimenti.

Dial, R. B. (1971). A probabilistic multi-path traffic assignment model which obviates path enumeration. *Transportation Research, 5*, 83–111.

Dial, R. B. (1996). Bicriterion traffic assignment: basic theory and elementary algorithms. *Transportation Science, 30*(2), 93–111.

Dodgson, T., Spackman, M., Pearman, A., & Phillips, L. (2000). *Multi-criteria analysis: a manual*. London: Department of the Environment, Transport and the Regions, HMSO.

Domencich, T. A., & McFadden, D. (1975). *Urban travel demand: a behavioural analysis*. New York: Elsevier.

Drake, Shofer, & May (1967). *A statistical analysis of speed-density hypotheses* (Highway Research Record 154). Transportation Research Board.

Drezner, Z., & Wesolowsky, G. (2003). Network design: selection and design of links and facility location. *Transportation Research A, 37*, 241–256.

Dyer, J. S. (2005). MAUT: Multiattribute utility theory. In J. Figueria, S. Greco, & M. Ehrgott (Eds.). *Multiple criteria decision analysis: state-of-the-art surveys* (pp. 265–285). Boston: Springer.

ECMT (2001). http://www.cemt.org.

Edwards, W. (1971). Social utilities. *Engineering Economist, 6*, 119–129.

Edwards, W., & Barron, F. H. (1994). SMARTS and SMARTER: improved simple methods for multiattribute utility measurement. *Organizational Behavior and Human Decision Process, 60*, 306–325.

Environmental Protection Agency (EPA) (1996). *Travel survey manual*. Washington: Department of Transportation.

Eurostat (2003). http://epp.eurostat.cec.eu.int.

Fernandez, J. E., & Friesz, T. L. (1983). Equilibrium predictions in transportation markets: the state of the art. *Transportation Research B, 17*, 155–172.

Ferrari, P. (1988). *Manuale dell'ingegneria civile: Vol. IV. Assegnazione del traffico alle reti di trasporto*. Firenze: Ed. Cremonese.

Ferrari, P. (1995). Road pricing and network equilibrium. *Transportation Research B, 29*, 357–372.

Ferrari, P. (1996). *Appunti di pianificazione dei trasporti* (2nd ed.). Internal report, Università di Pisa.

Ferrari, P. (1997). The meaning of capacity constraint multipliers in the theory of road network equilibrium. *Rendiconti del Circolo Matematico di Palermo, Serie II*, Suppl. *48*, 107–120.

Ferrari, P. (2003). A method for choosing from among alternative transportation projects. *European Journal of Operational Research, 150*, 194–203.

Ferrari, P., & Giannini, F. (1987). *Ingegneria stradale*. Vol. I. Milano: ISEDI.

Festa, D. C. (1997). *Modelli e metodi per l'ingegneria del traffico. Sviluppo e validazione di un sistema di modelli per la simulazione dell'inquinamento atmosferico prodotto dal traffico stradale*. Milano: Franco Angeli ed.

Festa, D. C., & Nuzzolo, A. (1989). Analisi sperimentale delle relazioni velocità-flusso per le strade urbane. *Le Strade, 1226*, 459–464.

Fisk, C. (1980). Some developments in equilibrium traffic assignment methodology, *Transportation Research B*, 243–255.

Fisk, C. (1984). Game theory and transportation systems modelling. *Transportation Research B, 18*, 301–313.

Fisk, S. C., & Nguyen, S. (1982). Solution algorithms for network equilibrium models with asymmetric user costs. *Transportation Science, 16*, 361–381.

Fleming, A. E., & Giugale, M. M. (2001). Financial systems in transition: a flow of funds analysis of financial evolution in Eastern Europe and Central Asia. *Journal of Comparative Economics, 29*, 586–587.

Florian, M., & Chen, Y. (1995). A coordinate descent method for the bi-level O-D matrix adjustment problem. *International Transactions Operations Research, 2*, 165–179.

Florian, M., Gaudry, M., & Lardinois, P. (1988). A two-dimensional framework for the understanding of transportation planning models. *Transportation Research B, 22*, 411–419.

Florian, M., & Hearn, D. (1995). Networks equilibrium models and algorithms. In M. O. Ball et al. (Eds.), *Handbooks in OR and MS*, Vol. 8. Amsterdam: Elsevier.
Florian, M., & Nguyen, S. (1976). An application and validation of equilibrium trip assignment methods. *Transportation Science*, *10*, 374–390.
Florian, M., & Spiess, H. (1982). The convergence of diagonalization algorithms for asymmetric network equilibrium problems. *Transportation Research B*, *16*, 477–483.
Florian, M., & Spiess, H. (1983). Transport network in practice. In *Proceedings of AIRO conference*, Napoli.
Friedrich, M., & Wekech, S. (2004). A schedule-based transit assignment model addressing the passengers' choice among competing connections. In N. H. M. Wilson & A. Nuzzolo (Eds.), *Schedule-based dynamic transit modeling. Theory and applications* (pp. 159–172). Dordrecht: Kluwer Academic.
Friesz, T. L. (1985). Transportation network equilibrium design and aggregation: key developments and research opportunities. *Transportation Research A*, *19*, 413–427.
Friesz, T. L., Bernstein, D., Smith, T. E., Tobin, R. L., & Wie, B. W. (1993). A variational inequality formulation of the dynamic network users equilibrium problem. *Operations Research*, *41*, 179–191.
Friesz, T. L., Cho, H. J., Mehta, N. J., Tobib, R. L., & Anandalingam, G. (1992). A simulated annealing approach to the network design problem with variational inequality constraints. *Transportation Science*, *26*, 18–26.
Friesz, T. L., Luque, J., Tobin, R. L., & Wie, B. Y. (1989). Dynamic network traffic assignment considered as a continuous time optimal control problem. *Operations Research*, *37*, 893–901.
Friesz, T. L., Tobin, R. L., & Harker, P. T. (1983). The state of the art in predictive freight network models. *Transportation Research A*, *17*, 409–417.
Fukushima, M. (1984). A modified Frank–Wolfe algorithm for solving the traffic assignment problem. *Transportation Research B*, *18*, 169–178.
Fusco Girard, L., & Nijkamp, P. (2005). *Energia, bellezza, partecipazione: la sfida della sostenibilità. Valutazioni integrate tra conservazione e sviluppo*. Milano: Franco Angeli ed.
Gallo, M. (2002). Un approccio meta-euristico alla progettazione dei parametri topologici delle reti di trasporto urbano. *Sistemi di Trasporto*, *1*, 15–33.
Gallo, M., D'Acierno, L., & Montella, B. (2009). A meta-heuristic approach for solving the Urban Network Design Problem. *European Journal of Operational Research*, in press.
Gallo, G., Longo, G., Nguyen, S., & Pallottino, S. (1993). Directed hypergraphs and applications. *Discrete Applied Mathematics*, *2*, 177–201.
Gallo, G., & Pallottino, S. (1982). Introduction and recent advances in shortest path methods. In *Proceedings of the 15th course on transportation planning models*. Amalfi: ICTS-CNR.
Gallo, G., & Pallottino, S. (1988). Shortest path algorithms. *Annals of Operational Research*, *13*, 3–79.
Gentile, G., Meschini, L., & Papola, N. (2004). Fast heuristics for continuous dynamic shortest paths and all-or-nothing assignment. Presented at AIRO 2004. Lecce: Italy.
Gentile, G., Meschini, L., & Papola, N. (2005). Macroscopic arc performance models with capacity constraints for within-day dynamic traffic assignment. *Transportation Research Part B*, *39*, 319–338.
Gentile, G., Meschini, L., & Papola, N. (2007). Spillback congestion in dynamic traffic assignment: a macroscopic flow model with time-varying bottlenecks. *Transportation Research Part B*, *41*, 1114–1138.
Geweke, J. (1991). Efficient simulation from the multivariate normal and Student-t distributions subject to linear constraints. In E.M. Keramidas (Ed.), *Computer science and statistics, proceedings of 23rd symposium on interface*.
Ghali, M. O., & Smith, M. J. (1993). Traffic assignment, traffic control and road pricing. In C. F. Daganzo (Ed.), *Transportation and traffic theory* (pp. 147–169). Amsterdam: Elsevier.
Golden, B., Wasil, E., & Harker, P. (Eds.), (1989). *The analytic hierarchy process: applications and studies*. New York: Springer.
Golob, T. F., & McNally, M. G. (1997). A model of activity participation and travel interactions between household heads. *Transportation Research B*, *31*(3), 177–194.

Greenshields, B. (1934). A study of traffic capacity. In *Proceedings of the highway research board 14*. Transportation Research Board.

Guihaire, V., & Hao, J.-K. (2008). Transit network design and scheduling: a global review. *Transportation Research A*, *42*, 1251–1273.

Gunn, H., & Bates, J. (1982). Statistical aspects of travel demand modelling. *Transportation Research A*, *16*, 371–382.

Haimes, Y. Y., & Chankong, V. (1985). *Decision making with multiple objectives*. Berlin: Springer.

Hajivassiliou, V., & McFadden, D. (1998). The method of simulated scores for the estimation of LDV models. *Econometrica*, *66*, 863–896.

Han, B. (1996). Optimising traffic signal settings for periods of time-varying demand. *Transportation Research A*, *30*, 207–230.

Harker, P. T. (1985). *Spatial price equilibrium: advances in theory, computation and application*. Heidelberg: Springer.

Harker, P. T. (1987). *Predicting intercity freight flows*. Topics in Transportation. Utrecht: VNU Science Press.

Hearn, D. W. (1997). Toll pricing models for traffic networks. In *Preprints of the 8th IFAC/IFIP/IFORS symposium*, Chania.

Hearn, D. W., Lawphongpanich, S., & Nguyen, S. (1984). Convex programming formulations of the asymmetric traffic assignment problem. *Transportation Research B*, *18*, 357–365.

Hensher, D. A., Barnard, P. O., & Truong, T. P. (1988). The role of stated preference methods in studies of travel choice. *Journal of Transport Economics and Policy*, *22*, 45–58.

Heydecker, B. G. (1996). A decomposition approach for signal optimisation in road networks. *Transportation Research B*, *30*, 99–114.

Heydecker, B. G., & Addison, J. D. (1998). Analysis of traffic models for dynamic equilibrium traffic assignment. In M. G. H. Bell (Ed.), *Transportation networks: recent methodological advances* (pp. 35–49). Oxford, UK: Pergamon.

Hickman, M. D., & Bernstein, D. H. (1997). Transit service and path choice models in stochastic and time-dependent networks. *Transportation Science*, *31*, 129–146.

Hickman, M. D., & Wilson, N. H. M. (1995). Passenger travel time and path choice implications of real-time transit information. *Transportation Research*, *4*, 211–226.

Hinloopen, E., & Nijkamp, P. (1990). Qualitative multiple criteria choice analysis. *Quality and Quantity*, *24*, 37–56.

Horowitz, J. L. (1981). Identification and diagnosis of specification errors in the multinominal logit model. *Transportation Research B*, *15*, 345–360.

Horowitz, J. L. (1982). Evaluation of usefulness of two standard goodness of fit indicators for comparing non-nested random utility models. *Transportation Research Records*, *874*, 19–25.

Horowitz, J. L. (1984). The stability of stochastic equilibrium in a two link transportation network. *Transportation Research B*, *18*, 13–28.

Horowitz, J. L. (1985). Travel and location behaviour: state of the art and research opportunities. *Transportation Research A*, *19*, 441–454.

Horowitz, J. L., Spermann, J. M., & Daganzo, C. F. (1982). An investigation on the accuracy of the Clark approximation for the multinominal probit model. *Transportation Science*, *16*, 382–401.

Hu, T. C. (1970). *Integer programming and network flows*. Reading: Addison–Wesley.

Hu, T. Y., & Mahmassani, H. S. (1997). Day-to-day evolution of network flows under real-time information and reactive signal control. *Transportation Research C*, *5*, 51–69.

Hurdle, V. F. (1984). *Signalized intersection delay models: a primer for the uninitiated*. Transportation Research Record 971.

Hutchinson, G. (1974). *Principles of urban transportation systems planning*. New York: McGraw–Hill.

Iman, M. O. (1998). Optimal design of public bus service with demand equilibrium. *Journal of Transportation Engineering*, *124*, 431–436.

Institute of Transportation Engineers (1982). *Transportation and traffic engineering handbook*. Englewood Cliffs: Prentice Hall.

Izard, W. (1951). Interregional and regional input–output analysis: a model of a space-economy. *The Review of Economics and Statistics*, *33*, 318–328.

Janson, B. N. (1989). Dynamic traffic assignment for urban road network. *Transportation Research B*, *25*, 143–161.

Jara-Diaz, S. R., & Friesz, T. (1982). Measuring the benefits of a transportation investment. *Transportation Research B*, *16*, 57–77.

Jayakrishnan, R., Mahamassani, H., & Hu, T. (1994). An evaluation tool for advanced traffic information and management system in urban networks. *Transportation Research C*, *2*, 129–147.

Jha, M., Madanat, S., & Peeta, S. (1998). Perception updating and day-to-day travel choice dynamics in traffic networks with information provision. *Transportation Research C*, *6*, 189–212.

Jolliffe, J. K., & Hutchinson, T. P. (1975). A behavioural explanation of the association between bus and passenger arrival times at a bus stop. *Transportation Science*, *9*, 248–282.

Jones, P. M., Koppelman, F., & Orfeuil, J. P. (1990). Activity analysis: state-of-the-art and future directions. In P. Jones (Ed.). *Developments in dynamic and activity-based approaches to travel analysis*. Gower: Aldershot.

Keane, M. (1994). A computationally practical simulation estimation for panel data. *Econometrica*, *62*, 95–116.

Keeney, R. L., & Raiffa, H. (1993). *Decision with multiple objectives*. Cambridge: Cambridge University Press.

Kimber, R. M., & Hollis, E. M. (1978). Peak period delays at road junctions and other bottlenecks. *Traffic Engineering and Control*, *19*, 442–446.

Kimber, R. M., Marlow, M., & Hollis, E. M. (1977). Flow-delay relationships at major-minor junctions. *Traffic Engineering and Control*, *18*, 516–520.

Kitamura, R., Kostyniuc, L., & Ting, K. L. (1979). Aggregation in spatial choice modelling. *Transportation Science*, *13*, 325–342.

Kleinrock, L. (1975). *Queuing system*, Vols. I, II. New York: Wiley.

Koppelman, F. S. (1989). Multidimensional model system for intercity travel choice behaviour. *Transportation Research Records*, *1241*, 1–8.

Koppelman, F. S., & Hauser, J. (1978). Destination choice behavior for non-grocery-shopping trips. *Transportation Research Records*, *673*, 157–165.

Koppelman, F. S., & Wen, C. H. (2000). The paired combinatorial logit model: properties, estimation and application. *Transportation Research B*, *34*, 75–89.

Langdon, M. G. (1984). Improved algorithms for estimating choice probabilities in the multinomial probit model. *Transportation Science*, *18*, 267–299.

Le Blanc, L. J. (1975). An algorithm for the discrete network design problem. *Transportation Science*, *9*, 183–199.

Le Blanc, L. J., & Boyce, D. E. (1986). A bilevel programming algorithm for exact solution of the network design problem with user optimal flows. *Transportation Research B*, *20*, 259–265.

Le Blanc, L. J., Morlook, E. K., & Pierskalla, W. P. (1975). An efficient approach to solving the road network equilibrium traffic assignment problem. *Transportation Research*, *9*, 309–318.

Lee, M., & McNally, M. G. (2006). An empirical investigation on the dynamic processes of activity scheduling and trip chaining. *Transportation*, *33*(6), 553–565.

Leontief, W., & Costa, P. (1987). *Il trasporto merci e l'economia italiana. Scenari di interazione al 2000 e al 2015. Sistemi operativi*. New York: Venezia.

Leontief, W., & Strout, A. (1963). *Multiregional input–output analysis*. Teoria economica delle interdipendenze settoriali, Etas/Kompass ed.

Leurent, F. (1993). Cost versus time equilibrium over a network. *European Journal of Operational Research*, *71*, 205–221.

Leurent, F. (1995). The practice of a dual criteria assignment model with continuously distributed values-of-time. In *Proceedings of the 23rd European transport forum: transportation planning methods, E* (pp. 117–128). London: PTRC.

Leurent, F. (1996). The theory and practice of a dual criteria assignment model with a continuously distributed value-of-time. In J. B. Lesort (Ed.), *Transportation and traffic theory* (pp. 455–477). Exeter: Pergamon.

Lighthill, M. J., & Witham, G. B. (1955). On kinematic waves. II. Theory of traffic flows on long crowded roads. *Proceedings of the Royal Society of London Series A*, *229*, 317–345.

Lindveld, K. (2003). *Dynamic O-D matrix estimation*. Ph.D. thesis.

Lo, K. H., Chang, E., & Chan, Y. C. (2001). Dynamic network traffic control. *Transportation Research A*, *35*, 721–744.

Los, M., & Lardinois, C. (1982). Combinatorial programming, statistical optimization and the optimal transportation network problem. *Transportation Research B*, *16*, 89–124.

Louviere, J. J. (1988). Conjoint analysis modelling of stated preferences. *Journal of Transport Economics and Policy*, *22*, 93–120.

Louviere, J. J., Hensher, D., & Swait, J. (2000). *Stated choice methods: analysis and application*. Cambridge: Cambridge University Press.

Lozano, A., & Storchi, G. (2001). Shortest viable path algorithm in multimodal networks. *Transportation Research A*, *35*, 225–241.

Lupi, M. (1986). Convergence of the Frank–Wolf algorithm in transportation networks. *Civil Engineering Systems*, *3*, 7–15.

Lupi, M. (1996). Determinazione del livello di servizio di una intersezione urbana: alcune osservazioni sulla modellizzazione del ritardo. In *Proceedings of SIDT conference*, Napoli.

Mahamassani, H., & Liu, Y. H. (1999). Dynamics of commuting decision behaviour under advanced traveller information systems. *Transportation Research C*, *7*, 91–107.

Maher, M. (1997). A probit-based stochastic user equilibrium assignment model. *Transportation Research B*, *31*, 341–355.

Maher, M. (1998). Algorithms for logit-based stochastic user equilibrium assignment. *Transportation Research B*, *32*, 539–549.

Maher, M., Stewart, K., & Rosa, A. (2005). Stochastic social optimum traffic assignment. *Transportation Research B*, *39*, 753–767.

Maher, M. J. (1983). Inference on trip matrices from observations on link volumes. A Bayesian statistical approach. *Transportation Research B*, *17*, 435–447.

Maher, M. J., & Hughes, P. C. (1997). An algorithm for SUEED – stochastic user equilibrium with elastic demand. In *Preprints of the 8th IFAC/IFIP/IFORS symposium* (pp. 1299–1304), Chania.

Maher, M. J., & Hughes, P. C. (1998). Recent developments in stochastic assignment modelling. *Traffic Engineering and Control*, *39*, 174–179.

Manheim, M. (1979). *Fundamentals of transportation systems analysis*. Cambridge: MIT Press.

Manski, C. (1977). The structure of random utility models. *Theory and Decision*, *8*, 229–254.

Manski, C. F., & McFadden, D. (1981). *Alternative estimators and sample designs for discrete choice analysis*. In *Structural analysis of discrete data with econometric applications*. Cambridge: MIT Press.

Manski, C. F., & Lerman, S. R. (1977). The estimation of probabilities from choice-based samples. *Econometrica*, *45*, 1977–1988.

Marcotte, P. (1983). Network optimization with continuous control parameters. *Transportation Science*, *17*, 181–197.

Marcotte, P., Nguyen, S., & Tanguay, K. (1996). Implementation of an efficient algorithm for the multiclass traffic assignment problem. In J. B. Lesort (Ed.), *Transportation and traffic theory* (pp. 455–477). Exeter: Pergamon.

Marcotte, P., & Zhu, D. (1996). An efficient algorithm for a bicriterion traffic assignment problem. In L. Bianco & P. Toth (Eds.), *Advanced methods in transportation analysis* (pp. 63–73). Berlin: Springer.

Martins, C. L., & Pato, M. V. (1998). Search strategies for the feeder bus network design problem. *European Journal of Operational Research*, *106*, 425–440.

Marzano, V., & Papola, A. (2004a). A link-based path multilevel logit model for route choice which allows implicit path enumeration. In *Proceedings of 2004 ETC conference*, Strasbourg.

Marzano, V., & Papola, A. (2004b). Modelling freight demand at a national level: theoretical developments and application to Italian demand. In *Proceedings of 2004 ETC conference*, Strasbourg.

Marzano, V., & Papola, A. (2008). On the covariance structure of the cross nested logit model. *Transportation Research B*, *42*(2), 83–98.

Marzano, V., Papola, A., & Simonelli, F. (2008). Investigating the effectiveness of the O-D matrix correction procedure using traffic counts. In *Proceedings of the 87th TRB meeting*, Washington.
May, A. D. (1990). *Traffic flow fundamentals*. Englewood Cliffs: Prentice Hall.
Mazzarino, D. (1997). Modeling freight transport demand: a survey. *Trasporti Europei*, 5.
McFadden, D. (1978). Conditional logit analysis of qualitative choice behaviour. In P. Zarembka (Ed.), *Frontiers of econometrics* (pp. 105–142). New York: Academic Press.
McNally, M. G. (2000). *The activity-based approach*. Irvine: Institute of Transportation Studies and Department of Civil & Environmental Engineering University of California.
McShane, W. R., & Roess, R. P. (1990). *Traffic engineering*. Englewood Cliffs: Prentice Hall.
McFadden, D. (1978). Modelling the choice of residential location. In *Spatial interaction theory and planning models* (pp. 75–96). Amsterdam.
Melkote, S., & Daskin, M. (2001). An integrated model of facility location and transportation network design. *Transportation Research A*, *35*, 515–538.
Meneguzzer, C. (1995). An equilibrium route choice model with explicit treatment of the effect of intersections. *Transportation Research B*, *29*, 329–356.
Meng, Q., Yang, H., & Bell, M. (2001). An equivalent continuously differentiable model and a locally convergent algorithm for the continuous network design problem. *Transportation Research B*, *35*, 83–105.
Menichini, F. (2003). Una analisi critica dei metodi di scelta fra progetti alternativi. *Sistemi di Trasporto*, *2*, 14–29.
Merchant, D. K., & Nemhauser, G. L. (1978). A model and an algorithm for the dynamic traffic assignment problem. *Transportation Science*, *12*, 183–199.
Messmer, A., & Papageorgiou, M. (1990). Metanet: a macroscopic simulation program for motorway networks. *Traffic Engineering and Control*, *31*, 466–470.
Meyer, M. D., & Miller, E. J. (2001). *Urban transportation planning: a decision oriented approach*. New York: McGraw–Hill.
Miller, R. E., & Blair, P. D. (1985). *Input output analysis: foundations and extensions*. Englewood Cliffs: Prentice Hall.
Mirchandani, P., & Head, L. (2001). A real-time traffic signal control system: architecture, algorithms, and analysis. *Transportation Research C*, *9*, 415–432.
Mirchandani, P., & Soroush, H. (1987). Generalized traffic equilibrium with probabilistic travel times and perceptions. *Transportation Science*, *21*, 133–152.
Mishan, E. J. (1974). *Analisi costi-benefici*. Milano: Etas Libri.
Modenese Vieira, L. F. (1992). *The value of service in freight transportation*. Ph.D. dissertation. Boston: MIT.
Montella, B. (1996). *Pianificazione e controllo del traffico urbano: modelli e metodi*. Napoli: Ed. Cuen.
Montella, B., & Cascetta, E. (1978). Tempo di attesa alle fermate di un servizio di trasporti collettivo urbano. *Ingegneria Ferroviaria*, *9*, 1–8.
Montella, B., D'Acierno, L., & Gallo, M. (2007). Transportation network design methods under the assumption of elastic demand. In *Proceedings of 11th world conference on transport research*. Berkeley, USA: University of California.
Montella, B., & Gallo, M. (2002). Un modello per la progettazione dei sistemi di trasporto collettivo in ambito urbano. In *Rilievi, modellizzazione e controllo del traffico veicolare* (pp. 182–201). Milano: Franco Angeli ed.
Montella, B., Gallo, M., & D'Acierno, L. (2000). Multimodal network design problems. In L. J. Sucharov (Ed.), *Urban transport V – urban transport and the environment for the 21st century* (pp. 405–414). Southampton: WIT Press.
Moses, L. N. (1955). The stability of interregional trading patterns and input–output analysis. *American Economic Review*, *45*, 803–832.
Munda, G. (1995). *Multicriteria evaluation in a fuzzy environment: theory and applications in ecological economics*. Heidelberg: Physica-Verlag.
Newell, G. F. (1971). *Application of queuing theory*. London: Chapman and Hall LTD.
Newell, G. F. (1980). *Traffic flows in transportation networks*. Cambridge: MIT Press.

Newell, G. F. (1993). A simplified theory of kinematic waves in highway traffic, part I: general theory; part II: queuing at freeway bottlenecks; part III: multi-destination flows. *Transportation Research Part B*, *27*, 281–313.

Nguyen, S. (1976). A unified approach to equilibrium methods for traffic assignments. In M. Florian (Ed.), *Lecture notes in economics and mathematical systems: Vol. 118. Traffic equilibrium methods* (pp. 148–182). New York: Springer.

Nguyen, S., & Dupuis, C. (1984). An efficient method for computing traffic equilibria in a network with asymmetric transportation costs. *Transportation Science*, *18*, 185–202.

Nguyen, S., Morello, E., & Pallottino, S. (1989). Discrete time dynamic estimation model for passenger origin–destination matrices on transit networks. *Transportation Research B*, *22*, 251–260.

Nguyen, S., & Pallottino, S. (1988). Equilibrium traffic assignment for large scale transit networks. *European Journal of Operational Research*, *37*, 176–186.

Nguyen, S., Pallottino, A., & Gendreau, M. (1993). *Implicit enumeration of hyperpaths in a logit model for transit networks*. CRT Publication Nr. 84.

Nielsen, O. A. (1997). On the distributions of the stochastic components in sue traffic assignment models. In *Proceedings of 25th European transport forum annual meeting, seminar F on transportation planning methods*, Vol. II.

Nielsen, O. A. (2000). A stochastic transit assignment model considering differences in passengers utility functions. *Transportation Research B*, *34*, 377–402.

Nielsen, O. A. (2004). A large scale stochastic multi-class schedule-based transit model with random coefficients. In N. H. M. Wilson & A. Nuzzolo (Eds.), *Schedule-based dynamic transit modeling. Theory and applications* (pp. 53–73). Dordrecht: Kluwer Academic.

Nielsen, O. A., Daly, A., & Frederiksen, R. D. (2002). A stochastic route choice model for car travellers in the Copenhagen region. *Networks and Spatial Economics*, *2*, 327–346.

Nielsen, O. A., Frederiksen, R. D., & Simonsen, N. (1998). Stochastic user equilibrium traffic assignment with turn–delays in intersections. *IFORS*, *5*, 555–568.

Nijkamp, P., Rietveld, P., & Voogd, H. (1990). *Multicriteria evaluation in physical planning*. Amsterdam: Elsevier.

Nijkamp, P., & Van Delft, A. (1977). *Multi-criteria analysis and regional decision-making*. Leiden: Martinus Nijhoff.

Nuzzolo, A., & Crisalli, U. (2004). The schedule-based approach in dynamic transit modelling: a general overview. In N. H. M. Wilson & A. Nuzzolo (Eds.), *Schedule-based dynamic transit modeling. Theory and applications* (pp. 1–24). Dordrecht: Kluwer Academic.

Nuzzolo, A., Crisalli, U., & Gangemi, F. (2000). A behavioural choice model for the evaluation of railway supply and pricing policies. *Transportation Research A*, *35*, 211–226.

Nuzzolo, A., & Russo, F. (1993). Un modello di rete diacronica per l'assegnazione dinamica al trasporto collettivo extraurbano. *Ricerca Operativa*, *67*, 37–56.

Nuzzolo, A., & Russo, F. (1995). A disaggregate freight modal choice model. In *Proceedings of 7th WCTR*, Sydney.

Nuzzolo, A., & Russo, F. (1996). Stochastic assignment models for transit low frequency services: some theoretical and operative aspects. In *Advanced methods in transportation analysis* (pp. 321–339). New York: Springer.

Nuzzolo, A., & Russo, F. (1997). *Modelli per l'analisi e la simulazione dei sistemi di trasporto collettivo*. Milano: Franco Angeli ed.

Nuzzolo, A., Russo, F., & Crisalli, U. (1999). A doubly dynamic assignment model for congested urban transit networks. In *Proceedings of 27th European transport forum, seminar F* (pp. 185–196), Cambridge, England.

Nuzzolo, A., Russo, F., & Crisalli, U. (2001). A doubly dynamic schedule-based assignment model for transit networks. *Transportation Science*, *35*, 268–285.

Nuzzolo, A., Russo, F., & Crisalli, U. (2003). *Transit network modelling. The schedule-based dynamic approach*. Milano: Franco Angeli ed.

OCED (2001). http://www.oecd.org.

Oi, K. I. Y., & Shuldiner, P. W. (1962). *An analysis of urban travel demands*. Evanston: Northwestern University Press.

Okutani, I., & Stephanades, Y. (1984). Dynamic prediction of traffic volume through Kalman filtering theory. *Transportation Research B*, *18*, 1–11.
Olaru, D., & Smith, B. (2005). Modelling behavioural rules for daily activity scheduling using fuzzy logic. *Transportation*, *32*(4), 423–441.
Ortuzar, J. de D. (1992). Stated preferences in travel demand modelling. In *Proceedings of world conference on transportation research*. Lyon.
Ortuzar, J. de D., & Willumsen, L. G. (2001). *Modelling transport* (3rd edn.). New York: Wiley.
Pallottino, S., & Scutellà, M. G. (1997). Shortest path algorithms in transportation models: classical and innovative aspects. In P. Marcotte & S. Nguyen (Eds.), *Proceedings of the equilibrium and advanced transportation modelling colloquium* (pp. 245–281). Amsterdam: Kluwer Academic.
Papageorgiou, M. (1991). *Concise encyclopedia of traffic and transportation systems*. Oxford: Pergamon Press.
Papola, A. (1996). I modelli di Valore Estremo Generalizzato (GEV) per la simulazione della domanda di trasporto (Internal report). Dipartimento di Ingegneria dei Trasporti, Università di Napoli "Federico II".
Papola, A. (2004). Some developments on the cross nested logit model. *Transportation Research B*, *38*, 833–851.
Pasquill, F., & Smith, F. B. (1983). *Atmospheric diffusion* (3rd edn.). Chichester: Ellis Horwood.
Patriksson, M. (1994). *The traffic assignment problem: models and methods. Topics in transportation*. Utrecht: VNU Science Press.
Pattnaik, S. B., Mohan, S., & Tom, V. M. (1998). Urban transit route network design using genetic algorithm. *Journal of Transportation Engineering*, *124*, 368–375.
Pearmin, D., Swanson, J., Kroes, E., & Bradley, M. (1991). *Stated preferences techniques: a guide to practice*. London: Steer Davies Gleave and Hague Consulting Group.
Picard, G., & Nguyen, S. (1987). Estimation of interregional freight flows using input–output analysis. In L. Bianco & A. La (Eds.), *Freight transport planning and logistics* (pp. 176–209). Berlin: Springer.
Pignataro, L. J. (1973). *Traffic engineering, theory and practice*. Englewood Cliffs: Prentice Hall.
Poorzahedy, H., & Turnquist, M. A. (1982). Approximate algorithms for the discrete network design problem. *Transportation Research B*, *16*, 45–55.
Potts, R. B., & Oliver, R. M. (1972). *Flows in transportation networks*. New York: Academic Press.
Powell, W. B., & Sheffi, Y. (1982). The convergence of equilibrium algorithms with predetermined step sizes. *Transportation Science*, *16*, 45–55.
Ran, B., & Boyce, D. E. (1994). *Dynamic urban transportation network models: theory and implications for intelligent vehicle-highway systems*. New York: Springer.
Ran, B., Rouphail, N. M., Tarko, A., & Boyce, D. E. (1997). Toward a class of link travel time functions for dynamic assignment models on signalised networks. *Transportation Research B*, *31*, 277–290.
Rama Moorthy, N. V. (1997). Planning of integrated transit network for bus and LRT. *Journal of Advanced Transportation*, *31*, 283–309.
Regan, A., & Garrido, R. A. (2001). Freight demand and shipper behavior modeling: state of the art, directions for the future. In D. A. Hensher & J. King (Eds.), *Travel behavior research, the leading edge* (pp. 185–216). Oxford: Pergamon Press.
Richards, P. I. (1956). Shock waves on the highway. *Operations Research*, *4*, 42–51.
Richards, M., & Ben-Akiva, M. (1975). *A disaggregate travel demand model*. Lexington: D.C. Heath.
Road Research Laboratory (1965). *Research on road traffic*. London: RRL Publication.
Robertson, D. I. (1979). Traffic models and optimum strategies of control: a review. In W. S. Homburger & L. Steinman (Eds.), *Proceedings of the international symposium on traffic control systems* (Vol. 1, pp. 262–288). Berkeley: University of California.
Robertson, H. D., Hummer, J. E., & Nelson, D. C. (1994). *Manual of transportation engineering studies*. Englewood Cliffs: Prentice Hall.
Roson, R. (1993). A multiregional network general equilibrium model. In *Proceedings of transportation and general equilibrium models workshop*, Venezia.

Rostirolla, P. (1998). *La fattibilità economico-finanziaria: metodi ed applicazioni*. Napoli: Liguori ed.

Roy, B., & Vanderpooten, D. (1996). The European school of MCDA: emergence, basic features and current works. *Journal of Multi-Criteria Decision Analysis*, *5*, 22–37.

Russo, F. (1995). Un modello per la progettazione delle frequenze. In *Sviluppi della ricerca sui sistemi di trasporto* (pp. 361–379). Milano: Franco Angeli ed.

Russo, F., & Cartenì, A. (2005). Application of a tour-based model to simulate freight distribution in a large urbanized area. In *Proceedings of 4th international conference on city logistics*, Langkawi.

Russo, F., & Vitetta, A. (1995). Networks and assignment models for the Italian national transportation systems. In *Proceeding of the 7th WCTR*, Sidney.

Russo, F., & Vitetta, A. (2003). An assignment model with modified logit, which obviates enumeration and overlapping problems. *Transportation*, *30*, 177–201.

Saaty, T. L. (1990). *Multicriteria decision making – the analytic hierarchy process–planning, priority, setting, resource allocation*. Pittsburgh: RWS.

Savage, L. J. (1954). *The foundations of statistics*. New York: Wiley.

Sayers, T. M., Jessop, A. T., & Hills, P. J. (2003). Multicriteria evaluation of transport options, flexible, transport and user friendly? *Transport Policy*, *10*, 95–106.

Seddon, P. A., & Day, M. P. (1974). Bus passenger waiting time in Greater Manchester. *Traffic Engineering and Control*, *15*, 442–445.

Sheffi, Y. (1985). *Urban transportation networks*. Englewood Cliffs: Prentice Hall.

Sheffi, Y., & Powell, W. B. (1983). Optimal signal settings over transportation networks. *Journal of Transportation Engineering*, *109*, 824–839.

Sheffi, Y., & Powell, W. B. (1981). A comparison of stochastic and deterministic traffic assignment over congested networks. *Transportation Research B*, *15*, 53–64.

Sheffi, Y., & Powell, W. B. (1982). An algorithm for the equilibrium assignment problem with random link times. *Networks*, *12*, 191–207.

Shiftan, Y., Ben-Akiva, M., de Jong, G., Hallkert, S., & Simmonds, D. (2002). Evaluation of externalities in transport projects. *European Journal of Transport and Infrastructure Research*, *2*, 285–303.

Shih, M. C., & Mahmassani, H. S. (1991). Vehicle sizing model for bus transit network. *Transportation Research Record*, *1452*, 35–41.

Shim, J. P. (1989). Bibliographical research on the analytic hierarchy process (AHP). *Socio-Economic Planning Sciences*, *23*, 161–167.

Small, K. A. (1982). The scheduling of commuter activities: work trips. *The American Economic Review*, *72*, 467–479.

Small, K. A. (1987). A discrete choice model for ordered alternatives. *Econometrica*, *55*, 409–424.

Smith, M. E. (1979). Design of small sample home interview travel surveys. *Transportation Research Records*, *701*, 29–35.

Smith, M. J. (1979). The existence, uniqueness and stability of traffic equilibria. *Transportation Research B*, *13*, 295–304.

Smith, M. J., & Van Vuren, T. (1993). Traffic equilibrium with responsive traffic control. *Transportation Science*, *27*, 118–132.

Soehodo, S., & Koshi, M. (1999). Design of public transit network in urban area with elastic demand. *Journal of Advanced Transportation*, *33*, 335–369.

Solanki, R. S., Gorti, J. K., & Southworth, F. (1998). The highway network design problem. *Transportation Research B*, *32*, 127–140.

Spiess, H., & Florian, M. (1989). Optimal strategies: a new assignment model for transit networks. *Transportation Research B*, *23*, 82–102.

Stopher, P. R., & Meybourg, A. H. (1975). *Urban transportation modelling and planning*. Lexington: Lexington Books.

Stopher, P. R., & Meybourg, A. H. (1976). *Transportation systems evaluation*. Lexington: Lexington Books.

Suwansirikul, C., Friesz, T. L., & Tobin, V. (1987). Equilibrium decomposed optimization: a heuristic for the continuous equilibrium network design problem. *Transportation Science*, *21*, 254–263.
Thomas, R. (1991). *Traffic assignment techniques*. Aldershot: Avebury Technical.
Tong, C. O., & Wong, S. C. (2000). A predictive dynamic traffic assignment model in congested capacity-constrained road networks. *Transportation Research B*, *34*, 625–644.
Train, K. (2003). *Discrete choice methods with simulation*. Cambridge: Cambridge University Press.
Transportation Research Board (2000). *Highway capacity manual 2000*. Washington: National Research Council.
Trezza, B., Moesch, G., & Rostirolla, P. (1978). *Economia pubblica: investimenti e tariffe*. Milano: Franco Angeli ed.
Tversky, A. (1972). Elimination by aspect: a theory of choice. *Psychological Review*, *79*, 281–299.
Van Vliet, D. (1981). Selected node-pair analysis in Dial's assignment algorithms. *Transportation Research B*, *15*, 65–68.
Van Vliet, D. (1987). The Frank–Wolfe algorithm for equilibrium traffic assignment viewed as a variational inequality. *Transportation Research*, *21*, 87–89.
Van Zuylen, J. H., & Willumsen, L. G. (1980). The most likely trip matrix estimated from traffic counts. *Transportation Research B*, *14*, 281–293.
Vinckie, P. (1992). *Multicriteria decision aid*. Chichester: Wiley.
Von Neumann, J., & Morgenstern, O. (1947). *Theory games and economic behavior*. Princeton: Princeton University Press.
Voogd, H. (1983). *Multi-criteria evaluation for urban and regional planning*. London: Pion.
Vovsha, P. (1997). Cross-nested logit model: an application to mode choice in the Tel-Aviv metropolitan area. *Transportation Research Record*, *1607*, 6–15.
Vovsha, P., & Bekhor, S. (1998). The link-nested logit model of route choice: overcoming the route overlapping problem. *Transportation Research Record*, *1645*, 133–142.
Vytoulkas, P. K. (1990). A dynamic stochastic assignment model for the analysis of general networks. *Transportation Research B*, *24*, 453–469.
Wachs, M. (1985). Planning, organizations and decision-making: a research agenda. *Transportation Research A*, *19*, 521–532.
Walker, J. (2001). *Extended discrete choice models: integrated framework, flexible error structures and latent variables*. Ph.D. dissertation, Massachusetts Institute of Technology.
Wardrop, J. G. (1952). Some theoretical aspects of road traffic research. In *Proceedings of the institution of civil engineers*, Part II (vol. 1, pp. 325–378).
Watling, D. P. (1996). Asymmetric problems and stochastic process models of traffic assignment. *Transportation Research B*, *30*, 339–357.
Watling, D. P. (1999). Stability of the stochastic equilibrium assignment problem: a dynamical systems approach. *Transportation Research B*, *33*, 281–312.
Webster, F. W. (1958). Traffic signal settings (Technical Paper Nr. 39). London: Transportation Road Research Laboratory.
Webster, F. W., & Cobbe, B. M. (1966). Traffic signals (Technical Paper Nr. 56). London: Ministry of Transport, Road Research.
Wen, C. H., & Koppelman, F. S. (2001). The generalized nested logit model. *Transportation Research B*, *35*, 627–641.
Wie, B. W., Friesz, T. L., & Tobin, T. L. (1990). Dynamic user optimal traffic assignment on congested multi-destination networks. *Transportation Research B*, *24*, 431–442.
Williams, H. C. W. L. (1977). On the formation of travel demand models and economic evaluation measures of user benefit. *Environment and Planning A*, *9*, 285–344.
Williams, H. C. W. L., & Ortùzar, J. de D. (1982). Behavioural theories of dispersion and the mis-specification of travel demand models. *Transportation Research B*, *16*, 167–219.
Willumsen, L. G., & Tamin, O. Z. (1989). Transport demand model estimation from traffic counts. *Transportation*, *16*, 3–26.
Wilson, A. G. (1974). *Urban and regional models in geography and planning*. New York: McGraw–Hill.

Wilson, N. H. M., & Nuzzolo, A. (2004). *Schedule-based dynamic transit modeling. Theory and applications*. Dordrecht: Kluwer Academic.

Wilson, N. H. M., & Nuzzolo, A. (2008). *Schedule-based modeling of transportation networks. Theory and applications*. Berlin: Springer. ISBN: 978-0-387-84811-2.

Winston, C. (1983). The demand for freight transportation: models and applications. *Transportation Research A*, *17*, 419–427.

Wohl, M., & Martin, B. V. (1967). *Traffic system analysis for engineers and planners*. New York: McGraw–Hill.

Wong, S. C., Wong, W. T., Leung, C. M., & Tong, C. O. (2002). Group-based optimization of a time-dependent transyt traffic model for area traffic control. *Transportation Research B*, *36*, 291–312.

Wong, C. S., & Yang, C. (1999). An iterative group-based signal optimization scheme for traffic equilibrium networks. *Journal of Advanced Transportation*, *33*, 201–217.

Wong, C. S., & Yang, H. (1997). Reserve capacity of a signal-controlled road network. *Transportation Research B*, *31*, 397–402.

Wu, J. H., Chen, Y., & Florian, M. (1998). The continuous dynamic network loading problem a mathematical formulation and solution method. *Transportation Research B*, *32*, 173–187.

Wu, J. H., Florian, M., & Marcotte, P. (1994). Transit equilibrium assignment: a model and solution algorithms. *Transportation Science*, *28*, 193–203.

Xu, Y. W., Wu, J. H., Florian, M., Marcotte, P., & Zhu, D. L. (1999). New advances in the continuous dynamic network loading problem. *Transportation Science*, *33*, 341–353.

Yang, H. (1995). Heuristic algorithms for the bi-level origin/destination matrix estimation problem. *Transportation Research B*, *29*, 231–242.

Yang, H., & Yagar, S. (1995). Traffic assignment and signal control in saturated road networks. *Transportation Research A*, *29*, 125–139.

Yang, Q., & Koutsopoulos, H. (1996). A microscopic traffic simulator for evaluation of dynamic traffic management systems. *Transportation Research C*, *4*, 113–129.

Yates, F. (1981). *Sampling methods for censuses and surveys*. London: Griffin.

Zadeh, L. A. (1976). A fuzzy-algorithmic approach to the definition of complex or imprecise concepts. *International Man-Machine Studies*, *8*, 249–291.

Zahedi, F. (1986). The analytic hierarchy process: a survey of the method and its applications. *Interfaces*, *16*, 96–108.

Zhao, Y., & Kockelman, K. M. (2003). The random utility based multiregional input–output model: solution existence and uniqueness. In *Proceedings of 82nd annual meeting of transportation research board*.

Zimmerman, G. (1995). *Multicriteria evaluation in a fuzzy environment, theory and applications in ecological economics*. Berlin: Physica-Verlag.

Ziyou, G., & Yifan, S. (2002). A reserve capacity model of optimal signal control with user-equilibrium route choice. *Transportation Research B*, *36*, 313–323.

Zlatoper, T., & Austrian, Z. (1989). Freight transport demand: a survey of recent econometric studies. *Transportation*, *16*, 27–46.

GPSR Compliance
The European Union's (EU) General Product Safety Regulation (GPSR) is a set of rules that requires consumer products to be safe and our obligations to ensure this.

If you have any concerns about our products, you can contact us on

ProductSafety@springernature.com

In case Publisher is established outside the EU, the EU authorized representative is:

Springer Nature Customer Service Center GmbH
Europaplatz 3
69115 Heidelberg, Germany

www.ingramcontent.com/pod-product-compliance
Lightning Source LLC
Chambersburg PA
CBHW051922150825
PP16919100001B/22

* 9 7 8 0 3 8 7 7 5 8 5 6 5 *